Journal of Homotopy and Related Structures

Volume 4(1), 2009

Aims and Scope

Homotopy is a basic discipline of mathematics having fundamental and various applications to important fields of mathematics.

The Journal has a wide scope which ranges from homotopical algebra and algebraic topology to algebraic number theory and functional analysis. Diverse algebraic, geometric, topological and categorical structures are closely related to mathematics such as general algebra, algebraic topology, algebraic geometry, category theory, differential geometry, computer science, K-theory, functional analysis, Galois theory and in physical sciences as well.

Journal of Homotopy and Related Structures intends to develop its vision on the determining role of homotopy in mathematics. The aim of the Journal is to show the importance, merit and diversity of homotopy in mathematical sciences.

Journal of Homotopy and Related Structures is primarily concerned with publishing carefully refereed significant and original research papers. However, a limited number of carefully selected survey and expository papers are also included, and special issues devoted to proceedings of meetings in the field and to Festschrifts will also be published.

Published jointly by College Publications and Tbilisi Center for Mathematical Sciences

ISBN 978-1-904987-69-7

College Publications
Scientific Director: Dov Gabbay
Managing Director: Jane Spurr
Department of Computer Science
King's College London, Strand, London WC2R 2LS, UK

http://www.collegepublications.co.uk

Original cover design by Laraine Welch
Printed by Lightning Source, Milton Keynes, UK

Journal of Homotopy and Related Structures

Volume **4**(1), 2009

Table of Contents

Journal of Homotopy and Related Structures, vol. 4(1), 2009, pp.1–5

REPARAMETRIZATIONS WITH GIVEN STOP DATA

MARTIN RAUSSEN

(*communicated by Ronnald Brown*)

Abstract

In [**1**], we performed a systematic investigation of reparametrizations of continuous paths in a Hausdorff space that relies crucially on a proper understanding of stop data of a (weakly increasing) reparametrization of the unit interval. I am grateful to Marco Grandis (Genova) for pointing out to me that the proof of Proposition 3.7 in [**1**] is wrong. Fortunately, the statement of that Proposition and the results depending on it stay correct. It is the purpose of this note to provide correct proofs.

1. Reparametrizations with given stop maps

To make this note self-contained, we need to include some of the basic definitions from [**1**]. The set of all (nondegenerate) closed subintervals of the unit interval $I = [0, 1]$ will be denoted by $\mathcal{P}_{[\,]}(I) = \{[a, b] \mid 0 \leqslant a < b \leqslant 1\}$.

Definition 1.1.
- A *reparametrization* of the unit interval I is a weakly increasing continuous self-map $\varphi : I \to I$ preserving the end points.
- A *non-degenerate* interval $J \subset I$ is a *φ-stop interval* if there exists a value $t \in I$ such that $\varphi^{-1}(t) = J$. The value $t = \varphi(J) \in I$ is called a *φ-stop value.*
- The set of all φ-stop intervals will be denoted as $\Delta_\varphi \subseteq \mathcal{P}_{[\,]}(I)$. Remark that the intervals in Δ_φ are disjoint and that Δ_φ carries a natural total order. We let $D_\varphi := \bigcup_{J \in \Delta_\varphi} J \subset I$ denote the *stop set* of φ; and $C_\varphi \subset I$ the set of all stop values.
- The *φ-stop map* $F_\varphi : \Delta_\varphi \to C_\varphi$ corresponding to a reparametrization φ is given by $F_\varphi(J) = \varphi(J)$.

It is shown in [**1**] that F_φ is an *order-preserving bijection* between (at most) *countable sets.* It is natural to ask (and important for some of the results in [**1**]) which order-preserving bijections between such sets arise from some reparametrization:

To this end, let

- $\Delta \subseteq \mathcal{P}_{[\,]}(I)$ denote a subset of *disjoint closed* sub-intervals – equipped with the natural total order;

Received August 19, 2008, revised September 2, 2008; published on February 11, 2009.
2000 Mathematics Subject Classification: 55,68
Key words and phrases: path, regular path, reparametrization, stop map.

- $C \subseteq I$ denote a subset with the same cardinality as Δ;
- $F : \Delta \to C$ denote an order-preserving bijection.

I am grateful to the referee for pointing out the following lemma and its proof:

Lemma 1.2. *A subset $\Delta \subseteq \mathcal{P}_{[\,]}(I)$ of disjoint closed intervals is countable.*

Proof. Given a set Δ of disjoint non-degenerate closed sub-intervals of the unit interval I, each will contain rational numbers by density. By the axiom of choice, choose for each disjoint sub-interval a specific rational number contained in that sub-interval. The chosen set $\Delta' \subset \mathbf{Q}$ of rationals is countable as a subset of $\mathbf{Q}$. Combining an enumeration of Δ' with the bijection between Δ' and Δ mapping each interval to its chosen rational yields an enumeration of Δ. □

Proposition 1.3. *There exists a reparametrization φ with $F_\varphi = F$ if and only if conditions* (1) - (8) *below are satisfied for intervals contained in Δ and for all* $0 < z < 1$*:*

1. $\min J = \sup_{J'<J} \max J' \Rightarrow F(J) = \sup_{J'<J} F(J')$;
2. $\max J = \inf_{J<J'} \min J' \Rightarrow F(J) = \inf_{J<J'} F(J')$;
3. $\sup_{J'<z} \max J' = \inf_{z<J''} \min J'' \Rightarrow \sup_{J'<z} F(J') = \inf_{z<J''} F(J'')$;
4. $\sup_{J'<z} \max J' < \inf_{z<J''} \min J'' \Rightarrow \sup_{J'<z} F(J') < \inf_{z<J''} F(J'')$;
5. $\inf_{0<J} \min J = 0 \Rightarrow \inf_{0<J} F(J) = 0$;
6. $\inf_{0<J} \min J > 0 \Rightarrow \inf_{0<J} F(J) > 0$;
7. $\sup_{J<1} \max J = 1 \Rightarrow \sup_{J<1} F(J) = 1$;
8. $\sup_{J<1} \max J < 1 \Rightarrow \sup_{J<1} F(J) < 1$.

Proof. Conditions (1) – (3), (5) and (7) are necessary for the stop data of a *continuous* reparametrization φ; (4), (6) and (8) are necessary to avoid further stop intervals.

Given a stop map satisfying conditions (1) – (8), we construct a reparametrization φ_F with $F(\varphi_F) = F$ as follows: For $t \in D = \bigcup_{J\in\Delta} J$, one has to define: $\varphi(t) = F(J)$ with $t \in J$. This defines a weakly increasing function φ_F on D. Conditions (1) and (2) make sure that this function is continuous (on D). Condition (3) makes it possible to extend φ_F uniquely to a weakly increasing continuous function on the closure $\bar{D}$: $\varphi_F(z)$ is defined as $\sup_{J'<z} F(J')$ for $z = \sup_{J'<z} \max J'$ and/or as $\inf_{z<J''} F(J'')$ for $z = \inf_{z<J''} \min J$. By (5) and (7), $\varphi_F(0) = 0$ and $\varphi_F(1) = 1$ if $0, 1 \in \bar{D}$; if not, we have to take these as a definition.

The complement $O = I \setminus \bar{D}$ is an open (possibly empty) subspace of I, hence a union of at most countably many open subintervals $J = [a_-^J, a_+^J]$ with boundary in $\partial D \cup \{0, 1\}$. Condition (4), (6) and (8) make sure, that $\varphi_F(a_-^J) < \varphi_F(a_+^J)$. Hence, every collection of strictly increasing homeomorphisms between $[a_-^J, a_+^J]$ and $[\varphi_F(a_-^J), \varphi_F(a_+^J)]$ – preserving endpoints – extends φ_F to a continuous increasing map $\varphi_F : I \to I$ with $\Delta_{\varphi_F} = \Delta, C_{\varphi_F} = C$ and $F_{\varphi_F} = F$. □

It is natural to ask, whether

- every at most countable subset $C \subset I$ occurs as set of stop values of some reparametrization: This is answered affirmatively in [1], Lemma 2.10;
- every set $\{I\} \neq \Delta \subset \mathcal{P}_{[\,]}(I)$ of closed disjoint intervals arises as set of stop intervals of a reparametrization:

Proposition 1.4. *For every $\{I\} \neq \Delta$ of closed disjoint sub-intervals in the unit interval I, there exists a reparametrization φ with $\Delta_\varphi = \Delta$.*

Proof. We use Lemma 1.2 to provide us with an enumeration j of the totally ordered set Δ (defined either on $\mathbf{N}$ or on a finite integer interval $[1, n]$). Using j, we are going to construct a reparametrization φ with stop value set C_φ included in the set $I[\frac{1}{2}] = \{0 \leqslant \frac{l}{2^k} \leqslant 1\}$ of rational numbers with denominators a power of 2. To this end, we will associate to every number $z \in I[\frac{1}{2}]$ either an interval in Δ or a degenerate one point interval; we end up with an ordered bijection beween $I[\frac{1}{2}]$ and a superset of Δ; all excess intervals will be degenerate one-point sets.

To get started, let I_0 denote either *the* interval in Δ containing 0 or, if no such interval exists, the degenerate interval $[0, 0] = \{0\}$; likewise define I_1. Every number $z \in I[\frac{1}{2}]$ apart from 0 and 1 has a unique representation $z = \frac{l}{2^k}$ with l *odd*, $0 < l < 2^k$. The construction proceeds by induction on k using the enumeration j.

Assume for a given $k \geqslant 1$, I_z and thus the map $I : z \mapsto I_z$ defined for all $z = \frac{l}{2^{k-1}}$, $0 \leqslant l \leqslant 2^{k-1}$ as an ordered map. For $0 < z = \frac{l}{2^k} < 1$ and l odd, both $z_\pm = z \pm \frac{1}{2^k}$ have a representation as fraction with denominator 2^{k-1} and thus $I_{z_-} < I_{z_+}$ are already defined. Let $I_z = j(m)$ with m minimal such that $I_{z_-} < j(m) < I_{z_+}$ if such an m exists; if not, then I_z is defined as the degenerate interval containing the single element $\frac{1}{2}(\max I_{z_-} + \min I_{z_+})$. The map $I : z \mapsto I_z$ thus constructed on $I[\frac{1}{2}]$ is order-preserving. Moreover, this map is onto, since – by an induction over k – $I_{j(k)}$ occurs as I_z with some z of the form $\frac{l}{2^k}$. Hence, there is an order-preserving inverse map $I^{-1} : I_z \mapsto z$.

For $k \geqslant 0$, let φ_k denote the piecewise linear reparametrization that has constant value z on I_z for $z = \frac{l}{2^k}$, $0 \leqslant l \leqslant 2^k$ and that is linear inbetween these intervals. Remark that $\varphi_{k+1} = \varphi_k$ on all I_z with $z = \frac{l}{2^k}$ including all occuring degenerate intervals. As a consequence, $\| \varphi_k - \varphi_{k+1} \| < \frac{1}{2^k}$, and hence for all $l > k$, $\| \varphi_k - \varphi_l \| < \frac{1}{2^{k-1}}$. Hence, the sequence $(\varphi_k)_{k \in \mathbf{N}}$ converges uniformly to a continuous reparametrization φ.

By construction, the resulting reparametrization φ is constant on all intervals in Δ; on every open interval between these stop intervals, it is linear and strictly increasing. In particular, $\Delta_\varphi = \Delta$. □

Remark 1.5. I was first tempted to prove Proposition 1.4 by taking some integral of the characteristic function of the complement of D and to normalize the resulting function. But in general, this does not work out since, as already remarked in [**1**], it may well be that $\bar{D} = I$!

2. Concluding remarks

Remark 2.1. 1. Instead of constructing the reparametrization φ in Proposition 1.4, it is also possible to apply the criteria in Proposition 1.3 to the restriction $I_{|\Delta}$ of the map I from the proof above.

2. Proposition 1.3 replaces Proposition 2.13 in [**1**]. To get sufficiency, requirements (1) and (2) had to be added to those mentioned in [**1**] in order to make sure that the map φ_F is continuous on D. Moreover, (6) and (8) had to be added to avoid stop intervals containing 0, resp. 1 in case Δ does not contain such intervals.
In particular, the midpoint map m that associates to every interval in Δ its midpoint satisfies the criteria given in [**1**], Proposition 2.13, but if fails in general to satisfy conditions (1) and (2) in Proposition 1.3 in this note; in particular, the map φ_m will in general not be continuous, as remarked by M. Grandis. The midpoint map m was used in the flawed proof of [**1**], Proposition 3.7 – stated as Proposition 2.2 below.

The main focus in [**1**] is on reparametrizations of continuous paths $p : I \to X$ into a Hausdorff space X. A continuous path q is called *regular* if it is constant or if the restriction $q|_J$ to every non-degenerate sub-interval $J \subseteq I$ is *non-constant.*

Proposition 2.2. (Proposition 3.7 in [**1**])
For every path $p : I \to X$, there exists a regular path q and a reparametrization such that $p = q \circ \varphi$.

Proof. A non-constant path p gives rise to the set of all (closed disjoint) *stop intervals* $\Delta_p \subset \mathcal{P}_{[\,]}(I)$, consisting of the maximal subintervals $J \subset I$ on which p is constant. Proposition 1.4 yields a reparametrization φ with $\Delta_\varphi = \Delta_p$ and thus a set-theoretic factorization

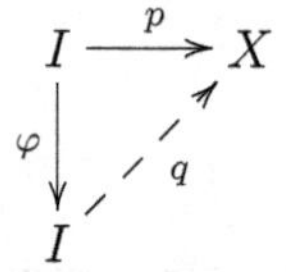

through a map $q : I \to X$ that is not constant on any non-degenerate subinterval $J \subseteq I$. The continuity of q follows as in the remaining lines of the proof in [**1**]. □

References

[**1**] U. Fahrenberg and M. Raussen *Reparametrizations of continuous paths*, J. Homotopy Relat. Struct. **2** (2007), no.2, 93 – 117.

See also the references in [**1**].

This article may be accessed via WWW at `http://jhrs.rmi.acnet.ge`

Martin Raussen
`raussen@math.aau.dk`
`www.math.aau.dk/~raussen`

Department of Mathematical Sciences
Aalborg University
Denmark
Fredrik Bajersvej 7G
DK-9220 Aalborg Øst

Journal of Homotopy and Related Structures, vol. 4(1), 2009, pp.7–38

NONISOMORPHIC VERDIER OCTAHEDRA ON THE SAME BASE

MATTHIAS KÜNZER

(*communicated by Antonio Cegarra*)

Abstract

We show by an example that in a Verdier triangulated category, there may exist two mutually nonisomorphic Verdier octahedra containing the same commutative triangle.

Contents

0. Introduction

0.1. Is being a 3-triangle characterised by 2-triangles?

VERDIER (implicitly) defined a Verdier octahedron to be a diagram in a triangulated category in the shape of an octahedron, four of whose triangles are distinguished, the four others commutative [**5,** Def. 1-1]; cf. also [**1,** 1.1.6]. It arises as follows.

Received September 7, 2008, revised September 29, 2008; published on March 16, 2009.
2000 Mathematics Subject Classification: 18E30.
Key words and phrases: triangulated categories, octahedra.

To a morphism in a triangulated category, we can attach an object, called its *cone*. The morphism we start with and its cone are contained in a distinguished triangle. To the morphism we started with, we refer as the *base* of this distinguished triangle.

Now given a commutative triangle, we can form the cone on the first morphism, on the second morphism and on their composite, yielding three distinguished triangles. These three cones in turn are contained in a fourth distinguished triangle. The whole diagram obtained by this construction is a Verdier octahedron. We shall refer to the commutative triangle we started with as the *base* of this Verdier octahedron.

A distinguished triangle has the property of being determined up to isomorphism by its base. Moreover, any morphism between the bases of two distinguished triangles can be extended to a morphism between the whole distinguished triangles.

We shall show that the analogous assertion is not true for Verdier octahedra. In §2, we give an example of two nonisomorphic Verdier octahedra on the same base. In particular, the identity morphism between the bases cannot be prolonged to a morphism between the whole Verdier octahedra.

In the terminology of Heller triangulated categories, a Verdier octahedron is a periodic 3-pretriangle X such that $Xd^{\#}$ is a 2-triangle (i.e. a distinguished triangle) for all injective periodic monotone maps $\bar{\Delta}_3 \xleftarrow{d} \bar{\Delta}_2$.

One of the two Verdier octahedra in our example will be a 3-triangle in the sense of [**4,** Def. 1.5], i.e. a "distinguished octahedron", whereas the other will not.

Note that unlike a Verdier octahedron, a 3-triangle is uniquely determined up to isomorphism by its base in the Heller triangulated context; cf. [**4,** Lem. 3.4.(6)].

0.2. Is being an n-triangle characterised by $(n-1)$-triangles?

The situation of §0.1 can be generalised in the following manner. Suppose given a closed Heller triangulated category $(\mathcal{C}, \mathsf{T}, \vartheta)$; cf. [**4,** Def. 1.5], Definition 13.

The Heller triangulation $\vartheta = (\vartheta_n)_{n \geqslant 0}$ on $(\mathcal{C}, \mathsf{T})$ can be viewed as a means to distinguish certain periodic n-pretriangles as n-triangles. Namely, a periodic n-pretriangle X is, by definition, an n-triangle if $X\vartheta_n = 1$; cf. [**4,** Def. 1.5.(ii.2)]. For instance, 2-triangles are distinguished triangles in the sense of Verdier; 3-triangles are particular, "distinguished" Verdier octahedra.

0.2.1. The example

Let $n \geqslant 3$. Let X be a periodic n-pretriangle. Suppose that $Xd^{\#}$ is an $(n-1)$-triangle for all injective periodic monotone maps $\bar{\Delta}_n \xleftarrow{d} \bar{\Delta}_{n-1}$. One might ask whether X is an n-triangle.

We shall show in §3 by an example that this is, in general, not the case.

0.2.2. Consequences

Suppose given $n \geqslant 3$ and a subset of the set of periodic n-pretriangles. We shall say for the moment that *determination* holds for this subset if for X and $\tilde{X}$ out of this subset, $X|_{\dot{\Delta}_n} \simeq \tilde{X}|_{\dot{\Delta}_n}$ implies that there is a periodic isomorphism $X \simeq \tilde{X}$. We shall say that *prolongation* holds for this subset, if for X and $\tilde{X}$ out of this subset and a morphism $X|_{\dot{\Delta}_n} \longrightarrow \tilde{X}|_{\dot{\Delta}_n}$, there exists a periodic morphism $X \longrightarrow \tilde{X}$ that restricts on $\dot{\Delta}_n$ to that given morphism. If prolongation holds, then determination holds.

- Consider the subset of periodic n-pretriangles X such that $Xd^\#$ is an $(n-1)$-triangle for all injective periodic monotone maps $\bar{\Delta}_n \xleftarrow{d} \bar{\Delta}_{n-1}$. Our example shows that in general, determination and prolongation do not hold for this subset. In fact, if X is such an n-pretriangle, but not an n-triangle, then the n-triangle on the base $X|_{\dot{\Delta}_n}$ is not isomorphic to X; cf. [**4,** Lem. 3.4.(1, 4)]. In particular, the reader familiar with Heller triangulated categories can skip §1.3, §1.4 and §2, where the case $n = 3$ is treated in the Verdier context.
- Bernstein, Beilinson and Deligne considered the subset of periodic n-pretriangles X such that $Xd^\#$ is a 2-triangle (i.e. a distinguished triangle) for all injective periodic monotone maps $\bar{\Delta}_n \xleftarrow{d} \bar{\Delta}_2$ [**1,** 1.1.14]. Given $1 \leqslant i \leqslant j \leqslant n+1$, we write $X_{(j-1)/(i-1)}$ for the object written $X([i,j[)$ in loc. cit., and $((j-1)/(i-1))^{+1} = (i-1)^{+1}/(j-1)$ for what is written $[i,j[^* = [j, i+n[$ there; condition (c) of loc. cit. translates into the requirement that $Xd^\#$ be a 2-triangle for any d – which implies that the whole commutative diagram is in fact an n-pretriangle by [**4,** Lem. A.17].
  Now our example shows that in general, determination and prolongation do not hold for this subset. In fact, this subset contains the previously described subset.

In both of the cases above, if $n = 3$, then the condition singles out the subset of Verdier octahedra; cf. §1.4.

- By [**4,** Lem. 3.4.(6); Lem. 3.2], determination and prolongation hold for the set of n-triangles.

So morally, our example shows that it makes sense to let the Heller triangulation ϑ distinguish n-triangles for all $n \geqslant 0$. There is no "sufficiently large" n we could be content with.

0.3. An appendix on transport of structure

Suppose given a Frobenius category $\mathcal{E}$; that is, an exact category with enough bijective objects (relative to pure short exact sequences). Let $\mathcal{B} \subseteq \mathcal{E}$ denote the full subcategory of bijective objects.

There are two variants of the stable category of $\mathcal{E}$. First, there is the *classical stable category* $\underline{\mathcal{E}}$, defined as the quotient of $\mathcal{E}$ modulo $\mathcal{B}$. Second, there is the *stable category* $\underline{\underline{\mathcal{E}}}$, defined as the quotient of the category of purely acyclic complexes with entries in $\mathcal{B}$ modulo the category of split acyclic complexes with entries in $\mathcal{B}$. The categories $\underline{\mathcal{E}}$ and $\underline{\underline{\mathcal{E}}}$ are equivalent. The advantage of the variant $\underline{\underline{\mathcal{E}}}$ is that it carries a shift automorphism, whereas $\underline{\mathcal{E}}$ carries a shift autoequivalence.

In [**4,** Cor. 4.7], we have endowed $\underline{\underline{\mathcal{E}}}$ with a Heller triangulation. Now in our particular situation, also $\underline{\mathcal{E}}$ carries a shift automorphism. Since $\underline{\mathcal{E}}$ is better suited for calculations within that category, the question arises whether the equivalence $\underline{\underline{\mathcal{E}}} \simeq \underline{\mathcal{E}}$ can be used to transport the structure of a Heller triangulated category from $\underline{\underline{\mathcal{E}}}$ to $\underline{\mathcal{E}}$. This is indeed the case; cf. Proposition 23.(1). Moreover, we give recipes how to detect and how to construct n-triangles in $\underline{\mathcal{E}}$; cf. Propositions 23.(2, 3), 26.

Roughly put, the variant $\underline{\underline{\mathcal{E}}}$ is rather suited for theoretical purposes, the variant $\underline{\mathcal{E}}$ is rather suited for practical purposes, and we had to pass a result from $\underline{\underline{\mathcal{E}}}$ to $\underline{\mathcal{E}}$. Not surprisingly, to do so, we had to grapple with the various equivalences and isomorphisms involved.

0.4. Acknowledgements

I thank AMNON NEEMAN for pointing out, years ago, why a counterexample as in §3 should exist, contrary to what I had believed.

This example has been found using the computer algebra system MAGMA [**2**]. I thank MARKUS KIRSCHMER for help with a Magma program.

I thank the referee for helpful comments.

0.5. Notations and conventions

We use the conventions listed in [**4**, §0.6]. In addition, we use the following conventions.

(i) If x and y are elements of a set, we let $\partial_{x,y} := 1$ if $x = y$, and we let $\partial_{x,y} := 0$ if $x \neq y$.

(ii) Given $a \in \mathbf{Z}$, we write $\mathbf{Z}/a := \mathbf{Z}/a\mathbf{Z}$.

(iii) Given a ring R and R-modules X and Y, we write, by choice, ${}_R(X,Y) = {}_{R\text{-Mod}}(X,Y) = \mathrm{Hom}_R(X,Y)$. Moreover, given $k \geqslant 0$, we write $X^{\oplus k} := \bigoplus_{i\in[1,k]} X$.

(iv) An *automorphism* T of a category $\mathcal{C}$ is an endofunctor on $\mathcal{C}$ for which there exists an endofunctor S such that $ST = 1_{\mathcal{C}}$ and $TS = 1_{\mathcal{C}}$. An *autoequivalence* T of a category $\mathcal{C}$ is an endofunctor on $\mathcal{C}$ for which there exists an endofunctor S such that $ST \simeq 1_{\mathcal{C}}$ and $TS \simeq 1_{\mathcal{C}}$.

(v) Let $n \geqslant 0$. Recall that $\bar{\Delta}_n^{\Delta\nabla} = \{\beta/\alpha \in \bar{\Delta}_n^{\#} : 0 \leqslant \alpha \leqslant \beta \leqslant 0^{+1}\} \subseteq \bar{\Delta}_n^{\#}$. We will often display an n-triangle or a periodic n-pretriangle in a Heller triangulated category $\mathcal{C}$ by showing its restriction to $\bar{\Delta}_n^{\Delta\nabla} \smallsetminus (\{\alpha/\alpha \;:\; 0 \leqslant \alpha \leqslant 0^{+1}\} \cup \{0^{+1}/0\})$. This is possible without loss of information, for we can reconstruct the whole diagram by adding zeroes on α/α for $0 \leqslant \alpha \leqslant 0^{+1}$ and on $0^{+1}/0$, and then by periodic prolongation.

(vi) Suppose given a Heller triangulated category $\mathcal{C}$. A *Verdier octahedron* in $\mathcal{C}$ is a periodic 3-pretriangle $X \in \mathrm{Ob}\,\mathcal{C}^{+,\,\text{periodic}}(\bar{\Delta}_3^{\#})$ such that $Xd^{\#} \in \mathrm{Ob}\,\mathcal{C}^{+,\,\text{periodic}}(\bar{\Delta}_2^{\#})$ is a 2-triangle for all injective periodic monotone maps $\bar{\Delta}_3 \xleftarrow{d} \bar{\Delta}_2$.

Henceforth, let $p \geqslant 2$ be a prime.

1. The classical stable category of $(\mathbf{Z}/p^m)$-mod

1.1. The category $(\mathbf{Z}/p^m)$-mod

Let $m \geqslant 0$. By $\mathcal{E} := (\mathbf{Z}/p^m)$-mod we understand the following category.

The objects are indexed by tuples $(a_i)_{i\in[0,m]}$ with $a_i \in \mathbf{Z}_{\geqslant 0}$. To such an index, we attach the object

$$\bigoplus_{i\in[0,m]} (\mathbf{Z}/p^i)^{\oplus a_i} .$$

As morphisms, we take $\mathbf{Z}/p^m$-linear maps.

Note that we have not chosen a skeleton. The trick here is to pick several zero objects.

The duality contrafunctor ${}_{\mathbf{Z}/p^m}(-,\mathbf{Z}/p^m)$ on $\mathcal{E}$, which sends $\mathbf{Z}/p^i$ to $\mathbf{Z}/p^i$ for $i \in [1,m]$, shows that an object in this category is injective if and only if it is projective. An object of $\mathcal{E}$ is bijective if and only if it is isomorphic to a finite direct sum of copies of $\mathbf{Z}/p^m$. The category $\mathcal{E}$ is an abelian Frobenius category, with all short exact sequences stipulated to be pure; cf. e.g. [**4**, Def. A.5.(2)].

1.2. The shift on $\underline{(\mathbf{Z}/p^m)\text{-mod}}$

To define a shift automorphism on the classical stable category $\underline{\mathcal{E}} = \underline{(\mathbf{Z}/p^m)\text{-mod}}$, we shall distinguish certain (pure) short exact sequences in $\mathcal{E}$; cf. §A.4.2.1, [**4**, Def. A.7].

Let $\mathrm{E}_k := \begin{pmatrix} 1 & & \\ & \ddots & \\ & & 1 \end{pmatrix}$ denote the unit matrix of size $k \times k$; let $\mathrm{E}'_k := \begin{pmatrix} & & 1 \\ & \cdot^{\cdot^{\cdot}} & \\ 1 & & \end{pmatrix}$ denote the reversed unit matrix of size $k \times k$.

As distinguished (pure) short exact sequences we take those of the form

$$\bigoplus_{i\in[0,m]} (\mathbf{Z}/p^i)^{\oplus a_i} \xrightarrow[\bullet]{\begin{pmatrix} p^m\mathrm{E}_{a_0} & & & \\ & p^{m-1}\mathrm{E}_{a_1} & & \\ & & \ddots & \\ & & & p^0\mathrm{E}_{a_m} \end{pmatrix}} (\mathbf{Z}/p^m)^{\oplus \sum_{i\in[0,m]} a_i}$$

$$\xrightarrow[+]{\begin{pmatrix} & & & \mathrm{E}'_{a_0} \\ & & \mathrm{E}'_{a_1} & \\ & \cdot^{\cdot^{\cdot}} & & \\ \mathrm{E}'_{a_m} & & & \end{pmatrix}} \bigoplus_{i\in[0,m]} (\mathbf{Z}/p^i)^{\oplus a_{m-i}}$$

So roughly speaking, distinguished short exact sequences are direct sums of those of the form

$$\mathbf{Z}/p^i \overset{p^{m-i}}{\longrightarrow\!\!\bullet} \mathbf{Z}/p^m \overset{1}{\longrightarrow\!\!+} \mathbf{Z}/p^{m-i},$$

where $i \in [0, m]$; we reorder the summands the cokernel term consists of.

With this choice, conditions (i, ii, iii) of §A.4.2.1 are satisfied.

On indecomposable objects and morphisms between them, the shift automorphism induced on $\underline{\mathcal{E}}$ by our set of distinguished short exact sequences is given by

$$(\mathbf{Z}/p^i \xrightarrow{a} \mathbf{Z}/p^j)^{+1} = (\mathbf{Z}/p^{m-i} \xrightarrow{p^{i-j}a} \mathbf{Z}/p^{m-j}),$$

where $i, j \in [0, m]$, and where a is a representative in $\mathbf{Z}$. Note that if $i < j$, then a is divisible by p^{j-i}.

Note that $\mathbf{Z}/p^i \xrightarrow{a} \mathbf{Z}/p^j$ represents zero in $\underline{\mathcal{E}}$ if and only if a is divisible by $p^{\min(m-i,\,j)}$.

1.3. A Verdier triangulation on $\underline{(\mathbf{Z}/p^m)\text{-mod}}$

By [**3**, Th. 2.6], $\underline{\mathcal{E}} = \underline{(\mathbf{Z}/p^m)\text{-mod}}$ is a Verdier triangulated category, i.e. a triangulated category in the sense of VERDIER [**5**, Def. 1-1].

This also follows by Remark 2 below and by [**4**, Prop. 3.6], which says that any Heller triangulated category in which idempotents split is also Verdier triangulated. The 2-triangles in the Heller context are the distinguished triangles in the Verdier context.

Given a morphism $X \xrightarrow{f} Y$ in $\mathcal{E}$, using the distinguished short exact sequence $X \rightarrowtail B \twoheadrightarrow X^{+1}$, where B is bijective, we can form the morphism

$$\begin{array}{ccccc} Y & \rightarrowtail & Z & \twoheadrightarrow & X^{+1} \\ {\scriptstyle f}\uparrow & & \uparrow & & \| \\ X & \rightarrowtail & B & \twoheadrightarrow & X^{+1} \end{array}$$

of short exact sequences, from which the sequence

$$X \xrightarrow{f} Y \rightarrowtail Z \twoheadrightarrow X^{+1}$$

represents a distinguished triangle in the Verdier triangulated category $\underline{\mathcal{E}}$.

1.4. Displaying Verdier octahedra

A Verdier octahedron in a Verdier triangulated category that has, as in [**1**, 1.1.6], as upper cap the diagram

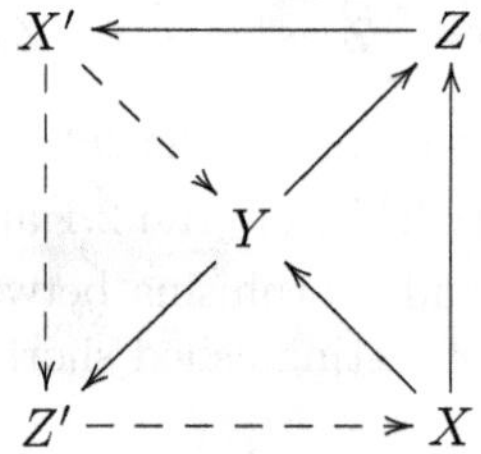

and as lower cap the diagram

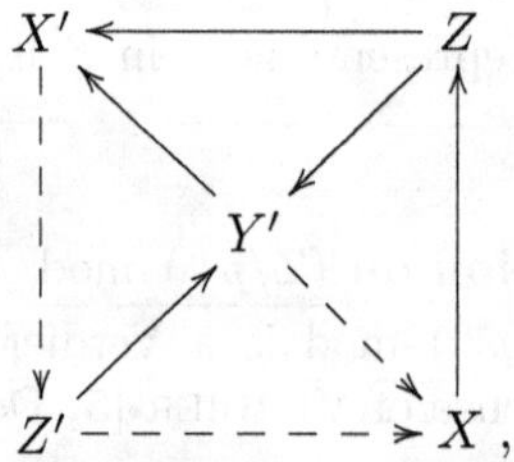

where the triangles (X, Y, Z'), (Y, Z, X'), (X, Z, Y') and (Z', Y', X') are distinguished and the remaining ones commutative, and where the dashed arrows represent morphisms of degree 1 (cf. [**1**, 1.1.1]), we will display, following the suggestion in [**1**, 1.1.14], as

$$\begin{array}{ccccccc}
 & & & & & & Z^{+1}\\
 & & & & & & \uparrow\\
 & & & & X' & \longrightarrow & Y^{+1}\\
 & & & & \uparrow & & \uparrow\\
 & & Z' & \longrightarrow & Y' & \longrightarrow & X^{+1}\\
 & & \uparrow & & \uparrow & & \\
X & \longrightarrow & Y & \longrightarrow & Z & & .
\end{array}$$

Note that the outer morphisms of the upper and lower cap are composites of the inner ones.

We parametrise such an octahedron by the indexing symbols

$$\begin{array}{ccccccc}
 & & & & & & 0^{+1}/3\\
 & & & & & & \uparrow\\
 & & & & 3/2 & \longrightarrow & 0^{+1}/2\\
 & & & & \uparrow & & \uparrow\\
 & & 2/1 & \longrightarrow & 3/1 & \longrightarrow & 0^{+1}/1\\
 & & \uparrow & & \uparrow & & \\
1/0 & \longrightarrow & 2/0 & \longrightarrow & 3/0 & & .
\end{array}$$

The operation $\mathrm{d}_0^{\#}$ maps this octahedron to the distinguished triangle (Z', Y', X'), the operation $\mathrm{d}_1^{\#}$ maps it to the distinguished triangle (Y, Z, X'), the operation $\mathrm{d}_2^{\#}$ maps it to the distinguished triangle (X, Z, Y'), and the operation $\mathrm{d}_3^{\#}$ maps it to the distinguished triangle (X, Y, Z').

1.5. A Heller triangulation on $(\mathbf{Z}/p^m)$-mod

Concerning the notation $\mathcal{E}^{\square}(\bar{\Delta}_n^{\vartriangle\triangledown})$, cf. §A.3. Given $n \geqslant 0$ and $X \in \operatorname{Ob}\mathcal{E}^{\square}(\bar{\Delta}_n^{\vartriangle\triangledown})$, we form $X^{\tau} \in \operatorname{Ob}\underline{\mathcal{E}}^{+,\,\mathrm{periodic}}(\bar{\Delta}_n^{\vartriangle\triangledown})$ with respect to the set of distinguished short exact sequences of §1.2 as described in §A.4.2.3; cf. Remark 25. That is, we replace the rightmost column of X by the column obtained using distinguished short exact sequences, so that $(X^{\tau})_{0^{+1}/*} = ((X^{\tau})_{*/0})^{+1} = (X_{*/0})^{+1}$; cf. §A.4.2.3.

Remark 1. *If the short exact sequences*

$$X_{\alpha/0} \overset{(x\ x)}{\rightarrowtail} X_{\alpha/\alpha} \oplus X_{0^{+1}/0} \overset{\binom{x}{-x}}{\twoheadrightarrow} X_{0^{+1}/\alpha}$$

appearing in the diagram X for $1 \leqslant \alpha \leqslant n$ already are distinguished, then the image of X in $\operatorname{Ob}\underline{\mathcal{E}}^{+}(\bar{\Delta}_n^{\vartriangle\triangledown})$ equals X^{τ}.

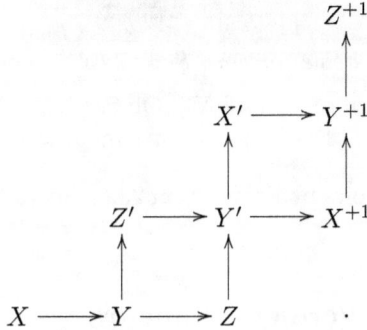

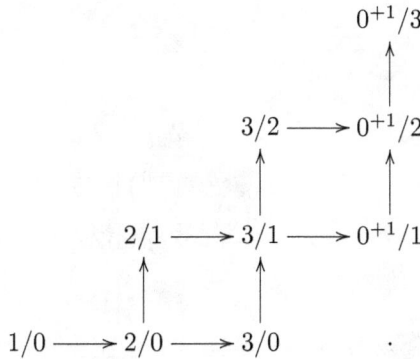

Concerning the notion of a closed Heller triangulated category, cf. Definition 13 in §A.2.

Remark 2. *The classical stable category* $\underline{\mathcal{E}} = (\mathbf{Z}/p^m)\text{-mod}$ *carries a closed Heller triangulation such that given* $n \geqslant 0$ *and* $X \in \mathrm{Ob}\,\mathcal{E}^{\square}(\bar{\Delta}_n^{\Delta\nabla})$, *the periodic prolongation of* X^τ *to an object of* $\underline{\mathcal{E}}^{+,\,\mathrm{periodic}}(\bar{\Delta}_n^{\#})$ *is an* n-*triangle.*

Proof. The assertion follows by Proposition 23.(1) in §A.4.2.1 and Proposition 26 in §A.4.2.4. $\qquad\qquad\square$

2. Nonisomorphic Verdier octahedra

Let $\mathcal{C} := (\mathbf{Z}/p^6)\text{-mod}$, and let it be endowed with a shift automorphism as in §1.2 and a Verdier triangulation as in §1.3.

Let the diagram X be given by

$$
\begin{array}{ccccc}
 & & & & \mathbf{Z}/p^3 \\
 & & & & \Big\uparrow{\scriptstyle p} \\
 & & & \mathbf{Z}/p^1\oplus\mathbf{Z}/p^5 \xrightarrow{\left(\begin{smallmatrix}-p^2\\1\end{smallmatrix}\right)} & \mathbf{Z}/p^3 \\
 & & & \Big\uparrow{\scriptstyle\left(\begin{smallmatrix}1&0\\0&p\end{smallmatrix}\right)} & \Big\uparrow{\scriptstyle p} \\
 & \mathbf{Z}/p^1\oplus\mathbf{Z}/p^5 \xrightarrow{\left(\begin{smallmatrix}p&0\\0&1\end{smallmatrix}\right)} & \mathbf{Z}/p^2\oplus\mathbf{Z}/p^4 \xrightarrow{\left(\begin{smallmatrix}-p\\1\end{smallmatrix}\right)} & \mathbf{Z}/p^3 \\
 & \Big\uparrow{\scriptstyle(1\ p^2)} & \Big\uparrow{\scriptstyle(1\ p)} & \\
 \mathbf{Z}/p^3 \xrightarrow{\ p\ } \mathbf{Z}/p^3 \xrightarrow{\ p\ } & \mathbf{Z}/p^3 & & .
\end{array}
$$

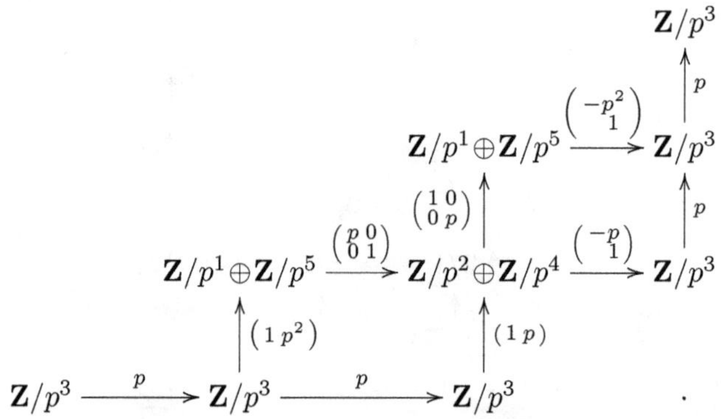

Let the diagram $\tilde{X}$ be given by

$$
\begin{array}{ccccc}
 & & & & \mathbf{Z}/p^3 \\
 & & & & \Big\uparrow{\scriptstyle p} \\
 & & & \mathbf{Z}/p^1\oplus\mathbf{Z}/p^5 \xrightarrow{\left(\begin{smallmatrix}-p^2\\1\end{smallmatrix}\right)} & \mathbf{Z}/p^3 \\
 & & & \Big\uparrow{\scriptstyle\left(\begin{smallmatrix}1&0\\-1&p\end{smallmatrix}\right)} & \Big\uparrow{\scriptstyle p} \\
 & \mathbf{Z}/p^1\oplus\mathbf{Z}/p^5 \xrightarrow{\left(\begin{smallmatrix}p&0\\1&1\end{smallmatrix}\right)} & \mathbf{Z}/p^2\oplus\mathbf{Z}/p^4 \xrightarrow{\left(\begin{smallmatrix}-p\\1\end{smallmatrix}\right)} & \mathbf{Z}/p^3 \\
 & \Big\uparrow{\scriptstyle(1\ p^2)} & \Big\uparrow{\scriptstyle(1\ p)} & \\
 \mathbf{Z}/p^3 \xrightarrow{\ p\ } \mathbf{Z}/p^3 \xrightarrow{\ p\ } & \mathbf{Z}/p^3 & & .
\end{array}
$$

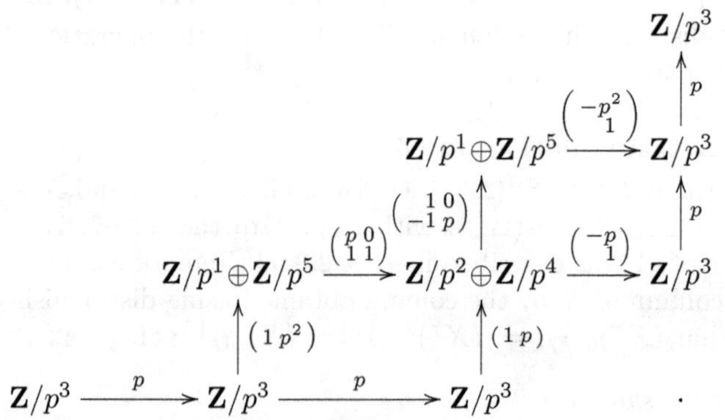

Lemma 3. *Both X and $\tilde{X}$ are Verdier octahedra.*

In contrast to the procedure in §3 below, to prove this, we will not make use of the folding operation.

Proof. We obtain $X\mathrm{d}_3^\# = \tilde{X}\mathrm{d}_3^\#$, horizontally displayed as

$$\mathbf{Z}/p^3 \xrightarrow{p} \mathbf{Z}/p^3 \xrightarrow{(1\ p^2)} \mathbf{Z}/p \oplus \mathbf{Z}/p^5 \xrightarrow{\binom{-p^2}{1}} \mathbf{Z}/p^3 \,.$$

The following morphism of short exact sequences in $(\mathbf{Z}/p^6)$-mod shows $X\mathrm{d}_3^\#$ to be a distinguished triangle.

$$\begin{array}{ccccc} \mathbf{Z}/p^3 & \xrightarrow{(1\ p^2)} & \mathbf{Z}/p \oplus \mathbf{Z}/p^5 & \xrightarrow{\binom{-p^2}{1}} & \mathbf{Z}/p^3 \\ {\scriptstyle p}\uparrow & & \uparrow{\scriptstyle (0\ 1)} & & \| \\ \mathbf{Z}/p^3 & \xrightarrow{p^3} & \mathbf{Z}/p^6 & \xrightarrow{1} & \mathbf{Z}/p^3 \end{array}$$

We obtain the distinguished triangle $X\mathrm{d}_1^\# = \tilde{X}\mathrm{d}_1^\# = X\mathrm{d}_3^\#$ again.
We obtain the diagram $X\mathrm{d}_2^\# = \tilde{X}\mathrm{d}_2^\#$, horizontally displayed as

$$\mathbf{Z}/p^3 \xrightarrow{p^2} \mathbf{Z}/p^3 \xrightarrow{(1\ p)} \mathbf{Z}/p^2 \oplus \mathbf{Z}/p^4 \xrightarrow{\binom{-p}{1}} \mathbf{Z}/p^3 \,.$$

The following morphism of short exact sequences in $(\mathbf{Z}/p^6)$-mod shows $X\mathrm{d}_2^\#$ to be a distinguished triangle.

$$\begin{array}{ccccc} \mathbf{Z}/p^3 & \xrightarrow{(1\ p)} & \mathbf{Z}/p^2 \oplus \mathbf{Z}/p^4 & \xrightarrow{\binom{-p}{1}} & \mathbf{Z}/p^3 \\ {\scriptstyle p^2}\uparrow & & \uparrow{\scriptstyle (0\ 1)} & & \| \\ \mathbf{Z}/p^3 & \xrightarrow{p^3} & \mathbf{Z}/p^6 & \xrightarrow{1} & \mathbf{Z}/p^3 \end{array}$$

We obtain the periodic isomorphism $X\mathrm{d}_0^\# \xrightarrow{\sim} \tilde{X}\mathrm{d}_0^\#$, horizontally displayed as

$$\begin{array}{ccccccc} \mathbf{Z}/p \oplus \mathbf{Z}/p^5 & \xrightarrow{\left(\begin{smallmatrix} p & 0 \\ 0 & 1 \end{smallmatrix}\right)} & \mathbf{Z}/p^2 \oplus \mathbf{Z}/p^4 & \xrightarrow{\left(\begin{smallmatrix} 1 & 0 \\ 0 & p \end{smallmatrix}\right)} & \mathbf{Z}/p \oplus \mathbf{Z}/p^5 & \xrightarrow{\left(\begin{smallmatrix} 0 & -p^4 \\ 1 & 0 \end{smallmatrix}\right)} & \mathbf{Z}/p \oplus \mathbf{Z}/p^5 \\ \| & & {\scriptstyle \wr}\downarrow{\scriptstyle \left(\begin{smallmatrix} 1 & 0 \\ 1 & 1 \end{smallmatrix}\right)} & & \| & & \| \\ \mathbf{Z}/p \oplus \mathbf{Z}/p^5 & \xrightarrow{\left(\begin{smallmatrix} p & 0 \\ 1 & 1 \end{smallmatrix}\right)} & \mathbf{Z}/p^2 \oplus \mathbf{Z}/p^4 & \xrightarrow{\left(\begin{smallmatrix} 1 & 0 \\ -1 & p \end{smallmatrix}\right)} & \mathbf{Z}/p \oplus \mathbf{Z}/p^5 & \xrightarrow{\left(\begin{smallmatrix} 0 & -p^4 \\ 1 & 0 \end{smallmatrix}\right)} & \mathbf{Z}/p \oplus \mathbf{Z}/p^5 \,. \end{array}$$

So we are reduced to show that $X\mathrm{d}_0^\#$ is a distinguished triangle, which it is as a direct sum of two distinguished triangles, as the following morphisms of short exact sequences in $(\mathbf{Z}/p^6)$-mod show.

$$\begin{array}{ccccc} \mathbf{Z}/p & \xrightarrow{p} & \mathbf{Z}/p^2 & \xrightarrow{1} & \mathbf{Z}/p \\ {\scriptstyle 1}\uparrow & & \uparrow{\scriptstyle 1} & & \| \\ \mathbf{Z}/p^5 & \xrightarrow{p} & \mathbf{Z}/p^6 & \xrightarrow{1} & \mathbf{Z}/p \end{array} \qquad \begin{array}{ccccc} \mathbf{Z}/p^4 & \xrightarrow{p} & \mathbf{Z}/p^5 & \xrightarrow{1} & \mathbf{Z}/p \\ {\scriptstyle 1}\uparrow & & \uparrow{\scriptstyle 1} & & \| \\ \mathbf{Z}/p^5 & \xrightarrow{p} & \mathbf{Z}/p^6 & \xrightarrow{1} & \mathbf{Z}/p \end{array}$$

□

By an isomorphism between Verdier octahedra, we shall understand an isomorphism of diagrams such that its entries on the rightmost vertical column arise by an application of the shift functor of $\mathcal{C}$ to its entries on the lower row.

Lemma 4. *The Verdier octahedra X and $\tilde{X}$ are not isomorphic.*

We will not use the fact that X is a 3-triangle, which in conjunction with [**4**, 3.4.(4, 6)] would permit us to restrict ourselves to consider isomorphisms that are identical on the lower row and the rightmost vertical column; cf. Lemma 7 in §3.2 below.

Proof. We *assume* the contrary. Consider an isomorphism $X \xrightarrow{\sim} \tilde{X}$, depicted as

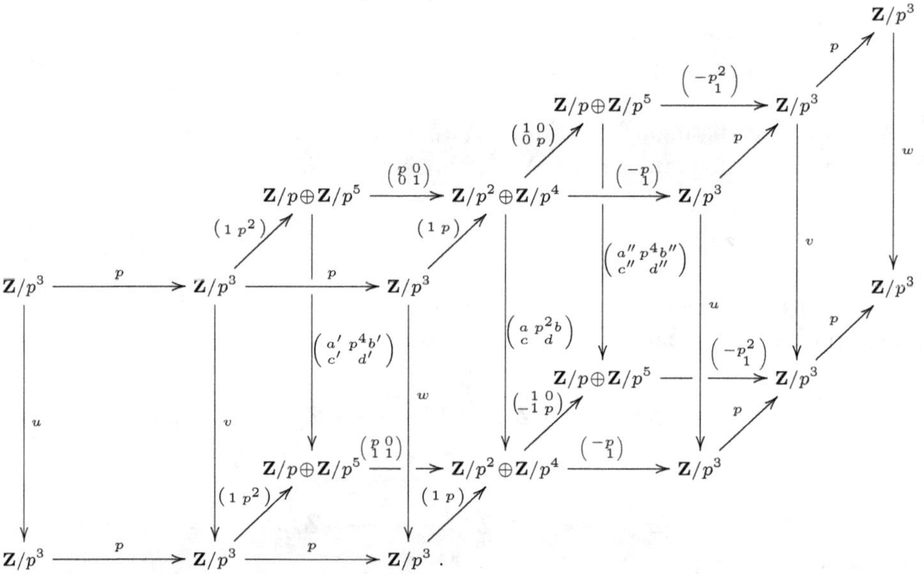

Note that all vertical quadrangles commute in $\mathcal{C}$.

The commutative quadrangles on $1/0 \longrightarrow 2/0 \longrightarrow 3/0$ yield $u \equiv_{p^2} v \equiv_{p^2} w$.
The commutative quadrangle on $3/0 \longrightarrow 3/1$ yields $pb + d \equiv_{p^2} w$.
The commutative quadrangle on $3/1 \longrightarrow 0^{+1}/1$ yields $-pc + d \equiv_{p^2} u$.
The commutative quadrangle on $3/1 \longrightarrow 3/2$ yields $b \equiv_p 0$ and $c \equiv_p d$.
Altogether, we have

$$u \equiv_{p^2} w \equiv_{p^2} pb + d \equiv_{p^2} d \equiv_{p^2} u + pc \equiv_{p^2} u + pd ,$$

whence

$$0 \equiv_p d \equiv_p w .$$

Since $\mathbf{Z}/p^3 \xrightarrow{w} \mathbf{Z}/p^3$ is an isomorphism in $\mathcal{C}$, we have $w \not\equiv_p 0$. This is *absurd*. □

In [**1**, 1.1.13], it is described how an octahedron gives rise to two "extra" triangles. Namely, as cone of the diagonal of a quadrangle appearing in an octahedron (cf. §1.4), we take the direct sum of the non-diagonal terms of the subsequent quadrangle, the morphisms being taken from the octahedron, with one minus sign inserted to ensure that the composition of two morphisms in the constructed triangle vanishes.

Lemma 5. *The triangles arising from X and from $\tilde{X}$ as described in* [**1**, 1.1.13] *are distinguished.*

Proof.[1] The morphism of short exact sequences in $(\mathbf{Z}/p^6)$-mod

$$\begin{array}{ccccc}
\mathbf{Z}/p^3 & \xrightarrow[\bullet]{\left(\begin{smallmatrix}1 & p^2 & -p\end{smallmatrix}\right)} & \mathbf{Z}/p\oplus\mathbf{Z}/p^5\oplus\mathbf{Z}/p^3 & \xrightarrow[+]{\left(\begin{smallmatrix}p & 0\\ 0 & 1\\ 1 & p\end{smallmatrix}\right)} & \mathbf{Z}/p^2\oplus\mathbf{Z}/p^4 \\
\uparrow{\scriptstyle\left(\begin{smallmatrix}-p\\ p^2\end{smallmatrix}\right)} & & \uparrow{\scriptstyle\left(\begin{smallmatrix}0 & -p & 1\\ 0 & 1 & 0\end{smallmatrix}\right)} & & \| \\
\mathbf{Z}/p^4\oplus\mathbf{Z}/p^2 & \xrightarrow[\bullet]{\left(\begin{smallmatrix}p^2 & 0\\ 0 & p^4\end{smallmatrix}\right)} & \mathbf{Z}/p^6\oplus\mathbf{Z}/p^6 & \xrightarrow[+]{\left(\begin{smallmatrix}1 & 0\\ 0 & 1\end{smallmatrix}\right)} & \mathbf{Z}/p^2\oplus\mathbf{Z}/p^4
\end{array}$$

and the isomorphism of diagrams with coefficients in $\mathcal{C}$

$$\begin{array}{ccccccc}
\mathbf{Z}/p^3 & \xrightarrow{\left(\begin{smallmatrix}1 & p^2 & -p\end{smallmatrix}\right)} & \mathbf{Z}/p\oplus\mathbf{Z}/p^5\oplus\mathbf{Z}/p^3 & \xrightarrow{\left(\begin{smallmatrix}p & 0\\ 0 & 1\\ 1 & p\end{smallmatrix}\right)} & \mathbf{Z}/p^2\oplus\mathbf{Z}/p^4 & \xrightarrow{\left(\begin{smallmatrix}-p^2\\ p\end{smallmatrix}\right)} & \mathbf{Z}/p^3 \\
\| & & \| & & \wr\downarrow{\scriptstyle\left(\begin{smallmatrix}1-p & 0\\ 1 & 1\end{smallmatrix}\right)} & & \| \\
\mathbf{Z}/p^3 & \xrightarrow{\left(\begin{smallmatrix}1 & p^2 & -p\end{smallmatrix}\right)} & (\mathbf{Z}/p\oplus\mathbf{Z}/p^5)\oplus\mathbf{Z}/p^3 & \xrightarrow{\left(\begin{smallmatrix}p & 0\\ 1 & 1\\ 1 & p\end{smallmatrix}\right)} & \mathbf{Z}/p^2\oplus\mathbf{Z}/p^4 & \xrightarrow{\left(\begin{smallmatrix}-p^2\\ p\end{smallmatrix}\right)} & \mathbf{Z}/p^3
\end{array}$$

show one of the triangles mentioned in loc. cit. to be distinguished in X and in $\tilde{X}$.

The morphism of short exact sequences in $(\mathbf{Z}/p^6)$-mod

$$\begin{array}{ccccc}
\mathbf{Z}/p^2\oplus\mathbf{Z}/p^4 & \xrightarrow[\bullet]{\left(\begin{smallmatrix}1 & 0 & p\\ 0 & p & -1\end{smallmatrix}\right)} & \mathbf{Z}/p\oplus\mathbf{Z}/p^5\oplus\mathbf{Z}/p^3 & \xrightarrow[+]{\left(\begin{smallmatrix}-p^2\\ 1\\ p\end{smallmatrix}\right)} & \mathbf{Z}/p^3 \\
\uparrow{\scriptstyle\left(\begin{smallmatrix}p & p^2\end{smallmatrix}\right)} & & \uparrow{\scriptstyle\left(\begin{smallmatrix}0 & 1 & 0\end{smallmatrix}\right)} & & \| \\
\mathbf{Z}/p^3 & \xrightarrow[\bullet]{p^3} & \mathbf{Z}/p^6 & \xrightarrow[+]{1} & \mathbf{Z}/p^3
\end{array}$$

and the isomorphism of diagrams with coefficients in $\mathcal{C}$

$$\begin{array}{ccccccc}
\mathbf{Z}/p^3 & \xrightarrow{\left(\begin{smallmatrix}p & p^2\end{smallmatrix}\right)} & \mathbf{Z}/p^2\oplus\mathbf{Z}/p^4 & \xrightarrow{\left(\begin{smallmatrix}1 & 0 & p\\ 0 & p & -1\end{smallmatrix}\right)} & (\mathbf{Z}/p\oplus\mathbf{Z}/p^5)\oplus\mathbf{Z}/p^3 & \xrightarrow{\left(\begin{smallmatrix}-p^2\\ 1\\ p\end{smallmatrix}\right)} & \mathbf{Z}/p^3 \\
\| & & \| & & \wr\downarrow{\scriptstyle\left(\begin{smallmatrix}1 & 0 & -p^2\\ 0 & 1 & 1\\ 1 & 0 & 1+p\end{smallmatrix}\right)} & & \| \\
\mathbf{Z}/p^3 & \xrightarrow{\left(\begin{smallmatrix}p & p^2\end{smallmatrix}\right)} & \mathbf{Z}/p^2\oplus\mathbf{Z}/p^4 & \xrightarrow{\left(\begin{smallmatrix}1 & 0 & p\\ -1 & p & -1\end{smallmatrix}\right)} & (\mathbf{Z}/p\oplus\mathbf{Z}/p^5)\oplus\mathbf{Z}/p^3 & \xrightarrow{\left(\begin{smallmatrix}-p^2\\ 1\\ p\end{smallmatrix}\right)} & \mathbf{Z}/p^3
\end{array}$$

show the other of the triangles mentioned in loc. cit. to be distinguished in X and in $\tilde{X}$. □

3. Nonisomorphic periodic n-pretriangles

More specifically, we give an example of two nonisomorphic periodic n-pretriangles whose periodic $(n-1)$-pretriangles are all $(n-1)$-triangles.

---

[1] Strictly speaking, we should reorder summands in the diagrams that follow; cf. §1.1. But then the proof would be more difficult to read.

Note that §2 is a particular case of §3, where we have given a somewhat longer alternative argument independent of [4].

Let $n \geqslant 3$. Let $\mathcal{C} := \underline{(\mathbf{Z}/p^{2n})}\text{-mod}$, and let it be endowed with a shift automorphism as in §1.2 and a $\overline{\text{Heller}}$ triangulation as in §1.5.

3.1. A $(2n-1)$-triangle

Let Y be the $(2n-1)$-triangle in $\mathcal{C}$ displayed as

$$
\begin{array}{c}
& & & & & & & & \mathbf{Z}/p^1 \\
& & & & & & & & \uparrow{\scriptstyle 1} \\
& & & & & & \mathbf{Z}/p^1 \xrightarrow{-p} \mathbf{Z}/p^2 \\
& & & & & & \uparrow{\scriptstyle 1} \qquad \uparrow{\scriptstyle 1} \\
& & & & \mathbf{Z}/p^1 \xrightarrow{p} \mathbf{Z}/p^2 \xrightarrow{-p} \mathbf{Z}/p^3 \\
& & & & \uparrow{\scriptstyle 1} \qquad \uparrow{\scriptstyle 1} \qquad \uparrow{\scriptstyle 1} \\
& & & & \vdots \qquad \vdots \qquad \vdots \\
& & & & \uparrow{\scriptstyle 1} \quad \uparrow{\scriptstyle 1} \quad \uparrow{\scriptstyle 1} \\
& \mathbf{Z}/p^1 \xrightarrow{p} \cdots \xrightarrow{p} \mathbf{Z}/p^{2n-4} \xrightarrow{p} \mathbf{Z}/p^{2n-3} \xrightarrow{-p} \mathbf{Z}/p^{2n-2} \\
& \uparrow{\scriptstyle 1} \qquad \uparrow{\scriptstyle 1} \qquad \uparrow{\scriptstyle 1} \qquad \uparrow{\scriptstyle 1} \\
\mathbf{Z}/p^1 \xrightarrow{p} \mathbf{Z}/p^2 \xrightarrow{p} \cdots \xrightarrow{p} \mathbf{Z}/p^{2n-3} \xrightarrow{p} \mathbf{Z}/p^{2n-2} \xrightarrow{-p} \mathbf{Z}/p^{2n-1} \\
\uparrow{\scriptstyle 1} \qquad \uparrow{\scriptstyle 1} \qquad \uparrow{\scriptstyle 1} \qquad \uparrow{\scriptstyle 1} \\
\mathbf{Z}/p^1 \xrightarrow{p} \mathbf{Z}/p^2 \xrightarrow{p} \mathbf{Z}/p^3 \xrightarrow{p} \cdots \xrightarrow{p} \mathbf{Z}/p^{2n-2} \xrightarrow{p} \mathbf{Z}/p^{2n-1}
\end{array}
$$

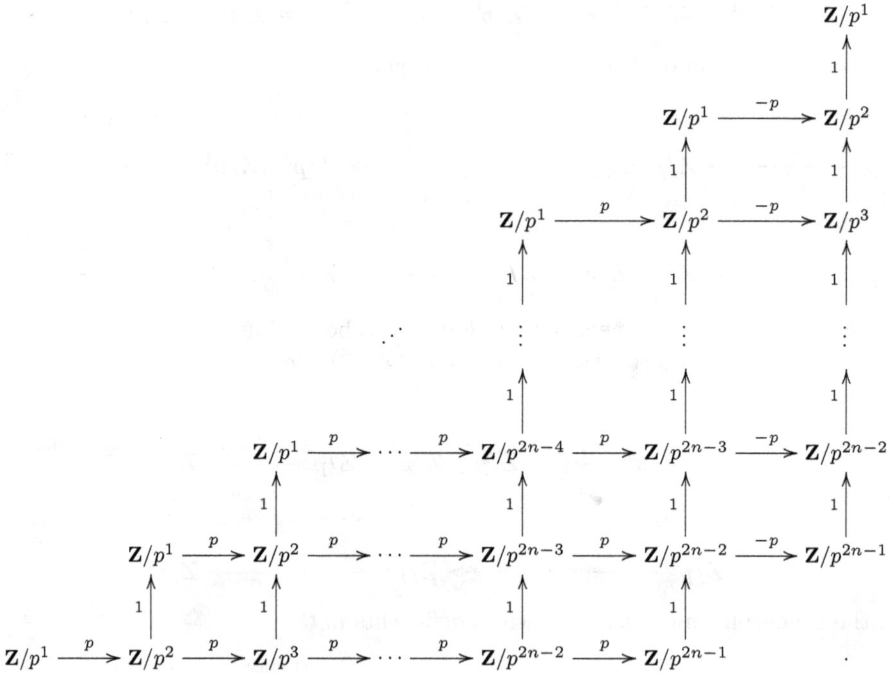

Here we have made use of the convention from §0.5 that we display of Y only its restriction to the subposet $\{\beta/\alpha \in \bar{\Delta}^{\#}_{2n-1} \;:\; 0 \leqslant \alpha < \beta \leqslant 0^{+1},\; \beta/\alpha \neq 0^{+1}/0\}$, which is possible without loss of information. Similarly below.

It arises from a diagram on $\bar{\Delta}^{\Delta\nabla}_n$ with values in $(\mathbf{Z}/p^{2n})$-mod that consists of squares, has entry $\mathbf{Z}/p^{2n}$ at position $0^{+1}/0$, and has the quadrangle

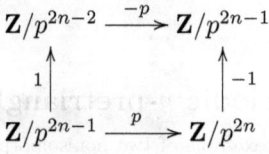

$$
\begin{array}{ccc}
\mathbf{Z}/p^{2n-2} & \xrightarrow{-p} & \mathbf{Z}/p^{2n-1} \\
\uparrow{\scriptstyle 1} & & \uparrow{\scriptstyle -1} \\
\mathbf{Z}/p^{2n-1} & \xrightarrow{p} & \mathbf{Z}/p^{2n}
\end{array}
$$

in its lower right corner. This diagram contains the necessary distinguished short exact sequences with the necessary signs inserted for Y to be in fact a $(2n-1)$-triangle; cf. Remarks 1, 2.

3.2. An n-triangle and a periodic n-pretriangle

We apply the folding operator $\mathfrak{f}_{n-1}$ to the $(2n-1)$-triangle Y obtained in §3.1, yielding the n-triangle $Y\mathfrak{f}_{n-1}$, which we shall display now; cf. [**4**, Lem. 3.4.(2), §1.2.2.3].

$$
\begin{array}{c}
\mathbf{Z}/p^n \\
\uparrow p \\
\mathbf{Z}/p^1\oplus\mathbf{Z}/p^{2n-1} \xrightarrow{\binom{-p^{n-1}}{-1}} \mathbf{Z}/p^n \\
\binom{1\,0}{0\,p}\uparrow \qquad \binom{1\,0}{0\,p}\uparrow \qquad \uparrow p \\
\mathbf{Z}/p^1\oplus\mathbf{Z}/p^{2n-1} \xrightarrow{\binom{p\,0}{0\,1}} \mathbf{Z}/p^2\oplus\mathbf{Z}/p^{2n-2} \xrightarrow{\binom{-p^{n-2}}{-1}} \mathbf{Z}/p^n \\
\binom{1\,0}{0\,p}\uparrow \qquad \binom{1\,0}{0\,p}\uparrow \qquad \uparrow p \\
\vdots \qquad\qquad \vdots \qquad\qquad \vdots \\
\binom{1\,0}{0\,p}\uparrow \qquad \binom{1\,0}{0\,p}\uparrow \qquad \uparrow p \\
\mathbf{Z}/p^1\oplus\mathbf{Z}/p^{2n-1} \xrightarrow{\binom{p\,0}{0\,1}} \cdots \xrightarrow{\binom{p\,0}{0\,1}} \mathbf{Z}/p^{n-3}\oplus\mathbf{Z}/p^{n+3} \xrightarrow{\binom{p\,0}{0\,1}} \mathbf{Z}/p^{n-2}\oplus\mathbf{Z}/p^{n+2} \xrightarrow{\binom{-p^2}{-1}} \mathbf{Z}/p^n \\
\binom{1\,0}{0\,p}\uparrow \qquad \binom{1\,0}{0\,p}\uparrow \qquad \uparrow p \\
\mathbf{Z}/p^1\oplus\mathbf{Z}/p^{2n-1} \xrightarrow{\binom{p\,0}{0\,1}} \mathbf{Z}/p^2\oplus\mathbf{Z}/p^{2n-2} \xrightarrow{\binom{p\,0}{0\,1}} \cdots \xrightarrow{\binom{p\,0}{0\,1}} \mathbf{Z}/p^{n-2}\oplus\mathbf{Z}/p^{n+2} \xrightarrow{\binom{p\,0}{0\,1}} \mathbf{Z}/p^{n-1}\oplus\mathbf{Z}/p^{n+1} \xrightarrow{\binom{-p}{-1}} \mathbf{Z}/p^n \\
\uparrow(1\,-p^{n-1}) \qquad \uparrow(1\,-p^{n-2}) \qquad \uparrow(1\,-p^2) \qquad \uparrow(1\,-p) \\
\mathbf{Z}/p^n \xrightarrow{p} \mathbf{Z}/p^n \xrightarrow{p} \mathbf{Z}/p^n \xrightarrow{p} \cdots \xrightarrow{p} \mathbf{Z}/p^n \xrightarrow{p} \mathbf{Z}/p^n
\end{array}
$$

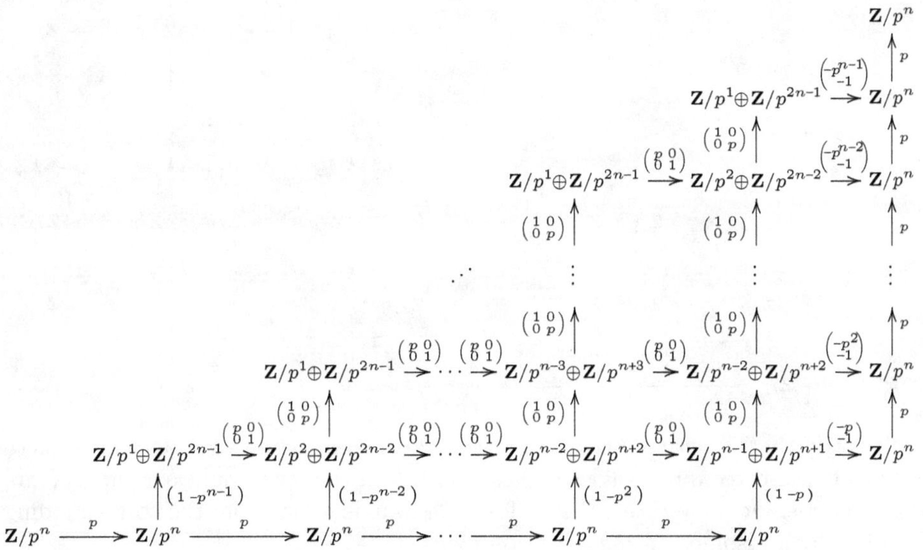

Let X be the n-triangle obtained from $Y\mathfrak{f}_{n-1}$ by isomorphic substitution along $\binom{1\ \ 0}{0\ -1}$ on all terms consisting of two summands; cf. [**4**, Lem. 3.4.(4)]. So X can be displayed as follows.

$$
\begin{array}{c}
\mathbf{Z}/p^n \\
\uparrow p \\
\mathbf{Z}/p^1\oplus\mathbf{Z}/p^{2n-1} \xrightarrow{\binom{-p^{n-1}}{1}} \mathbf{Z}/p^n \\
\binom{1\,0}{0\,p}\uparrow \qquad \uparrow p \\
\mathbf{Z}/p^1\oplus\mathbf{Z}/p^{2n-1} \xrightarrow{\binom{p\,0}{0\,1}} \mathbf{Z}/p^2\oplus\mathbf{Z}/p^{2n-2} \xrightarrow{\binom{-p^{n-2}}{1}} \mathbf{Z}/p^n \\
\binom{1\,0}{0\,p}\uparrow \qquad \binom{1\,0}{0\,p}\uparrow \qquad \uparrow p \\
\vdots \qquad\qquad \vdots \qquad\qquad \vdots \\
\binom{1\,0}{0\,p}\uparrow \qquad \binom{1\,0}{0\,p}\uparrow \qquad \uparrow p \\
\mathbf{Z}/p^1\oplus\mathbf{Z}/p^{2n-1} \xrightarrow{\binom{p\,0}{0\,1}} \cdots \xrightarrow{\binom{p\,0}{0\,1}} \mathbf{Z}/p^{n-3}\oplus\mathbf{Z}/p^{n+3} \xrightarrow{\binom{p\,0}{0\,1}} \mathbf{Z}/p^{n-2}\oplus\mathbf{Z}/p^{n+2} \xrightarrow{\binom{-p^2}{1}} \mathbf{Z}/p^n \\
\binom{1\,0}{0\,p}\uparrow \qquad \binom{1\,0}{0\,p}\uparrow \qquad \uparrow p \\
\mathbf{Z}/p^1\oplus\mathbf{Z}/p^{2n-1} \xrightarrow{\binom{p\,0}{0\,1}} \mathbf{Z}/p^2\oplus\mathbf{Z}/p^{2n-2} \xrightarrow{\binom{p\,0}{0\,1}} \cdots \xrightarrow{\binom{p\,0}{0\,1}} \mathbf{Z}/p^{n-2}\oplus\mathbf{Z}/p^{n+2} \xrightarrow{\binom{p\,0}{0\,1}} \mathbf{Z}/p^{n-1}\oplus\mathbf{Z}/p^{n+1} \xrightarrow{\binom{-p}{1}} \mathbf{Z}/p^n \\
\uparrow(1\,p^{n-1}) \qquad \uparrow(1\,p^{n-2}) \qquad \uparrow(1\,p^2) \qquad \uparrow(1\,p) \\
\mathbf{Z}/p^n \xrightarrow{p} \mathbf{Z}/p^n \xrightarrow{p} \mathbf{Z}/p^n \xrightarrow{p} \cdots \xrightarrow{p} \mathbf{Z}/p^n \xrightarrow{p} \mathbf{Z}/p^n
\end{array}
$$

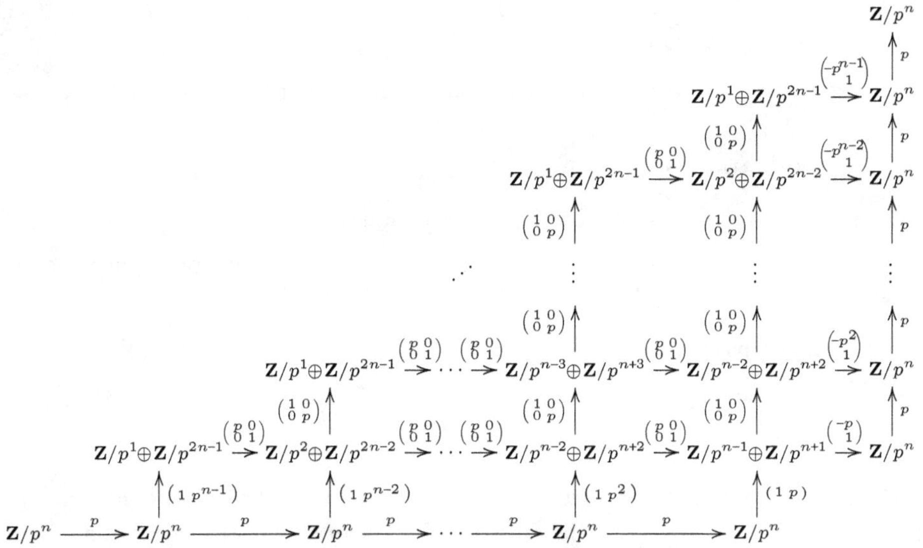

Let $\tilde{X}$ be the periodic n-pretriangle

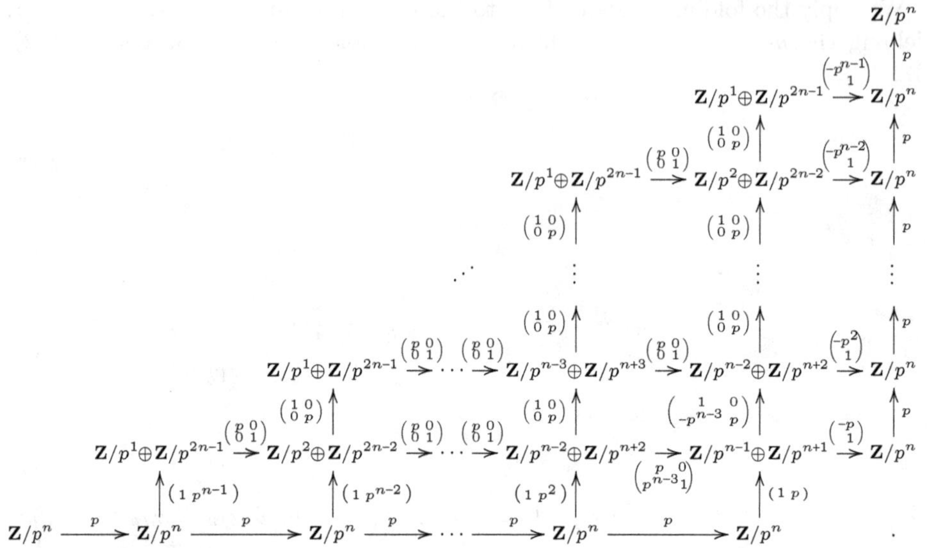

To verify that $\tilde{X}$ actually is an n-pretriangle, a comparison with X reduces us to show that the three quadrangles depicted in full in the lower right corner of $\tilde{X}$ are weak squares. Of these three, the middle quadrangle arises from the corresponding one of X by an isomorphic substitution along

$$\mathbf{Z}/p^{n-1} \oplus \mathbf{Z}/p^{n+1} \xrightarrow[\sim]{\left(\begin{smallmatrix} 1 & 0 \\ p^{n-3} & 1 \end{smallmatrix}\right)} \mathbf{Z}/p^{n-1} \oplus \mathbf{Z}/p^{n+1} ,$$

and thus is a weak square. For the lower one, we may apply [**4**, Lem. A.17] to the diagram

$$(\tilde{X}_{1/0}, \tilde{X}_{n-1/0}, \tilde{X}_{n/0}, \tilde{X}_{0^{+1}/0}, \tilde{X}_{1/1}, \tilde{X}_{n-1/1}, \tilde{X}_{n/1}, \tilde{X}_{0^{+1}/1})$$

and compare with X to show that it is a weak square. For the right hand side one, we may apply [**4**, Lem. A.17] to the diagram

$$(\tilde{X}_{n/0}, \tilde{X}_{n/1}, \tilde{X}_{n/2}, \tilde{X}_{n/n}, \tilde{X}_{0^{+1}/0}, \tilde{X}_{0^{+1}/1}, \tilde{X}_{0^{+1}/2}, \tilde{X}_{0^{+1}/n})$$

and compare with X to show that it is a weak square.

Given $k \in [0, n]$, we let $\bar{\Delta}_n \xleftarrow{\mathrm{d}_k} \bar{\Delta}_{n-1}$ be the periodic monotone map determined by $[0, n-1]\mathrm{d}_k = [0, n] \smallsetminus \{k\}$.

Lemma 6. *Suppose given* $k \in [0, n]$.

(1) *The diagram* $\tilde{X}\mathrm{d}_k^{\#}$ *is an* $(n-1)$*-triangle.*

(2) *We have* $X\mathrm{d}_k^{\#} \simeq \tilde{X}\mathrm{d}_k^{\#}$ *in* $\mathcal{C}^{+,\,\mathrm{periodic}}(\bar{\Delta}_{n-1}^{\#})$.

Proof. Since X is an n-triangle, $X\mathrm{d}_k^{\#}$ is an $(n-1)$-triangle; cf. [**4**, Lem. 3.4.(1)]. Since $X\mathrm{d}_k^{\#}|_{\dot{\Delta}_{n-1}} = \tilde{X}\mathrm{d}_k^{\#}|_{\dot{\Delta}_{n-1}}$, the diagram $\tilde{X}\mathrm{d}_k^{\#}$ is an $(n-1)$-triangle if and only if it is isomorphic to $X\mathrm{d}_k^\#$ in $\mathcal{C}^{+,\,\mathrm{periodic}}(\bar{\Delta}_{n-1}^\#)$; cf. [**4**, Lem. 3.4.(4, 6)]. So assertions (1) and (2) are equivalent. We will prove (2).

When referring to an object on a certain position in the diagram $X\mathrm{d}_k^\#$ resp. $\tilde{X}\mathrm{d}_k^\#$, we shall also mention in parentheses its position as an object in the diagram X resp. $\tilde{X}$ for ease of orientation.

When constructing a morphism in $\mathcal{C}^{+,\,\mathrm{periodic}}(\bar{\Delta}_{n-1}^\#)$, we will give its components on $\{j/i\,:\, 0 \leqslant i \leqslant j \leqslant n-1\} \subseteq \bar{\Delta}_{n-1}^\#$; the remaining components result thereof by periodic repetition.

Case $k \in \{1, n\}$. We have $X\mathrm{d}_1^\# = \tilde{X}\mathrm{d}_1^\#$ and $X\mathrm{d}_n^\# = \tilde{X}\mathrm{d}_n^\#$.

Case $k = 0$. We *claim* that $X\mathrm{d}_0^\#$ is isomorphic to $\tilde{X}\mathrm{d}_0^\#$ in $\mathcal{C}^{+,\,\mathrm{periodic}}(\bar{\Delta}_{n-1}^\#)$. In fact, an isomorphism $X\mathrm{d}_0^\# \xrightarrow{\sim} \tilde{X}\mathrm{d}_0^\#$ is given by

$$\mathbf{Z}/p^{n-1} \oplus \mathbf{Z}/p^{n+1} \xrightarrow[\sim]{\begin{pmatrix} 1 & 0 \\ p^{n-3} & 1 \end{pmatrix}} \mathbf{Z}/p^{n-1} \oplus \mathbf{Z}/p^{n+1}$$

at position $(n-1)/0$ (position $n/1$ in X resp. $\tilde{X}$), and by the identity elsewhere. This proves the *claim.*

Case $k \in [2, n-1]$. We *claim* that $X\mathrm{d}_k^\#$ is isomorphic to $\tilde{X}\mathrm{d}_k^\#$ in $\mathcal{C}^{+,\,\mathrm{periodic}}(\bar{\Delta}_{n-1}^\#)$. In fact, an isomorphism $X\mathrm{d}_k^\# \xrightarrow{\sim} \tilde{X}\mathrm{d}_k^\#$ is given as follows.

At position $j/0$ for $j \in [1, n-1]$ (position $j/0$ if $j \leqslant k-1$ and $(j+1)/0$ if $j \geqslant k$ in X resp. $\tilde{X}$), it is given by the identity on $\mathbf{Z}/p^n$.

At position j/i for $i, j \in [1, k-1]$ such that $i < j$ (position j/i in X resp. $\tilde{X}$), it is given by the identity on $\mathbf{Z}/p^{j-i} \oplus \mathbf{Z}/p^{2n-j+i}$.

At position j/i for $i, j \in [k, n-1]$ such that $i < j$ (position $(j+1)/(i+1)$ in X resp. $\tilde{X}$), it is given by the identity on $\mathbf{Z}/p^{j-i} \oplus \mathbf{Z}/p^{2n-j+i}$.

At position j/i for $i \in [1, k-1]$ and $j \in [k, n-1]$ such that $j/i \neq (n-1)/1$ (position $(j+1)/i$ in X resp. $\tilde{X}$), it is given by

$$\mathbf{Z}/p^{j+1-i} \oplus \mathbf{Z}/p^{2n-j-1+i} \xrightarrow[\sim]{\begin{pmatrix} 1 & 0 \\ -p^{j-1-i} & 1 \end{pmatrix}} \mathbf{Z}/p^{j+1-i} \oplus \mathbf{Z}/p^{2n-j-1+i}$$

At position $(n-1)/1$ (position $n/1$ in X resp. $\tilde{X}$), it is given by the identity on $\mathbf{Z}/p^{n-1} \oplus \mathbf{Z}/p^{n+1}$.

This proves the *claim.* □

Lemma 7. *X is not isomorphic to $\tilde{X}$ in $\mathcal{C}^{+,\,\mathrm{periodic}}(\bar{\Delta}_n^\#)$.*

In particular, $\tilde{X}$ is not an n-triangle; cf. [**4**, Lem. 3.4.(6)].

Proof. We *assume* the contrary. By [**4**, 3.4.(4)], X and $\tilde{X}$ are n-triangles. Thus, by [**4**, 3.4.(6)], there is an isomorphism $X \xrightarrow{\sim} \tilde{X}$ that is identical at $i/0$ and at $0^{+1}/i$ for $i \in [1, n]$. Let

$$\mathbf{Z}/p^{\ell-k} \oplus \mathbf{Z}/p^{2n-\ell+k} \xrightarrow{\begin{pmatrix} a_{\ell/k} & p^{2n-2\ell+2k} b_{\ell/k} \\ c_{\ell/k} & d_{\ell/k} \end{pmatrix}} \mathbf{Z}/p^{\ell-k} \oplus \mathbf{Z}/p^{2n-\ell+k}$$

denote the entry of this isomorphism at ℓ/k, where $1 \leqslant k < \ell \leqslant n$.

If $\ell - k \geqslant 2$, then we have on $\ell/k \longrightarrow \ell/(k+1)$ the commutative quadrangle

$$\begin{array}{ccc}
\mathbf{Z}/p^{\ell-k} \oplus \mathbf{Z}/p^{2n-\ell+k} & \xrightarrow{\left(\begin{smallmatrix}1 & 0\\ 0 & p\end{smallmatrix}\right)} & \mathbf{Z}/p^{\ell-k-1} \oplus \mathbf{Z}/p^{2n-\ell+k+1} \\
{\scriptstyle\left(\begin{smallmatrix}a_{\ell/k} & p^{2n-2\ell+2k} b_{\ell/k}\\ c_{\ell/k} & d_{\ell/k}\end{smallmatrix}\right)}\Big\downarrow & & \Big\downarrow{\scriptstyle\left(\begin{smallmatrix}a_{\ell/(k+1)} & p^{2n-2\ell+2k+2} b_{\ell/(k+1)}\\ c_{\ell/(k+1)} & d_{\ell/(k+1)}\end{smallmatrix}\right)} \\
\mathbf{Z}/p^{\ell-k} \oplus \mathbf{Z}/p^{2n-\ell+k} & \xrightarrow[\left(\begin{smallmatrix}1 & 0\\ 0 & p\end{smallmatrix}\right) - \partial_{\ell/k,\,n/1}\left(\begin{smallmatrix}0 & 0\\ p^{n-3} & 0\end{smallmatrix}\right)]{} & \mathbf{Z}/p^{\ell-k-1} \oplus \mathbf{Z}/p^{2n-\ell+k+1}
\end{array}$$

in $\mathcal{C}$. We read off the congruences

$$c_{\ell/k} - \partial_{\ell/k,\,n/1} p^{n-3} d_{\ell/k} \equiv_{p^{\ell-k-1}} p c_{\ell/(k+1)} \qquad \text{(i)}$$

$$b_{\ell/k} \equiv_{p^{\ell-k-1}} p b_{\ell/(k+1)} \,. \qquad \text{(ii)}$$

From (i) we infer

$$\begin{aligned} c_{n/1} - p^{n-3} d_{n/1} &\equiv_{p^{n-2}} p^1 c_{n/2} \equiv_{p^{n-2}} p^2 c_{n/3} \equiv_{p^{n-2}} \dots \\ &\dots \equiv_{p^{n-2}} p^{n-2} c_{n/(n-1)} \equiv_{p^{n-2}} 0 \,. \end{aligned} \qquad \text{(iii)}$$

From (ii) we infer

$$b_{n/1} \equiv_{p^{n-2}} p b_{n/2} \equiv_{p^{n-2}} p^2 b_{n/3} \equiv_{p^{n-2}} \dots \equiv_{p^{n-2}} p^{n-2} b_{n/(n-1)} \equiv_{p^{n-2}} 0 \,. \qquad \text{(iv)}$$

On $n/1 \longrightarrow 0^{+1}/1$, we have the commutative quadrangle

$$\begin{array}{ccc}
\mathbf{Z}/p^{n-1} \oplus \mathbf{Z}/p^{n+1} & \xrightarrow{\left(\begin{smallmatrix}-p\\ 1\end{smallmatrix}\right)} & \mathbf{Z}/p^{n} \\
{\scriptstyle\left(\begin{smallmatrix}a_{n/1} & p^2 b_{n/1}\\ c_{n/1} & d_{n/1}\end{smallmatrix}\right)}\Big\downarrow & & \Big\| \\
\mathbf{Z}/p^{n-1} \oplus \mathbf{Z}/p^{n+1} & \xrightarrow[\left(\begin{smallmatrix}-p\\ 1\end{smallmatrix}\right)]{} & \mathbf{Z}/p^{n}
\end{array}$$

in $\mathcal{C}$. We read off the congruence

$$-p c_{n/1} + d_{n/1} \equiv_{p^{n-1}} 1 \,. \qquad \text{(v)}$$

On $n/0 \longrightarrow n/1$, we have the commutative quadrangle

$$\begin{array}{ccc}
\mathbf{Z}/p^{n} & \xrightarrow{(1\;p)} & \mathbf{Z}/p^{n-1} \oplus \mathbf{Z}/p^{n+1} \\
\Big\| & & \Big\downarrow{\scriptstyle\left(\begin{smallmatrix}a_{n/1} & p^2 b_{n/1}\\ c_{n/1} & d_{n/1}\end{smallmatrix}\right)} \\
\mathbf{Z}/p^{n} & \xrightarrow[(1\;p)]{} & \mathbf{Z}/p^{n-1} \oplus \mathbf{Z}/p^{n+1}
\end{array}$$

in $\mathcal{C}$. We read off the congruence

$$p b_{n/1} + d_{n/1} \equiv_{p^{n-1}} 1 \,. \qquad \text{(vi)}$$

By (iii) resp. (iv) we conclude from (v) resp. (vi) that

$$(1 - p^{n-2})d_{n/1} \equiv_{p^{n-1}} 1 \qquad \text{(v}'\text{)}$$
$$d_{n/1} \equiv_{p^{n-1}} 1\,. \qquad \text{(vi}'\text{)}$$

Substituting (vi′) into (v′), we obtain

$$1 - p^{n-2} \equiv_{p^{n-1}} 1\,,$$

which is *absurd.* □

A. Appendix: Transport of structure

We use the notation of [**4**, §1, §2].

A.1. Transport of a Heller triangulation

Concerning weakly abelian categories, see e.g. [**4**, §A.6.3]. Recall that an additive functor between weakly abelian categories is called subexact if it induces an exact functor on the Freyd categories; cf. [**4**, §1.2.1.3]. For instance, an equivalence is subexact.

Set-up 8. *Suppose given a Heller triangulated category* $(\mathcal{C}, \mathsf{T}, \vartheta)$*; cf.* [**4**, Def. 1.5]. *Suppose given a weakly abelian category* $\mathcal{C}'$ *and an automorphism* T' *on* $\mathcal{C}'$*, called* shift*; cf.* [**4**, Def. A.26].

Assume given subexact functors $\mathcal{C} \xrightarrow{F} \mathcal{C}'$ *and* $\mathcal{C}' \xrightarrow{G} \mathcal{C}$*, and isotransformations* $1_{\mathcal{C}'} \underset{\sim}{\xrightarrow{\varepsilon}} GF$ *and* $\mathsf{T}'G \underset{\sim}{\xrightarrow{\sigma}} G\mathsf{T}$.

Suppose given $n \geqslant 0$. By abuse of notation, we write $F := \underline{F^+(\bar{\Delta}_n^\#)} : \underline{\mathcal{C}^+(\bar{\Delta}_n^\#)} \longrightarrow \underline{\mathcal{C}'^+(\bar{\Delta}_n^\#)}$ for the functor obtained by pointwise application of F.

Similarly, we write $\varepsilon := \underline{\varepsilon^+(\bar{\Delta}_n^\#)} : \underline{(1_{\mathcal{C}'})^+(\bar{\Delta}_n^\#)} \xrightarrow{\sim} \underline{(GF)^+(\bar{\Delta}_n^\#)}$ for the isotransformation obtained by pointwise application of ε.

More generally speaking, for notational convenience, induced functors of type $\underline{A^+(\bar{\Delta}_n^\#)}$ will often be abbreviated by A, and induced transformations of type $\underline{\alpha^+(\bar{\Delta}_n^\#)}$ will often be abbreviated by α. For instance, given $X \in \operatorname{Ob} \underline{\mathcal{C}^+(\bar{\Delta}_n^\#)}$, we will allow ourselves to write $X\,\mathsf{T} = X\underline{\mathsf{T}^+(\bar{\Delta}_n^\#)}$ $(= [X^{+1}])$.

Given $X' \in \operatorname{Ob} \underline{\mathcal{C}'^+(\bar{\Delta}_n^\#)} = \operatorname{Ob} \mathcal{C}'^+(\bar{\Delta}_n^\#)$, we define the isomorphism $[X']^{+1} \underset{\sim}{\xrightarrow{X'\vartheta'_n}} [X'^{+1}]$ in $\underline{\mathcal{C}'^+(\bar{\Delta}_n^\#)}$ by the following commutative diagram.

$$\begin{array}{ccc}
[X']^{+1} & \underset{\sim}{\xrightarrow{X'\vartheta'_n}} & [X'^{+1}] \\
{\scriptstyle [X'\varepsilon]^{+1}}\downarrow\wr & & \wr\downarrow{\scriptstyle [X'^{+1}]\varepsilon} \\
[X'GF]^{+1} & & [X'^{+1}]GF \\
\| & & \wr\downarrow{\scriptstyle X'\sigma F} \\
[X'G]^{+1}F & \underset{\sim}{\xrightarrow{X'G\vartheta_n F}} & [(X'G)^{+1}]F
\end{array}$$

In other words, we let

$$X'\vartheta'_n := ([X'\varepsilon]^{+1})(X'G\vartheta_n F)(X'\sigma^- F)([X'^{+1}]\varepsilon^-)\,.$$

As a composite of isotransformations, $(X'\vartheta'_n)_{X' \in \operatorname{Ob} \underline{\mathcal{C}'^{+}(\bar{\Delta}_n^{\#})}}$ is an isotransformation. Finally, let $\vartheta' := (\vartheta'_n)_{n \geqslant 0}$.

Lemma 9. *The triple $(\mathcal{C}', \mathsf{T}', \vartheta')$ is a Heller triangulated category.*

Cf. **[4,** Def. 1.5**]**. We will say that ϑ' is *transported from* $(\mathcal{C}, \mathsf{T}, \vartheta)$ *via* F *and* G. Strictly speaking, we should mention ε and σ here as well.

Proof. Suppose given $m,\, n \geqslant 0$, a periodic monotone map $\bar{\Delta}_n \xleftarrow{q} \bar{\Delta}_m$ and an n-pretriangle $X' \in \operatorname{Ob} \underline{\mathcal{C}'^{+}(\bar{\Delta}_n^{\#})}$. We *claim* that $X'\underline{q}^{\#}\vartheta'_m = X'\vartheta'_n\underline{q}^{\#}$. We have

$$\begin{array}{rcl} X'\underline{q}^{\#}\vartheta'_m & = & ([X'\underline{q}^{\#}\varepsilon]^{+1})(X'\underline{q}^{\#}G\vartheta_m F)(X'\underline{q}^{\#}\sigma^{-}F)([X'\underline{q}^{\#+1}]\varepsilon^{-})\,, \\ X'\vartheta'_n\underline{q}^{\#} & = & ([X'\varepsilon]^{+1}\underline{q}^{\#})(X'G\vartheta_n F\underline{q}^{\#})(X'\sigma^{-}F\underline{q}^{\#})([X'^{+1}]\varepsilon^{-}\underline{q}^{\#})\,. \end{array}$$

By respective pointwise definition, we have $[X'\underline{q}^{\#}\varepsilon]^{+1} = [X'\varepsilon]^{+1}\underline{q}^{\#}$ (using that q is periodic), $X'\underline{q}^{\#}\sigma^{-}F = X'\sigma^{-}F\underline{q}^{\#}$ and $[X'\underline{q}^{\#+1}]\varepsilon^{-} = [X'^{+1}]\varepsilon^{-}\underline{q}^{\#}$. Moreover, since $(\mathcal{C}, \mathsf{T}, \vartheta)$ is Heller triangulated, we get

$$X'\underline{q}^{\#}G\vartheta_m F \;=\; X'G\underline{q}^{\#}\vartheta_m F \;=\; X'G\vartheta_n\underline{q}^{\#}F \;=\; X'G\vartheta_n F\underline{q}^{\#}\,.$$

This proves the *claim.*

Suppose given $n \geqslant 0$ and $X' \in \operatorname{Ob} \underline{\mathcal{C}'^{+}(\bar{\Delta}_{2n+1}^{\#})}$. We *claim* that $X'\mathfrak{f}_{\underline{n}}\vartheta'_{n+1} = X'\vartheta'_{2n+1}\mathfrak{f}_{\underline{n}}$. We have

$$\begin{array}{rcl} X'\mathfrak{f}_{\underline{n}}\vartheta'_{n+1} & = & ([X'\mathfrak{f}_{\underline{n}}\varepsilon]^{+1})(X'\mathfrak{f}_{\underline{n}}G\vartheta_{n+1}F)(X'\mathfrak{f}_{\underline{n}}\sigma^{-}F)([X'\mathfrak{f}_{\underline{n}}{}^{+1}]\varepsilon^{-})\,, \\ X'\vartheta'_{2n+1}\mathfrak{f}_{\underline{n}} & = & ([X'\varepsilon]^{+1}\mathfrak{f}_{\underline{n}})(X'G\vartheta_{2n+1}F\mathfrak{f}_{\underline{n}})(X'\sigma^{-}F\mathfrak{f}_{\underline{n}})([X'^{+1}]\varepsilon^{-}\mathfrak{f}_{\underline{n}})\,. \end{array}$$

By additivity of F, G and T' and by respective pointwise definition, we have $[X'\mathfrak{f}_{\underline{n}}\varepsilon]^{+1} = [X'\varepsilon]^{+1}\mathfrak{f}_{\underline{n}}$ (using shiftcompatibility of $\mathfrak{f}_{\underline{n}}$), $X'\mathfrak{f}_{\underline{n}}\sigma^{-}F = X'\sigma^{-}F\mathfrak{f}_{\underline{n}}$ and $[X'\mathfrak{f}_{\underline{n}}{}^{+1}]\varepsilon^{-} = [X'^{+1}]\varepsilon^{-}\mathfrak{f}_{\underline{n}}$. Moreover, since $(\mathcal{C}, \mathsf{T}, \vartheta)$ is Heller triangulated, we get

$$X'\mathfrak{f}_{\underline{n}}G\vartheta_{n+1}F \;=\; X'G\mathfrak{f}_{\underline{n}}\vartheta_{n+1}F \;=\; X'G\vartheta_{2n+1}\mathfrak{f}_{\underline{n}}F \;=\; X'G\vartheta_{2n+1}F\mathfrak{f}_{\underline{n}}\,.$$

This proves the *claim.* □

A.2. Detecting n-triangles

Set-up 10. *Suppose given a Heller triangulated category $(\mathcal{C}, \mathsf{T}, \vartheta)$; cf.* **[4,** Def. 1.5**]**. *Suppose given an additive category $\mathcal{C}'$ and an automorphism T' on $\mathcal{C}'$, called shift.*

Suppose given mutually inverse equivalences $\mathcal{C} \xrightarrow{F} \mathcal{C}'$ and $\mathcal{C}' \xrightarrow{G} \mathcal{C}$. Note that $G \dashv F$, whence there exist isotransformations $1_{\mathcal{C}'} \xrightarrow[\sim]{\varepsilon} GF$ and $FG \xrightarrow[\sim]{\eta} 1_{\mathcal{C}}$ such that both $(F\varepsilon)(\eta F) = 1_F$ and $(\varepsilon G)(G\eta) = 1_G$ hold. We fix such ε and η.

Suppose given an isotransformation $\mathsf{T}'G \xrightarrow[\sim]{\sigma} G\mathsf{T}$.

Note that $\mathcal{C}'$ is weakly abelian, being equivalent to the weakly abelian category $\mathcal{C}$.

Let ϑ' be transported from $(\mathcal{C}, \mathsf{T}, \vartheta)$ via F and G. That is, we let

$$X'\vartheta'_n \;:=\; ([X'\varepsilon]^{+1})(X'G\vartheta_n F)(X'\sigma^{-}F)([X'^{+1}]\varepsilon^{-})$$

for $n \geqslant 0$ and $X' \in \operatorname{Ob} \underline{\mathcal{C}'^{+}(\bar{\Delta}_n^{\#})}$, defining $\vartheta' := (\vartheta'_n)_{n \geqslant 0}$. By Lemma 9, the triple $(\mathcal{C}', \mathsf{T}', \vartheta')$ is a Heller triangulated category. Moreover, let

$$(\mathsf{T}F \xrightarrow[\sim]{\rho} F\mathsf{T}') \;=\; (\mathsf{T}F \xrightarrow[\sim]{\eta^{-}\mathsf{T}F} FG\mathsf{T}F \xrightarrow[\sim]{F\sigma^{-}F} F\mathsf{T}'GF \xrightarrow[\sim]{F\mathsf{T}'\varepsilon^{-}} F\mathsf{T}')\,.$$

Notation 11. Suppose given $n \geqslant 0$. Concerning the full subposet

$$\bar{\Delta}_n^{\scriptscriptstyle\Delta\nabla} \;=\; \{\beta/\alpha \in \Delta_n^{\#} : 0 \leqslant \alpha \leqslant \beta \leqslant 0^{+1}\} \;\subseteq\; \bar{\Delta}_n^{\#}\,,$$

cf. **[4,** §2.5.1**]**.

(1) Suppose given $X' \in \operatorname{Ob} \underline{\mathcal{C}'^{+,\,\text{periodic}}(\bar{\Delta}_n^{\#})}$, where periodic means $[X']^{+1} = [X'^{+1}]$; cf. **[4,** §2.5.3]. Consider the diagram $X'G|_{\bar{\Delta}_n^{\Delta\nabla}}$. Denote by $X'G|^{\sigma}_{\bar{\Delta}_n^{\Delta\nabla}} \in \operatorname{Ob} \underline{\mathcal{C}'^{+}(\bar{\Delta}_n^{\#})}$ the diagram $X'G|_{\bar{\Delta}_n^{\Delta\nabla}}$ with $(X'G)_{0^{+1}/i} = X'_{0^{+1}/i}G = X'_{i/0}\,\mathsf{T}'\,G$ isomorphically replaced via $X'_{i/0}\,\sigma$ by $X'_{i/0}G\,\mathsf{T}$. Denote by $X'G^{\sigma} \in \operatorname{Ob} \underline{\mathcal{C}^{+,\,\text{periodic}}(\bar{\Delta}_n^{\#})}$ its periodic prolongation, characterised by $X'G^{\sigma}|_{\bar{\Delta}_n^{\Delta\nabla}} = X'G|^{\sigma}_{\bar{\Delta}_n^{\Delta\nabla}}$; cf. **[4,** §2.5.3]. Using, for $k \geqslant 0$,

$$\Big([X'G]^{+k} \xrightarrow{\sim} [X'G^{+1}]^{+(k-1)} \xrightarrow{\sim} \cdots \xrightarrow{\sim} [X'G^{+k}]\Big)\Big|_{\bar{\Delta}_n^{\Delta\nabla}},$$

given by

$$X'_{(j/i)^{+k}}G = X'_{j/i}\,\mathsf{T}'^{k}\,G\,\mathsf{T}^{0} \xrightarrow[\sim]{\mathsf{T}'^{k-1}\,\sigma\,\mathsf{T}^{0}} X'_{j/i}\,\mathsf{T}'^{k-1}\,G\,\mathsf{T}^{1} \xrightarrow[\sim]{\mathsf{T}'^{k-2}\,\sigma\,\mathsf{T}^{1}} \cdots \xrightarrow[\sim]{\mathsf{T}'^{0}\,\sigma\,\mathsf{T}^{k-1}} X'_{j/i}\,\mathsf{T}'^{0}\,G\,\mathsf{T}^{k}$$

at j/i for $0 \leqslant i \leqslant j \leqslant 0^{+1}$, and similarly for $k \leqslant 0$, using $\mathsf{T}'^{-}\,G \xrightarrow[\sim]{\mathsf{T}'^{-}\,\sigma^{-}\,\mathsf{T}^{-}} G\,\mathsf{T}^{-}$, we obtain an isomorphism $X'G \xrightarrow[\sim]{\phi} X'G^{\sigma}$ in $\underline{\mathcal{C}^{+}(\bar{\Delta}_n^{\#})}$ such that $\phi_{i/0} = 1_{X'_{i/0}G}$ and $\phi_{0^{+1}/i} = X'_{i/0}\,\sigma$ for $i \in [1,n]$.

(2) Suppose given $X \in \operatorname{Ob} \underline{\mathcal{C}^{+,\,\text{periodic}}(\bar{\Delta}_n^{\#})}$. Consider the diagram $XF|_{\bar{\Delta}_n^{\Delta\nabla}}$. Denote by $XF|^{\rho}_{\bar{\Delta}_n^{\Delta\nabla}} \in \operatorname{Ob} \underline{\mathcal{C}^{+}(\bar{\Delta}_n^{\#})}$ the diagram $XF|_{\bar{\Delta}_n^{\Delta\nabla}}$ with $(XF)_{0^{+1}/i} = X_{0^{+1}/i}F = X_{i/0}\,\mathsf{T}\,F$ isomorphically replaced via $X_{i/0}\,\rho$ by $X_{i/0}F\,\mathsf{T}'$. Denote by $XF^{\rho} \in \operatorname{Ob} \underline{\mathcal{C}'^{+,\,\text{periodic}}(\bar{\Delta}_n^{\#})}$ its periodic prolongation, characterised by $XF^{\rho}|_{\bar{\Delta}_n^{\Delta\nabla}} = XF|^{\rho}_{\bar{\Delta}_n^{\Delta\nabla}}$; cf. **[4,** §2.5.3].

Similarly as in (1), we have an isomorphism $XF \xrightarrow[\sim]{\psi} XF^{\rho}$ in $\underline{\mathcal{C}'^{+}(\bar{\Delta}_n^{\#})}$ such that $\psi_{i/0} = 1_{X_{i/0}F}$ and $\psi_{0^{+1}/i} = X_{i/0}\,\rho$ for $i \in [1,n]$.

Lemma 12. *Suppose given $n \geqslant 0$.*

(1) *Suppose given $X' \in \operatorname{Ob} \underline{\mathcal{C}'^{+,\,\text{periodic}}(\bar{\Delta}_n^{\#})}$. Then X' is an n-triangle if and only if $X'G^{\sigma}$ is an n-triangle.*

(2) *Suppose given $X \in \operatorname{Ob} \underline{\mathcal{C}^{+,\,\text{periodic}}(\bar{\Delta}_n^{\#})}$. Then X is an n-triangle if and only if XF^{ρ} is an n-triangle.*

Cf. **[4,** Def. 1.5.(ii.2)].

Proof. Ad (1). Since ϑ_n is a transformation, there exists a commutative quadrangle

$$\begin{array}{ccc}
[X'G^{\sigma}]^{+1} & \xrightarrow[\sim]{X'G^{\sigma}\vartheta_n} & [(X'G^{\sigma})^{+1}] \\
\wr\uparrow [\phi]^{+1} & & \wr\uparrow [\phi^{+1}] \\
[X'G]^{+1} & \xrightarrow[\sim]{X'G\vartheta_n} & [(X'G)^{+1}]
\end{array}$$

in $\underline{\mathcal{C}^{+}(\bar{\Delta}_n^{\#})}$. Therefore, $X'G^{\sigma}$ is an n-triangle if and only if

$$[\phi]^{+1} = (X'G\vartheta_n)[\phi^{+1}]\,.$$

By **[4,** Prop. 2.6], this equation is equivalent to $[\phi]^{+1}|_{\dot{\Delta}_n} = (X'G\vartheta_n)|_{\dot{\Delta}_n}[\phi^{+1}]|_{\dot{\Delta}_n}$; cf. **[4,** §2.1.1]; in other words, to

$$X'\sigma|_{\dot{\Delta}_n} = X'G\vartheta_n|_{\dot{\Delta}_n}$$

as morphisms from $X'\,\mathsf{T}'\,G|_{\dot{\Delta}_n}$ to $X'G\,\mathsf{T}|_{\dot{\Delta}_n}$ in $\underline{\mathcal{C}(\dot{\Delta}_n)}$. This, in turn, is equivalent to $X'\sigma = X'G\vartheta_n$ as morphisms from $X'\,\mathsf{T}'\,G$ to $X'G\,\mathsf{T}$ in $\underline{\mathcal{C}^{+}(\bar{\Delta}_n^{\#})}$ by **[4,** Prop. 2.6].

Now X' being an n-triangle is equivalent to $X'\vartheta'_n = 1$; i.e. to

$$([X'\varepsilon]^{+1})(X'G\vartheta_n F)(X'\sigma^{-}F)([X'^{+1}]\varepsilon^{-}) = 1\,.$$

Since $[X']^{+1} = [X'^{+1}]$, we have $[X'\varepsilon]^{+1} = [X']^{+1}\varepsilon = [X'^{+1}]\varepsilon$, whence this equation is equivalent to $(X'G\vartheta_n F)(X'\sigma^- F) = 1$. Since F is an equivalence, this amounts to $(X'G\vartheta_n)(X'\sigma^-) = 1$, as was to be shown.

Ad (2). Since ϑ'_n is a transformation, there exists a commutative quadrangle

$$\begin{array}{ccc} [XF^\rho]^{+1} & \xrightarrow[\sim]{XF^\rho\vartheta'_n} & [(XF^\rho)^{+1}] \\ \uparrow{\scriptstyle\wr}\,[\psi]^{+1} & & \uparrow{\scriptstyle\wr}\,[\psi^{+1}] \\ [XF]^{+1} & \xrightarrow[\sim]{XF\vartheta'_n} & [(XF)^{+1}] \end{array}$$

in $\underline{\mathcal{C}'^{+}(\bar{\Delta}_n^{\#})}$. Therefore, XF^ρ is an n-triangle if and only if $[\psi]^{+1} = (XF\vartheta'_n)[\psi^{+1}]$. By [**4**, Prop. 2.6], this equation is equivalent to $[\psi]^{+1}|_{\dot{\Delta}_n} = (XF\vartheta'_n)|_{\dot{\Delta}_n}[\psi^{+1}]|_{\dot{\Delta}_n}$; in other words, to

$$X\rho|_{\dot{\Delta}_n} = XF\vartheta'_n|_{\dot{\Delta}_n}$$

as morphisms from $X\,\mathsf{T}\,F|_{\dot{\Delta}_n}$ to $XF\,\mathsf{T}'\,|_{\dot{\Delta}_n}$ in $\underline{\mathcal{C}'(\dot{\Delta}_n)}$. This, in turn, is equivalent to $X\rho = XF\vartheta'_n$ as morphisms from $X\,\mathsf{T}\,F$ to $XF\,\mathsf{T}'$ in $\underline{\mathcal{C}'^{+}(\bar{\Delta}_n^{\#})}$ by [**4**, Prop. 2.6]. Which amounts to

$$(X\eta^-\,\mathsf{T}\,F)(XF\sigma^-F)(XF\,\mathsf{T}'\,\varepsilon^-) = ([XF\varepsilon]^{+1})(XFG\vartheta_n F)(XF\sigma^- F)([XF^{+1}]\varepsilon^-)\,;$$

i.e. to

$$X\eta^-\,\mathsf{T}\,F = ([X\eta^- F]^{+1})(XFG\vartheta_n F)\,.$$

Since $[X\eta^- F]^{+1} = [X\eta^-]^{+1}F$ and since ϑ_n is a transformation, the right hand side equals $(X\vartheta_n F)([(X\eta^-)^{+1}]F)$, and therefore we can continue the string of equivalent assertions with

$$X\eta^-\,\mathsf{T}\,F = (X\vartheta_n F)([(X\eta^-)^{+1}]F)\,;$$

i.e. with $X\vartheta_n F = 1$; i.e. with $X\vartheta_n = 1$; i.e. with X being an n-triangle. □

Definition 13. A Heller triangulated category $(\mathcal{C}, \mathsf{T}, \vartheta)$ is said to be *closed* if every morphism $X \xrightarrow{f} Y$ therein can be completed to a 2-triangle; i.e. if for all morphisms $X \xrightarrow{f} Y$ in $\mathcal{C}$, there exists $U \in \operatorname{Ob} \mathcal{C}^{+,\vartheta=1}(\bar{\Delta}_2^{\#})$ with $(X \xrightarrow{f} Y) = (U_{1/0} \xrightarrow{u} U_{2/0})$. If this is the case, also the Heller triangulation ϑ is called *closed*.

For instance, by [**4**, Prop. 3.6], a Heller triangulated category whose idempotents split is closed. Recall that ϑ' is transported from $(\mathcal{C}, \mathsf{T}, \vartheta)$ via F and G.

Lemma 14. *If $(\mathcal{C}, \mathsf{T}, \vartheta)$ is a closed Heller triangulated category, then $(\mathcal{C}', \mathsf{T}', \vartheta')$ is a closed Heller triangulated category.*

Proof. By Lemma 9, it remains to prove closedness of $(\mathcal{C}', \mathsf{T}', \vartheta')$. Suppose given $X' \xrightarrow{u'} Y'$ in $\mathcal{C}'$. We have to prove that it can be prolonged to a 2-triangle. Using closedness of $(\mathcal{C}, \mathsf{T}, \vartheta)$, we find a 2-triangle

$$X'G \xrightarrow{u'G} Y'G \xrightarrow{v} Z \xrightarrow{w} X'G\,\mathsf{T}\,.$$

We *claim* that

$$M' := (X' \xrightarrow{u'} Y' \xrightarrow{(Y'\varepsilon)(vF)} ZF \xrightarrow{(wF)(X'\sigma^- F)(X'\,\mathsf{T}'\,\varepsilon^-)} X'\,\mathsf{T}')$$

is a 2-triangle in $\mathcal{C}'$. By Lemma 12.(1), it suffices to show that $M'G^\sigma$ is a 2-triangle in $\mathcal{C}$. Consider the periodic isomorphism with upper row $M'G^\sigma$ and lower row a 2-triangle

$$\begin{array}{ccccccc}
X'G & \xrightarrow{u'G} & Y'G & \xrightarrow{(Y'\varepsilon G)(vFG)} & ZFG & \xrightarrow{(wFG)(X'\sigma^- FG)(X'\,\mathsf{T}'\,\varepsilon^- G)(X'\sigma)} & X'G\mathsf{T} \\
\| & & \| & & \wr\uparrow Z\eta^- & & \| \\
X'G & \xrightarrow{u'G} & Y'G & \xrightarrow{v} & Z & \xrightarrow{w} & X'G\mathsf{T}\,.
\end{array}$$

In fact, we have $(Y'\varepsilon G)(vFG) = (Y'G\eta^-)(vFG) = v(Z\eta^-)$ and

$$\begin{aligned}
(Z\eta^-)(wFG)(X'\sigma^- FG)(X'\,\mathsf{T}'\,\varepsilon^- G)(X'\sigma) &= w(X'G\,\mathsf{T}\,\eta^-)(X'\sigma^- FG)(X'\,\mathsf{T}'\,\varepsilon^- G)(X'\sigma) \\
&= w(X'\sigma^-)(X'\,\mathsf{T}'\,G\eta^-)(X'\,\mathsf{T}'\,\varepsilon^- G)(X'\sigma) \\
&= w(X'\sigma^-)(X'\sigma) \\
&= w\,.
\end{aligned}$$

This shows that $M'G^\sigma$ is a 2-triangle; cf. **[4,** Lem. 3.4.(4)]. This proves the *claim*. □

Remark 15. *Suppose given* $n \geqslant 0$.

(1) *Given* $X \in \operatorname{Ob} \mathcal{C}^+(\bar{\Delta}_n^\#)$, *we have* $(X\vartheta_n F)(X\rho) = XF\vartheta'_n$ *in* $\underline{\mathcal{C}'^+(\bar{\Delta}_n^\#)}$.
(2) *Given* $X' \in \operatorname{Ob} \mathcal{C}'^+(\bar{\Delta}_n^\#)$, *we have* $(X'\vartheta'_n G)(X'\sigma) = X'G\vartheta_n$ *in* $\underline{\mathcal{C}^+(\bar{\Delta}_n^\#)}$.

Proof. Ad (1). We have

$$\begin{aligned}
(X\vartheta_n F)(X\rho) &= (X\vartheta_n F)(X\eta^-\,\mathsf{T}\,F)(XF\sigma^- F)(XF\,\mathsf{T}'\,\varepsilon^-) \\
&= (X\vartheta_n F)([(X\eta^-)^{+1}]F)(XF\sigma^- F)([XF^{+1}]\varepsilon^-) \\
&= ([X\eta^-]^{+1}F)(XFG\vartheta_n F)(XF\sigma^- F)([XF^{+1}]\varepsilon^-) \\
&= ([X\eta^- F]^{+1})(XFG\vartheta_n F)(XF\sigma^- F)([XF^{+1}]\varepsilon^-) \\
&= ([XF\varepsilon]^{+1})(XFG\vartheta_n F)(XF\sigma^- F)([XF^{+1}]\varepsilon^-) \\
&= XF\vartheta'_n\,.
\end{aligned}$$

Ad (2). We have

$$\begin{aligned}
(X'\vartheta'_n G)(X'\sigma) &= ([X'\varepsilon]^{+1}G)(X'G\vartheta_n FG)(X'\sigma^- FG)([X'^{+1}]\varepsilon^- G)(X'\sigma) \\
&= ([X'\varepsilon]^{+1}G)(X'G\vartheta_n FG)(X'\sigma^- FG)([X'^{+1}]G\eta)(X'\sigma) \\
&= ([X'\varepsilon]^{+1}G)(X'G\vartheta_n FG)([X'G^{+1}]\eta)(X'\sigma^-)(X'\sigma) \\
&= ([X'\varepsilon G]^{+1})(X'G\vartheta_n FG)([X'G^{+1}]\eta) \\
&= ([X'G\eta^-]^{+1})(X'G\vartheta_n FG)([X'G^{+1}]\eta) \\
&= ([X'G]^{+1}\eta^-)(X'G\vartheta_n FG)([X'G^{+1}]\eta) \\
&= (X'G\vartheta_n)([X'G^{+1}]\eta^-)([X'G^{+1}]\eta) \\
&= X'G\vartheta_n\,.
\end{aligned}$$

□

A.3. Some lemmata

Let $\mathcal{E}$ be a Frobenius category, let $\mathcal{B} \subseteq \mathcal{E}$ be its full subcategory of bijective objects; cf. e.g. **[4,** Def. A.5]. We use the notations and conventions of **[4,** §A.2.3], in particular those of **[4,** Ex. A.6.(2)].

Let $n \geqslant 0$. Let $E \subseteq \bar{\Delta}_n^\#$ be a convex full subposet, i.e. whenever $\xi, \zeta \in E$ and $\lambda \in \bar{\Delta}_n^\#$ such that $\xi \leqslant \lambda \leqslant \zeta$, then $\lambda \in E$; cf. **[4,** §2.2.2.1]. For instance, $\Delta_n^{\Delta\nabla} \subseteq \bar{\Delta}_n^\#$ is such a convex full subposet; cf. Notation 11. A *pure square* in $\mathcal{E}$ is a commutative quadrangle (A, B, C, D) with pure short exact diagonal sequence $(A, B \oplus C, D)$; cf. **[4,** §A.4].

Denote by $\mathcal{E}^{\square}(E) \subseteq \mathcal{E}(E)$ the full subcategory determined by

$$\operatorname{Ob}\mathcal{E}^{\square}(E) := \left\{ X \in \operatorname{Ob}\mathcal{E}(E) \quad : \quad \begin{array}{ll} 1) & X_{\alpha/\alpha} \text{ is in } \operatorname{Ob}\mathcal{B} \text{ for all } \alpha \in \bar{\Delta}_n \\ & \text{such that } \alpha/\alpha \in E, \text{ and} \\ & X_{\alpha^{+1}/\alpha} \text{ is in } \operatorname{Ob}\mathcal{B} \text{ for all } \alpha \in \bar{\Delta}_n \\ & \text{such that } \alpha^{+1}/\alpha \in E. \\ 2) & \text{For all } \delta^{-1} \leqslant \alpha \leqslant \beta \leqslant \gamma \leqslant \delta \leqslant \alpha^{+1} \text{ in } \bar{\Delta}_n \\ & \text{such that } \gamma/\alpha,\ \gamma/\beta,\ \delta/\alpha \text{ and } \delta/\beta \text{ are in } E, \\ & \text{the quadrangle} \\ & \begin{array}{ccc} X_{\gamma/\beta} & \xrightarrow{x} & X_{\delta/\beta} \\ {\scriptstyle x}\uparrow & \square & \uparrow{\scriptstyle x} \\ X_{\gamma/\alpha} & \xrightarrow[x]{} & X_{\delta/\alpha} \end{array} \\ & \text{is a pure square.} \end{array} \right\} .$$

A particular case of this definition has been considered in **[4,** §4.1].

A.3.1. Cleaning the diagonal

Lemma 16. *Suppose given $X \in \operatorname{Ob}\mathcal{E}^{\square}(\bar{\Delta}_n^{\Delta\nabla})$. Suppose given $\beta \in \bar{\Delta}_n$ such that $0 \leqslant \beta \leqslant 0^{+1}$. There exists $\tilde{X} \in \operatorname{Ob}\mathcal{E}^{\square}(\bar{\Delta}_n^{\Delta\nabla})$ such that the following conditions* (1a, 1b, 2) *hold.*

(1a) *We have $\tilde{X}_{\alpha/\alpha} = X_{\alpha/\alpha}$ for $0 \leqslant \alpha \leqslant 0^{+1}$ such that $\alpha \neq \beta$.*
(1b) *We have $\tilde{X}_{\beta/\beta} = 0$.*
(2) *There exists an isomorphism $\tilde{X} \xrightarrow{\sim} X$ in $\underline{\mathcal{E}}(\bar{\Delta}_n^{\Delta\nabla})$.*

Proof. Pars pro toto, we consider the case $n = 4$ and $\beta = 2$. Given $X \in \operatorname{Ob}\mathcal{E}^{\square}(\bar{\Delta}_n^{\Delta\nabla})$, i.e. given

$$\begin{array}{ccccccccccc}
 & & & & & & & & & & X_{0^{+1}/0^{+1}} \\
 & & & & & & & & & & \uparrow{\scriptstyle x} \\
 & & & & & & & & X_{4/4} & \xrightarrow{x} & X_{0^{+1}/4} \\
 & & & & & & & & {\scriptstyle x}\uparrow & \square & \uparrow{\scriptstyle x} \\
 & & & & & & X_{3/3} & \xrightarrow{x} & X_{4/3} & \xrightarrow{x} & X_{0^{+1}/3} \\
 & & & & & & {\scriptstyle x}\uparrow & \square & \uparrow{\scriptstyle x} & \square & \uparrow{\scriptstyle x} \\
 & & & & X_{2/2} & \xrightarrow{x} & X_{3/2} & \xrightarrow{x} & X_{4/2} & \xrightarrow{x} & X_{0^{+1}/2} \\
 & & & & {\scriptstyle x}\uparrow & \square & \uparrow{\scriptstyle x} & \square & \uparrow{\scriptstyle x} & \square & \uparrow{\scriptstyle x} \\
 & & X_{1/1} & \xrightarrow{x} & X_{2/1} & \xrightarrow{x} & X_{3/1} & \xrightarrow{x} & X_{4/1} & \xrightarrow{x} & X_{0^{+1}/1} \\
 & & {\scriptstyle x}\uparrow & \square & \uparrow{\scriptstyle x} & \square & \uparrow{\scriptstyle x} & \square & \uparrow{\scriptstyle x} & \square & \uparrow{\scriptstyle x} \\
X_{0/0} & \xrightarrow{x} & X_{1/0} & \xrightarrow{x} & X_{2/0} & \xrightarrow{x} & X_{3/0} & \xrightarrow{x} & X_{4/0} & \xrightarrow{x} & X_{0^{+1}/0} \,,
\end{array}$$

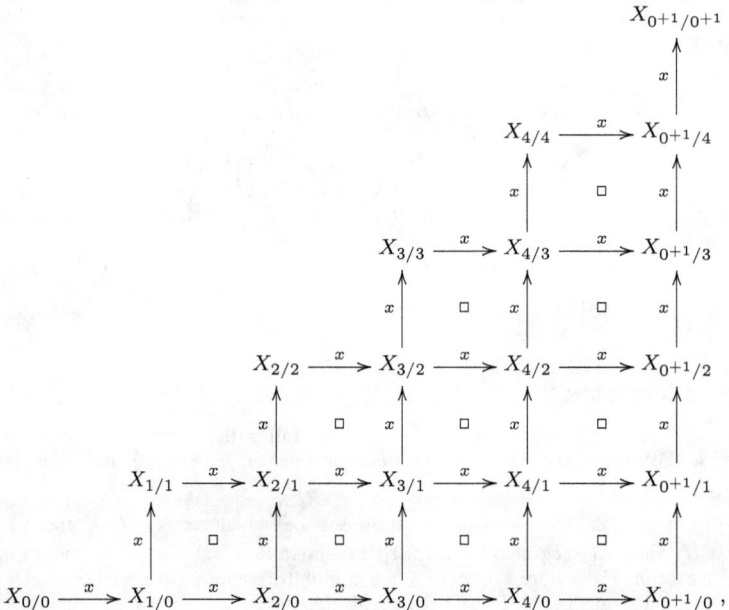

we define the diagram $\tilde{X}$ to be

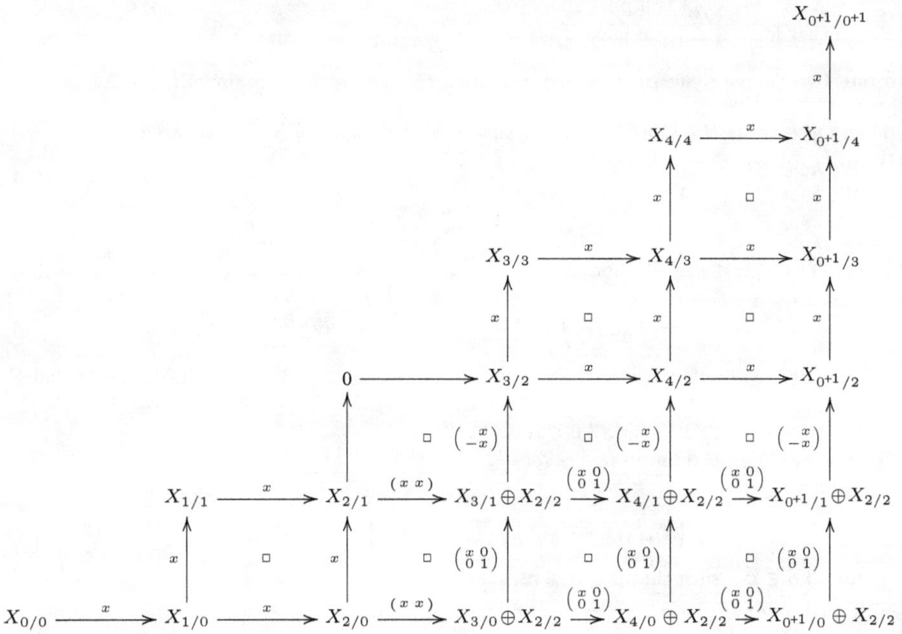

Using the Gabriel-Quillen-Laumon embedding theorem, we see that $\tilde{X}$ is actually an object of $\mathcal{E}^{\Box}(\bar{\Delta}_n^{\Delta\nabla})$; cf. [**4**, §A.2.2; Lem. A.11].

Since $X_{2/2}$ is bijective, inserting the zero morphism on all copies of $X_{2/2}$ and the identity on all other summands yields an isomorphism $\tilde{X} \xrightarrow{\sim} X$ in $\underline{\mathcal{E}}(\bar{\Delta}_n^{\Delta\nabla})$. □

Lemma 17. *Suppose given* $X \in \operatorname{Ob} \mathcal{E}^{\Box}(\bar{\Delta}_n^{\Delta\nabla})$.

There exists $X' \in \operatorname{Ob} \mathcal{E}^{\Box}(\bar{\Delta}_n^{\Delta\nabla})$ *such that the following conditions* (1, 2) *hold.*

(1) *We have* $X'_{\alpha/\alpha} = 0$ *for all* $\alpha \in \bar{\Delta}_n$ *such that* $0 \leqslant \alpha \leqslant 0^{+1}$.

(2) *There exists an isomorphism* $X' \xrightarrow{\sim} X$ *in* $\underline{\mathcal{E}}(\bar{\Delta}_n^{\Delta\nabla})$.

Proof. This follows by application of Lemma 16 consecutively for $\beta = 0, \beta = 1, \ldots, \beta = 0^{+1}$. □

A.3.2. Horseshoe lemma

Recall that $\mathcal{B}^{\mathrm{ac}}$ denotes the category of purely acyclic complexes with entries in $\mathcal{B}$, i.e. of complexes with entries in $\mathcal{B}$ that decompose into pure short exact sequences in $\mathcal{E}$; cf. [**4**, §A.2.3].

Suppose given $Y \in \operatorname{Ob}\mathcal{E}$. An object B of $\mathcal{B}^{\mathrm{ac}}$ is called a *(both-sided) bijective resolution* of Y if Y is isomorphic to $\operatorname{Im}(B^0 \longrightarrow B^1)$. Note that a bijective resolution of a bijective object is split acyclic.

We have a full and dense functor[2]

$$\begin{array}{ccc} \mathcal{B}^{\mathrm{ac}} & \xrightarrow{\hat{F}} & \mathcal{E} \\ B & \longmapsto & \operatorname{Im}(B^0 \longrightarrow B^1) \end{array}$$

We make the additional convention that if the image factorisation of a morphism d is chosen to be $d = \bar{d}\dot{d}$, then we choose the image factorisation $-d = \bar{d}(-\dot{d})$ over the same image object.

Pointwise application yields a functor $\mathcal{B}^{\mathrm{ac}}(\bar{\Delta}_n^{\Delta\nabla}) \xrightarrow{\hat{F}} \mathcal{E}(\bar{\Delta}_n^{\Delta\nabla})$, which is an abuse of notation.

---

[2]A functor induced by $\hat{F}$ will play the role of F of Set-up 10; cf. §A.4.2.2 below.

Suppose given $X \in \operatorname{Ob}\mathcal{E}^{\square}(\bar{\Delta}_n^{\Delta\nabla})$ such that $X_{\alpha/\alpha} = 0$ for all $0 \leqslant \alpha \leqslant 0^{+1}$. In particular, $X_{\beta/\alpha} \twoheadrightarrow X_{\gamma/\alpha} \twoheadrightarrow X_{\gamma/\beta}$ is a pure short exact sequence whenever given $0 \leqslant \alpha \leqslant \beta \leqslant \gamma \leqslant 0^{+1}$.

Recall that for $n \in \bar{\Delta}_n$, we have $n + 1 = 0^{+1}$; cf. [**4,** §1.1].

Lemma 18. *Suppose given a bijective resolution $C_{\alpha+1/\alpha}$ of $X_{\alpha+1/\alpha}$ for all $\alpha \in \bar{\Delta}_n$ such that $0 \leqslant \alpha \leqslant n$.*

Then there exists $B \in \operatorname{Ob}(\mathcal{B}^{\mathrm{ac}})^{\square}(\bar{\Delta}_n^{\Delta\nabla})$ such that the following conditions hold.

(1) *We have $B\hat{F} \simeq X$ in $\mathcal{E}^{\square}(\bar{\Delta}_n^{\#})$.*
(2) *We have $B_{\alpha/\alpha} = 0$ for all $0 \leqslant \alpha \leqslant 0^{+1}$.*
(3) *We have $B_{\alpha+1/\alpha} = C_{\alpha+1/\alpha}$ for all $\alpha \in \bar{\Delta}_n$ such that $0 \leqslant \alpha \leqslant n$.*

Proof. For $0 \leqslant \alpha \leqslant n$, we denote

$$\left(C^0_{\alpha+1/\alpha} \xrightarrow{\bar{d}} X_{\alpha+1/\alpha}\right) := \left(C^0_{\alpha+1/\alpha} \longrightarrow C_{\alpha+1/\alpha}\hat{F} \xrightarrow{\sim} X_{\alpha+1/\alpha}\right).$$

By duality and by induction, it suffices to find a morphism $Y \longrightarrow X$ in $\mathcal{E}^{\square}(\bar{\Delta}_n^{\Delta\nabla})$ such that the following conditions hold.

(i) We have $Y_{\beta/\alpha} \in \operatorname{Ob}\mathcal{B}$ for all $0 \leqslant \alpha \leqslant \beta \leqslant 0^{+1}$.
(ii) We have $Y_{\alpha/\alpha} = 0$ for all $0 \leqslant \alpha \leqslant 0^{+1}$.
(iii) We have

$$\left(Y_{\alpha+1/\alpha} \longrightarrow X_{\alpha+1/\alpha}\right) = \left(C^0_{\alpha+1/\alpha} \xrightarrow{\bar{d}} X_{\alpha+1/\alpha}\right)$$

for all $\alpha \in \bar{\Delta}_n$ such that $0 \leqslant \alpha \leqslant n$.

Note that any morphism $Y \longrightarrow X$ fulfilling (i, ii, iii) consists pointwise of pure epimorphisms, and that the kernel of such a morphism $Y \longrightarrow X$ taken in $\mathcal{E}(\bar{\Delta}_n^{\Delta\nabla})$ is in $\operatorname{Ob}\mathcal{E}^{\square}(\bar{\Delta}_n^{\Delta\nabla})$.

To construct $Y \longrightarrow X$, we let

$$Y_{\gamma/\alpha} := \bigoplus_{\beta \in \bar{\Delta}_n,\ \alpha \leqslant \beta < \gamma} C^0_{\beta+1/\beta}$$

for $0 \leqslant \alpha \leqslant \gamma \leqslant 0^{+1}$. For $\gamma/\alpha \leqslant \gamma'/\alpha'$, the diagram morphism $Y_{\gamma/\alpha} \longrightarrow Y_{\gamma'/\alpha'}$ is stipulated to be identical on the summands $C^0_{\beta+1/\beta}$ with $\alpha' \leqslant \beta < \gamma$ and zero elsewhere. This yields $Y \in \operatorname{Ob}\mathcal{E}^{\square}(\bar{\Delta}_n^{\Delta\nabla})$.

Given $0 \leqslant \alpha \leqslant \gamma \leqslant 0^{+1}$, we let $Y_{\gamma/\alpha} \longrightarrow X_{\gamma/\alpha}$ be defined as follows. For $0 \leqslant \beta \leqslant n$, we choose $Y_{\beta+1/\beta} \xrightarrow{e} X_{\beta+1/0}$ such that

$$(Y_{\beta+1/\beta} \xrightarrow{e} X_{\beta+1/0} \xrightarrow{x} X_{\beta+1/\beta}) = (Y_{\beta+1/\beta} \xrightarrow{\bar{d}} X_{\beta+1/\beta}).$$

The component of the morphism

$$(Y_{\gamma/\alpha} \longrightarrow X_{\gamma/\alpha}) = \left(\bigoplus_{\beta \in \bar{\Delta}_n,\ \alpha \leqslant \beta < \gamma} C^0_{\beta+1/\beta} \longrightarrow X_{\gamma/\alpha}\right)$$

at β is defined to be the composite

$$(C^0_{\beta+1/\beta} \longrightarrow X_{\gamma/\alpha}) := (C^0_{\beta+1/\beta} \xrightarrow{e} X_{\beta+1/0} \xrightarrow{x} X_{\gamma/\alpha}).$$

□

Remark 19. If $n = 2$, and if we restrict to $\{1/0,\ 2/0,\ 2/1\} \subseteq \bar{\Delta}_2^{\Delta\nabla}$, Lemma 18 yields the classical horseshoe lemma in its bothsided Frobenius category variant.

A.3.3. Applying $\hat{F}$ to a standard pure short exact sequence

Recall that $\mathcal{B} \subseteq \mathcal{E}$ is the full subcategory of bijective objects in $\mathcal{E}$. Recall that for $X \in \operatorname{Ob}\mathcal{B}^{\mathrm{ac}}$, we have chosen, in a functorial manner, a pure short exact sequence $X \rightarrowtail X\mathsf{I} \twoheadrightarrow X\mathsf{T}$ with a bijective middle term $X\mathsf{I}$, where the letter I stands for "injective"; cf. [**4,** §A.2.3].

Lemma 20. *Suppose given* $X \in \mathrm{Ob}\,\mathcal{B}^{\mathrm{ac}}$. *There exists an isomorphism of pure short exact sequences in* $\mathcal{E}$ *as follows.*

$$
\begin{array}{ccccc}
X\hat{F} & \xrightarrow{\;\;\bullet\;\;} & X|\hat{F} & \xrightarrow{\;\;+\;\;} & X\mathsf{T}\hat{F} \\
\| & & \wr\downarrow & & \| \\
X\hat{F} & \xrightarrow{\;\;\bullet\;\;} & X^1 & \xrightarrow{\;\;+\;\;} & X\mathsf{T}\hat{F}
\end{array}
$$

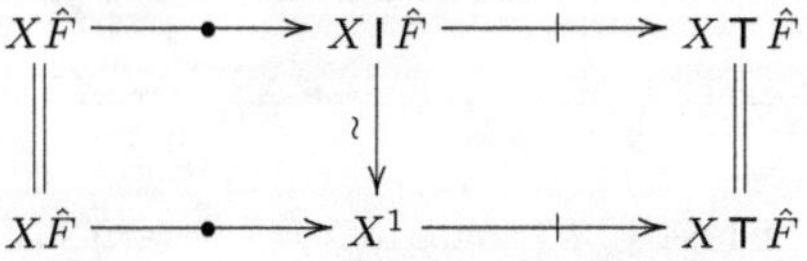

Therein, the upper sequence results from an application of $\hat{F}$ *to the pure short exact sequence* $X \xrightarrow{\bullet} X| \xrightarrow{+} X\mathsf{T}$ *in* $\mathcal{B}^{\mathrm{ac}}$. *The lower sequence is taken from the purely acyclic complex* X.

Proof. Concerning pure short exact sequences in $\mathcal{B}^{\mathrm{ac}}$, cf. [4, Ex. A.6]. Consider the following part of the pure short exact sequence $X \xrightarrow{\bullet} X| \xrightarrow{+} X\mathsf{T}$ in $\mathcal{B}^{\mathrm{ac}}$.

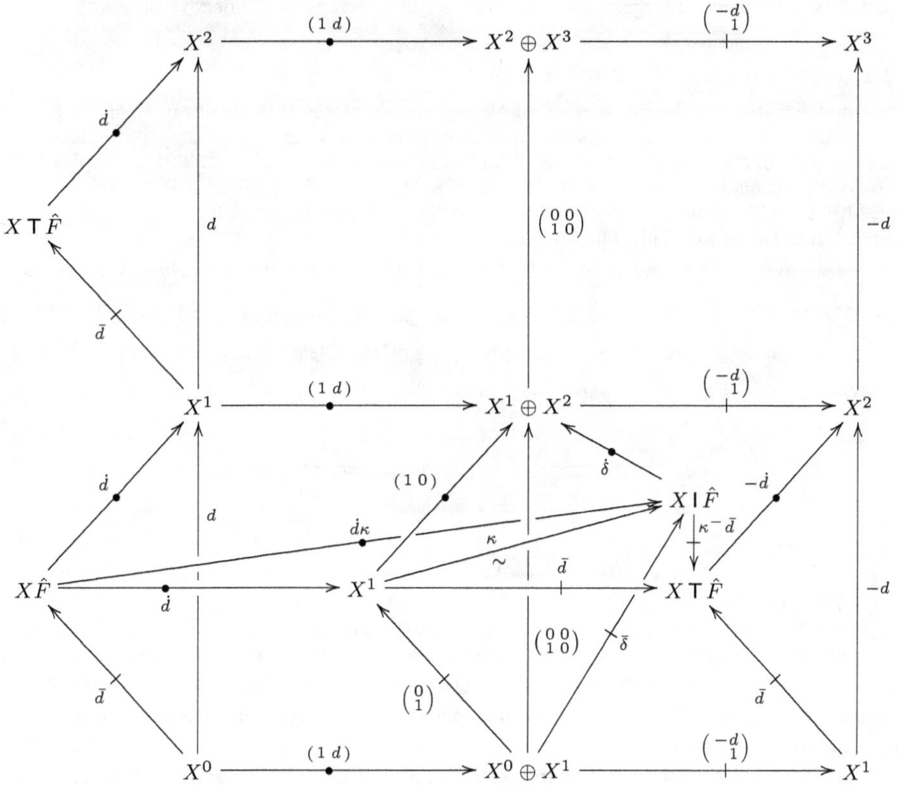

We have added the image factorisations

$$
X^0 \xrightarrow{\;\bar{d}\;+\;} X\hat{F} \xrightarrow{\;d\;\bullet\;} X^1
$$

and

$$
X^0 \oplus X^1 \xrightarrow{\;\bar{\delta}\;+\;} X|\hat{F} \xrightarrow{\;\dot{\delta}\;\bullet\;} X^1 \oplus X^2
$$

of the respective differentials, resulting from an application of $\hat{F}$. Factoring the differential of $X\mathsf{T}$ as

$$
\left(X^1 \xrightarrow{\;-d\;} X^2 \right) = \left(X^1 \xrightarrow{\;\bar{d}\;+\;} X\mathsf{T}\hat{F} \xrightarrow{\;-\dot{d}\;\bullet\;} X^2 \right)
$$

follows the additional convention made above.

Moreover, we have added the image factorisation $X^0 \oplus X^1 \xrightarrow{\binom{0}{1}} X^1 \xrightarrow{(1\,0)} X^1 \oplus X^2$ and, accordingly, the isomorphism $X^1 \xrightarrow[\sim]{\kappa} X \,\mathsf{I}\, \hat{F}$ that satisfies $\kappa\dot{\delta} = (1\,0)$ and $\binom{0}{1}\kappa = \bar{\delta}$.

The horizontal pure short exact sequence $X\hat{F} \xrightarrow{\dot{d}} X^1 \xrightarrow{\bar{d}} X \,\mathsf{T}\, \hat{F}$ lets all four arising parallelograms commute.

Now the sequence $X \to X\,\mathsf{I} \to X\,\mathsf{T}$ maps to $X\hat{F} \xrightarrow{\dot{d}\kappa} X\,\mathsf{I}\,\hat{F} \xrightarrow{\kappa^- \bar{d}} X\,\mathsf{T}\,\hat{F}$, for the commutativities $(\dot{d}\kappa)\dot{\delta} = \dot{d}\,(1\ d)$ and $\bar{\delta}(\kappa^- \bar{d}) = \binom{-d}{1}\,\bar{d}$ hold.

In particular, the sequence $(X \to X\,\mathsf{I} \to X\,\mathsf{T})\hat{F}$ actually is purely short exact. □

A.4. Stable vs. classically stable

Let $\mathcal{E}$ be a Frobenius category, let $\mathcal{B} \subseteq \mathcal{E}$ be its full subcategory of bijective objects.

A.4.1. n-triangles in the stable category

Recall that $\mathcal{B}^{\mathrm{ac}}$ denotes the category of purely acyclic complexes with entries in $\mathcal{B}$; cf. §A.3.2. Let $\mathcal{B}^{\mathrm{sp\,ac}} \subseteq \mathcal{B}^{\mathrm{ac}}$ denote the subcategory of split acyclic complexes. Let $\underline{\mathcal{E}} = \mathcal{B}^{\mathrm{ac}}/\mathcal{B}^{\mathrm{sp\,ac}}$ denote the stable category of $\mathcal{E}$; cf. [**4**, Def. A.7]. Let T be the automorphism on $\underline{\mathcal{E}}$ that shifts a complex to the left by one position, inserting signs; cf. [**4**, Ex. A.6.(1)]. Then $(\underline{\mathcal{E}}, \mathsf{T})$ carries a Heller triangulation ϑ; cf. [**4**, Cor. 4.7]. In fact, we may, and will, choose the tuple of isotransformations $\vartheta = (\vartheta_n)_{n\geqslant 0}$ constructed in the proof of [**4**, Th. 4.6].

Suppose given $n \geqslant 0$ and $X \in \mathrm{Ob}(\mathcal{B}^{\mathrm{ac}})^{\square}(\bar{\Delta}_n^{\#})$; cf. [**4**, §4.1] or §A.3. Now X maps to an object $X \in \mathrm{Ob}\,\underline{\underline{\mathcal{E}}}^{+}(\bar{\Delta}_n^{\#})$; cf. [**4**, Lem. A.29]. Thus we have an isomorphism $[X]^{+1} \xrightarrow{X\vartheta_n} [X^{+1}]$ in $\underline{\underline{\mathcal{E}}}^{+}(\bar{\Delta}_n^{\#})$. By the construction in the proof of [**4**, Th. 4.6], there is a representative $[X]^{+1} \xrightarrow{X\theta} [X^{+1}]$ in $\underline{\mathcal{E}}^{+}(\bar{\Delta}_n^{\#})$ of $X\vartheta_n$ such that in particular, there exists a morphism of pure short exact sequences

$$\begin{array}{ccccc}
X_{i/0} & \xrightarrow{(x\ x)} & X_{i/i} \oplus X_{0+1/0} & \xrightarrow{\binom{x}{-x}} & X_{0+1/i} \\
\| & & \downarrow & & \downarrow {\scriptstyle X\hat{\theta}_{i/0}} \\
X_{i/0} & \longrightarrow & X_{i/0}\,\mathsf{I} & \longrightarrow & X^{+1}_{i/0}
\end{array}$$

for each $i \in [1,n]$; where $X\hat{\theta}_{i/0}$ is a representative in $\mathcal{B}^{\mathrm{ac}}$ for the morphism $X\theta_{i/0}$ in $\underline{\mathcal{E}}$; where the upper pure short exact sequence stems from the diagram X; and where the lower pure short exact sequence is the standard one as in [**4**, Ex. A.6.(1)]. In particular, $X\theta_{i/0}$ is an isomorphism in $\underline{\mathcal{E}}$.

Let $X^{\vartheta} \in \mathrm{Ob}\,\underline{\underline{\mathcal{E}}}^{+,\,\mathrm{periodic}}(\bar{\Delta}_n^{\#})$ be defined as periodic prolongation of the image of the diagram $X|_{\bar{\Delta}_n^{\Delta\nabla}}$ in $\mathrm{Ob}\,\underline{\underline{\mathcal{E}}}^{+}(\bar{\Delta}_n^{\Delta\nabla})$ with $X_{0+1/i}$ isomorphically replaced via $X\theta_{i/0}$ by $X^{+1}_{i/0}$ for all $i \in [1,n]$. For short, the rightmost column of the image of $X|_{\bar{\Delta}_n^{\Delta\nabla}}$ becomes standardised; cf. [**4**, §2.1.3]. Using

$$\Big([X]^{+k} \xrightarrow{\sim} [X^{+1}]^{+(k-1)} \xrightarrow{\sim} \cdots \xrightarrow{\sim} [X^{+k}]\Big)\Big|_{\bar{\Delta}_n^{\Delta\nabla}}$$

for $k \geqslant 0$, and similarly for $k \leqslant 0$, we obtain an isomorphism $X \xrightarrow{\omega} X^{\vartheta}$ in $\underline{\underline{\mathcal{E}}}^{+}(\bar{\Delta}_n^{\#})$ such that $\omega_{i/0} = 1_{X_{i/0}}$ and $\omega_{0+1/i} = X\theta_{i/0}$ for $i \in [1,n]$; cf. Notation 11.(1).

Lemma 21. *Given $n \geqslant 0$ and $X \in \mathrm{Ob}(\mathcal{B}^{\mathrm{ac}})^{\square}(\bar{\Delta}_n^{\#})$, the periodic n-pretriangle X^{ϑ} is an n-triangle.*

The following proof is similar to the proof of Lemma 12.

Proof. We have to show that $X^{\vartheta}\vartheta_n = 1$; cf. [**4**, Def. 1.5.(ii.2)]. Since ϑ_n is a transformation, we have a commutative quadrangle

$$\begin{array}{ccc} [X^\vartheta]^{+1} & \xrightarrow[\sim]{X^\vartheta \vartheta_n} & [(X^\vartheta)^{+1}] \\ \uparrow \wr\, [\omega]^{+1} & & \uparrow \wr\, [\omega^{+1}] \\ [X]^{+1} & \xrightarrow[\sim]{X\vartheta_n} & [X^{+1}] \end{array}$$

in $\underline{\mathcal{E}}^+(\bar{\Delta}_n^\#)$. So we have to show that $(X\vartheta_n)[\omega^{+1}] = [\omega]^{+1}$. By [4, Prop. 2.6], it suffices to show that $(X\vartheta_n)|_{\dot{\Delta}_n}[\omega^{+1}]|_{\dot{\Delta}_n} = [\omega]^{+1}|_{\dot{\Delta}_n}$. Now $[\omega^{+1}]|_{\dot{\Delta}_n} = 1_{[X^{+1}]|_{\dot{\Delta}_n}}$ and, by construction, $(X\vartheta_n)|_{\dot{\Delta}_n} = [\omega]^{+1}|_{\dot{\Delta}_n}$. □

Corollary 22. *The Heller triangulated category $(\underline{\mathcal{E}}, \mathsf{T}, \vartheta)$ is closed.*

Cf. Definition 13.

Proof. We can extend any morphism $X_{1/0} \longrightarrow X_{2/0}$ of $\mathcal{B}^{\mathrm{ac}}$ to an object of $(\mathcal{B}^{\mathrm{ac}})^{\square}(\bar{\Delta}_2^\#)$ by choosing $X_{1/0} \rightarrowtail X_{1/1}$ with $X_{1/1}$ bijective and by choosing $X_{0+1/0} = 0$, then forming pushouts, then choosing $X_{2/1} \rightarrowtail X_{2/2}$ with $X_{2/2}$ bijective, etc. Dually in the other direction. Then we apply Lemma 21. □

A.4.2. The classical stable category under an additional hypothesis

A.4.2.1. The hypothesis

Let $\underline{\mathcal{E}} := \mathcal{E}/\mathcal{B}$ denote the classical stable category of $\mathcal{E}$.

Suppose given a set $\mathcal{D}$ of *distinguished* pure short exact sequences in $\mathcal{E}$ such that the following conditions hold.

(i) The middle term of each distinguished pure short exact sequence is bijective.
(ii) For all $X \in \mathrm{Ob}\,\mathcal{E}$, there exists a unique distinguished pure short exact sequence with kernel term X.
(iii) For all $X \in \mathrm{Ob}\,\mathcal{E}$, there exists a unique distinguished pure short exact sequence with cokernel term X.

A.4.2.2. Consequences

We shall define an endofunctor T' of $\underline{\mathcal{E}}$.

On objects. Given $X \in \mathrm{Ob}\,\underline{\mathcal{E}} = \mathrm{Ob}\,\mathcal{E}$, there exists a unique distinguished pure short exact sequence with kernel term X. Let $X\,\mathsf{T}'$ be the cokernel term of this sequence.

On morphisms. The image under T' of the residue class in $\underline{\mathcal{E}}$ of a morphism $X \xrightarrow{f} Y$ in $\mathcal{E}$ is represented by the morphism $X\,\mathsf{T}' \xrightarrow{g} Y\,\mathsf{T}'$ in $\mathcal{E}$ if there exists a morphism of distinguished pure short exact sequences as follows.

$$\begin{array}{ccccc} X & \rightarrowtail & B & \twoheadrightarrow & X\,\mathsf{T}' \\ \downarrow f & & \downarrow & & \downarrow g \\ Y & \rightarrowtail & C & \twoheadrightarrow & Y\,\mathsf{T}' \end{array}$$

Then T' is an automorphism of $\underline{\mathcal{E}}$; i.e. there exists an inverse T'^{-}, constructed dually, such that $\mathsf{T}'\,\mathsf{T}'^{-} = 1_{\underline{\mathcal{E}}}$ and $\mathsf{T}'^{-}\,\mathsf{T}' = 1_{\underline{\mathcal{E}}}$. As usual, we shall write $X^{+1} := X\,\mathsf{T}'$ for $X \in \mathrm{Ob}\,\underline{\mathcal{E}}$; etc.

The functor

$$\begin{array}{ccc} \underline{\underline{\mathcal{E}}} & \xrightarrow{F} & \underline{\mathcal{E}} \\ B & \longmapsto & \mathrm{Im}(B^0 \longrightarrow B^1) \end{array}$$

induced by $\hat{F}$ is an equivalence; cf. §A.3.2, [4, Lem. A.1]. Splicing purely acyclic complexes from distinguished pure short exact sequences, we obtain an inverse equivalence $\underline{\underline{\mathcal{E}}} \xleftarrow{G} \underline{\mathcal{E}}$. Define

$\mathsf{T}'G \xrightarrow{\sigma} G\mathsf{T}$ at $Y \in \operatorname{Ob}\underline{\mathcal{E}} = \operatorname{Ob}\mathcal{E}$ by letting

$$(Y\sigma^i) := (-1_{(YG)^{i+1}})^i .$$

Note that $Y\sigma F = 1_{Y\,\mathsf{T}'\,GF} = 1_{YG\,\mathsf{T}\,F}$; cf. §A.3.2.

Suppose given $Y \in \operatorname{Ob}\mathcal{E}$. We have a commutative diagram

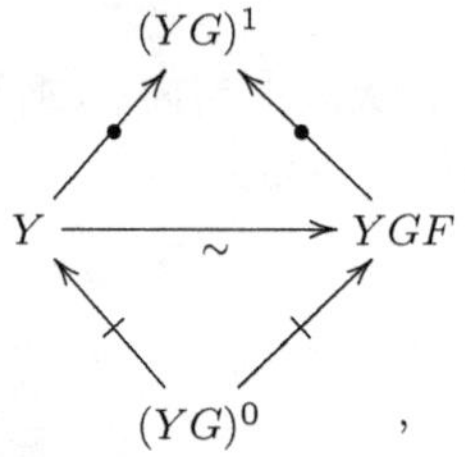

,

consisting of two image factorisations of the differential $(YG)^0 \longrightarrow (YG)^1$ and the induced isomorphism between the images $Y \xrightarrow{\sim} YGF$ that makes the upper and the lower triangle in this diagram commute.

The residue class in $\underline{\mathcal{E}}$ of this induced morphism $Y \xrightarrow{\sim} YGF$ shall be denoted by $Y \xrightarrow[\sim]{Y\varepsilon} YGF$. Letting Y vary, this gives rise to an isotransformation $1_{\underline{\mathcal{E}}} \xrightarrow[\sim]{\varepsilon} GF$. Since ε is a transformation, we have $(Y\varepsilon)(YGF\varepsilon) = (Y\varepsilon)(Y\varepsilon GF)$, whence $GF\varepsilon = \varepsilon GF$. Thus there is an isotransformation $FG \xrightarrow[\sim]{\eta} 1_{\underline{\mathcal{E}}}$ such that $(\varepsilon G)(G\eta) = 1_G$ and $(F\varepsilon)(\eta F) = 1_F$. Namely, for η we may take the inverse image under F of $F\varepsilon^-$.

So we are in the situation of Set-up 10 of §A.2. Define $(\mathsf{T}\,F \xrightarrow{\rho} F\,\mathsf{T}')$ as in §A.2.

Proposition 23.

(1) *By transport from $(\underline{\mathcal{E}}, \mathsf{T}, \vartheta)$ via F and G, we obtain a closed Heller triangulation ϑ' on $(\underline{\mathcal{E}}, \mathsf{T}')$.*

(2) *Suppose given $X' \in \operatorname{Ob}\underline{\mathcal{E}}^{+,\,\text{periodic}}(\bar{\Delta}_n^{\#})$. Then X' is an n-triangle if and only if $X'G^\sigma$ is an n-triangle.*

(3) *Suppose given $X \in \operatorname{Ob}\underline{\mathcal{E}}^{+,\,\text{periodic}}(\bar{\Delta}_n^{\#})$. Then X is an n-triangle if and only if XF^ρ is an n-triangle.*

Proof. Assertion (1) follows by Lemmata 9 and 14; cf. Corollary 22. Assertions (2, 3) follow by Lemma 12. □

Recall that $XF = X\hat{F}$ in $\operatorname{Ob}\underline{\mathcal{E}} = \operatorname{Ob}\mathcal{E}$ for $X \in \operatorname{Ob}\underline{\underline{\mathcal{E}}} = \operatorname{Ob}\mathcal{B}^{\mathrm{ac}}$.

Lemma 24. *Suppose given $X \in \operatorname{Ob}\mathcal{B}^{\mathrm{ac}}$. We have a morphism of pure short exact sequences*

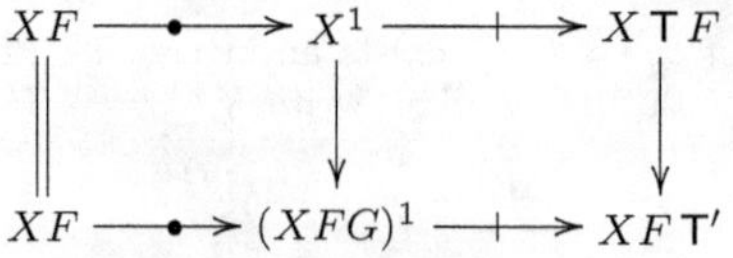

in $\mathcal{E}$ such that its morphism $X\,\mathsf{T}\,F \longrightarrow XF\,\mathsf{T}'$ represents $X\rho$ in $\underline{\mathcal{E}}$. Here, the upper pure short exact sequence is taken from the purely acyclic complex X; the lower pure short exact sequence is distinguished.

Proof. Given $X \in \operatorname{Ob} \mathcal{B}^{\mathrm{ac}}$, we can form a commutative diagram

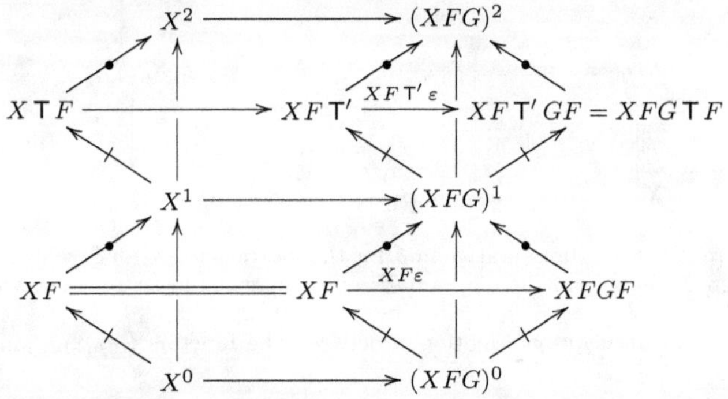

in $\mathcal{E}$. The pure monomorphisms $XF \rightarrowtail (XFG)^1$ and $XF\mathrm{T}' \rightarrowtail (XFG)^2$, and the pure epimorphisms $(XFG)^0 \twoheadrightarrow XF$ and $(XFG)^1 \twoheadrightarrow XF\mathrm{T}'$ appear in distinguished pure short exact sequences. Moreover, by abuse of notation, we have written $XF\varepsilon$ resp. $XF\mathrm{T}'\varepsilon$ for representatives in $\mathcal{E}$ of the respective morphisms in $\underline{\underline{\mathcal{E}}}$.

The partially displayed morphism of complexes $X \to XFG$ represents $X \xrightarrow{X\eta^-} XFG$ in $\underline{\underline{\mathcal{E}}}$, for F maps the morphism represented by $X \longrightarrow XFG$ to $XF\varepsilon = X\eta^- F$.

Therefore, the composite morphism

$$\Big(X\mathrm{T}F \longrightarrow XF\mathrm{T}' \xrightarrow{\quad XF\mathrm{T}'\varepsilon \quad} XF\mathrm{T}'GF = XFG\mathrm{T}F\Big)$$

from this diagram represents

$$\Big(X \xrightarrow{X\eta^-} XFG\Big)\mathrm{T}F\,;$$

note that there are no signs to be inserted at the respective pure epimorphisms of the image factorisation chosen by F; cf. §A.3.2. Thus the morphism $X\mathrm{T}F \to XF\mathrm{T}'$ from this diagram represents

$$\begin{aligned}&\Big(X\mathrm{T}F \xrightarrow{\quad X\eta^-\mathrm{T}F \quad} XFG\mathrm{T}F = XF\mathrm{T}'GF \xrightarrow{\quad XF\mathrm{T}'\varepsilon^- \quad} XF\mathrm{T}'\Big)\\ =\;&\Big(X\mathrm{T}F \xrightarrow{X\rho} XF\mathrm{T}'\Big)\,.\end{aligned}$$

□

A.4.2.3. Standardisation by substitution of the rightmost column

We mimic the construction $X \mapsto X^{\vartheta}$ made in §A.4.1, now for $\mathcal{E}$ instead of $\mathcal{B}^{\mathrm{ac}}$.

Denote by $\mathcal{E}^{\square}(\bar{\Delta}_n^{\Delta\nabla}) \xrightarrow{M'} \underline{\mathcal{E}}^{+}(\bar{\Delta}_n^{\Delta\nabla})$ the residue class functor; cf. §A.3, [**4,** §2.1.3, §4.1]. Denote by $\underline{\mathcal{E}}^{+,\,(\square)}(\bar{\Delta}_n^{\Delta\nabla})$ the full subcategory of $\underline{\mathcal{E}}^{+}(\bar{\Delta}_n^{\Delta\nabla})$ whose set of objects is given by

$$\operatorname{Ob}\underline{\mathcal{E}}^{+,\,(\square)}(\bar{\Delta}_n^{\Delta\nabla}) \;:=\; \big(\operatorname{Ob}\mathcal{E}^{\square}(\bar{\Delta}_n^{\Delta\nabla})\big)M'\,.$$

So $\underline{\mathcal{E}}^{+,\,(\square)}(\bar{\Delta}_n^{\Delta\nabla})$ is defined to be the "full image" in $\underline{\mathcal{E}}^{+}(\bar{\Delta}_n^{\Delta\nabla})$ of the residue class functor M'.

Suppose given $n \geqslant 0$ and $X \in \operatorname{Ob}\mathcal{E}^{\square}(\bar{\Delta}_n^{\Delta\nabla})$. Recall that $\dot{\Delta}_n = [1,n]$ is identified with

$$\{i/0 \,:\, i \in [1,n]\} \;\subseteq\; \bar{\Delta}_n^{\Delta\nabla}\,.$$

Write $X_{*/0} := X|_{\dot{\Delta}_n} = (i \mapsto X_{i/0}) \in \operatorname{Ob}\mathcal{E}(\dot{\Delta}_n)$ and $X_{0^{+1}/*} := (i \mapsto X_{0^{+1}/i}) \in \operatorname{Ob}\mathcal{E}(\dot{\Delta}_n)$, analogously for morphisms; analogously for objects in $\underline{\mathcal{E}}^{+}(\bar{\Delta}_n^{\Delta\nabla})$ and their morphisms.

Let the isomorphism $X_{0^{+1}/*} \xrightarrow[\sim]{X\tau} X^{+1}_{*/0}$ in $\underline{\mathcal{E}}(\dot{\Delta}_n)$ be defined by morphisms of pure short exact sequences

$$\begin{array}{ccccc} X_{i/0} & \xrightarrow{(x\ x)} & X_{i/i} \oplus X_{0^{+1}/0} & \xrightarrow{\binom{x}{-x}} & X_{0^{+1}/i} \\ \Vert & & \downarrow & & \downarrow{\scriptstyle X\hat{\tau}_i} \\ X_{i/0} & \longrightarrow & B_{i/0} & \longrightarrow & X^{+1}_{i/0} \end{array}$$

for $i \in [1, n]$; where $X\hat{\tau}_i$ is a representative in $\mathcal{E}$ for the morphism $X\tau_i$ in $\underline{\mathcal{E}}$; where the upper pure short exact sequence stems from the diagram X; and where the lower pure short exact sequence is distinguished.

In this way, we get an isotransformation τ between the functors $(-)_{0^{+1}/*}$ and $(-)^{+1}_{*/0}$ from $\underline{\mathcal{E}}^{+,(\square)}(\bar{\Delta}_n^{\Delta\nabla})$ to $\underline{\mathcal{E}}(\dot{\Delta}_n)$.

Let $\underline{\mathcal{E}}^{+,\,\text{periodic}}(\bar{\Delta}_n^{\Delta\nabla})$ be the (in general not full) subcategory of $\underline{\mathcal{E}}^{+}(\bar{\Delta}_n^{\Delta\nabla})$ given by the set of objects

$$\operatorname{Ob}\underline{\mathcal{E}}^{+,\,\text{periodic}}(\bar{\Delta}_n^{\Delta\nabla}) := \left\{ Y \in \operatorname{Ob}\underline{\mathcal{E}}^{+}(\bar{\Delta}_n^{\Delta\nabla}) \;:\; Y_{0^{+1}/*} = Y^{+1}_{*/0} \text{ in } \operatorname{Ob}\underline{\mathcal{E}}(\dot{\Delta}_n) \right\},$$

and by the set of morphisms

$${}_{\underline{\mathcal{E}}^{+,\,\text{periodic}}(\bar{\Delta}_n^{\Delta\nabla})}(Y,Y') = \{ f \in {}_{\underline{\mathcal{E}}^{+}(\bar{\Delta}_n^{\Delta\nabla})}(Y,Y') \;:\; f_{0^{+1}/*} = f^{+1}_{*/0} \text{ in } \underline{\mathcal{E}}(\dot{\Delta}_n) \},$$

for $Y, Y' \in \operatorname{Ob}\underline{\mathcal{E}}^{+,\,\text{periodic}}(\bar{\Delta}_n^{\Delta\nabla})$.

Given $X \in \operatorname{Ob}\mathcal{E}^{\square}(\bar{\Delta}_n^{\Delta\nabla})$, we let $X^{\tau} \in \operatorname{Ob}\underline{\mathcal{E}}^{+,\,\text{periodic}}(\bar{\Delta}_n^{\Delta\nabla})$ be defined as the diagram X with $X_{0^{+1}/*}$ isomorphically replaced via $X\tau$ by $X^{+1}_{*/0}$. For short, the rightmost column of X becomes standardised to obtain X^{τ}.

Given $X, X' \in \operatorname{Ob}\mathcal{E}^{\square}(\bar{\Delta}_n^{\Delta\nabla})$, a morphism $X \xrightarrow{f} X'$ in $\underline{\mathcal{E}}^{+}(\bar{\Delta}_n^{\Delta\nabla})$ induces a morphism $X^{\tau} \xrightarrow{f^{\tau}} X'^{\tau}$ in $\underline{\mathcal{E}}^{+,\,\text{periodic}}(\bar{\Delta}_n^{\Delta\nabla})$. Namely, we let $f^{\tau}_{\beta/\alpha} := f_{\beta/\alpha}$ for $0 \leqslant \alpha \leqslant \beta \leqslant n$, and we let $f^{\tau}_{0^{+1}/*}$ be characterised by the commutative quadrangle

$$\begin{array}{ccc} X_{0^{+1}/*} & \xrightarrow[\sim]{X\tau} & X^{+1}_{*/0} \\ {\scriptstyle f_{0^{+1}/*}}\downarrow & & \downarrow{\scriptstyle f^{\tau}_{0^{+1}/*}} \\ X'_{0^{+1}/*} & \xrightarrow[\sim]{X'\tau} & X'^{+1}_{*/0} \end{array}$$

in $\underline{\mathcal{E}}(\dot{\Delta}_n)$. In particular, since τ is an isotransformation, we have $f^{\tau}_{0^{+1}/*} = (f^{\tau}_{*/0})^{+1}$.

Remark 25. *The constructions made above define a functor*

$$\begin{array}{ccc} \underline{\mathcal{E}}^{+,(\square)}(\bar{\Delta}_n^{\Delta\nabla}) & \xrightarrow{(-)^{\tau}} & \underline{\mathcal{E}}^{+,\,\text{periodic}}(\bar{\Delta}_n^{\Delta\nabla}) \\ X & \longmapsto & X^{\tau}\,. \end{array}$$

A.4.2.4. n-triangles in the classical stable category

Proposition 26. *Suppose given $n \geqslant 0$ and $X \in \operatorname{Ob}\mathcal{E}^{\square}(\bar{\Delta}_n^{\Delta\nabla})$.*

The periodic prolongation of $X^{\tau} \in \operatorname{Ob}\underline{\mathcal{E}}^{+,\,\text{periodic}}(\bar{\Delta}_n^{\Delta\nabla})$ to an object of $\underline{\mathcal{E}}^{+,\,\text{periodic}}(\bar{\Delta}_n^{\#})$ is an n-triangle with respect to the triangulation ϑ' on $(\underline{\mathcal{E}}, \mathsf{T}')$ obtained as in Proposition 23.

Proof. By Lemma 17, there exists $X' \in \operatorname{Ob}\mathcal{E}^{\square}(\bar{\Delta}_n^{\Delta\nabla})$ such that $X'_{\alpha/\alpha} = 0$ for all $0 \leqslant \alpha \leqslant 0^{+1}$ and such that X is isomorphic to X' in $\underline{\mathcal{E}}^{+}(\bar{\Delta}_n^{\Delta\nabla})$. By Remark 25, the object X^{τ} is isomorphic to X'^{τ} in $\underline{\mathcal{E}}^{+,\,\text{periodic}}(\bar{\Delta}_n^{\Delta\nabla})$. Thus the periodic prolongation of X^{τ} is an n-triangle if and only if that of X'^{τ} is; cf. [**4**, Lem. 3.4.(4)].

Therefore, we may assume that $X_{\alpha/\alpha} \simeq 0$ for all $0 \leqslant \alpha \leqslant 0^{+1}$.

Let $\tilde{X} \in \mathrm{Ob}(\mathcal{B}^{\mathrm{ac}})^{\Box}(\bar{\Delta}_n^{\Delta\nabla})$ be such that there exists an isomorphism $X \xrightarrow[\sim]{\hat{a}} \tilde{X}\hat{F}$ in $\mathcal{E}^{\Box}(\bar{\Delta}_n^{\Delta\nabla})$ and such that $\tilde{X}_{\alpha/\alpha} = 0$ for all $0 \leqslant \alpha \leqslant 0^{+1}$; cf. Lemma 18. Denote by $X \xrightarrow[\sim]{a} \tilde{X}F$ the isomorphism in $\underline{\mathcal{E}}^{+}(\bar{\Delta}_n^{\Delta\nabla})$ represented by $X \xrightarrow[\sim]{\hat{a}} \tilde{X}\hat{F}$.

Let $\tilde{\tilde{X}} \in \mathrm{Ob}(\mathcal{B}^{\mathrm{ac}})^{\Box}(\bar{\Delta}_n^{\#})$ be such that $\tilde{\tilde{X}}|_{\bar{\Delta}_n^{\Delta\nabla}} = \tilde{X}$. By Lemma 21, the periodic n-pretriangle $\tilde{\tilde{X}}^{\vartheta} \in \mathrm{Ob}\,\underline{\underline{\mathcal{E}}}^{+,\,\mathrm{periodic}}(\bar{\Delta}_n^{\#})$ is an n-triangle. Note that $\tilde{\tilde{X}}^{\vartheta}$ depends only on $\tilde{X}$, not on the choice of $\tilde{\tilde{X}}$.

Thus, by Proposition 23.(3), $\tilde{\tilde{X}}^{\vartheta} F^{\rho} \in \mathrm{Ob}\,\underline{\mathcal{E}}^{+,\,\mathrm{periodic}}(\bar{\Delta}_n^{\#})$ is an n-triangle. Therefore, it suffices to show that X^{τ} and $\tilde{\tilde{X}}^{\vartheta} F^{\rho}|_{\bar{\Delta}_n^{\Delta\nabla}}$ are isomorphic in $\underline{\mathcal{E}}^{+,\,\mathrm{periodic}}(\bar{\Delta}_n^{\Delta\nabla})$, for then their periodic prolongations are isomorphic in $\underline{\mathcal{E}}^{+,\,\mathrm{periodic}}(\bar{\Delta}_n^{\#})$, which in turn shows the periodic prolongation of X^{τ} to be an n-triangle; cf. [**4**, Lem. 3.4.(4)].

We have a composite isomorphism

$$X^{\tau} \xleftarrow{\sim} X \xrightarrow[\sim]{a} \tilde{X}F = \tilde{\tilde{X}}F|_{\bar{\Delta}_n^{\Delta\nabla}} \xrightarrow{\sim} \tilde{\tilde{X}}^{\vartheta}F|_{\bar{\Delta}_n^{\Delta\nabla}} \xrightarrow{\sim} \tilde{\tilde{X}}^{\vartheta}F^{\rho}|_{\bar{\Delta}_n^{\Delta\nabla}}$$

in $\underline{\mathcal{E}}^{+}(\bar{\Delta}_n^{\Delta\nabla})$. We *claim* that it lies in $\underline{\mathcal{E}}^{+,\,\mathrm{periodic}}(\bar{\Delta}_n^{\Delta\nabla})$.

Suppose given $i \in [1, n]$. On $i/0$, this composite equals $a_{i/0}$. Thus we have to show that on $0^{+1}/i$, this composite equals $a_{i/0}^{+1} = a_{i/0}\,\mathsf{T}'$. Consider, to this end, the morphisms

$$\begin{array}{ccccc}
X_{i/0} & \xrightarrow{\bullet} & (X_{i/0}G)^1 & \xrightarrow{+} & X_{i/0}\,\mathsf{T}' \\
\| & & \uparrow & & \uparrow {\scriptstyle X\hat{\tau}_i} \\
X_{i/0} & \xrightarrow[\bullet]{x} & X_{0^{+1}/0} & \xrightarrow[+]{-x} & X_{0^{+1}/i} \\
\wr\downarrow {\scriptstyle \hat{a}_{i/0}} & & \wr\downarrow {\scriptstyle \hat{a}_{0^{+1}/0}} & & \wr\downarrow {\scriptstyle \hat{a}_{0^{+1}/i}} \\
\tilde{X}_{i/0}\hat{F} & \xrightarrow[\bullet]{\tilde{x}\hat{F}} & \tilde{X}_{0^{+1}/0}\hat{F} & \xrightarrow[+]{-\tilde{x}\hat{F}} & \tilde{X}_{0^{+1}/i}\hat{F} \\
\| & & \downarrow & & \downarrow {\scriptstyle X\hat{\theta}_{i/0}\hat{F}} \\
\tilde{X}_{i/0}\hat{F} & \xrightarrow{\bullet} & \tilde{X}_{i/0}\,\mathsf{I}\,\hat{F} & \xrightarrow{+} & \tilde{X}_{i/0}\,\mathsf{T}\,\hat{F} \\
\| & & \downarrow & & \| \\
\tilde{X}_{i/0}\hat{F} & \xrightarrow{\bullet} & (\tilde{X}_{i/0})^1 & \xrightarrow{+} & \tilde{X}_{i/0}\,\mathsf{T}\,\hat{F} \\
\| & & \downarrow & & \downarrow \\
\tilde{X}_{i/0}\hat{F} & \xrightarrow{\bullet} & (\tilde{X}_{i/0}\hat{F}G)^1 & \xrightarrow{+} & \tilde{X}_{i/0}\hat{F}\,\mathsf{T}'
\end{array}$$

of pure short exact sequences in $\mathcal{E}$. The fourth sequence is purely short exact by Lemma 20.

The first morphism from above arises by definition of $\hat{\tau}_i$; cf. §A.4.2.3. The second morphism is taken from $\hat{a}$. The third morphism arises by definition of $\hat{\theta}_{i/0}$ and an application of $\hat{F}$; cf. §A.4.1. The fourth morphism is given by Lemma 20. The fifth morphism is given by Lemma 24.

The first and the sixth pure short exact sequence are distinguished, and so the *claim* and hence the proposition follow.

References

[1] BERNSTEIN, J.; BEILINSON, A.A.; DELIGNE, P., *Faisceaux pervers,* Astérisque 100, 1982.

[2] BOSMA, W.; CANNON, J.; PLAYOUST, C., *The Magma algebra system. I. The user language,* J. Symbolic Comput. 24 (3-4), p. 235–265, 1997 (cf. magma.maths.usyd.edu.au).

[3] HAPPEL, D., *Triangulated Categories in the Representation Theory of Finite Dimensional Algebras,* LMS LN 119, 1988.

[4] KÜNZER, M., *Heller triangulated categories,* Homol. Homot. Appl. 9 (2), p. 233–320, 2007.

[5] VERDIER, J.L., *Catégories Derivées,* published in SGA 4 1/2, SLN 569, p. 262–311, 1977 (written 1963).

This article may be accessed via WWW at `http://www.rmi.acnet.ge/jhrs/`

Matthias Künzer
kuenzer@math.rwth-aachen.de
www.math.rwth-aachen.de/~kuenzer

Lehrstuhl D für Mathematik
RWTH Aachen
Templergraben 64
D-52062 Aachen
Germany

Journal of Homotopy and Related Structures, vol. 4(1), 2009, pp.39–68

COBORDISM CATEGORY OF PLUMBED 3-MANIFOLDS AND INTERSECTION PRODUCT STRUCTURES

YOSHIHIRO FUKUMOTO

(*communicated by James Stasheff*)

Abstract

In this paper, we introduce a category of graded commutative rings with certain algebraic morphisms, to investigate the cobordism category of plumbed 3-manifolds. In particular, we define a non-associative distributive algebra that gives necessary conditions for an abstract morphism between the homologies of two plumbed 3-manifolds to be realized geometrically by a cobordism. Here we also consider the homology cobordism monoid, and give a necessary condition using w-invariants for the homology 3-spheres to belong to the inertia group associated to some homology 3-spheres.

1. Introduction

In this paper, we introduce a category of graded commutative rings with certain algebraic morphisms in order to investigate the cobordism category of plumbed 3-manifolds. In fact, we define a non-associative distributive algebra, which we use to give necessary conditions for algebraic morphisms between the homologies of two plumbed 3-manifolds to be realized geometrically by cobordism. This paper is a generalization of the paper [**2**] on homology cobordisms, to general cobordisms between closed 3-manifolds. More precisely, we state the problem as follows. Let $\mathcal{C}_3$ be the cobordism category of 3-manifolds whose objects $M \in \mathrm{ob}(\mathcal{C}_3)$ are closed oriented 3-manifolds and whose morphisms $W \in \mathcal{C}_3(M, M')$ between two 3-manifolds M and M' are cobordisms $(W; M, M')$. On the other hand, let $\mathcal{L}_3$ be a category whose objects $(H_*, \bullet) \in \mathrm{ob}(\mathcal{L}_3)$ are graded commutative rings of dimension 3 and whose morphisms $(L_*; i, i', \bullet) \in \mathcal{L}_3((H_*, \bullet), (H'_*, \bullet))$ between two objects $(H_*, \bullet)$ and $(H'_*, \bullet)$ are composed of graded modules L_* of dimension 4 with certain product structures $\bullet$ and homomorphisms $H_* \xrightarrow{i} L_* \xleftarrow{i'} H'_*$ satisfying compatibility conditions on products. Note that there exists a functor $H_* : \mathcal{C}_3 \to \mathcal{L}_3$ given by $M \longmapsto (H_*(M; \mathbb{Z}), \bullet)$ and $(W; M, M') \longmapsto (L_*(W; \mathbb{Z}); i_*, i'_*, \bullet)$, where i_* and i'_*

Research supported by MEXT Grant-in-Aid for Scientific Research (18740039). The author would like to thank Professor Mikio Furuta, Professor Masaaki Ue, Professor Mikiya Masuda and Professor Yukio Kametani for helpful suggestions.
Received August 11, 2008, revised December 29, 2008; published on April 17, 2009.
2000 Mathematics Subject Classification: 57Q20, 55N45, 57R57, 58J20, 20M50, 55S20.
Key words and phrases: Cobordism category, cohomology ring, Seiberg-Witten theory, index theorem, V-manifolds.

are the induced homomorphisms

$$H_*(M;\mathbb{Z}) \xrightarrow{i_*} L_*(W;\mathbb{Z}) \xleftarrow{i'_*} H_*(M';\mathbb{Z}),$$

to the $\mathbb{Z}$-module $L_*(W;\mathbb{Z})$ defined by

$$L_k(W;\mathbb{Z}) = \operatorname{Im}\left(H_k(M;\mathbb{Z}) \oplus H_k(M';\mathbb{Z}) \stackrel{i_*+i'_*}{\to} H_k(W;\mathbb{Z})\right).$$

Let $R_*(W)$ be a $\mathbb{Z}$-module defined by

$$R_k(W;\mathbb{Z}) = \operatorname{Ker}\left(H_{k-1}(M;\mathbb{Z}) \oplus H_{k-1}(M';\mathbb{Z}) \stackrel{i_*+i'_*}{\to} H_{k-1}(W;\mathbb{Z})\right)$$

and $\partial_* \oplus \partial'_* : R_k(W;\mathbb{Z}) \to H_{k-1}(M;\mathbb{Z}) \oplus H_{k-1}(M';\mathbb{Z})$ be the induced homomorphism. Then the problem can be stated as follows.

Problem 1. *Let $M, M' \in \mathrm{ob}(\mathcal{C}_3)$ be two closed oriented 3-manifolds. Let $\phi = (L_*; i, i', \bullet) \in \mathcal{L}_3((H_*(M;\mathbb{Z}), \bullet), (H_*(M';\mathbb{Z}), \bullet))$ be an algebraic morphism of homology rings. Then does there exist a cobordism $(W; M, M') \in \mathcal{C}_3(M, M')$ such that $(L_*(W;\mathbb{Z}); i_*, i'_*, \bullet) \cong \phi = (L_*; i, i', \bullet)$.*

Note that by the exact sequence of the pair $(W, M \sqcup M')$, the cobordism $(W; M, M')$ preserves the intersection product structures, that is, the induced morphism $(L_*(W;\mathbb{Z}); i_*, i'_*, \bullet) \in \mathcal{L}_3((H_*(M;\mathbb{Z}), \bullet), (H_*(M';\mathbb{Z}), \bullet))$ must be composed of ring homomorphisms i_*, i'_* with respect to the intersection pairings.

Lemma 2. *Let $(W; M, M')$ be a cobordism between two closed oriented 3-manifolds M and M'. Let $(L_*(W;\mathbb{Z}); i_*, i'_*, \bullet)$ be the induced homomorphism $H_*(M;\mathbb{Z}) \xrightarrow{i_*} L_*(W;\mathbb{Z}) \xleftarrow{i'_*} H_*(M';\mathbb{Z})$. If we denote the intersection parings on M and W by $\bullet : H_k(M;\mathbb{Z}) \otimes H_\ell(M;\mathbb{Z}) \to H_{k+\ell-3}(M;\mathbb{Z})$ and $\bullet : R_{k+1}(W;\mathbb{Z}) \otimes L_\ell(W;\mathbb{Z}) \to L_{k+\ell-3}(W;\mathbb{Z})$, respectively, for any non-negative integers k, ℓ with $k + \ell \geqslant 3$, then we have $i_*(\partial_*(\eta) \cdot \theta) = \eta \cdot i_*(\theta)$ for any $\eta \in R_{k+1}(W;\mathbb{Z})$, $\theta \in L_\ell(W;\mathbb{Z})$.*

In fact, this lemma provides a necessary condition for the existence of cobordisms described above. However, the following is an example to which Lemma 2 cannot be applied.

Example 3. *Let (Γ, ω), (Γ', ω') be two Seifert graphs defined by*

$$\Gamma = (V, E),\ V = \{1, 2\},\ E = \{(1,2), (2,1)\}$$
$$\begin{cases} \omega_1 = \{2; (5,3), (5,3), (5,4)\}, \\ \omega_2 = \{2; (9,1), (9,1), (9,4)\}. \end{cases}$$

$$\Gamma' = (V', E'),\ V' = \{1, 2\},\ E' = \{(1,2), (2,1)\}$$
$$\begin{cases} \omega'_1 = \{2; (5,3), (5,3), (5,4)\}, \\ \omega'_2 = \{1; (9,2), (9,2), (9,2)\}. \end{cases}$$

Note that the associated plumbed 3-manifold $M(\Gamma)$ can be obtained by plumbing of S^1-V-bundles $E_v \to \Sigma_v$, $v \in V$ over the closed oriented V-surfaces Σ_v with Seifert *invariants $\omega(v)$ according to the graphs Γ, see Appendix 6 or [3]. This plumbed V-manifold can also be described by using the decorated plumbing graphs of N. Saveliev [12]. Then the homology groups of $M(\Gamma)$, $M(\Gamma')$ are calculated as follows.*

$$H_1(M(\Gamma);\mathbb{Z}) \cong \mathbb{Z}^8 \oplus \mathbb{Z}/45 \oplus \mathbb{Z}/675, \quad H_2(M(\Gamma);\mathbb{Z}) \cong \mathbb{Z}^8,$$
$$H_1(M(\Gamma');\mathbb{Z}) \cong \mathbb{Z}^6 \oplus \mathbb{Z}/45 \oplus \mathbb{Z}/675, \quad H_2(M(\Gamma');\mathbb{Z}) \cong \mathbb{Z}^6.$$

Let $\{\alpha_{vj}\}_{v\in\{1,2\},j\in\{1,2,\dots,2g_v\}} \subset H_1(\bar{\Sigma}_v;\mathbb{Z})$ be a system of α,β-cycles on the underlying topological space $\bar{\Sigma}_v$ of the V-surface Σ_v of genus g_v, so that $\alpha_{vj}\cdot\alpha_{v'j'} = \delta_{vv'}\varepsilon_{jj'}$ for $v,v' \in \{1,2\}$ and $j,j' \in \{1,2,\dots,2g_v\}$, where

$$\varepsilon_{jj'} = \begin{cases} 1 & (j = 2j''-1,\ j' = 2j'',\ j'' \in \{1,2,\dots,g_v\}) \\ -1 & (j = 2j'',\ j' = 2j''-1,\ j'' \in \{1,2,\dots,g_v\}) \\ 0 & (otherwise) \end{cases}.$$

We denote the same symbol $\{\alpha_{vj}\}_{v\in\{1,2\},j\in\{1,2,\dots,2g_v\}} \subset H_1(M(\Gamma);\mathbb{Z})$ to be the corresponding homology classes in $M(\Gamma)$, and fix an isomorphism on the free part $H_1(M(\Gamma);\mathbb{Z})/\mathrm{Tor} \cong \mathbb{Z}^8$ by using this $\{\alpha_{vj}\}$. For each $\alpha \in H_1(\bar{\Sigma}_v;\mathbb{Z})$, let $\theta_\alpha \in H_2(M(\Gamma);\mathbb{Z})$ be the corresponding 2-cycle obtained by using the natural homomorphism defined in Lemma 30.1. Then $\{\theta_{\alpha_{vj}}\}_{v\in\{1,2\},j\in\{1,2,\dots,2g_v\}}$ $\subset H_2(M(\Gamma);\mathbb{Z})$ form a basis, and we can fix an isomorphism $H_2(M(\Gamma);\mathbb{Z}) \cong \mathbb{Z}^6$ by using $\{\theta_{\alpha_{vj}}\}$. Let $L_* = \bigoplus_{k=0}^4 L_k$ be the graded $\mathbb{Z}$-module defined by

$$L_0 = \mathbb{Z},\ L_1 = \mathbb{Z}^8 \oplus \mathbb{Z}/45 \oplus \mathbb{Z}/675,\ L_2 = \mathbb{Z}^6,\ L_3 = \mathbb{Z},\ L_4 = 0.$$

Let $H_*(M(\Gamma)) \xrightarrow{i} L_* \xleftarrow{i'} H_*(M(\Gamma'))$ be two graded homomorphisms defined by

$$\begin{array}{ccccc} H_1(M(\Gamma);\mathbb{Z}) & \xrightarrow{i} & L_1 & \xleftarrow{i'} & H_1(M(\Gamma');\mathbb{Z}) \\ \downarrow\cong & & \downarrow\cong & & \downarrow\cong \\ \mathbb{Z}^8 & \xrightarrow{\Phi_1} & \mathbb{Z}^8 & \xleftarrow{\Phi_1'} & \mathbb{Z}^6 \\ \oplus & & \oplus & & \oplus \\ \mathbb{Z}/45 & \xrightarrow{1} & \mathbb{Z}/45 & \xleftarrow{1} & \mathbb{Z}/45 \\ \oplus & & \oplus & & \oplus \\ \mathbb{Z}/675 & \xrightarrow{1} & \mathbb{Z}/675 & \xleftarrow{1} & \mathbb{Z}/675 \end{array},$$

where

$$\Phi_1 = \begin{pmatrix} I_2 & O & O & O \\ O & I_2 & O & O \\ O & O & O & O \\ O & O & O & I_2 \end{pmatrix}, \quad \Phi_1' = \begin{pmatrix} I_2 & O & O \\ O & O & I_2 \\ O & I_2 & O \\ O & O & O \end{pmatrix},$$
$$I_2 = \begin{pmatrix} 1 & 0 \\ 0 & 1 \end{pmatrix},\ O = \begin{pmatrix} 0 & 0 \\ 0 & 0 \end{pmatrix},$$

$$\begin{array}{ccccc} H_2(M(\Gamma);\mathbb{Z}) & \xrightarrow{i} & L_2 & \xleftarrow{i'} & H_2(M(\Gamma');\mathbb{Z}) \\ \downarrow\cong & & \downarrow\cong & & \downarrow\cong \\ \mathbb{Z}^8 & \xrightarrow{\Phi_2} & \mathbb{Z}^6 & \xleftarrow{\Phi_2'} & \mathbb{Z}^6 \end{array},$$

and where

$$\Phi_2 = \begin{pmatrix} I_2 & O & O & O \\ O & I_2 & O & O \\ O & O & I_2 & O \end{pmatrix}, \quad \Phi_2' = \begin{pmatrix} I_2 & O & O \\ O & O & I_2 \\ O & O & O \end{pmatrix},$$

and

$$\begin{array}{ccccc} H_3(M(\Gamma);\mathbb{Z}) & \stackrel{i}{\to} & L_3 & \stackrel{i'}{\leftarrow} & H_3(M(\Gamma');\mathbb{Z}) \\ \downarrow\cong & & \downarrow\cong & & \downarrow\cong \\ \mathbb{Z} & \stackrel{1}{\to} & \mathbb{Z} & \stackrel{1}{\leftarrow} & \mathbb{Z} \end{array}.$$

On the other hand, let $R_* = \bigoplus_{k=0}^4 R_k$ *be the graded* $\mathbb{Z}$*-module defined by*

$$R_k = \operatorname{Ker}\left(H_{k-1} \oplus H'_{k-1} \stackrel{i-i'}{\to} L_{k-1}\right),$$

then we have

$$R_0 = 0,\ R_1 = \mathbb{Z},\ R_2 = \mathbb{Z}^6,\ R_3 = \mathbb{Z}^8,\ R_4 = \mathbb{Z}.$$

Let $\bullet : R_2 \otimes L_2 \to L_0 = \mathbb{Z}$ *and* $\bullet : R_3 \otimes L_2 \to L_1$ *be bilinear pairings defined by*

$$\begin{array}{c} L_2 \\ R_2\begin{pmatrix} \varepsilon_2 & O & O \\ O & \varepsilon_2 & O \\ O & O & \varepsilon_2 \end{pmatrix} \end{array}, \quad \begin{array}{c} L_1 \\ R_3\begin{pmatrix} -\varepsilon_2 & O & O & O \\ O & -\varepsilon_2 & O & O \\ O & O & O & -\varepsilon_2 \\ O & O & -\varepsilon_2 & O \end{pmatrix} \end{array}, \quad \begin{array}{c} L_2 \\ R_3\begin{pmatrix} 45\varepsilon_2\delta & O & O \\ O & 45\varepsilon_2\delta & O \\ O & O & O \\ O & O & O \end{pmatrix} \end{array},$$

$$\varepsilon_2 = \begin{pmatrix} 0 & 1 \\ -1 & 0 \end{pmatrix}, \quad \delta = 1 \in \mathbb{Z}/675 \subset L_1.$$

We fix spin structures,

$$c = ((1,1,0),(1,1,0)),\ c' = ((1,1,0),(0,0,0)),$$

on $M(\Gamma)$, $M(\Gamma')$, *respectively, where* $0, 1$ *are determined by the spin structures on punctured Riemann surfaces around* V*-singular points, see [**2**]. Then there exists no (spin) cobordism*

$$((W,\tilde{c});(M,c),(M',c'))$$

such that $(L_*(W;\mathbb{Z});i_*,i'_*,\bullet) \cong \phi = (L_*;i,i',\bullet)$ *and* $L_*(W;\mathbb{Z}) = H_*(W;\mathbb{Z})$.

Remark 4. *Note that the generators of* $R_2 = \mathbb{Z}^6$ *and* $R_3 = \mathbb{Z}^8$ *correspond to*

$$R_2 = \left\langle \overline{\alpha_{11}\alpha'_{11}}, \overline{\alpha_{12}\alpha'_{12}}, \overline{\alpha_{13}\alpha'_{21}}, \overline{\alpha_{14}\alpha'_{22}}, \overline{\alpha_{21}0}, \overline{\alpha_{22}0} \right\rangle,$$

$$R_3 = \left\langle \overline{\theta_{\alpha_{11}}\theta'_{\alpha'_{11}}}, \overline{\theta_{\alpha_{12}}\theta'_{\alpha'_{12}}}, \overline{\theta_{\alpha_{13}}\theta'_{\alpha'_{21}}}, \overline{\theta_{\alpha_{14}}\theta'_{\alpha'_{22}}}, \overline{\theta_{\alpha_{23}}0}, \overline{\theta_{\alpha_{24}}0}, \overline{0\theta'_{\alpha'_{13}}}, \overline{0\theta'_{\alpha'_{14}}} \right\rangle,$$

where we denote by $\overline{\alpha\alpha'}$ *(resp.* $\overline{\theta_\alpha\theta'_{\alpha'}}$*) the* 2*(resp.* 3*)-cycles corresponding to*

$$(-\alpha) \oplus \alpha' \in \operatorname{Ker}\left(H_1 \oplus H'_1 \stackrel{i+i'}{\to} L_1\right) \quad \textit{(resp. } (-\theta_\alpha) \oplus \theta'_{\alpha'} \in \operatorname{Ker}\left(H_1 \oplus H'_1 \stackrel{i+i'}{\to} L_1\right)\textit{)}$$

with $k = 2$ *(resp.* $k = 3$*). Take a generator* $\delta_1, \ldots, \delta_8, \bar{\varepsilon}, \bar{\delta}$ *of* L_1,

$$\langle \delta_1, \ldots, \delta_8 \rangle \oplus \langle \bar{\varepsilon} \rangle \oplus \langle \bar{\delta} \rangle \cong \mathbb{Z}^8 \oplus \mathbb{Z}/45 \oplus \mathbb{Z}/675 \cong L_1,$$

and a generator $\rho_1, \ldots, \rho_6$ *of* L_2,

$$\langle \rho_1, \ldots, \rho_6 \rangle \cong \mathbb{Z}^6 \cong L_2.$$

Then the pairings $\bullet : R_2 \otimes L_2 \to L_0 = \mathbb{Z}$ *and* $\bullet : R_3 \otimes L_2 \to L_1$ *are described as follows.*

$$\bullet : R_2 \otimes L_2 \to L_0 = \mathbb{Z}$$

$$\begin{aligned}
\overline{\alpha_{11}\alpha'_{11}} \cdot \rho_2 &= 1, \\
\overline{\alpha_{12}\alpha'_{12}} \cdot \rho_1 &= -1, \\
\overline{\alpha_{13}\alpha'_{21}} \cdot \rho_4 &= 1, \\
\overline{\alpha_{14}\alpha'_{22}} \cdot \rho_3 &= -1, \\
\overline{\alpha_{21}0} \cdot \rho_6 &= 1, \\
\overline{\alpha_{22}0} \cdot \rho_5 &= -1 \\
x \cdot y &= 0 \quad \textit{(otherwise)}.
\end{aligned}$$

$$\bullet : R_3 \otimes L_1 \to L_0 = \mathbb{Z}$$

$$\begin{aligned}
\overline{\theta_{\alpha_{11}}\theta_{\alpha'_{11}}} \cdot \delta_2 &= -1, \\
\overline{\theta_{\alpha_{12}}\theta_{\alpha'_{12}}} \cdot \delta_1 &= 1, \\
\overline{\theta_{\alpha_{13}}\theta_{\alpha'_{21}}} \cdot \delta_4 &= -1, \\
\overline{\theta_{\alpha_{14}}\theta_{\alpha'_{22}}} \cdot \delta_3 &= 1, \\
\overline{\theta_{\alpha_{23}}0} \cdot \delta_8 &= -1, \\
\overline{\theta_{\alpha_{24}}0} \cdot \delta_7 &= 1, \\
\overline{0\theta_{\alpha'_{13}}} \cdot \delta_6 &= -1, \\
\overline{0\theta_{\alpha'_{14}}} \cdot \delta_5 &= 1, \\
x \cdot y &= 0 \quad \textit{(otherwise)}.
\end{aligned}$$

$$\bullet : R_3 \otimes L_2 \to L_1$$

$$\begin{aligned}
\overline{\theta_{\alpha_{11}}\theta_{\alpha'_{11}}} \cdot \rho_2 &= -\bar{s}, \quad \bar{s} = 45\bar{\delta}, \\
\overline{\theta_{\alpha_{12}}\theta_{\alpha'_{12}}} \cdot \rho_1 &= \bar{s}, \\
\overline{\theta_{\alpha_{13}}\theta_{\alpha'_{21}}} \cdot \rho_4 &= -\bar{s}, \\
\overline{\theta_{\alpha_{14}}\theta_{\alpha'_{22}}} \cdot \rho_3 &= \bar{s}, \\
x \cdot y &= 0 \quad \textit{(otherwise)}.
\end{aligned}$$

As in the paper [**2**] on the homology cobordisms of plumbed 3-manifolds, we show the above statements by two approaches. Let $M(\Gamma)$ (resp. $M(\Gamma')$) be a plumbed

3-manifold associated to the tree Seifert graph Γ (resp. Γ') satisfying a certain condition (Ndeg) (Definition 6), and let $\phi = (L_*; i, i', \bullet)$ be an algebraic morphism between $H_*(M(\Gamma); \mathbb{Z})$ and $H_*(M(\Gamma'); \mathbb{Z})$. Then we construct a distributive algebra $\mathcal{R}_*(\Gamma, \Gamma', \phi)$ over $\mathbb{Q}$ by using the data (Γ, Γ', ϕ) as in Definition 6.

1. The 10/8-inequality.
   We apply a V-manifold version of the extended Furuta-Kametani-10/8-inequality for closed spin 4-manifolds with $b_1 > 0$. The 10/8-inequality contains terms depending on the quadruple cup product of the first cohomology on closed 4-V-manifolds. The quadruple products are calculated by using the algebra $\mathcal{R}_*(\Gamma, \Gamma', \phi)$ as a map $q(\Gamma, \Gamma', \phi) : \mathcal{R}_*(\Gamma, \Gamma', \phi)^{\otimes 4} \to \mathbb{Q}$.
2. The associativity of cup products.
   The distributive algebra $\mathcal{R}_*(\Gamma, \Gamma', \phi)$ is not necessarily associative by definition. However, if there exists a cobordism W between $M(\Gamma)$ and $M(\Gamma')$ realizing the algebraic morphism ϕ, then we see that there exists an injective ring homomorphism from $\mathcal{R}_*(\Gamma, \Gamma', \phi)$ to the homology ring $H_*(Z; \mathbb{Q})$ of a closed 4-V-manifold Z obtained by gluing $P(\Gamma)$, $P(\Gamma')$ and W along the boundaries $M(\Gamma)$ and $M(\Gamma')$, and hence that $\mathcal{R}_*(\Gamma, \Gamma', \phi)$ must be associative. Therefore, the associativity of the distributive algebra $\mathcal{R}_*(\Gamma, \Gamma', \phi)$ gives an obstruction to the existence of cobordisms W realizing ϕ.

Remark 5. *As in paper [2], the author does not know examples that can be detected by using the gauge theory in Approach 1 but cannot be detected by using the associativity of cup products in Approach 2.*

The homology cobordism category is defined to be the category whose objects are closed 3-manifolds and whose morphisms are homology cobordisms. If we take the quotient of the set of objects by the homology cobordism relation, then we obtain the homology cobordism monoid. In particular, we give a necessary condition, using w-invariants, for the homology 3-spheres to belong to the inertia group associated to some homology 3-spheres.

The organization of this paper is as follows. In Section 2, we introduce a distributive algebra $\mathcal{R}_*(\Gamma, \Gamma', \phi)$ and state the main theorems and their applications in Approach 1 using the 10/8-inequality, and in Approach 2 using the associativity of cup products. In Section 3, we recall the definition of cobordism the category of 3-manifolds and introduce a category of graded commutative rings with certain algebraic morphisms, which models that of the homology rings. In Section 5, we recall the definition of the w-invariants, which are integral lifts of the Rochlin invariants, and give several properties of the invariants under cobordisms and connected sum operations. Here we also give a necessary condition for the homology 3-spheres to belong to the inertia group associated to some 3-manifolds in the homology cobordism monoid. Finally in Section 6, to prove Lemma 32, we consider a cobordism of two plumbed 3-manifolds and calculate the intersection pairings of 3-cycles on closed 4-V-manifolds obtained by gluing 4-V-manifolds along the boundaries.

2. Main theorems and applications

2.1. A distributive algebra constructed from algebraic morphisms

Motivated by the above Lemma 32 below in Section 6, we introduce a distributive algebra $\mathcal{R}_*(\Gamma,\Gamma',\phi)$ using the data (Γ,Γ',ϕ).

Definition 6. *Let $\Gamma=(V,E,\omega)$, $\Gamma'=(V',E',\omega')$ be two tree Seifert graphs satisfying the condition (Ndeg) :*

1. *the intersection matrices $A(\Gamma)$, $A(\Gamma')$ of the plumbed V-manifold $P(\Gamma)$, $P(\Gamma')$ with respect to the standard basis are non-singular,*
2. *all the Euler numbers are non-zero $e(\omega_v)\neq 0$, $e(\omega'_{v'})\neq 0$ for all $v\in V$ and $v'\in V'$.*

Let $\phi=(L_*;i,i',\bullet)$ be an algebraic morphism between $(H_*(M(\Gamma);\mathbb{Z}),\bullet)$ and $(H_*(M(\Gamma');\mathbb{Z}),\bullet)$ in the $\mathcal{L}_3$ category . Then we define a graded commutative distributive algebra $\mathcal{R}_*(\Gamma,\Gamma',\phi)$ over $\mathbb{Q}$ to be

$$\begin{cases} \mathcal{R}_4(\Gamma,\Gamma',\phi)=\mathbb{Q}\,Z \\ \mathcal{R}_3(\Gamma,\Gamma',\phi)=R(\Gamma,\Gamma',\phi)\otimes\mathbb{Q} \\ \mathcal{R}_2(\Gamma,\Gamma',\phi)=\bigoplus_{v\in V}\mathbb{Q}\,\Sigma_v\ \oplus\bigoplus_{v'\in V'}\mathbb{Q}\,\Sigma'_{v'} \\ \mathcal{R}_1(\Gamma,\Gamma',\phi)=L(\Gamma,\Gamma',\phi)\otimes\mathbb{Q} \\ \mathcal{R}_0(\Gamma,\Gamma',\phi)=\mathbb{Q}\,pt \end{cases},$$

where

$$R(\Gamma,\Gamma',\phi)=\mathrm{Ker}\left(H(\Gamma)\oplus H(\Gamma')\overset{\theta\oplus\theta'}{\to}H_2(M(\Gamma);\mathbb{Z})\oplus H_2(M(\Gamma');\mathbb{Z})\overset{i+i'}{\to}L_2\right),$$

$$L(\Gamma,\Gamma',\phi)=\mathrm{Im}\left(H(\Gamma)\oplus H(\Gamma')\overset{\lambda\oplus\lambda'}{\to}H_1(M(\Gamma);\mathbb{Z})\oplus H_1(M(\Gamma');\mathbb{Z})\overset{i+i'}{\to}L_1\right),$$

as in Lemma 30 in Section 6. If we denote the elements corresponding to $(-\alpha)\oplus\alpha'\in R(\Gamma,\Gamma',\phi)$ formally by $\widetilde{\theta_\alpha\theta'_{\alpha'}}$ then the product structure on $\mathcal{R}_*(\Gamma,\Gamma',\phi)$ is given by

$$\begin{cases} Z\cdot x=x\quad (x\in\mathcal{R}_*(\Gamma,\Gamma',\phi)) \\ \widetilde{\theta_\alpha\theta'_{\alpha'}}\cdot\widetilde{\theta_\beta\theta'_{\beta'}}=-\sum_{v,v'\in V}A(\Gamma)^{vv'}(\alpha_{v'}\cdot\beta_{v'})\,\Sigma_v+\sum_{v,v'\in V'}A(\Gamma')^{vv'}(\alpha'_{v'}\cdot\beta'_{v'})\,\Sigma'_v, \\ \widetilde{\theta_\alpha\theta'_{\alpha'}}\cdot\Sigma_v=\Sigma_v\cdot\widetilde{\theta_\alpha\theta'_{\alpha'}}=-i(\alpha_v),\ \widetilde{\theta_\alpha\theta'_{\alpha'}}\cdot\Sigma'_v=\Sigma'_v\cdot\widetilde{\theta_\alpha\theta'_{\alpha'}}=i'(\alpha'_v), \\ i(\alpha_v)\cdot\widetilde{\theta_\beta\theta'_{\beta'}}=-\widetilde{\theta_\beta\theta'_{\beta'}}\cdot i(\alpha_v)=-\alpha_v\cdot\beta_v,\ i'(\alpha'_v)\cdot\widetilde{\theta_\beta\theta'_{\beta'}}=-\widetilde{\theta_\beta\theta'_{\beta'}}\cdot i'(\alpha'_v)=\alpha'_v\cdot\beta'_v, \\ \qquad\text{for any } \alpha,\alpha'\in H(\Gamma),\ \beta,\beta'\in H(\Gamma'),\ \widetilde{\theta_\alpha\theta'_{\alpha'}},\widetilde{\theta_\beta\theta'_{\beta'}}\in R(\Gamma,\Gamma',\phi) \\ \Sigma_v\cdot\Sigma_{v'}=-A(\Gamma)_{vv'}\,pt,\ \Sigma'_v\cdot\Sigma'_{v'}=A(\Gamma')_{vv'}\,pt, \\ x\cdot y=0\quad (x,y:\ \text{otherwise}). \end{cases}$$

Note that $\mathcal{R}_*(\Gamma,\Gamma',\phi)$ is not necessarily associative in general.

By Lemma 29, 30 in Section 6 below, we have the following

Theorem 7. *Let $\Gamma=(V,E,\omega)$, $\Gamma'=(V',E',\omega')$ be two tree Seifert graphs satisfying the condition (Ndeg), and let $\phi=(L_*;i,i',\bullet)$ be a morphism between homology rings $(H_*(M(\Gamma);\mathbb{Z}),\bullet)$, $(H_*(M(\Gamma');\mathbb{Z}),\bullet)$ in the $\mathcal{L}_3$ category. If there exists a cobordism $(W;M(\Gamma),M(\Gamma'))$ realizing $\phi=(L_*;i,i',\bullet)$, then there exists an injective ring homomorphism $\mathcal{R}_*(\Gamma,\Gamma',\phi)\to H_*(Z;\mathbb{Q})$, and hence $\mathcal{R}_*(\Gamma,\Gamma',\phi)$ must be an associative ring.*

Remark 8. *The V-manifold Z can be regarded as a rational homology manifold, and hence the homology ring $H_*(Z;\mathbb{Q})$ can be defined over the rationals $\mathbb{Q}$.*

2.2. The 10/8-inequality and the quadruple cup products

The 11/8-conjecture, due to Y. Matsumoto, states that for any closed spin 4-manifold the second Betti number is greater than or equal to 11/8 times the absolute value of the signature. A weaker inequality, called the 10/8-inequality, was first proved by M. Furuta using a technique based on the finite-dimensional approximation of the Seiberg-Witten equation. This inequality was proved under the assumption that the first Betti number is zero, but this condition can always be realized by surgeries. However, by dealing with the first cohomology of closed 4-manifolds, M. Furuta and Y. Kametani improved the 10/8-inequality for closed spin 4-manifolds with positive first Betti numbers by considering $Pin(2)$-equivariant maps between sphere bundles over the Jacobi tori of 4-manifolds constructed from the finite-dimensional approximation of the Seiberg-Witten equation [**9**]. Their result is based on the joint work of by M. Furuta, Y. Kametani, H. Matsue, and N. Minami [**8**] on the stable homotopy version [**7**] of the Seiberg-Witten invariants. The improved inequality contains terms that come from the quadruple cup product structures on closed spin 4-manifolds.

In the paper [**2**], we extended the Furuta-Kametani-10/8-inequality to the case of V-manifolds. For various definitions concerning V-manifolds, see [**11**].

Let $((M,c),(X,\hat{c}))$ be a pair consisting of a closed 3-manifold M with spin structure c and a closed spin 4-V-manifold X with V-spin structure $\hat{c}$ satisfying $\partial(X,\hat{c}) = (M,c)$. Then we can define an integral lift of the Rochlin invariant, which we call the w-invariant (Definition 19),

$$w((M,c),(X,\hat{c})) \equiv -\mu(M,c) \mod 16.$$

The w-invariant was defined in joint work with M. Furuta on applications of the 10/8-inequality [**4**]. In fact, we proved the vanishing of the V-indices of the Dirac operators on closed 4-V-manifolds X with $b_2^{\pm}(X) \leqslant 2$, which implies the homology cobordism invariance of w in a certain class of homology 3-spheres. In joint work with M. Furuta and M. Ue [**5**], and in the extensive work of N. Saveliev [**12**], it is shown that the Neumann-Siebenmann invariant for plumbed homology 3-spheres is equal to the w-invariant for some auxiliary plumbed V-manifold and its V-spin structure. Recently, M. Ue proved that the Ozsváth-Szabó correction term is equal to the Neumann-Siebenmann invariant (and hence the w-invariant) for a large class of plumbed rational homology 3-spheres [**15**].

We introduce the following quadruple product structure.

Lemma 9. *Let $\Gamma = (V,E,\omega)$, $\Gamma' = (V',E',\omega)$ be two tree Seifert graphs and let $\phi = (L_*; i, i', \bullet)$ be a morphism between the homology rings $(H_*(M(\Gamma)),\bullet)$, $(H_*(M(\Gamma')),\bullet)$ in the $\mathcal{L}_3$ category. Then a quadruple product $q(\Gamma,\Gamma',\phi)$:*

$\mathcal{R}_3(\Gamma, \Gamma', \phi)^{\otimes 4} \to \mathbb{Q}$ is calculated to be

$$\widetilde{\theta_\alpha \theta'_{\alpha'}} \cdot \widetilde{\theta_\beta \theta'_{\beta'}} \cdot \widetilde{\theta_\gamma \theta'_{\gamma'}} \cdot \widetilde{\theta_\delta \theta'_{\delta'}}$$
$$= - \sum_{v,v' \in V} A(\Gamma)^{vv'} (\alpha_v \cdot \beta_v)(\gamma_{v'} \cdot \delta_{v'}) + \sum_{v,v' \in V'} A(\Gamma')^{vv'} (\alpha'_v \cdot \beta'_v)(\gamma'_{v'} \cdot \delta'_{v'})$$

By Lemma 32, we generalize the Theorem in [**2**] to obtain

Theorem 10. *Let $\Gamma = (V, E, \omega)$, $\Gamma' = (V', E', \omega')$ be two tree Seifert graphs such that*

1. *$b^{\pm}(\Gamma) + b^{\mp}(\Gamma') \leqslant 2m + 2$,*
2. *the intersection matrices $A(\Gamma)$, $A(\Gamma')$ of the plumbed V-manifold $P(\Gamma)$, $P(\Gamma')$ with respect to the standard basis are non-singular, and*
3. *the Euler numbers are non-zero $e(\omega_v) \neq 0$, $e(\omega'_{v'}) \neq 0$ for all $v \in V$ and $v' \in V'$.*

Let $\phi = (L_*; i, i', \bullet)$ be a morphism between the homology rings $(H_*(M(\Gamma); \mathbb{Z}), \bullet)$, $(H_*(M(\Gamma'); \mathbb{Z}), \bullet)$ in the $\mathcal{L}_3$ category. Suppose that the associated plumbed 3-manifolds with spin structures $(M(\Gamma), c)$, $(M(\Gamma'), c')$ are spin cobordant $(M(\Gamma), c) \simeq^{\phi}_{(W,\tilde{c})} (M(\Gamma'), c')$ for some compact spin 4-manifold $(W, \tilde{c})$ inducing an algebraic morphism $\phi = (L_*; i, i', \bullet) \cong (L_*(W; \mathbb{Z}); i_*, i'_*, \bullet)$ such that $L_*(W; \mathbb{Z}) = H_*(W; \mathbb{Z})$. If there exists an injective homomorphism

$$h : H^1(\sharp_{i=1}^m T_i^4; \mathbb{Z}) \to R(\Gamma, \Gamma', \phi) \subset \mathcal{R}_3(\Gamma, \Gamma', \phi)$$

such that

$$h(x) \cdot h(y) \cdot h(z) \cdot h(w) \equiv \langle x \cup y \cup z \cup w, [\sharp_{i=1}^m T_i^4] \rangle \mod 2,$$

for any $x, y, z, w \in H^1(\sharp_{i=1}^m T_i^4; \mathbb{Z})$, where the T_i^4's are m-copies of the 4-torus T^4. Then we have

$$w((M(\Gamma), c), (P(\Gamma), \hat{c})) = w((M(\Gamma'), c'), (P(\Gamma), \hat{c}')).$$

Example 11. *Let Γ and Γ' be two Seifert graphs in the above Example 3. Let $\tilde{\Gamma} = \Gamma \sharp s \cdot (\{0\}, \emptyset)$ be the Seifert graph consisting of the disjoint union of Γ and s-vertices for $s \leqslant 2$, with no edges and with Seifert invariants*

$$\tilde{\omega}(v) = \omega(v), \ v \in V,$$
$$\tilde{\omega}_\ell(0) = \omega = \{0; (a_1, b_1), \ldots, (a_n, b_n)\}, \ 1 \leqslant r \leqslant s.$$

We also define $\tilde{\Gamma}' = \Gamma' \sharp s \cdot (\{0\}, \emptyset)$ and the Seifert invariants $\tilde{\omega}'$ by $\tilde{\omega}'(v) = \omega'(v)$ for $v \in V'$ and $\tilde{\omega}'(0) = \omega$. Suppose that one or more of the a_i's are even for $\tilde{\omega}(0)$, so that the associated disk V-bundle admits a V-spin structure. Then the plumbed 3-manifold $M(\tilde{\Gamma})$ is the connected sum $(M(\Gamma), c) \sharp s \cdot (\Sigma, c_\Sigma)$ of the plumbed 3-manifold $(M(\Gamma), c)$ and s-copies of the Seifert rational homology 3-sphere (Σ, c_Σ) of Seifert invariant $\tilde{\omega}(0)$. The plumbed 4-V-manifold $P(\tilde{\Gamma})$ is the boundary connected sum $(P(\Gamma), \hat{c}) \natural s \cdot (E, c_E)$, where E is the associated disk V-bundle of the S^1-fibration Σ. Suppose the Euler number $e(\omega)$ is positive then we have $b_2^{\pm}(P(\Gamma) \natural s \cdot E) +$

$b_2^{\mp}(P(\Gamma')\natural\, s\cdot E)\leqslant 2+s$. *Now the quadruple product associated to the 3-cocycles* $\widetilde{\theta_{\alpha_{11}}\theta'_{\alpha'_{11}}},\widetilde{\theta_{\alpha_{12}}\theta'_{\alpha'_{12}}},\widetilde{\theta_{\alpha_{13}}\theta'_{\alpha'_{21}}},\widetilde{\theta_{\alpha_{14}}\theta'_{\alpha'_{22}}}\in R(\Gamma,\Gamma',\phi)$ *satisfies*

$$\widetilde{\theta_{\alpha_{11}}\theta'_{\alpha'_{11}}}\cdot\widetilde{\theta_{\alpha_{12}}\theta'_{\alpha'_{12}}}\cdot\widetilde{\theta_{\alpha_{13}}\theta'_{\alpha'_{21}}}\cdot\widetilde{\theta_{\alpha_{14}}\theta'_{\alpha'_{22}}}=-5\equiv 1 \bmod 2$$

In fact, by noting that

$$i(\theta_{\alpha_{11}})=i'(\theta'_{\alpha'_{11}}),\ i(\theta_{\alpha_{12}})=i'(\theta'_{\alpha'_{12}}),\ i(\theta_{\alpha_{13}})=i'(\theta'_{\alpha'_{21}}),\ i(\theta_{\alpha_{14}})=i'(\theta'_{\alpha'_{22}}),$$

and

$$A(\Gamma)^{-1}=A(\Gamma')^{-1}=\begin{pmatrix}2 & -3\\ -3 & 6\end{pmatrix},$$

and applying Lemma 32, we obtain

$$\begin{aligned}&q(\Gamma,\Gamma',\phi)\left(\widetilde{\theta_{\alpha_{11}}\theta'_{\alpha'_{11}}}\otimes\widetilde{\theta_{\alpha_{12}}\theta'_{\alpha'_{12}}}\otimes\widetilde{\theta_{\alpha_{13}}\theta'_{\alpha'_{21}}}\otimes\widetilde{\theta_{\alpha_{14}}\theta'_{\alpha'_{22}}}\right)\\&=-A(\Gamma)^{11}(\alpha_{11}\cdot\alpha_{12})(\alpha_{13}\cdot\alpha_{14})+A(\Gamma')^{12}(\alpha'_{11}\cdot\alpha'_{12})(\alpha'_{21}\cdot\alpha'_{22})\\&=-2\cdot 1\cdot 1+(-3)\cdot 1\cdot 1\\&=-5\equiv 1 \bmod 2.\end{aligned}$$

On the other hand, the w-invariants are calculated to be

$$\begin{aligned}&w((M(\Gamma),c)\,\sharp\, s\cdot(\Sigma,c_\Sigma),(P(\Gamma),\hat{c})\,\natural\, s\cdot(E,c_E))\\&-w((M(\Gamma'),c')\,\sharp\, s\cdot(\Sigma,c_\Sigma),(P(\Gamma'),\hat{c}')\,\natural\, s\cdot(E,c_E))\\&=w((M(\Gamma),c),(P(\Gamma),\hat{c}))-w((M(\Gamma'),c'),(P(\Gamma'),\hat{c}'))\\&=12-(-4)=16\neq 0\end{aligned}$$

and $2+s\leqslant 2+2$. *Hence by Theorem 10, we see that there exists no cobordism* $(W,\tilde{c})$ *between* $(M(\Gamma),c)\,\sharp\, s\cdot(\Sigma,c_\Sigma)$ *and* $(M(\Gamma'),c')\,\sharp\, s\cdot(\Sigma,c_\Sigma)$ *inducing* $\phi=(L_*;i,i',\bullet)$ *for* $s\leqslant 2$ *such that* $L_*(W;\mathbb{Z})=H_*(W;\mathbb{Z})$. *Note that the difference is divisible by* 16 *and hence this cannot be detected by using the Rohlin invariant.*

2.3. Associativity of intersection products on homology

On the other hand, we can prove the above statement in Approach 2. Motivated by Lemma 32, we introduce the following triple product.

Definition 12.
Let $\Gamma=(V,E)$, $\Gamma'=(V',E')$ *be two Seifert graphs and let* $\phi=(K_*;i,i',\bullet)$ *be a morphism between homology rings* $(H_*(M(\Gamma);\mathbb{Z}),\bullet)$, $(H_*(M(\Gamma');\mathbb{Z}),\bullet)$ *in the category* $\mathcal{L}_3$. *Then a triple product* $t(\Gamma,\Gamma',\phi):\mathcal{R}_3(\Gamma,\Gamma',\phi)^{\otimes 3}\to\mathcal{R}_1(\Gamma,\Gamma',\phi)$ *is*

defined and calculated to be

$$\begin{aligned}
& t(\Gamma,\Gamma',\phi)\left(\widetilde{\theta_\alpha\theta'_{\alpha'}}\otimes\widetilde{\theta_\beta\theta'_{\beta'}}\otimes\widetilde{\theta_\gamma\theta'_{\gamma'}}\right)\\
&\equiv\left(\widetilde{\theta_\alpha\theta'_{\alpha'}}\cdot\widetilde{\theta_\beta\theta'_{\beta'}}\right)\cdot\widetilde{\theta_\gamma\theta'_{\gamma'}}-\widetilde{\theta_\alpha\theta'_{\alpha'}}\cdot\left(\widetilde{\theta_\beta\theta'_{\beta'}}\cdot\widetilde{\theta_\gamma\theta'_{\gamma'}}\right)\\
&=\sum_{v,v'\in V}A(\Gamma)^{vv'}(\alpha_v\cdot\beta_v)\,i(\gamma_{v'})-\sum_{v,v'\in V'}A(\Gamma')^{vv'}(\alpha'_v\cdot\beta'_v)\,i'(\gamma'_{v'})\\
&-\sum_{v,v'\in V}A(\Gamma)^{vv'}(\beta_v\cdot\gamma_v)\,i(\alpha_{v'})+\sum_{v,v'\in V'}A(\Gamma')^{vv'}(\beta'_v\cdot\gamma'_v)\,i'(\alpha'_{v'})\\
&\in\mathcal{R}_1(\Gamma,\Gamma',\phi).
\end{aligned}$$

In fact, we obtain the following criterion by Lemma 32.

Theorem 13. *Let $\Gamma=(V,E,\omega)$, $\Gamma'=(V',E',\omega')$ be two tree Seifert graphs satisfying the condition (Ndeg).*

Let $M(\Gamma)$ and $M(\Gamma')$ be the associated plumbed 3-manifolds, and $\phi=(L_*;i,i',\bullet)$ be a morphism between homology rings $(H_*(M(\Gamma)),\bullet)$, $(H_*(M(\Gamma')),\bullet)$ in the $\mathcal{L}_3$ category. If the triple product is not zero,

$$t(\Gamma,\Gamma',\phi)\left(\widetilde{\theta_\alpha\theta'_{\alpha'}}\otimes\widetilde{\theta_\beta\theta'_{\beta'}}\otimes\widetilde{\theta_\gamma\theta'_{\gamma'}}\right)\not\equiv 0\in\mathcal{R}_1(\Gamma,\Gamma',\phi)$$

for some triple $\widetilde{\theta_\alpha\theta'_{\alpha'}},\widetilde{\theta_\beta\theta'_{\beta'}},\widetilde{\theta_\gamma\theta'_{\gamma'}}\in R(\Gamma,\Gamma',\phi)$, then there exists no cobordism W between $M(\Gamma)$ and $M(\Gamma')$ inducing ϕ such that $L_*(W;\mathbb{Z})=H_*(W;\mathbb{Z})$.

Example 14. *In the above Example 3, we obtain*

$$t(\Gamma,\Gamma',\phi)\left(\widetilde{\theta_{\alpha_{11}}\theta'_{\alpha'_{11}}}\otimes\widetilde{\theta_{\alpha_{12}}\theta'_{\alpha'_{12}}}\otimes\widetilde{\theta_{\alpha_{13}}\theta'_{\alpha'_{21}}}\right)=5\delta_3\neq 0.$$

In fact, by Definition 12, we have

$$\begin{aligned}
& t(\Gamma,\Gamma',\phi)\left(\widetilde{\theta_{\alpha_{11}}\theta'_{\alpha'_{11}}}\otimes\widetilde{\theta_{\alpha_{12}}\theta'_{\alpha'_{12}}}\otimes\widetilde{\theta_{\alpha_{13}}\theta'_{\alpha'_{21}}}\right)\\
&=A(\Gamma)^{11}(\alpha_{11}\cdot\alpha_{12})\,i(\alpha_{13})-A(\Gamma')^{12}(\alpha'_{11}\cdot\alpha'_{12})\,i'(\alpha'_{21})\\
&-A(\Gamma)^{11}(\alpha_{12}\cdot\alpha_{13})\,i(\alpha_{11})+A(\Gamma')^{11}(\alpha'_{12}\cdot 0)\,i'(\alpha'_{11})+A(\Gamma')^{21}(0\cdot\alpha'_{21})\,i'(\alpha'_{11})\\
&=2\cdot 1\cdot i(\alpha_{13})-(-3)\cdot 1\cdot i'(\alpha'_{21})=5i(\alpha_{13})\\
&=5\delta_3\neq 0.
\end{aligned}$$

Hence by Theorem 13 we see that there exists no cobordism W between $M(\Gamma)\ \sharp\ s\cdot\Sigma$ and $M(\Gamma')\ \sharp\ s\cdot\Sigma$ inducing $\phi=(L_*;i,i')$ such that $L(W;\mathbb{Z})=H_*(W;\mathbb{Z})$ for any rational homology 3-sphere Σ and non-negative integer s.

3. Cobordism category of 3-manifolds and an abstract category of graded commutative rings

In this section, we give a precise definition of the $\mathcal{L}_3$ category, a category of graded commutative ring with certain algebraic morphisms which was introduced in order to investigate the cobordism category $\mathcal{C}_3$ of 3-manifolds.

3.1. Cobordism category of 3-manifolds

Let $(\mathcal{C}_3, \sqcup, \emptyset)$ be the category whose set of objects $\mathrm{ob}(\mathcal{C}_3)$ is the set of all disjoint unions M of closed oriented 3-manifolds and whose set of morphisms $\mathcal{C}_3(M, M')$ between $M \in \mathrm{ob}(\mathcal{C}_3)$ and $M' \in \mathrm{ob}(\mathcal{C}_3)$ is the set of all cobordisms $(W; M, M')$ between M and M', that is, W is a compact oriented smooth 4-manifold with boundary $\partial W \cong M \sqcup (-M')$. If M, M', and M'' are three objects in $\mathcal{C}_3$, then the composite operation on morphisms $\mathcal{C}_3(M, M') \times \mathcal{C}_3(M', M'') \to \mathcal{C}_3(M, M'')$, $(W, W') \longmapsto W \cup_{M'} W'$ is defined by gluing the 4-manifolds W and W' along the boundary component M'. There exists a bifunctor $\sqcup : \mathcal{C}_3 \times \mathcal{C}_3 \to \mathcal{C}_3$ defined by the disjoint union $\sqcup : \mathrm{ob}(\mathcal{C}_3) \times \mathrm{ob}(\mathcal{C}_3) \to \mathrm{ob}(\mathcal{C}_3)$, $(M_1, M_2) \longmapsto M_1 \sqcup M_2$ and $\sqcup : \mathcal{C}_3(M_1, M_1') \times \mathcal{C}_3(M_2, M_2') \to \mathcal{C}_3(M_1 \sqcup M_2, M_1' \sqcup M_2')$, $(W_1, W_2) \longmapsto W_1 \sqcup W_2$, and the empty set $\emptyset \in \mathrm{ob}(\mathcal{C}_3)$ defines the unit element. Hence $(\mathcal{C}_3, \sqcup, \emptyset)$ defines a monoidal category.

Let $\mathcal{C}_3^{\mathrm{spin}}$ be the category whose set of objects $\mathrm{ob}(\mathcal{C}_3^{\mathrm{spin}})$ is the set of all disjoint unions (M, c) of closed oriented 3-manifolds with spin structures c and whose set of morphisms $\mathcal{C}_3^{\mathrm{spin}}((M, c), (M', c'))$ between $(M, c) \in \mathrm{ob}(\mathcal{C}_3^{\mathrm{spin}})$ and $(M', c') \in \mathrm{ob}(\mathcal{C}_3^{\mathrm{spin}})$ is the set $((W, \tilde{c}); (M, c), (M', c'))$ of all spin cobordisms between (M, c) and (M', c'), that is, $(W, \tilde{c})$ is a compact spin smooth 4-manifold with boundary $\partial(W, \tilde{c}) \cong (M, c) \sqcup (-(M', c'))$. Then the monoidal category of 3-manifolds with spin structures $(\mathcal{C}_3^{\mathrm{spin}}, \sqcup, \emptyset)$ is defined similarly.

3.2. A category of graded commutative rings

Let $(\mathcal{L}_3, \oplus, 0)$ be a category whose set of objects $\mathrm{ob}(\mathcal{L}_3)$ is the set of all pairs $(H_*, \bullet)$ composed of

1. a graded $\mathbb{Z}$-module $H_* = \bigoplus_{k=0}^{3} H_k$ of dimension 3 such that $H_0 \cong \mathbb{Z}^c$ for some $c \in \mathbb{Z}_{\geqslant 0}$,
2. a graded commutative product $\bullet : H_k \otimes H_\ell \to H_{k+\ell-3}$, i.e. $\alpha_k \cdot \beta_\ell = (-1)^{(3-k)(3-\ell)} \beta_\ell \cdot \alpha_k$ for $\alpha_k \in H_k$, $\beta_\ell \in H_\ell$, satisfying the following conditions:
   (a) $\bullet : H_k \otimes H_{3-k} \to H_0 \cong \mathbb{Z}^c \xrightarrow{\varepsilon} \mathbb{Z}$ induces an isomorphism $H_{3-k} \otimes \mathbb{Q} \cong (H_k \otimes \mathbb{Q})^*$, where $\varepsilon : \mathbb{Z}^c \to \mathbb{Z}$ is given by $\varepsilon(\oplus_{i=1}^{c} m_i) = \sum_{i=1}^{c} m_i$,
   (b) $H_3 \cong \mathbb{Z}^c$ and the element $\mu \in H_3 \cong \mathbb{Z}^c$ corresponding to $\oplus_{i=1}^{c} 1 \in \mathbb{Z}^c$ is a unit element with respect to the product $\bullet : H_k \otimes H_3 \to H_k$,

and whose set of morphisms $\mathcal{L}_3((H_*, \bullet), (H'_*, \bullet))$ between two objects $(H_*, \bullet)$ and $(H'_*, \bullet)$ is the set of all quadruples $(L_*; i, i', \bullet)$ composed of

1. a graded $\mathbb{Z}$-module $L_* = \bigoplus_{k=0}^{4} L_*$ of dimension 4 such that $L_0 \cong \mathbb{Z}^d$ for some $d \in \mathbb{Z}_{\geqslant 0}$,
2. two homomorphisms $H_* \xrightarrow{i} L_* \xleftarrow{i'} H'_*$ such that
   (a) $L_k = \mathrm{Im}\left(H_k \oplus H'_k \xrightarrow{i+i'} L_k\right)$,
   (b) if R_k are $\mathbb{Z}$-modules defined by
   $$R_k = \mathrm{Ker}\left(H_{k-1} \oplus H'_{k-1} \xrightarrow{i+i'} L_{k-1}\right),$$

then the bilinear pairings $\bullet : R_k \otimes L_\ell \to L_{k+\ell-4}$, $\bullet : L_\ell \otimes R_k \to L_{k+\ell-4}$ induced by $\bullet : H_k \otimes H_\ell \to H_{k+\ell-3}$ and $\bullet : H'_k \otimes H'_\ell \to H'_{k+\ell-3}$ are graded-commutative with each other in the sense that $\alpha_k \cdot \beta_\ell = (-1)^{(4-k)(4-\ell)} \beta_\ell \cdot \alpha_k$ for $\alpha_k \in R_k$, $\beta_\ell \in L_\ell$, and the pairing $\bullet$ satisfies the following conditions:

i. the following commutative diagram holds,

$$\begin{array}{ccccc}
H_k & \otimes & H_\ell & \stackrel{\bullet}{\to} & H_{k+\ell-3} \\
{\scriptstyle \partial\otimes \mathrm{id}_{H_\ell}} & \uparrow & & & \\
R_{k+1} & \otimes & H_\ell & & \downarrow i \\
{\scriptstyle \mathrm{id}_{R_{k+1}}\otimes i} & \downarrow & & & \\
R_{k+1} & \otimes & L_\ell & \stackrel{\bullet}{\to} & L_{k+\ell-3} \\
{\scriptstyle \mathrm{id}_{R_{k+1}}\otimes i'} & \uparrow & & & \\
R_{k+1} & \otimes & H'_\ell & & \uparrow i' \\
{\scriptstyle \partial'\otimes \mathrm{id}_{H'_\ell}} & \downarrow & & & \\
H'_k & \otimes & H'_\ell & \stackrel{\bullet}{\to} & H'_{k+\ell-3}
\end{array},$$

where ∂, ∂' are the natural projections,

$$\partial \oplus \partial' : R_k = \mathrm{Ker}\left(H_{k-1} \oplus H'_{k-1} \stackrel{i+i'}{\to} L_{k-1}\right) \to H_{k-1} \oplus H'_{k-1}.$$

ii. $\bullet : R_k \otimes L_{4-k} \to L_0 \cong \mathbb{Z}^d \stackrel{\varepsilon}{\to} \mathbb{Z}$ induces an isomorphism $L_{4-k} \otimes \mathbb{Q} \cong (R_k \otimes \mathbb{Q})^*$, where $\varepsilon : \mathbb{Z}^d \to \mathbb{Z}$ is given by $\varepsilon\left(\oplus_{i=1}^d m_i\right) = \sum_{i=1}^d m_i$;

iii. $R_4 \cong \mathbb{Z}^d$ and the element $\nu \in R_4 \cong \mathbb{Z}^d$ corresponding to $\oplus_{i=1}^d 1 \in \mathbb{Z}^d$ is a unit element with respect to the product $\bullet : R_4 \otimes L_k \to L_k$.

Let $(H, \bullet)$, $(H', \bullet)$, and $(H'', \bullet)$ be three objects in $\mathcal{L}_3$. Then there exists a composite operation on morphisms

$$\begin{aligned}
\mathcal{L}_3\left((H,\bullet),(H',\bullet)\right) \times \mathcal{L}_3\left((H',\bullet),(H'',\bullet)\right) &\to \mathcal{L}_3\left((H,\bullet),(H'',\bullet)\right), \\
\left((L_*; i, i', \bullet), (L'_*; i'', i''', \bullet)\right) &\longmapsto \left((L \circ L')_*; i, i'', \bullet\right),
\end{aligned}$$

where $(L \circ L')_* = \bigoplus_{k=0}^3 (L \circ L')_k$ is the graded $\mathbb{Z}$-module defined by

$$(L \circ L')_k \cong \mathrm{Im}\left(H_k \oplus H''_k \stackrel{i'+i'''}{\to} \mathrm{Coker}\left(H'_k \stackrel{i'\oplus(-i'')}{\to} L_k \oplus L'_k\right)\right),$$

and the inclusions $H_k \stackrel{i}{\to} (L \circ L')_k \stackrel{i'''}{\leftarrow} H''_k$ are defined by the natural map

$$i \oplus i''' : H_k \oplus H''_k \stackrel{i\oplus i'''}{\to} L_k \oplus L'_k \to \mathrm{Coker}\left(H'_k \stackrel{i'\oplus(-i'')}{\to} L_k \oplus L'_k\right).$$

We define $R \circ R'$ in the same way as R is defined by L:

$$\begin{aligned}(R \circ R')_k &= \operatorname{Ker}\left(H_{k-1} \oplus H''_{k-1} \stackrel{i'+i'''}{\to} (L \circ L')_k\right) \\ &\cong \operatorname{Ker}\left(H_{k-1} \oplus H''_{k-1} \stackrel{i'+i'''}{\to} \frac{L_{k-1} \oplus L'_{k-1}}{(i' \oplus (-i''))\, H'_{k-1}}\right) \\ &\cong \operatorname{Im}\left(\operatorname{Ker}\left(R_k \oplus R'_k \stackrel{\partial'-\partial''}{\to} H'_{k-1}\right) \stackrel{\partial+\partial'''}{\to} H_{k-1} \oplus H''_{k-1}\right).\end{aligned}$$

Then the product $\bullet : (R \circ R')_k \otimes (L \circ L')_\ell \to (L \circ L')_{k+\ell-4}$ is induced by the natural product

$$\begin{aligned}\bullet\ : &\operatorname{Ker}\left(R_k \oplus R'_k \stackrel{\partial'-\partial''}{\to} H'_{k-1}\right) \otimes \operatorname{Coker}\left(H'_\ell \stackrel{i'\oplus(-i'')}{\to} L_\ell \oplus L'_\ell\right) \\ &\to \operatorname{Coker}\left(H'_{k+\ell-1} \stackrel{i'\oplus(-i'')}{\to} L_{k+\ell-5} \oplus L'_{k+\ell-5}\right)\end{aligned}$$

which is defined by using the products $\bullet : R_k \otimes L_\ell \to L_{k+\ell-4}$ and $\bullet : R'_k \otimes L'_\ell \to L'_{k+\ell-4}$. Then we have the following

Proposition 15. *The product* $\bullet : (R \circ R')_k \otimes (L \circ L')_{4-k} \to (L \circ L')_0 \cong \mathbb{Z}^{d''} \stackrel{\varepsilon}{\to} \mathbb{Z}$ *induces an isomorphism* $(R \circ R')_k \otimes \mathbb{Q} \cong ((L \circ L')_{4-k} \otimes \mathbb{Q})^*$.

Proof. Note that if $D : R_{k+1} \otimes \mathbb{Q} \to (L_{3-k} \otimes \mathbb{Q})^*$ and $D : H_k \otimes \mathbb{Q} \to (H_{3-k} \otimes \mathbb{Q})^*$ are the duality isomorphisms, then $D \circ \partial = i^* \circ D$, so we have

$$\begin{aligned}&\left(\operatorname{Ker}\left(R_k \oplus R'_k \stackrel{\partial'-\partial''}{\to} H'_{k-1}\right) \otimes \mathbb{Q}\right)^* \\ &\cong \operatorname{Coker}\left((R_k \otimes \mathbb{Q})^* \oplus (R'_k \otimes \mathbb{Q})^* \stackrel{(\partial'-\partial'')^*}{\leftarrow} (H'_{k-1} \otimes \mathbb{Q})^*\right) \\ &\cong \operatorname{Coker}\left(L_{4-k} \oplus L'_{4-k} \stackrel{i'\oplus(-i'')}{\leftarrow} H'_{4-k}\right) \otimes \mathbb{Q}.\end{aligned}$$

Note also that the identity map $\mathrm{id} : H_{3-k} \oplus H''_{3-k} \to H_{3-k} \oplus H''_{3-k}$ induces a natural isomorphism

$$\begin{aligned}&\operatorname{Ker}\left(H_{3-k} \oplus H''_{3-k} \stackrel{i+i'''}{\to} \operatorname{Coker}\left(H'_{3-k} \stackrel{i'\oplus(-i'')}{\to} L_{3-k} \oplus L'_{3-k}\right)\right) \\ &\cong \operatorname{Im}\left(H_{3-k} \oplus H''_{3-k} \stackrel{\partial+\partial'''}{\leftarrow} \operatorname{Ker}\left(R_{4-k} \oplus R'_{4-k} \stackrel{\partial'-\partial''}{\to} H'_{3-k}\right)\right).\end{aligned}$$

Then, using these isomorphisms and the fact that

$$\operatorname{Im}(f : V \to W)^* \cong \operatorname{Im}(f^* : V^* \leftarrow W^*)$$

for any homomorphism $f : V \to W$ of $\mathbb{Q}$-vector spaces, we have

$$((L \circ L')_k \otimes \mathbb{Q})^*$$

$$= \left(\mathrm{Im}\left(H_k \oplus H''_k \overset{i+i'''}{\to} \mathrm{Coker}\left(H'_k \overset{i' \oplus (-i'')}{\to} L_k \oplus L'_k\right)\right) \otimes \mathbb{Q}\right)^*$$

$$\cong \mathrm{Im}\left((H_{3-k} \otimes \mathbb{Q}) \oplus (H''_{3-k} \otimes \mathbb{Q}) \overset{\partial + \partial'''}{\leftarrow} \mathrm{Ker}\left(R_{4-k} \oplus R'_{4-k} \overset{\partial' - \partial''}{\to} H'_{3-k}\right) \otimes \mathbb{Q}\right)$$

$$\cong \mathrm{Ker}\left((H_{3-k} \otimes \mathbb{Q}) \oplus (H''_{3-k} \otimes \mathbb{Q}) \overset{i+i'''}{\to} \mathrm{Coker}\left(H'_{3-k} \overset{i' \oplus (-i'')}{\to} L_{3-k} \oplus L'_{3-k}\right) \otimes \mathbb{Q}\right)$$

$$= (R \circ R')_{4-k} \otimes \mathbb{Q}.$$

□

Remark 16. *The composition*

$$\mathcal{L}_3\left((H_*, \bullet), (H'_*, \bullet)\right) \times \mathcal{L}_3\left((H'_*, \bullet), (H''_*, \bullet)\right) \to \mathcal{L}_3\left((H_*, \bullet), (H''_*, \bullet)\right),$$
$$\left((L_*; i, i', \bullet), (L'_*; i'', i''', \bullet)\right) \longmapsto \left((L \circ L')_*; i, i'', \bullet\right)$$

satisfies the associativity law, and the unit morphism is defined by $(H_* \overset{1}{\to} H_* \overset{1}{\leftarrow} H_*) \in \mathcal{L}_3\left((H_*, \bullet), (H_*, \bullet)\right)$. *Hence* $\mathcal{L}_3$ *defines a category.*

Remark 17. *The condition of the Poincaré duality should be replaced with the condition that the product* $\bullet : R_k \otimes L_{4-k} \to L_0 \cong \mathbb{Z}^d \overset{\varepsilon}{\to} \mathbb{Z}$ *induces an isomorphism* $\overline{L}_{4-k} \cong \mathrm{Hom}\left(\overline{R}_k, \mathbb{Z}\right)$ *on the free parts in the integral coefficients. But this condition may not be preserved under composition, and we need to impose certain torsion-free conditions on the boundary. For example, we may introduce a category* $\mathcal{L}_3^0$ *whose objects are* $(H_*, \bullet, F_*)$ *with additional submodules* $F_* \subset H_*$ *and whose morphisms between* $(H_*, \bullet, F_*)$ *and* $(H'_*, \bullet, F'_*)$ *are* $(L_*; i, i', \bullet)$, *with the following additional torsion-free conditions:*

1. *there exists a lifting* $\sigma : \overline{R}_k \to R_k$ *such that* $\partial \circ \sigma\left(\overline{R}_k\right) \subset H_{k-1}$ *and* $\partial' \circ \sigma\left(\overline{R}_k\right) \subset H'_{k-1}$ *are torsion-free,*
2. F_* *and* F'_* *satisfies the following condition:*
   (a) $L_k / \left(i\left(H_k\right) + i'\left(F'_k\right)\right)$ *and* $L_k / \left(i\left(H'_k\right) + i'\left(F_k\right)\right)$ *are torsion-free,*
   (b) $F_k = \partial R_{k+1}$ *and* $F'_k = \partial' R_{k+1}$.

But these conditions may cause extra complications, which are not essential for our discussion. So we will discuss this matter elsewhere.

Note that there exists a functor $H_* : \mathcal{C}_3 \to \mathcal{L}_3$ given by $M \longmapsto (H_*(M; \mathbb{Z}), \bullet)$ and $(W; M, M')$
$\longmapsto (L_*(W; \mathbb{Z}); i_*, i'_*, \bullet)$, where i_* and i'_* are the induced homomorphisms

$$H_*(M; \mathbb{Z}) \overset{i_*}{\to} L_*(W; \mathbb{Z}) \overset{i'_*}{\leftarrow} H_*(M'; \mathbb{Z}),$$

to the $\mathbb{Z}$-module $L_*(W; \mathbb{Z})$ defined by

$$L_*(W; \mathbb{Z}) = \mathrm{Im}\left(H_*(M; \mathbb{Z}) \oplus H_*(M'; \mathbb{Z}) \overset{i_* + i'_*}{\to} H_*(W; \mathbb{Z})\right).$$

Remark 18. *If we consider the category $\mathcal{L}_3^0$ instead of $\mathcal{L}_3$, we need to replace the category $\mathcal{C}_3$ with a category $\mathcal{C}_3^0$ to define a functor $H_* : \mathcal{C}_3^0 \to \mathcal{L}_3^0$. The objects of $\mathcal{C}_3^0$ are 3-manifolds M with an additional submodule- $F_* \subset H_*(M;\mathbb{Z})$, and the morphisms between (M, F_*) and (M', F'_*) are cobordisms $(W; M, M')$ with the following additional torsion-free conditions:*

1. $\overline{H}_k(W;\mathbb{Z}) / (i'_* + i''_*)\left(\overline{H}_k(M;\mathbb{Z}) \oplus \overline{H}_k(M';\mathbb{Z})\right)$ *is torsion-free,*
2. *there exists a lifting* $\sigma : \overline{R}_k(W;\mathbb{Z}) \to R_k(W;\mathbb{Z})$ *such that* $\partial_* \circ \sigma\left(\overline{R}_k(W;\mathbb{Z})\right) \subset H_{k-1}(M;\mathbb{Z})$ *and* $\partial'_* \circ \sigma\left(\overline{R}_k(W;\mathbb{Z})\right) \subset H_{k-1}(M';\mathbb{Z})$ *are torsion-free,*
3. F_* *and* F'_* *satisfies the following conditions:*
   - *(a)* $L_k(W;\mathbb{Z}) / (i_*(H_k(M;\mathbb{Z})) + i'_*(F'_k))$ *and* $L_k(W;\mathbb{Z}) / (i_*(F_k) + i'(H_k(M';\mathbb{Z})))$ *are tor-sion-free,*
   - *(b)* $F_k = \partial_* R_{k+1}(W;\mathbb{Z})$ *and* $F'_k = \partial'_* R_{k+1}(W;\mathbb{Z})$.

There exists a bifunctor $\oplus : \mathcal{L}_3 \times \mathcal{L}_3 \to \mathcal{L}_3$ defined by the direct sum

$$\oplus : \mathrm{ob}(\mathcal{L}_3) \times \mathrm{ob}(\mathcal{L}_3) \to \mathrm{ob}(\mathcal{L}_3), ((H_{1*}, \bullet), (H_{2*}, \bullet)) \longmapsto (H_{1*}, \bullet) \oplus (H_{2*}, \bullet)$$
$$\oplus : \mathcal{L}_3((H_{1*}, \bullet), (H'_{1*}, \bullet)) \times \mathcal{L}_3((H_{2*}, \bullet), (H'_{2*}, \bullet)) \to$$
$$\mathcal{L}_3((H_{1*}, \bullet) \oplus (H_{2*}, \bullet), (H'_{1*}, \bullet) \oplus (H'_{2*}, \bullet)),$$
$$((L_{1*}; i_1, i'_1, \bullet), (L_{2*}; i_2, i'_2, \bullet)) \longmapsto (L_{1*} \oplus L_{2*}; i_1 \oplus i_2, i'_1 \oplus i'_2, \bullet),$$

and the zero $\mathbb{Z}$-module $0 \in \mathrm{ob}(\mathcal{L}_3)$ defines the unit element. Hence $(\mathcal{L}_3, \oplus, 0)$ defines a monoidal category.

Similarly, a functor $H_*^{\mathrm{spin}} : \mathcal{C}_3^{\mathrm{spin}} \to \mathcal{L}_3$ is defined by $(M, c) \longmapsto (H_*(M;\mathbb{Z}), \bullet)$ and

$$((W, \tilde{c}); (M, c), (M', c')) \longmapsto (L_*(W;\mathbb{Z}); i_*, i'_*, \bullet).$$

We call two objects $(H_{1*}, \bullet)$ and $(H_{2*}, \bullet)$ equivalent if and only if there exists a graded ring isomorphism $f = \{f_k\} : H_{1*} \to H_{2*}$, i.e. $f_k : H_{1,k} \to H_{2,k}$ are $\mathbb{Z}$-module isomorphisms such that $f_{k+\ell-3}(x \cdot y) = f_k(x) \cdot f_\ell(y)$ for any $x \in H_{1k}$, $y \in H_{2\ell}$. We also call two morphisms $\phi_1 = (L_{1*}; i_1, i'_1, \bullet)$ (resp. $\phi_2 = (L_{2*}; i_2, i'_2, \bullet)$) between H_{1*} and H'_{1*} (resp. H_{2*} and H'_{2*}) are equivalent if and only if there exist two ring isomorphisms $f : H_{1*} \to H_{2*}$, $f' : H'_{1*} \to H'_{2*}$ and a graded $\mathbb{Z}$-module isomorphism $g : L_{1*} \to L_{2*}$ such that the following diagrams commute.

$$\begin{array}{ccccc} H_{1*} & \xrightarrow{i_1} & L_{1*} & \xleftarrow{i'_1} & H'_{1*} \\ f \downarrow & & g \downarrow & & f' \downarrow \\ H_{2*} & \xrightarrow{i_2} & L_{2*} & \xleftarrow{i'_2} & H'_{2*} \end{array},$$

$$\begin{array}{ccccc} R_{1,k+1} & \otimes & L_{1,\ell} & \xrightarrow{\bullet} & L_{1,k+\ell-3} \\ (f_{k+1} \oplus f'_{k+1}) \otimes g_\ell & \downarrow & & & g_{k+\ell-3} \downarrow \\ R_{2,k+1} & \otimes & L_{2,\ell} & \xrightarrow{\bullet} & L_{2,k+\ell-3} \end{array}.$$

4. Homology cobordism category of 3-manifolds and a category of isomorphisms of graded commutative rings

In this section, we consider the homology cobordism category $\mathcal{C}_3^H$ of 3-manifolds and then reduce the category $\mathcal{L}_3$ to a category $\mathcal{L}_3^H$ of isomorphisms of graded commutative rings.

4.1. Homology cobordism category of 3-manifolds

Let $(\mathcal{C}_3^H, \sharp, S^3)$ be the category whose set of objects $\mathrm{ob}\,(\mathcal{C}_3^H)$ is the set of all closed oriented 3-manifolds M and whose set of morphisms $\mathcal{C}_3^H(M, M')$ between $M \in \mathrm{ob}\,(\mathcal{C}_3^H)$ and $M' \in \mathrm{ob}\,(\mathcal{C}_3^H)$ is the set of all homology cobordisms $(W; M, M')$ between M and M'. If M, M', and M'' are three objects in $\mathcal{C}_3^H$, then the composite operation on morphisms $\mathcal{C}_3^H(M, M') \times \mathcal{C}_3^H(M', M'') \to \mathcal{C}_3^H(M, M'')$, $(W, W') \longmapsto W \cup_{M'} W'$ is defined by gluing 4-manifolds W and W' along the boundary component M'. There exists a bifunctor $\sharp : \mathcal{C}_3^H \times \mathcal{C}_3^H \to \mathcal{C}_3^H$ defined by the connected sum $\sharp : \mathrm{ob}\,(\mathcal{C}_3^H) \times \mathrm{ob}\,(\mathcal{C}_3^H) \to \mathrm{ob}\,(\mathcal{C}_3^H)$, $(M_1, M_2) \longmapsto M_1 \sharp M_2$ and the boundary connected sum $\natural : \mathcal{C}_3^H(M_1, M_1') \times \mathcal{C}_3^H(M_2, M_2') \to \mathcal{C}_3^H(M_1 \sharp M_2, M_1' \sharp M_2')$, $(W_1, W_2) \longmapsto W_1 \natural W_2$, and the 3-sphere $S^3 \in \mathrm{ob}\,(\mathcal{C}_3^H)$ defines the unit element. Hence $(\mathcal{C}_3^H, \sharp, S^3)$ defines a monoidal category. The monoidal category $(\mathcal{C}_3^{H,\mathrm{spin}}, \sharp, S^3)$, whose objects (M, c) are 3-manifolds M with spin structures c and morphisms between (M, c) and (M', c') are homology spin cobordisms $((W, \tilde{c}); (M, c), (M', c'))$, can be defined similarly. Note that there exists a natural functor $\mathcal{C}_3^H \to \mathcal{C}_3$, but the monoidal operations $\sharp$ and $\sqcup$ are not compatible. Let $(\mathcal{S}_3^H, \sharp, S^3)$ be the subcategory of $(\mathcal{C}_3^H, \sharp, S^3)$ generated by the set $\mathrm{ob}\,(\mathcal{S}_3^H)$ of all objects $\Sigma \in \mathrm{ob}\,(\mathcal{C}_3^H)$ such that $H_*(\Sigma; \mathbb{Z}) \cong H_*(S^3; \mathbb{Z})$. In particular, there exists a monoidal operation $\sharp : \mathcal{S}_3^H \times \mathcal{C}_3^H \to \mathcal{C}_3^H$. Since spin structures on homology 3-spheres are unique, the corresponding subcategory $(\mathcal{S}_3^{H,\mathrm{spin}}, \sharp, S^3)$ of $(\mathcal{C}_3^{H,\mathrm{spin}}, \sharp, S^3)$ is equivalent to $(\mathcal{S}_3^H, \sharp, S^3)$.

Fix $H_* \in \mathrm{ob}\,(\mathcal{L}_3)$ and let $\mathcal{C}_3^H(H_*)$ be the subcategory of $\mathcal{C}_3^H$ generated by the set $\mathrm{ob}\,(\mathcal{C}_3^H(H_*))$ of all objects $M \in \mathrm{ob}\,(\mathcal{C}_3^H)$ such that $H_*(M; \mathbb{Z}) \cong H_*$. Then $\mathcal{C}_3^H = \bigcup_{H_* \in \mathrm{ob}(\mathcal{L}_3)} \mathcal{C}_3^H(H_*)$ and the monoidal operation $\sharp : \mathcal{S}_3^H \times \mathcal{C}_3^H \to \mathcal{C}_3^H$ induces $\sharp : \mathcal{S}_3^H \times \mathcal{C}_3^H(H_*) \to \mathcal{C}_3^H(H_*)$. Similarly, let $\mathcal{C}_3^{H,\mathrm{spin}}(H_*)$ be the subcategory of $\mathcal{C}_3^{H,\mathrm{spin}}$ generated by the set $\mathrm{ob}(\mathcal{C}_3^{H,\mathrm{spin}}(H_*))$ of all objects $(M, c) \in \mathrm{ob}(\mathcal{C}_3^{H,\mathrm{spin}})$ such that $H_*(M) \cong H_*$. Then we have $\mathcal{C}_3^{H,\mathrm{spin}} = \bigcup_{H_* \in \mathrm{ob}(\mathcal{L}_3)} \mathcal{C}_3^{H,\mathrm{spin}}(H_*)$ and $\sharp : \mathcal{S}_3^H \times \mathcal{C}_3^{H,\mathrm{spin}}(H_*) \to \mathcal{C}_3^{H,\mathrm{spin}}(H_*)$.

4.2. A category of isomorphisms of graded commutative rings

Let $(\mathcal{L}_3^H, \sharp, S)$ be a category whose objects $\mathrm{ob}\,(\mathcal{L}_3^H)$ are the same as $\mathrm{ob}\,(\mathcal{L}_3)$, except that the object $(H_*, \bullet)$ satisfies $H_0 = \mathbb{Z}$ and whose morphisms $\mathcal{L}_3^H((H_*, \bullet), (H_*', \bullet))$ between two objects $(H_*, \bullet)$ and $(H_*', \bullet)$ are the same as $\mathcal{L}_3((H_*, \bullet), (H_*', \bullet))$, except that the two homomorphisms $H_* \xrightarrow{i} L_* \xleftarrow{i'} H_*'$ are isomorphisms. This implies that $L_k \cong H_k \cong H'_k \cong R_{k+1}$ and the following diagram

$$\begin{array}{ccccc} H_k & \otimes & H_\ell & \overset{\bullet}{\to} & H_{k+\ell-3} \\ {\scriptstyle \partial\otimes \mathrm{id}_{H_\ell}} & \uparrow\cong & & & \\ R_{k+1} & \otimes & H_\ell & & \downarrow\cong i \\ {\scriptstyle \mathrm{id}_{R_{k+1}}\otimes i} & \downarrow\cong & & & \\ R_{k+1} & \otimes & L_\ell & \overset{\bullet}{\to} & L_{k+\ell-3} \\ {\scriptstyle \mathrm{id}_{R_{k+1}}\otimes i'} & \uparrow\cong & & & \\ R_{k+1} & \otimes & H'_\ell & & \uparrow\cong i' \\ {\scriptstyle \partial'\otimes \mathrm{id}_{H'_\ell}} & \downarrow\cong & & & \\ H'_k & \otimes & H'_\ell & \overset{\bullet}{\to} & H'_{k+\ell-3} \end{array}$$

commutes, and hence $H_* \overset{i}{\to} L_* \overset{i'}{\leftarrow} H'_*$ are in fact ring isomorphisms. Therefore we may define $\mathcal{L}_3^H((H_*,\bullet),(H'_*,\bullet))$ to be the set of all graded ring isomorphisms $\phi : H_* = \bigoplus_{k=0}^3 H_k \to H'_* = \bigoplus_{k=0}^3 H'_k$, $\phi(x\cdot y) = \phi(x)\cdot\phi(y)$ for any $x, y \in H_*$. If $(H,\bullet)$, $(H',\bullet)$, and $(H'',\bullet)$ are three objects in $\mathcal{L}_3^H$, then the composite operation on morphisms is defined as follows:

$$\begin{array}{ccc} \mathcal{L}_3^H((H_*,\bullet),(H'_*,\bullet)) \times \mathcal{L}_3^H((H'_*,\bullet),(H''_*,\bullet)) & \to & \mathcal{L}_3^H((H_*,\bullet),(H''_*,\bullet)) \\ (\phi_1,\phi_2) & \longmapsto & \phi_2\circ\phi_1 \end{array}$$

Note that there exists a functor $H_* : \mathcal{C}_3^H \to \mathcal{L}_3^H$ given by $M \longmapsto (H_*(M;\mathbb{Z}),\bullet)$ and $(W; M, M') \longmapsto \phi_W = i_*^{-1}\circ i'_*$.

There exists a bifunctor $\sharp : \mathcal{L}_3^H \times \mathcal{L}_3^H \to \mathcal{L}_3^H$ defined by the "connected sum",

$$\sharp : \mathrm{ob}\left(\mathcal{L}_3^H\right)\times \mathrm{ob}\left(\mathcal{L}_3^H\right) \to \mathrm{ob}\left(\mathcal{L}_3^H\right), ((H_{1*},\bullet),(H_{2*},\bullet)) \longmapsto (H_{1*}\sharp\, H_{2*},\bullet)$$

$$\begin{array}{ccc} \natural : \mathcal{L}_3^H((H_{1*},\bullet),(H'_{1*},\bullet))\times\mathcal{L}_3^H((H_{2*},\bullet),(H'_{2*},\bullet)) & \to & \mathcal{L}_3^H((H_{1*}\sharp\, H_{2*},\bullet),(H'_{1*}\sharp\, H'_{2*},\bullet)) \\ (\phi_1,\phi_2) & \longmapsto & \phi_1\natural\,\phi_2 \end{array}$$

where

$$(H_{1*}\sharp\, H_{2*})_k = \begin{cases} \mathbb{Z} & (k=0,3) \\ H_{1k}\oplus H_{2k} & (k=1,2) \end{cases},$$

with the product structure $\bullet$ defined naturally, and

$$(\phi_1\natural\,\phi_2)_k = \begin{cases} \mathrm{id}_{\mathbb{Z}} & (k=0,3) \\ \phi_{1k}\oplus\phi_{2k} & (k=1,2) \end{cases} : H_{1k}\sharp\, H_{2k} \to H'_{1k}\sharp\, H'_{2k}.$$

Note that $1 \in \mathbb{Z} = (H_{1*}\sharp\, H_{2*})_3$ satisfies $1\cdot x = x$ for any $x \in H_{1*}\sharp\, H_{2*}$. The "3-sphere" $S \in \mathrm{ob}\left(\mathcal{L}_3^H\right)$,

$$S_k = \begin{cases} \mathbb{Z} & (k=0,3) \\ 0 & (k=1,2) \end{cases}$$

defines the unit element. Hence $\left(\mathcal{L}_3^H, \sharp, S\right)$ defines a monoidal category.

4.3. Homology cobordism monoid

Let $(\mathcal{C}_3^H, \sharp, S^3)$ be the homology cobordism monoidal category. We define an equivalence relation $M \sim^H M'$, $M, M' \in \mathrm{ob}(\mathcal{C}_3^H)$ if and only if $\mathcal{C}_3^H(M, M') \neq \emptyset$. Let C_3^H be the abelian monoid defined by the quotient of $(\mathcal{C}_3^H, \sharp, S^3)$ by the equivalence relation $\sim^H$, and we call C_3^H the homology cobordism monoid. Note that the monoid corresponding to the subcategory $(\mathcal{S}_3^H, \sharp, S^3) \subset (\mathcal{C}_3^H, \sharp, S^3)$ is exactly the homology cobordism group Θ_3^H of homology 3-spheres. Then the monoidal operation $\sharp : \mathcal{S}_3^H \times \mathcal{C}_3^H \to \mathcal{C}_3^H$ induces the action $\sharp : \Theta_3^H \times C_3^H \to C_3^H$, and hence the homology cobordism monoid C_3^H is a Θ_3^H-space. Then for any $M \in C_3^H$, the inertia group $(\Theta_3^H)_M$,

$$(\Theta_3^H)_M = \{\Sigma \in \Theta_3^H \mid \Sigma \sharp M \sim^H M\}$$

is well-defined. Similarly, let $C_3^{H,\mathrm{spin}}$ be the abelian monoid obtained as the quotient of $(\mathcal{C}_3^{H,\mathrm{spin}}, \sharp, S^3)$ by the equivalence relation $\sim^{H,\mathrm{spin}}$ of homology spin cobordism. Then for any $(M, c) \in C_3^{H,\mathrm{spin}}$, the inertia group $(\Theta_3^H)^{\mathrm{spin}}_{(M,c)}$,

$$(\Theta_3^H)^{\mathrm{spin}}_{(M,c)} = \{\Sigma \in \Theta_3^H \mid \Sigma \sharp (M, c) \sim^{H,\mathrm{spin}} (M, c)\}$$

can be defined.

5. w-invariants

Let (M, c) be a closed oriented 3-manifold M with spin structure c, and let $(X, \hat{c})$ be a compact oriented 4-V-manifold with V-spin structure $\hat{c}$, satisfying $\partial(X, \hat{c}) = (M, c)$. Since the 3-dimensional spin cobordism group Ω_3^{spin} is zero, we can take a compact oriented 4-manifold W with a spin structure $\tilde{c}$, satisfying $\partial(W, \tilde{c}) = (-M, -c)$. Then we glue them along the boundary and obtain a closed oriented 4-V-manifold $Z = X \cup_M W$ with spin structure $\hat{c} = \hat{c} \cup_c \tilde{c}$. We fix a Riemannian V-metric on Z, and let $\mathcal{D}(Z)$ be the Dirac operator on Z associated with the V-spin structure $\hat{c}$. Then we define an invariant for the pair $((M, c), (X, \hat{c}))$ as follows. This invariant is an extension of the definition of the w-invariant [**4**], [**5**] for homology 3-spheres to the case of closed oriented 3-manifolds with $b_1 > 0$.

Definition 19.

$$w((M, c), (X, \hat{c})) = 8\,\mathrm{ind}_V \mathcal{D}(Z) + \mathrm{Sign}\,(W) \in \mathbb{Z}.$$

Remark 20. *By the excision property of the indices of the Dirac operators, and the Novikov additivity of the signature, this invariant does not depend on the choice of* (W, c_W).

Since the V-index of the Dirac operator is always divisible by 4, we see that the following proposition holds.

Proposition 21. *Let $\mu(M,c) \in \mathbb{Z}/16$ be the Rochlin invariant; then we have*

$$w((M,c),(X,\hat{c})) \equiv -\mu(M,c) \bmod 16.$$

By the excision properties of the indices of the Dirac operators, and the vanishing of the kernel of the Dirac operator on a round sphere, this invariant is additive under connected sums.

Proposition 22.

$$\begin{aligned} &w((M_1,c_1)\sharp(M_2,c_2),(X_1,\hat{c}_1)\natural(X_2,\hat{c}_2)) \\ &= w((M_1,c_1),(X_1,\hat{c}_1)) + w((M_2,c_2),(X_2,\hat{c}_2)) \end{aligned}$$

To state some properties of the invariant, we first recall some notation [**2**].

Definition 23. *Let k^+, k^- and r be non-negative integers. We define the set $\mathcal{X}(k^+,k^-;r)$ of all pairs $((M,c),(X,\hat{c}))$ composed of*

1. *(M,c) : a closed 3-manifold with a spin structure, and*
2. *$(X,\hat{c})$: a compact oriented 4-V-manifold with spin structure satisfying*
   (a) *$\partial(X,\hat{c}) = (M,c)$,*
   (b) *$b_2^+(X) \leqslant k^+$, $b_2^-(X) \leqslant k^-$, and*
   (c) *rank $\mathrm{Ker}\,(i_* : H_1(M;\mathbb{Q}) \to H_1(X;\mathbb{Q})) \leqslant r$.*

Then we define a set of 3-manifolds as follows.

$$\mathcal{Y}(k^+,k^-;r) = \{(M,c) | ((M,c),(X,\hat{c})) \in \mathcal{X}(k^+,k^-;r) \text{ for somc } (X,\hat{c})\}.$$

Remark 24. *$\mathcal{Y}(k^+,k^-;r)$ is not closed under connected sums. In fact, the connected sum defines a map*

$$\sharp : \mathcal{Y}(k_1^+,k_1^-;r_1) \times \mathcal{Y}(k_2^+,k_2^-;r_2) \to \mathcal{Y}(k_1^+ + k_2^+, k_1^- + k_2^-; \min(r_1,r_2)).$$

Then we have the following theorem [**2**].

Theorem 25. *Let k^+, k^- and r be non-negative integers satisfying $k^+ + k^- + r \leqslant 2$. Then the map*

$$w(k^+,k^-;r) : \mathcal{Y}(k^+,k^-;r) \ni (M,c) \longmapsto w((M,c),(X,\hat{c})) \in \mathbb{Z}$$

gives a homology spin cobordism invariant.

Theorem 26. *Suppose $(M,c) \in \mathrm{ob}(\mathcal{C}_3^{H,\mathrm{spin}})$ belongs to the class $\mathcal{Y}(k^+,k^-;r)$. If $\Sigma \in \left(\Theta_3^H\right)_M^{\mathrm{spin}}$ is in the class $\mathcal{Y}(l^+,l^-;0)$ with $k^+ + l^+ + k^- + l^- + r \leqslant 2$ then $w(l^+,l^-,0)(\Sigma) = 0$.*

Proof. Let (M,c) and Σ be as above. Then (M,c) and $\Sigma \sharp (M,c)$ belong to the class

$$\mathcal{Y}\left(k^+ + l^+, k^- + l^-; r\right).$$

Since $k^+ + l^+ + k^- + l^- + r \leqslant 2$, we apply Theorem 25 to $w\left(k^+ + l^+, k^- + l^-; r\right)$ and obtain

$$w\left(k^+ + l^+, k^- + l^-; r\right)(\Sigma \sharp (M,c)) = w\left(k^+ + l^+, k^- + l^-; r\right)(M,c).$$

By the additivity formula 22, we have

$$w\left(l^{+},l^{-};0\right)(\Sigma)+w\left(k^{+},k^{-};r\right)(M,c)=w\left(k^{+},k^{-};r\right)(M,c)$$

and therefore $w\left(l^{+},l^{-};0\right)(\Sigma)=0$. □

Example 27. *If one of a,b,c is even, then the Brieskorn homology 3-sphere $\Sigma(a,b,c)$ bounds a spin D^2-V-bundle X of Euler number $e=-1/(abc)$ associated to the S^1-fibration of $\Sigma(a,b,c)$ over a 2-sphere S^2. On the other hand, if all of a,b,c are odd, then $\Sigma(a,b,c)$ bounds a spin 4-V-manifold X with $b_2^{\pm}(X)=1$, constructed by using a "4-dimensional Seifert fibration," as in our joint work with M. Furuta and M. Ue [**5**]. Hence the pair $((\Sigma(a,b,c),c),(X,\hat{c}))$ belongs to $\mathcal{Y}(1,1;0)$. Therefore, if $(\Sigma(a,b,c),c)\in\left(\Theta_3^H\right)_M^{\mathrm{spin}}$, then $w(1,1;0)(\Sigma(a,b,c))=0$. Note that it is known that the w-invariant is equal to (-8)-times the Neumann-Siebenmann $\bar{\mu}$-invariant, $w((\Sigma(a,b,c),c),(X,\hat{c}))=-8\bar{\mu}(\Sigma(a,b,c))$, [**12**], [**5**]. Several sequence of the Brieskorn homology 3-spheres are known to bound contractible 4-manifolds due to the work of A. Casson and J. Harer [**1**].*

6. Plumbed 3-manifolds

Let $\Gamma=(V,E,\omega)$ be a Seifert graph. For simplicity, we assume that Γ is a tree graph. Let $P(\Gamma)$ be the plumbed 4-V-manifold with boundary obtained by plumbing according to Γ. For any vertices $v\in V$, we take the disk V-bundle $E_v\to\Sigma_v$ of Seifert invariant

$$\omega(v)=\left\{g_v;(a_{v1},b_{v1}),\ldots,(a_{vn_v},b_{vn_v})\right\},$$

where Σ_v is a closed V-surface of genus g_v, and if two vertices v and v' are connected by an edge $(v,v')\in E$ then we take a sufficiently small neighborhood $D^2\cong U_{v,v'}\subset S_v$ away from the singularity and glue the two disk V-bundles $E_v\to\Sigma_v$ by the map,

$$\phi_e:E_v|_{U_{v,v'}}\cong U_{v,v'}\times D^2\ni(z,w)\longmapsto(w,z)\in U_{v',v}\times D^2\cong E_{v'}|_{U_{v',v}}.$$

Remark 28. *We can also consider plumbing at singular points. In fact, this extension of the notion of plumbing is generalized to the notion of plumbed V-manifolds associated to decorated graphs by N. Saveliev [**12**].*

Then the surfaces Σ_v form a basis for the second homology $H_2(P(\Gamma),\mathbb{Q})$, and the intersection matrix $A(\Gamma)$ is given as follows:

$$A(\Gamma)_{vv'}=\begin{cases}e_v & v=v',\\ 1 & (v,v')\in E,\\ 0 & \text{otherwise},\end{cases}$$

where $e_v=\sum_{i=1}^{n_v}\frac{b_{vi}}{a_{vi}}$ is the Euler number of the disk V-bundle $E_v\to\Sigma_v$. The boundary $M(\Gamma)=\partial P(\Gamma)$ is a smooth 3-manifold, and $M(\Gamma)$ is a homology 3-sphere if and only if the following conditions hold:

$$\begin{cases} \Gamma : \text{a tree graph}, \\ g_v = 0 \text{ for any } v \in V, \\ a_{v1}, \ldots, a_{vn_v} : \text{pairwise coprime, and} \\ \det A(\Gamma) = \pm \dfrac{1}{\prod_{v\in V} \alpha_v}, \quad \alpha_v = \prod_{i=1}^{n_v} a_{vi}. \end{cases} \cdots (HS)$$

If Γ satisfies the condition

$$\begin{cases} \exists a_{vi} : \text{even or} \\ \forall a_{vi} : \text{odd and } \sum_{i=1}^{n_v} b_{vi} : \text{even} \end{cases} \quad \text{for any } v \in V \cdots (SP),$$

then $P(\Gamma)$ is a spin 4-V-manifold.

Let Γ, Γ' be two tree graphs with Seifert invariants and let $M(\Gamma)$, $M(\Gamma')$ be the corresponding plumbed 3-manifolds. Let $\phi = (L_*; i, i', \bullet)$ be an algebraic morphism between the homology rings $(H_*(M(\Gamma); \mathbb{Z}), \bullet)$, $(H_*(M(\Gamma'); \mathbb{Z}), \bullet)$ of the corresponding 3-manifolds $M(\Gamma)$, $M(\Gamma')$. We assume that Γ, Γ' satisfy the condition (Ndeg). Suppose that there exists a cobordism $(W; M(\Gamma), M(\Gamma'))$ inducing $\phi = (L_*; i, i', \bullet)$. Then $L_* \cong \operatorname{Im}\left(H_2(M(\Gamma); \mathbb{Z}) \oplus H_2(M(\Gamma'); \mathbb{Z}) \overset{i_* + i'_*}{\to} H_*(W; \mathbb{Z}) \right)$. Let Z be a 4-V-manifold obtained by gluing the 4-V-manifolds $P(\Gamma)$, $P(\Gamma')$ along their boundaries $M(\Gamma)$, $M(\Gamma')$, respectively. Then we have the following lemma [**2**].

Lemma 29. *The second homology group $H_2(Z; \mathbb{Q})$ is isomorphic to*

$$H_2(P(\Gamma); \mathbb{Q}) \oplus H_2(P(\Gamma'); \mathbb{Q}) \oplus \operatorname{Coker}\left(H_2(M(\Gamma); \mathbb{Z}) \oplus H_2(M(\Gamma'); \mathbb{Z}) \overset{i_* + i'_*}{\to} H_2(W; \mathbb{Z}) \right) \otimes \mathbb{Q}.$$

Let Z^0 be a smooth manifold obtained by removing the interiors of neighborhoods of the singularities. Then the boundary $\partial P(\Gamma)^0$ of $P(\Gamma)^0$ is composed of the disjoint union of the plumbed 3-manifold $M(\Gamma)$ and a disjoint union L of lens spaces. Then we have the following Lemma.

Lemma 30. *1. Let Γ be a Seifert graph and set $H(\Gamma) = \bigoplus_{v\in V} H_1(\bar{\Sigma}_v; \mathbb{Z})$.*

(a) There exists an injective homomorphism

$$H(\Gamma) \overset{\lambda}{\to} H_1(M(\Gamma); \mathbb{Z}),$$

(b) There exists the following natural commutative diagram,

$$\begin{array}{ccc} H(\Gamma) & \overset{\bar{\theta}}{\to} & H_3(P(\Gamma)^0, M(\Gamma) \sqcup L; \mathbb{Z}) \\ & {}_{\theta}\searrow & \downarrow \\ & & H_2(M(\Gamma); \mathbb{Z}) \end{array}.$$

2. Let Z^0 be the smooth 4-manifold obtained by removing the interiors of neighborhoods of singularities of Z.

(a) Set

$$L(\Gamma, \Gamma'; \phi) = \{ i_*\lambda(\alpha) + i'_*\lambda'(\alpha') \in H_1(W; \mathbb{Z}) \,|\, \alpha \oplus \alpha' \in H(\Gamma) \oplus H(\Gamma') \}.$$

Then there exists an injective homomorphism

$$\begin{array}{ccc} L(\Gamma,\Gamma';\phi) & \to & H_1(Z^0,L;\mathbb{Z}) \\ \beta & \longmapsto & k_*(\beta) \end{array}$$

for the inclusion $k: W \hookrightarrow Z^0$.

(b) *Set*

$$R(\Gamma,\Gamma';\phi) = \{\alpha \oplus \alpha' \in H(\Gamma) \oplus H(\Gamma') \,|\, i_*\theta_\alpha + i'_*\theta'_{\alpha'} = 0 \in H_2(W;\mathbb{Z})\}.$$

Then there exists an injective homomorphism

$$\begin{array}{cccc} \widetilde{\theta\theta'}: & R(\Gamma,\Gamma';\phi) & \to & H_3(Z^0,L;\mathbb{Z}) \\ & (-\alpha)\oplus\alpha' & \longmapsto & \widetilde{\theta_\alpha\theta'_{\alpha'}} \end{array}$$

such that for any pair $(-\alpha)\oplus\alpha' \in R(\Gamma,\Gamma';\phi)$, *the corresponding 3-cycle* $\widetilde{\theta_\alpha\theta'_{\alpha'}} \in H_3(Z^0,\partial Z^0;\mathbb{Z})$ *defines* $-\theta_\alpha + \theta'_{\alpha'}$ *on* $H_2(M(\Gamma) \sqcup M(\Gamma'))$.

Proof. 1. Let V be the set of vertices in Γ. For a V-manifold X, we denote X^0 be the manifold with boundary obtained by removing the interior of a sufficiently small neighborhood of the singularity of X.

(a) There exists an isomorphism

$$\iota: H(\Gamma) = \bigoplus_{v\in V} H_1(\bar{\Sigma}_v;\mathbb{Z}) \to H_1(\overline{P(\Gamma)};\mathbb{Z}) \cong \bigoplus_{v\in V} H_1(\bar{P}_v;\mathbb{Z}),$$

and by the exact sequence of relative homology of the pair $(\overline{P(\Gamma)}, M(\Gamma))$, the induced homomorphism $H_1(M(\Gamma);\mathbb{Z}) \xrightarrow{j} H_1(\overline{P(\Gamma)};\mathbb{Z})$ is onto and we can take a splitting homomorphism $H_1(M(\Gamma);\mathbb{Z}) \xleftarrow{\sigma} H_1(\overline{P(\Gamma)};\mathbb{Z})$, which establishes an injective homomorphism $\lambda: H(\Gamma) \xrightarrow{\iota} H_1(\overline{P(\Gamma)};\mathbb{Z}) \xrightarrow{\sigma} H_1(M(\Gamma);\mathbb{Z})$.

(b) Now for each 1-cycle $\alpha \in \bigoplus_{v\in V} H_1(\bar{\Sigma}_v;\mathbb{Z})$, we can associate a relative 3-cycle $\bar{\theta}_\alpha \in H_3(P(\Gamma)^0, M(\Gamma)\sqcup L;\mathbb{Z})$ as follows, where L is a disjoint union of lens spaces such that $\partial P(\Gamma)^0 \cong M(\Gamma)\sqcup L$. Let $M_v \to \Sigma_v$ be the Seifert fibration of Seifert invariant $\omega(v)$, and let $P_v \to \Sigma_v$ be the associated disk V-bundle. Since $\bar{P}_v$ is deformation retract to $\bar{\Sigma}_v$, $H^1(\bar{P}_v;\mathbb{Z}) \cong H^1(\bar{\Sigma}_v;\mathbb{Z})$. By the Meyer-Vietoris sequence for $\bar{P}_v = P_v^0 \cup \text{cone } L_v$ and Poincaré duality, we have $H^1(\bar{P}_v;\mathbb{Z}) \cong H^1(P_v^0;\mathbb{Z}) \cong H_3(P_v^0, M_v\sqcup L_v;\mathbb{Z})$, where L_v is a disjoint union of lens spaces such that $\partial P_v^0 \cong M_v \sqcup L_v$ and cone L_v is the disjoint union of cones over the lens spaces L_v. On the other hand, we have $H^1(\bar{\Sigma}_v;\mathbb{Z}) \cong \mathrm{Ker}\left(H^1(\Sigma_v^0;\mathbb{Z}) \to H^1(\partial\Sigma_v^0;\mathbb{Z})\right)$, and by Poincaré duality and the exact sequence of relative homology, this is isomorphic to $\mathrm{Im}\left(H_1(\Sigma_v^0;\mathbb{Z}) \to H_1(\Sigma_v^0,\partial\Sigma_v^0;\mathbb{Z})\right) \cong H_1(\bar{\Sigma}_v;\mathbb{Z})$. Therefore, we have $H_1(\bar{\Sigma}_v;\mathbb{Z}) \cong H_3(P_v^0, M_v\sqcup L_v;\mathbb{Z})$. Let Γ' be the graph obtained by removing a terminal vertex in Γ and the edge adjacent to it. Then the plumbed V-manifold $P(\Gamma)$ is obtained by gluing $P(\Gamma')$ and P_v along a local trivialization $D^2\times D^2 \subset P(\Gamma')$. Then $\partial P(\Gamma')^0$ is a disjoint union of the plumbed 3-manifold $M(\Gamma')$ and a disjoint union

L' of lens spaces. Set $M(\Gamma')^0 = M(\Gamma') - D^2 \times \partial D^2$. Then by the exact sequence for triples $(P(\Gamma')^0, M(\Gamma') \sqcup L', M(\Gamma')^0 \sqcup L')$, we see that $H_3(P(\Gamma')^0, M(\Gamma') \sqcup L'; \mathbb{Z}) \cong H_3(P(\Gamma')^0, M(\Gamma')^0 \sqcup L'; \mathbb{Z})$, and similarly $H_3(P_v^0, M_v \sqcup L_v; \mathbb{Z}) \cong H_3(P_v^0, M_v^0 \sqcup L_v; \mathbb{Z})$. Now $P(\Gamma)^0 = P(\Gamma')^0 \cup P_v^0$, $M(\Gamma) \sqcup L = M(\Gamma')^0 \sqcup L' \cup M_v^0 \sqcup L_v$, $P(\Gamma')^0 \cap P_v^0 = D \cong D^2 \times D^2$, and $M(\Gamma')^0 \sqcup L' \cap M_v^0 \sqcup L_v = T \cong S^1 \times S^1$. Note that $H_3(D, T; \mathbb{Z}) \to H_3(P(\Gamma')^0, M(\Gamma')^0 \sqcup L'; \mathbb{Z}) \oplus H_3(P_v^0, M_v^0 \sqcup L_v; \mathbb{Z})$ is the zero map and $H_2(D, T; \mathbb{Z}) \to H_2(P(\Gamma')^0, M(\Gamma')^0 \sqcup L'; \mathbb{Z}) \oplus H_2(P_v^0, M_v^0 \sqcup L_v; \mathbb{Z})$ is injective. Then by the Meyer-Vietoris sequence we have

$$H_3(P(\Gamma)^0, M(\Gamma) \sqcup L; \mathbb{Z}) \cong H_3(P(\Gamma')^0, M(\Gamma') \sqcup L'; \mathbb{Z}) \oplus H_3(P_v^0, M_v^0 \sqcup L_v; \mathbb{Z}).$$

Therefore, by induction on the number of vertices, we obtain

$$H_3(P(\Gamma)^0, M(\Gamma) \sqcup L; \mathbb{Z}) \cong \bigoplus_{v \in V} H_3(P_v^0, M_v \sqcup L_v; \mathbb{Z}).$$

Hence we have the natural isomorphism,

$$\bar{\theta} : \bigoplus_{v \in V} H_1(\bar{\Sigma}_v; \mathbb{Z}) \cong \bigoplus_{v \in V} H_3(P_v^0, M_v \sqcup L_v; \mathbb{Z}) \cong H_3(P(\Gamma)^0, M(\Gamma) \sqcup L; \mathbb{Z}).$$

Combining with the boundary connecting homomorphisms $\partial_* : H_3(P_v^0, M_v \sqcup L_v; \mathbb{Z}) \to H_2(M_v; \mathbb{Z})$ and $\partial_*: H_3(P(\Gamma)^0, M(\Gamma) \sqcup L; \mathbb{Z}) \to H_2(M(\Gamma); \mathbb{Z})$, which are isomorphisms, we have a natural isomorphism

$$\theta : \bigoplus_{v \in V} H_1(\bar{\Sigma}_v; \mathbb{Z}) \to \bigoplus_{v \in V} H_2(M_v; \mathbb{Z}) \cong H_2(M(\Gamma); \mathbb{Z}).$$

For $\alpha = \sum_{v \in V} \alpha_v \in \bigoplus_{v \in V} H_1(\bar{\Sigma}_v; \mathbb{Z})$, we denote the decomposition of relative 3-cycles $\bar{\theta}_\alpha \in H_3(P(\Gamma)^0, M(\Gamma) \sqcup L; \mathbb{Z}) \cong \bigoplus_{v \in V} H_3(P_v^0, M_v \sqcup L_v; \mathbb{Z})$, and denote it by $\bar{\theta}_\alpha = \sum_{v \in V} \bar{\theta}_{\alpha_v}$ with $\bar{\theta}_{\alpha_v} \in H_3(P_v^0, M_v \sqcup L_v; \mathbb{Z})$ and also denote that of the corresponding 2-cycles by $\theta_\alpha = \sum_{v \in V} \theta_{\alpha_v} \in H_2(M(\Gamma); \mathbb{Z})$ with $\theta_{\alpha_v} \in H_2(M_v; \mathbb{Z})$.

2. We have the following homomorphisms on homology ϕ.

$$\phi : \quad H_*(M(\Gamma); \mathbb{Z}) \quad \stackrel{i_*}{\to} \quad H_1(W; \mathbb{Z}) \quad \stackrel{i'_*}{\leftarrow} \quad H_*(M(\Gamma'); \mathbb{Z})$$

   (a) By the Meyer-Vietoris sequence

$$H_1(M(\Gamma) \sqcup M(\Gamma'); \mathbb{Z}) \stackrel{j_* \sqcup j'_* \oplus (i_* + i'_*)}{\longrightarrow} H_1(\overline{P(\Gamma)} \sqcup \overline{P(\Gamma')}; \mathbb{Z}) \oplus H_1(W; \mathbb{Z}) \to H_1(\bar{Z}; \mathbb{Z}) \to 0,$$

and the surjectivity of $H_1(M(\Gamma); \mathbb{Z}) \stackrel{j_*}{\to} H_1(\overline{P(\Gamma)}; \mathbb{Z})$, $H_1(\bar{Z}; \mathbb{Z})$ is isomorphic to

$$\left(H_1(\overline{P(\Gamma)} \sqcup \overline{P(\Gamma')}; \mathbb{Z}) \oplus H_1(W; \mathbb{Z}) \right) \Big/ \mathrm{Im}\,(j_* \sqcup j'_* \oplus (i_* + i'_*))$$

$$\cong \mathrm{Coker} \left(\mathrm{Ker}\, j_* \oplus \mathrm{Ker}\, j'_* \stackrel{i_* + i'_*}{\to} H_1(W; \mathbb{Z}) \right).$$

Since there exists an injective homomorphism $H(\Gamma) \overset{\lambda}{\to} H_1(M(\Gamma);\mathbb{Z})$ which factors injective homomorphism $H(\Gamma) \overset{\iota}{\to} H_1(\overline{P(\Gamma)};\mathbb{Z})$, there exists an injective homomorphism from

$$L(\Gamma,\Gamma';\phi) =$$

$$\mathrm{Im}\left(H(\Gamma) \oplus H(\Gamma') \overset{\lambda\oplus\lambda'}{\to} H_1(M(\Gamma);\mathbb{Z}) \oplus H_1(M(\Gamma);\mathbb{Z}) \overset{i_*+i'_*}{\to} H_1(W;\mathbb{Z})\right)$$

to $\mathrm{Coker}\left(\mathrm{Ker}\ j_* \oplus \mathrm{Ker}\ j'_* \overset{i_*+i'_*}{\to} H_1(W;\mathbb{Z})\right) \cong H_1(\bar{Z};\mathbb{Z})$. Note that by the Meyer-Vietoris sequence,

$$H_1(L;\mathbb{Z}) \overset{i_*\oplus(-j_*)}{\to} H_1(Z^0;\mathbb{Z}) \oplus H_1(V;\mathbb{Z}) \to H_1(\bar{Z};\mathbb{Z}) \to 0$$

and since $H_1(V) = 0$, $H_1(\bar{Z}) \cong H_1(Z^0)/i_*H_1(L)$. On the other hand, by the exact sequence of the pair (Z^0, L)

$$\begin{array}{l} H_1(L;\mathbb{Z}) \overset{i_*}{\to} H_1(Z^0;\mathbb{Z}) \overset{j_*}{\to} H_1(Z^0,L;\mathbb{Z}) \to \\ H_0(L;\mathbb{Z}) \overset{i_*}{\to} H_0(Z^0;\mathbb{Z}) \overset{j_*}{\to} H_0(Z^0,L;\mathbb{Z}) \to 0 \end{array}$$

and hence there exists a natural injective homomorphism $H_1(\bar{Z};\mathbb{Z}) \to H_1(Z^0,L;\mathbb{Z})$.

(b) Let us define $R(\Gamma,\Gamma';\phi) = \{\alpha \oplus \alpha' \in H(\Gamma) \oplus H(\Gamma') \mid i_*\theta_\alpha + i'_*\theta'_{\alpha'} = 0 \in H_2(W;\mathbb{Z})\}$. By using this ϕ, we can construct a 3-cycle $\widetilde{\theta_\alpha\theta'_{\alpha'}}$ as follows. Here we denote $X = P(\Gamma)$, $M = M(\Gamma)$, and $L = \partial Z^0$.

$$\begin{array}{ccccc} H_3(X^0 \sqcup X'^0, L;\mathbb{Z}) & \to & \underset{\widetilde{\theta_\alpha\theta'_{\alpha'}}}{H_3(Z^0,L;\mathbb{Z})} & \to & H_3(Z^0, X^0 \sqcup X'^0;\mathbb{Z}) \\ & & & & \longmapsto \\ & \overset{\cong}{\to} & \underset{\overline{\theta_\alpha\theta'_{\alpha'}}}{H_3(W, M \sqcup M';\mathbb{Z})} & \underset{\longmapsto}{\to} & \underset{-\theta_\alpha + \theta'_{\alpha'}}{H_2(M \sqcup M';\mathbb{Z})} \end{array}$$

Note that $i_*\theta_\alpha + i'_*\theta'_{\alpha'} = 0 \in H_2(W;\mathbb{Z})$. Then by the exact sequence of the pair $(W, M \sqcup M')$ there exists a relative homology class $\overline{\theta_\alpha\theta'_{\alpha'}} \in H_3(W, M \sqcup M';\mathbb{Z})$ that maps to $-\theta_\alpha + \theta'_{\alpha'}$. Note that $\overline{\theta_\alpha\theta'_{\alpha'}}$ is only determined up to the image of $H_3(W;\mathbb{Z}) \to H_3(W, M \sqcup M';\mathbb{Z})$. By the excision property $H_3(Z^0, X^0 \sqcup X^0;\mathbb{Z}) \cong H_3(W, M \sqcup M';\mathbb{Z})$, the surjectivity of the map $H_3(Z^0, L;\mathbb{Z}) \to H_3(Z^0, X^0 \sqcup X'^0;\mathbb{Z})$ (since $H_2(X^0 \sqcup X'^0, L;\mathbb{Z}) \to H_2(Z^0, L;\mathbb{Z})$ is injective), and $H_3(X^0 \sqcup X^0, L;\mathbb{Z}) \cong 0$, there exists a unique three-cycle $\widetilde{\theta_\alpha\theta'_{\alpha'}}$ in $H_3(Z^0, L, \mathbb{Z})$ which maps to $\overline{\theta_\alpha\theta'_{\alpha'}}$. This establishes a map $\widetilde{\theta\theta'} : R(\Gamma,\Gamma';\phi) \to H_3(Z^0, L;\mathbb{Z})$.

□

Remark 31. *Note that we can define the intersection pairing*

$$H_3(Z^0, L; \mathbb{Z}) \otimes H_3(Z^0, L; \mathbb{Z}) \to H_2(Z^0; \mathbb{Z})$$

by using

$$H_3(Z^0; \mathbb{Z}) \otimes H_3(Z^0, L; \mathbb{Z}) \to H_2(Z^0; \mathbb{Z})$$

since $H_3(Z^0; \mathbb{Z}) \to H_3(Z^0, L; \mathbb{Z}) \to H_2(L; \mathbb{Z}) = 0$, *and we can take a lift to* $H_3(Z^0; \mathbb{Z})$ *up to the image of* $H_3(L; \mathbb{Z})$. *Note that the pairings of elements in* $H_3(Z^0; \mathbb{Z})$ *and that of* $H_3(L; \mathbb{Z})$ *are trivial.*

As a generalization of Lemma in [**2**], we have the following

Lemma 32. *Let* $\Gamma = (V, E, \omega)$, $\Gamma' = (V', E', \omega')$ *be two tree Seifert graphs satisfying the condition (Ndeg) and* $\phi = (L_*; i, i', \bullet)$ *be a morphism between* $(H_*(M(\Gamma); \mathbb{Z}), \bullet)$ *and* $(H_*(M(\Gamma'); \mathbb{Z}), \bullet)$. *Suppose that there exists a cobordism* $(W; M(\Gamma), M(\Gamma'))$ *between* $M(\Gamma)$ *and* $M(\Gamma')$ *such that* $(L_*(W; \mathbb{Z}); i_*, i'_*, \bullet) \cong \phi$ *and* $L_*(W; \mathbb{Z}) = H_*(W; \mathbb{Z})$. *Let* Z *be the closed 4-*V*-manifold obtained by gluing* $P(\Gamma)$, $P(\Gamma')$ *and* W *along the boundaries* $M(\Gamma)$, $M(\Gamma')$, *and let* Z_0 *be the 4-manifold obtained by removing the interiors of sufficiently small regular neighborhoods of the singularity in* Z. *For a pair of 1-cycles* $(\alpha, \alpha') \in L(\Gamma, \Gamma', \phi)$, *we can define a 3-cycle* $\widetilde{\theta_\alpha \theta'_{\alpha'}} = \sum_{v \in V, v' \in V'} \widetilde{\theta_{\alpha_v} \theta'_{\alpha'_{v'}}} \in H_3(Z_0; \mathbb{Z})$, *and the intersection products among these 3-cycles can be calculated by using the intersection pairings among the closed* V*-surfaces* Σ_v *in* $P(\Gamma)$ *and curves on* Σ_v*'s as follows.*

$$\begin{aligned}
&\widetilde{\theta_\alpha \theta'_{\alpha'}} \cdot \widetilde{\theta_\beta \theta'_{\beta'}} \\
&= -\sum_{v, v' \in V} A(\Gamma)^{vv'} (\alpha_v \cdot \beta_v) \Sigma_{v'} + \sum_{v, v' \in V'} A(\Gamma')^{vv'} (\alpha'_v \cdot \beta'_v) \Sigma'_{v'} \in H_2(Z; \mathbb{Q}),
\end{aligned}$$

$$\begin{aligned}
&\left(\widetilde{\theta_\alpha \theta'_{\alpha'}} \cdot \widetilde{\theta_\beta \theta'_{\beta'}}\right) \cdot \widetilde{\theta_\gamma \theta'_{\gamma'}} \\
&= \sum_{v, v' \in V} A(\Gamma)^{vv'} (\alpha_v \cdot \beta_v) i(\gamma_{v'}) - \sum_{v, v' \in V'} A(\Gamma')^{vv'} (\alpha'_v \cdot \beta'_v) i'(\gamma'_{v'}) \in H_1(Z; \mathbb{Q}),
\end{aligned}$$

$$\begin{aligned}
&\widetilde{\theta_\alpha \theta'_{\alpha'}} \cdot \left(\widetilde{\theta_\beta \theta'_{\beta'}} \cdot \widetilde{\theta_\gamma \theta'_{\gamma'}}\right) \\
&= \sum_{v, v' \in V} A(\Gamma)^{vv'} (\beta_v \cdot \gamma_v) i(\alpha_{v'}) - \sum_{v, v' \in V'} A(\Gamma')^{vv'} (\beta'_v \cdot \gamma'_v) i'(\alpha'_{v'}) \in H_1(Z; \mathbb{Q}),
\end{aligned}$$

$$\begin{aligned}
&\left(\left(\widetilde{\theta_\alpha \theta'_{\alpha'}} \cdot \widetilde{\theta_\beta \theta'_{\beta'}}\right) \cdot \widetilde{\theta_\gamma \theta'_{\gamma'}}\right) \cdot \widetilde{\theta_\delta \theta'_{\delta'}} \\
&= -\sum_{v, v' \in V} A(\Gamma)^{vv'} (\alpha_v \cdot \beta_v)(\gamma_{v'} \cdot \delta_{v'}) + \sum_{v, v' \in V'} A(\Gamma')^{vv'} (\alpha'_v \cdot \beta'_v)(\gamma'_{v'} \cdot \delta'_{v'}) \\
&\in H_0(Z^0; \mathbb{Z}),
\end{aligned}$$

where $A(\Gamma_\ell)^{vv'}$ *are the inverses of the intersection matrices* $A(\Gamma)_{vv'} = \Sigma_v \cdot \Sigma_{v'}$, $v, v' \in V$ *in* $H_2(P(\Gamma); \mathbb{Q})$.

Proof. First we prove that the intersection pairing $\widetilde{\theta_\alpha\theta'_{\alpha'}} \cdot \widetilde{\theta_\beta\theta'_{\beta'}}$ is given by

$$\widetilde{\theta_\alpha\theta'_{\alpha'}} \cdot \widetilde{\theta_\beta\theta'_{\beta'}} = -\sum_{v,v'\in V} A(\Gamma)^{vv'}\left(\alpha_{v'}\cdot\beta_{v'}\right)\Sigma_v + \sum_{v,v'\in V'} A(\Gamma')^{vv'}\left(\alpha'_{v'}\cdot\beta'_{v'}\right)\Sigma'_v,$$

where $\alpha\alpha', \beta\beta' \in L(\Gamma, \Gamma'; \phi)$.

By Lemma 29 we can write

$$\widetilde{\theta_\alpha\theta'_{\alpha'}} \cdot \widetilde{\theta_\beta\theta'_{\beta'}} = \sum_{v\in V} c^v_{\alpha\beta}\Sigma_v + \sum_{v\in V'} c'^v_{\alpha\beta}\Sigma'_v$$

for some $c^v_{\alpha\beta}, c'^v_{\alpha\beta} \in \mathbb{Q}$. Now we multiply $\Sigma_{v'}$ from the left, we have

$$\Sigma_{v'}\cdot\left(\widetilde{\theta_\alpha\theta'_{\alpha'}} \cdot \widetilde{\theta_\beta\theta'_{\beta'}}\right) = \sum_{v\in V} c^v_{\alpha\beta}\Sigma_{v'}\cdot\Sigma_v = \sum_{v\in V} c^v_{\alpha\beta}\left(-A(\Gamma)_{v'v}\right)$$

On the other hand, by the associativity of the intersection pairing, we have

$$\begin{aligned}\Sigma_{v'}\cdot\left(\widetilde{\theta_\alpha\theta'_{\alpha'}} \cdot \widetilde{\theta_\beta\theta'_{\beta'}}\right) &= \left(\Sigma_{v'}\cdot\widetilde{\theta_\alpha\theta'_{\alpha'}}\right)\cdot\widetilde{\theta_\beta\theta'_{\beta'}}\\ &= \left(\Sigma_{v'}\cdot\bar{\theta}_\alpha\right)\cdot\widetilde{\theta_\beta\theta'_{\beta'}}\\ &= \left(-i\left(\alpha_{v'}\right)\right)\cdot\widetilde{\theta_\beta\theta'_{\beta'}}\\ &= \left(-\alpha_{v'}\right)\cdot\bar{\theta}_\beta = \alpha_{v'}\cdot\beta_{v'}.\end{aligned}$$

Therefore we have

$$c^{v''}_{\alpha\beta} = -\sum_{v'\in V}\left(\alpha_{v'}\cdot\beta_{v'}\right)A\left(\Gamma\right)^{v'v''}.$$

Similarly, if we multiply $\Sigma'_{v'}$ from the left, we have

$$\Sigma_{v'}\cdot\left(\widetilde{\theta_\alpha\theta'_{\alpha'}} \cdot \widetilde{\theta_\beta\theta'_{\beta'}}\right) = \sum_{v\in V'} c'^v_{\alpha\beta}\Sigma'_{v'}\cdot\Sigma'_v = \sum_{v\in V} c'^v_{\alpha\beta}A\left(\Gamma'\right)_{v'v}$$

On the other hand, by the associativity of intersection pairings, we have

$$\begin{aligned}\Sigma'_{v'}\cdot\left(\widetilde{\theta_\alpha\theta'_{\alpha'}} \cdot \widetilde{\theta_\beta\theta'_{\beta'}}\right) &= \left(\Sigma'_{v'}\cdot\widetilde{\theta_\alpha\theta'_{\alpha'}}\right)\cdot\widetilde{\theta_\beta\theta'_{\beta'}} = \left(\Sigma'_{v'}\cdot\left(-\bar{\theta}'_{\alpha'}\right)\right)\cdot\widetilde{\theta_\beta\theta'_{\beta'}}\\ &= -i'\left(\alpha'_{v'}\right)\cdot\widetilde{\theta_\beta\theta'_{\beta'}} = -\alpha'_{v'}\cdot\left(-\bar{\theta}_{\beta'}\right) = \alpha'_{v'}\cdot\beta'_{v'}.\end{aligned}$$

Therefore we have

$$c'^{v''}_{\alpha\beta} = \sum_{v'\in V}\left(\alpha'_{v'}\cdot\beta'_{v'}\right)A\left(\Gamma'\right)^{v'v''}.$$

Hence the assertion on the double products follows.

Next we calculate the triple products. Note that the intersections of Σ_v, Σ'_v and $\widetilde{\theta_\gamma\theta'_{\gamma'}}$ are calculated to be

$$\Sigma_v \cdot \widetilde{\theta_\gamma\theta'_{\gamma'}} = \Sigma_v \cdot \bar{\theta}_\gamma = -i(\gamma_v), \quad \Sigma'_v \cdot \widetilde{\theta_\gamma\theta'_{\gamma'}} = \Sigma'_v \cdot (-\bar{\theta}'_{\gamma'}) = -i'(\gamma'_v).$$

Then we can calculate $\left(\widetilde{\theta_\alpha\theta'_{\alpha'}} \cdot \widetilde{\theta_\beta\theta'_{\beta'}}\right) \cdot \widetilde{\theta_\gamma\theta'_{\gamma'}}$ and $\widetilde{\theta_\alpha\theta'_{\alpha'}} \cdot \left(\widetilde{\theta_\beta\theta'_{\beta'}} \cdot \widetilde{\theta_\gamma\theta'_{\gamma'}}\right)$ as follows.

$$\begin{aligned}
&\left(\widetilde{\theta_\alpha\theta'_{\alpha'}} \cdot \widetilde{\theta_\beta\theta'_{\beta'}}\right) \cdot \widetilde{\theta_\gamma\theta'_{\gamma'}} \\
&= \sum_{v\in V} c^v_{\alpha\beta}\Sigma_v \cdot \widetilde{\theta_\gamma\theta'_{\gamma'}} + \sum_{v\in V'} c'^v_{\alpha\beta}\Sigma'_v \cdot \widetilde{\theta_\gamma\theta'_{\gamma'}} \\
&= \sum_{v\in V} c^v_{\alpha\beta}(-i(\gamma_v)) + \sum_{v\in V'} c'^v_{\alpha\beta}(-i'(\gamma'_v)) \\
&= -\sum_{v\in V}\left(-\sum_{v'\in V}(\alpha_{v'}\cdot\beta_{v'})\,A(\Gamma)^{v'v}\right) i(\gamma_v) \\
&\quad + \sum_{v\in V}\left(\sum_{v'\in V}(\alpha'_{v'}\cdot\beta'_{v'})\,A(\Gamma')^{v'v}\right)(-i'(\gamma'_v)) \\
&= \sum_{v,v'\in V} A(\Gamma)^{v'v}(\alpha_{v'}\cdot\beta_{v'})\,i(\gamma_v) - \sum_{v,v'\in V'} A(\Gamma')^{v'v}(\alpha'_{v'}\cdot\beta'_{v'})\,i'(\gamma'_v).
\end{aligned}$$

$$\begin{aligned}
&\widetilde{\theta_\alpha\theta'_{\alpha'}} \cdot \left(\widetilde{\theta_\beta\theta'_{\beta'}} \cdot \widetilde{\theta_\gamma\theta'_{\gamma'}}\right) \\
&= \sum_{v\in V} c^v_{\beta\gamma}\widetilde{\theta_\alpha\theta'_{\alpha'}} \cdot \Sigma_v + \sum_{v\in V'} c'^v_{\beta\gamma}\widetilde{\theta_\alpha\theta'_{\alpha'}} \cdot \Sigma'_v \\
&= \sum_{v\in V} c^v_{\beta\gamma}(-i(\alpha_v)) + \sum_{v\in V'} c'^v_{\beta\gamma}(-i'(\alpha'_v)) \\
&= -\sum_{v\in V}\left(-\sum_{v'\in V}(\beta_{v'}\cdot\gamma_{v'})\,A(\Gamma)^{v'v}\right) i(\alpha_v) \\
&\quad + \sum_{v\in V'}\left(\sum_{v'\in V'}(\beta'_{v'}\cdot\gamma'_{v'})\,A(\Gamma')^{v'v}\right)(-i'(\alpha'_v)) \\
&= \sum_{v,v'\in V} A(\Gamma)^{v'v}(\beta_{v'}\cdot\gamma_{v'})\,i(\alpha_v) - \sum_{v,v'\in V'} A(\Gamma')^{v'v}(\beta'_{v'}\cdot\gamma'_{v'})\,i'(\alpha'_v).
\end{aligned}$$

The quadruple products can be calculated as follows.

$$\begin{aligned}
&\widetilde{\theta_\alpha\theta'_{\alpha'}}\cdot\widetilde{\theta_\beta\theta'_{\beta'}}\cdot\widetilde{\theta_\gamma\theta'_{\gamma'}}\cdot\widetilde{\theta_\delta\theta'_{\delta'}}\\
&=\left(\left(\widetilde{\theta_\alpha\theta'_{\alpha'}}\cdot\widetilde{\theta_\beta\theta'_{\beta'}}\right)\cdot\widetilde{\theta_\gamma\theta'_{\gamma'}}\right)\cdot\widetilde{\theta_\delta\theta'_{\delta'}}\\
&=\sum_{v\in V}\sum_{v'\in V}A(\Gamma)^{v'v}(\alpha_{v'}\cdot\beta_{v'})\,i(\gamma_v)\cdot\widetilde{\theta_\delta\theta'_{\delta'}}\\
&\quad-\sum_{v\in V'}\sum_{v'\in V'}A(\Gamma')^{v'v}(\alpha'_{v'}\cdot\beta'_{v'})\,i'(\gamma'_v)\cdot\widetilde{\theta_\delta\theta'_{\delta'}}\\
&=\sum_{v\in V}\sum_{v'\in V}A(\Gamma)^{v'v}(\alpha_{v'}\cdot\beta_{v'})(-\gamma_v\cdot\delta_v)\\
&-\sum_{v\in V'}\sum_{v'\in V'}A(\Gamma')^{v'v}(\alpha'_{v'}\cdot\beta'_{v'})(\gamma'_v\cdot(-\delta'_v))\\
&=-\sum_{v,v'\in V}A(\Gamma)^{v'v}(\alpha_{v'}\cdot\beta_{v'})(\gamma_v\cdot\delta_v)+\sum_{v,v'\in V'}A(\Gamma')^{v'v}(\alpha'_{v'}\cdot\beta'_{v'})(\gamma'_v\cdot\delta'_v).
\end{aligned}$$

□

Remark 33. *The quadruple product can be calculated in different ways. If we denote the above formula for the quadruple product* $\left(\left(\widetilde{\theta_\alpha\theta'_{\alpha'}}^\phi\cdot\widetilde{\theta_\beta\theta'_{\beta'}}^\phi\right)\cdot\widetilde{\theta_\gamma\theta'_{\gamma'}}^\phi\right)\cdot\widetilde{\theta_\delta\theta'_{\delta'}}^\phi$ *by* $\tilde\theta^\phi_{\alpha\alpha'\beta\beta'\gamma\gamma'\delta\delta'}$, *then we obtain the following formula.*

$$\begin{aligned}
\left(\widetilde{\theta_\alpha\theta'_{\alpha'}}^\phi\cdot\left(\widetilde{\theta_\beta\theta'_{\beta'}}^\phi\cdot\widetilde{\theta_\gamma\theta'_{\gamma'}}^\phi\right)\right)\cdot\widetilde{\theta_\delta\theta'_{\delta'}}^\phi&=\tilde\theta^\phi_{\beta\beta'\gamma\gamma'\alpha\alpha'\delta\delta'},\\
\widetilde{\theta_\alpha\theta'_{\alpha'}}^\phi\cdot\left(\left(\widetilde{\theta_\beta\theta'_{\beta'}}^\phi\cdot\widetilde{\theta_\gamma\theta'_{\gamma'}}^\phi\right)\cdot\widetilde{\theta_\delta\theta'_{\delta'}}^\phi\right)&=\tilde\theta^\phi_{\beta\beta'\gamma\gamma'\alpha\alpha'\delta\delta'},\\
\widetilde{\theta_\alpha\theta'_{\alpha'}}^\phi\cdot\left(\widetilde{\theta_\beta\theta'_{\beta'}}^\phi\cdot\left(\widetilde{\theta_\gamma\theta'_{\gamma'}}^\phi\cdot\widetilde{\theta_\delta\theta'_{\delta'}}^\phi\right)\right)&=\tilde\theta^\phi_{\gamma\gamma'\delta\delta'\alpha\alpha'\beta\beta'},\\
\left(\widetilde{\theta_\alpha\theta'_{\alpha'}}^\phi\cdot\widetilde{\theta_\beta\theta'_{\beta'}}^\phi\right)\cdot\left(\widetilde{\theta_\gamma\theta'_{\gamma'}}^\phi\cdot\widetilde{\theta_\delta\theta'_{\delta'}}^\phi\right)&=\tilde\theta^\phi_{\alpha\alpha'\beta\beta'\gamma\gamma'\delta\delta'}.
\end{aligned}$$

These formulas can be used to check the associativity in Theorem 13.

References

[1] A. Casson, J. Harer, "Some homology lens spaces which bound rational homology balls," Pacific J. Math. 96 (1981), 23-36.

[2] Y. Fukumoto, Homology spin cobordism problem of plumbed 3-manifolds and the cup product structures, preprint.

[3] Y. Fukumoto, "Plumbed homology 3-spheres bounding acyclic 4-manifolds," J. Math. Kyoto Univ. 40 vol.4

[4] Y. Fukumoto and M. Furuta, Homology 3-spheres bounding acyclic 4-manifolds, Math. Res. Lett. 7 (2000) 757-766.

[5] Y. Fukumoto, M. Furuta and M. Ue, w-invariants and Neumann-Siebenmann invariants for Seifert homology 3-spheres, Topology and its Appl. 116 (2001) 333-369.

[6] M. Furuta, Monopole equation and the 11/8 conjecture, Math. Res. Lett.8 (2001) 279-291.

[7] M. Furuta, Stable homotopy version of Seiberg-Witten invariant, MPI 97-110 (1997)

[8] M. Furuta, Y. Kametani, H. Matsue and N. Minami, Stable-homotopy Seiberg-Witten invariants and Pin bordisms, UTMS 2000-46 Sept. 1 (2000)

[9] M. Furuta, Y. Kametani, Equivariant maps between spheres bundles over tori and KO^*-degree, arXiv:Math.GT/0502511 v2 14 Mar 2005.

[10] T. Kawasaki, The index of elliptic operators over V-manifolds, Nagoya Math. J. 84 (1981), 135-137.

[11] I. Satake, The Gauss-Bonnet theorem for V-manifolds, J. of the Math. Soc. of Japan, 9 (1957), 464-492.

[12] N. Saveliev, Fukumoto-Furuta invariants of plumbed homology 3-spheres, Pacific J. Math. 205 (2002), 465-490.

[13] M. Ue, On the intersection forms of spin 4-manifolds bounded by spherical 3-manifolds, Algebr. Geom. Topol. 1 (2001), 549-578.

[14] M. Ue, The Neumann-Siebenmann invariant and Seifert surgery, Math. Z. 250 (2005), 475-493.

[15] M. Ue, The Ozsváth-Szabó and the Neumann-Siebenmann invariants for certain plumbed 3-manifolds, preprint.

This article may be accessed via WWW at `http://www.rmi.acnet.ge/jhrs/`

Yoshihiro Fukumoto
`fukumoto@kankyo-u.ac.jp`

Tottori University of Environmental Studies,
Tottori, JAPAN

Journal of Homotopy and Related Structures, vol. 4(1), 2009, pp.69–82

ON THE CLASSIFICATION OF UNSTABLE $H^*V - A$-MODULES

DORRA BOURGUIBA

(*communicated by Lionel Schwartz*)

Abstract

In this work, we begin studying the classification, up to isomorphism, of unstable $\mathrm{H}^*V - A$-modules E such that $\mathbb{F}_2 \otimes_{\mathrm{H}^*V} E$ is isomorphic to a given unstable A-module M. In fact this classification depends on the structure of M as unstable A-module. In this paper, we are interested in the case M a nil-closed unstable A-module and the case M is isomorphic to $\sum^n \mathbb{F}_2$. We also study, for $V = \mathbb{Z}/2\mathbb{Z}$, the case M is the Brown-Gitler module J(2).

1. Introduction

Let V be an elementary abelian 2-group of rank d, that is a group isomorphic to $(\mathbb{Z}/2\mathbb{Z})^d$, $d \in \mathbb{N}$, BV be a classifying space for the group V and $\mathrm{H}^*V = H^*(BV; \mathbb{F}_2)$. We recall that H^*V is an $\mathbb{F}_2$-polynomial algebra $\mathbb{F}_2[t_1, \ldots, t_d]$ on d generators $t_i, 1 \leqslant i \leqslant d$, of degree one.

Let A be the mod.2 Steenrod algebra and $\mathcal{U}$ the category of unstable A-modules. We recall that $\mathrm{H}^*V - \mathcal{U}$ is the category whose objects are unstable $\mathrm{H}^*V - A$-modules and morphisms are H^*V-linear and A-linear maps of degree zero. For example, the mod.2 equivariant cohomology of a V-CW-complex, which is the cohomology of the Borel construction, is an unstable $\mathrm{H}^*V - A$-module.

Let E be an unstable $\mathrm{H}^*V - A$-module, we denote by $\overline{E}$ the unstable A-module $\mathbb{F}_2 \otimes_{\mathrm{H}^*V} E = E/\widetilde{\mathrm{H}^*}V.E$, where $\widetilde{\mathrm{H}^*}V$ denotes the augmentation ideal of H^*V .

We have the following problem:

$(\mathcal{P})$: Let M be an unstable A-module.
Classify, up to isomorphism, unstable $\mathrm{H}^*V - A$-modules such that $\overline{E} \cong M$ (as unstable A-modules).

It is clear that, for every subgroup W of V, the unstable $\mathrm{H}^*V - A$-module:

$$\mathrm{H}^*W \otimes M$$

I would like to thank Professor Jean Lannes and Professor Said Zarati for several useful discussions. I am grateful to the referee for his suggestions.
Received July 26, 2008, revised November 29, 2008; published on April 17, 2009.
2000 Mathematics Subject Classification: 55M35, 55N91, 55T10, 18G05.
Key words and phrases: Unstable H*V-A-module, H*V-A-module injective, equivariant cohomology, Smith theory.

is a solution for the problem $(\mathcal{P})$.

For $W = 0$, a solution of $(\mathcal{P})$ is given by the unstable $\mathrm{H}^*V - A$-module M which is trivial as an H^*V-module.

For $W = V$, a solution of $(\mathcal{P})$ is given by the unstable $\mathrm{H}^*V - A$-module $\mathrm{H}^*V \otimes M$ which is free as an H^*V-module.

If $V = \mathbb{Z}/2\mathbb{Z}$ and $M = \Sigma N$ a suspension of an unstable A-module N, then we have, at least, the following two solutions of the problem $(\mathcal{P})$ which are free as $H^*(\mathbb{Z}/2\mathbb{Z})$-modules:

1. $\Sigma(H^*(\mathbb{Z}/2\mathbb{Z}) \otimes N)$.
2. $((H^*(\mathbb{Z}/2\mathbb{Z})^{\geqslant 1}) \otimes N$.

These two solutions are different as unstable A-modules (here $H^*(\mathbb{Z}/2\mathbb{Z})^{\geqslant 1}$ is the sub-algebra of $H^*(\mathbb{Z}/2\mathbb{Z})$ of elements of degree bigger than or equal to one). This shows that the solutions of the problem $(\mathcal{P})$ i.e. the classification, up to isomorphism, of unstable $\mathrm{H}^*V - A$-modules such that $\overline{E} \cong M$ (as unstable A-modules), depends on the structure of E as an H^*V-module and on the structure of M as unstable A-module.

In this paper we will discuss the solutions of $(\mathcal{P})$ if M is a nil-closed unstable A-module and E is free as an H^*V-module and the solutions of $(\mathcal{P})$ if M is isomorphic to $\sum^n \mathbb{F}_2$ or to $\mathrm{J}(2)$ and E is free as an H^*V-module .

We begin by proving the following result (which is solution of $(\mathcal{P})$ when M is a nil-closed unstable A-module).

Theorem 1.1. *Let E be unstable $\mathrm{H}^*V - A$-module which is free as an H^*V-module. If $\overline{E}$ is a nil-closed unstable A-module, then there exists two reduced $\mathcal{U}$-injectives I_0, I_1 and an $\mathrm{H}^*V - A$-linear map*
$\varphi : \mathrm{H}^*V \otimes I_0 \to \mathrm{H}^*V \otimes I_1$ such that:

1. *$E \cong ker\varphi$*
2. *$\overline{E} \cong ker\overline{\varphi}$*

The proof of this result is based on the classification of $\mathrm{H}^*V - \mathcal{U}$-injectives and on some properties of the injective hull in the category $\mathrm{H}^*V - \mathcal{U}$.

Our work is naturally motivated by topology as shown in the study of homotopy fixed points of a $\mathbb{Z}/2$-action (see [**L1**]). Let X be a space equipped with an action of $\mathbb{Z}/2$ and $X^{\mathrm{h}\mathbb{Z}/2}$ denote the space of homotopy fixed points of this action. The problem of determining the mod. 2 cohomology of $X^{\mathrm{h}\mathbb{Z}/2}$ (we ignore deliberately the questions of 2-completion) involves two steps:

- determining the mod. 2 equivariant cohomology $\mathrm{H}^*_{\mathbb{Z}/2}X$;
- determining $\mathrm{Fix}_{\mathbb{Z}/2}\,\mathrm{H}^*_{\mathbb{Z}/2}X$ (for the definition of the functor $\mathrm{Fix}_{\mathbb{Z}/2}$ see section 2).

For the first step, see for example [**DL**], the main information one has about the $\mathbb{Z}/2$-space X is that the Serre spectral sequence, for mod . 2 cohomology, associated to the fibration

$$X \to X_{h\mathbb{Z}/2} \to \mathrm{B}\mathbb{Z}/2$$

collapses ($X_{h\mathbb{Z}/2}$ denotes the Borel construction $\mathrm{E}\mathbb{Z}/2 \times_{\mathbb{Z}/2} X$). This collapsing implies that $\mathrm{H}^*_{\mathbb{Z}/2}X$ is H-free and that $\overline{\mathrm{H}^*_{\mathbb{Z}/2}X}$ is canonically isomorphic to H^*X. This gives clearly a topological application of problem $(\mathcal{P})$.

We then prove the following results (related to the case $\overline{E}$ is $\sum^n \mathbb{F}_2$ and $\mathrm{J}(2)$).

Theorem 1.2. *Let E be unstable $\mathrm{H}^*V - A$-module which is free as an H^*V-module. If $\overline{E}$ is isomorphic to $\sum^n \mathbb{F}_2$, then there exists an element u in H^*V such that:*

1. $u = \prod_i \theta_i^{\alpha_i}$, *where* $\theta_i \in (\mathrm{H}^1V) \setminus \{0\}$ *and* $\alpha_i \in \mathbb{N}$
2. $E \cong \sum^d u\mathrm{H}^*V$ *with* $d + \sum_i \alpha_i = n$

Proposition 1.3. *Let E be an $\mathrm{H} - A$-module which is H-free and such that $\overline{E}$ is isomorphic to $\mathrm{J}(2)$ then:*
$E \cong \mathrm{H} \otimes \mathrm{J}(2)$
or
E is the sub-$\mathrm{H} - A$-module of $\mathrm{H} \oplus \sum \mathrm{H}$ generated by $(t, \Sigma 1)$ and $(t^2, 0)$.

The proofs of these two results are based on Smith theory, some properties of the functor Fix and on a result of J.P. Serre.

The paper is structured as follows. In section 2, we introduce the definitions of reduced and nil-closed unstable A-modules. We give the classification of injective modules in the category $\mathcal{U}$ and in the category $H^*V - \mathcal{U}$. We also recall the algebraic Smith theory. In section 3, we establish some properties of E when $\overline{E}$ is a reduced unstable A-module. The results will be useful in section 4, where we give the solutions of the problem $(\mathcal{P})$ when E is free as an H^*V-module and $\overline{E}$ is nil-closed. In section 5, we give some topological applications. In section 6, we give the solutions of the problem $(\mathcal{P})$ when E is free as an H^*V-module and $\overline{E}$ is isomorphic to $\sum^n \mathbb{F}_2$, we also give a topological application. In section 7, we solve the problem $(\mathcal{P})$ when $\overline{E}$ is the Brown-Gitler module $\mathrm{J}(2)$ and V is $\mathbb{Z}/2\mathbb{Z}$.

2. Preliminaries on the categories $\mathcal{U}$ and $\mathrm{H}^*V - \mathcal{U}$

In this section, we will fix some notations, recall some definitions and results about the categories $\mathcal{U}$ and $\mathrm{H}^*V - \mathcal{U}$.

2.1. Nilpotent unstable A-modules

Let N be an unstable A-module. We denote by Sq_0 the $\mathbb{Z}/2\mathbb{Z}$-linear map:

$$Sq_0 : N \to N, \; x \mapsto Sq_0(x) = Sq^{|x|}x.$$

An unstable A-module N is called nilpotent if:

$$\forall\, x \in N,\ \exists\, n \in \mathbb{N};\ Sq_0^n x = 0.$$

For example, finite unstable A-modules and suspension of unstable A-modules are nilpotent. Let $Tor_1^{\mathrm{H}^*V}(\mathbb{F}_2, N)$ be the first derived functor of the functor $\mathbb{F}_2 \otimes_{\mathrm{H}^*V} - :$ $\mathrm{H}^*V - \mathcal{U} \to \mathcal{U}$, we have the following useful result.

Proposition 2.1.1. *([S] page 150) Let N be an unstable H^*V – A-module, then the unstable A-module $Tor_1^{\mathrm{H}^*V}(\mathbb{F}_2, N)$ is nilpotent.*

2.2. Reduced unstable A-modules

An unstable A-module M is called reduced if the $\mathbb{Z}/2\mathbb{Z}$-linear map:

$$Sq_0 : M \to M,\ x \mapsto Sq_0(x) = Sq^{|x|}x,$$

is an injection.

Another characterization of reduced unstable A-module in terms of nilpotent modules is the following.

Lemma 2.2.1. *([**LZ1**]) An unstable A-module is reduced if it does not contain a non-trivial nilpotent module.*

In particular, any A-linear map from a nilpotent A-module to a reduced one is trivial.

2.3. Nil-closed unstable A-modules

Let M be an unstable A-module. We denote by Sq_1 the $\mathbb{Z}/2\mathbb{Z}$-linear map:

$$Sq_1 : N \to N,\ x \mapsto Sq_1(x) = Sq^{|x|-1}x.$$

Definition 2.3.1. *([**EP**]) An unstable A-module M is called nil-closed if:*

1. *M is reduced.*
2. *$Ker(Sq_1) = Im(Sq_0)$.*

We have the following two characterizations of unstable nil-closed A-modules.

Lemma 2.3.2. *([**LZ1**]) Let M be an unstable A-module and $\mathcal{E}(M)$ be its injective hull. The unstable A-module M is nil-closed if and only if M and the quotient $\mathcal{E}(M)/M$ are reduced.*

Let $Ext^s_{\mathcal{U}}(-, M)$ be the s-th derived functor of the functor $\mathrm{H}om_{\mathcal{U}}(-, M)$.

Lemma 2.3.3. *([**LZ1**]) An unstable A-module M is nil-closed if and only if $Ext^s_{\mathcal{U}}(N, M) = 0$ for any nilpotent unstable A-module N and $s = 0, 1$.*

2.4. Injectives in the category $\mathcal{U}$

Let I be an unstable A-module, I is called an injective in the category $\mathcal{U}$ or $\mathcal{U}$-injective for short, if the functor $\mathrm{H}om_{\mathcal{U}}(-, I)$ is exact.

The classification of $\mathcal{U}$-injectives (see [**LZ1**], [**LS**]) is the following.

Let $\mathrm{J}(n)$, $n \in \mathbb{N}$, be the n-th Brown- Gitler module, characterized up to isomorphism, by the functorial bijection on the unstable A-module M:

$$\mathrm{Hom}_{\mathcal{U}}(M, \mathrm{J}(n)) \cong \mathrm{Hom}_{\mathbb{F}_2}(M^n, \mathbb{F}_2)$$

Clearly $\mathrm{J}(n)$ is an $\mathcal{U}$-injective and it is a finite module.
Let $\mathcal{L}$ be a set of representatives for $\mathcal{U}$-isomorphism classes of indecomposable direct factors of $\mathrm{H}^*(\mathbb{Z}/2\mathbb{Z})^m$, $m \in \mathbb{N}$ (each class is represented in $\mathcal{L}$ only once).
We have:

Theorem 2.4.1. *Let I be an $\mathcal{U}$-injective module. Then there exists a set of cardinals $a_{L,n}$, $(L, n) \in \mathcal{L} \times \mathbb{N}$, such that $I \cong \bigoplus_{(L,n)} (L \otimes \mathrm{J}(n))^{\oplus a_{L,n}}$.*
Conversely, any unstable A-module of that form is $\mathcal{U}$-injective.

Let's remark that H^*V is an $\mathcal{U}$-injective.

2.5. The injectives of the category $\mathrm{H}^*V - \mathcal{U}$

The classification of injectives of the category $\mathrm{H}^*V - \mathcal{U}$ ($\mathrm{H}^*V - \mathcal{U}$-injectives for short) is given by Lannes-Zarati [**LZ2**] as follows.
Let $\mathrm{J}_V(n)$, $n \in \mathbb{N}$, be the unstable $\mathrm{H}^*V - A$-module characterized, up to isomorphism, by the functorial bijection on the unstable $\mathrm{H}^*V - A$-module M:

$$\mathrm{Hom}_{\mathrm{H}^*V-\mathcal{U}}(M, \mathrm{J}_V(n)) \cong \mathrm{Hom}_{\mathbb{F}_2}(M^n, \mathbb{F}_2)$$

Clearly $\mathrm{J}_V(n)$ is an $\mathrm{H}^*V - \mathcal{U}$-injective.
Let $\mathcal{W}$ be the set of subgroups of V and let $(W, n) \in \mathcal{W} \times \mathbb{N}$, we write

$$E(V, W, n) = \mathrm{H}^*V \otimes_{\mathrm{H}^*V/W} \mathrm{J}_{V/W}(n)$$

(in this formula H^*V is an H^*V/W-module via the map induced in mod.2 cohomology by the canonical projection $V \to V/W$).

Theorem 2.5.1. *([**LZ2**]) If I is an injective of the category of $\mathrm{H}^*V - \mathcal{U}$, then $I \cong \bigoplus_{(L,W,n)\in\mathcal{L}\times\mathcal{W}\times\mathbb{N}} (E(V, W, n) \otimes_{\mathbb{F}_2} L)^{\oplus a_{L,W,n}}$.*
Conversely, each $\mathrm{H}^*V - A$-module of this form is an $\mathrm{H}^*V - \mathcal{U}$-injective.

Clearly H^*V is an $\mathrm{H}^*V - \mathcal{U}$-injective.

2.6. Algebraic Smith theory

2.6.1. The functors Fix

We introduce the functors Fix ([**L1**], [**LZ2**]). We denote by

$$Fix_V : \mathrm{H}^*V - \mathcal{U} \to \mathcal{U}$$

the left adjoint of the functor

$$\mathrm{H}^*V \otimes - : \mathcal{U} \to \mathrm{H}^*V - \mathcal{U}$$

We have the functorial bijection:

$$\mathrm{Hom}_{\mathrm{H}^*V-\mathcal{U}}(N,\ H^*V \otimes P) \cong \mathrm{Hom}_{\mathcal{U}}(Fix_V N,\ P)$$

for every unstable $\mathrm{H}^*V - A$-module N and every unstable A-module P.
The functor $\mathrm{F}ix_V$ has the following properties.
2.6.1.1. The functor $\mathrm{F}ix_V$ is an exact functor.
2.6.1.2. Let N be an unstable $\mathrm{H}^*V - A$-module and $\mathcal{E}(N)$ be its injective hull. Then, the module $\mathrm{F}ix_V\mathcal{E}(N)$ is the injective hull of $\mathrm{F}ix_V N$.

2.6.2.
Let N be an unstable $\mathrm{H}^*V - A$-module, we denote by

$$\eta_V : N \to \mathrm{H}^*V \otimes \mathrm{F}ix_V N$$

the adjoint of the identity of $Fix_V N$. We denote by $\mathrm{c}_V = \prod_{u \in \mathrm{H}^1V - \{0\}} u$ the top Dickson invariant, we have the following result (see [**LZ2**] corollary 2.3).

Proposition 2.6.1. *Let N be an unstable $\mathrm{H}^*V - A$-module. The localization of the map η_V*

$$\eta_V[\mathrm{c}_V^{-1}] : N[\mathrm{c}_V^{-1}] \to \mathrm{H}^*V[\mathrm{c}_v^{-1}] \otimes \mathrm{F}ix_V N$$

is an injection.

This shows in particular, that if N is torsion-free then the map η_V is an injection. The proposition 2.6.1 can be reformulated as follows.

Proposition 2.6.2. *Let N be an unstable $\mathrm{H}^*V - A$-module. If N is torsion-free then its injective hull in $\mathrm{H}^*V - \mathcal{U}$ is free as an H^*V-module and is isomorphic to* $\bigoplus_{(L,n) \in \mathcal{L} \times \mathbb{N}} (\mathrm{H}^*V \otimes \mathrm{J}(n)) \otimes L$

Proof. Since the module is torsion-free then the map $\eta_V : N \to \mathrm{H}^*V \otimes \mathrm{F}ix_V N$ adjoint of the identity of $\mathrm{F}ix_V N$ is an injection. So N is a sub-$\mathrm{H}^*V - A$-module of $\mathrm{H}^*V \otimes \mathrm{F}ix_V N$. By 2.6.1.1 and 2.6.1.2, we have that the injective hull of N is isomorphic to $\mathrm{H}^*V \otimes I$, where I is an $\mathcal{U}$-injective. □

Remark 2.6.3. As a consequence of proposition 2.6.2, we have that if E is an unstable $\mathrm{H}^*V - A$-module which is free as an H^*V-module then its injective hull (in the category $\mathrm{H}^*V - \mathcal{U}$) is also free as an H^*V-module.

Proposition 2.6.4. *[**LZ2**]. Let N be an unstable $\mathrm{H}^*V - A$-module which is of finite type as an H^*V-module. The localization of the map η_V*

$$\eta_V[\mathrm{c}_V^{-1}] : N[\mathrm{c}_V^{-1}] \to \mathrm{H}^*V[\mathrm{c}_V^{-1}] \otimes \mathrm{F}ix_V N$$

is an isomorphism.

In particular, the previous result shows that:

1. If N is free as an H^*V-module, then the map η_V is an injection.
2. The isomorphism of the proposition proves that $dim\overline{E} = dim\mathrm{F}ix_V E$ where dim is the total dimension (see [**LZ2**]).

3. Some properties of E when $\overline{E}$ is reduced

In this section we will prove some algebraic results which will be useful for section 4. In fact, we will analyze the relation between an unstable $\mathrm{H}^*V - A$-module E and its (associated) unstable A-module $\overline{E}$. For this, we will begin by giving some technical results.

3.1. Technical results

Lemma 3.1.1. *Let P and Q be unstable $\mathrm{H}^*V - A$-modules, free as H^*V-modules and $f : P \to Q$ an $\mathrm{H}^*V - A$-linear map. If the induced map $\overline{f} : \overline{P} \to \overline{Q}$ is an injection then f is also an injection.*

Proof. Let's denote by Imf the image of f, by $\widetilde{f} : P \to Imf$ the natural surjection and by $i : Imf \hookrightarrow Q$ the inclusion of Imf in Q. Since $\overline{f}$ is an injection so the induced map $(\overline{\widetilde{f}})$ is an isomorphism of unstable A-modules and then the induced map $\overline{i}$ is an injection. This shows that $\overline{Imf}$ is the image of $\overline{f}$. Since the module Imf is a sub-H^*V-module of the H^*V-free module Q and $\overline{i} : \overline{Imf} \hookrightarrow \overline{Q}$ is an injection, so Imf is free as an H^*V-module. In particular, we have that $Tor_1^{\mathrm{H}^*V}(\mathbb{F}_2, Imf)=0$ (see for example [**R**]). Let's denote by N the kernel of the map $\widetilde{f}$, so we have the following short exact sequence in $\mathrm{H}^*V - \mathcal{U}$:

$$0 \longrightarrow N \longrightarrow P \xrightarrow{\widetilde{f}} Imf \longrightarrow 0\ .$$

By applying the functor $(\mathbb{F}_2 \otimes_{\mathrm{H}^*V} -)$ to the previous sequence, we prove that $\overline{N}$ is trivial (since the map $(\overline{\widetilde{f}})$ is an isomorphism and Imf is free as an $\mathrm{H}^*V - A$-module). Hence the module N is trivial and the map f is an injection. □

The converse of this lemma is not true in general, but we have the following result:

Lemma 3.1.2. *Let P and Q be unstable $\mathrm{H}^*V - A$-modules, free as H^*V-modules and $f : P \to Q$ an $\mathrm{H}^*V - A$-linear map which is an injection. If $\overline{P}$ is a reduced unstable A-module, then the induced map $\overline{f} : \overline{P} \to \overline{Q}$ is an injection.*

Proof. We denote by C the quotient of Q by P, we have the following short exact sequence in $\mathrm{H}^*V - \mathcal{U}$:

$$0 \longrightarrow P \xrightarrow{f} Q \longrightarrow C \longrightarrow 0\ .$$

By applying the functor $(\mathbb{F}_2 \otimes_{\mathrm{H}^*V} -)$ to the previous sequence, we obtain an exact sequence in $\mathcal{U}$:

$$0 \longrightarrow Tor_1^{\mathrm{H}^*V}(\mathbb{F}_2, C) \longrightarrow \overline{P} \xrightarrow{\overline{f}} \overline{Q} \longrightarrow \overline{C} \longrightarrow 0\ .$$

Since $\overline{P}$ is reduced as unstable A-module and $Tor_1^{\mathrm{H}^*V}(\mathbb{F}_2, C)$ is nilpotent (see proposition 2.1.1), then the map $\overline{f}$ is an injection. □

3.2. Statement of some properties of E when $\overline{E}$ is reduced

The first result of this paragraph concerns the relation between the injective hull of E and the induced module $\overline{E}$.

Theorem 3.2.1. *Let E be an unstable $\mathrm{H}^*V - A$-module which is free as an H^*V-module and let $\mathcal{E}(E)$ be its injective hull (in the category $\mathrm{H}^*V - \mathcal{U}$). We suppose that $\overline{E}$ is reduced and let I be its injective hull in the category $\mathcal{U}$.*
Then $\mathcal{E}(E)$ is isomorphic, as an unstable $\mathrm{H}^*V - A$-module, to $\mathrm{H}^*V \otimes I$.

Proof. Since E is free as an H^*V-module, then $\mathcal{E}(E)$ is isomorphic, in the category $\mathrm{H}^*V - \mathcal{U}$, to $\mathrm{H}^*V \otimes J$, where J is an $\mathcal{U}$-injective (see proposition 2.6.2).
Let's denote by i the inclusion of E in $\mathcal{E}(E)$, we have, by lemma 3.1.2, that the induced map $\overline{i}$ is an injection. We will prove, by using the definition, that J is the injective hull of $\overline{E}$, in the category $\mathcal{U}$. Let P be a sub-A-module of J such that the A-module $(\overline{i})^{-1}(P)$ is trivial, we have to show that the unstable A-module P is trivial.
Since $(\overline{i})^{-1}(P)$ is trivial then the composition: $\pi \circ \overline{i} : \overline{E} \xrightarrow{\overline{i}} J \xrightarrow{\pi} J/P$ is an injection. By lemma 3.1.1, the following composition
$E \xrightarrow{i} \mathrm{H}^*V \otimes J \longrightarrow \mathrm{H}^*V \otimes (J/P)$ is an injection, which proves that the unstable $\mathrm{H}^*V - A$-module $i^{-1}(\mathrm{H}^*V \otimes P)$ is trivial. Since $\mathrm{H}^*V \otimes J$ is the injective hull of E so the unstable $\mathrm{H}^*V - A$-module $\mathrm{H}^*V \otimes P$ is trivial. □

Corollary 3.2.2. *Let E be an unstable $\mathrm{H}^*V - A$-module such that:*

1. *E is free as an H^*V-module.*
2. *$\overline{E}$ is reduced as unstable A-module.*

Then E is reduced as unstable A-module.

Proof. We have, by theorem 3.2.1, that the injective hull of E is $\mathrm{H}^*V \otimes I$, where I is the injective hull of $\overline{E}$ in $\mathcal{U}$. Since $\overline{E}$ is reduced, then I is a reduced $\mathcal{U}$-injective. This shows that E is reduced as an unstable A-module because its injective hull (in the category $\mathrm{H}^*V - \mathcal{U}$) is $\mathrm{H}^*V \otimes I$ which is reduced as unstable A-module. □

Remark 3.2.3. In the previous result the condition (1): E is free as an H^*V-module is necessary. In fact, the finite $\mathrm{H} - A$-module $\mathrm{J}_{\mathbb{Z}/2\mathbb{Z}}(1)$ is not free as an H-module and not reduced as an unstable A-module, however $\overline{\mathrm{J}_{\mathbb{Z}/2\mathbb{Z}}(1)} = \mathbb{F}_2$ is a reduced unstable A-module. Observe that $\mathrm{J}_{\mathbb{Z}/2\mathbb{Z}}(1)$ is isomorphic, as unstable A-module, to $\mathbb{F}_2 \oplus \sum \mathbb{F}_2$, the structure of H-module is given by: $t.\iota = \Sigma\iota$, where ι is the generator of $\mathbb{F}_2$ and t the generator of H.
Observe that the converse of corollary 3.2.2 is false. In fact, the $\mathrm{H} - A$-module $E = \mathrm{H}^{\geqslant 1}$ is reduced as unstable A-module however the unstable A-module $\overline{E} \cong \sum \mathbb{F}_2$ is not reduced.

4. Description of E when $\overline{E}$ is nil-closed

The main result of this paragraph concerns the relation between the two first terms of a (minimal) injective resolution of E and $\overline{E}$.

Theorem 4.1. *Let E be an unstable $\mathrm{H}^*V - A$-module which is free as an H^*V-module. We suppose that:*

1. *$\overline{E}$ is nil-closed.*
2. $0 \longrightarrow \overline{E} \longrightarrow I_0 \xrightarrow{i_1} I_1 \longrightarrow \cdots$ *is the beginning of a (minimal) $\mathcal{U}$- injective resolution of $\overline{E}$.*

Then there exists an $\mathrm{H}^*V - A$-linear map $\varphi : \mathrm{H}^*V \otimes I_0 \to \mathrm{H}^*V \otimes I_1$ such that:

1. $0 \longrightarrow E \longrightarrow \mathrm{H}^*V \otimes I_0 \xrightarrow{\varphi} \mathrm{H}^*V \otimes I_1 \longrightarrow \cdots$ *is the beginning of a (minimal) injective resolution of E (in the category $\mathrm{H}^*V - \mathcal{U}$).*
2. $\overline{\varphi} = i_1$

Proof. The unstable A-module $\overline{E}$ is nil-closed so is reduced, we have then, by theorem 3.2.1, that the injective hull of E is $\mathrm{H}^*V \otimes I_0$. We denote by C_0 the quotient of $\mathrm{H}^*V \otimes I_0$ by E. We have the following short exact sequence in $\mathrm{H}^*V - \mathcal{U}$:

$$0 \longrightarrow E \xrightarrow{i_0} \mathrm{H}^*V \otimes I_0 \longrightarrow C_0 \longrightarrow 0 .$$

Since the induced map $\overline{i_0}$ is an injection (see lemma 3.1.2), then the unstable A-module $Tor_1^{\mathrm{H}^*V}(\mathbb{F}_2, C_0)$ is trivial; this shows that the module C_0 is free as an H^*V-module (see for example [**NS**], proposition A.1.5).
We verify that the $\mathcal{U}$-injective hull of $\overline{C_0}$ is I_1 and that C_0 is reduced since $\overline{C_0}$ is reduced (see corollary 3.2.2). This implies, by theorem 3.2.1, that the $\mathrm{H}^*V - \mathcal{U}$-injective hull of C_0 is isomorphic to $\mathrm{H}^*V \otimes I_1$. □

Remark 4.2. let M be a nil-closed unstable A-module and
$0 \longrightarrow M \xrightarrow{i_0} I_0 \xrightarrow{i_1} I_1 \longrightarrow \cdots$ be the beginning of a (minimal) $\mathcal{U}$-injective resolution of M. We denote by

$$(Hom_{\mathrm{H}^*V-\mathcal{U}}(\mathrm{H}^*V \otimes I_0,\ \mathrm{H}^*V \otimes I_1))_{i_1}$$

the set of $\mathrm{H}^*V - A$-linear map $\varphi : \mathrm{H}^*V \otimes I_0 \to \mathrm{H}^*V \otimes I_1$ such that $\overline{\varphi} = i_1$.
Using Lannes T-functor (see [**L1**]) we have:

$$(Hom_{\mathrm{H}^*V-\mathcal{U}}(\mathrm{H}^*V \otimes I_0,\ \mathrm{H}^*V \otimes I_1))_{i_1} \cong (Hom_{\mathcal{U}}(T_V I_0,\ I_1))_{i_1}$$

where $(Hom_{\mathcal{U}}(T_V I_0,\ I_1))_{i_1}$ is the set of A-linear map $\psi : T_V I_0 \to I_1$ such that $\psi \circ i = i_1$, where $i : I_0 \hookrightarrow T_V I_0$ denotes the natural inclusion.
The kernel of any element $\psi \in (Hom_{\mathcal{U}}(T_V I_0,\ I_1))_{i_1}$, which is free as an H^*V-module, is an unstable $\mathrm{H}^*V - A$-module such that $\overline{ker\psi} \cong M$.

Remark 4.3. If $\overline{E}$ is an $\mathcal{U}$-injective then the only unstable free $\mathrm{H}^*V - A$-module, up to isomorphism, solution of the problem $(\mathcal{P})$ is $\mathrm{H}^*V \otimes \overline{E}$.
Let n be an even integer. The unstable free $\mathrm{H} - A$-modules, up to isomorphism, solution of the problem $(\mathcal{P})$ when M is $\mathrm{H}^*BSO(n)$ are $\mathrm{H}^*BO(n)$ and $\mathrm{H} \otimes \mathrm{H}^*BSO(n)$. We verify that these two $\mathrm{H} - A$-modules are not isomorphic in the category $\mathrm{H} - \mathcal{U}$ (since it does not exist an A-linear section of the projection $\mathrm{H}^*BO(n) \to \mathrm{H}^*BSO(n)$).

5. Applications

5.1.

Our first application concerns the determination of the mod. 2 cohomology of the mapping space $\mathbf{hom}(\mathrm{B}(\mathbb{Z}/2^n), Y)$ whose domain is a classifying space for the group $\mathbb{Z}/2^n$ and whose range is a space Y such that H^*Y is concentrated in even degrees.

We will just recall some facts, ignoring the p-completion problems. For further details see [**DL**].

One proceeds by induction on the integer n. Let us set

$$X = \mathbf{hom}(\mathrm{E}(\mathbb{Z}/2^n)/(\mathbb{Z}/2^{n-1}), Y) \quad .$$

The space X has the homotopy type of $\mathbf{hom}(\mathrm{B}(\mathbb{Z}/2^{n-1}), Y)$ and is equipped of an action $\mathbb{Z}/2$ such that one has a homotopy equivalence

$$\mathbf{hom}(\mathrm{B}(\mathbb{Z}/2^n), Y) \cong X^{\mathrm{h}\mathbb{Z}/2} \quad ,$$

$X^{\mathrm{h}\mathbb{Z}/2}$ denoting the homotopy fixed point space: $\mathbf{hom}_{\mathbb{Z}/2}(\mathrm{E}\mathbb{Z}/2, X)$. Using $\mathrm{Fix}_{\mathbb{Z}/2}$-theory [**L1**], one gets:

$$\mathrm{H}^*\mathbf{hom}(\mathrm{B}(\mathbb{Z}/2^n), Y) \cong \mathrm{Fix}_{\mathbb{Z}/2}\, \mathrm{H}^*_{\mathbb{Z}/2}X \quad .$$

Since the computation of the functor $\mathrm{Fix}_{\mathbb{Z}/2}$ on an unstable $\mathrm{H}-\mathrm{A}$-module is not difficult in general, the determination of the mod. 2 cohomology of the mapping space $\mathbf{hom}(\mathrm{B}(\mathbb{Z}/2^n), Y)$ is reduced to the determination of the unstable $\mathrm{H}-\mathrm{A}$-module $\mathrm{H}^*_{\mathbb{Z}/2}X$. As we are going to explain, this last point is closely related to problem $(\mathcal{P})$.

One knows by induction on n that the mod. 2 cohomology of the space X as the one of the space Y is concentrated in even degrees and one checks that the action of $\mathbb{Z}/2$ on $\mathrm{H}^*(Y;\mathbb{Z})$ is trivial. These two facts imply that the Serre spectral sequence, for mod. 2 cohomology, associated to the fibration

$$X \to X_{\mathrm{h}\mathbb{Z}/2} \to \mathrm{B}\mathbb{Z}/2$$

collapses ($X_{\mathrm{h}\mathbb{Z}/2}$ denotes the Borel construction $\mathrm{E}\mathbb{Z}/2 \times_{\mathbb{Z}/2} X$). This collapsing implies in turn that $\mathrm{H}^*_{\mathbb{Z}/2}X$ is H-free and that $\overline{\mathrm{H}^*_{\mathbb{Z}/2}X}$ is isomorphic to H^*X. So the determination of $\mathrm{H}^*\mathbf{hom}(\mathrm{B}(\mathbb{Z}/2^n), Y)$ is indeed reduced to the resolution of a problem $(\mathcal{P})$.

We conclude this subsection by a concrete example (we follow [**De**], section 6); we take $n = 2$ and $Y = \mathrm{BSU}(2)$. Using $\mathrm{T}_{\mathbb{Z}/2}$-computations one sees that X has the homotopy type of $\mathrm{BSU}(2) \coprod \mathrm{BSU}(2)$; one checks also that the $\mathbb{Z}/2$-action preserves the connected components. The $(\mathcal{P})$-problem asociated to the determination of the unstable $\mathrm{H}-\mathrm{A}$-module $\mathrm{H}^*_{\mathbb{Z}/2}X$ is the following one:

Find the unstable $\mathrm{H}-\mathrm{A}$-modules E such that

- E is H-free;
- the unstable A-module $\overline{E}$ is isomorphic to $\mathrm{H}^*\mathrm{BSU}(2)$.

Using the fact that the injective hull, in the category $\mathrm{H}-\mathcal{U}$, of E is $\mathrm{H}\otimes\mathrm{H}$ (see theorem 3.2), one checks that one has two possibilities:

- $E \cong \mathrm{H}\otimes \mathrm{H}^*\mathrm{BSU}(2)$;
- $E \cong \mathrm{H}\otimes_{\mathrm{H}^*\mathrm{BU}(1)} \mathrm{H}^*\mathrm{BU}(2)$ (the structures of unstable $\mathrm{H}^*\mathrm{BU}(1) - \mathrm{A}$-modules on $\mathrm{H} = \mathrm{H}^*\mathrm{BO}(1)$ and $\mathrm{H}^*\mathrm{BU}(2)$ are respectively induced by the inclusion of $\mathrm{O}(1)$ in $\mathrm{U}(1)$ and the determinant homomorphism from $\mathrm{U}(2)$ to $\mathrm{U}(1)$).

5.2.

The theorem 4.1 can be illustrated, topologically, as follows:

Proposition 5.2.1. *Let X be a CW-complex on which acts an elementary abelian group 2-group V. Suppose that:*

1. H^*X *is nil-closed*
2. $0 \longrightarrow \mathrm{H}^*X \longrightarrow I_0 \xrightarrow{\alpha} I_1 \longrightarrow \cdots$ *is the beginning of a (minimal) $\mathcal{U}$-injective resolution of* H^*X
3. $\mathrm{H}^*_V X$ *is free as an* H^*V*-module.*

Then there exists an $\mathrm{H}^*V - A$-linear map $\varphi: \mathrm{H}^*V\otimes I_0 \to \mathrm{H}^*V\otimes I_1$ such that:

1. $\mathrm{H}^*_V X \cong Ker(\varphi)$.
2. $0 \longrightarrow \mathrm{H}^*_V X \longrightarrow \mathrm{H}^*V\otimes I_0 \xrightarrow{\varphi} \mathrm{H}^*V\otimes I_1 \longrightarrow \cdots$ *is the beginning of a (minimal) injective resolution of $\mathrm{H}^*_V X$ (in the category $\mathrm{H}^*V - \mathcal{U}$).*
3. $\overline{\varphi} = \alpha : I_0 \to I_1$.

In particular, we have:

Corollary 5.2.2. *Let X be a CW-complex on which acts an elementary abelian group 2-group V. Suppose that:*

1. H^*X *is a reduced $\mathcal{U}$-injective,*
2. $\mathrm{H}^*_V X$ *is free as an* H^*V*-module.*

Then $\mathrm{H}^*_V X \cong \mathrm{H}^*V\otimes H^*X$.

6. Description of E when $\overline{E}$ is isomorphic to $\sum^n \mathbb{F}_2$

In this section, we prove the following result.

Theorem 6.1. *Let E be unstable $\mathrm{H}^*V - A$-module which is free as an H^*V-module. If $\overline{E}$ is isomorphic to $\sum^n \mathbb{F}_2$, then there exists an element u in H^*V such that:*

1. $u = \prod_i \theta_i^{\alpha_i}$, *where* $\theta_i \in (\mathrm{H}^1V)\setminus\{0\}$ *and* $\alpha_i \in \mathbb{N}$
2. $E \cong \sum^d u\mathrm{H}^*V$ *with* $d + \sum_i \alpha_i = n$.

Proof. Let N be an unstable A-module, we denote by $dimN$ the total dimension of N that is $dim\ N = \sum_i dim\ N^i$. We have the equality $dim\ \overline{E} = 1 = dim\ \mathrm{Fix}_V E$

(see [**LZ3**]), so we deduce that $Fix_V E = \sum^l \mathbb{F}_2$, where $l \in \mathbb{N}$. Let $\eta_V : E \to \mathrm{H}^*V \otimes Fix_V E$ be the adjoint of the identity of $Fix_V E$ (see [**LZ2**]). Since the map η_V is an injection, then the module E is a sub-H^*V – A-module of $\sum^l \mathrm{H}^*V$. Let's write $E = \sum^l E'$, where E' is sub-$\mathrm{H}^*V - A$-module of H^*V. By a result of J-P. Serre (see [**Se**]), there exists N such that: $c_V^N \mathrm{H}^*V \subset E' \subset \mathrm{H}^*V$. Since E' is free as an H^*V-module and of dimension one, then there exists $u \in \widetilde{\mathrm{H}}^*V$ such that $E' = u\mathrm{H}^*V$. The inclusion $c_V^N \mathrm{H}^*V \subset u\mathrm{H}^*V$ proves that $u = \prod_i \theta_i^{\alpha_i}$, where $\theta_i \in (\mathrm{H}^1V) \setminus \{0\}$ and $\alpha_i \in \mathbb{N}$. □

Remark 6.2. We remark that by the previous result, we can determinate E when $\overline{E}$ is isomorphic to $\mathbb{F}_2 \oplus \sum^n \mathbb{F}_2$. In this case, we verify that $E \cong \mathrm{H}^*V \oplus \sum^d u\mathrm{H}^*V$, where $u = \prod_i \theta_i^{\alpha_i}$, $\theta_i \in \mathrm{H}^*V \setminus \{0\}$, $\alpha_i \in \mathbb{N}$ and $d + \sum_i \alpha_i = n$. In fact, since the $\mathrm{H}^*V - \mathcal{U}$-injective module H^*V is a sub-H^*V-module of E, then $E \cong \mathrm{H}^*V \oplus E'$, where E' is an unstable H^*V – A-module, free as an H^*V-module and such that $\overline{E'} \cong \sum^n \mathbb{F}_2$. The result holds from theorem 6.1.

6.3 Example

We give an example showing how to realize topologically the cases of theorem 6.1 and remark 6.2.

Let $\rho : V \to \mathrm{O}(d)$ be a group homomorphism. ρ gives both an action of V on D^d, S^{d-1} and a d-dimensional orthogonal bundle whose mod.2 Euler class is denoted by $e(\rho)$.

The long exact sequence of the pair $(\mathrm{D}^d, \mathrm{S}^{d-1})$ and the Thom isomorphism give the long (Gysin) exact sequence (see for example [**Hu**]):

$$\cdots \longrightarrow \mathrm{H}^{*-1}V \longrightarrow \mathrm{H}_V^{*-1}\mathrm{S}^{d-1} \longrightarrow \Sigma^{-d}\mathrm{H}^*V \xrightarrow{\smile e(\rho)} \mathrm{H}^*V \longrightarrow \mathrm{H}_V^*\mathrm{S}^{d-1} \longrightarrow \cdots$$

The decomposition $\rho \cong \oplus_{i=1}^d \rho_i$ of the representation ρ into orthogonal representations of dimension 1 gives $e(\rho) = \prod_i e(\rho_i)$. We have now two cases.

- If none of the representations ρ_i is trivial then $e(\rho)$ is non zero and $\mathrm{H}_V^*(\mathrm{D}^d, \mathrm{S}^{d-1})$ is isomorphic to $e(\rho)\mathrm{H}^*V$ as an H^*V – A-module. This illustrates theorem 6.1.
- Otherwise, let's write $\rho = \sigma \oplus \tau$, σ (resp. τ) being the direct sum of the non trivial (resp. trivial) representations ρ_i. Then $\mathrm{H}_V^*\mathrm{S}^{d-1} \cong \mathrm{H}^*V \oplus \Sigma^{dim\tau}\ e(\sigma)\mathrm{H}^*V$ and $\mathrm{H}_V^*(\mathrm{S}^{d-1})$ is an illustration of the remark 6.2.

7. Determination of E when V is $\mathbb{Z}/2\mathbb{Z}$ and $\overline{E}$ is J(2)

In this section, we assume that V is $\mathbb{Z}/2\mathbb{Z}$ and $\overline{E}$ is the Brown-Gitler module J(2). We denote by $\mathrm{H} = \mathbb{F}_2[t]$ the cohomology of $\mathbb{Z}/2\mathbb{Z}$, where t is an element of H of degree one. We have the following result.

Proposition 7.1. *Let E be an* H – *A-module which is* H*-free and such that* $\overline{E}$ *is isomorphic to* J(2) *then:*

$E \cong \mathrm{H} \otimes \mathrm{J}(2)$

or

*E is the sub-*H – *A-module of* $\mathrm{H} \oplus \sum \mathrm{H}$ *generated by* $(t, \Sigma 1)$ *and* $(t^2, 0)$.

Proof. This proof uses the Smith theory (see [**DW**], [**LZ2**] theorem 2.1) which gives us an exact sequence (*) in $\mathrm{H}-\mathcal{U}$:

$$(*)\quad 0 \longrightarrow E \stackrel{\eta}{\longrightarrow} \mathrm{H}\otimes FixE \longrightarrow C \longrightarrow 0$$

where C the quotient of $\mathrm{H}\otimes FixE$ is finite and also $FixE$ is finite.
If the module C is trivial then E is isomorphic to $\mathrm{H}\otimes \mathrm{J}(2)$.
When C is a non trivial module. By applying the functor $\mathbb{F}_2\otimes_{\mathrm{H}} -$ to the exact sequence (*), we obtain:

$$0 \longrightarrow \sum \tau C \longrightarrow \overline{E} = \mathrm{J}(2) \longrightarrow FixE \longrightarrow \overline{C} \longrightarrow 0$$

where τC is the trivial part of C (see [**BHZ**]).
Let's denote by Q the quotient of $\overline{E}$ by $\sum \tau C$. By properties of the module $\mathrm{J}(2)$, we have that $\sum \tau C = \sum^2 \mathbb{F}_2$ and $Q = \sum \mathbb{F}_2$. The exact sequence:

$$0 \longrightarrow \sum \mathbb{F}_2 \longrightarrow FixE \longrightarrow \overline{C} \longrightarrow 0$$

gives that $FixE \cong \sum \mathbb{F}_2 \oplus \overline{C}$. One checks that the module $\overline{C}$ is either isomorphic to $\mathbb{F}_2$ or $\sum \mathbb{F}_2$. If $\overline{C} = \sum \mathbb{F}_2$ then $FixE \cong \sum \mathbb{F}_2 \oplus \sum \mathbb{F}_2$ as an unstable A-module, which implies that the module E is a suspension which is impossible because $\overline{E} = \mathrm{J}(2)$ is not a suspension. We conclude that $\overline{C} = \mathbb{F}_2$. Since $\tau C = \sum \mathbb{F}_2$ then we get C is isomorphic to $\mathrm{H}^{\leqslant 1}$, where $\mathrm{H}^{\leqslant 1}$ denotes the sub-$\mathrm{H}-A$-module of H consisting of elements of degree less or equal than 1. We have the following exact sequence in $\mathrm{H}-\mathcal{U}$:

$$0 \longrightarrow E \longrightarrow \mathrm{H}\oplus \sum \mathrm{H} \stackrel{\varphi}{\longrightarrow} \mathrm{H}^{\leqslant 1} \longrightarrow 0\,.$$

The module E, we are searching for, is the kernel of φ and we check that it is the sub-H – A-module of $\mathrm{H}\oplus\sum \mathrm{H}$ generated by the elements $(t, \Sigma 1)$ and $(t^2, 0)$. □

Remark 7.2. Let be $\mathbb{Z}/2\mathbb{Z}$ act on a real projective space $\mathbb{R}\mathrm{P}^2$; let x_0 be a fixed point of this action (the set of fixed point is not empty for example by an argument of Lefschetz number). We have:
- The Serre spectral sequence collapses to give that: $\mathrm{H}^*_V(\mathbb{R}\mathrm{P}^2, x_0)$ is H-free and $\overline{\mathrm{H}^*_V(\mathbb{R}\mathrm{P}^2, x_0)}$ is isomorphic to $\mathrm{J}(2)$.
- In [**DW**], Dwyer and Wilkerson have shown that $\mathrm{H}^*_V\mathbb{R}\mathrm{P}^2 = \mathbb{F}_2[t, y]/(f)$ where y restricts to x and $f = y^i(y+t)^j$ for $i + j = 3$. It is easy to check that this computation agrees with theorem 7.1.

References

[**BHZ**] D.Bourguiba, S.Hammouda, S.Zarati: Profondeur et cohomologie équivariante, African Diaspora Mathematics Research, Special Issue Vol 4 Number 3, 11-21.

[**De**] F.X.Dehon Cobordisme complexe des espaces profinis et foncteur T de Lannes, Mémoires de la Société Mathématique de France **98**, SMF 2004.

[DL] F.X.Dehon, J.Lannes: Sur les espaces fonctionnels dont la source est le classifiant d'un groupe de Lie compact, commutatif I.H.E.S. 89 (1999) 127-177.

[DW] W.G.Dwyer, C.W.Wilkerson: Smith theory revisited, Annals of Mathematics, 127(1988) 191-198.

[EP] M.J.Errockh, C.Peterson: Injective resolutions of unstable modules, Journal of Pure and Applied Algebra 97(1994) 37-50.

[Hu] D.Husemoller: Fibre bundles, McGraw-Hill, series in higher mathematics, 1966.

[L1] J.Lannes: Sur les espaces fonctionnels dont la source est le classifiant d'un p-groupe abélien élémentaire, Publ. I.H.E.S. 75 (1992) 135-224.

[LS] J.Lannes, L.Shwartz: Sur la structure des A-modules instables injectifs, Topology 28 (1989) 153-169.

[LZ1] J.Lannes, S.Zarati: Sur les $\mathcal{U}$-injectifs, Ann. Scient. Ec. Norm. Sup. 19 (1986) 1-31.

[LZ2] J.Lannes, S.Zarati: Théorie de Smith algébrique et classification des $\mathrm{H}^*V - \mathcal{U}$-injectifs, Bull. Soc. Math. France 123 (1995) 189-223.

[LZ3] J.Lannes, S.Zarati: Tor et Ext-dimensions des H^*V – A-modules instables qui sont de type fini comme H^*V-modules, Progress in Mathematics, Birkhäuser Verlag, vol 136 (1996) 241-253.

[NS] M.D.Neusel, L.Smith: Invariant theory of finite groups, volume 94 of Mathematical Surveys and Monographs. American Mathematical Society, Providence, RI, 2002.

[R] J.Rotman: An introduction to homological algebra, Academic Press, 1979.

[S] L.Schwartz: Unstable modules over the Steenrod algebra and Sullivan's fixed point set conjecture, University of Chicago Press, 1984.

[Se] J-P.Serre: Sur la dimension cohomologique des groupes profinis, Topology 3. (1965), 413-420.

This article may be accessed via WWW at `http://jhrs.rmi.acnet.ge`

Dorra Bourguiba
`dorra.bourguiba@fst.rnu.tn`

Faculté de Sciences–Mathématiques, Université de Tunis,
TN-1060 Tunis, Tunisie.

Journal of Homotopy and Related Structures, vol. 4(1), 2009, pp.83–109

PARTICLE CONFIGURATIONS AND COXETER OPERADS

SUZANNE M. ARMSTRONG, MICHAEL CARR,
SATYAN L. DEVADOSS, ERIC ENGLER, ANANDA LEININGER AND
MICHAEL MANAPAT

(*communicated by James Stasheff*)

Abstract

There exist natural generalizations of the real moduli space of Riemann spheres based on manipulations of Coxeter complexes. These novel spaces inherit a tiling by the graph-associahedra convex polytopes. We obtain explicit configuration space models for the classical infinite families of finite and affine Weyl groups using particles on lines and circles. A Fulton-MacPherson compactification of these spaces is described and this is used to define the Coxeter operad. A complete classification of the building sets of these complexes is also given, along with a computation of their Euler characteristics.

1. Motivation from Physics

1.1. Moduli spaces

A configuration space of n ordered, distinct particles on a variety V is

$$C_n(V) = V^n - \Delta, \quad \text{where } \Delta = \{(x_1, \ldots, x_n) \in V^n \mid \exists\, i, j,\ x_i = x_j\}.$$

Over the past decade, there has been an increased interest in the configuration space of n labeled particles on the projective line. The focus is on a quotient of this space by $\mathbb{P}\mathrm{Gl}_2(\mathbb{C})$, the affine automorphisms on $\mathbb{CP}^1$. The resulting variety is the moduli space of Riemann spheres with n punctures $\mathcal{M}_{0,n} = C_n(\mathbb{CP}^1)/\mathbb{P}\mathrm{Gl}_2(\mathbb{C})$. There is a compactification $\overline{\mathcal{M}}_{0,n}$ of this space, a smooth variety of complex dimension $n-3$, coming from Geometric Invariant Theory [**24**]. The space $\overline{\mathcal{M}}_{0,n}$ plays a crucial role as a fundamental building block in the theory of Gromov-Witten invariants, also appearing in symplectic geometry and quantum cohomology [**20**].

Our work is motivated by the *real* points $\overline{\mathcal{M}}_{0,n}(\mathbb{R})$ of this space, the set of points fixed under complex conjugation. These real moduli spaces have importance in their

---

All authors were partially supported by NSF grant DMS-9820570. Carr and Devadoss were also supported by NSF CARGO grant DMS-0310354.
We express our gratitude to Jim Stasheff and Vic Reiner for detailed comments on this work and to Alex Postnikov for working on the f-vectors of graph-associahedra. We also thank the referee for wonderful insight and careful analysis, especially with regards to the operad module structures.
Received November 21, 2008, revised January 28, 2009; published on May 29, 2009.
2000 Mathematics Subject Classification: Primary 18D50, Secondary 14H10, 52B11.
Key words and phrases: Operads, configuration spaces, compactifications, graph associahedra.

own right, appearing in areas such as ζ-motives of Goncharov and Manin [**17**] and Lagrangian Floer theory of Fukaya [**15**]. Indeed, $\overline{\mathcal{M}}_{0,n}(\mathbb{R})$ has even emerged in phylogenetic trees [**2**] and networks [**21**]. It was Kapranov [**19**] who first noticed a relationship between $\overline{\mathcal{M}}_{0,n}(\mathbb{R})$ and the braid arrangement of hyperplanes, associated to the Coxeter group of type A: Blow-ups of certain cells of the A_n Coxeter complex yield a space homeomorphic to a double cover of $\overline{\mathcal{M}}_{0,n+2}(\mathbb{R})$. This creates a natural tiling of $\overline{\mathcal{M}}_{0,n}(\mathbb{R})$ by associahedra, the combinatorics of which is discussed in [**12**]. Davis et. al have generalized this construction to all Coxeter groups, along with studying the fundamental groups of these blown-up spaces [**8**]. Carr and Devadoss [**6**] looked at the inherent tiling of these spaces by the convex polytopes *graph-associahedra.*

1.2. Overview

We begin with elementary results and notation: Section 2 provides the background of Coxeter groups and their associated Coxeter complexes and Section 3 constructs the appropriate configuration spaces of particles. Section 4 introduces the bracketing notation in order to visualize collisions in the configuration spaces, leading to viewing the hyperplanes of the Coxeter complexes in this new language. It is this notation that provides a transparent understanding of several results in this paper. In particular, it allows for a complete classification of the minimal building sets for the Coxeter complexes, along with their enumeration, as given in Tables 3 and 4. It is interesting to note that some configuration structures behave quite classically, whereas others (based on *thick* particles) are atypical.

The heart of the paper begins in Section 5 where the Fulton-MacPherson compactification [**16**] of these spaces is discussed and used to define the notion of a *Coxeter operad* in Definition 5.8, extending the mosaic operad of $\overline{\mathcal{M}}_{0,n}(\mathbb{R})$ [**10**]. Here, *nested* bracketings are used to describe the structure of the compactified spaces, enabling us to describe how the chambers of these spaces glue together. Section 6 ends with combinatorial results using the theory developed in [**6**]. For instance, the Euler characteristics of these Coxeter moduli spaces are given, exploiting the tilings by graph-associahedra.

As of writing this paper, the importance of the operadic structure of $\overline{\mathcal{M}}_{0,n}(\mathbb{R})$ has been brought further to light. For instance, Etingof et al. [**13**] have used the mosaic operad to compute the cohomology ring of $\overline{\mathcal{M}}_{0,n}(\mathbb{R})$. Recently, the Coxeter operad defined below appears in the work of E. Rains in computing the integral homology of these generalized Coxeter moduli spaces [**28**].

2. Spherical and Euclidean Complexes

2.1. Coxeter groups

In order to provide the construction of Coxeter moduli space generalizations of $\overline{\mathcal{M}}_{0,n}(\mathbb{R})$, as given in Section 5, we begin with standard facts and definitions about Coxeter systems. Most of the material here can be found in Bourbaki [**4**].

Definition 2.1. Given a finite set S, a *Coxeter group* W is given by the presentation

$$W = \langle\, s_i \in S \mid s_i^2 = 1,\ (s_i s_j)^{m_{ij}} = 1 \,\rangle,$$

where $m_{ij} = m_{ji}$ and $2 \leqslant m_{ij} \leqslant \infty$. The pair (W, S) is called a *Coxeter system.*

Associated to any Coxeter system (W, S) is its *Coxeter graph* Γ_W: Each node represents an element of S, where two nodes s_i, s_j determine an edge if and only if $m_{ij} \geqslant 3$. A Coxeter group is *irreducible* if its Coxeter graph is connected and it is *locally finite* if either W is finite or each proper subset of S generates a finite group. A Coxeter group is *simplicial* if it is irreducible and locally finite [**8**]. The classification of simplicial Coxeter groups and their graphs is well-known [**4**, Chapter 6].

We restrict our attention to infinite families of simplicial Coxeter groups which generalize to arbitrary number of generators. This will mimic configuration spaces of an arbitrary number of particles since our motivation comes from $\overline{\mathcal{M}}_{0,n}(\mathbb{R})$; it will also allow a well-defined construction of an operad. There are only seven such types of Coxeter groups: three spherical ones and four Euclidean ones. Figure 1 shows the Coxeter graphs associated to the Coxeter groups of interest; we label the edge with its order for $m_{ij} > 3$. The number of nodes of a graph is given by the subscript n for the spherical groups, whereas the number of nodes is $n+1$ for the Euclidean case.

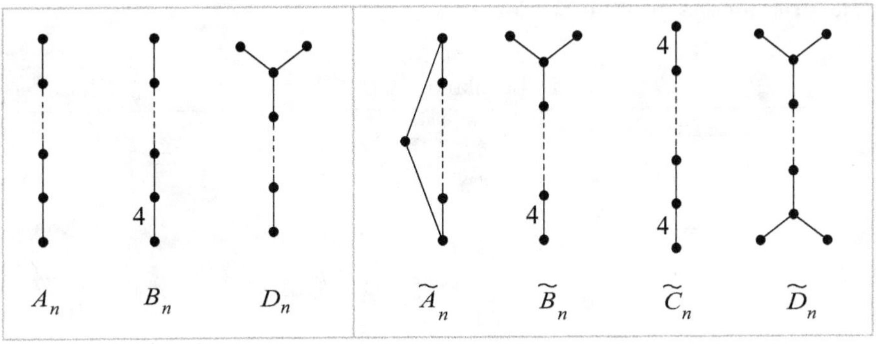

Figure 1: Coxeter graphs of spherical and Euclidean groups.

2.2. Coxeter complexes

Every spherical Coxeter group has an associated finite reflection group realized by reflections across linear hyperplanes on a sphere. Every conjugate of a generator s_i acts on the sphere as a reflection in some hyperplane, dividing the sphere into simplicial chambers. The sphere, along with its cellulation is the *Coxeter complex* corresponding to W, denoted $\mathcal{C}W$. The hyperplanes associated to each group given in Table 1 lie on the $(n-1)$ sphere. The W-action on the chambers of $\mathcal{C}W$ is simply transitive, and thus we may associate an element of W to each chamber. The number of chambers of $\mathcal{C}W$ comes from the order of the group.

W	Hyperplanes	# Chambers
A_n	$x_i = x_j$	$(n+1)!$
B_n	$x_i = 0,\ x_i = \pm x_j$	$2^n\ n!$
D_n	$x_i = \pm x_j$	$2^{n-1}\ n!$

Table 1: The spherical arrangements.

We move from spherical geometry coming from linear hyperplanes to Euclidean geometry arising from affine hyperplanes. Just as with the spherical case, each Euclidean Coxeter group has an associated Euclidean reflection group realized as reflections across affine hyperplanes in Euclidean space. Again, we focus on the infinite families of such Euclidean Coxeter groups which are $\widetilde{A}_n, \widetilde{B}_n, \widetilde{C}_n$, and $\widetilde{D}_n$. The hyperplanes associated to each group, given in Table 2, lie in $\mathbb{R}^n$.

We look at the quotient of the Euclidean space $\mathbb{R}^n$ by a group of translations, resulting in the n-torus $\mathbb{T}^n$. This is done for three reasons: First, the configuration space model is a more natural object after the quotient, resulting in particles on circles. Second, it is the correct generalization of the affine type A complex, as discussed in [**11**]. Third, and most importantly, it presents us with valid operad module structures as given in Section 5.

W	Hyperplanes ($k \in 2\mathbb{Z}$)	# Chambers
$\widetilde{A}_n$	$x_i = x_j + k$	$n!$
$\widetilde{B}_n$	$x_i = \pm x_j + k,\ x_i = 1 + k$	$2^{n-1}\ n!$
$\widetilde{C}_n$	$x_i = \pm x_j + k,\ x_i = 1 + k,\ x_i = 0 + k$	$2^n\ n!$
$\widetilde{D}_n$	$x_i = \pm x_j + k$	$2^{n-2}\ n!$

Table 2: The toroidal arrangements.

The translations for $\widetilde{A}_n$ are covered in [**11**, Section 2.3]. For the remaining cases, we choose a slightly non-standard collection of hyperplanes in order for the associated configuration spaces to be more canonical. This has the benefit of identifying the most ubiquitous set of hyperplanes $\{x_i = \pm x_j + k\}$, producing an arrangement that is familiar from the spherical cases. We refer to the quotient of the complex $\mathcal{C}W$ of Euclidean type as the *toroidal Coxeter complex*, denoted $\mathbb{T}\mathcal{C}W$.

Example 2.2. Figure 2(a) is $\mathcal{C}A_3$, the 2-sphere with hyperplane markings, and part (b) is the 2-sphere $\mathcal{C}B_3$. Figure 2(c) shows the hyperplanes of $\widetilde{C}_2$ in $\mathbb{R}^2$, whereas (d) shows the hyperplanes for $\widetilde{A}_2$. Part (e) is the cellulation of the toroidal complex $\mathbb{T}\mathcal{C}\widetilde{A}_2$.

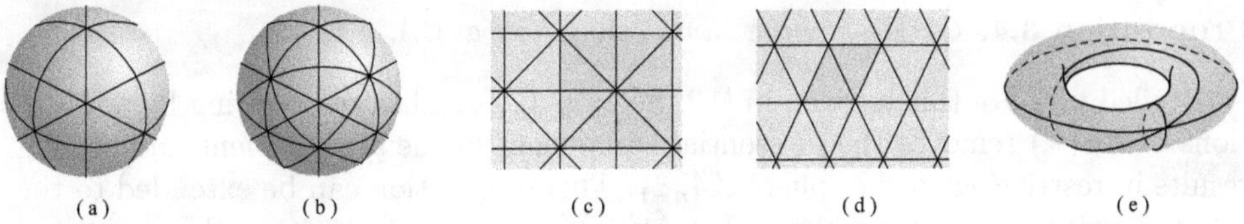

Figure 2: Coxeter complexes (a) $\mathcal{C}A_3$, (b) $\mathcal{C}B_3$, (c) $\mathcal{C}\widetilde{C}_2$, (d) $\mathcal{C}\widetilde{A}_2$ and (e) $\mathbb{T}\mathcal{C}\widetilde{A}_2$.

3. Configuration Spaces

3.1. Spherical complexes

We now give an explicit configuration space analog to each Coxeter complex above. These appear as (quotients of) configuration spaces of particles on the line $\mathbb{R}$ and the circle $\mathbb{S}$. The arguments used for the constructions below are elementary, immediately following from the hyperplane arrangements of the reflection groups. However, as shown in Section 5, the configuration space model we provide will enable us to elegantly capture the blow-ups of these Coxeter complexes.

Definition 3.1. Let $C_n(\mathbb{R}) = \mathbb{R}^n - \{(x_1, x_2, ..., x_n) \in \mathbb{R}^n \mid \exists\, i, j,\ x_i = x_j\}$ be the configuration space of n labeled particles on the real line $\mathbb{R}$. A generic point in $C_5(\mathbb{R})$ is $x_1 < x_2 < x_3 < x_4 < x_5$, which we notate (without labels) as ○-○-○-○-○.

Definition 3.2. Let $C_{\bar{n}}(\mathbb{R}_\bullet) = \mathbb{R}^n - \{(x_1, x_2, ..., x_n) \in \mathbb{R}^n \mid \exists\, i, j,\ x_i = \pm x_j \text{ or } x_i = 0\}$ be the space of n pairs of *symmetric* labeled particles (denoted $\bar{n}$) across the origin. A point in $C_{\bar{3}}(\mathbb{R}_\bullet)$ is $-x_3 < -x_2 < -x_1 < 0 < x_1 < x_2 < x_3$, which is depicted without labels as ○-○-○-●-○-○-○, where the black particle is fixed at the origin.

Definition 3.3. Let $C_{\bar{n}}(\mathbb{R}_\circ) = \mathbb{R}^n - \{(x_1, x_2, ..., x_n) \in \mathbb{R}^n \mid \exists\, i, j,\ x_i = \pm x_j\}$ be the space of $\bar{n}$ pairs of symmetric labeled particles across the origin, where the particle x_i and its symmetric partner $-x_i$ are both allowed to occupy the origin. A point in $C_{\bar{3}}(\mathbb{R}_\circ)$ is

$$-x_3 < -x_2 < -x_1, x_1 < x_2 < x_3 \tag{3.1}$$

drawn ○-○-○-+-○-○-○ without labels. Notice the mark at the origin where there is no fixed particle: The point $-x_3 < -x_2 < x_1, -x_1 < x_2 < x_3$ drawn as ○-○-○-+-○-○-○ lies in the same chamber of $C_{\bar{3}}(\mathbb{R}_\circ)$ as Eq.(3.1).

Let $\text{Aff}(\mathbb{R})$ be the group of affine transformations of $\mathbb{R}$ generated by translating and positive scaling. The action of $\text{Aff}(\mathbb{R})$ on $C_n(\mathbb{R})$ translates the leftmost of the n particles in $\mathbb{R}$ to -1 and the rightmost is scaled to 1. If we allow the particles in $C_n(\mathbb{R})/\text{Aff}(\mathbb{R})$ to *collide* (coincide with each other), the resulting space is denoted $C_n\langle\mathbb{R}\rangle$. In a sense, this includes the hyperplanes which were removed back into $\mathbb{R}^n$. The space $C_n\langle\mathbb{R}\rangle$ is sometimes referred to as the *naive compactification* of $C_n(\mathbb{R})/\text{Aff}(\mathbb{R})$.[1]

[1] The space $C_n\langle\mathbb{R}\rangle$ can also be thought of as the *closure* of $C_n(\mathbb{R})/\text{Aff}(\mathbb{R})$, though the space in

Proposition 3.4. *$C_n\langle\mathbb{R}\rangle$ has the same cellulation as $\mathcal{C}A_{n-1}$.*

A detailed proof of this is given in [**12**, Section 4]. Roughly, quotienting by translations of $\mathrm{Aff}(\mathbb{R})$ removes the inessential component of the arrangement and scaling results in restricting to the sphere $\mathcal{C}A_{n-1}$. This proposition can be extended to the other spherical Coxeter complexes. Let $\mathrm{Aff}(\bar{\mathbb{R}})$ be the transformations of $\mathbb{R}$ generated simply by positive scalings: The action of $\mathrm{Aff}(\bar{\mathbb{R}})$ scales the (symmetric) particles farthest from the origin to unit distance. Let $C_{\bar{n}}\langle\mathbb{R}_\bullet\rangle$ and $C_{\bar{n}}\langle\mathbb{R}_\circ\rangle$ denote spaces where particles of $C_{\bar{n}}(\mathbb{R}_\bullet)/\mathrm{Aff}(\bar{\mathbb{R}})$ and $C_{\bar{n}}(\mathbb{R}_\circ)/\mathrm{Aff}(\bar{\mathbb{R}})$ have collided, respectively.

Proposition 3.5. *$C_{\bar{n}}\langle\mathbb{R}_\bullet\rangle$ and $C_{\bar{n}}\langle\mathbb{R}_\circ\rangle$ have the same cellulation as $\mathcal{C}B_n$ and $\mathcal{C}D_n$ respectively.*

Proof. From the above definition above, it is clear $C_{\bar{n}}(\mathbb{R}_\bullet)$ and $C_{\bar{n}}(\mathbb{R}_\circ)$ are complements of the hyperplanes given in Table 1. Thus any collision of particles in the configuration space maps to a point on the hyperplanes defined by the associated finite reflection group. Quotienting by $\mathrm{Aff}(\bar{\mathbb{R}})$ allows choosing a particular representative for each fiber. Specifically, for the fiber containing $(x_1, x_2, ..., x_n)$, choose

$$(x_1, x_2, ..., x_n)/\sqrt{x_1^2 + x_2^2 + ... + x_n^2},$$

giving a map onto the unit sphere in $\mathbb{R}^n$. The cellulation of the sphere by these hyperplanes yields the desired Coxeter complex. □

3.2. Affine complexes

We move from the spherical to the affine (toroidal) complexes. However, the interest now is on configurations of particles on the circle $\mathbb{S}$. The group of rotations acts freely on $C_n(\mathbb{S})$, and its quotient is denoted by $C_n(\mathbb{S}')$; Figure 3(a) shows a point in $C_9(\mathbb{S}')$ drawn without labels.

Proposition 3.6. *$C_n\langle\mathbb{S}'\rangle$ has the same cellulation as $\mathbb{T}\mathcal{C}\widetilde{A}_{n-1}$.*

A proof of this is given in [**11**, Section 3]. A similar construction is produced below for the other three toroidal Coxeter complexes. Our focus now is on the circle $\mathbb{S}$ with the vertical line through its center as its axis of symmetry, where the two diametrically opposite points on the axis are labeled 0 and 1. The space of interest is the configuration space of pairs of *symmetric* labeled particles (again denoted $\bar{n}$) across this symmetric axis of the circle.

Definition 3.7. Let $C_{\bar{n}}(\mathbb{S}^\bullet_\circ) = \mathbb{T}^n - \{(x_1, x_2, ..., x_n) \in \mathbb{T}^n | \exists i, j,\ x_i = \pm x_j \text{ or } x_i = 1\}$ be the space of n pairs of symmetric labeled particles on $\mathbb{S}$ with a fixed particle at 1. Figure 3(b) shows a point in $C_{\bar{5}}(\mathbb{S}^\bullet_\circ)$.

Definition 3.8. Let $C_{\bar{n}}(\mathbb{S}^\bullet_\bullet) = \mathbb{T}^n - \{(x_1, x_2, ..., x_n) \in \mathbb{T}^n \mid \exists i, j,\ x_i = \pm x_j \text{ or } x_i = 1 \text{ or } x_i = 0\}$ be the space of n pairs of symmetric labeled particles on $\mathbb{S}$ with a fixed particle at 0 and 1. Figure 3(c) shows a point in $C_{\bar{5}}(\mathbb{S}^\bullet_\bullet)$.

---

which this closure is taken is non-trivial. An excellent treatment of these ideas in a general context is given by Sinha [**30**, Section 3].

Definition 3.9. Let $C_{\bar{n}}(\mathbb{S}^{\circ}_{\circ}) = \mathbb{T}^n - \{(x_1, x_2, ..., x_n) \in \mathbb{T}^n \mid \exists\, i, j,\ x_i = \pm x_j\}$ be the space of n pairs of symmetric labeled particles on $\mathbb{S}$ with no fixed particles. Figure 3(d) shows a point in $C_{\bar{5}}(\mathbb{S}^{\circ}_{\circ})$.

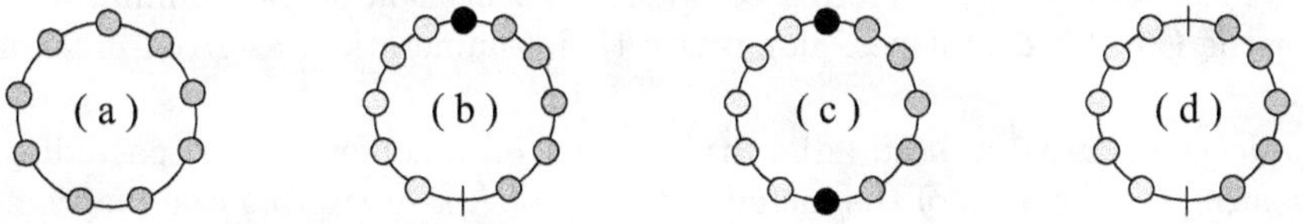

Figure 3: Configurations of particles (a) without and (b) - (d) with symmetry.

Proposition 3.10. *$C_{\bar{n}}\langle\mathbb{S}^{\bullet}_{\circ}\rangle$, $C_{\bar{n}}\langle\mathbb{S}^{\bullet}_{\bullet}\rangle$ and $C_{\bar{n}}\langle\mathbb{S}^{\circ}_{\circ}\rangle$ have the same cellulation as $\mathbb{T}\mathcal{C}\widetilde{B}_n$, $\mathbb{T}\mathcal{C}\widetilde{C}_n$ and $\mathbb{T}\mathcal{C}\widetilde{D}_n$ respectively.*

Proof. This is a direct consequence of the definitions of the configuration spaces, of the toroidal complexes, and their corresponding hyperplane arrangements given in Table 2. □

3.3. Group actions

The group of reflections W across the respective hyperplanes acts on the configuration space by permuting particles. The Coxeter group A_n (the symmetric group) is generated by transpositions s_{ij} which interchange the i-th and j-th particle. The Coxeter group B_n of *signed permutations* is generated by s_{ij} along with reflections $r_1, \ldots, r_n$, where r_i changes the sign of the i-th particle. Note that B_n is isomorphic to $\mathbb{Z}_2^n \rtimes \mathbb{S}_n$. The Coxeter group D_n is classically represented as the group of *even* signed permutations. Alternatively, D_n is isomorphic to the group $\mathbb{Z}_2^{n-1} \rtimes \mathbb{S}_n$, generated by transpositions s_{ij} along with reflections $r_2, \ldots, r_n$. The element r_1, which is present in B_n but not in D_n, corresponds to the reflection of the particle and its inverse that is closest to the origin.

Let $\sigma(W)$ denote the group acting *simply transitively* on the configuration spaces above. As mentioned above, for the spherical Coxeter groups, $\sigma(W)$ is isomorphic to W. However, the action of the affine groups is only transitive on the toroidal complexes. The simplest way to compute $\sigma(W)$ for the toroidal cases is from observing the diagrams given in Figure 3. Cutting the circle along a fixed point and "laying it flat" gives us the appropriate groups. Since a particle in $C_n(\mathbb{S}')$ is fixed by the group of rotations, then $\sigma(\widetilde{A}_n)$ is isomorphic to A_{n-1}. Similarly, $\sigma(\widetilde{B}_n)$ is isomorphic to D_n and $\sigma(\widetilde{C}_n)$ is isomorphic to B_n. The group $\sigma(\widetilde{D}_n)$ is isomorphic to $\mathbb{Z}_2^{n-2} \rtimes \mathbb{S}_n$, generated by transpositions s_{ij} along with reflections $r_2, \ldots, r_{n-1}$. The elements r_1 and r_n, which are not present in $\sigma(\widetilde{D}_n)$, correspond to the reflections of the particles and their inverses that are closest to the centrally symmetric axis.

4. Bracketings and Hyperplanes

4.1. Visualizing collisions

We introduce the bracket notation in order to visualize collisions in the configuration spaces, leading to a transparent understanding of our results below. In particular, Proposition 4.4 produces a complete classification of the minimal building sets for the Coxeter complexes, along with their enumeration, as given in Tables 3 and 4.

A *bracket* is drawn around adjacent particles on a configuration space diagram representing the collision of the included particles. A *k-bracketing* of a diagram is a set of k brackets representing multiple independent particle collisions. For example, the configuration

$$-x_4 = -x_3 < -x_2 < -x_1 = 0 = x_1 < x_2 < x_3 = x_4 \tag{4.1}$$

in $C_5\langle \mathbb{R}_\bullet \rangle$ corresponds to the bracketing . Each bracket on a configuration space diagram with symmetric particles will actually consist of two symmetric brackets, one on each side of the origin, with this symmetric pair counting as only one bracket. If this set includes the origin, we draw one symmetric bracket around the origin, which again counts as one bracket. Thus Eq.(4.1) is a 2-bracketing of its diagram.

Let α be an intersection of hyperplanes. We say that hyperplanes h_i *cellulate* α to mean the intersections $h_i \cap \alpha$ decompose α into cells. Denote by $\mathcal{H}_s\alpha$ the set of all hyperplanes that contain α, called the stabilizing hyperplanes of α. If reflections in these hyperplanes generate a finite reflection group, it is called the *stabilizer* of α. Note that in a simplicial Coxeter complex, the stabilizer exists for all intersections of hyperplanes.[2]

We define the *support* of a bracketing to be the configuration space associated to the bracketing diagram. That is, it is the subspace (of the configuration space) in which particles that share a bracket have collided. However, a set of collisions in a configuration space defines an intersection of hyperplanes. So, alternatively, the support of a bracketing is the smallest intersection of hyperplanes associated to the bracketing. The following is immediate:

Lemma 4.1. *If α is the support of a bracketing G, then for every pair of particles x_i and x_j that share a bracket in G, the hyperplane defined by $x_i = x_j$ is in $\mathcal{H}_s\alpha$.*

Example 4.2. Figure 4 shows part of the two-dimensional complexes $\mathcal{C}B_3$ and $\mathcal{C}D_3$, one with and one without a fixed particle at the axis of symmetry. As we move through the chambers, going from (a) through (g), a representative of each configuration is shown. Notice that since there is no fixed particle at the axis of symmetry for type D, there is no meaningful bracketing of the symmetric particles closest to the axis; they may pass each other freely *without* collision.

---

[2]By abuse of terminology, we also refer to the set $\mathcal{H}_s\alpha$ as the stabilizer of α.

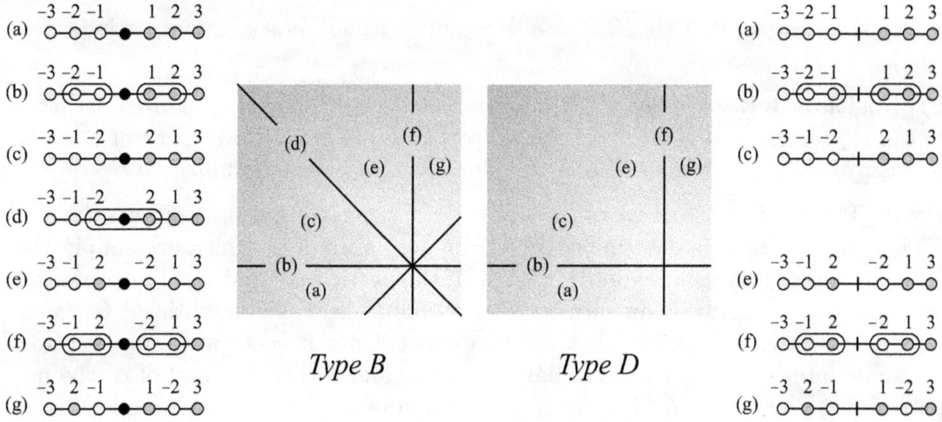

Figure 4: Local regions of $\mathcal{C}B_3$ and $\mathcal{C}D_3$.

4.2. Composition maps

There are natural composition maps on bracketed diagrams. These form the basic operations of our operads defined in Section 5.

Definition 4.3. There are three types of compositions:

1. Let H be a diagram of m particles of $C_k(\mathbb{R})$ or $C_k(\mathbb{S}')$, with one particle labeled i. Let G be a diagram of $C_k(\mathbb{R})$. The *composition* $H \circ_i G$ is the diagram of $m+k-1$ particles where the particle i is replaced by a bracket containing G.
2. Let H be a diagram of m paired particles of a configuration space, with one particle labeled i and its mirror image labeled $-i$. Let G be a diagram of $C_k(\mathbb{R})$. The *composition* $H \circ_i G$ is the diagram of $m+k-1$ paired particles, where the particle i is replaced by a bracket containing G and its pair $-i$ is replaced by a bracket containing the mirror image of G (left side of Figure 5).

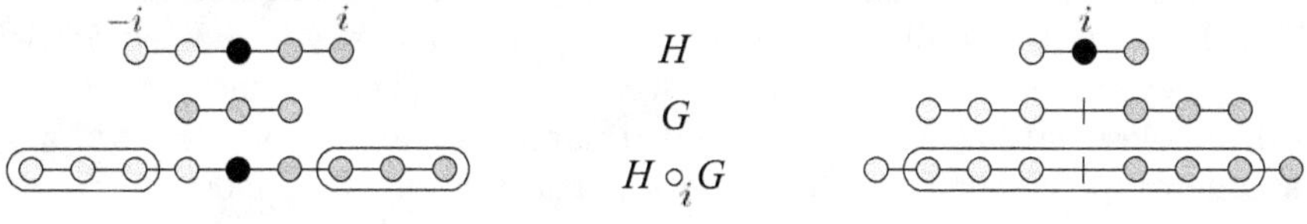

Figure 5: Composition operations on bracketings.

3. Let H be a diagram of m paired particles of $C_{\bar{m}}(\mathbb{R}_\bullet)$, $C_{\bar{m}}(\mathbb{S}^\bullet_\circ)$ or $C_{\bar{m}}(\mathbb{S}^\bullet_\bullet)$, with a fixed particle labeled i. Let G be a diagram of either $C_{\bar{k}}(\mathbb{R}_\bullet)$ or $C_{\bar{k}}(\mathbb{R}_\circ)$. The *composition* $H \circ_i G$ is the diagram of $m+k$ paired particles where the fixed particle i is replaced by a bracket containing G (right side of Figure 5).

Indeed any k-bracketing G can be represented as

$$G \;=\; H \circ_{i_1} G_1 \circ_{i_2} \cdots \circ_{i_k} G_k, \tag{4.2}$$

where the base H and the G_i's are diagrams without brackets and each i_j is a particle in H.

Proposition 4.4. *Let G be a k-bracketing as defined in Eq.(4.2). Moreover, let α be the support of G and let α_i be the support of G_i. Then the stabilizer of α is the product of the stabilizers of α_i and the cellulation of α is determined by H.*

Proof. The product structure of the stabilizer of α is a consequence of Lemmas 4.1. The cellulation of α is determined by the hyperplanes of α, which are simply the intersections of hyperplanes of $\mathcal{C}W$ with α. If the particles x_i and x_j share a bracket in α, then x_k cannot collide with x_j in α without also colliding with x_i. Geometrically, this property implies that the two hyperplanes $x_i = x_k$ and $x_j = x_k$ have the same intersection with α. Similarly, since x_i and x_j collide in all of α, the hyperplane $x_i = x_j$ plays no role in the cellulation of α. These two facts allow us to treat x_i and x_j as a single particle in G without changing the hyperplane arrangement. Repeating this process for all particles that share a bracket gives the desired result. □

4.3. Atypical complexes

It is easy to check that in most cases the cellulations of subspaces (intersections of hyperplanes) in Coxeter complexes are indeed other (smaller dimensional) Coxeter complexes. There are, however, three instances where this is not so, appearing as subspaces of the Coxeter complexes $\mathcal{C}D_n$, $\mathbb{T}\mathcal{C}\widetilde{B}_n$ and $\mathbb{T}\mathcal{C}\widetilde{D}_n$. In particular, they have cellulations combinatorially equivalent to Coxeter complexes with *additional* hyperplanes. We define these three atypical complexes below:

Definition 4.5. The complexes of interest are:

1. Let $\mathcal{C}D_{n,m}$ be $\mathcal{C}D_n$ with m additional hyperplanes $\{x_i = 0 \mid 1 \leqslant i \leqslant m\}$.
2. Let $\mathbb{T}\mathcal{C}\widetilde{B}_{n,m}$ be $\mathbb{T}\mathcal{C}\widetilde{B}_n$ with m additional hyperplanes $\{x_i = 0 \mid 1 \leqslant i \leqslant m\}$.
3. Let $\mathbb{T}\mathcal{C}\widetilde{D}_{n,m}$ be $\mathbb{T}\mathcal{C}\widetilde{D}_n$ with $2m$ additional hyperplanes $\{x_i = 0, 1 \mid 1 \leqslant i \leqslant m\}$.

The configuration space model provides intuition into how these cases arise naturally. Note how these are all complexes with associated configuration spaces on $\mathbb{R}_\circ$, $\mathbb{S}^\bullet_\circ$ and $\mathbb{S}^\circ_\circ$, where not all points along the axis of symmetry have fixed particles. The subspaces of these configuration spaces are those where some particles have collided. In these subspaces, sets of collided particles may be considered in aggregate as a new type of particle, called a *thick* particle. Figure 6(a) shows a bracketing and (b) its representation with thick particles. In general, thick particles allow us to represent any number of coincident particles by a single particle.

(a)

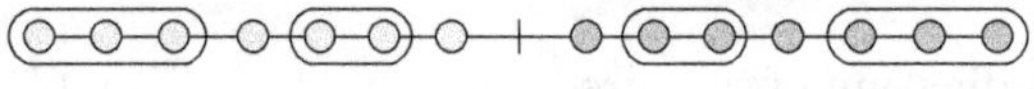

(b)

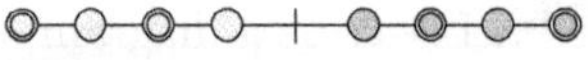

Figure 6: Bracketing and thick particles.

Recall that particles were defined such that they could occupy the same point as their inverse; that is, they do not form a collision with their inverse. Unlike (standard) particles, a thick particle and its inverse may not occupy the same point without collision. The reason comes from the hyperplane equations: In the subspace where x_i and x_j have collided, the hyperplane $x_i = -x_j$ represents the same configurations as $x_i = -x_i$ $(= 0)$. Thus, the m additional hyperplanes added to the complex correspond to the m thick particles in their configuration spaces. Then the diagram of Figure 6(b) is an element of $\mathcal{C}D_{4,2}$, sitting as a subspace of $\mathcal{C}D_7$ in Figure 6(a).

Remark. In the case of non-paired particles, the distinction between standard and thick particles is irrelevant, since no particle has an inverse to collide with. They are also inconsequential in configuration spaces that include a fixed particle wherever particles may meet their inverses.

5. Compactifications and Operads

5.1. Coxeter moduli spaces

Compactifying a configuration space $C_n(V)$ enables the points on V to collide and a *system* is introduced to record the *directions* points arrive at the collision. In the work of Fulton and MacPherson [**16**], this method is brought to rigor in the algebro-geometric context.[3] In [**9**, Section 4], De Concini and Procesi show that the *minimal blow-ups* of the Coxeter complexes $\mathcal{C}W$ are equivalent to the Fulton-MacPherson compactifications of their corresponding configuration spaces.

In order to describe these compactified Coxeter moduli spaces, we begin with definitions. The collection of hyperplanes $\{x_i = 0 \mid i = 1, \ldots, n\}$ of $\mathbb{R}^n$ generates the *coordinate* arrangement. A crossing of hyperplanes is *normal* if it is locally isomorphic to a coordinate arrangement. A construction which transforms any crossing into a normal crossing involves the algebro-geometric concept of a blow-up; a standard reference is [**18**].

Definition 5.1. The *blow-up* of a space V along a codimension k intersection α of hyperplanes is the closure of $\{(x, f(x)) \mid x \in V\}$ in $V \times \mathbb{P}^{k-1}$. That is, we replace α with the projective sphere bundle associated to the normal bundle of α.

A general collection of blow-ups is usually noncommutative in nature; in other words, the order in which spaces are blown up is important. For a given arrangement, De Concini and Procesi [**9**, Section 3] establish the existence and uniqueness of a *minimal building set*, a collection of subspaces for which blow-ups commute for a given dimension, and for which every crossing in the resulting space is normal. For a Coxeter complex $\mathcal{C}W$, we denote the minimal building set by $\mathrm{Min}(\mathcal{C}W)$.

Definition 5.2. The *Coxeter moduli space* $\mathcal{C}(W)_\#$ is the minimal blow-up of $\mathcal{C}W$, obtained by blowing up along elements of $\mathrm{Min}(\mathcal{C}W)$ in *increasing* order of dimension.

Remark. Kapranov showed the minimal blow-ups of $\mathcal{C}A_n$ yield a space homeomorphic to a double cover of the moduli space $\overline{\mathcal{M}}_{0,n+2}(\mathbb{R})$ [**19**, Proposition 4.8]. Thus,

---

[3] Axelrod and Singer [**1**] look at this compactification from a perspective of spherical blow-ups on real manifolds.

the Fulton-MacPherson compactifications of our configuration spaces yield generalizations of this moduli space.

Example 5.3. Figure 7(a) shows the blow-ups of the sphere $\mathcal{C}A_3$ of Figure 2(a) at nonnormal crossings. Each blown up point has become a hexagon with antipodal identification and the resulting manifold is the Coxeter moduli space $\mathcal{C}(A_3)_\#$, homeomorphic to the eight-fold connected sum of $\mathbb{RP}^2$ with itself.[4] Part (b) shows $\overline{\mathcal{M}}_{0,6}(\mathbb{R})$ coming from the iterated blow-ups of $\mathcal{C}A_4$. Figure 7(c) shows the minimal blow-up of $\mathcal{C}\widetilde{A}_2$ of Figure 2(e). Finally, part (d) is the cube with opposite facial identifications yielding the three-torus of $\mathcal{C}(\widetilde{A}_3)_\#$.

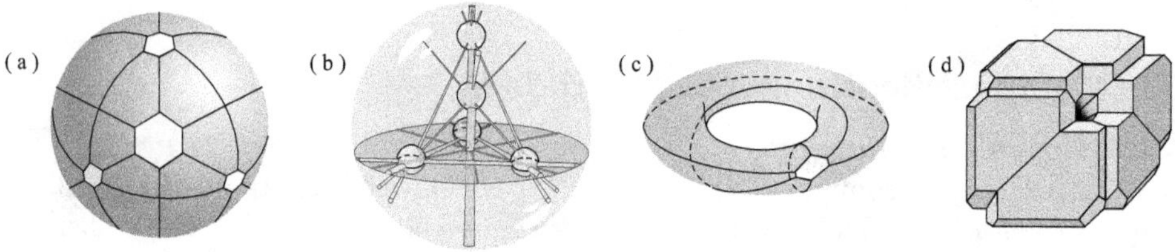

Figure 7: Coxeter moduli spaces (a) $\mathcal{C}(A_3)_\#$, (b) $\mathcal{C}(A_4)_\#$, (c) $\mathcal{C}(\widetilde{A}_2)_\#$ and (d) $\mathcal{C}(\widetilde{A}_3)_\#$.

The relationship between the set Min($\mathcal{C}W$) and the group W is given by the concept of reducibility. For intersections of hyperplanes α, β, and γ, the collection of hyperplanes $\mathcal{H}_s\alpha$ is *reducible* if $\mathcal{H}_s\alpha$ is a disjoint union $\mathcal{H}_s\beta \sqcup \mathcal{H}_s\gamma$, where $\alpha = \beta \cap \gamma$.

Lemma 5.4. *[8, Section 3] Let α be an intersection of hyperplanes of $\mathcal{C}W$. Then $\mathcal{H}_s\alpha$ is irreducible if and only if α is in Min($\mathcal{C}W$).*

This lemma can be rewritten in the language of bracketings:

Lemma 5.5. *Let α be an intersection of hyperplanes of $\mathcal{C}W$. Then α is in Min($\mathcal{C}W$) if and only if α is the support of a 1-bracketing.*

Proof. If α is the support of a 1-bracketing G, then $\mathcal{H}_s\alpha$ is determined by Lemma 4.1. Thus:

1. If G contains a fixed particle, then $\mathcal{H}_s\alpha \cong \mathcal{H}B_k$.
2. If G contains a particle and its inverse but no fixed particle, then $\mathcal{H}_s\alpha \cong \mathcal{H}D_k$.
3. If G does not contain a particle and its inverse, then $\mathcal{H}_s\alpha \cong \mathcal{H}A_k$.

All three of these hyperplane arrangements are irreducible, so α is in Min($\mathcal{C}W$).

Conversely, let α be the support of a k-bracketing $G = H \circ_{i_1} G_1 \circ_{i_2} \cdots \circ_{i_k} G_k$. Let β be the support of G_1 and γ be the product of the configuration spaces diagramed by $G_2, \cdots, G_k$. By the definition of reducibility, $\mathcal{H}_s\alpha = \mathcal{H}_s\beta \sqcup \mathcal{H}_s\gamma$, and thus α is not in Min($\mathcal{C}W$) by Lemma 5.4. □

[4] A projective version of this diagram is first found in a different context by Brahana and Coble in 1926 [5, Section 1] relating to possibilities of maps with twelve five-sided countries.

Remark. Table 3 itemizes the collection of elements in Min($\mathcal{CW}$) for the spherical cases and Table 4 for the Euclidean ones. In the tables, m represents the total number of thick particles and r the number of thick particles in the bracket (stabilizer) of the atypical complexes.

5.2. Nested bracketings

As bracketings encoded collisions of particles in configuration spaces, it is *nested* bracketings which encode the Fulton-MacPherson compactification of the configuration spaces [**9**, Section 2]. The FM compactification allows collisions of particles whose description comes from the repulsive potential observed by quantum physics: Pushing particles together creates a spherical bubble onto which the particles escape [**25**]. In other words, as particles try to collide, the result is a new bubble fused to the old at the point of collision, where the collided particles are now on the new bubble. The phenomena is dubbed as *bubbling*, with the resulting structure as a bubble-tree. Indeed, the nested bracketings are exactly the one-dimensional analogues of bubble-trees [**10**, Section 1].

Moreover, the codimension k faces of a chamber of a compactified configuration space are the nested k-bracketings on the configuration space diagrams. A diagram G denoted

$$G \;=\; H \circ_{i_1} G_1 \circ_{i_2} \cdots \circ_{i_k} G_k, \tag{5.1}$$

is a nested $(k+m_0+m_1+\cdots+m_k)$-bracketing, with each i_j a particle of H, where H is an m_0-bracketing and where G_i is a nested m_i-bracketing; see Figure 8. The composition maps are those in Definition 4.3.

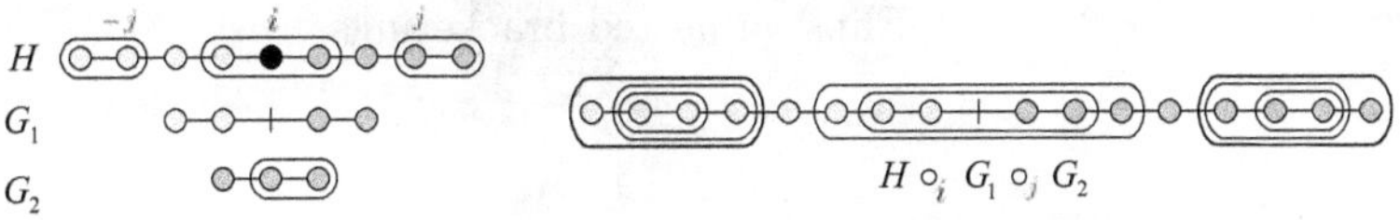

Figure 8: Composition operations on nested bracketings.

After compactification, different orderings of particles in a bracket do not necessarily represent the same cell. A different action is necessary to describe the identification of diagrams.

Definition 5.6. The *flip* $\hat{\sigma}(G)$ action on an unbracketed diagram G consists of the identity and the reflection (that reverses the order of the particles of G). On a nested bracketing diagram, the action of $\hat{\sigma}(G)$ acts independently on each bracketed component.

Theorem 5.7. *Let G be a nested bracketing where $G = H \circ_{i_1} G_1 \circ_{i_2} \cdots \circ_{i_k} G_k$. All bracketings in the image of G under $\hat{\sigma}(G_1) \times \cdots \times \hat{\sigma}(G_k)$ represent the same cell.*

Proof. By definition, blow-ups introduce a *projective* bundle around each subspace in the minimal building set. The analog in configuration spaces is an identification across each bracket: Flipping the positions of the particles in the bracket is defined to represent the same configuration. Thus the permutations that represent the same set of configurations as G are exactly the images of G under $\hat{\sigma}(G_1)\times\cdots\times\hat{\sigma}(G_k)$. □

Remark. This theorem gives us a gluing rule between the faces of two chambers. In other words, two nested k-bracketings G_1 and G_2 of a diagram (representing codimension k-faces) are identified if G_2 can be obtained from G_1 by flipping some of the brackets of G_2.

Figure 9 shows the permutations that preserve the cell represented by a particular configuration space diagram. Note that reflections commute with each other, and thus they generate a group which is isomorphic to $(\mathbb{Z}/2\mathbb{Z})^3$.

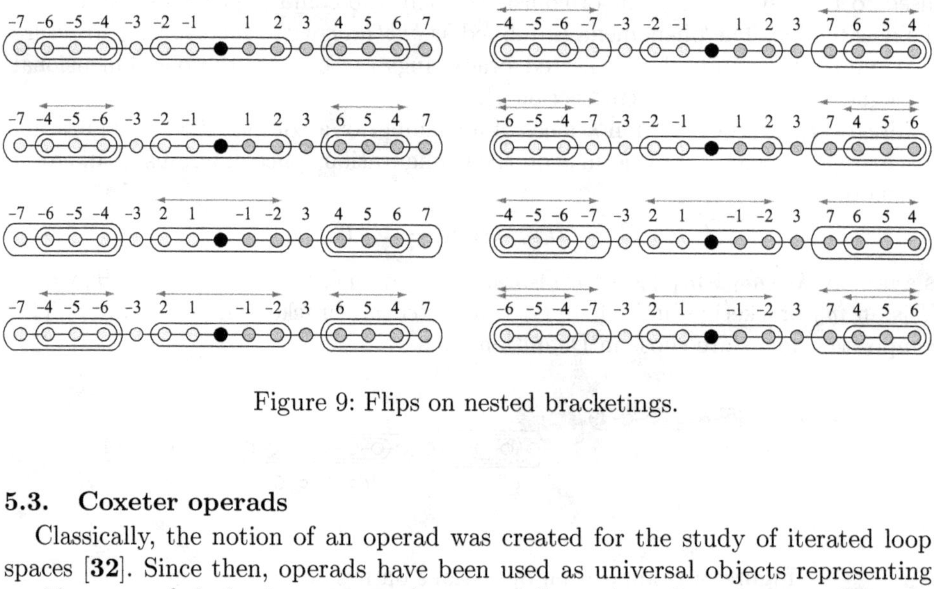

Figure 9: Flips on nested bracketings.

5.3. Coxeter operads

Classically, the notion of an operad was created for the study of iterated loop spaces [**32**]. Since then, operads have been used as universal objects representing a wide range of algebraic concepts. An *operad* $\mathcal{O}$ consists of a collection of objects $\{\mathcal{O}(n) \mid n \in \mathbb{N}\}$ in a monoidal category endowed with certain extra structures. Notably, $\mathcal{O}(n)$ carries an action by the symmetric group of n letters, and there are composition maps

$$\mathcal{O}(n)\otimes\mathcal{O}(k_1)\otimes\cdots\otimes\mathcal{O}(k_n)\rightarrow\mathcal{O}(k_1+\cdots+k_n)$$

which must be associative, unital, and equivariant; see [**22**, Chapter 1] for details.

One can view $\mathcal{O}(n)$ as objects consisting of n-ary operations, which yield an output given n inputs. We will be concerned mostly with operads in the context of topological spaces, where the objects $\mathcal{O}(n)$ will be equivalence classes of geometric objects. Classically, these objects can be pictured as in Figure 10. The composition $\mathcal{O}(i)\circ_k\mathcal{O}(j)$ is obtained by grafting the output of $\mathcal{O}(j)$ to the k-th input of $\mathcal{O}(i)$. The symmetric group acts by permuting the labeling of the inputs.

There are several variants and extensions of the operad definition above. A classic example is the *non-symmetric* version, which removes all references to the symmetric

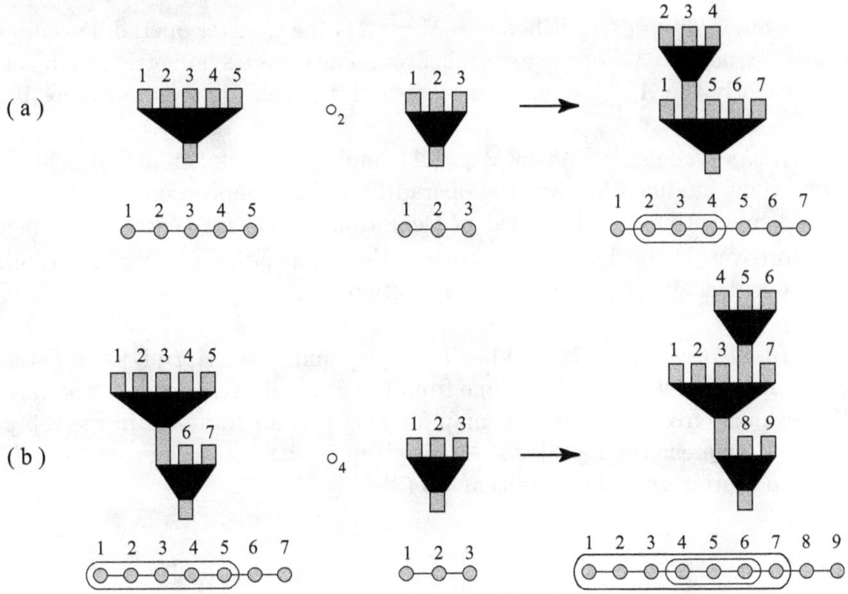

Figure 10: Examples of composition maps of an operad, along with dual figures related to (a) bracketing and (b) nested bracketing.

group in the definition above. A *right module* $\mathcal{O}_R$ over an operad $\mathcal{O}$ is a collection of objects $\{\mathcal{O}_R(n) \mid n \in \mathbb{N}\}$ with a set of composition maps

$$\mathcal{O}_R(n) \otimes \mathcal{O}(k_1) \otimes \cdots \otimes \mathcal{O}(k_n) \to \mathcal{O}_R(k_1 + \cdots + k_n).$$

A *bi-colored* operad is a multicategory with two objects [**23**, Section 2]: Intuitively, each of the inputs and output is given one of two colors. An element $\mathcal{O}(j)$ can be grafted into an input of $\mathcal{O}(i)$ if and only if the colors of the corresponding input and output match. Our version (the naming of which is credited to J. Stasheff) replaces the symmetric group in the classical definition with appropriate Coxeter groups instead. Let Ω be the collection $\{A, B, D, \widetilde{A}, \widetilde{B}, \widetilde{C}, \widetilde{D}\}$ of Coxeter groups from Figure 1.

Definition 5.8. For $W \in \Omega$, let $\mathcal{O}_W(n,k)$ be the collection of configuration spaces of nested k-bracketings with n particles associated to $\mathcal{C}(W)_{\#}$. If $\mathcal{U}$ is a subset of Ω, then let $\mathcal{O}_{\mathcal{U}} = \bigcup_{W \in \mathcal{U}} \mathcal{O}_W$. Then, the *Coxeter operad* is defined for each pair $\mathcal{U}, \mathcal{V}$ of (possibly empty) subsets of Ω for which the collection

$$\mathcal{O}_{\mathcal{U}}(n_H, k_H) \otimes \mathcal{O}_{\mathcal{V}}(n_1, k_1) \otimes \cdots \otimes \mathcal{O}_{\mathcal{V}}(n_m, k_m) \to \mathcal{O}_{\mathcal{U}}(n_*, k_*)$$

of composition maps exist, where $n_* = n_H - m + \sum n_i$ and $k_* = k_H + m + \sum k_i$.

We note several structures that appear based on this definition, giving a partial list:

5.3.0.1. *Classic Operads:* When $\mathcal{U} = \mathcal{V} = \{A\}$, the Coxeter operad becomes the A_∞ operad structure [**32**] of the associahedron. The classic symmetric group acting on $\mathcal{O}_A$ is exactly the A_n Coxeter group. Examples of this are seen in Figure 10.

5.3.0.2. *Right Modules:* When $\mathcal{V} = \{A\}$ and $\mathcal{U} = \{W\}$, for any $W \in \Omega \setminus A$, we have a right module $\mathcal{O}_W$ over the operad $\mathcal{O}_A$. The composition map (based on non-nested bracketings) is described in Definition 4.3 (2) for centrally symmetric spaces. Moreover, Definition 4.3 (1) provides the composition for $W = \widetilde{A}$, resulting in the cyclohedral structure given in [**22**, Section 4.4].

5.3.0.3. *Bi-Colored Operads:* When $\mathcal{U} = \{B\}$ and $\mathcal{V} = \{A, B, D\}$, we obtain a bi-colored operad. The two colors come from the centrally symmetric black particle and the ordinary free particles. The map for $D \in \mathcal{V}$ is diagrammed in the top part of Figure 11, whereas the map for $A \in \mathcal{V}$ is given in the bottom of the figure. This composition map is given in Definition 4.3 (3).

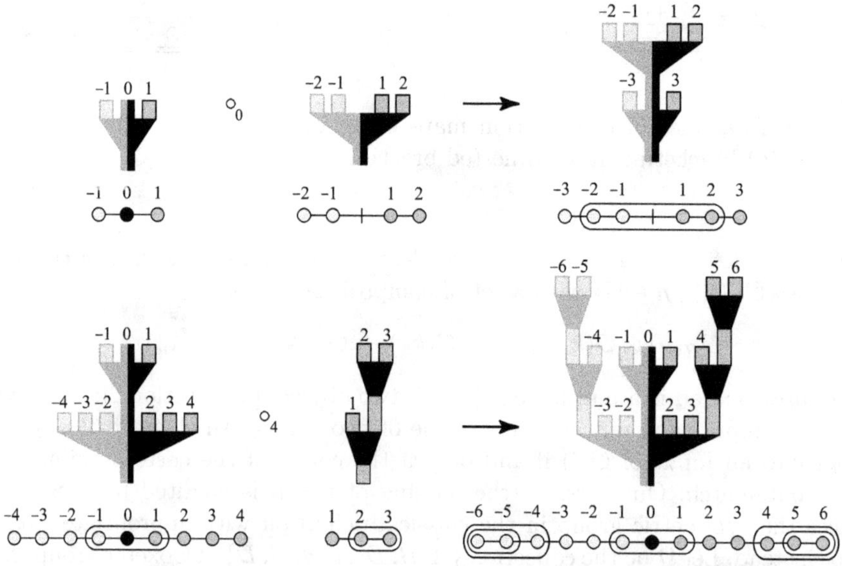

Figure 11: Examples of composition maps of the Coxeter operad, along with dual *tree* figures.

5.3.0.4. *Bi-Colored Right Modules:* The examples where $\mathcal{U} = \{\widetilde{B}, \widetilde{C}, \widetilde{D}\}$ and $\mathcal{V} = \{A, B, D\}$ result in the composition of bracketings on lines being glued onto bracketings on circles. This results in a bi-colored right module over a bi-colored operad. There are several options here which work for the different subsets of $\mathcal{U}$ and $\mathcal{V}$ given. For example, $\mathcal{O}_{\widetilde{D}}$ admits an action of $\mathcal{O}_A$ and $\mathcal{O}_{\{\widetilde{B},\widetilde{C}\}}$ admits an action of $\mathcal{O}_{\{B,D\}}$. From an operadic viewpoint, the affine Coxeter groups are analogous to the spherical ones shown, but an *unrooted tree*, rather than a rooted one, is used.

Remark. There exists a generalization of the classical operad to the *braid* operad, defined by Fiedorowicz, with the braid group playing the role of the symmetric group [**14**, Section 3]. Since the braid group is the Artin group of type A, it seems plausible that the Coxeter operads above can be extended to their corresponding Artin groups, yielding analogs to the braid operads.

6. Tiling by graph-associahedra

6.1. Definitions

This section uses the theory of graph-associahedra developed in [**6**]. As associahedra are related to the A_∞ operad, we show that graph-associahedra capture the structure of the Coxeter operad. Moreover, we provide combinatorial and enumerative results about the Coxeter moduli spaces. Notably, the Euler characteristics of these spaces are given.

Definition 6.1. Let Γ be a graph. A *tube* is a proper nonempty set of nodes of Γ whose induced graph is a proper, connected subgraph of Γ. There are three ways that two tubes t_1 and t_2 may interact on the graph.

1. Tubes are *nested* if $t_1 \subset t_2$.
2. Tubes *intersect* if $t_1 \cap t_2 \neq \emptyset$ and $t_1 \not\subset t_2$ and $t_2 \not\subset t_1$.
3. Tubes are *adjacent* if $t_1 \cap t_2 = \emptyset$ and $t_1 \bigcup t_2$ is a tube in Γ.

Tubes are *compatible* if they do not intersect and they are not adjacent. A *tubing* T of Γ is a set of tubes of Γ such that every pair of tubes in T is compatible. A k-*tubing* is a tubing with k tubes.

Theorem 6.2. *[**6**, Section 3] For a graph Γ with n nodes, the* graph-associahedron *$\mathcal{P}\Gamma$ is the convex polytope of dimension $n-1$ whose face poset is isomorphic to set of valid tubings of Γ, ordered such that $T \prec T'$ if T is obtained from T' by adding tubes.*

Figure 12 shows two examples of graph-associahedra, having underlying graphs as paths and cycles, respectively, with three nodes. These turn out to be the two-dimensional associahedron [**32**] and cyclohedron [**3**] polytopes.

Theorem 6.3. *[**6**, Section 4] Let W be a simplicial Coxeter group and Γ_W be its associated Coxeter graph. Then the W-action on $CW_\#$ has a fundamental domain combinatorially isomorphic to $\mathcal{P}\Gamma_W$.*

Notation. We write $\mathcal{P}W$ instead of $\mathcal{P}\Gamma_W$ when context makes it clear.

Theorem 6.4. *The tiling of the Coxeter moduli spaces are given as follows:*

1. *$\mathcal{P}A_n$ (the associahedron) tiles $\mathcal{C}(A_n)_\#$, $\mathcal{C}(B_n)_\#$ and $\mathbb{T}\mathcal{C}(\widetilde{C}_{n-1})_\#$.*
2. *$\mathcal{P}\widetilde{A}_n$ (the cyclohedron) tiles $\mathbb{T}\mathcal{C}(\widetilde{A}_n)_\#$.*
3. *$\mathcal{P}D_n$ tiles $\mathcal{C}(D_n)_\#$ and $\mathbb{T}\mathcal{C}(\widetilde{B}_{n-1})_\#$.*

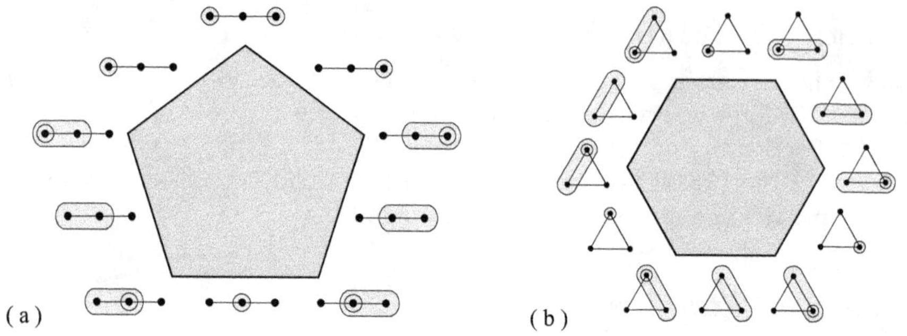

Figure 12: Graph-associahedra with a (a) path and (b) cycle as underlying graphs.

4. $\mathcal{P}\widetilde{D}_n$ *tiles* $\mathbb{TC}(\widetilde{D}_n)_{\#}$.

Proof. For a given graph Γ, the polytope $\mathcal{P}\Gamma_n$ depends only on the adjacency of nodes, not the label on the edges. □

Remark. There is a natural bijection from the set of all bracketings of a configuration space diagram to the set of all tubings of the associated Coxeter diagram. The bijection is such that two brackets intersect if and only if their images intersect or are adjacent as tubes. Thus the face poset of tubings is isomorphic to the face poset of bracketings, where k brackets correspond to a codimension k face. Figure 13 shows some examples of this bijection.

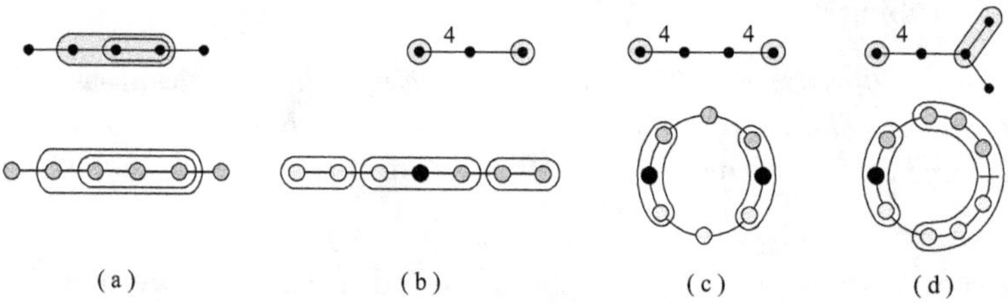

Figure 13: Examples of the bijection between tubings and nested bracketings.

6.2. Faces of polytopes

We analyze the structure of these tiling polyhedra $\mathcal{P}W$. For a given tube t and a graph Γ, let Γ_t denote the induced subgraph on the graph Γ. By abuse of notation, we sometimes refer to Γ_t as a tube.

Definition 6.5. Given a graph Γ and a tube t, construct a new graph Γ_t^* called the *reconnected complement*: If V is the set of nodes of Γ, then $V - t$ is the set of nodes of Γ_t^*. There is an edge between nodes a and b in Γ_t^* if either $\{a, b\}$ or $\{a, b\} \cup t$ is connected in Γ.

Theorem 6.6. *[6, Section 3] The facets of $\mathcal{P}\Gamma$ correspond to the set of 1-tubings on Γ. In particular, the facet associated to a 1-tubing $\{t\}$ is equivalent to $\mathcal{P}\Gamma_t \times \mathcal{P}\Gamma_t^*$.*

The facets of $\mathcal{P}W$ are of the form $\mathcal{P}\Gamma \times \mathcal{P}\Gamma^*$, which can be found by simple inspection. Using induction on each term of the product produces the following results:

Corollary 6.7. *[32] The faces of $\mathcal{P}A$ are of the form $\mathcal{P}A \times \cdots \times \mathcal{P}A$.*

Corollary 6.8. *[33] The faces of $\mathcal{P}\widetilde{A}$ are of the form $\mathcal{P}\widetilde{A} \times \mathcal{P}A \times \cdots \times \mathcal{P}A$.*

Before moving on to the other tiling polytopes, we need to look at some special graphs which appear as reconnected complements. They are displayed in the Figure 14 below, the subscript n denoting the number of vertices. Note that the polytope $\mathcal{P}X_4$ is the 3-dimensional *permutohedron.*

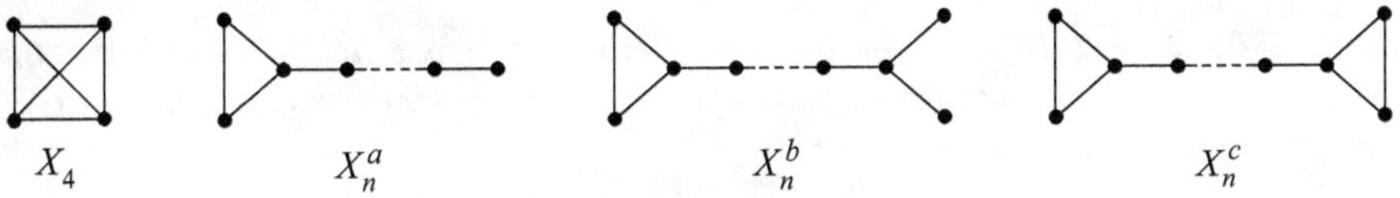

Figure 14: Special graphs appearing as reconnected complements.

Corollary 6.9. *The faces of $\mathcal{P}D$ are of the form*

1. $\mathcal{P}A \times \cdots \times \mathcal{P}A$
2. $\mathcal{P}D \times \mathcal{P}A \times \cdots \times \mathcal{P}A$
3. $\mathcal{P}X^a \times \mathcal{P}A \times \cdots \times \mathcal{P}A$.

Example 6.10. Figure 15 illustrates four different polyhedra. The first three are well-known objects: (a) the associahedron $\mathcal{P}A_4$, (b) cyclohedron $\mathcal{P}\widetilde{A}_4$ and (c) permutohedron $\mathcal{P}X_4$. The last one (d) is $\mathcal{P}D_4$ with six pentagons $\mathcal{P}A_3$, three squares $\mathcal{P}D_2 \times \mathcal{P}A_2$ and one hexagon $\mathcal{P}X_3^a$ for facets.

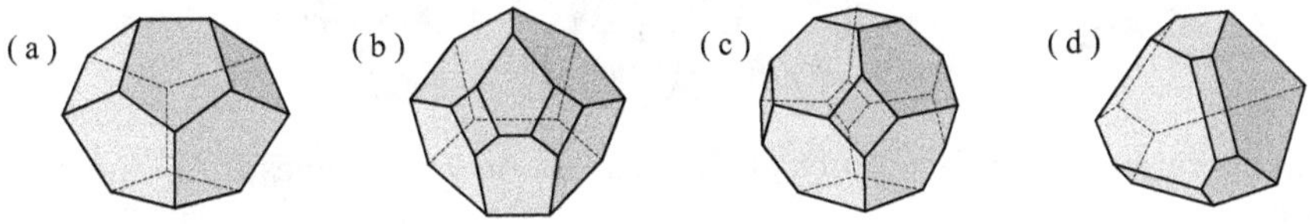

Figure 15: The 3-dimensional (a) associahedron $\mathcal{P}A_4$, (b) cyclohedron $\mathcal{P}\widetilde{A}_3$, (c) permutohedron $\mathcal{P}X_4$ and (d) $\mathcal{P}D_4$.

Corollary 6.11. *The faces of $\mathcal{P}\widetilde{D}$ are of the form*

(1) $\mathcal{P}A \times \cdots \times \mathcal{P}A$
(2) $\mathcal{P}D \times \mathcal{P}A \times \cdots \times \mathcal{P}A$
(3) $\mathcal{P}X^a \times \mathcal{P}A \times \cdots \times \mathcal{P}A$
(4) $\mathcal{P}X^a \times \mathcal{P}X^a \times \mathcal{P}A \times \cdots \times \mathcal{P}A$
(5) $\mathcal{P}D \times \mathcal{P}X^a \times \mathcal{P}A \times \cdots \times \mathcal{P}A$
(6) $\mathcal{P}D \times \mathcal{P}D \times \mathcal{P}A \times \cdots \times \mathcal{P}A$
(7) $\mathcal{P}X^b \times \mathcal{P}A \times \cdots \times \mathcal{P}A$
(8) $\mathcal{P}X^c \times \mathcal{P}A \times \cdots \times \mathcal{P}A$
(9) $\mathcal{P}\widetilde{D} \times \mathcal{P}A \times \cdots \times \mathcal{P}A$
(10) $\mathcal{P}X_4 \times \mathcal{P}A \times \cdots \times \mathcal{P}A$.

6.3. Euler characteristic

We compute the Euler characteristics of the Coxeter moduli spaces. From Theorem 6.6, we see that the number of codimension k faces of the polytope $\mathcal{P}W$ cellulating $\mathcal{C}W_\#$ is precisely the number of k-tubings of the associated Coxeter diagram Γ_W.

Theorem 6.12. *Let $f_k(\mathcal{P}W)$ be the number of k-dimensional faces of $\mathcal{P}W$, and let g be the number of chambers in the spherical or toroidal Coxeter complex $\mathcal{C}W$. If $\dim(\mathcal{P}W) = n$, then*

$$\chi(\mathcal{C}(W)_\#) = \sum_{k=0}^{n} (-1)^k \, \frac{g \cdot f_k}{2^{n-k}}.$$

Proof. In order to count the number of k-dimensional faces in the space $\mathcal{C}(W)_\#$, take the number of total chambers g and multiply it by the number of k-dimensional faces $f_k(\mathcal{P}W)$ for each tile $\mathcal{P}W$. The amount of overcounting is simply how different chambers of the complex meet at each k-dimensional face of a tile. Since all the crossings in $\mathcal{C}(W)_\#$ are normal, each k-dimensional face is identified with 2^{n-k} copies. □

Remark. The number of vertices in $\mathcal{P}A$ is the well-known Catalan number [**31**, Section 6.5]. The faces $f_k(W)$ of $\mathcal{P}W$ provide natural generalizations; see [**26**] for further exposition.

Theorem 6.4 shows only four types of graph-associahedra tiling the Coxeter moduli spaces: $\mathcal{P}A_n$ (the associahedron), $\mathcal{P}\widetilde{A}_n$ (the cyclohedron), $\mathcal{P}D_n$ and $\mathcal{P}\widetilde{D}_n$. The enumeration of the faces of the associahedra $\mathcal{P}A_n$ is a classic result of A. Cayley [**7**], obtained by just counting the number of n-gons with k non-intersecting diagonals:

$$f_k(\mathcal{P}A_n) = \frac{1}{n+1}\binom{n-1}{k}\binom{2n-k}{n}.$$

The enumeration of the face poset of the cylohedron $\mathcal{P}\widetilde{A}_n$ comes from Simion [**29**, Section 3]:

$$f_k(\mathcal{P}\widetilde{A}_n) = \binom{n}{k}\binom{2n-k}{n}.$$

In a recent paper, Postnikov provides a recursive formula for the generating function of the numbers f_k [**26**, Theorem 7.11]. Using this, a closed formulas for the graph-associahedra of types D_n and $\widetilde{D}_n$ can be found; see [**27**, Section 12]. We thank A. Postnikov for sharing the following result:

Proposition 6.13. *The face poset enumerations of types D_n and $\widetilde{D}_n$ are*

$$f_k(\mathcal{P}D_n) = 2f_k(\mathcal{P}A_n)-2f_k(\mathcal{P}A_{n-1})-f_k(\mathcal{P}A_{n-2})-f_{k-1}(\mathcal{P}A_{n-1})-f_{k-1}(\mathcal{P}A_{n-2}) \tag{6.1}$$

$$\begin{aligned} f_k(\mathcal{P}\widetilde{D}_n) &= 4f_k(\mathcal{P}A_{n+1}) - 8f_k(\mathcal{P}A_n) - 4f_{k-1}(\mathcal{P}A_n) + f_{k-2}(\mathcal{P}A_{n-1}) \\ &\quad + 4f_k(\mathcal{P}A_{n-2}) + 6f_{k-1}(\mathcal{P}A_{n-2}) + 2f_{k-2}(\mathcal{P}A_{n-2}) \\ &\quad + f_k(\mathcal{P}A_{n-3}) + 2f_{k-1}(\mathcal{P}A_{n-3}) + f_{k-2}(\mathcal{P}A_{n-3}). \end{aligned} \tag{6.2}$$

Theorem 6.14. *The Euler characteristics of the spherical blown-up Coxeter complexes are as follows: When n is even, the values are zero; when $n = 2m+1$ is odd,*

$$\chi(\mathcal{C}(A_n)_\#) = (-1)^m\, 2n\, ((n-2)!!)^2 \tag{6.3}$$

$$\chi(\mathcal{C}(B_n)_\#) = 2^n \frac{1}{(n+1)}\, \chi(\mathcal{C}(A_n)_\#) \tag{6.4}$$

$$\chi(\mathcal{C}(D_n)_\#) = 2^{n-3}\left[\frac{8}{n+1} - \frac{1}{n-2}\right] \chi(\mathcal{C}(A_n)_\#). \tag{6.5}$$

The Euler characteristics of the toroidal blown-up Coxeter complexes are as follows: When n is odd, the values are zero; when $n = 2m$ is even,

$$\chi(\mathbb{T}\mathcal{C}(\widetilde{A}_n)_\#) = (-1)^m\, ((n-1)!!)^2 \tag{6.6}$$

$$\chi(\mathbb{T}\mathcal{C}(\widetilde{B}_n)_\#) = \frac{1}{2(n+1)}\, \chi(\mathcal{C}(D_{n+1})_\#) \tag{6.7}$$

$$\chi(\mathbb{T}\mathcal{C}(\widetilde{C}_n)_\#) = 2^n \frac{1}{(n+2)(n+1)}\, \chi(\mathcal{C}(A_{n+1})_\#) \tag{6.8}$$

$$\chi(\mathbb{T}\mathcal{C}(\widetilde{D}_n)_\#) = 2^{n-6}\frac{1}{(n+1)}\left[\frac{64}{n+2} - \frac{15}{n-1}\right] \chi(\mathcal{C}(A_{n+1})_\#). \tag{6.9}$$

Proof. We use Theorem 6.12 to obtain a summation, using the values f_k given above along with the number of chambers provided by Tables 1 and 2. The values for Eqs.(6.3) and (6.6) have been previously calculated in [**10**, Section 3.2] and [**29**, Section 4.3] respectively. Equations (6.4), (6.7) and (6.8) are consequences of Theorem 6.4, where these spaces share the same tiling polytopes as previous calculations. From Eq. (6.1) and Theorem 6.12, we obtain a linear combination

$$\begin{aligned} \chi(\mathcal{C}(D_n)_\#) = &\frac{2^n}{(n+1)!}\, \chi(\mathcal{C}(A_n)_\#) - \frac{2^{n-1}}{n!}\, \chi(\mathcal{C}(A_{n-1})_\#) - \frac{2^{n-3}}{(n-1)!}\, \chi(\mathcal{C}(A_{n-2})_\#) \\ &+ \frac{2^{n-1}}{n!}\, \chi(\mathcal{C}(A_{n-1})_\#) + \frac{2^{n-2}}{(n-1)!}\, \chi(\mathcal{C}(A_{n-2})_\#). \end{aligned}$$

Algebraic manipulations result in Eq.(6.5). Similar calculations using Eq. (6.2) yield Eq. (6.9) after simplification. □

Remark. The reason there is a dimension shift between the spherical and toroidal cases is due to the convention of the affine case having $n + 1$ nodes in its Coxeter graph, compared to n nodes for the spherical.

6.4. Tiling atypical complexes

The polytopes tiling the Coxeter moduli spaces are given by Theorem 6.4. We now discuss the tiling of the atypical complexes, given in Definition 4.5, after minimal blow-ups. As in other (compactified) configuration spaces, the chambers of these complexes correspond to orderings of the particles. However, different orderings of particles may give different face posets to the chamber, since switching a standard and thick particle may change the valid bracketings of the diagram. Specifically, near an axis of symmetry with no fixed particles, having thick particles allows more collisions and hence more brackets, than having standard particles. It is here where the polytopes $\mathcal{P}X^a, \mathcal{P}X^b$, and $\mathcal{P}X^c$ based on Figure 14 appear.

Recall that the chambers of these complexes arise as subspaces of $\mathcal{C}(D_n)_\#$, $\mathbb{T}\mathcal{C}(\widetilde{B}_n)_\#$ or $\mathbb{T}\mathcal{C}(\widetilde{D}_n)_\#$. From Theorem 6.4, these chambers must be *faces* of either $\mathcal{P}D$ or $\mathcal{P}\widetilde{D}$. Converting bracketings to tubings allows us to compute the face poset of the chamber using Theorem 6.6. The following is an example of this method. Note how there is not simply one type of polytope tiling each blown-up atypical complex, as was the case with the Coxeter complexes.

Example 6.15. By taking different bracketings in configurations $C_{\bar{5}}(\mathbb{R}_\circ)$ associated to $\mathcal{C}(D_5)_\#$, we can produce the configuration space diagrams of different chambers of $\mathcal{C}(D_{4,1})_\#$, as in Figure 16. By converting bracketings to tubings using the bijection, each facet of the chamber corresponds to the appropriate reconnected complement.

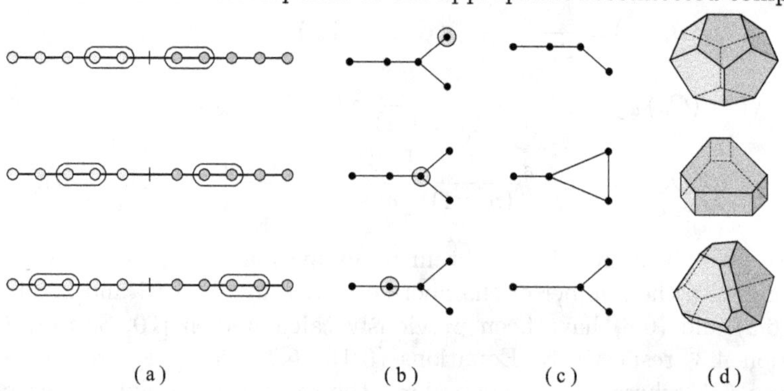

Figure 16: (a) Configuration diagram brackets, (b) associated tubing, (c) reconnected complements and (d) fundamental chambers.

The gluing rules for the compactified configuration spaces applies to these atypical models as well. The reflection action can change the ordering of standard and thick particles, and thus it encodes the manner in which polytopes of different types glue to tile the space.

Example 6.16. The illustration in Figure 17 of five chambers of $\mathcal{C}(D_{4,1})_\#$ shows how the configuration of three (standard) particles and 1 thick particle encodes gluing of faces among different types of chambers either across hyperplanes or antipodal maps. This is done by the flip action $\hat{\sigma}$ of Theorem 5.7. We see the gluing of a face of $\mathcal{P}D_4$ to a face of $\mathcal{P}X_4^a$ (a - b) and the corresponding labeled configuration spaces. Another face of this $\mathcal{P}X_4^a$ attaches to a face of $\mathcal{P}B_4$ (c - d) which glues to the face of another $\mathcal{P}B_4$ (e - f). This identification is across the hyperplane $x_i = 0$, where x_i is the label for the thick particle. Finally, this $\mathcal{P}B_4$ glues to another chamber of type $\mathcal{P}D_4$ (g - h) through an antipodal map.

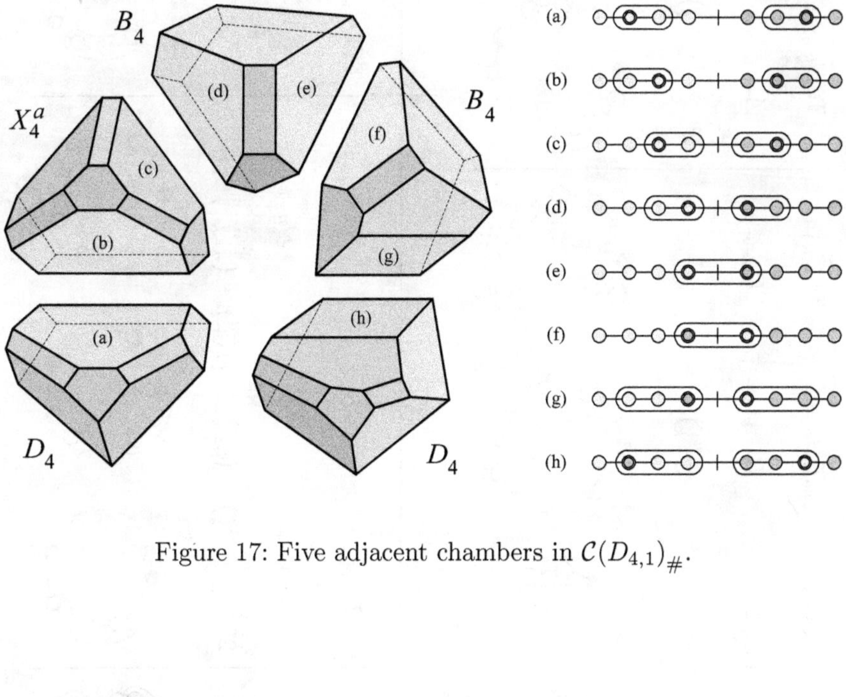

Figure 17: Five adjacent chambers in $\mathcal{C}(D_{4,1})_{\#}$.

$\mathcal{C}W$	Subspace	Stabilizer	Enumeration	Configuration
$\mathcal{C}A_n$	$\mathcal{C}A_{k+1}$	A_{n-k-1}	$\binom{n+1}{n-k}$	
$\mathcal{C}B_n$	$\mathcal{C}B_{k+1}$	B_{n-k-1}	$\binom{n}{n-k-1}$	
	$\mathcal{C}B_{k+1}$	A_{n-k-1}	$2^{n-k-1}\binom{n}{n-k}$	
$\mathcal{C}D_n$	$\mathcal{C}B_{k+1}$	D_{n-k-1}	$\binom{n}{n-k-1}$	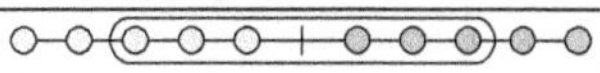
	$\mathcal{C}D_{k+1,1}$	A_{n-k-1}	$2^{n-k-1}\binom{n}{n-k}$	
$\mathcal{C}D_{n,m}$	$\mathcal{C}B_{k+1}$	$D_{n-k-1,r}$	$\binom{m}{r}\binom{n-m}{n-k-r-1}$	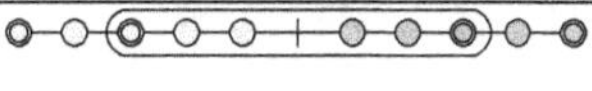
	$\mathcal{C}D_{k+1,m-r+1}$	A_{n-k-1}	$2^{n-k-1}\binom{n}{n-k}$	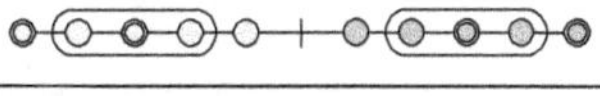

Table 3: Minimal building sets of dimension k for spherical complexes.

$\mathcal{CW}$	$\mathbb{T}\mathcal{C}\widetilde{A}_n$	$\mathbb{T}\mathcal{C}\widetilde{B}_n$			$\mathbb{T}\mathcal{C}\widetilde{C}_n$	
Subspace	$\mathbb{T}\mathcal{C}\widetilde{A}_{k+1}$	$\mathbb{T}\mathcal{C}\widetilde{B}_{k+1,1}$	$\mathbb{T}\mathcal{C}\widetilde{B}_{k+1}$	$\mathbb{T}\mathcal{C}\widetilde{C}_{k+1}$	$\mathbb{T}\mathcal{C}\widetilde{C}_{k+1}$	$\mathbb{T}\mathcal{C}\widetilde{C}_{k+1}$
Stabilizer	A_{n-k}	A_{n-k-1}	B_{n-k}	D_{n-k}	A_{n-k-1}	B_{n-k}
Enumeration	$\binom{n+1}{n+1-k}$	$2^{n-k-1}\binom{n}{n-k}$	$\binom{n}{n-k}$	$\binom{n}{n-k}$	$2^{n-k-1}\binom{n}{n-k}$	$2\binom{n}{n-k}$
Configuration	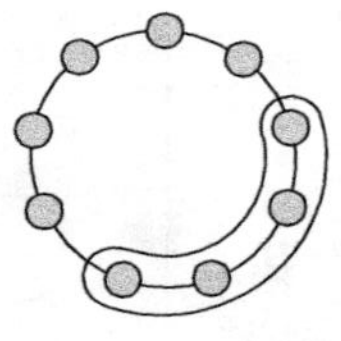	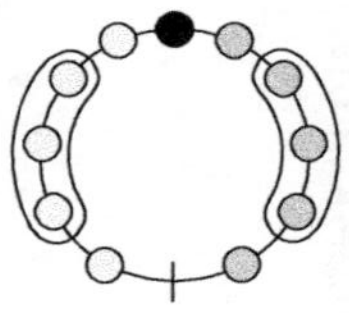	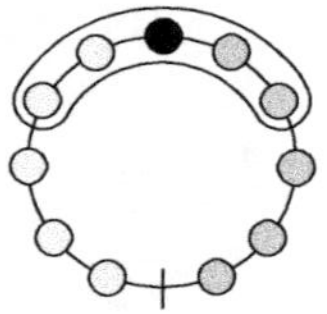	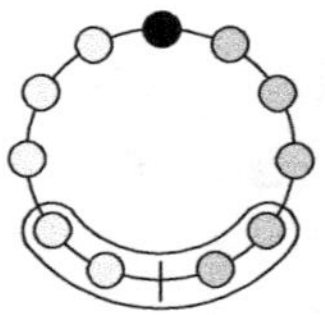	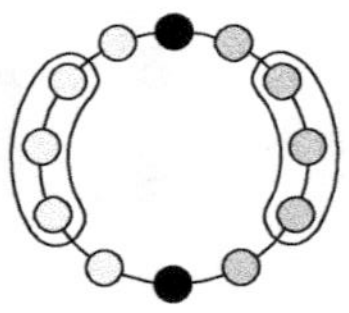	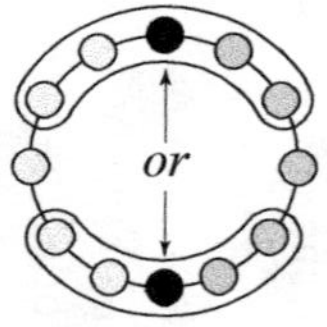

$\mathcal{CW}$	$\mathbb{T}\mathcal{C}\widetilde{D}_n$		$\mathbb{T}\mathcal{C}\widetilde{B}_{n,m}$			$\mathbb{T}\mathcal{C}\widetilde{D}_{n,m}$	
Subspace	$\mathbb{T}\mathcal{C}\widetilde{D}_{k+1,1}$	$\mathbb{T}\mathcal{C}\widetilde{B}_{k+1}$	$\mathbb{T}\mathcal{C}\widetilde{B}_{k,m-r+1}$	$\mathbb{T}\mathcal{C}\widetilde{B}_{k+1,r}$	$\mathbb{T}\mathcal{C}\widetilde{C}_{k+1}$	$\mathbb{T}\mathcal{C}\widetilde{D}_{k+1,m-r+1}$	$\mathbb{T}\mathcal{C}\widetilde{B}_{k+1,r}$
Stabilizer	A_{n-k-1}	D_{n-k}	A_{n-k-1}	B_{n-k-1}	$D_{n-k-1,r}$	A_{n-k-1}	$D_{n-k-1,r}$
Enumeration	$2^{n-k-1}\binom{n}{n-k}$	$2\binom{n}{n-k}$	$2^{n-k-1}\binom{n}{n-k}$	$\binom{n-m}{n-k-r-1}\binom{m}{r}$	$\binom{n-m}{n-k-r-1}\binom{m}{r}$	$2^{n-k-1}\binom{n}{n-k}$	$2\binom{n-m}{n-k-r-1}\binom{m}{r}$
Configuration	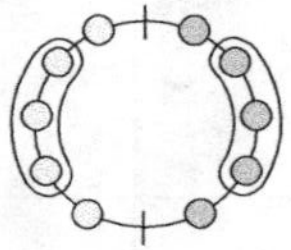	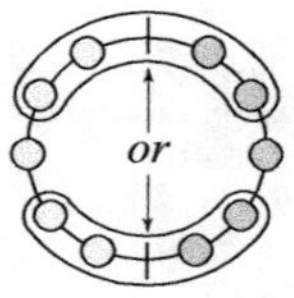	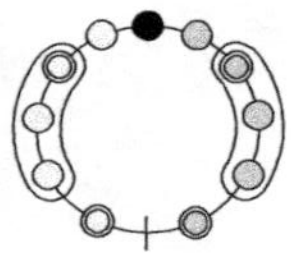	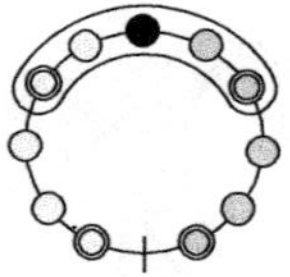	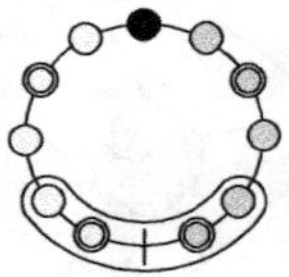	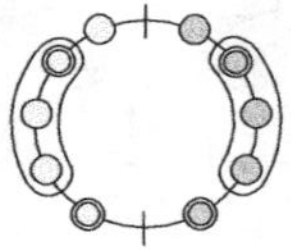	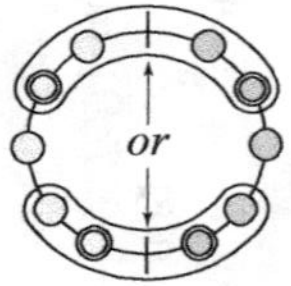

Table 4: Minimal building sets of dimension k for Euclidean complexes.

References

[1] S. Axelrod, I. M. Singer, Chern-Simons perturbation theory II, *J. Diff. Geom.* **39** (1994) 173-213.

[2] L. Billera, S. Holmes, K. Vogtmann, Geometry of the space of phylogenetic trees, *Adv. App. Math.* **27** (2001) 733-767.

[3] R. Bott and C. Taubes, On the self-linking of knots, *J. Math. Phys.* **35** (1994) 5247-5287.

[4] N. Bourbaki, *Lie Groups and Lie Algebras: Chapters 4-6,* Springer-Verlag, Berlin, 2002.

[5] H. R. Brahana and A. M. Coble, Maps of twelve countries with five sides with a group of order 120 containing an Ikosahedral subgroup, *Amer. J. Math.* **48** (1926) 1-20.

[6] M. Carr and S. L. Devadoss, Coxeter complexes and graph-associahedra, *Topology Appl.* **153** (2006) 2155-2168.

[7] A. Cayley, On the partitions of a polygon, *Proc. Lond. Math. Soc.* **22** (1890) 237-262.

[8] M. Davis, T. Januszkiewicz, R. Scott, Nonpositive curvature of blowups, *Selecta Math.* **4** (1998) 491 - 547.

[9] C. De Concini and C. Procesi, Wonderful models of subspace arrangements, *Selecta Math.* **1** (1995) 459-494.

[10] S. L. Devadoss, Tessellations of moduli spaces and the mosaic operad, *Contemp. Math.* **239** (1999) 91-114.

[11] S. L. Devadoss, A space of cyclohedra, *Disc. Comp. Geom.* **29** (2003) 61-75.

[12] S. L. Devadoss, Combinatorial equivalence of real moduli spaces, *Notices Amer. Math. Soc.* **51** (2004) 620-628.

[13] P. Etingof, A. Henriques, J. Kamnitzer, E. Rains, The cohomology ring of the real locus of the moduli space of stable curves of genus 0 with marked points, preprint math.AT/0507514.

[14] Z. Fiedorowicz, The symmetric bar construction, preprint.

[15] K. Fukaya, Y-G Oh, H. Ohta, K. Ono, Lagrangian intersection Floer theory: anomaly and obstruction. *Kyoto. Dept. Math.* 00-17.

[16] W. Fulton and R. MacPherson, A compactification of configuration spaces, *Ann. Math.* **139** (1994) 183-225.

[17] A. Goncharov and Y. Manin, Multiple ζ-motives and moduli spaces $\overline{\mathcal{M}}_{0,n}$, *Compos. Math.* **140** (2004) 1-14.

[18] R. Hartshorne, *Algebraic Geometry,* Springer-Verlag, New York, 1977.

[19] M. M. Kapranov, The permutoassociahedron, MacLane's coherence theorem, and asymptotic zones for the KZ equation, *J. Pure Appl. Alg.* **85** (1993) 119-142.

[20] M. Kontsevich and Y. Manin, Gromov-Witten classes, quantum cohomology, and enumerative geometry, *Comm. Math. Phys.* **164** (1994) 525-562.

[21] D. Levy and L. Pachter. The neighbor-net algorithm, preprint math.CO/0702515.

[22] M. Markl, S. Shnider, J. Stasheff, *Operads in Algebra, Topology and Physics,* Amer. Math. Soc., Rhode Island, 2002.

[23] M. Markl, Homotopy Algebras are Homotopy Algebras, *Forum Math.* **16** (2004) 129-160.

[24] D. Mumford, J. Fogarty, F. Kirwan, *Geometric Invariant Theory,* Springer-Verlag, New York, 1994.

[25] T. Parker and J. Wolfson, Pseudo-holomorphic maps and bubble trees, *J. Geom. Anal.* **3** (1993) 63-98.

[26] A. Postnikov, Permutohedra, associahedra, and beyond, preprint math.CO/0507163.

[27] A. Postnikov, V. Reiner, L. Williams, Faces of generalized permutohedra, preprint math.CO/0609184.

[28] E. Rains, The homology of real subspace arrangements, preprint math.AT/0610743.

[29] R. Simion, A type-B associahedron. *Adv. Appl. Math.* **30** (2003) 2-25.

[30] D. P. Sinha, Manifold-theoretic compactifications of configuration spaces, *Selecta Math.* **10** (2004) 391-428.

[31] R. P. Stanley, *Enumerative Combinatorics: Volume 2,* Cambridge University Press, New York, 1999.

[32] J. Stasheff, Homotopy associativity of H-spaces I, *Trans. Amer. Math. Soc.* **108** (1963) 275-292.

[33] J. Stasheff (Appendix B coauthored with S. Shnider), *Contemp. Math.* **202** (1997) 53-81.

This article may be accessed via WWW at `http://jhrs.rmi.acnet.ge`

Suzanne M. Armstrong
suzanne.m.armstrong@williams.edu

Department of Mathematics and Statistics
Williams College
Williamstown, MA 01267

Michael Carr
mpcarr@umich.edu

Department of Mathematics
University of Michigan
Ann Arbor, MI 48109

Satyan L. Devadoss
satyan.devadoss@williams.edu

Department of Mathematics and Statistics
Williams College
Williamstown, MA 01267

Eric Engler
eric.h.engler@williams.edu

Department of Mathematics and Statistics
Williams College
Williamstown, MA 01267

Ananda Leininger
anandal@mit.edu

Department of Mathematics
MIT
Cambridge, MA 02139

Michael Manapat
manapat@ocf.berkeley.edu

Department of Mathematics
University of California, Berkeley
Berkeley, CA 94720

Journal of Homotopy and Related Structures, vol. 4(1), 2009, pp.111–121

RATIONAL SELF-HOMOTOPY EQUIVALENCES AND WHITEHEAD EXACT SEQUENCE

MAHMOUD BENKHALIFA

(*communicated by James Stasheff*)

Abstract

For a simply connected CW-complex X, let $\mathcal{E}(X)$ denote the group of homotopy classes of self-homotopy equivalence of X and let $\mathcal{E}_\sharp(X)$ be its subgroup of homotopy classes which induce the identity on homotopy groups. As we know, the quotient group $\frac{\mathcal{E}(X)}{\mathcal{E}_\sharp(X)}$ can be identified with a subgroup of $Aut(\pi_*(X))$. The aim of this work is to determine this subgroup for rational spaces. We construct the Whitehead exact sequence associated with the minimal Sullivan model of X which allows us to define the subgroup $\text{Coh.Aut}\big(\text{Hom}(\pi_*(X),\mathbb{Q})\big)$ of self-coherent automorphisms of the graded vector space $\text{Hom}(\pi_*(X),\mathbb{Q})$. As a consequence we establish that $\frac{\mathcal{E}(X)}{\mathcal{E}_\sharp(X)} \cong \text{Coh.Aut}\big(\text{Hom}(\pi_*(X),\mathbb{Q})\big)$. In addition, by computing the group $\text{Coh.Aut}\big(\text{Hom}(\pi_*(X),\mathbb{Q})\big)$, we give examples of rational spaces that have few self-homotopy equivalences.

1. Introduction

If X is a simply connected CW-complex, let $\mathcal{E}(X)$ denote the set of homotopy classes of self-homotopy equivalence of X. Equipped with the composition of homotopy classes, $\mathcal{E}(X)$ is a group. Let $\mathcal{E}_\sharp(X)$ denote the subgroup of homotopy classes which induce the identity on homotopy groups. Clearly $\mathcal{E}_\sharp(X)$ is a normal subgroup of $\mathcal{E}(X)$. In this paper we study the quotient group $\frac{\mathcal{E}(X)}{\mathcal{E}_\sharp(X)}$ where X is a rational space. Recall that there exists a homomorphism $\mathcal{E}(X) \to Aut(\pi_*(X))$ whose kernel is $\mathcal{E}_\sharp(X)$, thus we can identify $\frac{\mathcal{E}(X)}{\mathcal{E}_\sharp(X)}$ with a subgroup G of $Aut(\pi_*(X))$

The aim of this paper is to determine the subgroup G when X is a rational space. Due to the theory elaborated by Sullivan [**5**], the homotopy theory of rational spaces is equivalent to the homotopy theory of minimal cochain commutative algebras over the rationals (mccas, for short). Because of this equivalence we can translate our problem to study the quotient group $\frac{\mathcal{E}(\Lambda V,\partial)}{\mathcal{E}_\sharp(\Lambda V,\partial)}$, where $(\Lambda V, \partial)$ is the mcca associated

Received December 19, 2008, revised April 23, 2009; published on June 9, 2009.
2000 Mathematics Subject Classification: Primary 55P62, 55Q05; Secondary 55S35.
Key words and phrases: Groups of self-homotopy equivalences, rational homotopy theory, Whitehead exact sequence, coherent morphisms.

with X (called the minimal Sullivan model of X), $\mathcal{E}(\Lambda V, \partial)$ denotes the group of self-homotopy equivalence of $(\Lambda V, \partial)$ and $\mathcal{E}_\sharp(\Lambda V, \partial)$ denotes the subgroup of $\mathcal{E}(\Lambda(V), \partial)$ consisting of the elements inducing the identity on the indecomposables.

For this purpose we associate with each mcca $(\Lambda V, \partial)$ an exact sequence, denoted by $WES(\Lambda V, \partial)$ and called the Whitehead exact sequence of $(\Lambda V, \partial)$. This sequence allows us to define the semigroup Coh.Mor(V^*) (respect. the group Coh.Aut(V^*)) of self-coherent homomorphisms (respect. self-coherent automorphisms) of the graded vector space V^* and to exhibit a short exact sequence of semigroups (with units):

$$\mathcal{E}_\sharp(\Lambda(V), \partial) \rightarrowtail [(\Lambda(V), \partial), (\Lambda(V), \partial)] \twoheadrightarrow \text{Coh.Mor}(V^*)$$

and a short exact sequence of groups:

$$\mathcal{E}_\sharp(\Lambda(V), \partial) \rightarrowtail \mathcal{E}(\Lambda(V), \partial) \twoheadrightarrow \text{Coh.Aut}(V^*),$$

where $[(\Lambda(V), \partial), (\Lambda(V), \partial)]$ denotes the semigroup of homotopy classes of cochain morphism from $(\Lambda V, \partial)$ to itself.

Because of the homotopy equivalence mentioned above, our main result says:

Theorem. *If X is a simply connected rational CW-complex, then:*
There exist a short exact sequence of semigroups:

$$\mathcal{E}_\sharp(X) \rightarrowtail [X, X] \twoheadrightarrow \text{Coh.Mor}\big(\text{Hom}(\pi_*(X), \mathbb{Q})\big).$$

There exist a short exact sequence of groups:

$$\mathcal{E}_\sharp(X) \rightarrowtail \mathcal{E}(X) \twoheadrightarrow \text{Coh.Aut}\big(\text{Hom}(\pi_*(X), \mathbb{Q})\big).$$

Here $[X, X]$ denotes the semigroup of self-homotopy maps of X.

In addition and by using techniques of rational homotopy theory, we compute the groups Coh.Aut$\big(\text{Hom}(\pi_*(X), \mathbb{Q})\big)$ for certain rational spaces via their minimal Sullivan models. For instance, we investigate the following question asked by M. Arkowitz and G Lupton in [**1**]: Which finite groups can be realized as the group of self-homotopy equivalence of a rational space? We show that the groups $\bigoplus_{2^n} \mathbb{Z}_2$ (2^n copies of $\mathbb{Z}_2$) are realizable for $n \leqslant 10$. At the end of this work we ask the following question: Is it true that the groups $\bigoplus_{2^n} \mathbb{Z}_2$ are always realizable for $n \geqslant 11$?

2. Coherent morphishms

2.1. Whitehead exact sequence of 1-connected mcca

Let $(\Lambda V, \partial)$ be a 1-connected mcca. As we have done in ([**3**], section 2) (respect. ([**4**], section 2)) for 1-connected minimal free cochain algebras (respect. free chain algebras) over a P.I.D, we can define the Whitehead exact sequence of $(\Lambda V, \partial)$ as follows:

First define the pair:

$$\left(\Lambda V^{\leqslant n+1}; \Lambda V^{\leqslant n-1}\right) = \frac{(\Lambda V^{\leqslant n+1}, \partial|_{\Lambda V^{\leqslant n+1}})}{(\Lambda V^{\leqslant n-1}, \partial|_{\Lambda V^{\leqslant n-1}})} \ , \ \forall n \geqslant 3$$

where $\partial|_{\Lambda V^{\leqslant n}}$ denotes the restriction of the differential ∂ to $\Lambda V^{\leqslant n}$.

To each pair $\left(\Lambda V^{\leqslant n+1};\Lambda V^{\leqslant n-1}\right)$ corresponds the following short exact sequence of cochain complexes:

$$(\Lambda V^{\leqslant n-1}),\partial|_{\Lambda V^{\leqslant n-1}})\rightarrowtail(\Lambda V^{\leqslant n+1},\partial|_{\Lambda V^{\leqslant n+1}})\twoheadrightarrow\left(\Lambda V^{\leqslant n+1};\Lambda V^{\leqslant n-1}\right)$$

which yields the following long exact cohomology sequence:

$$\begin{array}{ccc} \cdots\to V^n\cong H^n\left(\Lambda V^{\leqslant n+1});\Lambda V^{\leqslant n-1})\right) & \xrightarrow{b^n} & H^{n+1}(\Lambda V^{\leqslant n-1})) \\ & & \downarrow \\ \cdots\leftarrow H^{n+1}(\Lambda V^{\leqslant n+1}))\leftarrow V^{n+1}\cong H^{n+1}\left(\Lambda V^{\leqslant n+1});\Lambda V^{\leqslant n-1})\right) & \xleftarrow{j^{n+1}} & H^{n+1}(\Lambda V^{\leqslant n+1})) \end{array}$$

Consequently, if we combine the two long exact cohomology sequences associated with the two pairs $\left(\Lambda V^{\leqslant n+1};\Lambda V^{\leqslant n-1}\right)$ and $\left(\Lambda V^{\leqslant n+2};\Lambda V^{\leqslant n}\right)$ respectively, we get the following long exact sequence:

$$\to V^n\xrightarrow{b^n}H^{n+1}(\Lambda V^{\leqslant n-1})\to H^{n+1}(\Lambda V^{\leqslant n+1})\xrightarrow{j^{n+1}}V^{n+1}\xrightarrow{b^{n+1}}H^{n+2}(\Lambda V^{\leqslant n})\to \qquad (2.1)$$

where the homomorphisms b^n and j^{n+1} are defined as follows:

$$b^n(v_n)=[\partial^n(v_n)] \qquad\qquad j^{n+1}([v_{n+1}+q_{n+1}])=v_{n+1}. \qquad (2.2)$$

Here $[\partial^n(v_n)]$ and $[v_{n+1}+q_{n+1}]$ denote respectively the cohomology classes of $\partial^n(v_n)\in(\Lambda V^{\leqslant n-1})^{n+1}$ and $v_{n+1}+q_{n+1}\in(\Lambda V^{\leqslant n+1})^{n+1}$.

Since it is well-known that $H^{n+1}(\Lambda V^{\leqslant n+1})\cong H^{n+1}(\Lambda V)$, then from (2.1) we get the following long sequence:

$$\cdots\to V^n\xrightarrow{b^n}H^{n+1}(\Lambda V^{\leqslant n-1})\longrightarrow H^{n+1}(\Lambda V)\longrightarrow V^{n+1}\xrightarrow{b^{n+1}}\cdots$$

called the Whitehead exact sequence of $(\Lambda V,\partial)$.

This sequence is natural with respect to cochain morphisms. That is, if $\alpha:(\Lambda(V),\partial)\to(\Lambda(W),\delta)$ is a cochain morphism, then α induces the following commutative diagram:

$$(1)\qquad \begin{array}{ccccccccc} \cdots\to V^n & \xrightarrow{b^n} & H^{n+1}(\Lambda V^{\leqslant n-1}) & \longrightarrow & H^{n+1}(\Lambda V) & \longrightarrow & V^{n+1} & \xrightarrow{b^{n+1}} & \cdots \\ \downarrow{\scriptstyle\widetilde{\alpha}^n} & & \downarrow{\scriptstyle H^{n+1}(\alpha_{(n-1)})} & & \downarrow{\scriptstyle H^{n+1}(\alpha)} & & \downarrow{\scriptstyle\widetilde{\alpha}^{n+1}} & & \\ \cdots\to W^n & \xrightarrow{b'^n} & H^{n+1}(\Lambda W^{\leqslant n-1}) & \longrightarrow & H^{n+1}(\Lambda W) & \longrightarrow & W^{n+1} & \xrightarrow{b'^{n+1}} & \cdots \end{array}$$

where $\widetilde{\alpha}:V^*\to W^*$ is the graded homomorphism induced by α on the indecomposables and where $\alpha_{(n-1)}:(\Lambda V^{\leqslant n-1},\partial)\to(\Lambda W^{\leqslant n-1},\delta)$ is the restriction of α.

2.2. Coherent morphisms between Whitehead exact sequences

Let $(\Lambda V,\partial)$, $(\Lambda W,\delta)$ be two 1-connected mccas and let $\xi:V^*\to W^*$ be a given graded linear application. For every $n\geqslant 2$, let $\{\xi^{\leqslant n}\}$ denote the set of all cochain morphisms from $(\Lambda V^{\leqslant n}, \partial)$ to $(\Lambda W^{\leqslant n}, \delta)$ inducing $\xi^{\leqslant n}$ on the indecomposables.

Definition 2.1. Let $(\Lambda V, \partial)$ and $(\Lambda W, \delta)$ be two 1-connected mccas. A graded linear map $\xi^* : V^* \to W^*$ is called a coherent morphism if the following holds:
For every $n \geqslant 2$, if the set $\{\xi^{\leqslant n}\}$ is not empty, then it contains $\alpha_{(n)}$ making the following diagram commute:

$$
\begin{array}{ccc}
V^{n+1} & \xrightarrow{\quad \xi^{n+1} \quad} & W^{n+1} \\
\downarrow{\scriptstyle \flat^{n+1}} & & \downarrow{\scriptstyle \flat'^{n+1}} \\
H^{n+2}(\Lambda V^{\leqslant n}) & \xrightarrow{\quad H^{n+2}(\alpha_{(n)}) \quad} & H^{n+2}(\Lambda W^{\leqslant n})
\end{array}
\tag{2}
$$

Example 2.1. If $\alpha : (\Lambda(V), \partial) \to (\Lambda(W), \delta)$ is a cochain algebra morphism between two 1-connected mccas, then, according to diagram (1), the graded linear map $\widetilde{\alpha} : V^* \to W^*$ is a coherent morphism.

Example 2.2. It easy to see that if $(\Lambda(V), \partial)$ is a 1-connected mcca, then Id_{V^*} is a coherent morphism. Observe that in this case the set $\{Id_{V^*}\}$ of cochain morphisms from $(\Lambda(V), \partial)$ to it self inducing Id_{V^*} on the indecomposables is always not empty since it contains $Id_{(\Lambda(V),\partial)}$.

Now let $(\Lambda V, \partial)$, $(\Lambda W, \delta)$ be two 1-connected mccas and let $\text{Coh.Mor}(V^*, W^*)$ denote the set of all the coherent automorphisms from V^* to W^*. Example 2.1 allows us to define a map $\Phi : [(\Lambda V, \partial), (\Lambda W, \delta)] \to \text{Coh.Mor}(V^*, W^*)$ by setting $\Phi([\alpha]) = \widetilde{\alpha}$. Here $[(\Lambda V, \partial), (\Lambda W, \delta)]$ denote the set of homotopy classes from $(\Lambda V, \partial)$ to $(\Lambda W, \delta)$. Recall that there is a reasonable concept of "homotopy" among cochain morphisms (see for example [**5**] for details), analogous in many respects to the topological notion of homotopy.

Remark 2.1. It is well-known ([**5**] proposition 12.8) that if two cochain morphisms $\alpha, \alpha' : (\Lambda V, \partial) \to (\Lambda W, \delta)$ are homotopic, then they induce the same graded linear maps on the indecomposables i.e, $\widetilde{\alpha} = \widetilde{\alpha'}$. So the map Φ is well-defined.

Proposition 2.1. *The map Φ is surjective.*

Proof. Let $\xi \in \text{Coh.Mor}(V^*, W^*)$. Assume, by induction, that we have constructed a cochain morphism $\theta_{(n)} : (\Lambda V^{\leqslant n}, \partial) \to (\Lambda W^{\leqslant n}, \delta)$ such that $\Phi([\theta_{(n)}]) = \xi^{\leqslant n}$. This implies that the set $\{\xi^{\leqslant n}\}$ is not empty. Therefore, by definition 2.1, this set contains an element $\alpha_{(n)}$ making the diagram (2) commutes. Now choose $(v_\sigma)_{\sigma \in \Sigma}$ as a basis of V^{n+1}. Recall that, in this diagram, we have:

$$
\begin{aligned}
H^{n+2}(\alpha_{(n)}) \circ \flat^{n+1}(v_\sigma) &= \alpha_{(n)} \circ \partial^{n+1}(v_\sigma) + \operatorname{Im} \delta^{n+1} \\
\flat'^{n+1} \circ \xi^{n+1}(v_\sigma) &= \delta^{n+1} \circ \xi^{n+1}(v_\sigma) + \operatorname{Im} \delta^{n+1}
\end{aligned}
\tag{2.3}
$$

where $\delta^{n+1} : (\Lambda W^{\leqslant n})^{n+1} \to (\Lambda W^{\leqslant n})^{n+2}$. Since the diagram (2) commutes, the element $(\alpha_{(n)} \circ \partial^{n+1} - \delta^{n+1} \circ \xi^{n+1})(v_\sigma) \in \operatorname{Im} \delta^{n+1}$. As a consequence there exists $u_\sigma \in (\Lambda W^{\leqslant n})^{n+1}$ such that:

$$(\alpha_{(n)} \circ \partial^{n+1} - \delta^{n+1} \circ \xi^{n+1})(v_\sigma) = \delta^{n+1}(u_\sigma). \tag{2.4}$$

Thus we define $\theta_{(n+1)} : (\Lambda V^{\leqslant n+1}, \partial) \to (\Lambda W^{\leqslant n+1}, \delta)$ by setting:

$$\theta_{(n+1)}(v_\sigma) = \xi^{n+1}(v_\sigma) + u_\sigma \ , \ v_\sigma \in V^{n+1} \quad \text{and} \quad \theta = \alpha_{(n)} \text{ on } V^i \ , \ \forall i \leqslant n \tag{2.5}$$

As $\partial^{n+1}(v_\sigma) \in (\Lambda V^{\leqslant n})^{n+2}$ then, by (2.4), we get:

$$\begin{aligned}\delta^{n+1} \circ \theta_{(n+1)}(v_\sigma) = \delta^{n+1}(\xi^{n+1}(v_\sigma)) + \delta^{n+1}(u_\sigma) = \\ \alpha_{(n)} \circ \partial^{n+1}(v_\sigma) = \alpha_{(n)} \circ \partial^{n+1}(v_\sigma).\end{aligned}$$

So $\theta_{(n+1)}$ is a cochain morphism. Now due to the fact that $u_\sigma \in (\Lambda W^{\leqslant n})^{n+1}$, the homomorphism $\widetilde{\theta}^{n+1}_{(n+1)} : V^{n+1} \to W^{n+1}$ coincides with ξ^{n+1}. This implies that $\Phi([\theta_{(n+1)}]) = \xi^{\leqslant n+1}$ and the set $\{\xi^{\leqslant n+1}\}$ is not empty, completing the induction step. Finally the iteration of this process yields a cochain morphism $\theta : (\Lambda V, \partial) \to (\Lambda W, \delta)$ satisfying $\widetilde{\theta} = \xi$ □

Remark 2.2. If we assume that $(\Lambda W^{\leqslant n})^{n+1} = 0$ for all $n \leqslant k$, then the cochain morphism $\theta_{(n+1)}$ given in (2.5) will satisfy $\theta_{(n+1)} = \xi^{\leqslant n+1}$ and the set $\{\xi^{\leqslant n+1}\}$ contains just one element for all $n \leqslant k$.

Remark 2.3. It is well-known (see [**5**]) that any cochain morphism between two 1-connected mccas inducing a graded linear isomorphism on the indecomposables is an isomorphism. Consequently if the coherent morphism ξ is an isomorphism, then the cochain morphism $\alpha : (\Lambda V, \partial) \to (\Lambda W, \delta)$ constructed in the proof of proposition 2.1 is such that $\alpha_{(n)} : (\Lambda V^{\leqslant n}, \partial) \to (\Lambda W^{\leqslant n}, \delta)$ is a cochain isomorphism for every $n \geqslant 2$.

Now let us denote by $\mathcal{E}(\Lambda(V), \partial)$ the group of the self-homotopy equivalences of $(\Lambda(V), \partial)$ and by $\mathcal{E}_\sharp(\Lambda(V), \partial)$ the subgroup of $\mathcal{E}(\Lambda(V), \partial)$ consisting of the elements inducing the identity on the indecomposables. Also let $\mathrm{Coh.Aut}(V^*)$ denote the set of the self-coherent automorphisms of V^*.

Proposition 2.2. $\mathrm{Coh.Aut}(V^*)$ *is a subgroup of the group* $Aut(V^*)$.

Proof. Let $\xi, \xi' \in \mathrm{Coh.Aut}(V^*)$. By definition 2.1 to prove that $\xi' \circ \xi \in \mathrm{Coh.Aut}(V^*)$, we must show that, for every $n \geqslant 2$, if the set $\{(\xi' \circ \xi)^{\leqslant n}\}$ is not empty, then it contains an element $\lambda_{(n)}$ making making the following diagram commutes:

$$\begin{array}{ccc} V^{n+1} & \xrightarrow{\xi'^{n+1}\circ\xi^{n+1}} & V^{n+1} \\ \downarrow{\scriptstyle b^{n+1}} & & \downarrow{\scriptstyle b^{n+1}} \\ H^{n+2}(\Lambda V^{\leqslant n}) & \xrightarrow{H^{n+2}(\lambda_{(n)})} & H^{n+2}(\Lambda V^{\leqslant n}) \end{array}$$

Recall that $\{(\xi'\circ\xi)^{\leqslant n}\}$ is the set of all cochain morphisms from $(\Lambda V^{\leqslant n},\partial)$ to itself inducing $(\xi'\circ\xi)^{\leqslant n}$ on the indecomposables.

Indeed, if $\xi,\xi'\in \text{Coh.Aut}(V^*)$, then, according the proof of proposition 2.1, there exist two cochain isomorphisms $\alpha,\alpha':(\Lambda(V),\partial)\to(\Lambda(V),\partial)$ such that $\widetilde{\alpha}=\xi$, $\widetilde{\alpha}'=\xi'$ and satisfying:

$$b^{n+1}\circ\xi'^{n+1}=H^{n+2}(\alpha'_{(n)})\circ b^{n+1}\qquad b^{n+1}\circ\xi^{n+1}=H^{n+2}(\alpha_{(n)})\circ b^{n+1}\quad,\quad\forall n\geqslant 2$$

So, for all $n\geqslant 2$, the set $\{(\xi'\circ\xi)^{\leqslant n}\}$ contains $\lambda_{(n)}=\alpha'_{(n)}\circ\alpha_{(n)}$ and an easy computation shows that:

$$\begin{aligned} b^{n+1}\circ\xi'^{n+1}\circ\xi^{n+1} &= H^{n+2}(\alpha'_{(n)})\circ b^{n+1}\circ\xi^{n+1}=H^{n+2}(\alpha'_{(n)})\circ H^{n+2}(\alpha'_{(n)})\circ b^{n+1}\\ &= H^{n+2}(\alpha'_{(n)}\circ\alpha'_{(n)})\circ b^{n+1}=H^{n+2}(\lambda_{(n)})\circ b^{n+1} \end{aligned}\tag{2.6}$$

which implies that $\xi'\circ\xi$ is a coherent automorphism.

Now let $\xi\in\text{Coh.Aut}(V^*)$. By proposition 2.1 and remark 2.3 we get a cochain isomorphism $\alpha:(\Lambda V,\partial)\to(\Lambda V,\partial)$ satisfying $\widetilde{\alpha}=\xi$ and such that $\alpha_{(n)}$ is a cochain isomorphism for all $n\geqslant 2$. Consequently there exists $\alpha'_{(n)}$ such that $\alpha_{(n)}\circ\alpha'_{(n)}=\alpha'_{(n)}\circ\alpha_{(n)}=Id_{(\Lambda V^{\leqslant n},\partial)}$ which implies that $\widetilde{\alpha'}_{(n)}=(\xi^{-1})^n$ for all $n\geqslant 2$. So the set $\{(\xi^{-1})^{\leqslant n}\}$ contains $\alpha'_{(n)}$. Moreover as $\xi\in\text{Coh.Aut}(V^*)$ it satisfies $b^{n+1}\circ\xi^{n+1}=H^{n+2}(\alpha_{(n)})\circ b^{n+1}$ which implies that $H^{n+2}(\alpha'_{(n)})\circ b^{n+1}=b^{n+1}\circ(\xi^{n+1})^{-1}$. Hence $\xi^{-1}\in\text{Coh.Aut}(V^*)$ □

Theorem 2.1. *Let $(\Lambda(V),\partial)$ be a 1-connected mcca. There exists a short exact sequence of groups:*

$$\mathcal{E}_\sharp(\Lambda(V),\partial)\rightarrowtail\mathcal{E}(\Lambda(V),\partial)\overset{\Phi}{\twoheadrightarrow}\text{Coh.Aut}(V^*)\tag{2.7}$$

Proof. First we have $\Phi(\{\alpha\circ\alpha'\})=\widetilde{\alpha\circ\alpha'}=\widetilde{\alpha}\circ\widetilde{\alpha}'=\Phi(\{\alpha\})\circ\Phi(\{\alpha'\})$. Next the surjection of Φ is assured by proposition 2.1 and finally it is clear that $\ker\Phi=\mathcal{E}_\sharp(\Lambda V,\partial)$ □

The set $[(\Lambda(V),\partial),(\Lambda(V),\partial)]$ of self-homotopy classes of a 1-connected mcca $(\Lambda V,\partial)$, equipped with the composition of maps, is a semigroup with unit. So let Coh.Mor(V^*) denote the set of the self-coherent morphisms of V^*. From proposition 2.1 we deduce that Coh.Mor(V^*) is a semigroup with unit and the map Φ is a homomorphism of semigroups. Hence theorem 2.1 implies that:

Corollary 2.1. *Let $(\Lambda(V), \partial)$ be a 1-connected mcca. There exists a short exact sequence of semigroups:*

$$\mathcal{E}_\sharp(\Lambda(V), \partial) \rightarrowtail [(\Lambda(V), \partial), (\Lambda(V), \partial)] \overset{\Phi}{\twoheadrightarrow} \text{Coh.Mor}(V^*). \tag{2.8}$$

Because of the equivalence between the homotopy theory of rational spaces and the homotopy theory of mccas, we can construe the above results as follows. Let X be a simply connected rational CW-complex of finite type. By the properties of the Sullivan minimal model $(\Lambda(V), \partial)$ of X, we can identify $\mathcal{E}(X)$ with $\mathcal{E}(\Lambda(V), \partial)$ and $\mathcal{E}_\sharp(X)$ with $\mathcal{E}_\sharp(\Lambda(V), \partial)$. Moreover $WES(\Lambda(V), \partial)$ can be written as follows:

$$\cdots \to \text{Hom}(\pi_n(X), \mathbb{Q}) \overset{b_X^{n-1}}{\to} \Gamma_X^n \to H^n(X, \mathbb{Q}) \to \text{Hom}(\pi_n(X), \mathbb{Q}) \overset{b_X^n}{\to} \cdots$$

where $\Gamma_X^n = H^n(\Lambda(V^{\leqslant n-2}))$. We call this sequence the Whitehead exact sequence of X and we denote it by $WES(X)$. Clearly this sequence is an invariant of homotopy.

As a consequences of theorem 2.1 and corollary 2.1 we establish the following result:

Corollary 2.2. *There exist a short exact sequence of groups:*

$$\mathcal{E}_\sharp(X) \rightarrowtail \mathcal{E}(X) \overset{\Psi}{\twoheadrightarrow} \text{Coh.Aut}\big(\text{Hom}(\pi_*(X), \mathbb{Q})\big). \tag{2.9}$$

There exist a short exact sequence of semigroups:

$$\mathcal{E}_\sharp(X) \rightarrowtail [X, X] \overset{\Psi}{\twoheadrightarrow} \text{Coh.Mor}\big(\text{Hom}(\pi_*(X), \mathbb{Q})\big) \tag{2.10}$$

Here $[X, X]$ denotes the semigroup of the self-homotopy classes of X and the linear map $\Psi(\{\alpha\}) : \text{Hom}(\pi_*(X), \mathbb{Q}) \to \text{Hom}(\pi_*(X), \mathbb{Q})$ is defined as follows:

$$\Psi(\{\alpha\})(\eta) = \eta \circ \pi_*(\alpha) \ , \ \forall \eta \in \text{Hom}\big(\pi_*(X), \mathbb{Q}\big)$$

Recall that $\text{Coh.Aut}\big(\text{Hom}(\pi_*(X), \mathbb{Q})\big)$ (respect. $\text{Coh.Mor}\big(\text{Hom}(\pi_*(X), \mathbb{Q})\big)$ denotes the subgroup of the self-coherent automorphisms (respect. self-coherent morphisms) of $\text{Hom}(\pi_*(X), \mathbb{Q})$.

Definition 2.2. Let $(\Lambda V, \partial)$ and $(\Lambda W, \delta)$ be two 1-connected mccas. We say that $WES(\Lambda V, \partial)$ and $WES(\Lambda W, \delta)$ are coherently isomorphic if the set $\text{Coh.Iso}(V^*, W^*)$ of the coherent isomorphisms from V^* to W^* is not empty.

Corollary 2.3. *Two simply connected rational CW-complexes of finite type X and Y are homotopy equivalent if and only their $WES(X)$ and $WES(Y)$ are coherently isomorphic.*

Proof. Let $(\Lambda V, \partial)$ (respect. $(\Lambda W, \delta)$) the Sullivan minimal model of X (respect. of Y). First recall that $WES(X) = WES(\Lambda V, \partial)$ and $WES(Y) = WES(\Lambda W, \delta)$. Now if $WES(X)$ and $WES(Y)$ are coherently isomorphic, then there exists a coherent isomorphism $\xi : V^* \to W^*$. Now proposition 2.1 yields a cochain morphism

$\alpha : (\Lambda V, \partial) \to (\Lambda W, \delta)$ such that the map $\widetilde{\alpha}$, induced by α on the indecomposables, satisfies $\widetilde{\alpha} = \xi$. So $\widetilde{\alpha}$ is an isomorphism which means that the models $(\Lambda V, \partial)$ and $(\Lambda W, \delta)$ are isomorphic. Hence, by the properties of the Sullivan minimal model, we conclude that X and Y are homotopy equivalent □

3. Examples

In this section we give some examples showing how the group $\text{Coh.Aut}(\text{Hom}(\pi_*(X), \mathbb{Q}))$ can be used to compute the group $\mathcal{E}(X)$ when X is a simply connected rational CW-complex. First let us consider the following example which has already treated in ([**1**], example 5.3), where the authors have used another technique, which is radically different from our approach, to determine $\mathcal{E}(X)$.

Example 3.1. Let $\Lambda V = \Lambda(x_1, x_2, y_1, y_2, y_3, z)$ with $|x_1| = 10$, $|x_2| = 12$, $|y_1| = 41$, $|y_2| = 43$, $|y_3| = 45$ and $|z| = 119$. The differential is as follows:

$$\partial(x_1) = 0 \quad \partial(y_1) = x_1^3 x_2 \qquad\qquad \partial(y_3) = x_1 x_2^3$$
$$\partial(x_2) = 0 \quad \partial(y_2) = x_1^2 x_2^2 \quad \partial(z) = y_1 y_2 x_2^3 - y_1 y_3 x_1 x_2^2 + y_2 y_3 x_1^2 x_2 + x_1^{12} + x_2^{10}$$

An easy computation shows that:

$$H^{46}(\Lambda V^{\leqslant 44}) = \mathbb{Q}\{x_1 x_2^3\} \quad H^{44}(\Lambda V^{\leqslant 42}) = \mathbb{Q}\{x_1^2 x_2^2\} \quad H^{42}(\Lambda V^{\leqslant 40}) = \mathbb{Q}\{x_1^3 x_2\}$$
$$H^{120}(\Lambda V^{\leqslant 118}) = \mathbb{Q}\{y_1 y_2 x_2^3 - y_1 y_3 x_1 x_2^2 + y_2 y_3 x_1^2 x_2, x_1^{12}, x_2^{10}\}$$

First any linear map $\xi^i : V^i \to V^i$, where $i = 10, 12, 41, 43, 45, 119$, is multiplication with a rational number, so write $\xi^i = p_i$. Hence in this case any element of $\text{Coh.Mor}(V^*)$ can be identified with $(p_{10}, p_{12}, p_{41}, p_{43}, p_{45}, p_{119}) \in \mathbb{Q}^6$, therefore $\text{Coh.Mor}(V^*)$ can be regarded as a semigroup of $(\mathbb{Q}^6, \times)$.
Now define the cochain algebra morphism $\alpha_{(40)} : \Lambda V^{\leqslant 40} \to \Lambda V^{\leqslant 40}$ by $\alpha_{(40)}(x_1) = \xi^{10}(x_1)$ and $\alpha_{(40)}(x_2) = \xi^{12}(x_2)$. So the set of the cochain morphisms from $(\Lambda V^{\leqslant 40}, \partial)$ to itself inducing ξ^{10}, ξ^{12} on the indecomposables is not empty. To be a coherent morphism the linear map $\xi^{41} : V^{41} \to V^{41}$ must satisfy, according to the diagram (2), the relation:

$$b^{41} \circ \xi^{41} = H^{42}(\alpha_{(40)}) \circ b^{41} \tag{3.1}$$

where the linear map $b^{41} : V^{41} = \mathbb{Q}\{y_1\} \to H^{42}(\Lambda V^{\leqslant 40}) = \mathbb{Q}\{x_1^3 x_2\}$ can be regarded as multiplication with a rational. Now as $H^{42}(\alpha_{(40)})$ is identified with multiplication by $p_{10}^3 p_{12}$, the relation (3.1) implies the equation $p_{41} = p_{10}^3 p_{12}$. By going back to the proof of proposition 2.1, this equation allows us to extend $\alpha_{(40)}$ to a cochain morphism $\alpha_{(41)} : \Lambda V^{\leqslant 41} \to \Lambda V^{\leqslant 41}$. Because $(\Lambda V^{\leqslant 40})^{41} = 0$, the element $u_\sigma \in (\Lambda V^{\leqslant 40})^{41}$, given in (2.4), is zero. Consequently we have $\alpha_{(41)}(y_1) = \xi^{41}(y_1)$.
By using a similar argument in degree 43, 45, 119 we get the following equations:

$$p_{43} = p_{10}^2 p_{12}^2 \ , \ p_{45} = p_{10} p_{12}^3 \ , \ p_{119} = p_{41} p_{43} p_{10}^2 p_{12}^3 = p_{10}^{12} = p_{12}^{10}$$

which have 3 solutions $(0,0,0,0,0,0)$, $(1,1,1,1,1,1)$, $(1,-1,-1,1,-1,1)$. So we get 3 coherent homomorphisms.

Now by corollary 2.1 we have $\Phi^{-1}(Id_{V_*}) = \xi_\sharp(\Lambda V, \partial)$. Due to the fact that $(\Lambda V^{\leqslant n})^{n+1} = 0$ for $n = 10, 12, 41, 43, 45, 119$ and by using remark (2.2, we deduce that $\Phi^{-1}(Id_{V_*}) = \{Id_{(\Lambda V, \partial)}\}$ which implies that Φ, given in the short exact sequence (2.8), is an isomorphism of semigroups. Hence $[(\Lambda V, \partial), (\Lambda V, \partial)]$ has 3 elements and then $\mathcal{E}(\Lambda V, \partial)$ has 2 elements corresponding to the coherent automorphisms $(1, 1, 1, 1, 1, 1)$, $(1, -1, -1, 1, -1, 1)$.

Example 3.2. Let $(\Lambda W, \delta)$ be the mcca obtained from the graded algebra $(\Lambda V, \partial)$, given in example 3.1, by adding a new generator $|x_0|$, with $|x_0| = 2$ and where the differential is as follows, $\delta(x_0) = 0$, $\delta(z) = y_1y_2x_2^3 - y_1y_3x_1x_2^2 + y_2y_3x_1^2x_2 + x_1^{12} + x_2^{10} + x_0^{60}$ and $\delta = \partial$ on the other generators. In this case a simple computation shows that:

$$\operatorname{Im} b^{46} = \mathbb{Q}\{x_1x_2^3\}\ ,\ \operatorname{Im} b^{44} = \mathbb{Q}\{x_1^2x_2^2\}\ ,\ \operatorname{Im} b^{42} = \mathbb{Q}\{x_1^3x_2\}$$

$$\operatorname{Im} b^{120} = \mathbb{Q}\{y_1y_2x_2^3 - y_1y_3x_1x_2^2 + y_2y_3x_1^2x_2, x_1^{12}, x_2^{10}, x_0^{60}\}$$

Write $\xi^2 : W^2 = \mathbb{Q}\{x_0\} \to W^2 = \mathbb{Q}\{x_0\}$ as $\xi^2(x_0) = p_2x_0$ with $p_2 \in \mathbb{Q}$. By similar arguments as in example 3.1 we get the following equations:

$$p_{41} = p_{10}^3p_{12}\ ,\ p_{43} = p_{10}^2p_{12}^2\ ,\ p_{45} = p_{10}p_{12}^3\ ,\ p_{119} = p_{41}p_{43}p_{10}^2p_{12}^3 = p_{10}^{12} = p_{12}^{10} = p_2^{60}$$

which give 5 coherent homomorphisms:

$$(0,0,0,0,0,0,0), (1,1,1,1,1,1,1), (1,1,-1,-1,1,-1,1)$$

$$(-1,1,1,1,1,1,1), (-1,1,-1,-1,1,-1,1).$$

As in the example 3.1 we have $\Phi^{-1}(Id_{W_*}) = \{Id_{(\Lambda W, \delta)}\}$, so $[(\Lambda W, \delta), (\Lambda W, \delta)]$ has 5 elements and then $\mathcal{E}(\Lambda W, \delta)$ has 4 elements corresponding to the coherent automorphisms:

$$(1,1,1,1,1,1,1), (1,-1,-1,1,-1,1), (-1,1,1,1,1,1,1), (-1,1,-1,-1,1,-1,1).$$

As the last three elements are of order 2 we conclude that $\mathcal{E}(\Lambda W, \delta) \cong \mathbb{Z}_2 \oplus \mathbb{Z}_2$.

Remark 3.1. In [1] M. Arkowitz and G Lupton ended their work by the following question: Which finite groups can be realized as the group of self-homotopy equivalence of a rational space? Examples 3.1 and 3.2 show that the groups $\mathbb{Z}_2$ and $\mathbb{Z}_2 \oplus \mathbb{Z}_2$ are realizable.

Now let $(\Lambda U_1, \partial_1)$ be the mcca obtained from $(\Lambda W, \delta)$, given in example 3.2, by adding a new generator x_3 with $|x_3| = 3$ and where the differential is $\partial_1(x_3) = 0$, $\partial_1(z) = \delta(z) + x_3^{40}$ and $\partial_1 = \delta$ on the other generators. If we write $\xi^3 : U_1^3 = \mathbb{Q}\{x_3\} \to U_1^3 = \mathbb{Q}\{x_3\}$ as $\xi^3(x_3) = p_3x_3$ with $p_3 \in \mathbb{Q}$, then we will get the following equations:

$$p_{41} = p_{10}^3p_{12}\ ,\ p_{43} = p_{10}^2p_{12}^2\ ,\ p_{45} = p_{10}p_{12}^3\ ,$$

$$p_{119} = p_{41}p_{43}p_{10}^2p_{12}^3 = p_{10}^{12} = p_{12}^{10} = p_2^{60} = p_3^{40} \tag{3.2}$$

which have the following nontrivial solutions:

$$p_2 = p_3 = p_{12} = p_{41} = p_{45} = \pm 1 \quad , \quad p_{10} = p_{43} = p_{119} = 1.$$

which give 8 coherent automorphisms (seven of them are of order 2) which are:

$$(1,1,1,1,1,1,1,1),(1,1,1,1,1,1,1,-1),(1,1,-1,-1,1,-1,1,1),(1,1,-1,-1,1,-1,1,-1)$$
$$(-1,1,1,1,1,1,1,1),(-1,1,1,1,1,1,1,-1),(-1,1,-1,-1,1,-1,1,1),$$
$$(-1,1,-1,-1,1,-1,1,1,-1).$$

Hence $\mathcal{E}(\Lambda U_1,\delta_1)\cong\mathbb{Z}_2\oplus\mathbb{Z}_2\oplus\mathbb{Z}_2\oplus\mathbb{Z}_2$. Now let $(\Lambda U_2,\delta_2)$ be the mcca obtained from $(\Lambda U_1,\delta_1)$ by adding a new generator x_4 with $|x_4|=4$ and where the differential is $\delta_2(x_4)=0$, $\delta_2(z)=\delta_1(z)+x_4^{30}$ and $\delta_2=\delta_1$ on the other generators. If we write $\xi^4:U_2^4=\mathbb{Q}\{x_4\}\to U_2^4=\mathbb{Q}\{x_4\}$ as $\xi^4(x_4)=p_4x_4$ with $p_4\in\mathbb{Q}$. Then we find the same equations given in (3.2) and the relation $p_4^{30}=p_3^{40}$ which have the following nontrivial solutions:

$$p_2=p_3=p_4=p_{12}=p_{41}=p_{45}=\pm1\quad,\quad p_{10}=p_{43}=p_{119}=1.$$

Hence we get 16 coherent automorphisms of order 2. So $\mathcal{E}(\Lambda U_2,\delta_2)\cong\underset{2^4}{\oplus}\mathbb{Z}_2$ (2^4 copies of $\mathbb{Z}_2$).

Next $(\Lambda U_3,\delta_3)$ is the mcca obtained from $(\Lambda U_2,\delta_2)$ by adding a new generator x_5 with $|x_5|=5$ and where the differential is $\delta_3(x_5)=0$, $\delta_3(z)=\delta_2(z)+x_5^{20}$ and $\delta_3=\delta_2$ on the other generators. If we write $\xi^5:U_3^5=\mathbb{Q}\{x_5\}\to U_3^5=\mathbb{Q}\{x_5\}$ as $\xi^5(x_5)=p_5x_5$ with $p_5\in\mathbb{Q}$, Then we find the same equations given in (3.2) and the relations $p_5^{20}=p_4^{30}$ which have the following nontrivial solutions:

$$p_2=p_3=p_4=p_5=p_{12}=p_{41}=p_{45}=\pm1\quad,\quad p_{10}=p_{43}=p_{119}=1.$$

So we get 32 coherent automorphisms of order 2 and $\mathcal{E}(\Lambda U_3,\delta_3)\cong\underset{2^5}{\oplus}\mathbb{Z}_2$.

Now define the following mccas:

$(\Lambda U_4,\delta_4)$ is the mcca obtained from $(\Lambda U_3,\delta_3)$ by adding a new generator x_6 with $|x_6|=6$ and where the differential is $\delta_4(x_6)=0$, $\delta_4(z)=\delta_3(z)+x_6^{20}$ and $\delta_4=\delta_3$ on the other generators.

$(\Lambda U_5,\delta_5)$ is the mcca obtained from $(\Lambda U_4,\delta_4)$ by adding a new generator x_{15} with $|x_{15}|=15$ and where the differential is $\delta_5(x_{15})=0$, $\delta_5(z)=\delta_4(z)+x_{15}^{8}$ and $\delta_5=\delta_4$ on the other generators.

$(\Lambda U_6,\delta_6)$ is the mcca obtained from $(\Lambda U_5,\delta_5)$ by adding a new generator x_{20} with $|x_{20}|=20$ and where the differential is $\delta_6(x_{20})=0$, $\delta_6(z)=\delta_5(z)+x_{20}^{6}$ and $\delta_6=\delta_5$ on the other generators.

$(\Lambda U_7,\delta_7)$ is the mcca obtained from $(\Lambda U_6,\delta_6)$ by adding a new generator x_{30} with $|x_{30}|=30$ and where the differential is $\delta_7(x_6)=0$, $\delta_7(z)=\delta_6(z)+x_{30}^{4}$ and $\delta_7=\delta_6$ on the other generators.

$(\Lambda U_8,\delta_8)$ is the mcca obtained from $(\Lambda U_7,\delta_7)$ by adding a new generator x_{60} with $|x_{60}|=60$ and where the differential is $\delta_7(x_{60})=0$, $\delta_8(z)=\delta_7(z)+x_{60}^{2}$ and $\delta_8=\delta_7$ on the other generators.

By the same arguments developed above we get:

$$\mathcal{E}(\Lambda U_4,\delta_4)\cong\underset{2^6}{\oplus}\mathbb{Z}_2\ ,\ \mathcal{E}(\Lambda U_5,\delta_5)\cong\underset{2^7}{\oplus}\mathbb{Z}_2\ ,\ \mathcal{E}(\Lambda U_6,\delta_6)\cong\underset{2^8}{\oplus}\mathbb{Z}_2$$

$$\mathcal{E}(\Lambda U_7, \delta_7) \cong \bigoplus_{2^9} \mathbb{Z}_2 \quad , \; \mathcal{E}(\Lambda U_8, \delta_8) \cong \bigoplus_{2^{10}} \mathbb{Z}_2.$$

Therefore the groups $\bigoplus_{2^n} \mathbb{Z}_2$ *are realizable for* $n \leqslant 10$.

Finally we end this work by asking the following question: Is it true that the groups $\bigoplus_{2^n} \mathbb{Z}_2$ are always realizable for $n \geqslant 11$?

References

[1] M. Arkowitz and G Lupton, Rational obstruction theory and rational homotopy sets, Math. Z, 235, N.3, 525-539, 2002.

[2] H.J. Baues, *Homotopy Type and Homology,* Oxford Mathematical Monographs, Oxford, 1996, 496p.

[3] M. Benkhalifa, On the classification problem of the quasi-isomorphism classes of 1-connected minimal free cochain algebras, Topology and its Applications 155, 1350-1370, 2008.

[4] M. Benkhalifa, On the classification problem of the quasi-isomorphism classes of free chain algebras, Journal of Pure and Applied Algebra 210, 343-362, 2007.

[5] Y. Felix, S. Halperin and J-C. Thomas *Rational homotopy theory,* Springer-Verlag, GTM 205, 2000

[6] J.H.C. Whitehead, A certain exact sequence, Ann. Math, 52:51-110, 1950.

This article may be accessed via WWW at `http://www.rmi.acnet.ge/jhrs/`

Mahmoud Benkhalifa
`makhalifa@uqu.edu.sa`

Department of Mathematics,
Faculty of Applied Sciences,
Umm Al-Qura University,
Mekka, Saudi Arabia

Journal of Homotopy and Related Structures, vol. 4(1), 2009, pp.123–151

OCHA AND THE SWISS-CHEESE OPERAD

EDUARDO HOEFEL

(*communicated by James Stasheff*)

Abstract

In this paper we show that the relation between Kajiura-Stasheff's OCHA and A. Voronov's swiss-cheese operad is analogous to the relation between SH Lie algebras and the little discs operad. More precisely, we show that the OCHA operad is quasi-isomorphic to the operad generated by the top-dimensional homology classes of the swiss-cheese operad.

Introduction

OCHA refers to the homotopy algebra of open and closed strings introduced by Kajiura and Stasheff [15] inspired by the work of Zwiebach on string field theory [35]. In [15] the *A_∞-algebras over L_∞-algebras* are also introduced, they are the strong homotopy version of $\mathfrak{g}$-algebras (or Leibniz pairs, see [8]). An OCHA is a structure obtained by adding other operations to an A_∞-algebra over an L_∞-algebra. The physical meaning of those additional operations is given by the "opening of a closed string into an open one".

Considering that its relevance to Physics is well acknowledged (see also [16]), in the present paper we further explore the *mathematical significance* of a full OCHA, not restricted to an A_∞-algebra over an L_∞-algebra. In [13] we have proven that any degree one coderivation $D \in \mathrm{Coder}(S^cL \otimes T^cA)$ such that $D^2 = 0$ defines an OCHA structure on the pair (L, A). In this work we study the relation between OCHA's and A. Voronov's swiss-cheese operad and show that it is analogous to the relation between SH Lie algebras and the little discs operad. A graded Lie algebra is part of the structure of a Gerstenhaber Algebra, which in turn is equivalent to an algebra over the homology little discs operad. The Lie part of a Gerstenhaber algebra is given by the top-dimensional homology classes of the little discs operad.

We will study the suboperad of the homology swiss-cheese operad generated by top-dimensional homology classes and show that it is quasi-isomorphic to the operad

---

The author wishes to thank Jim Stasheff and Murray Gerstenhaber for the kind hospitality during his stay as a visiting graduate student at the University of Pennsylvania (CNPq-Brasil grant SWE-201064/04). We are also grateful to J. Stasheff and H. Kajiura for valuable discussions. Some results of this paper consist of strengthened versions of some of the results present in the author's Ph.D. thesis [14] defended in 2006 at Unicamp under the supervision of A. Rigas and T. E. Barros.
Received February 05, 2008, revised January 01, 2009; published on June 15, 2009.
2000 Mathematics Subject Classification: 18G55, 18D50.
Key words and phrases: Coloured operads, Axelrod-Singer compactification, homotopical algebra, resolutions of operads.

whose algebras are OCHAs. The quasi-isomorphism, however, is not of operads but only a quasi-isomorphism of modules over the operad $\mathcal{L}_\infty$ of L_∞-algebras.

Let $\mathcal{OC}_\infty$ be the OCHA operad and let $\mathcal{OC}$ denote the suboperad of the homology swiss-cheese operad generated by top-dimensional homology classes. Our main result is the following.

Theorem. *There is a morphism of differential graded $\mathcal{L}_\infty$-modules $\mu : \mathcal{OC}_\infty \longrightarrow \mathcal{OC}$ which induces an isomorphism in cohomology.*

The paper is organized as follows. In section 1 we briefly review F. Cohen's theorem on the homology of the little discs operad and state it using trees. Section 2 reviews the analogous description of the homology swiss-cheese operad in terms of generators and relations given by trees. In section 3 we define OCHA in a grading and signs convention which is different from the original one in [15]. The definition given here is appropriate for studying its correspondence with the compactified configuration space of points on the closed upper half plane. We show that both definitions are equivalent through the (de)suspension operator. A definition of the OCHA operad $\mathcal{OC}_\infty$ is provided in section 4 using the partially planar trees, a type of tree which is defined in the same section. Section 5 reviews the construction of the compactified configuration spaces $\overline{C(p,q)}$ first introduced by Kontsevich in [19]. The combinatorial structure of its boundary strata is described in terms of partially planar trees and some examples are provided. The well known equivalence between $\overline{C(1,q)}$ and the cyclohedron W_{q+1} is explained in terms of those trees. In section 6 we prove the quasi-isomorphism between $\mathcal{OC}_\infty$ and $\mathcal{OC}$ viewed as modules over the operad $\mathcal{L}_\infty$ of L_∞ algebras. The main tool used in its proof is the spectral sequence of $\overline{C(p,q)}$ as a manifold with corners.

Notation and Conventions

Let us fix a field k of characteristic zero. In this paper, all vector spaces are over k and 'graded vector space' will always mean '$\mathbb{Z}$-graded vector space', unless otherwise stated. Let V be a graded vector space, we define a left action of the symmetric group S_n on $V^{\otimes n}$ in the following way: if $\tau \in S_2$ is a transposition, then the action is given by $x_1 \otimes x_2 \overset{\tau}{\mapsto} (-1)^{|x_1||x_2|} x_2 \otimes x_1$. Since any $\sigma \in S_n$ is a composition of transpositions, the sign of the action of σ on $V^{\otimes n}$ is well defined:

$$x_1 \otimes \cdots \otimes x_n \overset{\sigma}{\mapsto} \epsilon(\sigma) x_{\sigma(1)} \otimes \cdots \otimes x_{\sigma(n)}. \tag{1}$$

We will refer to $\epsilon(\sigma)$ as the Koszul sign of the permutation. Let us define $\chi(\sigma) = (-1)^\sigma \epsilon(\sigma)$, where $(-1)^\sigma$ is the sign of the permutation.

Given two homogeneous maps $f, g : V \to W$ between graded vector spaces, according to the Koszul sign convention (which will be used throughout this work), we have:

$$(f \otimes g)(v_1 \otimes v_2) = (-1)^{|g||v_1|}(f(v_1) \otimes g(v_2)). \tag{2}$$

We will use the notation of Lada-Markl [20] for the suspension and desuspension operators: $\uparrow$ and $\downarrow$. Let $\uparrow V$ (resp. $\downarrow V$) denote the suspension (resp. desuspension) of the graded vector space V defined by: $(\uparrow V)^p = V^{p-1}$ (resp. $(\downarrow V)^p = V^{p+1}$). We thus have the natural maps $\uparrow : V \to \uparrow V$ of degree 1, and $\downarrow : V \to \downarrow V$ of degree -1. Let

$\uparrow^{\otimes n}$ denote $\bigotimes^n \uparrow\colon \bigotimes^n V \to \bigotimes^n \uparrow V$ and $\downarrow^{\otimes n}$ is defined analogously. The operators $\uparrow^{\otimes n}$ and $\downarrow^{\otimes n}$ transform symmetric operations into anti-symmetric ones. In fact, let E (resp. A) denote the symmetric (resp. anti-symmetric) left action of the group of permutations S_n on $V^{\otimes n}$:

$$E(\sigma)(x_1 \otimes \cdots \otimes x_n) = \epsilon(\sigma) x_{\sigma(1)} \otimes \cdots \otimes x_{\sigma(n)} \tag{3}$$

$$A(\sigma)(x_1 \otimes \cdots \otimes x_n) = \chi(\sigma) x_{\sigma(1)} \otimes \cdots \otimes x_{\sigma(n)} \tag{4}$$

Both actions are related by: $\uparrow^{\otimes n} E(\sigma) \downarrow^{\otimes n} = (-1)^{n(n-1)/2} A(\sigma)$, for any $\sigma \in S_n$. In particular, $\uparrow^{\otimes n} \circ \downarrow^{\otimes n} = (-1)^{n(n-1)/2} \cdot \mathbb{1}$. The sign $(-1)^{n(n-1)/2}$ is a consequence of the Koszul sign convention (2) defined above (see also [7]).

Let us now describe how the notation for operads and its related concepts (such as: representations, ideals and modules) will be used in this paper. Our description will not necessarily include precise definitions. Those can be found in [24]. An operad is any sequence $\mathcal{O} = \{\mathcal{O}(n)\}_{n \geqslant 1}$ of objects in a symmetric monoidal category (such as the category of topological spaces or the category of vector spaces) endowed with a right action of the symmetric group S_n on each $\mathcal{O}(n)$ and a composition law satisfying natural associative and equivariance conditions.

Given a graded vector space V, the endomorphism operad of V is defined as $\mathrm{End}_V(n) = \mathrm{Hom}(V^{\otimes n}, V)$. The composition law $\circ_i$ in End_V is defined by the usual composition in the ith variable of multilinear maps and the right action of S_n on $\mathrm{Hom}(V^{\otimes n}, V)$ is the composition with the symmetric left action E defined by (3). In particular, this means that for graded vector spaces, according to our conventions, 'symmetric' always mean 'graded symmetric'.

Among the standard examples of operads are those defined in terms of *trees*. In this paper, in accordance with [10], trees are oriented and not necessarily compact: an edge may be terminated by a vertex at only one end (or none). Such an edge is called *external*. An external edge oriented toward its vertex is called a *leaf*, otherwise it is called *the root*. Trees are assumed to have only one root. The leaves of each tree are labeled by natural numbers. The action of S_n on trees with n leaves is defined by permuting the labels. The composition law $\circ_i$ on operads defined in terms of trees is given by the grafting operation, i.e., the identification of the root of one tree with the leaf labelled i of the other tree.

We also need to mention the *coloured operads*, a concept that goes back to Boardman and Vogt [3]. Following the notation of Berger and Moerdijk [2], given a set of colours C, a C-coloured operad $\mathcal{P}$ is defined by assigning to each $(n+1)$-tuple of colours $(c_1, \ldots, c_n; c)$ an object

$$\mathcal{P}(c_1, \ldots, c_n; c) \qquad \text{in some monoidal category}$$

endowed with a composition law and a symmetric group action. The defining conditions for coloured operads are analogous to those for ordinary operads.

Given a family $A = \{A_c\}_{c \in C}$ of vector spaces indexed by C, the C-coloured operad $\mathrm{End}(A)$ is defined by:

$$\mathrm{End}(A)(c_1, \ldots, c_n; c) = \mathrm{Hom}(A_{c_1} \otimes \cdots \otimes A_{c_n}, A_c). \tag{5}$$

For coloured operads, the composition law is only defined when the colour of the output coincides with the colour of the input. Another example of a coloured operad is given by trees with coloured edges, i.e., trees such that for each edge is assigned a element in some set C. For trees with coloured edges, the grafting operation is only defined when the colour of the root coincides with the colour of the corresponding leaf. Coloured trees will be used throughout the present paper. In fact, we will use 2-coloured trees where the colours of the edges are wiggly or straight, according to the notation used in [15].

Let $\mathcal{P}$ be an operad and let $M = \{M(n)\}_{n\geqslant 1}$ be a sequence of objects where each $M(n)$ has a right S_n-action. We say that M is a left $\mathcal{P}$-module if it is endowed with a left 'operadic action' $\circ_i^\lambda$:

$$\circ_i^\lambda : \mathcal{P}(n) \otimes M(m) \to M(m+n-1) \tag{6}$$

which is equivariant and satisfies associative conditions analogous to those in the definition of operads. The definition of right modules is similar.

An *ideal* in an operad $\mathcal{P}$ is a sequence of objects $\mathcal{I} = \{I(n)\}_{n\geqslant 1}$ with $I(n) \subseteq \mathcal{P}(n)$, where each $I(n)$ is invariant under the action of S_n and $\mathcal{I}$ is a left and right $\mathcal{P}$-module. We refer the reader to [24] for the precise definitions and further details about these concepts.

1. The homology little disks operad

We begin by recalling the description of the homology little discs operad $H_\bullet(\mathcal{D})$ in terms of generators and relations. To keep our notation in accordance with [15], we will represent classes in $H_\bullet(\mathcal{D})$ by trees with wiggly edges. As usual in operad theory, all trees are assumed to be *rooted* and *oriented* toward the root. A tree with only one vertex and n incoming edges is called an n-corolla.

The little disks operad (also called little 2-disks operad) is a sequence $\mathcal{D} = \{\mathcal{D}(n)\}_{n\geqslant 1}$ of topological spaces $\mathcal{D}(n)$ defined as the space of all maps

$$d : \coprod_{1\leqslant s\leqslant n} D_s \to D$$

from the disjoint union of n numbered standard two dimensional disks $D_1, \ldots, D_n$ to D, where

$$D := \{(x_1, x_2) \in \mathbb{R}^2 \mid x_1^2 + x_2^2 \leqslant 1\},$$

such that d, when restricted to each disk, is a composition of translation and multiplication by a positive real number and the images of the interiors of the disks are disjoint. The symmetric group acts by renumbering the disks. We may interpret $d \in \mathcal{D}(n)$ as the standard disk D with n numbered disjoint circular holes. The operad composition $\gamma_{\mathcal{D}}(d; d_1, \ldots, d_n)$ is, intuitively speaking, given by gluing n disks $d_1, \ldots, d_n$ in the holes of the disk with holes d, and erasing the seams (see [24, 25] for more precise definitions).

Since $\mathcal{D}(2)$ is homotopy equivalent to S^1, its homology has two generators. The zero dimensional generator will be denoted by a 2-corolla with a "white" vertex: $\circ$, while the one dimensional generator will be denoted by a 2-corolla with a "black" vertex: $\bullet$. Let

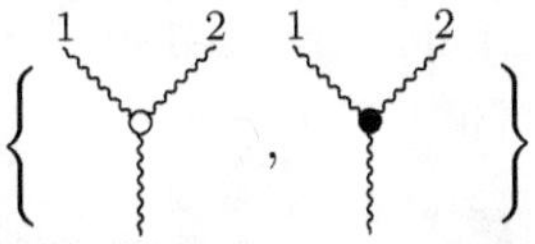

be a basis for $H_\bullet(\mathcal{D}(2))$, where the first generator has degree zero, the second has degree one and both are invariant under the action of the permutation group. An algebra over $H_\bullet(\mathcal{D})$ will thus have two graded symmetric operations. Notice however that the bracket defined below by (9) is *skew* graded symmetric, as required by the definition of a Gerstenhaber algebra. F. Cohen's famous result about the homology of $\mathcal{D}$ can be stated in the language of trees, as follows.

Theorem 1.1 (F. Cohen [5]). *The homology little disks operad $H_\bullet(\mathcal{D})$ is isomorphic, as an operad of $\mathbb{Z}$-graded vector spaces, to the operad generated by the above trees, subject to the following relations:*

i) Both generators are invariant under permutation of their labels;

ii) Jacobi identity:

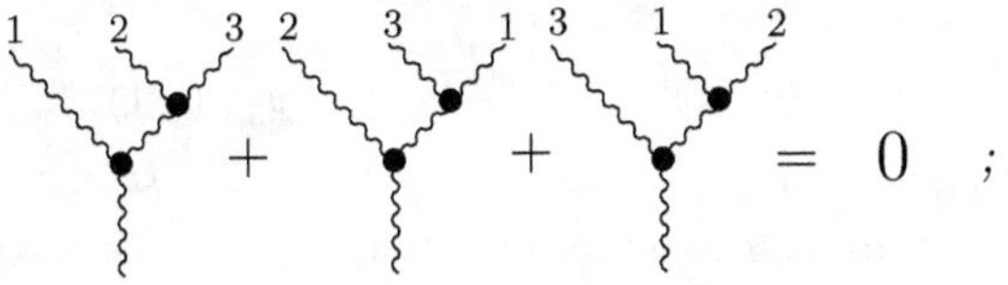

iii) Leibniz rule:

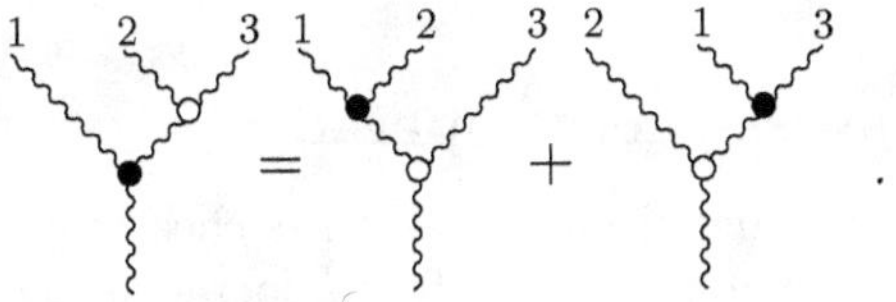

Let us now describe algebras over $H_\bullet(\mathcal{D})$. It is well known that algebras over $H_\bullet(\mathcal{D})$ are equivalent to Gerstenhaber algebras. However, a brief description of that equivalence will help clarify our exposition involving the swiss-cheese operad and its relations to OCHA.

Remember that V is an algebra over $H_\bullet(\mathcal{D})$ if there is a morphism of operads $\Phi : H_\bullet(\mathcal{D}) \to \mathrm{End}_V$ where $\mathrm{End}_V(n) = \{\mathrm{Hom}(V^{\otimes n}, V)\}$, is the endomorphism operad. Consequently, V is a graded vector space endowed with two graded symmetric operations $m_2 : V \otimes V \to V$ of degree $|m_2| = 0$, corresponding to the first generating tree, and $l_2 : V \otimes V \to V$ of degree $|l_2| = 1$ corresponding to the second generating tree.

The two relations presented above in terms of trees correspond to the equalities:

$$\sum_\sigma l_2 \circ (\mathbb{1} \otimes l_2) \circ E(\sigma) = 0 \tag{7}$$

and

$$l_2 \circ (\mathbb{1} \otimes m_2) = m_2 \circ (l_2 \otimes \mathbb{1}) + m_2 \circ (\mathbb{1} \otimes l_2) \circ E(\tau_{1,2}). \tag{8}$$

In the first equality, σ runs over all cyclic permutations and $E(\sigma)$ is defined by (3). In the second equality $\tau_{1,2}$ denotes the transposition (1 2). Notice that $E(\sigma)$ appears in both formulas above because, by definition, the right action of S_n on $\mathrm{Hom}(V^{\otimes n}, V)$ is given by composition with $E(\sigma)$. Given homogeneous elements $x, y, z \in V$, identities (7) and (8) are expressed by:

$$(-1)^{|x||z|} l_2(l_2(x,y),z) + (-1)^{|x||y|} l_2(l_2(y,z),x) + (-1)^{|y||z|} l_2(l_2(z,x),y) = 0$$

$$l_2(x, y \cdot z) = l_2(x,y) \cdot z + (-1)^{(|x|-1)|y|} y \cdot l_2(x,z)$$

where the dot product denotes m_2. Notice that the sign $(-1)^{(|x|-1)|y|}$ occurs in the second expression as a consequence of the transposition $\tau_{1,2}$ and the Koszul sign convention.

To see that an algebra over $H_\bullet(\mathcal{D})$ is equivalent to a Gerstenhaber algebra, we just need to define the bracket:

$$[x,y] := (-1)^{|x|} l_2(x,y) \tag{9}$$

it is not dificult to see that

$$[x,y] = -(-1)^{(|x|-1)(|y|-1)}[y,x].$$

and

$$(-1)^{(|x|-1)(|z|-1)}[[x,y],z] + (-1)^{(|x|-1)(|y|-1)}[[y,z],x] + (-1)^{(|y|-1)(|z|-1)}[[z,x],y] = 0.$$

This shows that the structure of an algebra over $H_\bullet(\mathcal{D})$ is equivalent to the structure of an Gertenhaber algebra on V with bracket defined by (9).

2. The Homology swiss-cheese operad

In this section we recall the definition of the swiss-cheese operad [32], denoted by $\mathcal{SC}$. We will present the homology swiss-cheese operad using 2-coloured trees. Harrelson [11] has presented similarly the homology of open-closed strings in the wider context of PROPs. The following presentation of the homology swiss-cheese operad is a particular case of Harrelson's open-closed homology PROP.

The swiss-cheese operad $\mathcal{SC}$ is a 2-coloured operad. We will use the initials of open and closed as our set of colours: $C = \{o, c\}$. For $m \geqslant 0, n \geqslant 0$ with $m + n \geqslant 1$, $\mathcal{SC}(m, n; o)$ is the configuration space of non-overlapping disks labeled 1 through m and upper semi-disks labeled 1 through n embedded by translations and dilations in the standard unit upper semi-disk so that the embedded semi-disks are all centered on the diameter of the big semi-disk.

For $m \geqslant 1$ and $n = 0$, $\mathcal{SC}(m, 0; c) = \mathcal{D}(m)$ is just the usual component of the little disks operad, and $\mathcal{SC}(m, n; c)$ is the empty set for $n \geqslant 1$.

Observation 2.1. *The components of the form $\mathcal{SC}(m, 0; o)$ were excluded in the original definition of the operad $\mathcal{SC}$, (see [32]), i.e., those components which have only discs as inputs and intervals as output were not to be considered in the original definition. Here, however, we shall keep the components $\mathcal{SC}(m, 0; o)$, $m \geqslant 1$, since they are crucial for the OCHA structure. In fact, as we will see in this paper, a* *zero dimensonal generator of the homology of $\mathcal{SC}(1,0;o)$ corresponds to the map $n_{1,0} : L \to A$. The physical meaning of $n_{1,0}$ being given by the "opening of a closed string into an open one", see* [15,16].

Let us now describe the homology swiss-cheese operad $H_\bullet(\mathcal{SC})$ in terms of generators and relations using trees. Since $H_\bullet(\mathcal{SC})$ is a 2-coloured operad, our trees must also be 2-coloured. The colours we use are wiggly and straight.

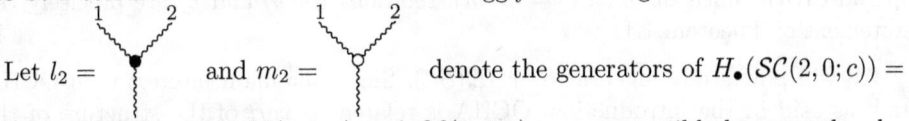

Let $l_2 =$ 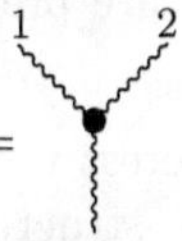 and $m_2 =$ 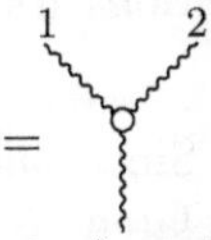 denote the generators of $H_\bullet(\mathcal{SC}(2,0;c)) = H_\bullet(\mathcal{D}(2))$. Both spaces $\mathcal{SC}(1,0;o)$ and $\mathcal{SC}(0,2;o)$ are contractible because the elements of $\mathcal{SC}(1,0;o)$ have only one interior disc while $\mathcal{SC}(0,2;o)$ is homeomorphic to $\mathcal{C}_1(2)$ (where $\mathcal{C}_1$ is the little intervals operad) which is well known to be contractible. Their degree zero homology generators will be denoted respectively by

$n_{1,0} =$  and $n_{0,2} =$ 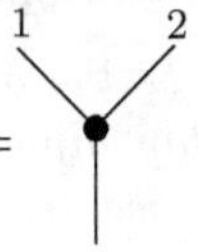 .

So, the degrees of the above generators are: $|l_2| = 1$ and $|m_2| = |n_{1,0}| = |n_{0,2}| = 0$. The analog of Cohen's theorem can be stated as follows.

Theorem 2.2. *The homology swiss-cheese operad $H_\bullet(\mathcal{SC})$ is isomorphic, as a $\mathbb{Z}$-graded 2-coloured operad, to the 2-coloured operad generated by: l_2, m_2, $n_{1,0}$ and $n_{0,2}$, satisfying the following relations:*

a) l_2 is invariant under permutation and satisfies the Jacobi identity;

b) m_2 is invariant under permutation and associative;

c) l_2 and m_2 satisfy the Leibniz rule;

d) $n_{0,2}$ is associative;

e) 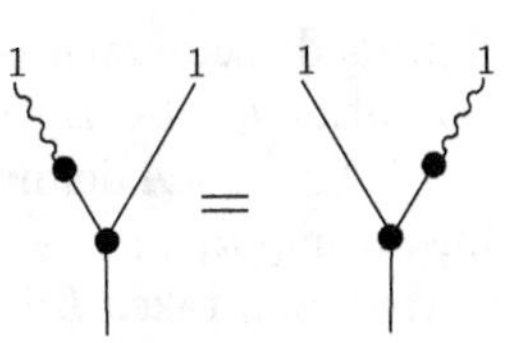 *and* 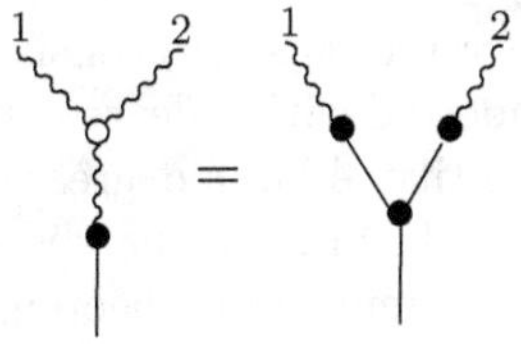 .

Observation 2.3. *Given a coloured tree, if it has k leaves of some colour c, then those leaves are labeled 1 to k. So, on the same tree we may have two leaves of different colours with the same label.*

Proof of Theorem 2.2. We first show that l_2, m_2, $n_{1,0}$ and $n_{0,2}$ generate the operad $H_\bullet(\mathcal{SC})$. In fact, $\mathcal{SC}(0,n;o)$ is homeomorphic to the little intervals operad $\mathcal{C}_1$ and hence $H_\bullet(\mathcal{SC}(0,n;o))$ is generated by $n_{0,2}$. The space $\mathcal{SC}(1,0;o)$ consists of configurations of one disk inside the standard semi disk and is also contractible, hence the need for the zero dimensional generator $n_{1,0}$. Finally observe that $\mathcal{SC}(m,n;o)$ is homotopy equivalent to $\mathcal{D}(m)\times S_n$, thus from Theorem 1.1 any class in $H_\bullet(\mathcal{SC}(m,n;o))$ is obtained from operadic composition of l_2, $n_{1,0}$ and $n_{0,2}$. In order to generate the full operad $H_\bullet(\mathcal{SC})$ we need to consider $\mathcal{SC}(m,0;c) = \mathcal{D}(m)$. But we already know that its homology is generated by l_2 and m_2 from Theorem 1.1.

We now show that the generators in fact satisfy the above relations. Relation *e)* involve only zero dimensional homology classes and one can easily check that the compositions indicated in *e)* belong to the same path component of the swiss-cheese. Item *d)* follows immediately from the fact that $\mathcal{SC}(0,n;o)$ is the little intervals operad $\mathcal{C}_1(n)$. Since $\mathcal{SC}(m,0;c) = \mathcal{D}(m)$, relations *a), b)* and *c)* are precisely the statement of Theorem 1.1. □

We will now study algebras over $H_\bullet(\mathcal{SC})$. Since our main interest is in OCHA and, as said in the introduction, OCHA is related to *part* of the structure of the homology swiss-cheese, let us define a suboperad of $H_\bullet(\mathcal{SC})$ containing the relevant structure.

Definition 2.4. *$\mathcal{OC}$ is the suboperad of $H_\bullet(\mathcal{SC})$ generated by l_2, $n_{1,0}$ and $n_{0,2}$. Algebras over $\mathcal{OC}$ will be called* open-closed algebras *or simply OC-*algebras.

Observation 2.5. *Let $\mathcal{L}$ be the operad defined by $\mathcal{L}(n) = H_{n-1}(\mathcal{D}(n))$ for $n \geqslant 1$, i.e., $\mathcal{L}$ is the operad of top dimensional homology classes of the little discs operad. From Theorem 1.1, we see that the operad $\mathcal{L}$ is generated by an element $l_2 \in H_1(\mathcal{D}(2))$ which is invariant under the symmetric group action and satisfies the Jacobi identity. Consequently, l_2 corresponds to a degree one graded commutative bilinear operation satisfying the Jacobi identity. Under operadic desuspension, the new generator will have degree zero, will be graded anti-commutative and will satisfy the Jacobi identity. In other words, the operadic desuspension $\mathfrak{s}^{-1}$ transforms $\mathcal{L}$ into the Lie operad: $\mathcal{L}ie = \mathfrak{s}^{-1}\mathcal{L}$ (see* [24] *for the definition of operadic (de)suspension). In this paper we refer to $\mathcal{L}$ as the* Lie operad *by "abus de langage".*

There is an analogy between the operads $\mathcal{OC}$ and $\mathcal{L}$ which is sumarized bellow:

$$\mathcal{OC} \iff \text{top-dimensional generators of } H_\bullet(\mathcal{SC})$$
$$\mathcal{L} \iff \text{top-dimensional generators of } H_\bullet(\mathcal{D}).$$

An algebra over $\mathcal{OC}$ (or OC-algebra) consists of a pair of $\mathbb{Z}$-graded vector spaces L and A such that L is endowed with a degree one symmetric operation $l_2 : L \otimes L \to L$ satisfying the Jacobi identity, A has a degree zero operation $m_2 : A \otimes A \to A$ defining a structure of associative algebra and there is a degree zero linear map $n_{1,0} : L \to A$. From the first identity in item e) of Theorem 2.2, it follows that $n_{1,0}$ takes L into the center of A.

3. Open-closed homotopy algebras

OCHA's were originally defined in a particular grading and signs convention where all multilinear maps have degree one and, after being lifted as a coderivation $D \in \mathrm{Coder}(S^c L \otimes T^c A)$, the OCHA axioms are translated into the single condition: $D^2 = 0$.

In order to study the relation between OCHA and the swiss-cheese operad, we need a definition where grading and signs are given by the corresponding compactified configuration space. More specifically, a definition where the degrees are equal to minus the dimension of the configuration space and the signs in the axioms are chosen so as to make them compatible with the boundary operator in the first row of the E^1 term of the spectral sequence of the compactified configuration space. In this section we present the definition in this geometrical setting. It is proven in the Appendix that both definitions are equivalent.

Let us first recall the definition of SH Lie [21] algebras in a grading and signs convention compatible with its compactified configuration space description (see [18, 31]).

Definition 3.1 (Strong Homotopy Lie algebra). *A strong homotopy Lie algebra (or L_∞-algebra) is a $\mathbb{Z}$-graded vector space V endowed with a collection of graded symmetric n-ary brackets $l_n : V^{\otimes n} \to V$, of degree $3-2n$ such that $l_1^2 = 0$ and for $n \geqslant 2$:*

$$\partial\, l_n(v_1, \dots, v_n) = \sum_{\substack{\sigma\in\Sigma_{k+l=n}\\ k\geqslant 2, l\geqslant 1}} \epsilon(\sigma)\, l_{1+l}(l_k(v_{\sigma(1)}, \dots, v_{\sigma(k)}), v_{\sigma(k+1)}, \dots, v_{\sigma(n)}) = 0 \tag{10}$$

where σ runs over all (k,l)-unshuffles, i.e., permutations $\sigma \in S_n$ such that $\sigma(i) < \sigma(j)$ for $1 \leqslant i < j \leqslant k$ and for $k+1 \leqslant i < j \leqslant k+l$.

Observation. *The operator ∂ in the above definition denotes the induced differential on the endomorphism complex, i.e.:*

$$\partial\, l_n = l_1 l_n + l_n(l_1 \otimes \mathbf{1} \otimes \cdots \otimes \mathbf{1} + \cdots + \mathbf{1} \otimes \cdots \otimes \mathbf{1} \otimes l_1).$$

Definition 3.2 (Open-Closed Homotopy Algebra – OCHA). *An OCHA consists of a 4-tuple $(L, A, \mathfrak{l}, \mathfrak{n})$ where L and A are $\mathbb{Z}$-graded vector spaces, $\mathfrak{l} = \{l_n : L^{\otimes n} \to L\}_{n\geqslant 1}$ and $\mathfrak{n} = \{n_{p,q} : L^{\otimes p} \otimes A^{\otimes q} \to A\}_{p+q\geqslant 1}$ are two families of multilinear maps where l_n has degree $3-2n$ and $n_{p,q}$ has degree $2-2p-q$, such that $(L, \mathfrak{l})$ is an L_∞-algebra and the two families satisfy the following compatibility condition:*

$$\begin{aligned}
&\partial\, n_{n,m}(v_1, \dots, v_n, a_1, \dots, a_m) = \\
&= \sum_{\sigma\in\Sigma_{p+r=n},\ p\geqslant 2} (-1)^{\epsilon(\sigma)} n_{1+r,m}(l_p(v_{\sigma(1)}, \dots, v_{\sigma(p)}), v_{\sigma(p+1)}, \dots, v_{\sigma(n)}, a_1, \dots, a_m) + \\
&\quad + \sum_{\substack{\sigma\in\Sigma_{p+r=n},\ i+j=m-s\\ (r,s)\neq(0,1),(n,m)}} (-1)^{\mu_{p,i}(\sigma)} n_{p,i+1+j}(v_{\sigma(1)}, .., v_{\sigma(p)}, a_1, .., a_i, n_{r,s}(v_{\sigma(p+1)}, .., v_{\sigma(n)}, \\
&\qquad\qquad a_{i+1}, .., a_{i+s}), a_{i+s+1}, .., a_m).
\end{aligned}$$

where $\mu_{p,i}(\sigma) = s + i + si + ms + \epsilon(\sigma) + s(v_{\sigma(1)} + \cdots + v_{\sigma(p)} + a_1 + \cdots + a_i) + (a_1 + \cdots + a_i)(v_{\sigma(i+1)}) + \cdots + v_{\sigma(n)})$.

Observation. *The operator ∂ in the above definition denotes the induced differential on the endomorphism complex, i.e.:*

$$\partial\, n_{n,m} = n_{0,1} n_{n,m} - (-1)^m\, n_{n,m}(d_{L^n} \otimes \mathbf{1}_A^{\otimes m} + \mathbf{1}_L^{\otimes n} \otimes d_{A^m})$$

where $d_{L^n} = l_1 \otimes \mathbf{1} \otimes \cdots \otimes \mathbf{1} + \cdots + \mathbf{1} \otimes \cdots \otimes \mathbf{1} \otimes l_1$ *and* $d_{A^m} = n_{0,1} \otimes \mathbf{1} \otimes \cdots \otimes \mathbf{1} + \cdots + \mathbf{1} \otimes \cdots \otimes \mathbf{1} \otimes n_{0,1}$.

It is convenient to have a shorthand expression for the OCHA relations:

$$\partial\, n_{n,m} = \sum_{\sigma\in\Sigma_{p+r=n},\ p\geqslant 2} n_{1+r,m}(l_p \otimes \mathbf{1}_L^{\otimes r} \otimes \mathbf{1}_A^{\otimes m})(E(\sigma)\otimes \mathbf{1}_A^{\otimes m}) +$$

$$+ \sum_{\substack{\sigma\in\Sigma_{p+r=n},\ i+j=m-s\\ (r,s)\neq(0,1),(n,m)}} (-1)^{s+i+si+ms} n_{p,i+1+j}(\mathbf{1}_L^{\otimes p}\otimes \mathbf{1}_A^{\otimes i}\otimes n_{r,s}\otimes \mathbf{1}_A^{\otimes j})(E(\sigma)\otimes \mathbf{1}_A^{\otimes m}) \quad (11)$$

where $E(\sigma)$ was defined by formula (3) on page 125. The complicated sign of the definition is absorbed in the above expression if we assume the following standard convention: given two maps $h_1, h_2 : V\otimes W \to U$, the tensor product $h_1\otimes h_2$ defined on $V^{\otimes^2}\otimes W^{\otimes^2}$ is given by: $(h_1\otimes h_2)((v_1\otimes v_2)\otimes(w_1\otimes w_2)) = (-1)^{|v_2||w_1|}h_1(v_1,w_1)\otimes h_2(v_2,w_2)$.

Example 3.3. *Here is a list of the first few OCHA relations:*

$$\partial\, n_{0,1} = 2\,(n_{0,1})^2 = 0 \quad (12)$$

$$\partial\, n_{1,1} = n_{0,2}(n_{1,0}\otimes \mathbb{1}_A) - n_{0,2}(\mathbb{1}_A\otimes n_{1,0}) \quad (13)$$

$$\partial\, n_{2,0} = n_{1,0}l_2 + n_{1,1}(\mathbb{1}_L\otimes n_{1,0}) + n_{1,1}(\mathbb{1}_L\otimes n_{1,0})E(\tau_{1,2}) \quad (14)$$

$$\partial\, n_{1,2} = n_{1,1}(\mathbf{1}_L\otimes n_{0,2}) - n_{0,2}(n_{1,1}\otimes \mathbf{1}_A) - n_{0,2}(\mathbf{1}_A\otimes n_{1,1}) +$$
$$+ n_{0,3}(n_{1,0}\otimes \mathbf{1}_A\otimes \mathbf{1}_A) - n_{0,3}(\mathbf{1}_A\otimes n_{1,0}\otimes \mathbf{1}_A) + n_{0,3}(\mathbf{1}_A\otimes \mathbf{1}_A\otimes n_{1,0}) \quad (15)$$

$$\partial\, n_{2,1} = n_{1,1}(l_2\otimes \mathbf{1}_A) + n_{1,1}(\mathbf{1}_L\otimes n_{1,1}) + n_{1,1}(\mathbf{1}_L\otimes n_{1,1})(E(1\ 2)\otimes \mathbf{1}_A) +$$
$$+ n_{0,2}(n_{2,0}\otimes \mathbf{1}_A) - n_{0,2}(\mathbf{1}_A\otimes n_{2,0}) + n_{1,2}(\mathbf{1}_L\otimes n_{1,0}\otimes \mathbf{1}_A) - n_{1,2}(\mathbf{1}_L\otimes \mathbf{1}_A\otimes n_{1,0}) +$$
$$+ n_{1,2}(\mathbf{1}_L\otimes n_{1,0}\otimes \mathbf{1}_A)(E(1\ 2)\otimes \mathbf{1}_A) - n_{1,2}(\mathbf{1}_L\otimes \mathbf{1}_A\otimes n_{1,0})(E(1\ 2)\otimes \mathbf{1}_A) \quad (16)$$

Relation (12) simply says that $n_{0,1}$ is a differential operator. On the other hand, (13) means that $n_{1,0}$ takes L into the homotopy center of A where $n_{1,1}$ is the homotopy operator. The configuration space corresponding to $n_{1,1}$ is the cyclohedron W_2 (see example 5.3). The configuration space corresponding to relation (14) is "The Eye" (Figure 4 pg.148). Relation (15) corresponds to the configuration space W_3 (Figure 3 pg.148). Finally, relation (16) corresponds to the configuration space illustrated by Figure 5 pg.149. If we consider an OCHA structure where the maps $n_{1,0}$ and $n_{2,0}$ are set equal to zero, then relations (15) and (16) together say that $n_{1,1} : L\otimes A \to A$ is a Lie algebra action by derivations up to homotopy.

It is a well known fact that A_∞ and L_∞ algebras can be defined both in the *geometrical setting* (where the degrees of the multilinear maps are minus the dimension of the corresponding configuration space) and the *algebraic setting* where all the maps have degree 1. It is also well known that both definitions are equivalent through the (de)suspension operator. The same is true for OCHA.

Proposition 3.4. *An OCHA structure* $(L, A, \mathfrak{l}, \mathfrak{n})$, *in the grading and signs conventions of defintion 3.2, is equivalent to a degree one coderivation* $D \in \mathrm{Coder}(S^c(\downarrow\downarrow L) \otimes T^c(\downarrow A))$ *such that* $D^2 = 0$.

The proof of this fact amounts to an appropriate use of the Lada-Markl notation for the suspension and desuspension operators $\uparrow$ and $\downarrow$ (see [20]) and is provided in the Appendix.

4. The OCHA operad $\mathcal{OC}_\infty$

In this section we study the operad $\mathcal{OC}_\infty$ whose algebras are precisely OCHA's as given in Definition 3.2. Our presentation of $\mathcal{OC}_\infty$ is slightly different from (but naturally equivalent to) the original definition in [15] because of the different conventions. We begin by defining the partially planar trees.

Definition 4.1. *A* partially planar tree *is an isotopy class of oriented rooted trees embedded in the euclidean 3 dimensional space* $\mathbb{R}^3$ *such that a fixed subset of edges is contained in the xy-plane. Planar edges will be denoted by straight lines, while spatial edges will be denoted by wiggly lines.*

Observation 4.2. *Partially planar trees have appeared in the work of Merkulov* [26]. *Merkulov, however, uses wiggly lines for planar edges and straight lines for spatial edges.*

The partially planar trees relevant for the definition of $\mathcal{OC}_\infty$ have a specific form we now begin to describe. We define l_n as the corolla which has n leaves and only spatial edges and $n_{p,q}$ as the corolla with planar root, p spatial leaves and q planar leaves. Leaves of different colours are labelled by different sets:

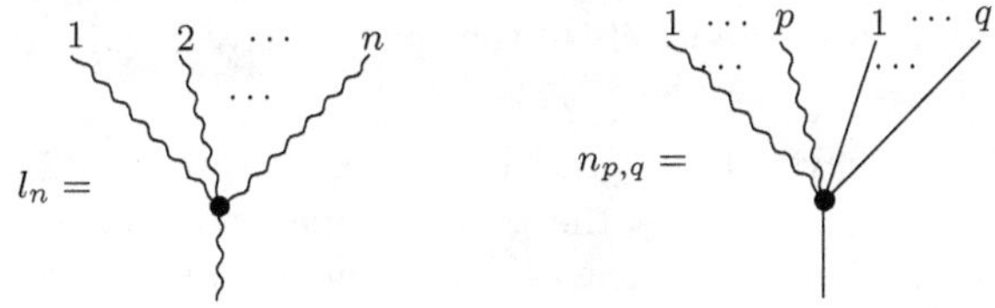

(17)

As mentioned in the introduction, the grafting operation of a tree T_2 on some leaf of a tree T_1 is only defined when the colour of the root of T_2 is equal to the colour of the corresponding leaf of T_1. The grafting of a tree T_2 with spatial root on the ith spatial leaf of some tree T_1 will be denoted by:

$$T_1 \circ_i T_2 \tag{18}$$

On the other hand, the grafting of a tree with planar root T_4 on the ith planar leaf of some tree T_3 will be denoted by:

$$T_3 \bullet_i T_4 \tag{19}$$

Consider the set of all corollas $n_{p,q}$ and l_n with $2p + q \geqslant 2$ and $n \geqslant 2$. Let $T(n)$ denote the set of all partially planar trees T with n leaves which can be obtained by grafting a finite number of corollas in the above set. Let $\mathcal{T}_o(p,q) \subseteq \mathcal{T}(p+q)$ denote the subset of trees with planar root having p spatial leaves and q planar leaves. Let $\mathcal{T}_c(n) \subseteq \mathcal{T}(n)$ be the subset of trees with spatial root.

Definition 4.3. *For $2p+q \geqslant 2$, we define $\mathcal{N}_\infty(p,q)$ as the vector space spanned by $\mathcal{T}_o(p,q)$ and for $n \geqslant 2$, $\mathcal{L}_\infty(n)$ is defined as the vector space spanned by $\mathcal{T}_c(n)$. The space $\mathcal{N}_\infty(0,1)$ is defined as the vector space spanned by the tree with only one planar edge and no vertices, while $\mathcal{L}_\infty(1)$ is defined similarly as the vector space spanned by the tree with only one spatial edge and no vertices.*

Observation 4.4. *Notice that if a tree in $\mathcal{T}(n)$ has a spatial root, then all of its edges must also be spatial because of the corollas we have chosen as generators.*

Let $|i(T)|$ be the number of internal edges of T, we define the degree of $T \in \mathcal{T}(n)$ as follows:

$$|T| = \begin{cases} |i(T)| + 2 - 2p - q, & \text{if } T \in \mathcal{T}_o(p,q) \\ |i(T)| + 3 - 2n, & \text{if } T \in \mathcal{T}_c(n) \end{cases} \tag{20}$$

in particular, $|n_{p,q}| = 2 - 2p - q$ and $|l_n| = 3 - 2n$. Now we can define the spaces: $\mathcal{N}_\infty = \bigoplus_{k+l\geqslant 1} \mathcal{N}(k,l)$ and $\mathcal{L}_\infty = \bigoplus_{n\geqslant 1} \mathcal{L}_\infty(n)$, and finally define:

$$\mathcal{OC}_\infty = \mathcal{L}_\infty \oplus \mathcal{N}_\infty. \tag{21}$$

There is a symmetric group action on spatial leaves by permuting the labels of the spatial leaves, and there is no symmetric group action on planar leaves. In other words, given a tree $T \in \mathcal{OC}_\infty$ with p spatial leaves and q planar leaves, the group S_p acts on T by permuting the labels of the spatial leaves, while the planar leaves remain fixed.

The space $\mathcal{OC}_\infty$ we have just defined has the structure of a 2-coloured operad of graded vector spaces defined by the grafting operations $\circ_i$ and $\bullet_i$ and by the symmetric group action on spatial leaves. Let us now define a differential operator $d : \mathcal{OC}_\infty \to \mathcal{OC}_\infty$. We proceed analogously to the definition given in [12] (see also [18]).

Let us first define the action of d on corollas l_n and $n_{p,q}$:

$$d\, l_n = \sum_{\substack{k+l=n+1 \\ k,l\geqslant 2}} \ \sum_{\substack{\text{unshuffles } \sigma: \\ \{1,2,\ldots,n\}=I_1\cup I_2 \\ \#I_1=k,\ \#I_2=l-1}} \tag{22}$$

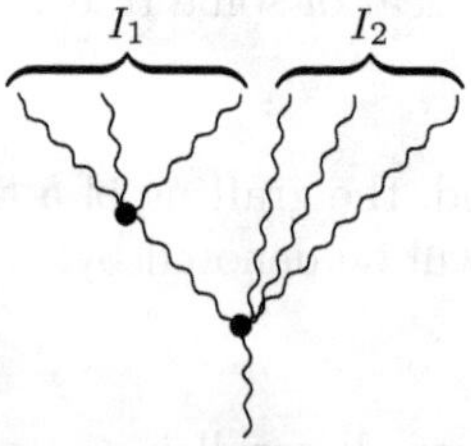

observing that an unshuffle σ is equivalent to a partition $(1,2,\ldots,n) = I_1 \cup I_2$ into two ordered subsets I_1 and I_2. On the other hand: $d\, n_{n,m} =$

$$= \sum_{\substack{k+l=n+1\\ k,l\geqslant 2}} \; \sum_{\substack{\text{unshuffles } \sigma:\\ \{1,2,\dots,n\}=I_1\cup I_2\\ \#I_1=k,\ \#I_2=l-1}} \Bigg($$

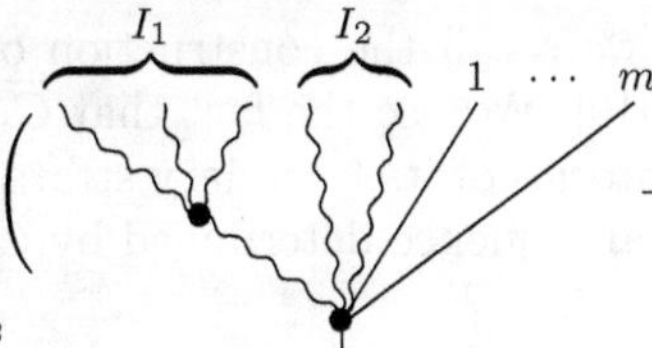

$$+$$

$$\sum_{0\leqslant i,s\leqslant m} (-1)^{s+i+si+ms}$$

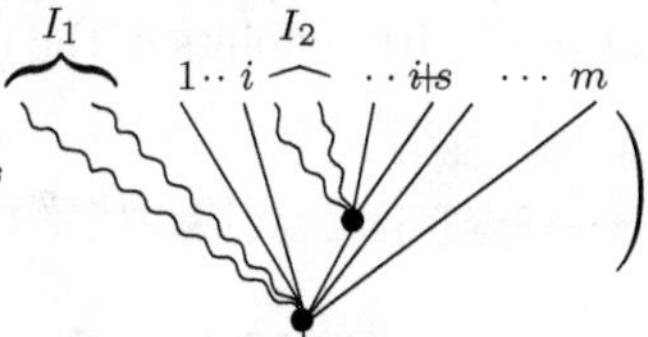

$$\Bigg). \qquad (23)$$

Once d is defined on the generators of $\mathcal{OC}_\infty$, it is extended to the whole operad by the Leibniz rule:

$$d\,(T\circ_i T_1) = d\,T\circ_i T_1 + (-1)^{|T|} T\circ_i d\,T_1 \qquad d\,(T\bullet_i T_2) = d\,T\bullet_i T_2 + (-1)^{|T|} T\ \bullet_i d\,T_2$$

where: T_1 is a tree with spatial root and T_2 is a tree with planar root. With the operator d, $\mathcal{OC}_\infty$ becomes a differential graded 2-coloured operad.

Observation 4.5. *For trees in $\mathcal{OC}_\infty$, let $T' \to T$ indicate that T is obtained from T' by contracting a spatial or planar internal edge. The above defined differential operator $d : \mathcal{OC}_\infty \to \mathcal{OC}_\infty$ is, up to sign, simply given by:*

$$d(T) = \sum_{T'\to T} \pm T',$$

the only difference between the above definition and the original one in [15] is the sign.

According to the grading defined by (20), the operator d has degree 1. So $\mathcal{OC}_\infty$ is an operad of cochain complexes.

Given two differential graded vector spaces L and A, we say that (L, A) is an algebra over $\mathcal{OC}_\infty$ if there is a morphism of differential graded 2-coloured operads:

$$\Psi : \mathcal{OC}_\infty \to \mathrm{End}_{L,A}$$

where $\mathrm{End}_{L,A}$ is the 2-coloured endomorphism operad of the pair (L, A), as described by (5). Since Ψ is a chain map and the differential operator ∂ on $\mathrm{End}_{L,A}$ is precisely the one used in formulas (10) and (11), it follows that (L, A) is an algebra over $\mathcal{OC}_\infty$ if, and only if, it admits the structure of an OCHA.

Observation 4.6. *By definition, $\mathcal{OC}_\infty = \mathcal{L}_\infty \oplus \mathcal{N}_\infty$, where $\mathcal{N}_\infty$ is spanned by trees with planar root. For trees in $\mathcal{OC}_\infty$, grafting two trees T_1 and T_2 where at least one of them has a planar root always results in a tree with a planar root. So, $\mathcal{N}_\infty$ is an ideal in $\mathcal{OC}_\infty$. On the other hand, $\mathcal{L}_\infty$ is a suboperad of $\mathcal{OC}_\infty$, since trees with spatial root can only have spatial edges. Finally, we observe that $\mathcal{OC}_\infty$ has a strucutre of module over $\mathcal{L}_\infty$ given by the grafting operation in $\mathcal{OC}_\infty$.*

5. The compactification $\overline{C(p,q)}$

In this section we recall the construction of the space $\overline{C(p,q)}$, first introduced by Kontsevich in [19]. We use the fact that $\overline{C(p,q)}$ is a manifold with corners and study the combinatorics of its boundary strata to show that the first row of the E^1 term of the spectral sequence determined by $\overline{C(p,q)}$ is isomorphic, as a differential complex, to $\mathcal{N}_\infty(p,q)$.

Let p, q be non-negative integers satisfying the inequality $2p+q \geqslant 2$. We denote by $\mathrm{Conf}(p,q)$ the configuration space of marked points on the upper half plane $H = \{z \in \mathbb{C} \mid \mathrm{Im}(z) \geqslant 0\}$ with p points in the interior and q points on the boundary (real line):

$$\mathrm{Conf}(p,q) = \{(z_1,\dots,z_p,x_1,\dots,x_q) \in H^{p+q} \mid z_{i_1} \neq z_{i_2},\, x_{j_1} \neq x_{j_2}\ \forall i_1 \neq i_2, j_1 \neq j_2 \\ \mathrm{Im}(z_i) > 0,\, \mathrm{Im}(x_j) = 0\ \forall i,j\}$$

The above configuration space $\mathrm{Conf}(p,q)$ is the cartesian product of an open subset of H^p and an open subset of $\mathbb{R}^q$ and, consequently, is a $2p+q$ dimensional smooth manifold. Let $C(p,q)$ be the quotient of $\mathrm{Conf}(p,q)$ by the action of the group of orientation preserving affine transformations that leaves the real line fixed:

$$C(p,q) = \mathrm{Conf}(p,q)\big/(z \mapsto az+b) \quad a,b \in \mathbb{R},\ a > 0.$$

The condition $2p+q \geqslant 2$ ensures that the action is free and thus $C(p,q)$ is a $2p+q-2$ dimensional smooth manifold.

Let $\mathrm{Conf}_n(\mathbb{C})$ be the configuration space of n points in the complex plane. We take the quotient by affine transformations $z \mapsto az+b$ where $a \in \mathbb{R}$, $a > 0$ and $b \in \mathbb{C}$ and define $C(n) := \mathrm{Conf}_n(\mathbb{C})/(z \mapsto az+b)$. Again $C(n)$ is a smooth manifold. The real version of the Fulton-MacPherson compactification $\overline{C(n)}$ is defined in the usual way, see [1,9,27]. Let ϕ be the embedding:

$$\phi : C(p,q) \longrightarrow C(2p+q) \tag{24}$$

defined by $\phi(z_1,\dots,z_p,x_1,\dots,x_q) = (z_1,\bar{z}_1,\dots,z_p,\bar{z}_p,x_1,\dots,x_q)$, where $\bar{z}$ denotes complex conjugation.

Definition 5.1. *The compactification of the configuration space $C(p,q)$ is defined as the closure in $\overline{C(2p+q)}$ of the image of ϕ. It will be denoted by $\overline{C(p,q)}$.*

The compactification $\overline{C(n)}$ of points in the plane can be intuitivelly described through "*bubbling offs*" on the sphere (the one point compactification of the plane). In the case of $\overline{C(p,q)}$, one can think of the closed disc as the one point compactification of the upper half plane and think of the embedding ϕ as taking the closed disc to the upper hemisphere of the above sphere. Punctures in the bulk of the disc are reflected through the equator. Points in $\overline{C(p,q)}$ can be intuitivelly described through "*bubbling offs*" on the disc. Those bubbling offs are pictured in the next figures.

5.1. The Stratification of $\overline{C(p,q)}$

The combinatorics of the compactification $\overline{C(n)}$ of the configuration space of points in the complex plane is well known to be described in terms of trees. In other

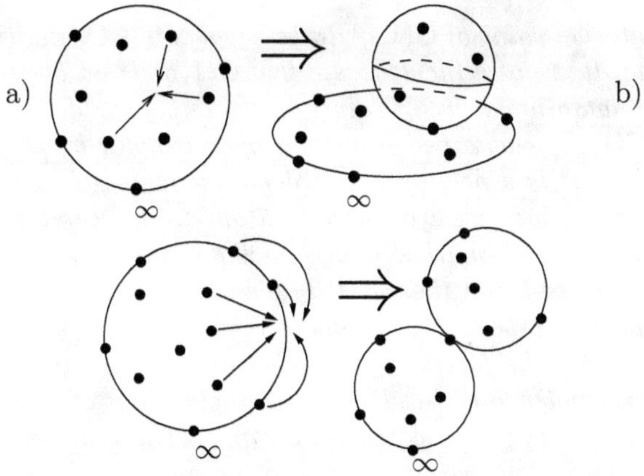

Figure 1: The two possible types of bubbling off on the closed disc.

words, its boundary strata can be labeled by trees (see: [1, 9, 18, 27, 33, 34]). Since $\overline{C(p,q)}$ was defined through the embedding $\phi : C(p,q) \to C(2p+q)$, it naturally inherits its combinatorics from that of $\overline{C(2p+q)}$. Leaves corresponding to the p points in the bulk of the upper half plane are spatial, while leaves corresponding to the q points on the boundary (real line) are planar. The combinatorics of $\overline{C(p,q)}$ is thus described by partially planar trees. We follow the notation of [18] and state this fact in the following theorem.

Theorem 5.2. *There is a stratification of* $\overline{C(p,q)}$ *such that:*

(1) $\overline{C(p,q)} = \coprod_{T \in \mathcal{T}_o(p,q)} S_T$. *Each stratum* S_T *is a smooth submanifold and* $\operatorname{codim}_{\mathbb{R}} S_T = |i(T)| =$ *number of internal edges of* T ;

(2) *there is a unique open stratum* $S_{n_{p,q}} = C(p,q) \quad 2p+q \geqslant 2$;

(3) *for each tree* $T \in \mathcal{T}_o(p,q)$ *we have the identity*

$$S_T = S_{n_{p_1,q_1}} \times S_{\delta_1} \times \cdots \times S_{\delta_n}$$

where each δ_i *is a corolla of the form* $n_{k,l}$ *or* l_k, *and* T *is obtained by grafting the corollas* $\delta_1, \ldots, \delta_n$ *to* n_{p_1,q_1}.

(4) *The boundary of the closure* $\overline{S_T}$ *of each stratum is given by* $\partial\overline{S_T} = \bigcup_{T' \to T} \overline{S_{T'}}$, *where* $T' \to T$ *means that* T *is obtained from* T' *by contracting a internal edge.*

In case $p = 0$, the space $\overline{C(0,q)}$ is the associahedron K_q [28, 29]. The labeling of the boundary strata of $\overline{C(0,q)}$, in this case, reduces to the well known labeling of the facets of K_q by planar trees.

Example 5.3 ($\overline{C(1,q)}$ is the cyclohedron W_{q+1}). *The cyclohedron was introduced by Bott and Taubes [4] and received its name from Stasheff [30]. It is defined as the Fulton-MacPherson compactification of the configuration space of points on the circle* S^1 *modded out by the group of rotations* $SO(2) = S^1$. *The equivalence between* $\overline{C(1,q)}$ *and* W_{q+1} *will be described below in terms of partially planar trees using the above theorem.*

By fixing the interior point of $C(1,q)$ to be equal to $i \in \mathbb{C}$, the remaining points are on the real line. It is not difficult to see that $C(1,q)$ is an open simplex homeomorphic to the configuration space of points on the circle modded out by $SO(2)$. The compactification $\overline{C(1,q)}$ is obtained from that open simplex by performing iterated blow ups. Hence $\overline{C(1,q)}$ is a polytope. In order to show that $\overline{C(1,q)}$ and W_{q+1} are equivalent polytopes, we just need to establish a one-to-one correspondence between bracketings around the $q+1$ marked points on the circle (q points on the real line plus one point marked ∞) and the partially planar trees in $\mathcal{T}_o(1,q)$, showing also that the correspondence respects the incidence relations.

In fact, for any $q \geqslant 0$ the facets of the cyclohedron W_{q+1} are labeled by (i.e. are in one-to-one correspondence with) all the meaningful ways of inserting brackets in an expression of $q+1$ letters disposed on a circle. The codimension of the facet corresponding to a given bracketing is equal to the number of brackets inserted, as illustrated by Figure 3 *(see also Devadoss' paper* [6] *on the cyclohedra.).*

First recall that $\overline{C(1,q)}$ can be described as the compactified configuration space of points on the closed disc (the one point compactification of the upper half plane) with 1 point in the interior of the disc, q points on the boundary of the disc plus one boundary point marked as ∞. From the "bubbling off" description of points in the compactification, we know that each facet of codimension k in $\overline{C(1,q)}$ corresponds to $k+1$ discs joined at points in the boundary such that exactly one of those discs contains one point in the bulk while the others contain only points in the boundary.

Let us exhibit the correspondence between 'circular bracketings' and 'bubbling offs'. Consider a point in $\overline{C(1,q)}$ in the bubbling off description, as a number of discs joined at "double points". There is only one of these discs which contains the interior point, the remaining ones only contain points in the boundary. The correspondence goes as follows (see Figure 2):

1. *the disc containing the interior point corresponds to the circle;*
2. *points on the boundary of the disc containing the interior point correspond to points on the circle which are not inside any bracket;*
3. *a disc joined to the disc containing the interior point corresponds to a bracketing; two discs joined correspond to a bracketing inside another bracketing or to two disjoint bracketings, and so on.*

In order to get a tree from the bubbling off, we associate to the discs their dual graphs. According to the usual procedure, each disc correspond to a vertex; the point marked ∞ corresponds to the root; the double points correspond to the edges and the remaining marked points correspond to the leaves. Since the correspondence between bracketings on S^1 and joining discs is established, there follows the correspondence between bracketings on S^1 and trees in $\mathcal{T}_o(1,q)$ (see Figure 2). Since the facets of both polytopes are in a one-to-one correspondence compatible with their corresponding boundaries, it follows that $W_q = \overline{C(1,q)}$.

Figures 3 and 4 on page 148 illustrate the spaces $\overline{C(1,2)}$ and $\overline{C(2,0)}$. Figure 5 on page 149 is a portrait of $\overline{C(2,1)}$ due to S. Devadoss where we have included the partially planar trees corresponding to its codimension one boundary strata (see also [17]). The OCHA relations corresponding to the spaces $\overline{C(2,0)}$, $\overline{C(1,2)}$ and

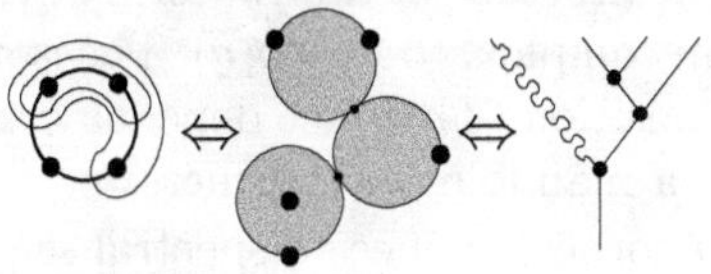

Figure 2: Example of the correspondence between circular bracketings and trees.

$\overline{C(2,1)}$ are given in formulas (14), (15) and (16).

5.2. The space $\overline{C(p)}$ as a deformation retract of $\overline{C(p,q)}$

We close the present section by pointing out a fact that will play a crucial role in the proof of our main theorem (Theorem 6.6). There is a stratum S_T in $\overline{C(p,q)}$ which is homeomorphic to $C(p)$, where T is the following tree:

$T =$ 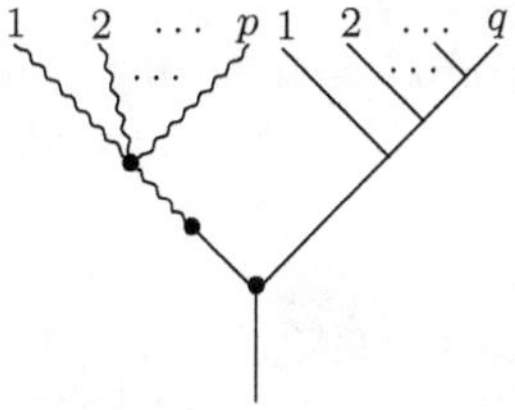 , (25)

moreover, S_T is a deformation retract of $\overline{C(p,q)}$.

In fact, by putting a collar neighborhood along the boundary, we see that there is a deformation retraction of $\overline{C(p,q)}$ onto $C(p,q) = \overline{C(p,q)} \setminus \partial\,\overline{C(p,q)}$. This last space was defined as the quotient of $\mathrm{Conf}(p,q)$ by the group of affine transformations $z \mapsto az + b$, where $a, b \in \mathbb{R}$. Since that group is contractible, $C(p,q)$ is homotopy equivalent to $\mathrm{Conf}(p,q)$. Now, $\mathrm{Conf}(p,q)$ is homeomorphic to $\mathrm{Conf}_{\mathbb{C}}(p) \times \mathrm{Conf}_{\mathbb{R}}(q)$ and $\mathrm{Conf}_{\mathbb{R}}(q)$ is well known to be contractible. Thus, by composing all those contractions and homotopy equivalences, we get the claimed deformation retraction of $\overline{C(p,q)}$ onto S_T. To see that S_T is in fact homeomorphic to $C(p)$, just notice that T is obtained by grafting the trees l_p, $n_{1,0}$ and a binary planar tree. The configuration space corresponding to $n_{1,0}$ and to any binary planar tree is just one point, the homeomorphism thus follows from Theorem 5.2. Notice that the retraction is essentialy determined by the contraction of $\mathrm{Conf}_{\mathbb{R}}(q)$ to a single point. This means that the configuration of the interior points are unaffected during the contraction of $\overline{C(p,q)}$ onto $\overline{S_T} = \overline{C(p)}$.

Those facts will be used in the next section along with the fact, proven by P. May in [25], that $C(p)$ is S_p-equivariantly homotopy equivalent to $\mathcal{D}(p)$, where $\mathcal{D}$ denotes the little discs operad.

6. OCHA and the spectral sequence of $\overline{C(p,q)}$

In this section we show that the first row of the E^1 term of the spectral sequence of $\overline{C(p,q)}$ is isomorphic, as a chain complex, to $\mathcal{N}_\infty(p,q)$. The isomorphism is natural with respect to the operad composition. This fact depends crucially on the study of the stratification of $\overline{C(p,q)}$ as a manifold with corners.

Every compact manifold with corners induces a spectral sequence converging to its homology. In fact, the boundary strata of the manifold induces a natural filtration on its singular chain complex which ensures the existence of the spectral sequence. Since the boundary filtration is finite, the spectral sequence is convergent.

Let us study the spectral sequence in the case of the manifold $\overline{C(p,q)}$. Consider the topological filtration of $\overline{C(p,q)}$:

$$F^i\overline{C(p,q)} = \{\text{closure of the union } \coprod_T S_T \text{ of strata of dimension } i\} =$$

$$\bigcup \{\overline{S_T} \mid \dim S_T = i\}$$

We will denote $F^i\overline{C(p,q)}$ more simply by F^i, with $2p+q-2 \geqslant i \geqslant 0$, remembering that the dimension of $C(p,q)$ is $2p+q-2$. The topological filtration induces a filtration on the singular chain complex of $\overline{C(p,q)}$ and we have the spectral sequence.

Theorem 6.1. *There is a spectral sequence $E^r_{m,n}$ converging to $H_*(C(p,q))$. Its E^1 term has the form $E^1_{m,n} = H_{m+n}(F^m, F^{m-1})$ and, for $n=0$, the complex*

$$0 \to E^1_{2p+q-2,0} \to \cdots \to E^1_{m,0} \to E^1_{m-1,0} \to \cdots \to E^1_{0,0} \to 0$$

is isomorphic to the p,q component $\mathcal{N}_\infty(p,q)$ of the ideal $\mathcal{N}_\infty \triangleleft \mathcal{OC}_\infty$.

Proof. From Theorem 5.2, we have: $F^m \setminus F^{m-1} = \coprod_{|T|=-m} S_T$, where $|T|$ is the degree of T defined by (20). The equality $|T| = -m$ means that the number of internal edges of T is equal to the codimension of its corresponding submanifold S_T. Using the Lefshetz duality theorem:

$$E^1_{m,0} = H_m(F^m, F^{m-1}) = H^0(F^m \setminus F^{m-1}) = H^0(\coprod_{|T|=-m} S_T) = \bigoplus_{|T|=-m} k. \quad (26)$$

As a vector space, $E^1_{m,0}$ is thus exactly the vector space generated by trees of degree $-m$ in $\mathcal{N}_\infty(p,q)$. The spectral sequence is a homology spectral sequence, hence $|d^1| = -1$. On the other hand, $\mathcal{N}_\infty$ is an ideal in $\mathcal{OC}_\infty$ which is an operad of cochain complex. So the diferential d in the ideal $\mathcal{N}_\infty$ has degree 1. This is consistent with the fact, used in (26), that the degree of a tree is minus the dimension of its corresponding relative homology class.

Now we need to check that the differential d^1 on the first row of the spectral sequence coincides with d defined by formulas (22) and (4).

In fact, by Theorem 5.2 and the Lefshetz duality, each relative class in $H_m(F^m, F^{m-1})$ is given by the closure $\overline{S_T}$ of a submanifold S_T. To see that d^1 coincides with d, recall that the operator d^1 is given by the relativization of the boundary operator ∂:

$$H_{m-1}(F^{m-1})\xrightarrow{j_{m-1}}$$

$$H_m(F^m, F^{m-1})\xrightarrow{\partial_m} \xrightarrow{d^1_m} H_{m-1}(F^{m-1}, F^{m-2}).$$

Item 4 of Thm. 5.2, says that: $\partial\overline{S_T} = \bigcup_{T'\to T}\overline{S_{T'}}$. Observation 4.5 (pg. 135) implies that $d^1 = d$. □

Observation 6.2. *It is still necessary to check that the signs given by the coboundary operator in the spectral sequence coincide with the signs given in formulas (22) and (4). That is a somewhat tedious exercise which consists of comparing the orientation induced on the product of two oriented manifolds with the orientation induced by the operadic embedding of that product manifold into the boundary strata of other oriented manifold.*

6.1. The Quasi-isomorphism of $\mathcal{L}_\infty$-modules

According to Definition 2.4, the open-closed operad $\mathcal{OC}$ is the operad generated by top-dimensional homology classes of the swiss-cheese operad, i.e., $\mathcal{OC}$ is the suboperad of $H_\bullet(\mathcal{SC})$ generated by $n_{1,0}$, l_2 and $n_{0,2}$ (see Definition 2.4). Recall $\mathcal{L}$ is the suboperad of $H_\bullet(\mathcal{D})$ defined by $\mathcal{L}(n) = H_{n-1}(\mathcal{D}(n))$ for $n \geqslant 1$. We refer to $\mathcal{L}$ as the Lie operad, since algebras over it are equivalent to Lie algebras (see Observation 2.5 pg. 130). From the tree description of $H_\bullet(\mathcal{SC})$ given in section 2, we see that $\mathcal{OC}$ is a suboperad of $\mathcal{OC}_\infty$.

We know that $\mathcal{OC}_\infty$ is an $\mathcal{L}_\infty$-module (see Observation 4.6) and that $\mathcal{OC}$ is an $\mathcal{L}$-module, since $\mathcal{L}$ is an suboperad of $\mathcal{OC}$. Consequently, $\mathcal{OC}$ has a natural structure of $\mathcal{L}_\infty$-module induced by the well known quasi-isomorphism of operads: $\mu : \mathcal{L}_\infty \to \mathcal{L}$ defined by $\mu(l_2) = l_2$ and $\mu(l_n) = 0$ for $n \geqslant 3$ (see [22–24] for details).

Proposition AppendixB.1 in the Appendix says that there is a morphism of differential graded $\mathcal{L}_\infty$-modules extending the identity on $\mathcal{OC}$:

$$\mu : \mathcal{OC}_\infty \to \mathcal{OC}. \tag{27}$$

The $\mathcal{L}_\infty$-morphism μ vanishes on the corollae that are not in $\mathcal{OC}$. The restriction of μ to $\mathcal{L}_\infty$ coincides with the above mentioned quasi-isomorphism between $\mathcal{L}_\infty$ and $\mathcal{L}$. Since $\mathcal{OC}_\infty = \mathcal{L}_\infty \oplus \mathcal{N}_\infty$, in order to prove that μ is a quasi-isomorphism of $\mathcal{L}_\infty$-modules, we need to study the cohomology of the ideal $\mathcal{N}_\infty$(see corollary 6.5).

Let us begin by showing that, for any p, q such that $2p + q \geqslant 2$, the cohomology of $\mathcal{N}_\infty(p, q)$ is isomorphic, as $\mathbb{Z}$-graded vector spaces, to $H_\bullet(\mathcal{D}(p))$ (where $\mathcal{D}$ is the little discs operad). From Theorem 6.1, $\mathcal{N}_\infty(p, q)$ is isomorphic to the complex given by:

$$0 \longrightarrow H_{2p+q-2}(F^{2p+q-2}, F^{2p+q-3}) \longrightarrow \cdots$$

$$\cdots \longrightarrow H_m(F^m, F^{m-1}) \longrightarrow \cdots \longrightarrow H_0(F^0) \longrightarrow 0$$

where each F^i is the closure of the disjoint union of the i-dimensional strata of $\overline{C(p,q)}$.

Notice that $E^r_{m,n}$ is a homology spectral sequence converging to $H_*(C(p,q))$. In the proof of Theorem 6.1, we have seen that the above complex is isomorphic to a complex generated by trees whose degree is minus the degree of their corresponding relative homology classes. So, it is a cochain complex generated by trees, precisely: $\mathcal{N}_\infty(p,q)$. In what follows, we use this to establish a relation between the *cohomology* of $\mathcal{N}_\infty$ and the *homology* of the little disks operad.

Since each stratum is a smooth submanifold, it follows that F^i has the homotopy type of a CW complex. The manifold $\overline{C(p,q)}$ has thus the homotopy type of a CW complex X such that each skeleton X^i is homotopy equivalent to F^i. It is well known that for any CW complex, the map $H_n(X^n) \to H_n(X^{n+1}) \simeq H_n(X)$, induced by the inclusion $X^n \hookrightarrow X^{n+1}$, is surjective for all n. Consequently, the map

$$H_n(F^n) \to H_n(F^{n+1}) \simeq H_n(\overline{C(p,q)})$$

is also surjective. For the same reason, we have: $H_n(F^{n-1}) \simeq H_n(X^{n-1}) = 0$. Now, consider the usual commutative diagram:

$$\begin{array}{ccccccc}
 & 0 \longrightarrow & & H_n(F^{n+1}) \simeq H_n(\overline{C(p,q)}) & & & \\
 & & H_n(F^n) \xrightarrow{j_n} & & & & \\
\cdots \longrightarrow & H_{n+1}(F^{n+1}, F^n) & \xrightarrow[\partial_{n+1}]{d_{n+1}} & H_n(F^n, F^{n-1}) & \xrightarrow[\partial_n]{d_n} & H_{n-1}(F^{n-1}, F^{n-2}) & \longrightarrow \cdots, \\
 & & & & H_{n-1}(F^{n-1}) \xrightarrow{j_{n-1}} & & \\
 & & & 0 \longrightarrow & & &
\end{array}$$

since $H_n(F^n) \to H_n(F^{n+1}) \simeq H_n(\overline{C(p,q)})$ is surjective and $H_n(F^n) \xrightarrow{j_n} H_n(F^n, F^{n-1})$ is injective, from the exactness of the sequence of the pair (F^n, F^{n-1}) one can see that the nth cohomology group of the complex $\mathcal{N}_\infty(p,q)$ is isomorphic to $H_n(\overline{C(p,q)})$. As observed before, $\overline{C(p,q)}$ is homotopy equivalent to $\mathcal{D}(p)$, so the following lemma is proved.

Lemma 6.3. *$H^k(\mathcal{N}_\infty(p,q)) \simeq H_k(\mathcal{D}(p))$ for every $k \geqslant 0$ and p, q such that $2p+q \geqslant 2$.*

For any $q \geqslant 0$, consider the following sequence of vector spaces:

$$H^\bullet(\mathcal{N}_\infty(\_, q)) := \{H^\bullet(\mathcal{N}_\infty(p,q))\}_{p \geqslant 1}.$$

Since $\mathcal{L}$ is just the operad generated by a binary tree l_2 of degree 1 which is invariant under the action of the symmetric group S_2 and satisfies the Jacobi identity, there is a natural injection $\mathcal{L} \hookrightarrow H^\bullet(\mathcal{OC}_\infty)$. Since $\mathcal{N}_\infty$ is an ideal in $\mathcal{OC}_\infty$, it follows that $H^\bullet(\mathcal{N}_\infty)$ is an ideal in $H^\bullet(\mathcal{OC}_\infty)$. Consequently, for any $q \geqslant 0$ we have a structure of $\mathcal{L}$-module on $H^\bullet(\mathcal{N}_\infty(\_, q))$. Since $\mathcal{L} = \{H_{n-1}(\mathcal{D}(n))\}_{n \geqslant 1}$ is a suboperad of $H_\bullet(\mathcal{D})$, we also have a natural structure of $\mathcal{L}$-module on $H_\bullet(\mathcal{D})$. The next proposition is a stronger version of the above lemma.

Proposition 6.4. *For any $q \geqslant 0$, $H^\bullet(\mathcal{N}_\infty(\_, q))$ and $H_\bullet(\mathcal{D})$ are isomorphic as $\mathcal{L}$-modules.*

Proof. At the end of section 5 we observed that $\overline{C(p,q)}$ deformation retracts to a stratum S_T which is homeomorphic to $C(p)$. That deformation retract takes each stratum of dimension m (represented by a partially planar tree of degree $-m$) in $\overline{C(p,q)}$ to an m-dimensional singular chain in $\overline{S_T} = \overline{C(p)}$.

In fact, following the notation of Theorem 5.2, let S_U be a stratum of dimension m corresponding to a tree U of degree $|U| = -m$. Its closure $\overline{S_U}$ is a connected smooth oriented manifold with corners (topologically it is a manifold with boundary). Let $[\overline{S_U}]$ be the relative fundamental class in $H_m(\overline{S_U}, \partial\overline{S_U})$. For each stratum S_U, take a singular chain in $\overline{C(p,q)}$ representing the fundamental class $[\overline{S_U}]$. By composing with the contraction $\overline{C}(p,q) \to \overline{C}(p)$, we see that those singular chains are taken to singular chains in $\overline{C}(p)$. Hence we have a chain map:

$$\psi_p : \mathcal{N}_\infty(p,q) \longrightarrow C_*(\overline{C(p)}) \tag{28}$$

and an induced map in homology

$$\Psi_p : H^\bullet(\mathcal{N}_\infty(p,q)) \longrightarrow H_\bullet(C(p)) \simeq H_\bullet(\mathcal{D}(p)), \qquad \text{for each } p \geqslant 1. \tag{29}$$

Since the contraction leaves the configuration of the interior points unaffected (see subsection 5.2), classes representred by trees with only spatial edges will also be unaffected. It follows that the class $[S_{\delta_k \circ_i U}] = [S_{\delta_k}] \times [S_U]$ will be taken to the class $[S_{\delta_k}] \times \psi_p([S_U])$ for any spatial corolla $\delta_k \in \mathcal{L}_\infty$. Consequently, the sequence of maps $\{\Psi_p\}$ define a morphism of $\mathcal{L}$-modules:

$$\Psi : H^\bullet(\mathcal{N}_\infty(\_,q)) \longrightarrow H_\bullet(\mathcal{D}).$$

In order to show that Ψ is an isomorphism, let us now construct a map from $H_\bullet(\mathcal{D}(p))$ to $H^\bullet(\mathcal{N}_\infty(p,0))$, for each $p \geqslant 1$. In case $p = 1$, define the map:

$$\Phi_1 : H_\bullet(\mathcal{D}(1)) \longrightarrow H^\bullet(\mathcal{N}_\infty(1,0))$$

by taking the identity in $e \in H_\bullet(\mathcal{D}(1))$ into $n_{1,0} =$ .

When $p = 2$, the map $\Phi_2 : H_\bullet(\mathcal{D}(2)) \longrightarrow H^\bullet(\mathcal{N}_\infty(2,0))$ is defined by:

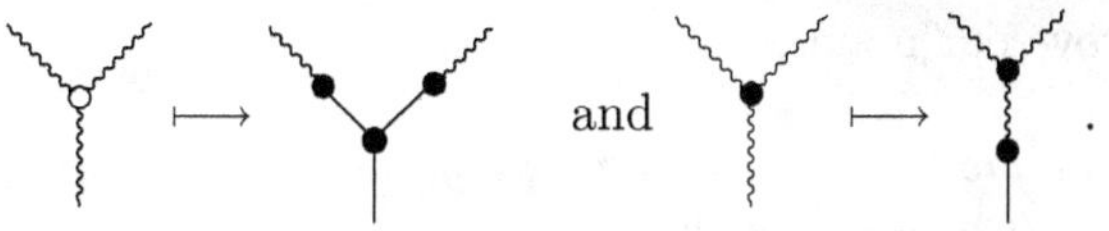

Now that we have defined our maps on the operad generators of $H_\bullet(\mathcal{D})$, we define the map $\Phi_p : H_\bullet(\mathcal{D}(p)) \to H^\bullet(\mathcal{N}_\infty(p,0))$, for any p, in the following way:

i) *if $T \in H_\bullet(\mathcal{D}(p))$ has only white vertices (i.e., corresponds to a zero dimensional homology class), then $\Phi_p(T)$ is defined by grafting $n_{1,0}$ to all the leaves of the tree obtained from T by making all vertices black and all edges straight;*

ii) *extend Φ_p to the whole $H_\bullet(\mathcal{D}(p))$ so that the resulting map*

$$\Phi : H_\bullet(\mathcal{D}) \to H^\bullet(\mathcal{N}_\infty(\_,0)) \tag{30}$$

becomes a morphism of left modules over $\mathcal{L} = \{H_{n-1}(\mathcal{D}(n))\}_{n\geqslant 1}$, i.e., such that

$$\Phi(T \circ_i l_2) = \Phi(T) \circ_i l_2, \ \textit{for any } T \in H_\bullet(\mathcal{D}).$$

In conclusion, we have defined another morphism of $\mathcal{L}$-modules

$$\Phi : H_\bullet(\mathcal{D}) \to H^\bullet(\mathcal{N}(\_,0)).$$

To see that $\Phi : H_\bullet(\mathcal{D}) \to H^\bullet(\mathcal{N}_\infty(\_,0))$ is an isomorphism, let us show that the composition $\Psi \circ \Phi$ is the identity in $H_\bullet(\mathcal{D})$. In fact: since both Ψ and Φ are morphisms of $\mathcal{L}$-modules, we need only to check that on generators. Observe that  correspond to a zero dimensional component of the boundary strata of $\overline{C(2,0)}$ and is taken to the zero dimensional generator  $\in H_0(\mathcal{D}(2))$ under the deformation retraction $\overline{C(2,0)} \to \overline{C(2)} \cong \mathcal{D}(2)$ used to define Ψ. On the other hand,  is homemorphic to S^1 (see Figure 4 pg. 148) and is naturally taken to  $\in H_1(\mathcal{D}(2))$ under the same deformation retraction, so $\Psi \circ \Phi = Id$. From lemma 6.3, we know that the vector spaces $H_\bullet(\mathcal{D}(p))$ and $H^\bullet(\mathcal{N}_\infty(p,0))$ have the same dimension for each $p \geqslant 1$. It follows that Φ is in fact a bijection.

Finally we just need to observe that $H^\bullet(\mathcal{N}_\infty(\_,0))$ is naturally isomorphic as an $\mathcal{L}$-module to $H^\bullet(\mathcal{N}_\infty(\_,q))$ for any $q \geqslant 0$. The isomorphism being induced by the grafting operation with some fixed binary planar tree T with $q+1$ leaves. □

Corollary 6.5. *The cohomology $H^\bullet(\mathcal{N}_\infty)$ is the ideal of $H^\bullet(\mathcal{OC}_\infty)$ generated by $n_{1,0}$ and $n_{0,2}$.*

Proof. It is immediate from the explicit definition of the $\mathcal{L}$-isomorphism Φ that any class in $H^\bullet(\mathcal{N}_\infty(p,q))$ can be obtained by grafting a finite number of trees of the form $n_{1,0}$ and $n_{0,2}$ followed by grafting a finite number of the form l_2, i.e., by the action of $\mathcal{L}$ on $H^\bullet(\mathcal{N}(\_,q))$. □

We can now prove our main result.

Theorem 6.6. *The morphism of differential graded $\mathcal{L}_\infty$-modules $\mu : \mathcal{OC}_\infty \longrightarrow \mathcal{OC}$ induces an isomorphism in cohomology.*

Proof. It is sufficient to show that the cohomology OCHA operad $H^\bullet(\mathcal{OC}_\infty)$ and $\mathcal{OC}$ are isomorphic as operads of graded vector spaces. Let us first recall that the operad $\mathcal{OC}_\infty$ is decomposed as a direct sum: $\mathcal{OC}_\infty = \mathcal{L}_\infty \oplus \mathcal{N}_\infty$, where $\mathcal{L}_\infty$ is the operad of L_∞-algebras and $\mathcal{N}_\infty$ is the is the ideal of partially planar trees with planar root. Since the differential operator d respects the direct sum decomposition, the homology of $\mathcal{OC}_\infty$ is a direct sum: $H^\bullet(\mathcal{OC}_\infty) = \mathcal{L} \oplus H^\bullet(\mathcal{N}_\infty)$. Now we just observe that $\mathcal{L}$ is the operad generated by l_2 and, from Corollary 6.5, $H^\bullet(\mathcal{N}_\infty)$ is generated by $n_{1,0}$ and $n_{0,2}$. The relations listed in the statement of Theorem 2.2 are naturally satisfied in $H^\bullet(\mathcal{OC}_\infty)$ since they are just the homology version of the OCHA axioms. □

Considering that $\mathcal{OC}$ is a suboperad of $H_\bullet(\mathcal{SC})$, an interesting problem that might be pursued in a sequel to the present paper is to extend our results to the whole operad $H_\bullet(\mathcal{SC})$. That would involve the entire spectral sequence of $\overline{C(p,q)}$ (see also the comments at the end of [32]).

AppendixA. OCHA as a Coderivation Differential

We say that a coderivation $\phi \in \mathrm{Coder}(S^c(U) \otimes T^c(V))$ is in **OCHA form** if it can be written as a sumation

$$\phi = \sum_{n\geqslant 1} \tilde{g}_n + \sum_{p+q\geqslant 1} \tilde{f}_{p,q},$$

where $\tilde{g}_n$ and $\tilde{f}_{p,q}$ denote the lifting as a coderivation of some maps: $g_n : U^{\wedge n} \to U$ and $f_{p,q} : U^{\wedge p} \otimes V^{\otimes q} \to V$. In [13] we have proven that all coderivations in $\mathrm{Coder}(S^c(U) \otimes T^c(V))$ are in OCHA form for any vector spaces U and V over a field k of characteristic zero.

Proposition 3.4. *An OCHA structure $(L, A, \mathfrak{l}, \mathfrak{n})$, in the grading and signs conventions of defintion 3.2, is equivalent to a degree one coderivation $D \in \mathrm{Coder}(S^c(\downdownarrows L) \otimes T^c(\downarrow A))$ such that $D^2 = 0$.*

Proof. Let us begin by defining: $\tilde{l}_1 = -l_1$ and $\tilde{n}_{0,1} = -n_{0,1}$ as the differential operators respectively on $\downdownarrows L$ and on $\downarrow A$. Let $D \in \mathrm{Coder}(S^c(\downdownarrows L) \otimes T^c(\downarrow A))$ be any degree one coderivation such that $D^2 = 0$. Since any coderivation in $\mathrm{Coder}(S^c(\downdownarrows L) \otimes T^c(\downarrow A))$ is in OCHA form, D is obtained by lifting maps $\tilde{l}_n : (\downdownarrows L)^{\otimes n} \to \downdownarrows L$ for $n \geqslant 1$ and $\tilde{n}_{p,q} : (\downdownarrows L)^{\otimes p} \otimes (\downarrow A)^{\otimes q} \to \downarrow A$ for $p+q \geqslant 1$, where all the maps $\tilde{l}_n$ and $\tilde{n}_{p,q}$ have degree one.

Equation $D^2 = 0$ holds if and only if $\{\tilde{l}_n\}_{n\geqslant 1}$ satisfies the conditions of an L_∞ algebra and $\{\tilde{n}_{p,q}\}_{p+q\geqslant 1}$ satisfies the conditions of an OCHA as originally defined in [15]:

$$0 = \sum_{\sigma\in\Sigma_{p+r=n}} \Big(\tilde{n}_{1+r,m}(\tilde{l}_p \otimes \mathbf{1}_L^{\otimes r} \otimes \mathbf{1}_A^{\otimes m}) +$$

$$\sum_{i+j+s=m} \tilde{n}_{p,i+1+j}(\mathbf{1}_L^{\otimes p} \otimes \mathbf{1}_A^{\otimes i} \otimes \tilde{n}_{r,s} \otimes \mathbf{1}_A^{\otimes j})\Big)(E(\sigma) \otimes \mathbf{1}_A^{\otimes m}). \tag{31}$$

Now define maps $l_n : L^{\otimes n} \to L$ and $n_{p,q} : L^{\otimes p} \otimes A^{\otimes q} \to A$, with $\deg(l_n) = 3-2n$ and $\deg(n_{p,q}) = 2-2p-q$ such that: $\tilde{l}_p = \downdownarrows l_p(\upuparrows)^{\otimes p}$ and $\tilde{n}_{p,q} = \downarrow n_{p,q}(\upuparrows^{\otimes p} \otimes \uparrow^{\otimes q})$.

Thus:

$$\tilde{n}_{1+r,m}(\tilde{l}_p \otimes \mathbf{1}_L^{\otimes r} \otimes \mathbf{1}_A^{\otimes m}) + \sum_{i+j+s=m} \tilde{n}_{p,i+1+j}(\mathbf{1}_L^{\otimes p} \otimes \mathbf{1}_A^{\otimes i} \otimes \tilde{n}_{r,s} \otimes \mathbf{1}_A^{\otimes j}) =$$

$$= \downarrow n_{1+r,m}(\Uparrow^{\otimes 1+r} \otimes \uparrow^{\otimes m})(\Downarrow l_p(\Uparrow)^{\otimes p} \otimes \mathbf{1}_L^{\otimes r} \otimes \mathbf{1}_A^{\otimes m}) +$$

$$+ \sum_{i+j+s=m} \downarrow n_{p,i+1+j}(\Uparrow^{\otimes p} \otimes \uparrow^{\otimes i+1+j})(\mathbf{1}_L^{\otimes p} \otimes \mathbf{1}_A^{\otimes i} \otimes \downarrow n_{r,s}(\Uparrow^{\otimes r} \otimes \uparrow^{\otimes s}) \otimes \mathbf{1}_A^{\otimes j}) =$$

$$= \downarrow n_{1+r,m}(l_p \otimes \mathbf{1}_L^{\otimes r} \otimes \mathbf{1}_A^{\otimes m})(\Uparrow^{\otimes n} \otimes \uparrow^{\otimes m}) +$$

$$+ \sum_{i+j+s=m} (-1)^{js+i} \downarrow n_{p,i+1+j}(\mathbf{1}_L^{\otimes p} \otimes \mathbf{1}_A^{\otimes i} \otimes n_{r,s} \otimes \mathbf{1}_A^{\otimes j})(\Uparrow^{\otimes n} \otimes \uparrow^{\otimes m}) =$$

$$= \downarrow \Big(n_{1+r,m}(l_p \otimes \mathbf{1}_L^{\otimes r} \otimes \mathbf{1}_A^{\otimes m})$$

$$+ \sum_{i+j+s=m} (-1)^{js+i} n_{p,i+1+j}(\mathbf{1}_L^{\otimes p} \otimes \mathbf{1}_A^{\otimes i} \otimes n_{r,s} \otimes \mathbf{1}_A^{\otimes j}) \Big)(\Uparrow^{\otimes n} \otimes \uparrow^{\otimes m}),$$

where the sign $(-1)^{js+i}$ comes from the Koszul sign convention. Observing that $\tilde{l}_1 = -l_1$, $\tilde{n}_{0,1} = -n_{0,1}$ and $(-1)^{js+i} = (-1)^{s+i+si+ms}$, we obtain formula (11) from formula (AppendixA). □

AppendixB. Existence of the DG $\mathcal{L}_\infty$-module morphism

Proposition AppendixB.1. *There is a morphism of differential graded $\mathcal{L}_\infty$-modules $\mu : \mathcal{OC}_\infty \to \mathcal{OC}$ extending the identity on $\mathcal{OC}$, i.e., such that the following diagram is commutative:*

$$\begin{array}{ccc} \mathcal{OC}_\infty & \xrightarrow{\ \mu\ } & \\ \uparrow & & \\ \mathcal{OC} & \xrightarrow{\ id\ } & \mathcal{OC}\,. \end{array}$$

Proof. In this proof we shall omit the labels on trees because they are not crucial in the argument.

The open-closed operad $\mathcal{OC}$ is a differential graded operad where the differential operator δ is trivial: $\delta \equiv 0$. On the other hand, the differential operator d of the OCHA operad $\mathcal{OC}_\infty$ is defined by formulas (22) and (4). We will exhibit a chain map $\mu : \mathcal{OC}_\infty \to \mathcal{OC}$ which is also a morphism of $\mathcal{L}_\infty$-modules. In other words, μ must satisfy two conditions:

$$\mu(dT) = 0, \quad \forall T \in \mathcal{OC}_\infty$$
$$\mu(l \circ_i T) = l \circ_i \mu(T), \quad \forall T \in \mathcal{OC}_\infty \text{ and } \forall l \in \mathcal{L}_\infty.$$

Let $\mathcal{E}$ be the $\mathcal{L}_\infty$-submodule of $\mathcal{OC}_\infty$ generated by $\mathcal{OC}$ and by 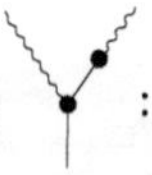 :

$$\mathcal{E} = \Big\langle \mathcal{OC},$$  $$\Big\rangle$$

On the generators of the submodule $\mathcal{E}$, the map μ is defined in the following way:

$$\mu(T) = T \quad \forall T \in \mathcal{OC} \quad \text{and} \quad \mu\Big($$ 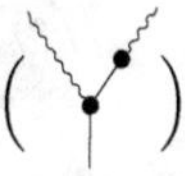 $$\Big) = -\frac{1}{2}$$ 

and it is extended to $\mathcal{E}$ as an $\mathcal{L}_\infty$-morphism. Finally, for any tree $T \in \mathcal{OC}_\infty$ such that $T \notin \mathcal{E}$, we define $\mu(T) = 0$. We thus have an $\mathcal{L}_\infty$-morphism:

$$\mu : \mathcal{OC}_\infty \to \mathcal{OC}.$$

It remains to show that μ is a chain map, i.e., that $\mu(dT) = 0$ for any tree $T \in \mathcal{OC}_\infty$. Given any tree $T \in \mathcal{OC}_\infty$, dT is a summation of trees. By the definition of μ, if T is such that dT has no components in $\mathcal{E}$, then $\mu(dT) = 0$. Hence, we just need to consider those elements $T \in \mathcal{OC}_\infty$ such that dT has some component in $\mathcal{E}$. Such elements form an $\mathcal{L}_\infty$-submodule of $\mathcal{OC}_\infty$ which will be denoted by $\mathcal{E}'$. More precisely:

$$\mathcal{E}' := \{T \in \mathcal{OC}_\infty : dT = T_1 + T_2, \quad T_1 \in \mathcal{E}, T_1 \neq 0\}.$$

Any tree T is obtained by grafting a finite number of corollae which we call the *irreducible components* of T. Recall that, for $n \geqslant 3$, the $\mathcal{L}_\infty$-module action of $l_n \in \mathcal{L}_\infty$ on any element of $\mathcal{OC}$ is zero since that action is defined through the quasi-ismorphism $\mu : \mathcal{L}_\infty \to \mathcal{L}$, and $\mu(l_n) = 0$ for $n \geqslant 3$. From the definition of $\mu : \mathcal{OC}_\infty \to \mathcal{OC}$ and the definition of the $\mathcal{L}_\infty$-module structure on $\mathcal{OC}$, one can see that the irreducible components of any tree $T \in \mathcal{E}'$ such that $\mu(dT) \neq 0$ could only be one of the following corollae:

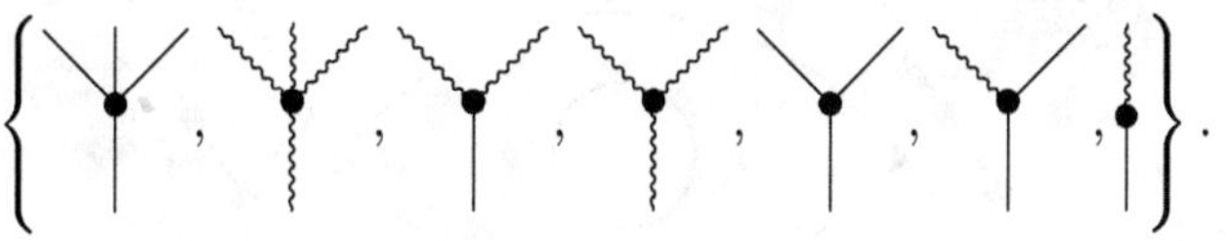

Consequently, we just need to check that $\mu(dT) = 0$ where T is any of the above corollae. In the case of $T =$  : 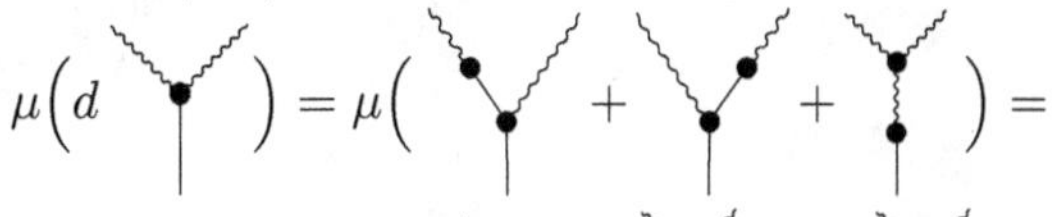

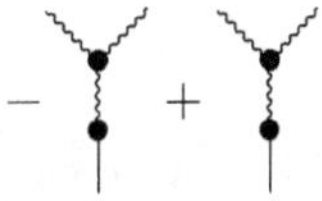 $= 0$, since by definition we have: 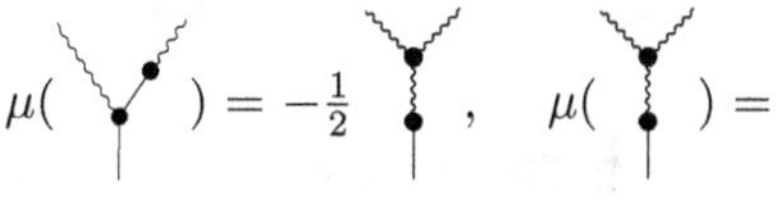

 and because the wiggly edges are spatial, we also have: 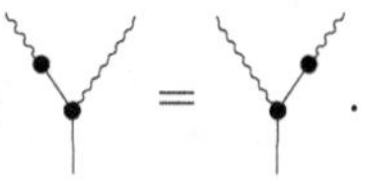 . The other corollae can be handled similarly. ☐

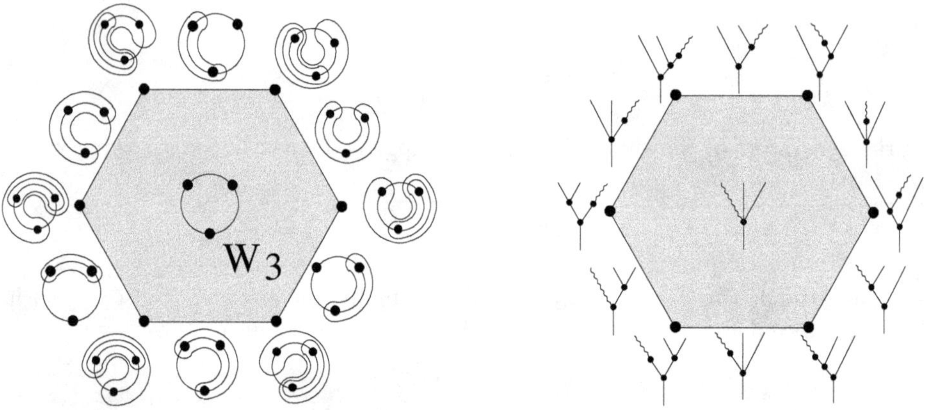

Figure 3: Cyclohedron $\overline{C(1,2)}$ and its cells labelled by circular bracketings and by trees.

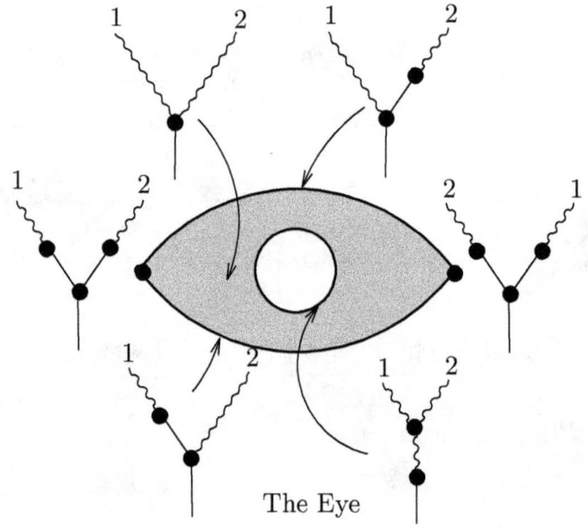

Figure 4: The space $\overline{C(2,0)}$ = "The Eye" and its boundary strata labelled by trees.

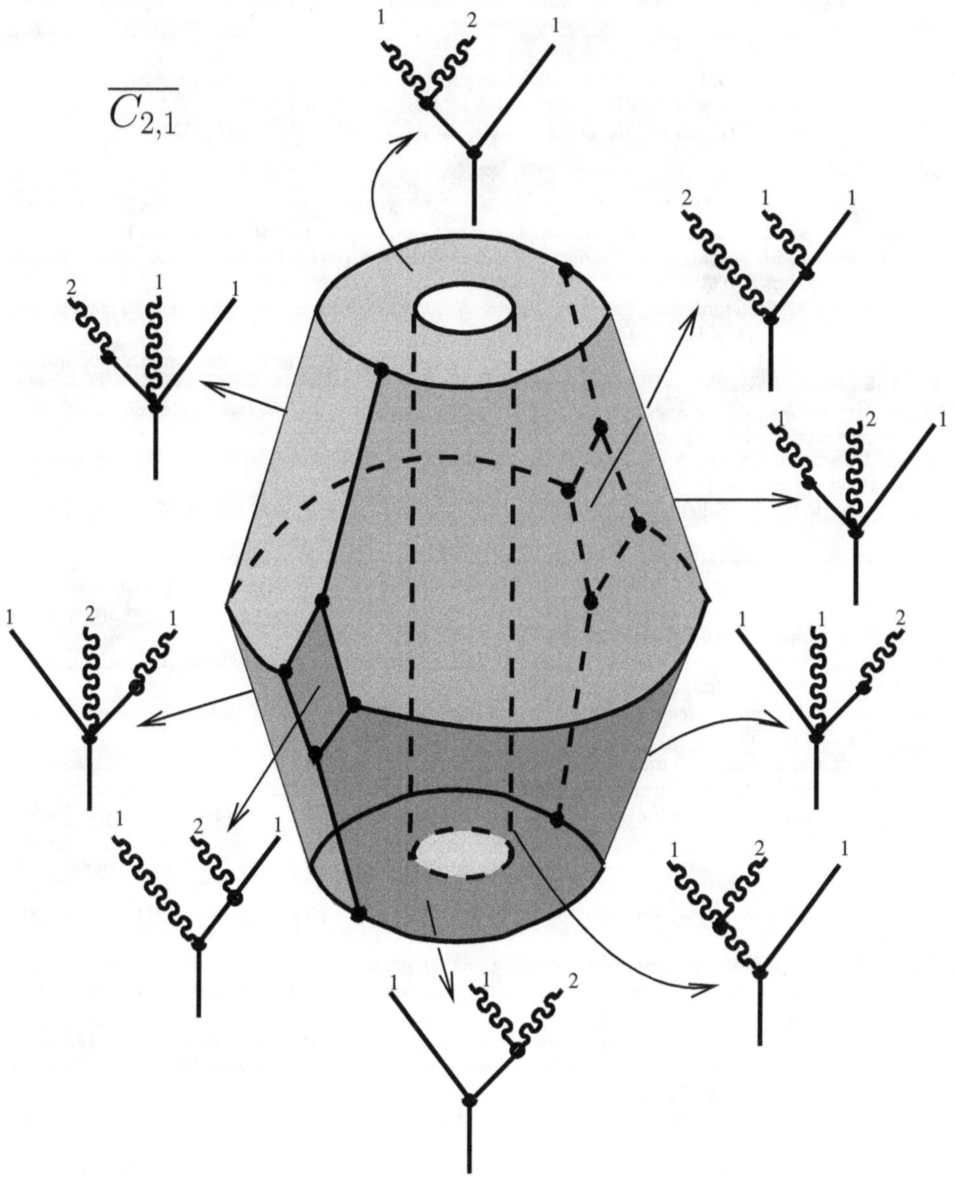

Figure 5: The space $\overline{C(2,1)}$, which is topologically equivalent to a solid torus, and its codimension 1 boundary components labelled by partially planar trees.

References

[1] S. Axelrod and I. M. Singer, *Chern-Simons perturbation theory II*, Perspectives in mathematical physics. Conf. Proc. Lecture Notes Math. Phys. **III** (1994), 17–49.

[2] C. Berger and I. Moerdijk, *Resolution of coloured operads and rectification of homotopy algebras*, Contemp. Math. **431** (2007), 31–58.

[3] J.M. Boardman and R.M. Vogt, *Homotopy invariant algebraic structures on topological spaces*, Lecture Notes in Mathematics, vol. 347. Springer-Verlag, 1973.

[4] R. Bott and C. Taubes, *On the self-linking of knots*, J. Math. Phys. **35** (1994), no. 10, 5247–5287.

[5] F.R. Cohen, *The homology of C_{n+1}-spaces*, The homology of iterated loop spaces (F.R. Cohen, T.J. Lada, and J.P. May, eds.), Lecture Notes in Mathematics, vol. 533. Springer-Verlag, 1976.

[6] S. L. Devadoss, *A space of cyclohedra*, Discrete Comput. Geom. **29** (2003), no. 1, 61–75.

[7] M. Doubek, M. Markl, and P. Zima, *Deformation theory (lecture notes)*, Arch. Math. (Brno) **43** (2007), no. 5, 333–371.

[8] M. Flato, M. Gerstenhaber, and A.A. Voronov, *Cohomology and deformation of Leibniz pairs*, Lett. Math. Phys. **34** (1995), no. 1, 77–90.

[9] W. Fulton and R. MacPherson, *A compactification of configuration spaces*, Ann. of Math. **139** (1994), 183–225.

[10] E. Getzler and J. D. S. Jones, *Operads, homotopy algebra and iterated integrals for double loop spaces*, available at `hep-th/9403055`.

[11] E. Harrelson, *On the homology of open-closed string field theory*, available at `arXiv:math/0412249v2[math.AT]`.

[12] V. Hinich and V. Schechtman, *Homotopy Lie algebras*, Advances in Soviet Math. **16** (1995), no. 2, 1–28.

[13] E. Hoefel, *On the coalgebra description of OCHA*, available at `math.QA/0607435`.

[14] ———, *Espaços de configurações e OCHA*, Ph.D Thesis - Unicamp, 2006 (Portuguese).

[15] H. Kajiura and J. Stasheff, *Homotopy algebras inspired by classical open-closed string field theory*, Comm. Math. Physics **263** (2006), no. 3, 553 –581.

[16] ———, *Open-closed homotopy algebra in mathematical physics*, Journal of Mathematical Physics **47** (2006), no. 2, 28p.

[17] ———, *Homotopy algebra of open-closed strings*, Geometry & Topology Monographs **13** (2008), 229–259.

[18] T. Kimura, J. Stasheff, and A. Voronov, *On operad structures of moduli spaces and string theory*, Comm. Math. Physics **171** (1995), no. 1, 1–25.

[19] M. Kontsevich, *Deformation quantization of Poisson manifolds*, Lett. Math. Phys. **66** (2003), no. 3, 157–216.

[20] T. Lada and M. Markl, *Strongly homotopy Lie algebras*, Communications in Algebra **23** (1995), 2147–2161.

[21] T. Lada and J. Stasheff, *Introduction to sh Lie algebras for physicists*, Int. J. Theo. Phys. **32** (1993), 1087–1103.

[22] M. Markl, *Homotopy algebras via resolutions of operads*, Proceedings of the 19th Winter School "Geometry and physics", Srní, Czech Republic, January 9-15, 1999. Supplem. Rend. Circ. Mat. Palermo, Ser. II. **63** (2000), 157–164.

[23] ———, *Homotopy algebras are homotopy algebras*, Forum Math. **16** (2004), no. 1, 129–160.

[24] M. Markl, S. Shnider, and J. Stasheff, *Operads in Algebra, Topology and Physics*, Mathematical Surveys and Monographs, 96. AMS, 2002.

[25] P. May, *The geometry of iterated loop spaces*, Lectures Notes in Mathematics, vol. 271. Springer-Verlag, 1972.

[26] S. Merkulov, *Operads, deformation theory and F-manifolds*, Frobenius manifolds. Quantum cohomology and singularities. Proceedings of the workshop, Bonn, Germany, July 8–19, 2002 (Hertling, Claus (ed.) et al.) Wiesbaden: Vieweg. Aspects of Mathematics E **36** (2004), 213–251.

[27] Dev P. Sinha, *Manifold-theoretic compactifications of configuration spaces*, Sel. Math., New Ser. **10** (2004), no. 3, 391–428.

[28] J. Stasheff, *On the homotopy associativity of H-spaces I*, Trans. AMS **108** (1963), 275–292.

[29] ———, *On the homotopy associativity of H-spaces II*, Trans. AMS **108** (1963), 293–312.

[30] ———, *From operads to "physically inspired" theories*, Contemp. Math. **202** (1997), 53–81.

[31] A. Voronov, *Topological field theories, string backgrounds and homotopy algebras*, Proceedings of the XXIInd international conference on differential geometric methods in theoretical physics. Universidad Nacional Autónoma de México (J. Keller and Z. Oziewicz, eds.), Advances in Applied Clifford Algebras. **4** (1994), no. S1, 167–178.

[32] ______, *The swiss-cheese operad*, Contemporary Math. **239** (1999), 365–373.

[33] ______, *Homotopy Gerstenhaber algebras*, Conférence Moshé Flato 1999: Quantization, deformations, and symmetries, Dijon, France (Dito, Giuseppe (ed.) et al.) Volume II. Dordrecht: Kluwer Academic Publishers. Math. Phys. Stud. **22** (2000), 307–331.

[34] ______, *Topics in mathematical physics (lecture notes)*, University of Minnesota, 2001.

[35] B. Zwiebach, *Oriented open-closed string theory revisited*, Annals Phys. **267** (1998), no. 193, 33–152.

This article may be accessed via WWW at `http://www.rmi.acnet.ge/jhrs/`

Eduardo Hoefel
hoefel@ufpr.br

Dep. de Matemática
Universidade Federal do Paraná
Brasil
Curitiba
c.p. 019081 cep: 81531-990

Journal of Homotopy and Related Structures, vol. 4(1), 2009, pp.153–179

THE TRANSFER IN MOD-P GROUP COHOMOLOGY BETWEEN $\Sigma_P \int \Sigma_{PN-1}$, $\Sigma_{PN-1} \int \Sigma_P$ AND Σ_{PN}

NONDAS E. KECHAGIAS

(*communicated by Frederick Cohen*)

Abstract

In this work we compute the induced transfer map:

$$\bar{\tau}^* : \mathrm{Im}\left(res^* : H^*\left(G\right) \to H^*\left(V\right)\right) \to \mathrm{Im}\left(res^* : H^*\left(\Sigma_{p^n}\right) \to H^*\left(V\right)\right)$$

in mod p-cohomology. Here Σ_{p^n} is the symmetric group acting on an n-dimensional $\mathbb{F}_p$ vector space V, $G = \Sigma_{p^n,p}$ a p-Sylow subgroup, $\Sigma_{p^{n-1}} \int \Sigma_p$, or $\Sigma_p \int \Sigma_{p^{n-1}}$. Some answers are given by natural invariants which are related to certain parabolic subgroups. We also compute a free module basis for certain rings of invariants over the classical Dickson algebra. This provides a computation of the image of the appropriate restriction map. Finally, if $\xi : \mathrm{Im}\left(res^* : H^*\left(G\right) \to H^*\left(V\right)\right) \to \mathrm{Im}\left(res^* : H^*\left(\Sigma_{p^n}\right) \to H^*\left(V\right)\right)$ is the natural epimorphism, then we prove that $\bar{\tau}^* = \xi$ in the ideal generated by the top Dickson algebra generator.

1. Introduction-Results

Let H be a subgroup of a finite group G. There are two important maps in group cohomology going in the opposite direction: the restriction and transfer. The Weyl subgroup acts on the right in group cohomology and the inclusion $H \hookrightarrow G$ induces a map

$$\left(res_H^G\right)^* : H^*\left(G\right) \to H^*\left(H\right)^{W_G(H)}$$

In other words the image of the restriction map is contained in the $W_G\left(V\right)$-invariants. The role of classical invariant theory in determining and analyzing cohomology of finite groups is important.

The inclusion $H \hookrightarrow G$ also induces a transfer map

$$tr^* : H^*\left(H\right) \to H^*\left(G\right)$$

The transfer map plays a fundamental role in group cohomology.

We thank the referee and N. Kuhn very much for their suggestions regarding the exposition of this work.

Received October 10, 2008, revised March 9, 2009; published on June 18, 2009.

2000 Mathematics Subject Classification: Primary 20J05, 18G10, 55S10; Secondary 13F20, 13A50.

Key words and phrases: Restriction map, Transfer map, Cohomology of symmetric groups, Parabolic invariants, Dickson algebra, Steenrod algebra action, Free modules over the Dickson algebra.

In this work we compute the maps above for particular cases. Some answers are given by particular invariants which are of the form: a free module basis over the fundamental object in modular invariant theory, i.e. the Dickson algebra.

We studied the case $G = \mathbb{Z}/p \wr ... \wr \mathbb{Z}/p$ in [**5**]. We extend those results for $G = \Sigma_p \wr \Sigma_{p^{n-1}}$ and $\Sigma_{p^{n-1}} \wr \Sigma_p$. The methods applied in [**5**] can not be applied in this case. We compute the image of the restriction map in Theorem 17 for $\Sigma_{p^{n_l}} \wr ... \wr \Sigma_{p^{n_1}}$. To compute the transfer, we need to express the previous ring as a module over the Dickson algebra. We do so in proposition 23 and Theorem 32. Finally, we show that the induced transfer coincides with the natural, so called, epimorphism on a certain ideal in Theorems 41 and 43.

Let $V \cong \mathbb{F}_p^n$ be an n-dimensional $\mathbb{F}_p$ vector space. Let Σ_{p^n} denote the permutations on V. Now V has a left action on itself and defines an inclusion: $V \hookrightarrow \Sigma_{p^n}$. Let $\Sigma_p \wr \Sigma_{p^{n-1}}$ denote the semidirect product of Σ_p with $\left(\Sigma_{p^{n-1}}\right)^p$ with Σ_p acting by permuting factors. And for $\Sigma_{p^{n-1}} \wr \Sigma_p$ respectively. Let

$$\Sigma_{p^n,p} := \mathbb{Z}/p \wr ... \wr \mathbb{Z}/p$$

which is a p-Sylow subgroup of $\Sigma_{p^n,p}$ containing V. The maximal elementary abelian p-subgroup V is contained by both $\Sigma_p \wr \Sigma_{p^{n-1}}$ and $\Sigma_{p^{n-1}} \wr \Sigma_p$.

Simple coefficients are taken in $\mathbb{F}_p \cong \mathbb{Z}/p$ where p is an odd prime. For $p = 2$ minor modifications are needed and left to the interested reader. Hence $H^*(G)$ stands for $H^*(G, \mathbb{Z}/p)$.

It is known that

$$H^*(V) \cong \begin{cases} \mathbb{F}_p[y_1, \cdots, y_n], \text{ for } p = 2 \\ E_{\mathbb{F}_p}(x_1, \cdots x_n) \otimes \mathbb{F}_p[y_1, \cdots, y_n] \end{cases}$$

It is known that the Weyl subgroups $W_{\Sigma_{p^n}}(V)$, $W_{\Sigma_{p^n,p}}(V)$, $W_{\Sigma_p \wr \Sigma_{p^{n-1}}}(V)$ and $W_{\Sigma_{p^{n-1}} \wr \Sigma_p}(V)$ are the general linear group $GL(n, \mathbb{F}_p)$, the upper triangular subgroup U_n, and the parabolic subgroups $P(1, n-1)$ and $P(n-1, 1)$ respectively. Here

$$P(k, n-k) = \left\{ \begin{pmatrix} A & C \\ \mathbf{0} & B \end{pmatrix} \mid A \in GL(k), B \in GL(n-k)) \right\}$$

Kuhn ([**8**]) proved that the following diagram is commutative and this is the key point for our study:

$$\begin{array}{ccc} H^*\left(\Sigma_p \wr \Sigma_{p^{n-1}}\right) & \xrightarrow{tr^*} & H^*\left(\Sigma_{p^n}\right) \\ \left(res_V^{\Sigma_p \wr \Sigma_{p^{n-1}}}\right)^* \downarrow & & \downarrow \left(res_V^{\Sigma_{p^n}}\right)^* \\ H^*(V)^{W_{\Sigma_p \wr \Sigma_{p^{n-1}}}(V)} & \xrightarrow{\tau^*} & H^*(V)^{W_{\Sigma_{p^n}}(V)} \end{array}$$

In this work we investigate the induced transfer homomorphisms:

$$\mathrm{Im}\left(res^* : H^*(G) \to H^*(V)\right) \xrightarrow{\overline{\tau}^*} \mathrm{Im}\left(res^* : H^*\left(\Sigma_{p^n}\right) \to H^*(V)\right)$$

For $G = \Sigma_{p^n,p}$, $\Sigma_p \wr \Sigma_{p^{n-1}}$ and $\Sigma_{p^{n-1}} \wr \Sigma_p$. The problem reduces to find free module bases for certain algebras of modular invariants. This is a hard problem for a general parabolic subgroup.

The restriction map is not an onto map and our first task is to compute its image. Please note that for $p = 2$ the restriction map is onto. We give an invariant theoretic proof of the following Theorem first proved by Mui ([**11**]) using cohomological methods in section 3. It requires technical results from group cohomology and invariant theory.

Theorem 15 ([11]). *The image* $\mathrm{Im}\left(res^* : H^*\left(\Sigma_{p^n,p}\right) \to H^*\left(V\right)\right)$ *is isomorphic with the tensor product between an exterior and a polynomial algebra*

$$E_{\mathbb{F}_p}\left(\hat{M}_{1,0}, \hat{M}_{2,1}\hat{L}_1^{(p-3)/2}, \cdots, \hat{M}_{n,n-1}\hat{L}_{n-1}^{(p-3)/2}\right) \otimes H_n^t.$$

Definitions and notation are given in section 2.

Since the transfer is an additive map (and the identity on the Dickson algebra), it is important to describe these images of the appropriate rings as modules over the Dickson algebra $\left(H^*\left(V\right)^{GL(n,\mathbb{F}_p)}\right)$. The bulk of this work is to that direction.

As an application of last Theorem we derive the next proposition in section 3. The image is given by natural invariants which have the following form.

Proposition 16. *The image* $Im\left(res^* : H^*\left(\Sigma_{p^n,p}\right) \to H^*\left(V\right)\right)$ *is isomorphic with*

$$H_n^t \bigoplus_i \bigoplus_{s_i} H_n^t \hat{M}_{i,s_1,\ldots,s_{k-1},i-1} \hat{L}_{i-1}^{(p-3)/2} \prod_1^{k-1} \hat{L}_{s_t}^{(p-3)/2} \prod_1^{k-1} \hat{L}_{s_t+1}^{(p-3)/2}$$

Here $k \leqslant i \leqslant n$ *and* $0 \leqslant s_1 < \ldots < s_{k-1} < i-1$.

Let $I = (n_l, \ldots, n_1)$ be a sequence of positive integers such that $\sum n_i = n$ and $P(I)$ the associated parabolic subgroup. We call

$$D_n := \left(\mathbb{F}_p[y_1, \ldots, y_n]\right)^{GL(n,\mathbb{F}_p)}$$

(the classical Dickson algebra) and

$$\mathbb{F}_p\left(I\right) := \left(\mathbb{F}_p[y_1, \cdots, y_n]\right)^{P(I)}$$

Implementing last Theorem and the ring $H^*\left(V\right)^{P(I)}$, we compute the image of the restriction map in section 3.

Theorem 17. *The image* $\mathrm{Im}\left(res^* : H^*\left(\Sigma_{p^{n_l}} \int \ldots \int \Sigma_{p^{n_1}}\right) \to H^*\left(V\right)\right)$ *is isomorphic to the subalgebra generated by*

$$\left\{ \begin{array}{c} \hat{d}_{\nu_i,\nu_i-k_i}, \hat{M}_{\nu_i,\nu_i-k_i}\left(\hat{L}_{\nu_i}\right)^{p-2}, \hat{M}_{\nu_i,\nu_i-k_j,\nu_i-k_i}\left(\hat{L}_{\nu_i}\right)^{p-2} \mid \\ 1 \leqslant i \leqslant \ell,\ 1 \leqslant k_i \leqslant n_i, k_i < k_j < \nu_i,\ \nu_i = \sum_{t=1}^{i} n_t \end{array} \right\}$$

along with certain relations.

For notation and relations between the generators please see Theorem 11 in section 2.

It is a hard problem to express the subalgebra above as a free module over the appropriate subalgebra of the Dickson algebra. Instead we study certain rings of invariants of parabolic subgroups.

It is known that $\mathbb{F}_p\left(I\right)$ is a finitely generated free module over D_n. In order to provide a free basis, we define a new generating set for $\mathbb{F}_p\left(1, n-1\right)$ and $\mathbb{F}_p\left(n-1, 1\right)$. There are two advantages for this new set. Mainly, it is closed under the action of

Steenrod's algebra and secondly the algebra generators for D_n can be decomposed with respect to the new ones. We prove the following proposition in section 4.

Proposition 23. *Let $I = (1, n-1)$, then*

$$\mathbb{F}_p(I) = \mathbb{F}_p[h_1^{p-1}, d_{n,i}(I) \mid 1 \leqslant i \leqslant n-1]$$

$$H^*(V)^{P(I)} \cong \mathbb{F}_p(I) \oplus \mathbb{F}_p(I)\left[M_{1,0}h_1^{p-2} \bigoplus_{t_i} M_{n,t_1,\ldots,t_k} L_n^{p-2}\right]$$

Here $1 \leqslant t_k$ and $0 \leqslant t_1 < \ldots < t_k \leqslant n-1$.

Kuhn and Mitchell described $\mathbb{F}_p(I)$ using appropriate Dickson algebra generators in [**9**]. Their set is elegant and more easily described than ours, but their set is not closed under the action of Steenrod's algebra, and their set is not as useful as ours is in computations.

The next Theorem provides a free module basis for $\mathbb{F}_p(n-1,1)$ over D_n proved in section 5.

For each t, $1 \leqslant t \leqslant n-1$, we define the set of all $(n-t)$-tuples

$$\mathcal{M}(n-2,t) = \{M = (p, m_t, \ldots, m_{n-2}) \mid 0 \leqslant m_i \leqslant p-1\}$$

and, for each $M \in \mathcal{M}(n-2,t)$ we define

$$d_{n-1}^M = d_{n-1,t-1}^p d_{n-1,t}^{m_t} \ldots d_{n-1,n-2}^{m_{n-2}}$$

Theorem 32. We have

$$B_{D_n}(\mathbb{F}_p(n-1,1)) = \bigcup_{t=1}^{n-1} \{d_{n-1}^M \mid M \in \mathcal{M}(n-2,t)\}$$

as a free module basis for $\mathbb{F}_p(n-1,1)$ over D_n.

The following corollary is the main result in this work.

Corollary 33. *i)* $\operatorname{Im}\left(res^* : H^*\left(\Sigma_p \int \Sigma_{p^{n-1}}\right) \to H^*(V)\right)$ *is isomorphic to a free module over D_n on*

$$\left\{ \begin{array}{c} \hat{M}_{1,0}\hat{L}_1^{(p-2)}\hat{h}_1^{(p-1)m}, \hat{M}_{n,s_1,\ldots,s_k}\hat{L}_n^{(p-2)} d_{n,0}^{\left(\left[\frac{k+1}{2}\right]-1\right)} \hat{h}_1^{(p-1)m} \mid \\ 0 \leqslant m < A_1, k \leqslant n, 1 \leqslant s_k, 0 \leqslant s_1 < \ldots < s_k \leqslant n-1 \end{array} \right\}$$

Here $A_1 = p^{n-1} + \ldots + p$.

ii) $\operatorname{Im}\left(res^* : H^*\left(\Sigma_{p^{n-1}} \int \Sigma_p\right) \to H^*(V)\right)$ *is isomorphic to a free module over D_n on*

$$\left\{ \begin{array}{c} \hat{M}_{n,n-1}\hat{L}_n^{(p-2)} f, \hat{M}_{n-1,s_1,\ldots,s_k}\hat{L}_{n-1}^{(p-2)} d_{n-1,0}^{\left(\left[\frac{k+1}{2}\right]-1\right)} g \mid \\ f, g \in B_{D_n}(\mathbb{F}_p(n-1,1)),\ k \leqslant n-1, 0 \leqslant s_1 < \ldots < s_k \leqslant n-1 \end{array} \right\}$$

Finally, the transfer map is studied in the last section. There is a natural description of $\mathbb{F}_p(1,n-1)$ or $\mathbb{F}_p(n-1,1)$ as a polynomial algebra (proposition **23** or as described in [**9**]). According to last corollary, there is an alternate description of it as a free module over the Dickson algebra. The natural epimorphisms

$$\xi : \mathbb{F}_p(1,n-1) \to D_n \text{ and } \xi : \mathbb{F}_p(n-1,1) \to D_n$$

which "rewrites" an element of the polynomial algebra in terms of the free module basis are shown to be equal with the induced transfer maps. Let us consider an example.

Example. Let $n = 3$ and $p = 2$. $\mathbb{F}_p(2,1) = \mathbb{F}_p[d_{2,0}, d_{2,1}, d_{3,2}]$ and the basis is $B = \left\{d_{2,0}^i d_{2,1}^j, d_{2,0}^2 d_{2,1}^j, d_{2,1}^2 | 0 \leqslant i,j \leqslant 1\right\}$. We need to describe the way in which the three generators of $\mathbb{F}_p(2,1)$ can be written in terms of B and D_3. Here is the way:

$$\begin{aligned} d_{2,0}d_{2,1}^2 &= d_{3,0} + d_{3,2}d_{2,0} \\ d_{2,1}^3 &= d_{3,1} + d_{3,2}d_{2,1} + d_{2,0}^2 \\ d_{2,0}^3 &= d_{3,1}d_{2,0} + d_{3,0}d_{2,1} \end{aligned}$$

Suppose we want to find $\xi\left(d_{2,0}^2 d_{2,1}^7\right)$. According to B and the relations above, this element "rewrites" as follows

$$d_{2,0}^2 d_{2,1}^7 = d_{3,0}^2 d_{3,1} + d_{3,0}^2 d_{3,2} d_{2,1} + d_{3,0}^2 d_{2,0}^2 + d_{3,2}^3 d_{2,0}^2 d_{2,1} + d_{3,0} d_{3,2}^2 d_{2,0} d_{2,1}$$

Thus $\xi\left(d_{2,0}^2 d_{2,1}^7\right) = d_{3,0}^2 d_{3,1}$.

Theorem 41. *Let $\xi : \mathbb{F}_p(n-1,1) \longrightarrow D_n$ be the natural epimorphism with respect to the given free module basis B and $\bar{\tau}^* : \mathbb{F}_p(n-1,1) \to D_n$ the transfer map. Then $\xi = \bar{\tau}^*$.*

The advantage of the map ξ is that it calculates $\bar{\tau}^*$.

Although the transfer map satisfies the nice property described in last Theorem for the polynomial part of the ring of invariants, it does not for the exterior part. Please see example 42. But the transfer coincides with the map ξ in the ideal generated by the top Dickson algebra generator.

Theorem 43. *Let $\xi, \bar{\tau}^* : \mathrm{Im}\left(res_V^{\Sigma_{p^n,p}}\right)^* \to Im\left(res_V^{\Sigma_{p^n}}\right)^*$ be the rewriting and the induced transfer maps. Then $\xi = \bar{\tau}^*$ in the ideal generated by $(d_{n,0})$.*

Our method strongly depends on the action of Steenrod's algebra on the rings of invariants. This action is the key ingredient in the proof of Theorem 15 which is the building block for the computation of the images of the appropriate restriction maps. This method was inspired by a similar method used by Adem and Milgram VI 1 in [**1**]. All background material can be found in this excellent account. For the computation of the free module bases, we follow Campbell and Hughes [**2**]. Taking into account proposition 16 which is a long and technical result, the familiar reader may proceed to sections 5 and 6.

2. The rings of invariants

Let us repeat some classical results from the literature. Let $G = GL(n, \mathbb{F}_p)$, B_n, or U_n be the general linear group, the Borel subgroup, and the upper triangular subgroup with 1's on the diagonal, respectively. G acts as usual on V. Let $I = (n_l, ..., n_1)$ be an ordered sequence of positive integers such that $\sum n_i = n$. We order such sequences as above by refinements: $I \leqslant I'$ if I is a refinement of I'. For example $(1, ..., 1) \leqslant (n_1, n_2) \leqslant (n)$. Given such a sequence I let $V^1 \subset ... \subset V^l = V$

be defined by

$$V^i =< e_1, e_2, ..., e_{(n_1+...+n_i)} >$$

This is called a flag by Kuhn [**8**]. It is well known that the set

$$P(I) := \{g \in GL(n, \mathbb{F}_p) \mid \forall i \; g(V^i) = V^i\}$$

$$P(I) = \left\{ \begin{pmatrix} GL_{n_1} & * & * \\ 0 & \ddots & * \\ 0 & 0 & GL_{n_\ell} \end{pmatrix} \right\} \leqslant GL(n, \mathbb{F}_p)$$

is a subgroup of $GL(n, \mathbb{F}_p)$ called a parabolic subgroup related to the partition I. Moreover, if G is a subgroup of $GL(n, \mathbb{F}_p)$ containing the Borel subgroup B_n, then $G = P(I)$ for some sequence I, ([**3**] page 112).

Since $H^*(V) = E_{\mathbb{F}_p}(x_1, \cdots x_n) \otimes \mathbb{F}_p[y_1, \cdots, y_n]$, the object of study is

$$\left(E_{\mathbb{F}_p}(x_1, \cdots x_n) \otimes \mathbb{F}_p[y_1, \cdots, y_n]\right)^{P(I)}$$

The classical Dickson algebra, $D_n = (\mathbb{F}_p[y_1, \cdots, y_n])^{GL(n,\mathbb{F}_p)}$, is described as follows. Let

$$h_i = \prod_{v \in V^{i-1}} (y_i - v) \text{ and } L_n = \prod_1^n h_i$$

Let $L_{n,i}$ be the determinant of the $n \times n$ matrix $\begin{pmatrix} y_1 & \cdots & y_n \\ \vdots & \cdots & \vdots \\ y_1^{p^n} & \cdots & y_n^{p^n} \end{pmatrix}$ where the $i+1$-row is missing, i.e. the row $\begin{pmatrix} y_1^{p^i}, & \cdots, & y_n^{p^i} \end{pmatrix}$. Moreover, $L_n = L_{n,n}$ and $L_{n,0} = L_n^p$.

Let $L_{n,i}(\hat{t}) = \det \begin{pmatrix} y_1 & \cdots & \hat{y}_t & \cdots & y_n \\ \vdots & \cdots & \vdots & \cdots & \vdots \\ y_1^{p^{n-1}} & \cdots & \hat{y}_t^{p^{n-1}} & \cdots & y_n^{p^{n-1}} \end{pmatrix}$ where the $i+1$-row is missing. Now the following formula holds:

$$L_n = (-1)^{t-1} \left[y_t L_{n,n-1}(\hat{t}) - y_t^p L_{n,1}(\hat{t}) + ... + (-1)^{n-1} y_t^{p^{n-1}} L_{n,n-1}(\hat{t})\right] \quad (1)$$

Finally, let

$$d_{n,i} = \frac{L_{n,i}}{L_n}$$

The degrees of the previous elements are $|h_i| = 2p^{i-1}$, $|L_n| = 2\frac{p^n-1}{p-1}$, and $|d_{n,i}| = 2(p^n - p^i)$.

We shall also need the matrix ω which consists of 1's along the antidiagonal for the transpose of these groups, please see remark 12.

Definition 1. *Let $f \in H^*(V)$, then $\hat{f}$ stands for ωf. In particular $\hat{h}_i = \omega h_i$ or $\hat{h}_i = \prod_{v \in \langle y_{n+2-i}, \cdots, y_n \rangle} (y_{n+1-i} - v)$.*

Theorem 2 (Dickson [4]). $D_n = \mathbb{F}_p[d_{n,0}, \cdots, d_{n,n-1}]$.

Theorem 3 (Mui [11]). *i)* $H_n := (\mathbb{F}_p[y_1, \cdots, y_n])^{U_n} = \mathbb{F}_p[h_n, \cdots, h_1]$ *and*

$$H_n^t := (\mathbb{F}_p[y_1, \cdots, y_n])^{U_n^t} = \mathbb{F}_p[\hat{h}_n, \cdots, \hat{h}_1]$$

ii) $(\mathbb{F}_p[y_1, \cdots, y_n])^{B_n} = \mathbb{F}_p[(h_n)^{p-1}, \cdots, (h_1)^{p-1}]$ *and*

$$(\mathbb{F}_p[y_1, \cdots, y_n])^{B_n^t} = \mathbb{F}_p[(\hat{h}_n)^{p-1}, \cdots, (\hat{h}_1)^{p-1}]$$

Relations between the generators of rings of invariants are given as follows:

Proposition 4 ([5]). $d_{n,n-i} = \sum_{1 \leqslant j_1 < \cdots < j_i \leqslant n} \prod_{s=1}^{i} \left(h_{j_s}^{p-1}\right)^{p^{n-i+s-j_s}}$.

Corollary 5. $d_{n,n-i} = d_{n-1,n-i} h_n^{p-1} + d_{n-1,n-i-1}^p$.

Theorem 6 (Kuhn and Mitchell [9]). *Let* $I = (n_l, \cdots, n_1)$.

i) $\mathbb{F}_p(I) := \mathbb{F}_p[d_{\nu_i, \nu_i - k_i} \mid 1 \leqslant i \leqslant \ell,\ 1 \leqslant k_i \leqslant n_i,\ \nu_i = \sum_{t=1}^{i} n_t]$.

ii) $\mathbb{F}_p(I)^t := \mathbb{F}_p[\hat{d}_{\nu_i, \nu_i - k_i} \mid 1 \leqslant i \leqslant \ell,\ 1 \leqslant k_i \leqslant n_i,\ \nu_i = \sum_{t=1}^{i} n_t]$.

All the rings of invariants considered in this work are algebras over the Steenrod algebra. The action of Steenrod's algebra on Dickson algebra elements has been completely computed in [6]. We repeat here the following Theorem applied several times in this work.

Theorem 7. *([6], page 170) i) Let* $q = \Sigma_1^{n-1} a_t p^{t+l}$ *such that* $p - 1 \geqslant a_t \geqslant a_{t-1} > a_{i-1} = 0$. *Then*

$$P^q d_{n,0}^{p^l} = d_{n,0}^{p^l} (-1)^{a_{n-1}} \Pi_i^{n-1} \binom{a_t}{a_{t-1}} d_{n,t}^{p^l (a_t - a_{t-1})}$$

Otherwise, $P^q d_{n,0}^{p^l} = 0$.

ii) Let $q = \Sigma_1^{n-1} a_t p^{t+l}$ *such that* $p - 1 \geqslant a_t \geqslant a_{t-1} > a_i = 0$ *and* $a_i + 1 \geqslant a_{i-1} \geqslant a_t \geqslant a_{t-1} \geqslant 0$. *Then*

$$P^q d_{n,i}^{p^l} =$$

$$d_{n,i}^{p^l} (-1)^{a_{n-1}} \left(\Pi_{i+1}^{n-1} \binom{a_t}{a_{t-1}} \right) \binom{a_i + 1}{a_{i-1}} \left(\Pi_s^{i-1} \binom{a_t}{a_{t-1}} \right) \Pi_s^{n-1} d_{n,t}^{p^l (a_t - a_{t-1})}$$

Here $a_{s-1} = 0$. *Otherwise,* $P^q d_{n,0}^{p^l} = 0$.

We need some technical results for the proof of Theorem 15. Let

$$h_i(\hat{\jmath}) := \prod_{v \in \langle y_1, \ldots, \hat{y}_j, \ldots, y_{i-1} \rangle} (y_i - v) \tag{2}$$

and $d_{n,t}(\hat{\jmath})$ be the Dickson algebra generator of degree $2\left(p^{n-1} - p^t\right)$ in

$$(\mathbb{F}_p[y_1, \cdots, \hat{y}_j, \ldots, y_n])^{GL(n-1, \mathbb{F}_p)}$$

Let $\delta_{i,j} \in GL(n, \mathbb{F}_p)$ such that it permutes only the i and j coordinates. Let

$$h_i(j) := \delta_{i,j} h_i = \prod_{v \in \langle y_1, \ldots, \hat{y}_j, \ldots, y_i \rangle} (y_j - v) \tag{3}$$

for $j \leqslant i$.

Lemma 8. $h_i = h_i^p(\hat{\jmath}) - h_i(\hat{\jmath})(h_{i-1}(j))^{p-1}$.

Proof.

$$h_i = \prod_a \prod_{v \in \langle y_2, \cdots, y_{i-1} \rangle} (y_i - a y_1 - v) =$$

$$\prod_a \sum_{t=0}^{i-2} (y_i + a y_1)^{p^{i-2-t}} (-1)^t d_{i-1,t}(\hat{1}) = \prod_a \left(h_i(\hat{1}) + a h_{i-1}(1)\right)$$

Since $\sum_a a \equiv 0 \bmod p$, $\sum_{a_{i_t} \neq a_{i_l}} \prod_{t=1}^{p-2} a_{i_t} \equiv 0 \bmod p$ and $\prod_{a \neq 0} a \equiv p-1 \bmod p$, $h_i = h_i^p(\hat{1}) - h_i(\hat{1})(h_{i-1}(1))^{p-1}$. Now applying $\delta_{1,j}$ the statement follows. $\square$

The Dickson's result was extended for $H^*(V)^{GL(2,\mathbb{F}_p)}$ by Cardenas and Mui for the general case. For full details please see [**11**].

In $E_{\mathbb{Z}}(x_1, \ldots, x_n) \otimes \mathbb{Z}[y_1, \ldots, y_n]$, let $M_{n,s_1,\ldots,s_k}$ be defined as

$$\frac{1}{k!} \det \begin{pmatrix} x_1 & \cdots & x_1 & y_1 & \cdots & \widehat{y_1^{p^{s_1}}} & \cdots & \widehat{y_1^{p^{s_k}}} & \cdots & y_1^{p^{n-1}} \\ \vdots & & \vdots & \vdots & & \vdots & & \vdots & & \vdots \\ x_n & \cdots & x_n & y_n & \cdots & \widehat{y_n^{p^{s_1}}} & \cdots & \widehat{y_n^{p^{s_k}}} & \cdots & y_n^{p^{n-1}} \end{pmatrix}$$

Here $0 \leqslant s_1 < \ldots < s_k \leqslant n-1$. The columns $\begin{pmatrix} \widehat{y_1^{p^{s_i}}} \\ \vdots \\ \widehat{y_n^{p^{s_i}}} \end{pmatrix}$ are missing and the matrix for the proceeding determinant is filed out with k columns of the form $\begin{pmatrix} x_1 \\ \vdots \\ x_n \end{pmatrix}$ to have n rows and columns. Let

$$M_{n,i}(\hat{t}) = Det \begin{pmatrix} x_1 & y_1 & \cdots & \widehat{y_1^{p^i}} & \cdots & y_1^{p^{n-2}} \\ \cdots & & & & & \cdots \\ x_n & y_n & \cdots & \widehat{y_n^{p^i}} & \cdots & y_n^{p^{n-2}} \end{pmatrix}$$

and the t-th row is missing i.e. $\left[x_t, y_t \ldots, y_t^{p^{n-2}}\right]$. Now the following formula is obvious:

$$M_{n,n-1} = (-1)^{t-1}\left[x_t L_{n,n-1}(\hat{t}) - y_t M_{n,0}(\hat{t}) + \ldots + (-1)^{n-1} y_t^{p^{n-2}} M_{n,n-2}(\hat{t})\right] \tag{4}$$

We recall that $\hat{M}_{m,s_1,\ldots,s_k} = \omega M_{m,s_1,\ldots,s_k}$ and $\hat{d}_{m,t} = \omega d_{m,t}$ for $1 \leqslant m \leqslant n$ and $\omega \in GL(n, \mathbb{F}_p)$.

Theorem 9 (Mui). *i)* $H^*(V)^{GL(n,\mathbb{F}_p)} \cong D_n \bigoplus_k \bigoplus_{s_i} D_n M_{n,s_1,\ldots,s_k} L_n^{p-2}$. *Here a double summation is taken over* $k = 1, \ldots, n$ *and* $0 \leqslant s_1 < \ldots < s_k \leqslant n-1$. *Furthermore the generators satisfy: 1)* $M_{n,s}^2 = 0$ *and*
2) $M_{n,s_1} \ldots M_{n,s_k} = (-1)^{k(k-1)/2} M_{n,s_1,\ldots,s_k} L_n^{k-1}$.

ii) $H^*(V)^{U_n^t} \cong H_n^t \bigoplus_i \bigoplus_{s_t} H_n^t \hat{M}_{i,s_1,\ldots,s_{k-1},i-1}$. *Here* $k \leqslant i \leqslant n$ *and* $0 \leqslant s_1 < \ldots < s_{k-1} < i-1$.

The next lemma describes relations between exterior and polynomial algebra generators.

Lemma 10. *i) Let* $0 \leqslant s_1 < \ldots < s_k \leqslant n-2$. *Then*

$$M_{n-1,s_1,\ldots,s_k} h_n =$$

$$M_{n,s_1,\ldots,s_k} - \sum_{(t_1,\ldots,t_k)>(s_k-k+1,\ldots,s_k)} (-1)^{k+i} M_{n,s_1,\ldots,\hat{s}_i,\ldots,s_k} d_{n-1,s_i}$$

ii) Let $0 \leqslant s_1 < \ldots < s_k \leqslant k-1$. *Then*

$$M_{l,s_1,\ldots,s_k} h_{l+1} \ldots h_n = M_{n,s_1,\ldots,s_k} + \sum_{(t_1,\ldots,t_k)>(s_k-k+1,\ldots,s_k)} M_{n,t_1,\ldots,t_k} f_{t_1,\ldots,t_k}$$

Here $f_{t_1,\ldots,t_k} \in H_n$.

The next Theorem is an extension of Mui's Theorem for parabolic subgroups ([**5**]).

Theorem 11 (Kechagias). *Let* $I = (n_l, \cdots, n_1)$ *be a sequence of non-negative integers such that* $\sum n_i = n$ *and* $P(I)$ *be the associated parabolic subgroup of* $GL(n, \mathbb{F}_p)$, *then*

$$H^*(V)^{P(I)} \cong \mathbb{F}_p(I) \bigoplus_i \bigoplus_k \bigoplus_{s_t} \mathbb{F}_p(I) M_{\nu_i,s_1,\ldots,s_k} L_{\nu_i}^{p-2}$$

Here $1 \leqslant i \leqslant \ell$, $\nu_i = \sum_{t=1}^{i} n_t$, $1 \leqslant k \leqslant \nu_i$, $\nu_{i-1} \leqslant s_k$ *and* $0 \leqslant s_1 < \ldots < s_k \leqslant \nu_i - 1$.

3. The restriction map

We remind the reader about a well known analogy between

$$U_n \leqslant B_n \leqslant P(I) \leqslant GL(n, \mathbb{F}_p)$$

and subgroups of the symmetric group Σ_{p^n}. There exists a regular embedding $V \hookrightarrow \Sigma_{p^n}$ which takes $u \in V$ to the permutation on V induced by $v \mapsto u + v$.

Let us recall that the wreath product between $H \leqslant \Sigma_l$ and $K \leqslant \Sigma_m$ is defined by

$$1 \to H^m \to K \int H \to K \to 1$$

and $K \int H \leqslant \Sigma_{ml}$.

Let $\Sigma_{p^n,p} := (\mathbb{Z}_p)_n \int \cdots \int (\mathbb{Z}_p)_1$ and $\Sigma(I) := \Sigma_{p^{n_l}} \int ... \int \Sigma_{p^{n_1}}$. Then $\Sigma_{p^n,p}$ is a p-Sylow subgroup of Σ_{p^n} and here is the analogy

$$\Sigma_{p^n,p} \leqslant \Sigma(1, ..., 1) \leqslant \Sigma(I) \leqslant \Sigma_{p^n}$$

Here the inclusion $V \hookrightarrow \Sigma_{p^n,p}$ factors as follows

$$V = \mathbb{Z}_p \times (\mathbb{Z}_p)^{n-1} \overset{1\times\Delta^p}{\to} \mathbb{Z}_p \int \Sigma_{p^{n-1},p} \to \Sigma_p \int \Sigma_{p^{n-1},p} \to \Sigma_{p^n}$$

Moreover, the Weyl subgroups of V in $\Sigma_{p^n,p}$, $\Sigma(I)$, and Σ_{p^n} are the upper triangular group U_n, $P(I)$ and the general linear group $GL(n, \mathbb{F}_p)$ respectively. Please see [**8**] Theorem 3.2.

Finally, $Aut(V) \cong GL(n, \mathbb{F}_p)$ and let

$$\rho : W_{\Sigma_{p^n}}(V) \hookrightarrow GL(n, \mathbb{F}_p)$$

be the regular representation. Now the contragredient representation ρ^* acts on $V^* \cong H^1(V)$. Here $\rho^*(g) = \rho\left(g^{-1}\right)^t$. Moreover the Weyl group, $W_{\Sigma_{p^n}}(V) \cong GL(n, \mathbb{F}_p)$, acts on V^* as follows:

$$(a_{i,j})x_k := \sum_i a_{i,k}x_i$$

Here, $V^* = \langle x_1, \cdots x_n \rangle$.

Let E_G and B_G denote the total and classifying spaces of a finite group G. Let $H \leqslant G$ be a subgroup, then E_G can also be a total space for H and $pt \times_H E_G$ is a model for B_G. Moreover,

$$G/H \to B_H \overset{\pi}{\to} B_G$$

is a fibration. The inclusion described above, $V \hookrightarrow G$, induces a map

$$\left(res_V^G\right)^* : H^*(G) \to H^*(V)^{W_G(V)}$$

Here $G = \Sigma(I)$ and $H^*(G) := H^*(B_G, \mathbb{Z}/p)$.

Since $H^1(V) \cong V^*$ and the Bockstein homomorphism is an isomorphism $\beta : H^1(V) \to H^2(V)$, let $y_i = \beta x_i$ for $1 \leqslant i \leqslant n$. Now

$$H^*(V) = E_{\mathbb{F}_p}(x_1, \cdots x_n) \otimes \mathbb{F}_p[y_1, \cdots, y_n]$$

and $H^*(V)^{GL(n,\mathbb{F}_p)}$ denotes the Dickson algebra.

Remark 12. *Note that*

$$\operatorname{Im}\left(res_V^G\right)^* \leqslant H^*(V)^{W_G(V)} = \left(E_{\mathbb{F}_p}(x_1, \cdots x_n) \otimes \mathbb{F}_p[y_1, \cdots, y_n]\right)^{W_G(V)^t}$$

In other words we consider the transposes of the groups described above.

The following important Theorem first proved by Cardenas for $n = 2$ and extended by Kuhn provides the effective tools for our calculations. Here we use a particular version of that Theorem. Please see VI, 1.6 in [**1**].

Theorem 13 (Cardenas, Mui, Kuhn). .
i) Let $res^* : H^*\left(\Sigma_p \int \Sigma_{p^{n-1}}\right) \to H^*(V)$, *then*

$$\operatorname{Im}(res^*) = H^*(V)^{P(1,n-1)} \cap \operatorname{Im}\left(res^* : H^*(\Sigma_{p^n,p}) \to H^*(V)\right)$$

ii) Let $res^* : H^*(\Sigma_{p^{n-1}} \int \Sigma_p) \to H^*(V)$, then

$$\mathrm{Im}(res^*) = H^*(V)^{P(n-1,1)} \cap \mathrm{Im}(res^* : H^*(\Sigma_{p^n,p}) \to H^*(V))$$

Our first task is to give an invariant theoretic description of $\mathrm{Im}(res^* : H^*(\Sigma_{p^n,p}) \to H^*(V))$. Using a Theorem of Steenrod and the action of the Steenrod algebra on upper triangular invariants we compute this ring. For completeness we repeat some well known facts on group cohomology. For full details please see VII in [**12**].

Let $H \triangleleft G$, then we have a fibering. An application of this fibering is the following:

$$(B_G)^p \xrightarrow{j} E_{\mathbb{Z}_p} \times_{\mathbb{Z}_p} (B_G)^p \xrightarrow{\pi} B_{\mathbb{Z}_p}$$

Here $G^p \triangleleft \mathbb{Z}_p \int G$ and $(B_G)^p \simeq B_{G^p}$. The last implies

$$H^*(G) \otimes ... \otimes H^*(G) \cong H^*(G^p)$$

Let $\Delta^p : B_G \to (B_G)^p$ be the diagonal and

$$1 \times \Delta^p : B_{\mathbb{Z}_p} \times B_G \to B(\mathbb{Z}_p \int G) \simeq E_{\mathbb{Z}_p} \times_{\mathbb{Z}_p} (B_G)^p$$

the induced map. The image of the restriction map is the image of $1 \times \Delta^p$. Now $H^*(\mathbb{Z}_p \int G)$ is an $H^*(\mathbb{Z}_p)$-module and $(\Delta^p)^*$ is an $H^*(\mathbb{Z}_p)$-module homomorphism. Moreover the map π^* is a monomorphism.

Let $\{u_j | j \in J\}$ be an $\mathbb{F}_p$ basis of $H^*(G)$. Then

$$M := \langle u_j \otimes ..... \otimes u_j | j \in J \rangle$$

is an $\mathbb{F}_p$-submodule of $H^*(G^p)$ and

$$F := \langle u_{j_1} \otimes ... \otimes u_{j_p} | j_1 \leqslant ... \leqslant j_p \; j_1 < j_p \rangle$$

is a free $\mathbb{F}_p$-submodule of $H^*(G^p)$. It is well known that

$$H^*(\mathbb{Z}_p \int G) \cong H^*(\mathbb{Z}_p; (H^*(G)^p)) \cong \mathbb{F}_p \otimes F^{\mathbb{Z}_p} \oplus H^*(\mathbb{Z}_p) \otimes M$$

Please see IV Theorem 1.7 in [**1**]. If $v \in H^*(\mathbb{Z}_p)$, then v acts on $H^*(\mathbb{Z}_p \int G)$ by $1^p \otimes v$.

Given a class $v \in H^*(G)$ we have a class $v \otimes ..... \otimes v \in H^*(G^p)$. Now $(\Delta^p)^*(v \otimes ... \otimes v) = v^p$ and Steenrod defined a map on the cochain level in order to compute the image of the restriction map

$$P : H^q(G) \to H^{pq}(\mathbb{Z}_p \int G)$$

such that Pv is the cohomology class $\varepsilon \otimes v^p$ where ε is the augmentation on the chain level. More precisely,

$$Pv = 1 \otimes v^p \in \mathbb{F}_p \otimes F^{\mathbb{Z}_p} \oplus \mathbb{F}_p \otimes M$$

Moreover, the Steenrod map satisfies

$$P(u \cup v) = (-1)^{p(p-1)/2|u||v|} Pu \cup Pv$$

Please see page 190 in [**1**]. Now $H^*(\mathbb{Z}_p) \otimes \mathrm{Im}\, P \cong H^*(\mathbb{Z}_p) \otimes M$ and $H^*(\mathbb{Z}_p) \otimes \mathrm{Im}(\Delta^p)^* P = \mathrm{Im}(\Delta^p)^*$.

Theorem 14 (Steenrod, May). *Let $v \in H^q(G)$, $\eta = (p-1)/2$ and $\mu(q) = (\eta!)^{-q}(-1)^{\eta(q^2+q)/2}$. Then*

$$(1\times\Delta^p)^* Pv = \mu(q)\left[\sum_i (-1)^i y^{(q-2i)\eta}\otimes P^i v + \sum_i (-1)^{i+q} x y^{(q-2i)\eta-1}\otimes \beta P^i v\right]$$

Here $H^*(\mathbb{Z}_p)\cong E_{\mathbb{F}_p}(x)\otimes\mathbb{F}_p[y]$.

Please see IV Theorem 4.1 in [**1**].

Now we are ready to prove the main Theorem of this section.

Theorem 15.

$$\operatorname{Im}(res^*: H^*(\Sigma_{p^n,p})\to H^*(V))\cong$$

$$E_{\mathbb{F}_p}\left(\hat{M}_{1,0},\hat{M}_{2,1}\hat{L}_1^{(p-3)/2},\cdots,\hat{M}_{n,n-1}\hat{L}_{n-1}^{(p-3)/2}\right)\otimes H_n^t$$

Proof. We apply induction on n. We shall prove

i) $(1\times\Delta^p)^* P\left(\hat{h}_i(\hat{n})\right) = c\hat{h}_i$ and

ii) $(1\times\Delta^p)^* P\left(\hat{M}_{i,i-1}(\hat{n})\hat{L}_{i-1}^{(p-3)/2}(\hat{n})\right) = c'\hat{M}_{i,i-1}\hat{L}_{i-1}^{(p-3)/2}$.

Here $c, c'\in(\mathbb{F}_p)^*$. Or equivalently,

$(1\times\Delta^p)^* P\left(h_i(\hat{1})\right) = ch_i$ and

$(1\times\Delta^p)^* P\left(M_{i,i-1}(\hat{1})L_{i-1}^{(p-3)/2}(\hat{1})\right) = c'M_{i,i-1}L_{i-1}^{(p-3)/2}$.

i) We apply Steenrod-May's formula.

$$(1\times\Delta^p)^* P\left(h_i(\hat{1})\right) = \mu\left(2p^{i-2}\right)\sum_m (-1)^m y_1^{(2p^{i-2}-2m)\eta}\otimes P^m h_i(\hat{1}) \tag{5}$$

We recall definitions 2, 3 and lemma 8:

$$h_i = h_i^p\left(\hat{1}\right) - h_i\left(\hat{1}\right)\left(h_{i-1}(1)\right)^{p-1} \tag{6}$$

The idea is to compare the coefficients of y_1^l for certain l's in the expressions (5) and (6).

We start with the action of Steenrod's algebra $P^m h_i(\hat{1})$. We apply Theorem 20 repeatedly.

If $m = p^{i-2}$, then $P^m h_i(\hat{1}) = h_i^p(\hat{1})$.

Now let $m = a_{i-3}p^{i-3}+...+a_s p^s$, then

$$P^m h_i(\hat{1}) = (-1)^{a_{i-3}} h_i(\hat{1}) d_{i-1,i-2}(\hat{1})\binom{a_{i-3}+1}{a_{i-4}}\prod_{t=s}^{i-4}\binom{a_{t+1}}{a_t}\prod_{t=s}^{i-4} d_{i-1,t}^{a_t-a_{t-1}}(\hat{1})$$

We recall definition 3:

$$\left(h_{i-1}(1)\right)^{p-1} = \left(y_1^{p^{i-2}} + \sum(-1)^t y_1^{p^{i-2-t}} d_{i-1,t}(\hat{1})\right)^{p-1}$$

Let $r\leqslant p-1$, $0\leqslant t_1<...<t_r\leqslant i-2$ and $\lambda_{t_1}+...+\lambda_{t_r} = p-1$. Then the coefficient of $y_1^{\Sigma\lambda_{t_i}p^{t_i}}$ in the last expression is given by

$$(-1)^{(p-1)(i-2)-\Sigma\lambda_{t_i}t_i}\left(\frac{(p-1)!}{\lambda_{t_1}!...\lambda_{t_r}!}\right)\Pi d_{i-1,t_i}^{\lambda_{t_i}}(\hat{1})$$

Here $(p-1)(i-2) - \Sigma\lambda_{t_i} t_i \equiv \Sigma\lambda_{t_i} t_i \mod 2$.

Next the corresponding coefficient of y_1 in (5) shall be considered.

Let $(p^{i-2} - m)(p-1) = \Sigma\lambda_{t_i} p^{t_i}$. Then

$$m(p-1) = p^{i-2}(p-1) - \Sigma\lambda_{t_i} p^{t_i} = (b_{t_r} - 1) p^{i-3} + b_{t_r}(p^{i-4} + ... + p^{t_r}) +$$

$$b_{t_{r-1}}(p^{t_r-1} + ... + p^{t_{r-1}}) + ... + b_{t_2}(p^{t_3-1} + ... + p^{t_2}) + b_{t_1}(p^{t_2-1} + ... + p^{t_1})$$

Here $b_{t_1} = \lambda_{t_1}$, $b_{t_2} - b_{t_1} = \lambda_{t_2}$, ..., $b_{t_r} - b_{t_{r-1}} = \lambda_{t_r}$ and $b_{t_r} = p-1$. Thus $b_{t_i} = a_{t_i} = ... = a_{t_{i+1}-1}$ for $i \leqslant r-1$ and $b_{t_r} = a_{t_r} = ... = a_{i-4} = a_{i-3} + 1$. It is an easy computation to prove that the exponents of (-1) are equal in both sides i.e. $(a_{i-3} + m) \equiv \Sigma\lambda_{t_i} t_i mod 2$.

We conclude $(1 \times \Delta^p)^* P(h_i(\hat{1})) \equiv -\mu(2p^{i-2}) h_i$.

ii) We shall prove that

$$(1 \times \Delta^p)^* P\left(M_{i,i-1}(\hat{1}) L_{i-1}^{(p-3)/2}(\hat{1})\right) = c' M_{i,i-1} L_{i-1}^{(p-3)/2} \tag{7}$$

by comparing the corresponding coefficients of powers of y_1. First we consider elements

$\beta P^m \left(M_{i,i-1}(\hat{1}) L_{i-1}^{(p-3)/2}(\hat{1})\right) \neq 0$. Please see proposition 21. This is equivalent with

$$m = p^{i-3} + ... + 1 + \Sigma^l a_{i_t}(p^{i-3} + ... + p^{i_t}) \text{ and } \Sigma^l a_{i_t} \leqslant \frac{p-3}{2} \tag{8}$$

In this case

$$\beta P^m \left(M_{i,i-1}(\hat{1}) L_{i-1}^{(p-3)/2}(\hat{1})\right) = (a_{i_1}, ..., a_{i_l}) L_i(\hat{1}) \left(\prod L_{i-1,i_t}^{a_t}(\hat{1})\right) L_{i-1}^{a_{l+1}}$$

Here $a_{l+1} = \left(\frac{p-3}{2} - \Sigma a_{i_t}\right)$ and

$$(a_{i_1}, ..., a_{i_l}) = ((p-3)/2)!/\Sigma a_{i_t}! \left(\frac{p-3}{2} - \Sigma a_{i_t}\right)!$$

In Steenrod-May's formula, the corresponding exponent of y_1 is

$$\frac{p-1}{2}\left(2p^{i-2} - 2(p^{i-3} + ... + 1) - 2\left(\Sigma a_{i_t}(p^{i-3} + ... + p^{i_t})\right)\right) - 1 =$$

$$\Sigma^l a_{i_t} p^{i_t} + a_{l+1} p^{i-2}$$

For each m satisfying condition (8),

$$(-1)^{m+p^{i-2}} \beta P^m \left(M_{i,i-1}(\hat{1}) L_{i-1}^{(p-3)/2}(\hat{1})\right) x_1 y_1^{\Sigma a_{i_t} p^{i_t} + a_{l+1} p^{i-2}} =$$

$$(-1)^{m+1}(a_{i_1}, ..., a_{i_l}) L_{i-1}(\hat{1}) \left(\prod L_{i-2,i_t}^{a_{i_t}}(\hat{1})\right) L_{i-2}^{a_{l+1}}(\hat{1}) x_1 y_1^{\Sigma a_{i_t} p^{i_t} + a_{l+1} p^{i-2}}$$

The corresponding coefficient of $x_1 y_1^{\Sigma a_{i_t} p^{i_t} + a_{l+1} p^{i-2}}$ in the decomposition of $M_{i,i-1} L_{i-1}^{(p-3)/2}$ (right hand side in (7)) with respect to $x_1 y_1$ (according to formulas 1 and 4) is

$$(-1)^{\Sigma_{t=1}^{l+1} a_{i_t} i_t}(a_{i_1}, ..., a_{i_l}) L_{i-1}(\hat{1}) \left(\prod L_{i-1,i_t}^{a_{i_t}}(\hat{1})\right) L_{i-1,i-2}^{a_{l+1}}(\hat{1})$$

Those two elements differ by $(-1)^{1+(i-2)(p-1)/2}$.

Next we consider elements of the form

$$(-1)^m P^m \left(M_{i,i-1}(\hat{1})L_{i-1}^{(p-3)/2}(\hat{1})\right) y_1^{\frac{p-1}{2}\left(2p^{i-2}-2m\right)}$$

in the left hand side of (7).

For non-zero elements we have $m = p^{i-3} + ... + p^k + m'$ with $m' = \Sigma^l a_{i_t}\left(p^{i-3} + ... + p^{i_t}\right)$ and $\Sigma^l a_{i_t} \leqslant \frac{p-3}{2}$. Replacing m in the exponent of y_1 it takes the form

$$\frac{p-1}{2}\left(2p^{i-2} - 2\left(p^{i-3} + ... + 1\right) - 2\left(\Sigma a_{i_t}\left(p^{i-3} + ... + p^{i_t}\right)\right)\right) + p^k - 1$$

As before the corresponding coefficients of y_1 to the particular exponent differ by

$$(-1)^{1+(i-2)(p-1)/2} : m - \left(k - 1 + \Sigma^{l+1} a_{i_t} i_t\right) \equiv 1 + (i-2)(p-1)/2 \quad mod \quad p$$

Now the proof is complete. □

Proposition 16. *The image* $\mathrm{Im}\left(res^* : H^*\left(\Sigma_{p^n,p}\right) \to H^*(V)\right)$ *is isomorphic with*

$$H_n^t \bigoplus_i \bigoplus_{s_t} H_n^t \hat{M}_{i,s_1,\ldots,s_{k-1},i-1} \hat{L}_{i-1}^{(p-3)/2} \prod_1^{k-1} \hat{L}_{s_t}^{(p-3)/2} \prod_1^{k-1} \hat{L}_{s_t+1}^{(p-3)/2}$$

Here $k \leqslant i \leqslant n$ *and* $0 \leqslant s_1 < ... < s_{k-1} < i-1$.

Proof. This is an application of Theorem 15 and lemma 10. □

The next Theorem is an application of last Theorem and Cardenas-Mui-Kuhn Theorem.

Theorem 17. $\mathrm{Im}\left(res^* : H^*\left(\Sigma_{p^{n_l}} \int ... \int \Sigma_{p^{n_1}}\right) \to H^*(V)\right)$ *is isomorphic to the subalgebra generated by*

$$\left\{\begin{array}{c} \hat{d}_{\nu_i,\nu_i-k_i}, \hat{M}_{\nu_i,\nu_i-k_i}\left(\hat{L}_{\nu_i}\right)^{p-2}, \hat{M}_{\nu_i,\nu_i-k_j,\nu_i-k_i}\left(\hat{L}_{\nu_i}\right)^{p-2} \mid \\ 1 \leqslant i \leqslant \ell,\ 1 \leqslant k_i \leqslant n_i, k_i < k_j < \nu_i,\ \nu_i = \sum_{t=1}^{i} n_t \end{array}\right\}$$

Subject to relations described in Theorem 9 and lemma 10.

4. Relations between parabolic and Dickson algebra generators

Since D_n is a subalgebra of $\mathbb{F}_p[V]^{P(I)}$, any Dickson algebra generator can be decomposed in terms of generators of the later algebra. We shall describe these relations in this section for $I = (n-1, 1)$ and $(1, n-1)$.

We recall that a Dickson algebra generator $d_{n,n-i}$ consists of the sum of all possible combinations of i elements from $\{h_1^{p-1}, \cdots, h_n^{p-1}\}$ in certain p-th exponents (proposition 4) and this might be more than what a $P(I)$-generator needs. For instance, we would like to replace $d_{n,n-1}$ by another element which is a $P(I)$-invariant but not a $GL(n, \mathbb{F}_p)$-one. An example is in order.

Example. Proposition 4 is applied.

a) Let $n = 4$ and $n_1 = 3$.

$$d_{4,3} = h_1^{(p-1)p^3} + h_2^{(p-1)p^2} + h_3^{(p-1)p} + h_4^{(p-1)} \Rightarrow$$

$$d_{4,3} - h_1^{(p-1)p^3} - h_2^{(p-1)p^2} - h_3^{(p-1)p} = h_4^{(p-1)}$$

And this polynomial is a $P(3,1)$-invariant.

b) Let $n = 4$ and $n_1 = 1$.

$$d_{4,3} = h_1^{(p-1)p^3} + h_2^{(p-1)p^2} + h_3^{(p-1)p} + h_4^{(p-1)} \Rightarrow$$

$$d_{4,3} - h_1^{(p-1)p^3} = h_2^{(p-1)p^2} + h_3^{(p-1)p} + h_4^{(p-1)}$$

And this polynomial in a $P(1,3)$-invariant. Let us call the last sum $d_{4,3}(I)$. Thus

$$d_{4,3}(I) = d_{4,3} - h_1^{(p-1)p^3}$$

Next we consider $d_{4,2}$:

$$d_{4,2} = h_1^{(p-1)p^2} h_2^{(p-1)p^2} + h_1^{(p-1)p} h_3^{(p-1)p^2} + h_1^{(p-1)} h_4^{(p-1)p^2} + h_2^{(p-1)p} h_3^{(p-1)p} +$$

$$h_2^{(p-1)} h_4^{(p-1)p} + h_3^{(p-1)} h_4^{(p-1)} \Rightarrow$$

$$d_{4,2} - (h_2^{(p-1)p^3} + h_3^{(p-1)p} + h_4^{(p-1)}) h_1^{(p-1)p^2} = h_3^{(p-1)} h_4^{(p-1)}$$

Let us call the last sum $d_{4,2}(I)$. Thus

$$d_{4,2}(I) = d_{4,2} - h_1^{(p-1)p^2} d_{4,3}(I)$$

Now $d_{4,1}$:

$$d_{4,1} = h_2^{(p-1)} h_3^{(p-1)} h_4^{(p-1)} + h_1^{(p-1)} h_3^{(p-1)} h_4^{(p-1)p} + h_1^{(p-1)p} h_2^{(p-1)p} h_4^{(p-1)p} +$$

$$h_1^{(p-1)p} h_2^{(p-1)p} h_3^{(p-1)p} \Rightarrow$$

$$d_{4,1} - (h_3^{(p-1)} h_4^{(p-1)p} + h_2^{(p-1)p} h_4^{(p-1)p} + h_2^{(p-1)p} h_3^{(p-1)p}) h_1^{(p-1)p} =$$

$$h_2^{(p-1)} h_3^{(p-1)} h_4^{(p-1)}$$

Thus

$$d_{4,1}(I) = d_{4,1} - h_1^{(p-1)p} d_{4,2}(I)$$

Finally $d_{4,0}$:

$$d_{4,0} = h_1^{(p-1)} h_2^{(p-1)} h_3^{(p-1)} h_4^{(p-1)} = h_1^{(p-1)} d_{4,1}(I)$$

Remark 18. *According to proposition 4 each Dickson algebra generator is a function on* $\left\{h_1^{p-1}, ..., h_n^{p-1}\right\}$*:* $F_{n,i}(h_1^{p-1}, ..., h_n^{p-1}) = d_{n,n-i}$*. Let* $I = (n_1, n - n_1)$ *we define*

$$d_{n,n-i}(I) = F_{n,i}(0, ..., 0, h_{i+1}^{p-1}, ..., h_n^{p-1})$$

for $n - n_1 \geqslant i$*.*

Let us note that $d_{n,n-i}(I)$ also depends on the value of n_1. Moreover, the new polynomial is a summand of $d_{n,n-i}$ and it will be expressed in terms of old generators. The following proposition is an application of corollary 5 and Theorem 11.

Proposition 19. *Let* $I = (n-1, 1)$*, then*

$$\mathbb{F}_p(I) = \mathbb{F}_p[d_{n-1,i}, h_n^{p-1} \mid 0 \leqslant i \leqslant n-2]$$

and

$$H^*(V)^{P(I)} \cong \mathbb{F}_p(I) \bigoplus_{s_i} \mathbb{F}_p(I) \left[M_{n-1,s_1,\ldots,s_k} L_{n-1}^{p-2} \bigoplus_{t_i} M_{n,t_1,\ldots,t_k,n-1} h_n^{p-2} \right]$$

Here $0 \leqslant s_1 < \ldots < s_k \leqslant n-2$ *and* $0 \leqslant t_1 < \ldots < t_k < n-1$.

Next we compute the action of Steenrod's algebra on an upper triangular generator.

Theorem 20. *i) Let* $m = \sum_{t=s}^{n-2} a_t p^t$ *and* $p-1 \geqslant a_{n-2}+1 \geqslant a_{n-3} \geqslant \ldots \geqslant a_s \geqslant 0$, *then*

$$P^m h_n = h_n \left(d_{n-1,n-2} (-1)^{a_{n-2}} \binom{a_{n-2}+1}{a_{n-3}} \prod_{t=s}^{n-3} \binom{a_t}{a_{t-1}} \prod_{t=s}^{n-2} d_{n-1,t}^{a_t - a_{t-1}} \right)$$

ii) Let $m = p^{n-1}$*, then*

$$P^m h_n = h_n^p$$

iii) For all other cases, $P^m h_n = 0$.

Proof. $P^m h_n = P^m \left(\sum_{t=0}^{n-1} (-1)^t y_n^{p^{n-1-t}} d_{n-1,t} \right) =$
$\sum_{t=0}^{n-1} (-1)^t \sum_{i+j=m} P^i y_n^{p^{n-1-t}} P^j d_{n-1,t}$. If $P^i y_n^{p^{n-1-t}} = 0$ for all t, then $P^m h_n = 0$. Thus $P^m h_n \neq 0$ implies $\exists t$ and i such that

$$P^i y_n^{p^{n-1-t}} = \begin{cases} y_n^{p^{n-1-t}} & \text{for } i = 0 \\ P^i y_n^{p^{n-t}} & \text{for } i = p^{n-1-t} \\ 0, \text{ otherwise} & \end{cases} \quad \text{and } P^j d_{n-1,t} \neq 0$$

Thus y_n divides $P^m h_n$. Since $P^m h_n \in H_n$, $P^m h_n = h_n f$ and $f \in H_n$.

Let $m < p^{n-1}$ and $P^m h_n \neq 0$, then $P^m y_n^{p^{n-1}} = 0$ and $P^i y_n^{p^{n-2}} \neq 0$. Otherwise, $P^i y_n^{p^l} = y_n^{p^{l+1}}$ for $l+1 < n-1$. In that case $P^m h_n = 0$. Thus $i = p^{n-2}$ and $m = p^{n-2} + j$.

According to Theorem 7, $P^j d_{n-1,n-2} \neq 0$ if and only if $j = \sum_{t=s}^{n-2} a_t p^t$ and $p-1 \geqslant a_{n-2}+1 \geqslant a_{n-3} \geqslant \ldots \geqslant a_s \geqslant 0$. Now the statement follows. □

Proposition 21 ([5]). *i)* $P^m M_{i,i-1} L_{i-1}^{(p-3)/2} =$

$$M_{i,i-1} P^m L_{i-1}^{(p-3)/2} + \Sigma_0^{i-1} M_{i,t} P^{m-(p^{i-1}+\ldots+p^t)} L_{i-1}^{(p-3)/2}$$

ii) If $m = p^{i-2} + ... + 1 + \Sigma a_{i_t}(p^{i-2} + ... + p^{i_t})$ *and* $\Sigma a_{i_t} \leqslant \frac{p-3}{2}$*, then*

$$\beta P^m M_{i,i-1} L_{i-1}^{(p-3)/2} = (a_{i_1}, ..., a_{i_l})\, L_i \left(\prod L_{i-1,i_t}^{a_t}\right) L_{i-1}^{(p-3)/2-\Sigma a_{i_t}}$$

And $\beta P^m M_{i,i-1} L_{i-1}^{(p-3)/2} = 0$*, otherwise.*

Here $(a_{i_1}, ..., a_{i_l}) = ((p-3)/2)!/\Sigma a_{i_t}!\,(\frac{p-3}{2} - \Sigma a_{i_t})!$.

According to Theorem 20 and proposition 21, the action of Steenrod's algebra is closed on the generating set above.

Next we proceed to the case $I = (1, n-1)$.

Proposition 22. *Let* $I = (1, n-1)$ *and* $n - 1 \geqslant i \geqslant 1$*, then* $d_{n,n-i}(I)$ *can be decomposed in terms of* $d_{n,n-t}$ *and vise versa for* $t < i$ *as follows*

$$d_{n,n-i} = h_1^{(p-1)p^{n-i}} d_{n,n-i+1}(I) + d_{n,n-i}(I)$$

$$d_{n,n-i}(I) = d_{n,n-i} - \sum_{t=n-i}^{n-1} (-1)^{t+i+1-n} h_1^{(p-1)(p^{n-i}+...+p^t)} d_{n,t+1}$$

Proof. We apply proposition 4 and induction on i. $d_{n,n-i}$ is a combination of i elements from $\{h_1^{(p-1)}, ..., h_n^{(p-1)}\}$ or 1 from $\{h_1^{(p-1)}\}$ and $i-1$ from $\{h_2^{(p-1)}, ..., h_n^{(p-1)}\}$ on certain powers:

$$d_{n,n-i} = h_1^{(p-1)p^{n-i}} d_{n,n-i+1}(I) + d_{n,n-i}(I)$$

Now the claim follows. □

Theorem 23. *Let* $I = (1, n-1)$*, then*

$$\mathbb{F}_p(I) = \mathbb{F}_p[h_1^{p-1}, d_{n,i}(I) \mid 1 \leqslant i \leqslant n-1]$$

and

$$H^*(V)^{P(I)} \cong \mathbb{F}_p(I) \oplus \mathbb{F}_p(I) \left[M_{1,0} h_1^{p-2} \bigoplus_{t_l} M_{n,t_1,...,t_k} L_n^{p-2} \right]$$

Here $1 \leqslant t_k$ *and* $0 \leqslant t_1 < ... < t_k \leqslant n-1$.

Proof. Because of last proposition the $d_{n,i}(I)$'s are invariants and consist a polynomial basis. The claim follows from Theorem 11. □

Proposition 24. *The action of Steenrod's algebra on the generating set* $\left\{h_1^{p-1}, d_{n,i}(I) \mid 1 \leqslant i \leqslant n-1\right\}$ *is closed.*

Proof. We need to evaluate the action of Steenrod's algebra for the Steenrod algebra generators P^{p^l} only. We recall Theorem 7.

$$P^{p^l} d_{n,i} = \begin{cases} d_{n,i-1}, \text{ for } l = i-1 \\ -d_{n,i} d_{n,n-1}, \text{ for } l = n-1 \quad \text{and} \\ 0, \text{ otherwise} \end{cases}$$

$$P^{p^l} h_1^{(p-1)p^k} = \begin{cases} -h_1^{p^{k+1}-(p-2)p^k}, \text{ for } l = k \\ 0, \text{otherwise} \end{cases}$$

Because of proposition 22, we have to consider $P^{p^{n-1}}, ..., P^{p^{n-i-1}}$ only. Let $n-i > 1$, then $P^{p^l} d_{n,i}(I)$ is a function on the set:

$$\left\{h_1^{p-1}, d_{n,i}(I) \mid 1 \leqslant i \leqslant n-1\right\}$$

Let $n - i = 1$, then we apply relation $d_{n,0} = h_1^{(p-1)} d_{n,1}(I)$ on $P^{p^l} d_{n,1}(I)$. □

For the general case please see [7].

5. $\mathbb{F}_p(n-1,1)$ and $\mathbb{F}_p(1,n-1)$ as free modules over D_n

D_n serves as a homogeneous system of parameters and in fact both $\mathbb{F}_p[V]^{U_n}$ and $\mathbb{F}_p(I)$ are free D_n-modules. A free basis has been given for $\mathbb{F}_p[V]^{U_n}$ as a module over D_n ([2] and [5]).

Since U_n is a p-Sylow subgroup of $GL(n, \mathbb{F}_p)$ and H_n is a polynomial algebra, $\mathbb{F}_p(I)$ is Cohen-Macaulay. Hence, $\mathbb{F}_p(I)$ is a free module over D_n.

Remark 25. *i) The rank of $\mathbb{F}_p(I)$ over D_n is $[GL(n, \mathbb{F}_p) : P(I)]$.*
ii) Let $P(G,t)$ denote the Poincaré series of $\mathbb{F}_p(n-1,1)$. Note that $|d_{n-1,i}| = p^{n-1} - p^i$ divides $|d_{n-1+1,i+1}|$ and hence
$P(D_n, t) / P(G, t) = \prod_i \left(1 + t^{|d_{n-1,i}|} + t^{2|d_{n-1,i}|} + \cdots + t^{(p-1)|d_{n-1,i}|}\right).$

Definition 26. *Let the symbol $B_A(A')$ stand for a free module basis of the algebra A' over the algebra A.*

Theorem 27 ([2])**.** $B_{D_n}(H_n) = \left\{h_1^{r_1}...h_n^{r_n} \mid 0 \leqslant r_i < p^{n-i+1} - 1\right\}$ *is a free module basis for H_n over D_n.*

Corollary 28. $\operatorname{Im}(res^* : H^*(\Sigma_{p^n,p}) \to H^*(V))$ *is isomorphic to the free module over D_n on*

$$\left\{ \begin{array}{c} \hat{M}_{i,s_1,\ldots,s_{k-1},i-1} \hat{L}_{i-1}^{(p-3)/2} \prod_2^k \hat{L}_{s_t}^{(p-3)/2} \prod_2^k \hat{L}_{s_t+1}^{(p-3)/2} \hat{h}_1^{r_1} ... \hat{h}_n^{r_n} \mid \\ 0 \leqslant r_i < p^{n-i+1} - 1, k \leqslant i \leqslant n, 0 \leqslant s_1 < ... < s_{k-1} < i-1 \end{array} \right\}$$

Proof. This is an application of last Theorem and proposition 16. □

Proposition 29. $B_{D_n}(\mathbb{F}_p(1,n-1)) = \left\{h_1^{(p-1)m} \mid 0 \leqslant m \leqslant A_1\right\}$ *is a free module basis for $\mathbb{F}_p(1,n-1)$ over D_n. Here $A_1 = p^{n-1} + ... + p$.*

Proof. Our statement follows directly from the following formulas:

$$d_{n,0} = d_{n,1}(I)\, h_1^{(p-1)}$$

$$d_{n,1} = d_{n,1}(I) + \sum_{t=1}^{n-1} (-1)^t d_{n,1+t} h_1^{(p-1)p^t + ... + p}$$

$$d_{n,0} = d_{n,1}h_1^{(p-1)} + \sum_{t=1}^{n-1}(-1)^t d_{n,1+t}h_1^{(p-1)p^t+p^{t-1}+...+1}$$

□

Corollary 30 ([5]).

$$B_{D_n}(\mathbb{F}_p(1,...,1)) = \left\{h_1^{(p-1)m_1}...h_{n-1}^{(p-1)m_{n-1}} \,|\, 0 \leqslant m_i \leqslant A_i\right\}$$

is a free module basis for $\mathbb{F}_p(1,...,1)$ *over* D_n. *Here* $A_i = p^{n-i} + ... + p$.

In the opposite direction as in the last proposition, we consider the analogue statement. Next lemma demonstrates our approach.

Lemma 31. $B_{D_4}(\mathbb{F}_p(3,1)) = \left\{d_{3,0}^i d_{3,1}^j d_{3,2}^k \mid 0 \leqslant i,j,k \leqslant p-1\right\} \cup$ $\left\{d_{3,0}^p d_{3,1}^i d_{3,2}^j \,|0 \leqslant i,j \leqslant p-1\right\} \cup \{d_{3,1}^p d_{3,2}^i \mid 0 \leqslant i \leqslant p-1\} \cup \{d_{3,2}^p\}$ *is a free module basis for* $\mathbb{F}_p(3,1)$ *over* D_4.

Proof. Because of remark 25, our statement follows directly from the following relations and induction on the total degree of $d_{3,2}^{m_2}d_{3,1}^{m_1}d_{3,0}^{m_0}$:
i) $d_{4,0} = d_{4,3}d_{3,0} - d_{3,0}d_{3,2}^p \Rightarrow d_{3,0}d_{3,2}^p = -d_{4,0} + d_{4,3}d_{3,0}$;
ii) $d_{4,1} = d_{4,3}d_{3,1} + d_{3,0}^p - d_{3,1}d_{3,2}^p \Rightarrow d_{3,1}d_{3,2}^p = -\,d_{4,1} + d_{4,3}d_{3,1} + d_{3,0}^p$;
iii) $d_{4,2} = d_{4,3}d_{3,2} + d_{3,1}^p - d_{3,2}^{p+1} \Rightarrow d_{3,2}^{p+1} = -d_{4,2} + d_{4,3}d_{3,2} + d_{3,1}^p$;
iv) $-d_{3,0}^{p+1} = d_{4,0}d_{3,1} - d_{4,1}d_{3,0}$;
v) $-d_{3,0}d_{3,1}^p = d_{4,0}d_{3,2} - d_{4,2}d_{3,0}$;
vi) $-d_{3,1}^{p+1} = d_{4,1}d_{3,2} - d_{4,2}d_{3,1} - d_{3,0}^p d_{3,2}$. □

For each t, $1 \leqslant t \leqslant n-1$, we define the set of all $(n-t)$-tuples

$$\mathcal{M}(n-2,t) = \{M = (p, m_t, ..., m_{n-2}) \mid 0 \leqslant m_i \leqslant p-1\}$$

and, for each $M \in \mathcal{M}(n-2,t)$ we define

$$d_{n-1}^M = d_{n-1,t-1}^p d_{n-1,t}^{m_t} ... d_{n-1,n-2}^{m_{n-2}}$$

Theorem 32. *We have*

$$B_{D_n}(\mathbb{F}_p(n-1,1)) = \bigcup_{t=1}^{n-1}\{d_{n-1}^M \,|\, M \in \mathcal{M}(n-2,t)\}$$

as a free module basis for $\mathbb{F}_p(n-1,1)$ *over* D_n.

Proof. Because of remark 25, we only have to prove that the given set is a generating set. We use induction on the total degree $|m| := \sum m_i$ of a typical monomial $d^m = \prod_{i=0}^{n-2} d_{n-1,i}^{m_i}$.
Let us recall our relations:
$d_{n,i} = d_{n,n-1}d_{n-1,i} - d_{n-1,i}d_{n-1,n-2}^p + d_{n-1,i-1}^p \Rightarrow$

$$d_{n-1,i}d_{n-1,n-2}^p = -d_{n,i} + d_{n,n-1}d_{n-1,i} + d_{n-1,i-1}^p \text{ for } 0 \leqslant i \leqslant n-2 \qquad (9)$$

$$d_{n,i}d_{n-1,j} - d_{n,j}d_{n-1,i} = d^p_{n-1,i-1}d_{n-1,j} - d_{n-1,i}d^p_{n-1,j-1} \Rightarrow$$

$$d_{n-1,i}d^p_{n-1,j-1} = -d_{n,i}d_{n-1,j} + d_{n,j}d_{n-1,i} + d^p_{n-1,i-1}d_{n-1,j} \text{ for } 0 \leqslant i \leqslant j-1. \quad (10)$$

$$d^{p+1}_{n-1,0} = d_{n,1}d_{n-1,0} - d_{n,0}d_{n-1,1} \quad (11)$$

Please note that relations (9) and (11) reduce the total degree. On the other hand, relation (10) does not, but it moves the same type of degree to the left with respect to index i.

It is obvious, because of the types of the relations above, that no other relation can be deduced from the ones given. Namely, any combination of these ends up to the one given.

Let d_i denote $d_{n-1,i}$ for simplicity.

Let $d^m = \prod_{i=1}^{l} d_{s_i}^{m_{s_i}}$, $0 \leqslant s_1 < ... < s_l \leqslant n-2$, and $0 < m_{s_i}$. Let $f = d^m/d_{s_1}$. Then $f = \sum d(i) f(i)$ where $d(i) \in D_n$ and $f(i)$ is a basis element by induction. Let $g = f(i)d_{s_1}$ and $f(i) = \prod d_{s'_i}^{m_{s'_i}}$. Here $m_{s'_1} \leqslant p$ and $m_{s'_i} < p$.

If $s_1 < s'_1$, then g is a basis element. If $s_1 = s'_1$ and $m_{s'_1} < p$, then g is a basis element.

Let $s_1 = s'_t$, $m_{s'_t} = p$ and t maximal. Thus $g = g'd^{p+1}_{s'_t}$.

i) If $s'_t = 0$ or $n-2$, then the total degree of the decomposition according to relations (9) and (11) is strictly less than that of g.

ii) Let $0 < s'_t < n-2$. According to relation (10), $d^{p+1}_{s'_t} = d^p_{s'_t-1}d_{s'_t+1} + others$.

(*) Now we consider $g'd^p_{s'_t-1}d_{s'_t+1}$. Again by relation (10), this element is either a basis element or decomposes to $g'd^p_{s'_t-2}d_{s'_t}d_{s'_t+1} + others$. After a finite number of steps either a basis element is obtained plus others or $d_0^{k_0}d_1^{k_1}...d_{s'_t+1}^{k_{s'_t+1}}$ such that $k_0 > p$ and $k_i < p$. Now relation (11) is in order.

Let $s_1 = s'_t$, $m_{s'_t} = p-1$ and $t > 1$. In this case g has the form $g = ...d_{s'_{t-1}}d^p_{s'_t}...$ and we proceed as in (*) above. □

Corollary 33. *i)* $\operatorname{Im}\left(res^* : H^*\left(\Sigma_p \int \Sigma_{p^{n-1}}\right) \to H^*(V)\right)$ *is isomorphic to a free module over* D_n *on*

$$\left\{ \begin{array}{c} \hat{M}_{1,0}\hat{L}_1^{(p-2)}\hat{h}_1^{(p-1)m}, \hat{M}_{n,s_1,\ldots,s_k}\hat{L}_n^{(p-2)}d_{n,0}^{([\frac{k+1}{2}]-1)}\hat{h}_1^{(p-1)m} \mid \\ 0 \leqslant m < A_1, k \leqslant n, 1 \leqslant s_k, 0 \leqslant s_1 < ... < s_k \leqslant n-1 \end{array} \right\}$$

Here $A_1 = p^{n-1} + ... + p$.

ii) $\operatorname{Im}\left(res^* : H^*\left(\Sigma_{p^{n-1}} \int \Sigma_p\right) \to H^*(V)\right)$ *is isomorphic to a free module over* D_n *on*

$$\left\{ \begin{array}{c} \hat{M}_{n,n-1}\hat{L}_n^{(p-2)}f, \hat{M}_{n-1,s_1,\ldots,s_k}\hat{L}_{n-1}^{(p-2)}d_{n-1,0}^{([\frac{k+1}{2}]-1)}g \mid \\ f,g \in B_{D_n}(\mathbb{F}_p(n-1,1)), k \leqslant n-1, 0 \leqslant s_1 < ... < s_k \leqslant n-1 \end{array} \right\}$$

Proof. This is an application of proposition 29, Theorem 32 and 11, corollary 28 and lemma 10. □

6. The transfer

In the opposite direction of the restriction map, a map is defined called the transfer for H a subgroup of finite index in G:

$$tr^* : H^*(H) \to H^*(G)$$

At the cochain level $tr^*(a)(\lambda) = \sum_{1}^{[G:H]} g_i a\left(g_i^{-1}\lambda\right)$. Here $a \in C_H^i = Hom_{\mathbb{Z}(H)}(C_i, \mathbb{F}_p)$, $\lambda \in C_i$ and $\{g_i\}$ is a set of left coset representatives ([**1**] page 71).

Let us recall from the introduction that the Weyl subgroups of V in $\Sigma_{p^{n_1}} \int \Sigma_{p^{n_2}}$ and Σ_{p^n} are $P(n_1, n_2)$ and the general linear group $GL(n, \mathbb{F}_p)$ respectively. The induced inclusion

$$W_{\Sigma_{p^{n_1}} \int \Sigma_{p^{n_2}}}(V) \to W_{\Sigma_{p^n}}(V)$$

induces

$$H^*(V)^{P(n_1,n_2)} \xrightarrow{\tau^*} H^*(V)^{GL(n,\mathbb{F}_p)}$$

given by $\tau^*(f) = \sum_{1}^{[G:H]} g_i f$. Here f is a $P(n_1, n_2)$-invariant polynomial. V is a $\mathbb{F}_p G$-module. In our case the transfer is surjective and $H^*(V)^{GL(n,\mathbb{F}_p)}$ is a direct summand. The following diagram is commutative, please see [**8**].

$$\begin{array}{ccc} H^*\left(\Sigma_{p^{n_1}} \int \Sigma_{p^{n_2}}\right) & \xrightarrow{tr^*} & H^*(\Sigma_{p^n}) \\ (res_V^{\Sigma_{p^{n_1}} \int \Sigma_{p^{n_2}}})^* \downarrow & & \downarrow (res_V^{\Sigma_{p^n}})^* \\ H^*(V)^{W_{\Sigma_{p^{n_1}} \int \Sigma_{p^{n_2}}}(V)} & \xrightarrow{\tau^*} & H^*(V)^{W_{\Sigma_{p^n}}(V)} \end{array}$$

Campbell and Hughes ([**2**]) have studied the transfer for the case:

$$\tau^* : H_n \to D_n$$

We extended the result above ([**5**]) for

$$H^*(V)^{U_n} \xrightarrow{\tau^*} H^*(V)^{GL(n,\mathbb{F}_p)}$$

In this work we consider the induced map

$$\bar{\tau}^* : \mathrm{Im}\left(res^* : H^*(G) \to H^*(V)\right) \to \mathrm{Im}\left(res^* : H^*(\Sigma_{p^n}) \to H^*(V)\right)$$

Here $G = \Sigma_{p^n,p}$, $\Sigma_p \int \Sigma_{p^{n-1}}$ and $\Sigma_{p^{n-1}} \int \Sigma_p$.

Next we define a set of coset representatives for the groups under consideration. We apply the method of Campbell and Hughes.

Let $\mathrm{Pr}_n(x) \in \mathbb{F}_p[x]$ be an irreducible polynomial of degree n and σ_n a root of $\mathrm{Pr}_n(x)$ in the $(p^n - 1)$-st cyclotomic field $\mathbb{F}_p^n$ over $\mathbb{F}_p$ ([**10**]).Let σ_n be a primitive root of unity and its minimal polynomial

$$\Pr_n(x) = c_0 + c_1 x + ... + c_{n-1}x^{n-1} + x^n$$

Here there exists j such that $c_j c_0 \neq 0$. Then

$$\sigma_n^n = -\left(c_0 + c_1\sigma_n + ... + c_{n-1}\sigma_n^{n-1}\right)$$

and the companion matrix of $\Pr_n(x)$ is

$$A_n = \begin{pmatrix} 0 & \dots & 0 & -c_0 \\ 1 & & & -c_1 \\ \vdots & \ddots & & \vdots \\ 0 & \dots & 1 & -c_{n-1} \end{pmatrix}$$

with $\Pr_n(A_n) = 0_{n\times n}$. So A_n is a representative for $\left(\mathbb{F}_p^n\right)^*$ and can be identified with σ_n.

$$\mathbb{F}_p^n = <\sigma_n^0, \sigma_n, ..., \sigma_n^{n-1}> = <\sigma_n, ..., \sigma_n^{p^n-1}>, <\sigma_n> = \left(\mathbb{F}_p^n\right)^*$$
$$<\sigma_n^{\frac{p^n-1}{p-1}}> = \mathbb{F}_p^*$$

A_n acts linearly on $\mathbb{F}_p^n$ and $A_n\left(c\sigma_n^i\right) = c\sigma_n^{i+1}$. Let us note that this action is compatible with the given action on the rings of invariants: let $c\sigma_n^i$ be represented by $(0,...,0,c,0,...,0)$ with respect to the given basis, then $A_n\left(c\sigma_n^i\right)$ is the matrix multiplication between $(0,...,0,c,0,...,0)$ and the $i+1$-th column of A_n.

Moreover, $\sigma_n^k = \sigma_n^{k-n+1}\sigma_n^{n-1}$ or $A_n^{k-n+1}\left(\sigma_n^{n-1}\right)$ for any k. Thus the last column of A_n^k can be any non-zero element of $\left(\mathbb{F}_p^n\right)^*$.

Let

$$\Phi_n :< \sigma_n^0, \sigma_n, ..., \sigma_n^{n-1} > \longrightarrow V^n$$

Then the map induced by $\Phi_n(\sigma_n^i) = y_{n-i}$ and linearity is an isomorphism. Moreover, A_n acts on V via Φ_n: $A_n y_i = \Phi_n\left(\sigma_n\sigma_n^{n-i}\right)$. Now, $\left(\mathbb{F}_p^n\right)^*$ can be viewed as a subset of the group of automorphisms $GL(n,\mathbb{F}_p)$.

Inductively we define σ_m such that $\langle\sigma_m\rangle \cong \langle y_{n-m+1},...,y_n\rangle^*$ and $\Phi_m :< \sigma_m^0, \sigma_m, ..., \sigma_m^{m-1} > \rightarrow \langle y_{n-m+1},...,y_n\rangle$. We consider $A_m \in GL(n,\mathbb{F}_p)$ such that $A_m(y_j) = y_j$ for $1 \leqslant j \leqslant n-m$ and $A_m(y_j) = \Phi_m\left(\sigma_m\sigma_m^{n-m+1-j}\right)$ for $1+n-m \leqslant j \leqslant n$.

Lemma 34. *i) Let* $Ç = \{\left(A_n^i\right)^{-1} \mid 0 \leqslant i \leqslant p^n-2\}/\sim$ *where* $A_n^i \sim A_n^j$, *if there exists* $c \in (\mathbb{F}_p)^*$ *such that* $A_n^i = cA_n^j$. *Then the set* $Ç$ *is a set of left coset representatives for* $GL(n,\mathbb{F}_p)$ *over* $P(1,n-1)$.

ii) Let $\acute{C} = \{\left(A_n^i\right)^t \mid 0 \leqslant i \leqslant p^n-2\}/\sim$ *where* $A_n^i \sim A_n^j$, *if there exists* $c \in \mathbb{F}_p^*$ *such that* $A_n^i = cA_n^j$. *Then the set* $\acute{C}$ *is a set of left coset representatives for* $GL(n,\mathbb{F}_p)$ *over* $P(n-1,1)$.

Proof. We recall that $|GL_n : P(1,n-1)| = \frac{p^n-1}{p-1} = |GL_n : P(n-1,1)| = |Ç| = |\acute{C}|$. The first column of $\left(A_n^k\right)^{-1}$ or the last row of $\left(A_n^k\right)^t$ can be any non-zero element of $\left(\mathbb{F}_p^n\right)^*$. Let $g, g' \in P(1,n-1)$ and $\left(A_n^k\right)^{-1} g = \left(A_n^l\right)^{-1} g'$, then $\left(A_n^{k-l}\right)^{-1} \in P(1,n-1)$ which is not the case for $k \neq l$. The same is true for $P(n-1,1)$. □

The following proposition has been proved by Campbell and Hughes in [**2**].

Proposition 35. *The set* $\left\{ \left(A_n^{i_n}\right)^{-1} ... \left(A_1^{i_1}\right)^{-1} \mid 0 \leqslant i_m \leqslant p^m - 2 \right\}$ *is a set of left coset representatives for* $GL(n, \mathbb{F}_p)$ *over* U_n.

Proof. We apply induction on n. □

Proposition 36. *Let* $\xi : \mathbb{F}_p(1, n-1) \longrightarrow D_n$ *be the natural epimorphism with respect to the given free module basis* B *and*

$$\tau^* : \mathbb{F}_p(1, n-1) \to D_n$$

the transfer map. Then $\xi = \tau^*$.

Proof. Let us recall that a free module basis consists of $\left(h_1^{p-1}\right)^m$ for $0 \leqslant m \leqslant \frac{p^n - p}{p-1} = p^{n-1} + ... + p$.

$$\tau^* \left(h_1^{p-1}\right)^m = \sum_i \left(A_n^i\right)^{-1} \left(h_1^{p-1}\right)^m = \sum_i \left(\left(A_n^i\right)^{-1} h_1\right)^{(p-1)m} =$$

$$\sum_{u \in V} (u)^{(p-1)m} = (p-1) \sum_{1 \leqslant i \leqslant n, v \in <y_1, ..., y_{i-1}>} (y_i + v)^{(p-1)m}$$

The last summand is a GL-invariant and so only $m\,(p-1) = p^n - p^k$ for $1 \leqslant k \leqslant n-1$ should be considered, i.e. $\tau^* \left(h_1^{p-1}\right)^m$ is a scalar multiple of $d_{n,k}$. Because of proposition 4, $d_{n,k}$ contains $\left(\prod_{t=1}^{n-k} y_{k+t}^{p^{k+t-1}}\right)^{p-1}$. Next we consider the coefficient of this monomial in $\tau^* \left(h_1^{p-1}\right)^m$ or in

$$\sum_{v \in <y_{k+1}, ..., y_{n-1}>, u \in <y_1, ..., y_k>} (y_n + v + u)^{(p-1)m}$$

This coefficient is $p^k \frac{(p^n - p^k)!}{\prod_{t=1}^{n-k} (p^{k+t-1}(p-1))!} \equiv 0 \mod p$. Thus $\tau^* \left(h_1^{p-1}\right)^m = 0$. □

Remark 37. *According to the last proof, if* $m\,(p-1) = p^n - 1$, *then*

$$\tau^* \left(h_1^{p-1}\right)^m = (p-1)\, d_{n,0}$$

Theorem 38 ([2]). *Let* $\xi : H_n \longrightarrow D_n$ *be the natural epimorphism with respect to the given basis* B *and* $\tau^* : H_n \to D_n$ *the transfer map. Then* $\xi = \tau^*$.

Proof. It is obvious that $\prod \left(A_s^{i_s}\right)^{-1} \left(\prod h_t^{r_t}\right) =$

$$\left(A_n^{i_n}\right)^{-1} \left(h_1^{r_1} \left(A_{n-1}^{i_{n-1}}\right)^{-1} \left(h_2^{r_2} ... \left(A_2^{i_2}\right)^{-1} \left(h_{n-1}^{r_{n-1}} \left(\left(A_1^{i_1}\right)^{-1} h_n^{r_n}\right)\right) ...\right)$$

Let $r_i < p^{n-i+1} - 1$, then the proof of last proposition for $n = n - i + 1$ implies that $\sum_{i_{n-i+1}} \left(A_{n-i+1}^{i_{n-i+1}}\right)^{-1} h_i^{r_i} = 0$. Now the statement follows. □

Corollary 39. *Let $\xi : \mathbb{F}_p(1, ..., 1) \longrightarrow D_n$ be the natural epimorphism with respect to the given basis B and $\tau^* : \mathbb{F}_p(1, ..., 1) \to D_n$ the transfer map. Then $\xi = \tau^*$.*

Next we consider $P(n-1, 1)$. In this case the use of the coset representatives arises technical problems. Instead, using degree arguments, we shall prove that only particular elements of the given basis might be expressed with respect to Dickson algebra generators. Then applying Steenrod operations on Dickson algebra generators, we shall prove that the transfer map $\tau^* : \mathbb{F}_p(n-1, 1) \to D_n$ coincides with the natural epimorphism $\xi : \mathbb{F}_p(n-1, 1) \longrightarrow D_n$ with respect to basis B.

The next technical lemma will be needed for the proof of our next Theorem.

Lemma 40. *Let m_i and m'_j be non-negative integers such that $0 \leqslant m_i \leqslant p-1$, $m'_j \geqslant 0$ and p a prime number.*

1) Let $0 \leqslant i \leqslant n-2$ and $0 \leqslant j \leqslant n-1$. Then the equation

$$\sum_0^{n-2} m_i \left(p^{n-1} - p^i\right) = \sum_0^{n-1} m'_j \left(p^n - p^j\right)$$

does not have an integral solution.

2) Let $i_0 < i \leqslant n-2$ and $0 \leqslant j \leqslant n-1$. Then the equation

$$\left(p^n - p^{i_0+1}\right) + \sum_{i_0+1}^{n-2} m_i \left(p^{n-1} - p^i\right) = \sum_0^{n-1} m'_j \left(p^n - p^j\right)$$

admits solutions of type $m_{i_0+1} = ... = m_k = p-1$, $m_i = 0$ for $k < i$ and $m'_{n-1} = k - i_0$, $m'_{k+1} = 1$ and zero otherwise. Here $i_0 < k \leqslant n-2$ and $m_i = 0$ for any $i > i_0$, $m'_{i_0+1} = 1$ and zero otherwise.

Theorem 41. *Let $\xi : \mathbb{F}_p(n-1, 1) \longrightarrow D_n$ be the natural epimorphism with respect to the given free module basis B and $\tau^* : \mathbb{F}_p(n-1, 1) \to D_n$ the transfer map. Then $\xi = \tau^*$.*

Proof. If we show that $\tau^*(d) = 0$ for all d in the basis, then $\xi = \tau^*$. Because of the statement in last lemma only the following cases should be considered: d^p_{n-1,i_0}, and $d^p_{n-1,i_0} d^{p-1}_{n-1,i_0+1} ... d^{p-1}_{n-1,k}$.

Let $\tau^*\left(d^p_{n-1,i_0}\right) = c d_{n,i_0+1}$. Applying $P^{p^{i_0}}$, we get $\tau^*\left(d^p_{n-1,i_0-1}\right) = c d_{n,i_0}$. Applying $P^p ... P^{p^{i_0}-1}$ on the previous element, we get $\tau^*\left(d^p_{n-1,0}\right) = c d_{n,1}$ But

$$P^1 \tau^*\left(d^p_{n-1,0}\right) = 0 \neq P^1\left(c d_{n,1}\right) = c d_{n,0}$$

Let $\tau^*\left(d^p_{n-1,i_0} d^{p-1}_{n-1,i_0+1} ... d^{p-1}_{n-1,k}\right) = c d_{n,k+1} d^{k-i_0}_{n,n-1}$.

Let $f = d^p_{n-1,i_0} d^{p-1}_{n-1,i_0+1} ... d^{p-1}_{n-1,k}$ and $g = d_{n,k+1} d^{k-i_0}_{n,n-1}$. We would like to apply P^{p^k}. Using Theorem 7 we show that no monomial of $P^{p^k} f \neq 0$ is in the ideal $(d_{n-1,k})$. Then $P^{p^k} f \in (d_{n-1,0}, ..., d_{n-1,k-1})$. Since

$$p^k = (p-1)\left(p^{k-1} + ... + p^{i_0}\right) + p^{i_0}$$

$P^{p^k} f$ contains the summand $d^p_{n-1,i_0-1} d^{p-1}_{n-1,i_0} ... d^{p-1}_{n-1,k-1}$. So $P^{p^k} f \neq 0$.

Claim: No monomial of $P^{p^k}f$ is in the ideal $(d_{n-1,k})$. We prove the claim by showing that there does not exist a solution of

$$p^k = \sum_{j=i_0}^{k-1}\sum_{i=0}^{j} a_{j,i}p^i$$

unless $a_{k-1,k-1} = p-1$. In that case

$$P^{p^k}f = \sum_{(m_0,\dots,m_{k-1})} c\left(m_0,\dots,m_{k-1}\right)\prod_{t=0}^{k-1} d_{n-1,t}^{m_t}$$

Here $0 \leqslant a_{j,i} \leqslant a_{j,i+1} \leqslant p-1$ for $i_0+1 \leqslant j \leqslant k-1$ and $0 \leqslant a_{i_0,i} \leqslant a_{i_0,i+1} \leqslant 1$. We consider the extreme cases and prove that there is no positive solution.

Let $a_{k-1,k-1} = p-2$, $a_{k-1,i} = p-2$, $a_{j,i} = p-1$ for $i_0+1 \leqslant j \leqslant k-2$ and $a_{i_0,t} = 1$. Then

$$\sum_{j=i_0}^{k-1}\sum_{i=0}^{j} a_{j,i}p^i = p^k - (k-i_0-3) < p^k$$

Now the claim follows.

It is obvious that if $a_{k-1,k-1} = p-1$, then no summand of $P^{p^k}f$ is in $(d_{n-1,k})$.

Applying $P^1 \dots P^{p^k}$, we get

$$P^1 \dots P^{p^k} f = 0 \text{ and } P^1 \dots P^{p^k}\left(d_{n,k}d_{n,n-1}^{k-i_0}\right) = d_{n,0}d_{n,n-1}^{k-i_0}$$

The last line is an application of Theorem 7. □

The next example is a counterexample to the statement of last Theorem in the case
$\operatorname{Im}\left(res_V^{\Sigma_{p^n},p}\right)^* \supsetneq H_n$.

Example 42. *Let* $p = 3$ *and* $n = 2$. *Then* $\operatorname{Pr}_2(x) = 2 + x + x^2$ *and* $\sigma_2^2 = 2\sigma_2 + 1$ *or* $A_2 = \begin{pmatrix} 0 & 1 \\ 1 & 2 \end{pmatrix}$, $A_2^{-1} = \begin{pmatrix} 1 & 1 \\ 1 & 0 \end{pmatrix}$. *A set of coset representatives for* $GL(2,3)$ *over* $P(1,1)$ *is given in proposition 35.*

By direct computation, $\tau^*\left(M_{1,0}h_1^{p^2-1-p}\right) = M_{2,1}L_2^{p-2} \neq 0$. *Let us note that* $M_{1,0}h_1^{p^2-1-p}$ *is a basis element.*

Theorem 43 ([5]). *Let* $\xi, \bar{\tau}^* : \operatorname{Im}\left(res_V^{\Sigma_{p^n},p}\right)^* \to \operatorname{Im}\left(res_V^{\Sigma_{p^n}}\right)^*$. *Then* $\xi = \bar{\tau}^*$ *in the ideal generated by* $(d_{n,0})$.

Proof. Let $f \in \operatorname{Im}\left(res_V^{\Sigma_{p^n},p}\right)^*$. Then $\bar{\tau}^*(fd_{n,0}) = \bar{\tau}^*(f)\,d_{n,0}$. But according to lemma 10:

$$\bar{\tau}^* (f d_{n,0}) = \bar{\tau}^* \left(\sum_J M_{n;J} L_n^{p-2} h^{I(J)} \right) = \sum_J M_{n;J} L_n^{p-2} \bar{\tau}^* \left(h^{I(J)} \right) =$$

$$\sum_J M_{n;J} L_n^{p-2} \xi \left(h^{I(J)} \right)$$

If $\xi \left(h^{I(J)} \right)$ is not divisible by $d_{n,0}$, then it must be zero. □

References

[1] Adem, A. and Milgram, R.J.: Cohomology of finite groups, Springer-Verlag, Heidelberg 1994.

[2] Campbell, H. E. A. and Hughes, I. P.: The ring of upper triangular invariants as a module over the Dickson algebra. Mathematishes Annalen. Volume 306 (1996), 429-443.

[3] Carter, R. W.: Simple Groups of Lie Type. Wiley, New York, 1972.

[4] Dickson, L. E.: A fundamental system of invariants of the general modular linear group with a solution of the form problem. Trans. A. M. S. 12 (1911), 75-98.

[5] Kechagias, Nondas: The transfer between rings of modular invariants of subgroups of $GL(n, \mathbb{F}_p)$. The Fields Institute Communications of the A. M. S.. Stable-Unstable Homotopy. S. Kochman and R. Selick, editors. Volume 19 (1998), pp. 165-180.

[6] Kechagias, Nondas: An invariant theoretic description of the primitive elements of the *modp* cohomology of a finite loop space which are annihilated by Steenrod Operations. "Proceedings of a Workshop on Invariant Theory, April 2002, Queen's University". C.R.M.- A.M.S. Proceedings and Lecture Notes Volume 35 (2003), 159-174.

[7] Kechagias, Nondas: A Note on Parabolic Subgroups and the Steenrod Algebra Action. Algebra Colloquium, Vol. 15, No. 4 (December 2008), 689 - 698.

[8] Kuhn, N. J.: Chevalley group theory and the transfer in the homology of symmetric groups. Topology Vol 24, No. 3. pp 247-264. 1985.

[9] Kuhn, N. J. and Mitchell, S. A.: The multiplicity of the Steinberg representation of $GL(n, \mathbb{F}_q)$ in the symmetric algebra. Proc. Amer. Math. Soc. 96 (1986), no. 1, 1–6.

[10] Lidl, R. and Niederreiter, H.: Finite fields. Second edition. Encyclopedia of Mathematics and its Applications, 20. Cambridge University Press, Cambridge, 1997.

[11] Huyhn Mui: Modular invariant theory and the cohomology algebras of the symmetric groups. J. Fac. Sci. Univ. Tokyo, IA (1975), 319-369.

[12] Steenrod, N. E. and Epstein, D. B.A.: Cohomology Operations. Ann. of Math. Studies No50, Princeton University Press 1962.

This article may be accessed via WWW at `http://www.rmi.acnet.ge/jhrs/`

Nondas E. Kechagias
`nkechag@uoi.gr`

Department of Mathematics
University of Ioannina
Ioannina 45110
Greece

Journal of Homotopy and Related Structures, vol. 4(1), 2009, pp.181–185

REIDEMEISTER TORSION AND ANALYTIC TORSION OF SPHERES

THIAGO DE MELO AND MAURO SPREAFICO

(*communicated by Jonathan Rosenberg*)

Abstract

We provide a simple topological derivation of a formula for the Reidemeister and the analytic torsion of spheres.

Weng and You gave in [**6**] a formula for the analytic torsion of a sphere with the standard metric. Their result is obtained by direct calculation applying the definition of analytic torsion, and is based on the explicit knowledge of the spectrum of the Laplacian on forms. Beside simplicity and generality, the result is not particularly illuminating, as stated in [**6**]. We prove in this note a different formula for the analytic torsion of a sphere, and we show that this new formula is equivalent to the one given in [**6**]. Our proof is based on purely geometric topological means, namely evaluation of the Reidemeister (R) torsion [**5**] and application of the Cheeger-Müller theorem [**1, 3**]. The main motivations, beside the different proof, are: from one side, the topological approach is natural and simpler, from the other, our formula provides a nice geometric interpretation of the result, that now we state, after introducing some notation. For a closed connected Riemannian manifold (W, g), with metric g, and an orthogonal representation ρ of the fundamental group of W, we denote by $T((W, g); \rho)$ the analytic torsion of (W, g) with respect to ρ (see [**4**] Definition 1.6), and by $\tau_{\mathrm{R}}((W, g); \rho)$ the Ray and Singer version of the Reidemeister torsion of (W, g) with respect to the same representation (see equation (3) below for the definition).

Theorem 1. *Let S^n_l be the sphere of radius l in $\mathbb{R}^{n+1}$ $(n > 0)$ with the standard Riemannian metric g_l induced by the immersion. Let ρ_0 be a trivial orthogonal representation of $\pi_1(S^n_l)$. Then,*

$$T((S^n_l, g_l); \rho_0) = \tau_{\mathrm{R}}((S^n_l, g_l); \rho_0) = \begin{cases} 1, & \text{if } n \text{ is even,} \\ (\mathrm{Vol}_{g_l}(S^n_l))^{\mathrm{rank}(\rho_0)}, & \text{if } n \text{ is odd.} \end{cases}$$

The rest of this note is dedicated to the proof of this theorem. We recall that, by the Cheeger-Müller theorem [**1, 3**], the analytic torsion $T((S^n_l, g_l); \rho)$ coincides with the R torsion $\tau_{\mathrm{R}}((S^n_l, g_l); \rho)$. Thus, it remains to evaluate the R torsion. For we first introduce some notation. Let

$$C: \qquad C_n \xrightarrow{\partial_n} C_{n-1} \xrightarrow{\partial_{n-1}} \cdots \xrightarrow{\partial_2} C_1 \xrightarrow{\partial_1} C_0,$$

M. Spreafico thanks M. Lesh for useful conversations.
Received October 29, 2008, revised May 27, 2009; published on July 5, 2009.
2000 Mathematics Subject Classification: 57Q10, 58J52.
Key words and phrases: Reidemeister-Franz torsion, analytic torsion.

be a chain complex of real vector spaces. Denote by $Z_q = \ker \partial_q$, by $B_q = \mathrm{Im}\partial_{q+1}$, and by $H_q(C) = Z_q/B_q$ as usual. For two bases $x = \{x_1, \dots, x_m\}$ and $y = \{y_1, \dots, y_m\}$ of a vector space V, denote by (y/x) the matrix defined by the change of base. For each q, fix a base c_q for C_q, and a base h_q for $H_q(C)$. Let b_q be a set of (independent) elements of C_q such that $\partial_q(b_q)$ is a base for B_{q-1}. Then the set of elements $\{\partial_{q+1}(b_{q+1}), h_q, b_q\}$ is a base for C_q. In this situation, the Reidemeister torsion of the complex C with respect to the graded base $h = \{h_q\}$ is the positive real number

$$\tau_{\mathrm{R}}(C; h) = \prod_{q=0}^{n} |\det(\partial_{q+1}(b_{q+1}), h_q, b_q/c_q)|^{(-1)^q}.$$

Let K be a connected finite cell complex of dimension n and $\tilde{K}$ its universal covering complex, and identify the fundamental group of K with the group of the covering transformations of $\tilde{K}$. Let $C(\tilde{K}; \mathbb{R})$ be the real chain complex of $\tilde{K}$. The action of the group of covering transformations makes each chain group $C_q(\tilde{K}; \mathbb{R})$ into a module over the group algebra $\mathbb{R}\pi_1(K)$, and each of these modules is $\mathbb{R}\pi_1(K)$-free and finitely generated by the natural choice of the q-cells of K. We have got the complex $C(\tilde{K}; \mathbb{R}\pi_1(K))$ of free finitely generated modules over $\mathbb{R}\pi_1(K)$. Let $\rho : \pi_1(K) \to O(m, \mathbb{R})$ be a representation of the fundamental group, and consider the twisted complex $C(K; \mathbb{R}^m_\rho)$. Assume $H_q(C(K; \mathbb{R}^m_\rho))$ are free finitely generated modules over $\mathbb{R}\pi_1(K)$. The Reidemeister torsion of K with respect to the representation ρ and to the graded base h is defined applying the previous construction to the twisted complex $C(K; \mathbb{R}^m_\rho)$, namely

$$\tau_{\mathrm{R}}(K; h, \rho) = \prod_{q=0}^{n} |\det \rho(\partial_{q+1}(b_{q+1}), h_q, b_q/c_q)|^{(-1)^q}, \tag{1}$$

in $\mathbb{R}^+$. If K is the cellular (or simplicial) decomposition of a space X, the Reidemeister torsion of X is defined accordingly, and denoted by $\tau_{\mathrm{R}}(X; h, \rho)$. It was proved in [**2**] that $\tau_{\mathrm{R}}(X; h, \rho)$ does not depend on the decomposition K.

Let W be a closed connected orientable Riemannian manifold of dimension n with Riemannian metric g. Then, all the previous assumptions are satisfied, and the R torsion $\tau_{\mathrm{R}}(W; h, \rho)$ is well defined for each fixed graded base h for the homology of W, and each representation ρ of the fundamental group. In this context, Ray and Singer suggest a natural geometric invariant object, by fixing an appropriate base h using the geometric structure, as follows. Let $E_\rho \to W$ be the real vector bundle associated to the representation $\rho : \pi_1(W) \to O(m, \mathbb{R})$. Let $\Omega(W, E_\rho)$ be the graded linear space of smooth forms on W with values in E_ρ. The exterior differential on W defines the exterior differential on $\Omega^q(W, E_\rho)$, $d : \Omega^q(W, E_\rho) \to \Omega^{q+1}(W, E_\rho)$. The metric g defines an Hodge operator on W and hence on $\Omega^q(W, E_\rho)$, $\star : \Omega^q(W, E_\rho) \to \Omega^{n-q}(W, E_\rho)$, and, using the inner product in E_ρ, an inner product on $\Omega^q(W, E_\rho)$. Let $\mathcal{H}^q$ be the space of the harmonic forms in $\Omega^q(W, E_\rho)$, and let $\mathcal{A}^q$ be the de Rham map $\mathcal{A}^q : \mathcal{H}^q \to C^q(W; E_\rho)$,

$$\mathcal{A}^q(\omega)(c \otimes_\rho v) = \int_c (\omega, v),$$

with $c \otimes_\rho v \in C_q(W; E_\rho)$, and where we identify the chain c with the q-cell that c represents, and $(\, , \,)$ is the inner product in $\mathbb{R}^m$. Following Ray and Singer, we define the following map, where $\hat{c}_{q,j}$ is the Poincaré dual cell to $c_{q,j}$,

$$\begin{aligned} \mathcal{A}_q : &\mathcal{H}^q \to C_q(W; E_\rho), \\ \mathcal{A}_q : &\omega \mapsto (-1)^{(n-1)q} \sum_{j,k} \int_{\hat{c}_{q,j}} (\star\omega, e_k)(c_{q,j} \otimes_\rho e_k). \end{aligned} \tag{2}$$

In this situation, let a be a graded orthonormal base for the space of the harmonic forms in $\Omega(W, E_\rho)$, then we call the positive real number

$$\tau_{\mathrm{R}}((W,g);\rho) = \tau_{\mathrm{R}}(W; \mathcal{A}(a), \rho) = \prod_{q=0}^{n} |\det \rho(\partial_{q+1}(b_{q+1}), \mathcal{A}_q(a_q), b_q/c_q)|^{(-1)^q}, \tag{3}$$

the R torsion of (W, g) with respect to the representation ρ. It can be proved that $\tau_{\mathrm{R}}((W,g);\rho)$ does not depend on the choice of the orthonormal base a.

We now compute the R torsion of the sphere S^n_l. Recall that the sphere $S^n_l = \{x \in \mathbb{R}^{n+1} \mid |x| = l\}$, is parameterized in spherical coordinates by

$$\left\{ \begin{array}{rcl} x_1 &=& l \sin\theta_n \sin\theta_{n-1} \cdots \sin\theta_3 \sin\theta_2 \sin\theta_1 \\ x_2 &=& l \sin\theta_n \sin\theta_{n-1} \cdots \sin\theta_3 \sin\theta_2 \cos\theta_1 \\ x_3 &=& l \sin\theta_n \sin\theta_{n-1} \cdots \sin\theta_3 \cos\theta_2 \\ x_4 &=& l \sin\theta_n \sin\theta_{n-1} \cdots \cos\theta_3 \\ &\vdots& \\ x_n &=& l \sin\theta_n \cos\theta_{n-1} \\ x_{n+1} &=& l \cos\theta_n \end{array} \right.$$

with $0 \leqslant \theta_1 \leqslant 2\pi$, $0 \leqslant \theta_2, \ldots, \theta_n \leqslant \pi$. The induced Riemannian metric is

$$g_l = l^2 \left((d\theta_n)^2 + \sin^2\theta_n (d\theta_{n-1})^2 + \cdots + \prod_{i=2}^{n} \sin^2\theta_i (d\theta_1)^2 \right) = l^2 g_1.$$

Let K be the standard cellular decomposition of S^n_l, with one top cell and one 0-cell. Let ρ_0 be the trivial representation when $n = 1$, of rank m. Then the relevant complex is

$$C: \quad 0 \longrightarrow \mathbb{R}^m[c^1_n] \longrightarrow 0 \longrightarrow \cdots \longrightarrow 0 \longrightarrow \mathbb{R}^m[c^1_0] \longrightarrow 0,$$

with preferred bases $c_0 = \{c^1_0\}$ and $c_n = \{c^1_n\}$. To fix the base for the homology, we need a graded orthonormal base a for the harmonic forms. This is $a_0 = \left\{ \frac{1}{\sqrt{\mathrm{Vol}_{g_l}(S^n_l)}} \right\}$ and $a_n = \left\{ \frac{1}{\sqrt{\mathrm{Vol}_{g_l}(S^n_l)}} \sqrt{|g_l|} d\theta_n \wedge \cdots \wedge d\theta_1 \right\}$. Applying the formula in equation (2) we obtain $h_0 = \{h_0^1\}$, $h_n = \{h_n^1\}$, with

$$h_0^1 = \mathcal{A}_0(a_0^1) = \frac{1}{\sqrt{\mathrm{Vol}_{g_l}(S_l^n)}}\int_{S_l^n} \sqrt{|g_l|}d\theta_n \wedge \cdots \wedge d\theta_1 c_0^1 = \sqrt{\mathrm{Vol}_{g_l}(S_l^n)}c_0^1,$$

$$h_n^1 = \mathcal{A}_n(a_n^1) = \frac{1}{\sqrt{\mathrm{Vol}_{g_l}(S_l^n)}}\int_{\mathrm{pt}} \star\sqrt{|g_l|}d\theta_n \wedge \cdots \wedge d\theta_1 c_n^1 = \frac{1}{\sqrt{\mathrm{Vol}_{g_l}(S_l^n)}}c_n^1.$$

As $b_q = \emptyset$, for all q, we have that

$$|\det \rho(h_0/c_0)| = \left(\sqrt{\mathrm{Vol}(S_l^n)}\right)^m, \qquad |\det \rho(h_n/c_n)| = \frac{1}{(\sqrt{\mathrm{Vol}(S_l^n)})^m}.$$

Applying the definition in equation (3), this gives

$$\tau_{\mathrm{R}}((S_l^n, g_l);\rho_0) = \left(\sqrt{\mathrm{Vol}_{g_l}(S_l^n)}\right)^m \left(\frac{1}{\sqrt{\mathrm{Vol}_{g_l}(S_l^n)}}\right)^{m(-1)^n},$$

and completes the proof of the theorem.

We conclude with some remarks.

- The formula in Theorem 1 is an expected result for spheres. In fact, it is well known that fixing non zero volume elements $u_q \in \Lambda^{d_q}(H^q(W;E_\rho))$, $d_q = \dim H^q(W;E_\rho)$, and hence fixing a graded base, say $h(u)$ for the homology, we have

$$\tau_{\mathrm{R}}(W;h(\alpha u)) = \frac{\alpha_0}{\alpha_1}\frac{\alpha_2}{\alpha_3}\cdots\tau_{\mathrm{R}}(W;h(u)),$$

  where $\alpha u = \{\alpha_q u_q\}$, $0 \neq \alpha_q \in \mathbb{R}$.

- If we use the cellular decomposition with two cells in each dimension, a similar calculation gives the same result, since the determinant in all dimensions q, with $0 < q < n$, is 1.

- Consider the product $S_a^n \times S_b^k$ with the product metric. Using the cellular decomposition with four cells: c_0^1, c_n^1, c_k^1, and c_{n+k}^1, we obtain

$$T(S_a^n \times S_b^k) = \begin{cases} \mathrm{Vol}(S_b^k)^{\chi(S_a^n)} & n \text{ even, } k \text{ odd,} \\ \mathrm{Vol}(S_a^n)^{\chi(S_b^k)} & k \text{ even, } n \text{ odd,} \\ 1 & n,k \text{ even or } n,k \text{ odd,} \end{cases}$$

  in agreement with one of the main properties of the torsion.

- We prove the equivalence of the formula given in Theorem 1 (in the case of $\mathrm{rank}(\rho_0) = 1$) with the formula given in [**6**]. This follows recalling the volume of the sphere S_l^n

$$\mathrm{Vol}_{g_l}(S_l^n) = \frac{2\pi^{\frac{n+1}{2}}l^n}{\Gamma\left(\frac{n+1}{2}\right)},$$

where $\Gamma(z)$ is the Euler Gamma function. This gives:

$$T((S_l^{2k+1}, g_l); \rho_0) = \frac{2\pi^{k+1} l^{2k+1}}{k!}.$$

References

[1] J. Cheeger, *Analytic torsion and the heat equation*, Ann. Math. **109** (1979) 259-322.

[2] J. Milnor, *Whitehead torsion*, Bull. AMS **72** (1966) 358-426.

[3] W. Müller, *Analytic torsion and R-torsion of Riemannian manifolds*, Adv. Math. **28** (1978) 233-305.

[4] D.B. Ray and I.M. Singer, *R-torsion and the Laplacian on Riemannian manifolds*, Adv. Math. **7** (1971) 145-210.

[5] K. Reidemeister, *Homotopieringe und Linseräume*, Hamburger Abhandl. **11** (1935) 102-109.

[6] L. Weng and Y. You, *Analytic torsions of spheres*, Int. J. Math. **7** (1996) 109-125.

This article may be accessed via WWW at `http://www.rmi.acnet.ge/jhrs/`

Thiago de Melo
tmelo.mat@gmail.com
Mauro Spreafico
mauros@icmc.usp.br

ICMC, Universidade de São Paulo, São Carlos, Brazil.

Journal of Homotopy and Related Structures, vol. 4(1), 2009, pp.187–244

PARALLEL TRANSPORT AND FUNCTORS

URS SCHREIBER AND KONRAD WALDORF

(*communicated by James Stasheff*)

Abstract

Parallel transport of a connection in a smooth fibre bundle yields a functor from the path groupoid of the base manifold into a category that describes the fibres of the bundle. We characterize functors obtained like this by two notions we introduce: local trivializations and smooth descent data. This provides a way to substitute categories of functors for categories of smooth fibre bundles with connection. We indicate that this concept can be generalized to connections in categorified bundles, and how this generalization improves the understanding of higher dimensional parallel transport.

Table of Contents

Received January 29, 2008, revised June 2, 2009; published on July 19, 2009.
2000 Mathematics Subject Classification: 53C05, 18B40, 55R10
Key words and phrases: Principal bundle, connection, diffeological space, groupoid

1. Introduction

Higher dimensional parallel transport generalizes parallel transport along curves to parallel transport along higher dimensional objects, for instance surfaces. One motivation to consider parallel transport along surfaces comes from two-dimensional conformal field theories, where so-called Wess-Zumino terms have been recognized as surface holonomies [**Gaw88, CJM02, SSW07**].

Several mathematical objects have have been used to define higher dimensional parallel transport, among them classes in Deligne cohomology [**Del91**], bundle gerbes with connection and curving [**Mur96**], or 2-bundles with 2-connections [**BS04, BS07**]. The development of such definitions often occurs in two steps: an appropriate definition of parallel transport along curves, followed by a generalization to higher dimensions. For instance, bundle gerbes with connection can be obtained as a generalization of principal bundles with connection. However, in the case of both bundle gerbes and Deligne classes one encounters the obstruction that the structure group has to be abelian. It is hence desirable to find a reformulation of fibre bundles with connection, that brings along a natural generalization for arbitrary structure group.

A candidate for such a reformulation are holonomy maps [**Bar91, CP94**]. These are group homomorphisms

$$\mathcal{H} : \pi_1^1(M, *) \longrightarrow G$$

from the group of thin homotopy classes of based loops in a smooth manifold M into a Lie group G. Any principal G-bundle with connection over M defines a group homomorphism $\mathcal{H}$, but the crucial point is to distinguish those from arbitrary ones. By imposing a certain smoothness condition on $\mathcal{H}$, these holonomy maps correspond – for connected manifolds – bijectively to principal G-bundles with connection [**Bar91, CP94**]. On the other hand, they have a natural generalization from loops to surfaces. However, the obstruction for M being connected becomes even stronger: only if the manifold M is connected and simply-connected, holonomy maps generalized to surfaces capture all aspects of surface holonomy [**MP02**]. Especially the second obstruction erases one of the most interesting of these aspects, see, for example, [**GR02**].

In order to obtain a formulation of parallel transport along curves without topological assumptions on the base manifold M, one considers functors

$$F : \mathcal{P}_1(M) \longrightarrow T$$

from the path groupoid $\mathcal{P}_1(M)$ of M into another category T [**Mac87, MP02**]. The set of objects of the path groupoid $\mathcal{P}_1(M)$ is the manifold M itself, and the set of morphisms between two points x and y is the set of thin homotopy classes of curves starting at x and ending at y. A functor $F : \mathcal{P}_1(M) \longrightarrow T$ is a generalization of a group homomorphism $\mathcal{H} : \pi_1^1(M, *) \longrightarrow G$, but it is not clear how the smoothness condition for holonomy maps has to be generalized to these functors.

Let us first review how a functor $F : \mathcal{P}_1(M) \longrightarrow T$ arises from parallel transport in a, say, principal G-bundle P with connection. In this case, the category T is the category G-Tor of smooth manifolds with smooth, free and transitive G-action from the right, and smooth equivariant maps between those. Now, the connection on P associates to any smooth curve $\gamma : [0,1] \longrightarrow M$ and any element in the fibre $P_{\gamma(0)}$ over the starting point, a unique horizontal lift $\tilde{\gamma} : [0,1] \longrightarrow P$. Evaluating this lift at its endpoint defines a smooth map

$$\tau_\gamma : P_{\gamma(0)} \longrightarrow P_{\gamma(1)},$$

the parallel transport in P along the curve γ. It is G-equivariant with respect to the G-action on the fibres of P, and it is invariant under thin homotopies. Moreover, it satisfies

$$\tau_{\mathrm{id}_x} = \mathrm{id}_{P_x} \quad \text{and} \quad \tau_{\gamma' \circ \gamma} = \tau_{\gamma'} \circ \tau_\gamma,$$

where id_x is the constant curve and γ and γ' are smoothly composable curves. These are the axioms of a functor

$$\mathrm{tra}_P : \mathcal{P}_1(M) \longrightarrow G\text{-Tor}$$

which sends an object x of $\mathcal{P}_1(M)$ to the object P_x of G-Tor and a morphism γ of $\mathcal{P}_1(M)$ to the morphism τ_γ of G-Tor. Summarizing, every principal G-bundle with connection over M defines a functor tra_P. Now the crucial point is to characterize these functors among all functors from $\mathcal{P}_1(M)$ to G-Tor.

In this article we describe such a characterization. For this purpose, we introduce, for general target categories T, the notion of a transport functor. These are certain functors

$$\mathrm{tra} : \mathcal{P}_1(M) \longrightarrow T,$$

such that the category they form is – in the case of $T = G$-Tor – equivalent to the category of principal G-bundles with connection.

The defining properties of a transport functor capture two important concepts: the existence of local trivializations and the smoothness of associated descent data.

Just as for fibre bundles, local trivializations are specified with respect to an open cover of the base manifold M and to a choice of a typical fibre. Here, we represent an open cover by a surjective submersion $\pi : Y \longrightarrow M$, and encode the typical fibre in the notion of a structure groupoid: this is a Lie groupoid Gr together with a functor

$$i : \mathrm{Gr} \longrightarrow T.$$

Now, a π-local i-trivialization of a functor $F : \mathcal{P}_1(M) \longrightarrow T$ is another functor

$$\mathrm{triv} : \mathcal{P}_1(Y) \longrightarrow \mathrm{Gr}$$

together with a natural equivalence

$$t : F \circ \pi_* \longrightarrow i \circ \mathrm{triv},$$

where $\pi_* : \mathcal{P}_1(Y) \longrightarrow \mathcal{P}_1(M)$ is the induced functor between path groupoids. In detail, the natural equivalence t gives for every point $y \in Y$ an isomorphism $F(\pi(y)) \cong i(\mathrm{triv}(y))$ that identifies the "fibre" $F(\pi(y))$ of F over $\pi(y)$ with the image of a "typical fibre" $\mathrm{triv}(y)$ under the functor i. In other words, a functor is π-locally i-trivializable, if its pullback to the cover Y factors through the functor i up to a natural equivalence. Functors with a chosen π-local i-trivialization (triv, t) form a category $\mathrm{Triv}^1_\pi(i)$.

The second concept we introduce is that of smooth descent data. Descent data is specified with respect to a surjective submersion π and a structure groupoid $i : \mathrm{Gr} \longrightarrow T$. While descent data for a fibre bundle with connection is a collection of transition functions and local connection 1-forms, descent data for a functor $F : \mathcal{P}_1(M) \longrightarrow T$ is a pair (triv, g) consisting of a functor $\mathrm{triv} : \mathcal{P}_1(Y) \longrightarrow \mathrm{Gr}$ like the one from a local trivializations and of a certain natural equivalence g that compares triv on the two-fold fibre product of Y with itself. Such pairs define a descent category $\mathfrak{Des}^1_\pi(i)$. The first result of this article (Theorem 2.9) is to prove the descent property: extracting descent data and, conversely, reconstructing a functor from descent data, are equivalences of categories

$$\mathrm{Triv}^1_\pi(i) \cong \mathfrak{Des}^1_\pi(i).$$

We introduce descent data because one can precisely decide whether a pair (triv, g) is smooth or not (Definition 3.1). The smoothness conditions we introduce can be expressed in basic terms of smooth maps between smooth manifolds, and arises from the theory of diffeological spaces [**Che77**]. The concept of smooth descent data is our generalization of the smoothness condition for holonomy maps to functors.

Combining both concepts we have introduced, we call a functor that allows – for some surjective submersion π – a π-local i-trivialization whose corresponding descend data is smooth, a transport functor on M in T with Gr-structure. The category formed by these transport functors is denoted by $\mathrm{Trans}^1_{\mathrm{Gr}}(M, T)$.

Let us return to the particular target category $T = G$-Tor. As described above, one obtains a functor $\mathrm{tra}_P : \mathcal{P}_1(M) \longrightarrow G$-Tor from any principal G-bundle P with connection. We consider the Lie groupoid $\mathrm{Gr} = \mathcal{B}G$, which has only one object, and where every group element $g \in G$ is an automorphism of this object. The notation indicates the fact that the geometric realization of the nerve of this category yields the classifying space BG of the group G. The Lie groupoid $\mathcal{B}G$ can be embedded in the category G-Tor via the functor $i_G : \mathcal{B}G \longrightarrow G\text{-Tor}$ which sends the object of $\mathcal{B}G$ to the group G regarded as a G-space, and a morphism $g \in G$ to the equivariant smooth map which multiplies with g from the left.

The descent category $\mathfrak{Des}^1_\pi(i_G)$ for the structure groupoid $\mathcal{B}G$ and some surjective submersion π is closely related to differential geometric objects: we derive a one-to-one correspondence between smooth functors $\mathrm{triv} : \mathcal{P}_1(Y) \longrightarrow \mathcal{B}G$, which are part of the objects of $\mathfrak{Des}^1_\pi(i_G)$, and 1-forms A on Y with values in the Lie algebra of G (Proposition 4.7). The correspondence can be symbolically expressed as the path-ordered exponential

$$\mathrm{triv}(\gamma) = \mathcal{P}\exp\left(\int_\gamma A\right)$$

for a path γ. Using this relation between smooth functors and differential forms, we show that a functor $\mathrm{tra}_P : \mathcal{P}_1(M) \longrightarrow G\text{-Tor}$ obtained from a principal G-bundle with connection, is a transport functor on M in G-Tor with $\mathcal{B}G$-structure. The main result of this article (Theorem 5.4) is that this establishes an equivalence of categories

$$\mathfrak{Bun}^\nabla_G(M) \cong \mathrm{Trans}^1_{\mathcal{B}G}(M, G\text{-Tor})$$

between the category of principal G-bundles with connection over M and the category of transport functors on M in G-Tor with $\mathcal{B}G$-structure. In other words, these transport functors provide a proper reformulation of principal bundles with connection, emphasizing the aspect of parallel transport.

This article is organized as follows. In Section 2 we review the path groupoid of a smooth manifold and describe some properties of functors defined on it. We introduce local trivializations for functors and the descent category $\mathfrak{Des}^1_\pi(i)$. In Section 3 we define the category $\mathrm{Trans}^1_{\mathrm{Gr}}(M,T)$ of transport functors on M in T with Gr-structure and discuss several properties. In Section 4 we derive the result that relates the descent category $\mathfrak{Des}^1_\pi(i_G)$ for the particular functor $i_G : \mathcal{B}G \longrightarrow G\text{-Tor}$ to differential forms. In Section 5 we provide examples that show that the theory of transport functors applies well to several situations: we prove our main result concerning principal G-bundles with connection, show a similar statement for vector bundles with connection, and also discuss holonomy maps. In Section 6 we discuss principal groupoid bundles and show how transport functors can be used to derive the definition of a connection on such groupoid bundles. Section 7 contains various directions in which the concept of transport functors can be generalized. In particular, we outline a possible generalization of transport functors to transport n-functors

$$\mathrm{tra} : \mathcal{P}_n(M) \longrightarrow T,$$

which provide an implementation for higher dimensional parallel transport. The discussion of the interesting case $n = 2$ is the subject of a separate publication [**SW08a**].

2. Functors and local Trivializations

We give the definition of the path groupoid of a smooth manifold and describe functors defined on it. We introduce local trivializations and descent data of such functors.

2.1. The Path Groupoid of a smooth Manifold

We start by setting up the basic definitions around the path groupoid of a smooth manifold M. We use the conventions of [**CP94, MP02**], generalized from loops to paths.

Definition 2.1. *A path $\gamma : x \longrightarrow y$ between two points $x, y \in M$ is a smooth map $\gamma : [0,1] \longrightarrow M$ which has a sitting instant: a number $0 < \epsilon < \frac{1}{2}$ such that $\gamma(t) = x$ for $0 \leqslant t < \epsilon$ and $\gamma(t) = y$ for $1 - \epsilon < t \leqslant 1$.*

Let us denote the set of such paths by PM. For example, for any point $x \in M$ there is the constant path id_x defined by $\mathrm{id}_x(t) := x$. Given a path $\gamma_1 : x \longrightarrow y$ and another path $\gamma_2 : y \longrightarrow z$ we define their composition to be the path $\gamma_2 \circ \gamma_1 : x \longrightarrow z$ defined by

$$(\gamma_2 \circ \gamma_1)(t) := \begin{cases} \gamma_1(2t) & \text{for } 0 \leqslant t \leqslant \frac{1}{2} \\ \gamma_2(2t-1) & \text{for } \frac{1}{2} \leqslant t \leqslant 1. \end{cases}$$

This gives a smooth map since γ_1 and γ_2 are both constant near the gluing point, due to their sitting instants. We also define the inverse $\gamma^{-1} : y \longrightarrow x$ of a path $\gamma : x \longrightarrow y$ by $\gamma^{-1}(t) := \gamma(1-t)$.

Definition 2.2. *Two paths $\gamma_1 : x \longrightarrow y$ and $\gamma_2 : x \longrightarrow y$ are called thin homotopy equivalent, if there exists a smooth map $h : [0,1] \times [0,1] \longrightarrow M$ such that*

1. *there exists a number $0 < \epsilon < \frac{1}{2}$ with*
   (a) $h(s,t) = x$ for $0 \leqslant t < \epsilon$ and $h(s,t) = y$ for $1 - \epsilon < t \leqslant 1$.
   (b) $h(s,t) = \gamma_1(t)$ for $0 \leqslant s < \epsilon$ and $h(s,t) = \gamma_2(t)$ for $1 - \epsilon < s \leqslant 1$.
2. *the differential of h has at most rank 1 everywhere, i.e.*

$$\mathrm{rank}(dh|_{(s,t)}) \leqslant 1$$

for all $(s,t) \in [0,1] \times [0,1]$.

Due to condition (b), thin homotopy defines an equivalence relation on PM. The set of thin homotopy classes of paths is denoted by P^1M, and the projection to classes is denoted by

$$\mathrm{pr} : PM \longrightarrow P^1M.$$

We denote a thin homotopy class of a path $\gamma : x \longrightarrow y$ by $\overline{\gamma} : x \longrightarrow y$. Notice that thin homotopies include the following type of reparameterizations: let $\beta : [0,1] \longrightarrow [0,1]$ be a path $\beta : 0 \longrightarrow 1$, in particular with $\beta(0) = 0$ and $\beta(1) = 1$. Then, for any path $\gamma : x \longrightarrow y$, also $\gamma \circ \beta : x \longrightarrow y$ is a path and

$$h(s,t) := \gamma(t\beta(1-s) + \beta(t)\beta(s))$$

defines a thin homotopy between them.

The composition of paths defined above on PM descends to P^1M due to condition (a), which admits a smooth composition of smooth homotopies. The composition of thin homotopy classes of paths obeys the following rules:

Lemma 2.3. *For any path* $\gamma : x \longrightarrow y$,

a) $\overline{\gamma} \circ \overline{\mathrm{id}_x} = \overline{\gamma} = \overline{\mathrm{id}_y} \circ \overline{\gamma}$,

b) For further paths $\gamma' : y \longrightarrow z$ *and* $\gamma'' : z \longrightarrow w$,

$$(\overline{\gamma''} \circ \overline{\gamma'}) \circ \overline{\gamma} = \overline{\gamma''} \circ (\overline{\gamma'} \circ \overline{\gamma}).$$

c) $\overline{\gamma} \circ \overline{\gamma^{-1}} = \overline{\mathrm{id}_y}$ *and* $\overline{\gamma^{-1}} \circ \overline{\gamma} = \overline{\mathrm{id}_x}$.

These three properties lead us to the following

Definition 2.4. *For a smooth manifold M, we consider the category whose set of objects is M, whose set of morphisms is P^1M, where a class $\overline{\gamma} : x \longrightarrow y$ is a morphism from x to y, and the composition is as described above. Lemma 2.3 a) and b) are the axioms of a category and c) says that every morphism is invertible. Hence we have defined a groupoid, called the path groupoid of M, and denoted by $\mathcal{P}_1(M)$.*

For a smooth map $f : M \longrightarrow N$, we denote by

$$f_* : \mathcal{P}_1(M) \longrightarrow \mathcal{P}_1(N)$$

the functor with $f_*(x) = f(x)$ and $(f_*)(\overline{\gamma}) := \overline{f \circ \gamma}$. The latter is well-defined, since a thin homotopy h between paths γ and γ' induces a thin homotopy $f \circ h$ between $f \circ \gamma$ and $f \circ \gamma'$.

In the following we consider functors

$$F : \mathcal{P}_1(M) \longrightarrow T \tag{2.1}$$

for some arbitrary category T. Such a functor sends each point $p \in M$ to an object $F(p)$ in T, and each thin homotopy class $\overline{\gamma} : x \longrightarrow y$ of paths to a morphism $F(\overline{\gamma}) : F(x) \longrightarrow F(y)$ in T. We use the following notation: we call M the *base space* of the functor F, and the object $F(p)$ the *fibre* of F over p. In the remainder of this section we give examples of natural constructions with functors (2.1).

ADDITIONAL STRUCTURE ON T. Any additional structure for the category T can be applied pointwise to functors into T, for instance,

a) if T has direct sums, we can take the direct sum $F_1 \oplus F_2$ of two functors.
b) if T is a monoidal category, we can take tensor products $F_1 \otimes F_2$ of functors.
c) if T is monoidal and has a duality regarded as a functor $d : T \longrightarrow T^{\mathrm{op}}$, we can form the dual $F^* := d \circ F$ of a functor F.

PULLBACK. If $f : M \longrightarrow N$ is a smooth map and $F : \mathcal{P}_1(N) \longrightarrow T$ is a functor, we define

$$f^*F := F \circ f_* : \mathcal{P}_1(M) \longrightarrow T$$

to be the pullback of F along f.

FLAT FUNCTORS. Instead of the path groupoid, one can also consider the fundamental groupoid $\Pi_1(M)$ of a smooth manifold M, whose objects are points in M, just like for $\mathcal{P}_1(M)$, but whose morphisms are smooth homotopy classes of paths (whose differential may have arbitrary rank). The projection from thin homotopy classes to smooth homotopy classes provides a functor

$$p : \mathcal{P}_1(M) \longrightarrow \Pi_1(M).$$

We call a functor $F : \mathcal{P}_1(M) \longrightarrow T$ *flat*, if there exists a functor $\tilde{F} : \Pi_1(M) \longrightarrow T$ with $F \cong \tilde{F} \circ p$. This is motivated by parallel transport in principal G-bundles: while it is invariant under *thin* homotopy, it is only homotopy invariant if the bundle is flat, i.e. has vanishing curvature. However, aside from Section 7.2 we will not discuss the flat case any further in this article.

RESTRICTION TO PATHS BETWEEN FIXED POINTS. Finally, let us consider the restriction of a functor $F : \mathcal{P}_1(M) \longrightarrow T$ to paths between two fixed points. This yields a map

$$F_{x,y} : \mathrm{Mor}_{\mathcal{P}_1(M)}(x,y) \longrightarrow \mathrm{Mor}_T(F(x), F(y)).$$

Of particular interest is the case $x = y$, in which $\mathrm{Mor}_{\mathcal{P}_1(M)}(x,x)$ forms a group under composition, which is called the thin homotopy group of M at x, and is denoted by $\pi_1^1(M,x)$ [**CP94, MP02**]. Even more particular, we consider the target category G-Tor: by choosing a diffeomorphism $F(x) \cong G$, we obtain an identification

$$\mathrm{Mor}_{G\text{-Tor}}(F(x), F(x)) = G,$$

and the restriction $F_{x,x}$ of a functor $F : \mathcal{P}_1(M) \longrightarrow G$-Tor to the thin homotopy group of M at x gives a group homomorphism

$$F_{x,x} : \pi_1^1(M,x) \longrightarrow G.$$

This way one obtains the setup of [**Bar91, CP94**] and [**MP02**] for the case $G = U(1)$ as a particular case of our setup. A further question is, whether the group homomorphism $F_{x,x}$ is smooth in the sense used in [**Bar91, CP94, MP02**]. An answer is given in Section 5.2.

2.2. Extracting Descent Data from a Functor

To define local trivializations of a functor $F : \mathcal{P}_1(M) \longrightarrow T$, we fix three attributes:

1. A surjective submersion $\pi : Y \longrightarrow M$. Compared to local trivializations of fibre bundles, the surjective submersion replaces an open cover of the manifold. Indeed, given an open cover $\{U_\alpha\}_{\alpha \in A}$ of M, one obtains a surjective submersion by taking Y to be the disjoint union of the U_α and $\pi : Y \longrightarrow M$ to be the union of the inclusions $U_\alpha \hookrightarrow M$.
2. A Lie groupoid Gr, i.e. a groupoid whose sets of objects and morphisms are smooth manifolds, whose source and target maps

$$s, t : \mathrm{Mor}(\mathrm{Gr}) \longrightarrow \mathrm{Obj}(\mathrm{Gr})$$

are surjective submersions, and whose composition

$$\circ : \mathrm{Mor}(\mathrm{Gr})\ {}_s\!\times_t\ \mathrm{Mor}(\mathrm{Gr}) \longrightarrow \mathrm{Mor}(\mathrm{Gr})$$

and the identity $\mathrm{id} : \mathrm{Obj}(\mathrm{Gr}) \longrightarrow \mathrm{Mor}(\mathrm{Gr})$ are smooth maps. The Lie groupoid Gr plays the role of the typical fibre of the functor F.

3. A functor $i : \mathrm{Gr} \longrightarrow T$, which relates the typical fibre Gr to the target category T of the functor F. In all of our examples, i will be an equivalence of categories. This is important for some results derived in Section 3.2.

Definition 2.5. *Given a Lie groupoid* Gr*, a functor* $i : \mathrm{Gr} \longrightarrow T$ *and a surjective submersion* $\pi : Y \longrightarrow M$*, a* *$\pi$-local i-trivialization* *of a functor*

$$F : \mathcal{P}_1(M) \longrightarrow T$$

is a pair (triv, t) *of a functor* $\mathrm{triv} : \mathcal{P}_1(Y) \longrightarrow \mathrm{Gr}$ *and a natural equivalence*

$$t : \pi^*F \longrightarrow i \circ \mathrm{triv}.$$

The natural equivalence t is also depicted by the diagram

$$\begin{array}{ccc} \mathcal{P}_1(Y) & \xrightarrow{\ \pi_*\ } & \mathcal{P}_1(M) \\ {\scriptstyle \mathrm{triv}}\downarrow & \overset{t}{\swarrow} & \downarrow{\scriptstyle F} \\ \mathrm{Gr} & \xrightarrow[\ i\ ]{} & T. \end{array}$$

To set up the familiar terminology, we call a functor *locally i-trivializable*, if it admits a π-local i-trivialization for some choice of π. We call a functor *i-trivial*, if it admits an id_M-local i-trivialization, i.e. if it is naturally equivalent to the functor $i \circ \mathrm{triv}$. To abbreviate the notation, we will often write triv_i instead of $i \circ \mathrm{triv}$.

Note that local trivializations can be pulled back: if $\zeta : Z \longrightarrow Y$ and $\pi : Y \longrightarrow M$ are surjective submersions, and (triv, t) is a π-local i-trivialization of a functor F, we obtain a $(\pi \circ \zeta)$-local i-trivialization $(\zeta^*\mathrm{triv}, \zeta^*t)$ of F. In terms of open covers, this corresponds to a refinement of the cover.

Definition 2.6. *Let* Gr *be a Lie groupoid and let* $i : \mathrm{Gr} \longrightarrow T$ *be a functor. The category* $\mathrm{Triv}^1_\pi(i)$ *of* *functors with π-local i-trivialization* *is defined as follows:*

1. *its objects are triples* (F, triv, t) *consisting of a functor* $F : \mathcal{P}_1(M) \longrightarrow T$ *and a π-local i-trivialization* (triv, t) *of* F.
2. *a morphism*

$$(F, \mathrm{triv}, t) \xrightarrow{\ \alpha\ } (F', \mathrm{triv}', t')$$

is a natural transformation $\alpha : F \longrightarrow F'$*. Composition of morphisms is simply composition of these natural transformations.*

Motivated by transition functions of fibre bundles, we extract a similar datum from a functor F with π-local i-trivialization (triv, t); this datum is a natural equivalence

$$g : \pi_1^*\mathrm{triv}_i \longrightarrow \pi_2^*\mathrm{triv}_i$$

between the two functors $\pi_1^*\mathrm{triv}_i$ and $\pi_2^*\mathrm{triv}_i$ from $\mathcal{P}_1(Y^{[2]})$ to T, where π_1 and π_2 are the projections from the two-fold fibre product $Y^{[2]} := Y \times_M Y$ of Y to the components. In the case that the surjective submersion comes from an open cover of M, $Y^{[2]}$ is the disjoint union of all two-fold intersections of open subsets. The natural equivalence g is defined by

$$g := \pi_2^* t \circ \pi_1^* t^{-1};$$

its component at a point $\alpha \in Y^{[2]}$ is the morphism $t(\pi_2(\alpha)) \circ t(\pi_1(\alpha))^{-1}$ in T. The composition is well-defined because $\pi \circ \pi_1 = \pi \circ \pi_2$.

Transition functions of fibre bundles satisfy a cocycle condition over three-fold intersections. The natural equivalence g has a similar property when pulled back to the three-fold fibre product $Y^{[3]} := Y \times_M Y \times_M Y$.

Proposition 2.7. *The diagram*

$$\begin{array}{ccccc} & & \pi_2^*\mathrm{triv}_i & & \\ & \overset{\pi_{12}^*g}{\nearrow} & & \overset{\pi_{23}^*g}{\searrow} & \\ \pi_1^*\mathrm{triv}_i & & \xrightarrow[\pi_{13}^*g]{} & & \pi_3^*\mathrm{triv}_i \end{array}$$

of natural equivalences between functors from $\mathcal{P}_1(Y^{[3]})$ to T is commutative.

Now that we have defined the data (triv, g) associated to an object (F, triv, t) in $\mathrm{Triv}^1_\pi(i)$, we consider a morphism

$$\alpha : (F, \mathrm{triv}, t) \longrightarrow (F', \mathrm{triv}', t')$$

between two functors with π-local i-trivializations, i.e. a natural transformation $\alpha : F \longrightarrow F'$. We define a natural transformation

$$h : \mathrm{triv}_i \longrightarrow \mathrm{triv}'_i$$

by $h := t' \circ \pi^*\alpha \circ t^{-1}$, whose component at $x \in Y$ is the morphism $t'(x) \circ \alpha(\pi(x)) \circ t(x)^{-1}$ in T. From the definitions of g, g' and h one obtains the commutative diagram

$$\begin{array}{ccc} \pi_1^*\mathrm{triv}_i & \xrightarrow{g} & \pi_2^*\mathrm{triv}_i \\ {\scriptstyle \pi_1^*h}\downarrow & & \downarrow{\scriptstyle \pi_2^*h} \\ \pi_1^*\mathrm{triv}'_i & \xrightarrow[g']{} & \pi_2^*\mathrm{triv}'_i . \end{array} \qquad (2.2)$$

The behaviour of the natural equivalences data g and h leads to the following definition of a category $\mathfrak{Des}^1_\pi(i)$ of descent data. This terminology will be explained in the next section.

Definition 2.8. *The category $\mathfrak{Des}^1_\pi(i)$ of* <u>*descent data*</u> *of π-locally i-trivialized functors is defined as follows:*

1. *its objects are pairs* (triv, g) *of a functor* triv : $\mathcal{P}_1(Y) \longrightarrow$ Gr *and a natural equivalence*

$$g : \pi_1^*\mathrm{triv}_i \longrightarrow \pi_2^*\mathrm{triv}_i,$$

such that the diagram

$$\begin{array}{ccccc} & & \pi_2^*\mathrm{triv}_i & & \\ & \overset{\pi_{12}^*g}{\nearrow} & & \overset{\pi_{23}^*g}{\searrow} & \\ \pi_1^*\mathrm{triv}_i & & \xrightarrow[\pi_{13}^*g]{} & & \pi_3^*\mathrm{triv}_i \end{array} \tag{2.3}$$

is commutative.

2. *a morphism* (triv, g) $\longrightarrow$ (triv$'$, g') *is a natural transformation*

$$h : \mathrm{triv}_i \longrightarrow \mathrm{triv}'_i$$

such that the diagram

$$\begin{array}{ccc} \pi_1^*\mathrm{triv}_i & \xrightarrow{g} & \pi_2^*\mathrm{triv}_i \\ {\scriptstyle \pi_1^*h}\downarrow & & \downarrow{\scriptstyle \pi_2^*h} \\ \pi_1^*\mathrm{triv}'_i & \xrightarrow[g']{} & \pi_2^*\mathrm{triv}'_i. \end{array} \tag{2.4}$$

is commutative. The composition is the composition of these natural transformations.

Summarizing, we have defined a functor

$$\mathrm{Ex}_\pi : \mathrm{Triv}_\pi^1(i) \longrightarrow \mathfrak{Des}_\pi^1(i), \tag{2.5}$$

that extracts descent data from functors with local trivialization and of morphisms of those in the way described above.

2.3. Reconstructing a Functor from Descent Data

In this section we show that extracting descent data from a functor F preserves all information about F. We also justify the terminology *descent data*, see Remark 2.10 below.

Theorem 2.9. *The functor*

$$\mathrm{Ex}_\pi : \mathrm{Triv}_\pi^1(i) \longrightarrow \mathfrak{Des}_\pi^1(i)$$

is an equivalence of categories.

For the proof we define a weak inverse functor

$$\mathrm{Rec}_\pi : \mathfrak{Des}_\pi^1(i) \longrightarrow \mathrm{Triv}_\pi^1(i) \tag{2.6}$$

that reconstructs a functor (and a π-local i-trivialization) from given descent data. The definition of Rec_π is given in three steps:

1. We construct a groupoid $\mathcal{P}_1^\pi(M)$ covering the path groupoid $\mathcal{P}_1(M)$ by means of a surjective functor $p^\pi : \mathcal{P}_1^\pi(M) \longrightarrow \mathcal{P}_1(M)$, and show that any object (triv, g) in $\mathfrak{Des}_\pi^1(i)$ gives rise to a functor
$$R_{(\mathrm{triv},g)} : \mathcal{P}_1^\pi(M) \longrightarrow T.$$
We enhance this to a functor
$$R : \mathfrak{Des}_\pi^1(i) \longrightarrow \mathrm{Funct}(\mathcal{P}_1^\pi(M), T), \tag{2.7}$$
where $\mathrm{Funct}(\mathcal{P}_1^\pi(M), T)$ is the category of functors from $\mathcal{P}_1^\pi(M)$ to T and natural transformations between those.
2. We show that the functor $p^\pi : \mathcal{P}_1^\pi(M) \longrightarrow \mathcal{P}_1(M)$ is an equivalence of categories and construct a weak inverse
$$s : \mathcal{P}_1(M) \longrightarrow \mathcal{P}_1^\pi(M).$$
The pullback along s is the functor
$$s^* : \mathrm{Funct}(\mathcal{P}_1^\pi(M), T) \longrightarrow \mathrm{Funct}(\mathcal{P}_1(M), T) \tag{2.8}$$
obtained by pre-composition with s.
3. By constructing canonical π-local i-trivializations of functors in the image of the composition $s^* \circ R$ of the functors (2.7) and (2.8), we extend this composition to a functor
$$\mathrm{Rec}_\pi := s^* \circ R : \mathfrak{Des}_\pi^1(i) \longrightarrow \mathrm{Triv}_\pi^1(i).$$
Finally, we give in Appendix B.1 the proof that Rec_π is a weak inverse of the functor Ex_π and thus show that Ex_π is an equivalence of categories.

Before we perform the steps 1 to 3, let us make the following remark about the nature of the category $\mathfrak{Des}_\pi^1(i)$ and the functor Rec_π.

Remark 2.10. We consider the case $i := \mathrm{id}_{\mathrm{Gr}}$. Now, the forgetful functor
$$v : \mathrm{Triv}_\pi^1(i) \longrightarrow \mathrm{Funct}(\mathcal{P}_1(M), \mathrm{Gr})$$
has a canonical weak inverse, which associates to a functor $F : \mathcal{P}_1(M) \longrightarrow \mathrm{Gr}$ the π-local i-trivialization $(\pi^*F, \mathrm{id}_{\pi^*F})$. Under this identification, $\mathfrak{Des}_\pi^1(i)$ is the descent category of the functor category $\mathrm{Funct}(M, \mathrm{Gr})$ with respect to π in the sense of a stack [**Moe02, Str04**]. The functor
$$\mathrm{Rec}_\pi : \mathfrak{Des}_\pi^1(i) \longrightarrow \mathrm{Funct}(\mathcal{P}_1(M), \mathrm{Gr})$$
realizes the descent.

STEP 1: THE GROUPOID $\mathcal{P}_1^\pi(M)$. The groupoid $\mathcal{P}_1^\pi(M)$ we introduce is the *universal path pushout* associated to the surjective submersion $\pi : Y \longrightarrow M$. Heuristically, $\mathcal{P}_1^\pi(M)$ is the path groupoid of the covering Y combined with "jumps" in the fibres of π. We explain its universality in Appendix A.1 for completeness and introduce here a concrete realization (see Lemma A.4).

Definition 2.11. *The groupoid $\mathcal{P}_1^\pi(M)$ is defined as follows. Its objects are points $x \in Y$ and its morphisms are formal (finite) compositions of two types of basic morphisms: thin homotopy classes $\overline{\gamma} : x \longrightarrow y$ of paths in Y, and points $\alpha \in Y^{[2]}$ regarded as morphisms $\alpha : \pi_1(\alpha) \longrightarrow \pi_2(\alpha)$. Among the morphisms, we impose three relations:*

(1) for any thin homotopy class $\overline{\Theta} : \alpha \longrightarrow \beta$ of paths in $Y^{[2]}$, we demand that the diagram

$$\begin{array}{ccc} \pi_1(\alpha) & \xrightarrow{\alpha} & \pi_2(\alpha) \\ {\scriptstyle (\pi_1)_*(\overline{\Theta})}\downarrow & & \downarrow{\scriptstyle (\pi_2)_*(\overline{\Theta})} \\ \pi_1(\beta) & \xrightarrow[\beta]{} & \pi_2(\beta). \end{array}$$

of morphisms in $\mathcal{P}_1^\pi(M)$ is commutative.

(2) for any point $\Xi \in Y^{[3]}$, we demand that the diagram

$$\begin{array}{ccccc} & & \pi_2(\Xi) & & \\ & {\scriptstyle \pi_{12}(\Xi)}\nearrow & & \searrow{\scriptstyle \pi_{23}(\Xi)} & \\ \pi_1(\Xi) & & \xrightarrow[\pi_{13}(\Xi)]{} & & \pi_3(\Xi) \end{array}$$

of morphisms in $\mathcal{P}_1^\pi(M)$ is commutative.

(3) we impose the equation $\overline{\mathrm{id}_x} = (x, x) \in Y^{[2]}$ for any $x \in Y$.

It is clear that this definition indeed gives a groupoid. It is important for us because it provides the two following natural definitions.

Definition 2.12. *For an object (triv, g) in $\mathfrak{Des}_\pi^1(i)$, we have a functor*

$$R_{(\mathrm{triv},g)} : \mathcal{P}_1^\pi(M) \longrightarrow T$$

that sends an object $x \in Y$ to $\mathrm{triv}_i(x)$, a basic morphism $\overline{\gamma} : x \longrightarrow y$ to $\mathrm{triv}_i(\overline{\gamma})$ and a basic morphism α to $g(\alpha)$.

The definition is well-defined since it respects the relations among the morphisms: (1) is respected due to the commutative diagram for the natural transformation g, (2) is the cocycle condition (2.3) for g and (3) follows from the latter since g is invertible.

Definition 2.13. *For a morphism $h : (\mathrm{triv}, g) \longrightarrow (\mathrm{triv}', g')$ in $\mathfrak{Des}_\pi^1(i)$ we have a natural transformation*

$$R_h : R_{(\mathrm{triv},g)} \longrightarrow R_{(\mathrm{triv}',g')}$$

that sends an object $x \in Y$ to the morphism $h(x)$ in T.

The commutative diagram for the natural transformation R_h for a basic morphism $\overline{\gamma} : x \longrightarrow y$ follows from the one of h, and for a basic morphism $\alpha \in Y^{[2]}$ from the condition (2.4) on the morphisms of $\mathfrak{Des}^1_\pi(i)$.

We explain in Appendix A.1 that Definitions (2.12) and (2.13) are consequences of the universal property of the groupoid $\mathcal{P}_1^\pi(M)$, as specified in Definition A.1 and calculated in Lemma A.4. Here we summarize the definitions above in the following way:

Lemma 2.14. *Definitions (2.12) and (2.13) yield a functor*

$$R : \mathfrak{Des}^1_\pi(i) \longrightarrow \mathrm{Funct}(\mathcal{P}_1^\pi(M), T). \tag{2.9}$$

STEP 2: PULLBACK TO M. To continue the reconstruction of a functor from given descent data let us introduce the projection functor

$$p^\pi : \mathcal{P}_1^\pi(M) \longrightarrow \mathcal{P}_1(M) \tag{2.10}$$

sending an object $x \in Y$ to $\pi(x)$, a basic morphism $\overline{\gamma} : x \longrightarrow y$ to $\pi_*(\overline{\gamma})$ and a basic morphism $\alpha \in Y^{[2]}$ to $\mathrm{id}_{\pi(\pi_1(\alpha))}$ $(= \mathrm{id}_{\pi(\pi_2(\alpha))})$. In other words, it is just the functor π_* and forgets the jumps in the fibres of π. More precisely,

$$p^\pi \circ \iota = \pi_*,$$

where $\iota : \mathcal{P}_1(Y) \longrightarrow \mathcal{P}_1^\pi(M)$ is the obvious inclusion functor.

Lemma 2.15. *The projection functor $p^\pi : \mathcal{P}_1^\pi(M) \longrightarrow \mathcal{P}_1(M)$ is a surjective equivalence of categories.*

Proof. Since $\pi : Y \longrightarrow M$ is surjective, it is clear that p^π is surjective on objects. It remains to show that the map

$$(p^\pi)_1 : \mathrm{Mor}_{\mathcal{P}_1^\pi(M)}(x, y) \longrightarrow \mathrm{Mor}_{\mathcal{P}_1(M)}(\pi(x), \pi(y)) \tag{2.11}$$

is bijective for all $x, y \in Y$. Let $\gamma : \pi(x) \longrightarrow \pi(y)$ be any path in M. Let $\{U_i\}_{i\in I}$ an open cover of M with sections $s_i : U_i \longrightarrow Y$. Since the image of $\gamma : [0,1] \longrightarrow M$ is compact, there exists a *finite* subset $J \subset I$ such that $\{U_i\}_{i\in J}$ covers the image. Let $\gamma = \gamma_n \circ ... \circ \gamma_1$ be a decomposition of γ such that $\gamma_i \in PU_{j(i)}$ for some assignment $j : \{1, ..., n\} \longrightarrow J$. Let $\tilde{\gamma}_i := (s_{j(i)})_*\gamma_i \in PY$ be lifts of the pieces, $\tilde{\gamma}_i : a_i \longrightarrow b_i$ with $a_i, b_i \in Y$. Now we consider the path

$$\tilde{\gamma} := (b_n, y) \circ \tilde{\gamma}_n \circ (b_{n-1}, a_n) \circ ... \circ \tilde{\gamma}_2 \circ (b_1, a_2) \circ \tilde{\gamma}_1 \circ (x, a_1),$$

whose thin homotopy class is evidently a preimage of the thin homotopy class of γ under $(p^\pi)_1$. The injectivity of (2.11) follows from the identifications (1), (2) and (3) of morphisms in the groupoid $\mathcal{P}_1^\pi(M)$. □

Since p^π is an equivalence of categories, there exists a (up to natural isomorphism) unique weak inverse functor $s : \mathcal{P}_1(M) \longrightarrow \mathcal{P}_1^\pi(M)$ together with natural equivalences $\lambda : s \circ p^\pi \longrightarrow \mathrm{id}_{\mathcal{P}_1^\pi(M)}$ and $\rho : p^\pi \circ s \longrightarrow \mathrm{id}_{\mathcal{P}_1(M)}$. The inverse functor s can be constructed explicitly: for a fixed choice of lifts $s(x) \in Y$ for every point $x \in M$, and a fixed choice of an open cover, each path can be lifted as described in the proof of Lemma 2.15. In this case we have $\rho = \mathrm{id}$, and the component of λ at $x \in Y$ is the morphism $(s(\pi(x)), x)$ in $\mathcal{P}_1^{\pi}(M)$. Now we have a canonical functor

$$s^* \circ R : \mathfrak{Des}_{\pi}^1(i) \longrightarrow \mathrm{Funct}(\mathcal{P}_1(M), T).$$

It reconstructs a functor $s^*R_{(\mathrm{triv},g)}$ from a given object (triv, g) in $\mathfrak{Des}_{\pi}^1(i)$ and a natural transformation s^*R_h from a given morphism h in $\mathfrak{Des}_{\pi}^1(i)$.

STEP 3: LOCAL TRIVIALIZATION. What remains to enhance the functor $s^* \circ R$ to a functor

$$\mathrm{Rec}_{\pi} : \mathfrak{Des}_{\pi}^1(i) \longrightarrow \mathrm{Triv}_{\pi}^1(i)$$

is finding a π-local i-trivialization (triv, t) of each reconstructed functor $s^*R_{(\mathrm{triv},g)}$. Of course the given functor $\mathrm{triv} : \mathcal{P}_1(Y) \longrightarrow \mathrm{Gr}$ serves as the first component of the trivialization, and it remains to define the natural equivalence

$$t : \pi^* s^* R_{(\mathrm{triv},g)} \longrightarrow \mathrm{triv}_i. \tag{2.12}$$

We use the natural equivalence $\lambda : s \circ p^{\pi} \longrightarrow \mathrm{id}_{\mathcal{P}_1^{\pi}(M)}$ associated to the functor s and obtain a natural equivalence

$$\iota^*\lambda : s \circ \pi_* \longrightarrow \iota$$

between functors from $\mathcal{P}_1(Y)$ to $\mathcal{P}_1^{\pi}(M)$. Its component at $x \in Y$ is the morphism $(s(\pi(x)), x)$ going from $s(\pi(x))$ to x. Using

$$\pi^* s^* R_{(\mathrm{triv},g)} = (s \circ \pi_*)^* R_{(\mathrm{triv},g)} \quad \text{and} \quad \mathrm{triv}_i = \iota^* R_{(\mathrm{triv},g)},$$

we define by

$$t := g \circ \iota^*\lambda$$

the natural equivalence (2.12). Indeed, its component at $x \in Y$ is the morphism $g((s(\pi(x)), x)) : \mathrm{triv}_i(s(\pi(x))) \longrightarrow \mathrm{triv}_i(x)$, these are natural in x and isomorphisms because g is one. Diagrammatically, it is

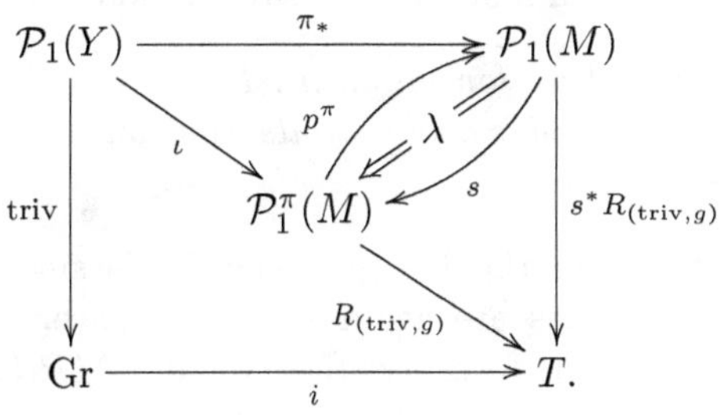

This shows

Lemma 2.16. *The pair* (triv, t) *is a* π*-local* i*-trivialization of the functor* $s^*R_{(\mathrm{triv},g)}$.

This finishes the definition of the reconstruction functor Rec_{π}. The remaining proof that Rec_{π} is a weak inverse of Ex_{π} is postponed to Appendix B.1.

3. Transport Functors

Transport functors are locally trivializable functors whose descent data is smooth. Wilson lines are restrictions of a functor to paths between two fixed points. We deduce a characterization of transport functors by the smoothness of their Wilson lines.

3.1. Smooth Descent Data

In this section we specify a subcategory $\mathfrak{Des}^1_\pi(i)^\infty$ of the category $\mathfrak{Des}^1_\pi(i)$ of descent data we have defined in the previous section. This subcategory is supposed to contain *smooth* descent data. The main issue is to decide, when a functor $F : \mathcal{P}_1(X) \longrightarrow \mathrm{Gr}$ is smooth: in contrast to the objects and the morphisms of the Lie groupoid Gr, the set P^1X of morphisms of $\mathcal{P}_1(X)$ is not a smooth manifold.

Definition 3.1. *Let* Gr *be a Lie groupoid and let X be a smooth manifold. A functor $F : \mathcal{P}_1(X) \longrightarrow \mathrm{Gr}$ is called* smooth, *if the following two conditions are satisfied:*

1. *On objects, $F : X \longrightarrow \mathrm{Obj}(\mathrm{Gr})$ is a smooth map.*
2. *For every $k \in \mathbb{N}_0$, every open subset $U \subset \mathbb{R}^k$ and every map $c : U \longrightarrow PX$ such that the composite*

$$U \times [0,1] \xrightarrow{c\times\mathrm{id}} PX \times [0,1] \xrightarrow{\mathrm{ev}} X \tag{3.1}$$

is smooth, also

$$U \xrightarrow{c} PX \xrightarrow{\mathrm{pr}} P^1X \xrightarrow{F} \mathrm{Mor}(\mathrm{Gr})$$

is smooth.

In (3.1), ev is the evaluation map $\mathrm{ev}(\gamma, t) := \gamma(t)$. Similar definitions of smooth maps defined on thin homotopy classes of paths have also been used in [**Bar91, CP94, MP02**]. We explain in Appendix A.2 how Definition 3.1 is motivated and how it arises from the general concept of diffeological spaces [**Che77**], a generalization of the concept of a smooth manifold, cf. Proposition A.7 i).

Definition 3.2. *A natural transformation $\eta : F \longrightarrow G$ between smooth functors $F, G : \mathcal{P}_1(X) \longrightarrow \mathrm{Gr}$ is called* smooth, *if its components form a smooth map $X \longrightarrow \mathrm{Mor}(\mathrm{Gr}) : X \longmapsto \eta(X)$.*

Because the composition in the Lie groupoid Gr is smooth, compositions of smooth natural transformations are again smooth. Hence, smooth functors and smooth natural transformations form a category $\mathrm{Funct}^\infty(\mathcal{P}_1(X), \mathrm{Gr})$. Notice that if $f : M \longrightarrow X$ is a smooth map, and $F : \mathcal{P}_1(X) \longrightarrow \mathrm{Gr}$ is a smooth functor, the pullback f^*F is also smooth. Similarly, pullbacks of smooth natural transformations are smooth.

Definition 3.3. *Let* Gr *be a Lie groupoid and let $i : \mathrm{Gr} \longrightarrow T$ be a functor. An object* (triv, g) *in $\mathfrak{Des}^1_\pi(i)$ is called* smooth, *if the following two conditions are satisfied:*

1. *The functor*

$$\mathrm{triv} : \mathcal{P}_1(Y) \longrightarrow \mathrm{Gr}$$

is smooth in the sense of Definition 3.1.

2. *The natural equivalence*

$$g : \pi_1^* \mathrm{triv}_i \longrightarrow \pi_2^* \mathrm{triv}_i$$

factors through i *by a natural equivalence* $\tilde{g} : \pi_1^* \mathrm{triv} \longrightarrow \pi_2^* \mathrm{triv}$ *which is smooth in the sense of Definition 3.2. For the components at a point* $\alpha \in Y^{[2]}$, *the factorization means* $g(\alpha) = i(\tilde{g}(\alpha))$.

In the same sense, a morphism

$$h : (\mathrm{triv}, g) \longrightarrow (\mathrm{triv}', g')$$

between smooth objects is called smooth, if it factors through i *by a smooth natural equivalence* $\tilde{h} : \mathrm{triv} \longrightarrow \mathrm{triv}'$.

Remark 3.4. If i is faithful, the natural equivalences $\tilde{g}$ and $\tilde{h}$ in Definition 3.3 are uniquely determined, provided that they exist. If i is additionally full, also the existence of g and h is guaranteed.

Smooth objects and morphisms in $\mathfrak{Des}^1_\pi(i)$ form the subcategory $\mathfrak{Des}^1_\pi(i)^\infty$. Using the equivalence Ex_π defined in Section 2.2, we obtain a subcategory $\mathrm{Triv}^1_\pi(i)^\infty$ of $\mathrm{Triv}^1_\pi(i)$ consisting of those objects (F, triv, t) for which $\mathrm{Ex}_\pi(F, \mathrm{triv}, t)$ is smooth and of those morphisms h for which $\mathrm{Ex}_\pi(h)$ is smooth.

Proposition 3.5. *The functor* $\mathrm{Rec}_\pi : \mathfrak{Des}^1_\pi(i) \longrightarrow \mathrm{Triv}^1_\pi(i)$ *restricts to an equivalence of categories*

$$\mathrm{Rec}_\pi : \mathfrak{Des}^1_\pi(i)^\infty \longrightarrow \mathrm{Triv}^1_\pi(i)^\infty.$$

Proof. This follows from the fact that $\mathrm{Ex}_\pi \circ \mathrm{Rec}_\pi = \mathrm{id}_{\mathfrak{Des}^1_\pi(i)}$, see the proof of Theorem 2.9 in Appendix B.1. □

Now we are ready to define transport functors.

Definition 3.6. *Let* M *be a smooth manifold,* T *a category,* Gr *a Lie groupoid and* $i : \mathrm{Gr} \longrightarrow T$ *a functor.*

1. *A transport functor on* M *in* T *with* Gr*-structure is a functor*

$$\mathrm{tra} : \mathcal{P}_1(M) \longrightarrow T$$

such that there exists a surjective submersion $\pi : Y \longrightarrow M$ *and a* π*-local* i*-trivialization* (triv, t), *such that* $\mathrm{Ex}_\pi(\mathrm{tra}, \mathrm{triv}, t)$ *is smooth.*

2. *A morphism between transport functors on* M *in* T *with* Gr*-structure is a natural equivalence* $\eta : \mathrm{tra} \longrightarrow \mathrm{tra}'$ *such that there exists a surjective submersion* $\pi : Y \longrightarrow M$ *together with* π*-local* i*-trivializations of* tra *and* tra′, *such that* $\mathrm{Ex}_\pi(\eta)$ *is smooth.*

It is clear that the identity natural transformation of a transport functor tra is a morphism in the above sense. To show that the composition of morphisms between transport functors is possible, note that if $\pi : Y \longrightarrow M$ is a surjective submersion for which $\mathrm{Ex}_\pi(\eta)$ is smooth, and $\zeta : Z \longrightarrow Y$ is another surjective submersion, then also $\mathrm{Ex}_{\pi\circ\zeta}(\eta)$ is smooth. If now $\eta : \mathrm{tra} \longrightarrow \mathrm{tra}'$ and $\eta' : \mathrm{tra}' \longrightarrow \mathrm{tra}''$ are morphisms of transport functors, and $\pi : Y \longrightarrow M$ and $\pi' : Y' \longrightarrow M$ are surjective submersions for which $\mathrm{Ex}_\pi(\eta)$ and $\mathrm{Ex}_{\pi'}(\eta')$ are smooth, the fibre product $\tilde{\pi} : Y \times_M Y' \longrightarrow M$ is a surjective submersion and factors through π and π' by surjective submersions. Hence, $\mathrm{Ex}_{\tilde{\pi}}(\eta' \circ \eta)$ is smooth.

Definition 3.7. *The category of all transport functors on M in T with* Gr*-structure and all morphisms between those is denoted by* $\mathrm{Trans}^1_{\mathrm{Gr}}(M,T)$.

From the definition of a transport functor with Gr-structure it is not clear that, for a fixed surjective submersion $\pi : Y \longrightarrow M$, all choices of π-local i-trivializations (triv, t) with smooth descent data give rise to isomorphic objects in $\mathfrak{Des}^1_\pi(i)^\infty$. This is at least true for full functors $i : \mathrm{Gr} \longrightarrow T$ and contractible surjective submersions: a surjective submersion $\pi : Y \longrightarrow M$ is called *contractible*, if there exists a smooth map $c : Y \times [0,1] \longrightarrow Y$ such that $c(y,0) = y$ for all $y \in Y$ and $c(y,1) = y_k$ for some fixed choice of $y_k \in Y_k$ for each connected component Y_k of Y. We may assume without loss of generality, that c has a sitting instant with respect to the second parameter, so that we can regard c also as a map $c : Y \longrightarrow PY$. For example, if Y is the disjoint union of the open sets of a good open cover of M, $\pi : Y \longrightarrow M$ is contractible.

Lemma 3.8. *Let $i : \mathrm{Gr} \longrightarrow T$ be a full functor, let $\pi : Y \longrightarrow M$ be a contractible surjective submersion and let (triv, t) and (triv', t') be two π-local i-trivializations of a transport functor $\mathrm{tra} : \mathcal{P}_1(M) \longrightarrow T$ with* Gr*-structure. Then, the identity natural transformation $\mathrm{id}_{\mathrm{tra}} : \mathrm{tra} \longrightarrow \mathrm{tra}$ defines a morphism*

$$\mathrm{id}_{\mathrm{tra}} : (\mathrm{tra}, \mathrm{triv}, t) \longrightarrow (\mathrm{tra}, \mathrm{triv}', t')$$

in $\mathrm{Triv}^1_\pi(i)^\infty$, in particular, $\mathrm{Ex}_\pi(\mathrm{tra}, \mathrm{triv}, t)$ and $\mathrm{Ex}_\pi(\mathrm{tra}, \mathrm{triv}', t')$ are isomorphic objects in $\mathfrak{Des}^1_\pi(i)^\infty$.

Proof. Let $c : Y \times [0,1] \longrightarrow Y$ be a smooth contraction, regarded as a map $c : Y \longrightarrow PY$. For each $y \in Y_k$ we have a path $c(y) : y \longrightarrow y_k$, and the commutative diagram for the natural transformation t gives

$$t(y) = \mathrm{triv}_i(\overline{c(y)})^{-1} \circ t(y_k) \circ \mathrm{tra}(\pi_*(\overline{c(y)})),$$

and analogously for t'. The descent datum of the natural equivalence $\mathrm{id}_{\mathrm{tra}}$ is the natural equivalence

$$h := \mathrm{Ex}_\pi(\mathrm{id}) = t' \circ t^{-1} : \mathrm{triv}_i \longrightarrow \mathrm{triv}'_i.$$

Its component at $y \in Y_k$ is the morphism

$$h(y) = \mathrm{triv}'_i(\overline{c(y)})^{-1} \circ t'(y_k) \circ t(y_k)^{-1} \circ \mathrm{triv}_i(\overline{c(y)}) : \mathrm{triv}_i(y) \longrightarrow \mathrm{triv}'_i(y) \quad (3.2)$$

in T. Since i is full, $t'(y_k) \circ t(y_k)^{-1} = i(\kappa_k)$ for some morphism

$$\kappa_k : \mathrm{triv}(y_k) \longrightarrow \mathrm{triv}'(y_k),$$

so that h factors through i by

$$\tilde{h}(y) := \mathrm{triv}'(\overline{c(y)})^{-1} \circ \kappa_k \circ \mathrm{triv}(\overline{c(y)}) \in \mathrm{Mor}(\mathrm{Gr}).$$

Since triv and triv$'$ are smooth functors, $\mathrm{triv} \circ \mathrm{pr} \circ c$ and $\mathrm{triv}' \circ \mathrm{pr} \circ c$ are smooth maps, so that the components of $\tilde{h}$ form a smooth map $Y \longrightarrow \mathrm{Mor}(\mathrm{Gr})$. Hence, h is a morphism in $\mathfrak{Des}^1_\pi(i)^\infty$. □

To keep track of all the categories we have defined, consider the following diagram of functors which is strictly commutative:

$$\begin{array}{ccccc} \mathfrak{Des}^1_\pi(i)^\infty & \xrightarrow{\mathrm{Rec}_\pi} & \mathrm{Triv}^1_\pi(i)^\infty & \xrightarrow{v^\infty} & \mathrm{Trans}^1_{\mathrm{Gr}}(M,T) \\ \downarrow & & \downarrow & & \downarrow \\ \mathfrak{Des}^1_\pi(i) & \xrightarrow[\mathrm{Rec}_\pi]{} & \mathrm{Triv}^1_\pi(i) & \xrightarrow[v]{} & \mathrm{Funct}(M,T) \end{array} \tag{3.3}$$

The vertical arrows are the inclusion functors, and v^∞ and v are forgetful functors. In the next subsection we show that the functor v^∞ is an equivalence of categories.

3.2. Wilson Lines of Transport Functors

We restrict functors to paths between two fixed points and study the smoothness of these restrictions. For this purpose we assume that the functor $i : \mathrm{Gr} \longrightarrow T$ is an equivalence of categories; this is the case in all examples of transport functors we give in Section 5.

Definition 3.9. *Let $F : \mathcal{P}_1(M) \longrightarrow T$ be a functor, let* Gr *be a Lie groupoid and let $i : \mathrm{Gr} \longrightarrow T$ be an equivalence of categories. Consider two points $x_1, x_2 \in M$ together with a choice of objects G_k in* Gr *and isomorphisms $t_k : F(x_k) \longrightarrow i(G_k)$ in T for $k = 1, 2$. Then, the map*

$$\mathcal{W}^{F,i}_{x_1,x_2} : \mathrm{Mor}_{\mathcal{P}_1(M)}(x,y) \longrightarrow \mathrm{Mor}_{\mathrm{Gr}}(G_1,G_2) : \overline{\gamma} \longmapsto i^{-1}(t_2 \circ F(\overline{\gamma}) \circ t_1^{-1})$$

is called the Wilson line of F from x_1 to x_2.

Note that because i is essentially surjective, the choices of objects G_k and morphisms $t_k : F(x_k) \longrightarrow G_k$ exist for all points $x_k \in M$. Because i is full and faithful, the morphism $t_2 \circ F(\overline{\gamma}) \circ t_1^{-1} : i(G_1) \longrightarrow i(G_2)$ has a unique preimage under i, which is the Wilson line. For a different choice $t'_k : F(x_k) \longrightarrow i(G'_k)$ of objects in Gr and isomorphisms in T the Wilson line changes like

$$\mathcal{W}^{F,i}_{x_1,x_2} \longmapsto \tau_2^{-1} \circ \mathcal{W}^{F,i}_{x_1,x_2} \circ \tau_1$$

for $\tau_k : G'_k \longrightarrow G_k$ defined by $i(\tau_k) = t_k \circ t'^{-1}_k$.

Definition 3.10. *A Wilson line $\mathcal{W}^{F,i}_{x_1,x_2}$ is called smooth, if for every $k \in \mathbb{N}_0$, every open subset $U \subset \mathbb{R}^k$ and every map $c : U \longrightarrow PM$ such that $c(u)(t) \in M$ is smooth on $U \times [0,1]$, $c(u,0) = x_1$ and $c(u,1) = x_2$ for all $u \in U$, also the map*

$$\mathcal{W}^{F,i}_{x_1,x_2} \circ \mathrm{pr} \circ c : U \longrightarrow \mathrm{Mor}_{\mathrm{Gr}}(G_1,G_2)$$

is smooth.

This definition of smoothness arises again from the context of diffeological spaces, see Proposition A.6 i) in Appendix A.2. Notice that if a Wilson line is smooth for some choice of objects G_k and isomorphisms t_k, it is smooth for any other choice. For this reason we have not labelled Wilson lines with additional indices G_1, G_2, t_1, t_2.

Lemma 3.11. *Let $i : \mathrm{Gr} \longrightarrow T$ be an equivalence of categories, let*

$$F : \mathcal{P}_1(M) \longrightarrow T$$

be a functor whose Wilson lines $\mathcal{W}^{F,i}_{x_1,x_2}$ are smooth for all points $x_1, x_2 \in M$, and let $\pi : Y \longrightarrow M$ be a contractible surjective submersion. Then, F admits a π-local i-trivialization (triv, t) *whose descent data* $\mathrm{Ex}_\pi(\mathrm{triv}, t)$ *is smooth.*

Proof. We choose a smooth contraction $r : Y \longrightarrow PY$ and make, for every connected component Y_k of Y, a choice of objects G_k in Gr and isomorphisms $t_k : F(\pi(y_k)) \longrightarrow i(G_k)$. First we set $\mathrm{triv}(y) := G_k$ for all $y \in Y_k$, and define morphisms

$$t(y) := t_k \circ F(\pi_*(\overline{r(y)})) : F(\pi(y)) \longrightarrow i(G_k)$$

in T. For a path $\gamma : y \longrightarrow y'$, we define the morphism

$$\mathrm{triv}(\overline{\gamma}) := i^{-1}(t(y') \circ F(\pi_*(\overline{\gamma})) \circ t(y)^{-1}) : G_k \longrightarrow G_k$$

in Gr. By construction, the morphisms $t(y)$ are the components of a natural equivalence $t : \pi^*F \longrightarrow \mathrm{triv}_i$, so that we have defined a π-local i-trivialization (triv, t) of F. Since triv is locally constant on objects, it satisfies condition 1 of Definition 3.1. To check condition 2, notice that, for any path $\gamma : y \longrightarrow y'$,

$$\mathrm{triv}(\overline{\gamma}) = \mathcal{W}^{F,i}_{y_k,y_k}(\pi_*(\overline{r(y')} \circ \overline{\gamma} \circ \overline{r(y)}^{-1})). \tag{3.4}$$

More generally, if $c : U \longrightarrow PY$ is a map, we have, for every $u \in U$, a path

$$\tilde{c}(u) := \pi_*(r(c(u)(1)) \circ c(u) \circ r(c(u)(0))^{-1})$$

in M. Then, equation (3.4) becomes

$$\mathrm{triv} \circ \mathrm{pr} \circ c = \mathcal{W}^{F,i}_{y_k,y_k} \circ \mathrm{pr} \circ \tilde{c}.$$

Since the right hand side is by assumption a smooth function; triv is a smooth functor. The component of the natural equivalence $g := \pi_2^* t \circ \pi_1^* t^{-1}$ at a point $\alpha = (y, y') \in Y^{[2]}$ with $y \in Y_k$ and $y' \in Y_l$ is the morphism

$$g(\alpha) = t_l \circ F(\pi(c(y'))) \circ F(\pi(c(y)))^{-1} \circ t_k^{-1} : i(G_k) \longrightarrow i(G_l),$$

and hence of the form $g(\alpha) = i(\tilde{g}(\alpha))$. Now consider a chart $\varphi : V \longrightarrow Y^{[2]}$ with an open subset $V \in \mathbb{R}^n$, and the path $c(u) := r(\pi_2(\varphi(u))) \circ r(\pi_1(\varphi(u)))^{-1}$ in Y. We find

$$\tilde{g} \circ \varphi = \mathcal{W}^{F,i}_{y_k,y_l} \circ \mathrm{pr} \circ c$$

as functions from U to $\mathrm{Mor}(G_k, G_l)$. Because the right hand side is by assumption a smooth function, $\tilde{g}$ is smooth on every chart, and hence also a smooth function. □

Theorem 3.12. *Let $i : \mathrm{Gr} \longrightarrow T$ be an equivalence of categories. A functor*

$$F : \mathcal{P}_1(M) \longrightarrow T$$

is a transport functor with Gr*-structure if and only if for every pair (x_1, x_2) of points in M the Wilson line $\mathcal{W}^{F,i}_{x_1,x_2}$ is smooth.*

Proof. One implication is shown by Lemma 3.11, using the fact that contractible surjective submersions always exist. To prove the other implication we express the Wilson line of the transport functor locally in terms of the functor $R_{(\mathrm{triv},g)}$: $\mathcal{P}_1^{\pi}(M) \longrightarrow T$ from Section 2.3. We postpone this construction to Appendix B.2. □

Theorem 3.12 makes it possible to check explicitly, whether a given functor F is a transport functor or not. Furthermore, because every transport functor has smooth Wilson lines, we can apply Lemma 3.11 and have

Corollary 3.13. *Every transport functor* tra : $\mathcal{P}_1(M) \longrightarrow T$ *with* Gr*-structure (with i : Gr $\longrightarrow T$ an equivalence of categories) admits a π-local i-trivialization with smooth descent data for any contractible surjective submersion π.*

This corollary can be understood analogously to the fact, that every fibre bundle over M is trivializable over every good open cover of M.

Proposition 3.14. *For an equivalence of categories $i : \mathrm{Gr} \longrightarrow T$ and a contractible surjective submersion $\pi : Y \longrightarrow M$, the forgetful functor*

$$v^{\infty} : \mathrm{Triv}^1_{\pi}(i)^{\infty} \longrightarrow \mathrm{Trans}^1_{\mathrm{Gr}}(M,T)$$

is a surjective equivalence of categories.

Proof. By Corollary 3.13 v^{∞} is surjective. Since it is certainly faithful, it remains to prove that it is full. Let η be a morphism of transport functors with π-local i-trivialization, i.e. there exists a surjective submersion $\pi' : Y' \longrightarrow M$ such that $\mathrm{Ex}_{\pi'}(\eta)$ is smooth. Going to a contractible surjective submersion $Z \longrightarrow Y \times_M Y'$ shows that also $\mathrm{Ex}_{\pi}(\eta)$ is smooth. □

Summarizing, we have for i an equivalence of categories and π a contractible surjective submersion, the following equivalences of categories:

$$\mathfrak{Des}^1_{\pi}(i)^{\infty} \underset{\mathrm{Ex}_{\pi}}{\overset{\mathrm{Rec}_{\pi}}{\rightleftarrows}} \mathrm{Triv}^1_{\pi}(i)^{\infty} \xrightarrow{\;v^{\infty}\;} \mathrm{Trans}^1_{\mathrm{Gr}}(M,T).$$

4. Differential Forms and smooth Functors

We establish a relation between smooth descent data we have defined in the previous section and more familiar geometric objects like differential forms, motivated by [**BS04**] and [**Bae07**]. The relation we find can be expressed as a path ordered exponential, understood as the solution of an initial value problem.

Lemma 4.1. *Let G be a Lie group with Lie algebra $\mathfrak{g}$. There is a canonical bijection between the set $\Omega^1(\mathbb{R},\mathfrak{g})$ of $\mathfrak{g}$-valued 1-forms on $\mathbb{R}$ and the set of smooth maps*

$$f:\mathbb{R}\times\mathbb{R}\longrightarrow G$$

satisfying the cocycle condition

$$f(y,z)\cdot f(x,y)=f(x,z). \tag{4.1}$$

Proof. The idea behind this bijection is that f is the path-ordered exponential of a 1-form A,

$$f(x,y)=\mathcal{P}\exp\left(\int_x^y A\right).$$

Let us explain in detail what that means. Given the 1-form A, we pose the initial value problem

$$\frac{\partial}{\partial t}u(t)=-\mathrm{d}r_{u(t)}|_1(A_t\left(\frac{\partial}{\partial t}\right))\quad\text{and}\quad u(t_0)=1 \tag{4.2}$$

for a smooth function $u:\mathbb{R}\longrightarrow G$ and a number $t_0\in\mathbb{R}$. Here, $r_{u(t)}$ is the right multiplication in G and $\mathrm{d}r_{u(t)}|_1:\mathfrak{g}\longrightarrow T_{u(t)}G$ is its differential evaluated at $1\in G$. The sign in (4.2) is a convention well-adapted to the examples in Section 5. Differential equations of this type have a unique solution $u(t)$ defined on all of $\mathbb{R}$, such that $f(t_0,t):=u(t)$ depends smoothly on both parameters. To see that f satisfies the cocycle condition (4.1), define for fixed $x,y\in\mathbb{R}$ the function $\Psi(t):=f(y,t)\cdot f(x,y)$. Its derivative is

$$\begin{aligned}\frac{\partial}{\partial t}\Psi(t)&=\mathrm{d}r_{f(x,y)}|_1\left(\frac{\partial}{\partial t}f(y,t)\right)\\&=-\mathrm{d}r_{f(x,y)}|_1(\mathrm{d}r_{f(y,t)}|_1(A_t\left(\frac{\partial}{\partial t}\right)))\\&=-\mathrm{d}r_{\Psi(t)}(A_t\left(\frac{\partial}{\partial t}\right))\end{aligned}$$

and furthermore $\Psi(y)=f(x,y)$. So, by uniqueness

$$f(y,t)\cdot f(x,y)=\Psi(t)=f(x,t).$$

Conversely, for a smooth function $f:\mathbb{R}\times\mathbb{R}\longrightarrow G$, let $u(t):=f(t_0,t)$ for some $t_0\in\mathbb{R}$, and define

$$A_t\left(\frac{\partial}{\partial t}\right):=-\mathrm{d}r_{u(t)}|_1^{-1}\frac{\partial}{\partial t}u(t), \tag{4.3}$$

which yields a 1-form on $\mathbb{R}$. If f satisfies the cocycle condition, this 1-form is independent of the choice of t_0. The definition of the 1-form A is obviously inverse to (4.2) and thus establishes the claimed bijection. □

We also need a relation between the functions f_A and $f_{A'}$ corresponding to 1-forms A and A', when A and A' are related by a gauge transformation. In the following we denote the left and right invariant Maurer-Cartan forms on G forms by θ and $\bar{\theta}$ respectively.

Lemma 4.2. *Let $A \in \Omega^1(\mathbb{R}, \mathfrak{g})$ be a $\mathfrak{g}$-valued 1-form on $\mathbb{R}$, let $g : \mathbb{R} \longrightarrow G$ be a smooth function and let $A' := \mathrm{Ad}_g(A) - g^*\bar{\theta}$. If f_A and $f_{A'}$ are the smooth functions corresponding to A and A' by Lemma 4.1, we have*

$$g(y) \cdot f_A(x, y) = f_{A'}(x, y) \cdot g(x).$$

Proof. By direct verification, the function $g(y) \cdot f_A(x, y) \cdot g(x)^{-1}$ solves the initial value problem (4.2) for the 1-form A'. Uniqueness gives the claimed equality. □

In the following we use the two lemmata above for 1-forms on $\mathbb{R}$ to obtain a similar correspondence between 1-forms on an arbitrary smooth manifold X and certain smooth functors defined on the path groupoid $\mathcal{P}_1(X)$. For a given 1-form $A \in \Omega^1(X, \mathfrak{g})$, we first define a map

$$k_A : PX \longrightarrow G$$

in the following way: a path $\gamma : x \longrightarrow y$ in X can be continued to a smooth function $\gamma : \mathbb{R} \longrightarrow X$ with $\gamma(t) = x$ for $t < 0$ and $\gamma(t) = y$ for $t > 1$, due to its sitting instants. Then, the pullback $\gamma^*A \in \Omega^1(\mathbb{R}, \mathfrak{g})$ corresponds by Lemma 4.1 to a smooth function $f_{\gamma^*A} : \mathbb{R} \times \mathbb{R} \longrightarrow G$. Now we define

$$k_A(\gamma) := f_{\gamma^*A}(0, 1).$$

The map k_A defined like this comes with the following properties:

a) For the constant path id_x we obtain the constant function $f_{\mathrm{id}_x^*A}(x, y) = 1$ and thus

$$k_A(\mathrm{id}_x) = 1. \tag{4.4}$$

b) For two paths $\gamma_1 : x \longrightarrow y$ and $\gamma_2 : y \longrightarrow z$, we have

$$f_{(\gamma_2 \circ \gamma_1)^*A}(0, 1) = f_{(\gamma_2 \circ \gamma_1)^*A}(\frac{1}{2}, 1) \cdot f_{(\gamma_2 \circ \gamma_1)^*A}(0, \frac{1}{2}) = f_{\gamma_1^*A}(0, 1) \cdot f_{\gamma_2^*A}(0, 1)$$

and thus

$$k_A(\gamma_2 \circ \gamma_1) = k_A(\gamma_2) \cdot k_A(\gamma_1). \tag{4.5}$$

c) If $g : X \longrightarrow G$ is a smooth function and $A' := \mathrm{Ad}_g(A) - g^*\bar{\theta}$,

$$g(y) \cdot k_A(\gamma) = k_{A'}(\gamma) \cdot g(x) \tag{4.6}$$

for any path $\gamma : x \longrightarrow y$.

The next proposition shows that the definition of $k_A(\gamma)$ depends only on the thin homotopy class of γ.

Proposition 4.3. *The map $k_A : PX \longrightarrow G$ factors in a unique way through the set P^1X of thin homotopy classes of paths, i.e. there is a unique map*

$$F_A : P^1X \longrightarrow G$$

such that $k_A = F_A \circ \mathrm{pr}$ *with* $\mathrm{pr} : PX \longrightarrow P^1X$ *the projection.*

Proof. If k_A factors through the surjective map $\mathrm{pr} : PX \longrightarrow P^1X$, the map F_A is determined uniquely. So we only have to show that two thin homotopy equivalent paths $\gamma_0 : x \longrightarrow y$ and $\gamma_1 : x \longrightarrow y$ are mapped to the same group element, $k_A(\gamma_0) = k_A(\gamma_1)$. We have moved this issue to Appendix B.3. □

In fact, the map $F_A : P^1X \longrightarrow G$ is not just a map. To understand it correctly, we need the following category:

Definition 4.4. *Let* G *be a Lie group. We denote by* $\mathcal{B}G$ *the following Lie groupoid: it has only one object, and* G *is its set of morphisms. The unit element* $1 \in G$ *is the identity morphism, and group multiplication is the composition, i.e.* $g_2 \circ g_1 := g_2 \cdot g_1$.

To understand the notation, notice that the geometric realization of the nerve of $\mathcal{B}G$ yields the classifying space of the group G, i.e. $|N(\mathcal{B}G)| = BG$. We claim that the map F_A defined by Proposition 4.3 defines a functor

$$F_A : \mathcal{P}_1(X) \longrightarrow \mathcal{B}G.$$

Indeed, since $\mathcal{B}G$ has only one object one only has to check that F_A respects the composition (which is shown by (4.4)) and the identity morphisms (shown in (4.5)).

Lemma 4.5. *The functor* F_A *is smooth in the sense of Definition 3.1.*

Proof. Let $U \subset \mathbb{R}^k$ be an open subset of some $\mathbb{R}^k$ and let $c : U \longrightarrow PX$ be a map such that $c(u)(t)$ is smooth on $U \times [0,1]$. We denote the path associated to a point $x \in U$ and extended smoothly to $\mathbb{R}$ by $\gamma_x := c(x) : \mathbb{R} \longrightarrow X$. This means that $U \longrightarrow \Omega^1(\mathbb{R}, \mathfrak{g}) : x \longmapsto \gamma_x^*A$ is a smooth family of $\mathfrak{g}$-valued 1-forms on $\mathbb{R}$. We recall that

$$(k_A \circ c)(x) = k_A(\gamma_x) = f_{\gamma_x^*A}(0,1)$$

is defined to be the solution of a differential equation, which now depends smoothly on x. Hence, $k_A \circ c = F_A \circ \mathrm{pr} \circ c : U \longrightarrow G$ is a smooth function. □

Let us summarize the correspondence between 1-forms on X and smooth functors developed in the Lemmata above in terms of an equivalence between categories. One category is a category $\mathrm{Funct}^\infty(\mathcal{P}_1(X), \mathcal{B}G)$ of smooth functors and smooth natural transformations. The second category is the category of differential G-cocycles on X:

Definition 4.6. *Let* X *be a smooth manifold and* G *be a Lie group with Lie algebra* $\mathfrak{g}$. *We consider the following category* $Z^1_X(G)^\infty$*: objects are all* $\mathfrak{g}$*-valued 1-forms* A *on* X*, and a morphism* $A \longrightarrow A'$ *is a smooth function* $g : X \longrightarrow G$ *such that*

$$A' = \mathrm{Ad}_g(A) - g^*\bar{\theta}.$$

The composition is the multiplication of functions, $g_2 \circ g_1 = g_2 g_1$.

We claim that the Lemmata above provide the structure of a functor

$$\mathcal{P} : Z^1_X(G)^\infty \longrightarrow \mathrm{Funct}^\infty(\mathcal{P}_1(X), \mathcal{B}G).$$

It sends a $\mathfrak{g}$-valued 1-form A on X to the functor F_A defined uniquely in Proposition 4.3 and which is shown by Lemma 4.5. It sends a function $g : X \longrightarrow G$ regarded as a morphism $A \longrightarrow A'$ to the smooth natural transformation $F_A \longrightarrow F_{A'}$ whose component at a point x is $g(x)$. This is natural in x due to (4.6).

Proposition 4.7. *The functor*

$$\mathcal{P} : Z^1_X(G)^\infty \longrightarrow \mathrm{Funct}^\infty(X, \mathcal{B}G).$$

is an isomorphism of categories, which reduces on the level of objects to a bijection

$$\Omega^1(X, \mathfrak{g}) \cong \{\textit{Smooth functors } F : \mathcal{P}_1(X) \longrightarrow \mathcal{B}G\}.$$

Proof. If A and A' are two $\mathfrak{g}$-valued 1-forms on X, the set of morphisms between them is the set of smooth functions $g : X \longrightarrow G$ satisfying the condition $A' = \mathrm{Ad}_g(A) - g^*\bar{\theta}$. The set of morphisms between the functors F_A and $F_{A'}$ are smooth natural transformations, i.e. smooth maps $g : X \longrightarrow G$, whose naturality square is equivalent to the same condition. So, the functor $\mathcal{P}$ is manifestly full and faithful. It remains to show that it is a bijection on the level of objects. This is done in Appendix B.4 by an explicit construction of a 1-form A to a given smooth functor F. □

One can also enhance the category $Z^1_X(G)^\infty$ in such a way that it becomes the familiar category of local data of principal G-bundles with connection.

Definition 4.8. *The category $Z^1_\pi(G)^\infty$ of differential G-cocycles of the surjective submersion π is the category whose objects are pairs (g, A) consisting of a 1-form $A \in \Omega^1(Y, \mathfrak{g})$ and a smooth function $g : Y^{[2]} \longrightarrow G$ such that*

$$\pi_{13}^* g = \pi_{23}^* g \cdot \pi_{12}^* g \quad \textit{and} \quad \pi_2^* A = \mathrm{Ad}_g(\pi_1^* A) - g^*\bar{\theta}.$$

A morphism

$$h : (g, A) \longrightarrow (g', A')$$

is a smooth function $h : Y \longrightarrow G$ such that

$$A' = \mathrm{Ad}_h(A) - h^*\bar{\theta} \quad \textit{and} \quad \pi_2^* h \cdot g = g' \cdot \pi_1^* h.$$

Composition of morphisms is given by the product of these functions, $h_2 \circ h_1 = h_2 h_1$.

To explain the notation, notice that for $\pi = \mathrm{id}_X$ we obtain $Z^1_X(G)^\infty = Z^1_\pi(G)^\infty$. As an example, we consider the group $G = U(1)$ and a surjective submersion $\pi : Y \longrightarrow M$ coming from a good open cover $\mathfrak{U}$ of M. Then, the group of isomorphism classes of $Z^1_\pi(U(1))^\infty$ is the Deligne hypercohomology group $H^1(\mathfrak{U}, \mathcal{D}(1))$, where $\mathcal{D}(1)$ is the Deligne sheaf complex $0 \longrightarrow \underline{U(1)} \longrightarrow \underline{\Omega}^1$.

Corollary 4.9. *The functor $\mathcal{P}$ extends to an equivalence of categories*

$$Z^1_\pi(G)^\infty \cong \mathfrak{Des}^1_\pi(i_G)^\infty,$$

where $i_G : \mathcal{B}G \longrightarrow G\text{-Tor}$ sends the object of $\mathcal{B}G$ to the group G regarded as a G-space, and a morphism $g \in G$ to the equivariant smooth map which multiplies with g from the left.

This corollary is an important step towards our main theorem, to which we come in the next section.

5. Examples

Various structures in the theory of bundles with connection are special cases of transport functors with Gr-structure for particular choices of the structure groupoid Gr. In this section we spell out some prominent examples.

5.1. Principal Bundles with Connection

In this section, we fix a Lie group G. Associated to this Lie group, we have the Lie groupoid $\mathcal{B}G$ from Definition 4.4, the category G-Tor of smooth manifolds with right G-action and G-equivariant smooth maps between those, and the functor $i_G : \mathcal{B}G \longrightarrow G\text{-Tor}$ that sends the object of $\mathcal{B}G$ to the G-space G and a morphism $g \in G$ to the G-equivariant diffeomorphism that multiplies from the left by g. The functor i_G is an equivalence of categories.

As we have outlined in the introduction, a principal G-bundle P with connection over M defines a functor

$$\mathrm{tra}_P : \mathcal{P}_1(M) \longrightarrow G\text{-Tor}.$$

Before we show that tra_P is a transport functor with $\mathcal{B}G$-structure, let us recall its definition in detail. To an object $x \in M$ it assigns the fibre P_x of the bundle P over the point x. To a path $\gamma : x \longrightarrow y$, it assigns the parallel transport map $\tau_\gamma : P_x \longrightarrow P_y$.

For preparation, we recall the basic definitions concerning local trivializations of principal bundles with connections. In the spirit of this article, we use surjective submersions instead of coverings by open sets. In this language, a local trivialization of the principal bundle P is a surjective submersion $\pi : Y \longrightarrow M$ together with a G-equivariant diffeomorphism

$$\phi : \pi^*P \longrightarrow Y \times G$$

that covers the identity on Y. Here, the fibre product $\pi^*P = Y \times_M P$ comes with the projection $p : \pi^*P \longrightarrow P$ on the second factor. It induces a section

$$s : Y \longrightarrow P : y \longmapsto p(\phi^{-1}(y, 1)).$$

The transition function $\tilde{g}_\phi : Y^{[2]} \longrightarrow G$ associated to the local trivialization ϕ is defined by

$$s(\pi_1(\alpha)) = s(\pi_2(\alpha)) \cdot \tilde{g}_\phi(\alpha) \tag{5.1}$$

for every point $\alpha \in Y^{[2]}$. A connection on P is a $\mathfrak{g}$-valued 1-form $\omega \in \Omega^1(P, \mathfrak{g})$ that obeys

$$\omega_{\rho g}\left(\frac{\mathrm{d}}{\mathrm{d}t}(\rho g)\right) = \mathrm{Ad}_g^{-1}\left(\omega_\rho\left(\frac{\mathrm{d}\rho}{\mathrm{d}t}\right)\right) + \theta_g\left(\frac{\mathrm{d}g}{\mathrm{d}t}\right) \tag{5.2}$$

for smooth maps $\rho : [0,1] \longrightarrow P$ and $g : [0,1] \longrightarrow G$. In this setup, a tangent vector $v \in T_pP$ is called horizontal, if it is in the kernel of ω.

Notice that all our conventions are chosen such that the transition function $\tilde{g}_\phi : Y^{[2]} \longrightarrow G$ and the local connection 1-form $\tilde{A}_\phi := s^*\omega \in \Omega^1(Y, \mathfrak{g})$ define an object in the category $Z^1_\pi(G)^\infty$ from Definition 4.8.

To define the parallel transport map τ_γ associated to a path $\gamma : x \longrightarrow y$ in M, we assume first that γ has a lift $\tilde{\gamma} : \tilde{x} \longrightarrow \tilde{y}$ in Y, that is, $\pi_*\tilde{\gamma} = \gamma$. Consider then the path $s_*\tilde{\gamma}$ in P, which can be modified by the pointwise action of a path g in G from the right, $(s_*\tilde{\gamma})g$. This modification has now to be chosen such that every tangent vector to $(s_*\tilde{\gamma})g$ is horizontal, i.e.

$$0 = \omega_{(s_*\tilde{\gamma})g}\left(\frac{\mathrm{d}}{\mathrm{d}t}((s_*\tilde{\gamma})g)\right) \overset{(5.2)}{=} \mathrm{Ad}_g^{-1}\left(\omega_{s_*\tilde{\gamma}}\left(\frac{\mathrm{d}(s_*\tilde{\gamma})}{\mathrm{d}t}\right)\right) + \theta_g\left(\frac{\mathrm{d}g}{\mathrm{d}t}\right)$$

This is a linear differential equation for g, which has together with the initial condition $g(0) = 1$ a unique solution $g = g(\tilde{\gamma})$. Then, for any $p \in P_x$,

$$\tau_\gamma(p) := s(y)(g(1) \cdot h), \tag{5.3}$$

where h is the unique group element with $s(x)h = p$. It is evidently smooth in p and G-equivariant. Paths γ in M which do not have a lift to Y have to be split up in pieces which admit lifts; τ_γ is then the composition of the parallel transport maps of those.

Lemma 5.1. *Let P be a principal G-bundle over M with connection $\omega \in \Omega^1(P, \mathfrak{g})$. For a surjective submersion $\pi : Y \longrightarrow M$ and a trivialization ϕ with associated section $s : Y \longrightarrow P$, we consider the smooth functor*

$$F_\omega := \mathcal{P}(s^*\omega) : \mathcal{P}_1(Y) \longrightarrow \mathcal{B}G$$

associated to the 1-form $s^*\omega \in \Omega^1(Y, \mathfrak{g})$ by Proposition 4.7. Then,

$$i_G(F_\omega(\overline{\gamma})) = \phi_y \circ \tau_{\pi_*\gamma} \circ \phi_x^{-1} \tag{5.4}$$

for any path $\gamma : x \longrightarrow y$ in PY.

Proof. Recall the definition of the functor F_ω: for a path $\gamma : x \longrightarrow y$, we have to consider the 1-form $\gamma^*s^*\omega \in \Omega^1(\mathbb{R}, \mathfrak{g})$, which defines a smooth function $f_\omega : \mathbb{R} \times \mathbb{R} \longrightarrow G$. Then, $F_\omega(\overline{\gamma}) := f_\omega(0, 1)$. We claim the equation

$$f_\omega(0, t) = g(t). \tag{5.5}$$

This comes from the fact that both functions are solutions of the same differential equation, with the same initial value for $t = 0$. Using (5.5),

$$i_G(F_\omega(\overline{\gamma}))(h) = F_\omega(\overline{\gamma}) \cdot h = g(1) \cdot h$$

for some $h \in G$. On the other hand,

$$\phi_y(\tau_{\pi_*\gamma}(\phi_x^{-1}(h))) = \phi_y(\tau_{\pi_*\gamma}(s(x)h)) \overset{(5.3)}{=} \phi_y(s(y)(g(1) \cdot h)) = g(1) \cdot h.$$

This proves equation (5.4). □

Now we are ready to formulate the basic relation between principal G-bundles with connection and transport functors with $\mathcal{B}G$-structure.

Proposition 5.2. *The functor*

$$\mathrm{tra}_P : \mathcal{P}_1(M) \longrightarrow G\text{-Tor}$$

obtained from parallel transport in a principal G-bundle P, is a transport functor with $\mathcal{B}G$-structure in the sense of Definition 3.6.

Proof. The essential ingredient is, that P is locally trivializable: we choose a surjective submersion $\pi : Y \longrightarrow M$ and a trivialization ϕ. The construction of a functor $\mathrm{triv}_\phi : \mathcal{P}_1(Y) \longrightarrow \mathcal{B}G$ and a natural equivalence

$$\begin{array}{ccc} \mathcal{P}_1(Y) & \xrightarrow{\pi_*} & \mathcal{P}_1(M) \\ {\scriptstyle \mathrm{triv}_\phi}\downarrow & \overset{t_\phi}{\swarrow\!\!\!\swarrow} & \downarrow{\scriptstyle \mathrm{tra}_P} \\ \mathcal{B}G & \xrightarrow[i_G]{} & G\text{-Tor} \end{array}$$

is as follows. We let $\mathrm{triv}_\phi := \mathcal{P}(s^*\omega)$ be the smooth functor associated to the 1-form $s^*\omega$ by Proposition 4.7. To define the natural equivalence t_ϕ, consider a point $x \in Y$. We find $\pi^*\mathrm{tra}_P(x) = P_{\pi(x)}$ and $(i_G \circ \mathrm{triv}_\phi)(x) = G$. So we define the component of t_ϕ at x by

$$t_\phi(x) := \phi_x : P_{\pi(x)} \longrightarrow G.$$

This is natural in x since the diagram

$$\begin{array}{ccc} P_{\pi(x)} & \xrightarrow{\phi_x} & G \\ {\scriptstyle \tau_{\pi_*\overline{\gamma}}}\downarrow & & \downarrow{\scriptstyle i_G(\mathrm{triv}_\phi(\overline{\gamma}))} \\ P_{\pi(y)} & \xrightarrow[\phi_y]{} & G \end{array}$$

is commutative by Lemma 5.1. Notice that the natural equivalence

$$g_\phi := \pi_2^* t_\phi \circ \pi_1^* t_\phi \tag{5.6}$$

factors through the smooth transition function $\tilde{g}_\phi$ from (5.1), i.e. $g_\phi = i_G(\tilde{g}_\phi)$. Hence, the pair $(\mathrm{triv}_\phi, g_\phi)$ is a smooth object in $\mathfrak{Des}^1_\pi(i)^\infty$. □

Now we consider the morphisms. Let $\varphi : P \longrightarrow P'$ be a morphism of principal G-bundles over M (covering the identity on M) which respects the connections, i.e. $\omega = \varphi^*\omega'$. For any point $p \in M$, its restriction $\varphi_x : P_x \longrightarrow P'_x$ is a smooth G-equivariant map. For any path $\gamma : x \longrightarrow y$, the parallel transport map satisfies

$$\varphi_y \circ \tau_\gamma = \tau'_\gamma \circ \varphi_x.$$

This is nothing but the commutative diagram for the components $\eta_\varphi(x) := \varphi_x$ natural transformation $\eta_\varphi : \mathrm{tra}_P \longrightarrow \mathrm{tra}_{P'}$.

Proposition 5.3. *The natural transformation*

$$\eta_\varphi : \mathrm{tra}_P \longrightarrow \mathrm{tra}_{P'}$$

obtained from a morphism $\varphi : P \longrightarrow P'$ of principal G-bundles, is a morphism of transport functors in the sense of Definition 3.6.

Proof. Consider a surjective submersion $\pi : Y \longrightarrow M$ such that π^*P and π^*P' are trivializable, and choose trivializations ϕ and ϕ'. The descent datum of η_φ is the natural equivalence $h := t'_\phi \circ \pi^*\eta_\varphi \circ t_\phi^{-1}$. Now define the map

$$\tilde{h} : Y \longrightarrow G : x \longmapsto p_G(\phi'(x, \varphi(s(x))))$$

where p_G is the projection to G. This map is smooth and satisfies $h = i_G(\tilde{h})$. Thus, η_φ is a morphism of transport functors. □

Taking the Propositions 5.2 and 5.3 together, we have defined a functor

$$\mathfrak{Bun}_G^\nabla(M) \longrightarrow \mathrm{Trans}^1_{\mathrm{Gr}}(M, G\text{-Tor}) \tag{5.7}$$

from the category of principal G-bundles over M with connection to the category of transport functors on M in G-Tor with $\mathcal{B}G$-structure. In particular, this functor provides us with lots of examples of transport functors.

Theorem 5.4. *The functor*

$$\mathfrak{Bun}_G^\nabla(M) \longrightarrow \mathrm{Trans}^1_{\mathcal{B}G}(M, G\text{-Tor}) \tag{5.8}$$

is an equivalence of categories.

We give two proofs of this Theorem: the first is short and the second is explicit.

First Proof. Let $\pi : Y \longrightarrow M$ be a contractible surjective submersion, over which every principal G-bundle is trivializable. Extracting a connection 1-form $\tilde{A}_\phi \in \Omega^1(Y, \mathfrak{g})$ and the transition function (5.1) yields a functor

$$\mathfrak{Bun}_G^\nabla(M) \longrightarrow Z^1_\pi(G)^\infty$$

to the category of differential G-cocycles for π, which is in fact an equivalence of categories. We claim that the composition of this equivalence with the sequence

$$Z^1_\pi(G)^\infty \xrightarrow{\ \mathcal{P}\ } \mathfrak{Des}^1_\pi(i)^\infty \xrightarrow{\ \mathrm{Rec}_\pi\ } \mathrm{Triv}^1_\pi(i)^\infty \xrightarrow{\ v^\infty\ } \mathrm{Trans}^1_{\mathrm{Gr}}(M, G\text{-Tor}) \tag{5.9}$$

of functors is naturally equivalent to the functor (5.8). By Corollary 4.9, Theorem 2.9 and Proposition 3.14 all functors in (5.9) are equivalences of categories, and so is (5.8). To show the claim recall that in the proof of Proposition 5.2 we have defined a local trivialization of tra_P, whose descent data $(\mathrm{triv}_\phi, g_\phi)$ is the image of the local data $(\tilde{A}_\phi, \tilde{g}_\phi)$ of the principal G-bundle under the functor $\mathcal{P}$. This reproduces exactly the steps in the sequence (5.9). □

Second proof. We show that the functor (5.8) is faithful, full and essentially surjective. In fact, this proof shows that it is even surjective. So let P and P' two principal G-bundles with connection over M, and let tra_P and $\mathrm{tra}_{P'}$ be the associated transport functors.

Faithfulness follows directly from the definition, so assume now that

$$\eta : \mathrm{tra}_P \longrightarrow \mathrm{tra}_{P'}$$

is a morphism of transport functors. We define a morphism $\varphi : P \longrightarrow P$ pointwise as $\varphi(x) := \eta(p(x))(x)$ for any $x \in P$, where $p : P \longrightarrow M$ is the projection of the bundle P. This is clearly a preimage of η under the functor (5.8), so that we only have to show that φ is a smooth map. We choose a surjective submersion such that

P and P' are trivializable and such that $h := \mathrm{Ex}_\pi(\eta) = t_{\phi'} \circ \pi^*\eta \circ t_\phi^{-1}$ is a smooth morphism in $\mathfrak{Des}^1_\pi(i)^\infty$. Hence it factors through a smooth map $\tilde{h} : Y \longrightarrow G$, and from the definitions of t_ϕ and $t_{\phi'}$ it follows that $\pi^*\varphi$ is the function

$$\pi^*\varphi : \pi^*P \longrightarrow \pi^*P' : (y,p) \longmapsto \phi'^{-1}(\phi(y,p)\tilde{h}(y)),$$

and thus smooth. Finally, since π is a surjective submersion, φ is smooth.

It remains to prove that the functor (5.8) is essentially surjective. First we construct, for a given transport functor $\mathrm{tra} : \mathcal{P}_1(M) \longrightarrow G\text{-Tor}$ a principal G-bundle P with connection over M, performing exactly the inverse steps of (5.9). We choose a surjective submersion $\pi : Y \longrightarrow M$ and a π-local i-trivialization (triv, t) of the transport functor tra. By construction, its descent data $(\mathrm{triv}, g) := \mathrm{Ex}_\pi(\mathrm{triv}, g)$ is an object in $\mathfrak{Des}^1_\pi(i)^\infty$. By Corollary 4.9, there exists a 1-form $A \in \Omega^1(Y, \mathfrak{g})$, and a smooth function $\tilde{g} : Y^{[2]} \longrightarrow G$, forming an object $(A, \tilde{g})$ in the category $Z^1_\pi(G)^\infty$ of differential cocycles such that

$$\mathcal{P}(A, \tilde{g}) = (\mathrm{triv}, g) \tag{5.10}$$

in $\mathfrak{Des}^1_\pi(i)^\infty$. In particular $g = i_G(\tilde{g})$. The pair $(A, \tilde{g})$ is local data for a principal G-bundle P with connection ω. The reconstructed bundle comes with a canonical trivialization $\phi : \pi^*P \longrightarrow Y \times G$, for which the associated section $s : Y \longrightarrow P$ is such that $A = s^*\omega$, and whose transition function is $\tilde{g}_\phi = \tilde{g}$.

Let us extract descent data of the transport functor tra_P of P: as described in the proof of Proposition 5.4, the trivialization ϕ of the bundle P gives rise to a π-local i_G-trivialization $(\mathrm{triv}_\phi, t_\phi)$ of the transport functor tra_P, namely

$$\mathrm{triv}_\phi := F_\omega := \mathcal{P}(s^*\omega) = \mathcal{P}(A) \tag{5.11}$$

and $t_\phi(x) := \phi_x$. Its natural equivalence g_ϕ from (5.6) is just $g_\phi = i_G(\tilde{g}_\phi)$.

Finally we construct an isomorphism $\eta : \mathrm{tra}_P \longrightarrow \mathrm{tra}$ of transport functors. Consider the natural equivalence

$$\zeta := t^{-1} \circ t_\phi : \pi^*\mathrm{tra}_P \longrightarrow \pi^*\mathrm{tra}.$$

From condition (2.4) it follows that $\zeta(\pi_1(\alpha)) = \zeta(\pi_2(\alpha))$ for every point $\alpha \in Y^{[2]}$. So ζ descends to a natural equivalence

$$\eta(x) := \zeta(\tilde{x})$$

for $x \in M$ and any $\tilde{x} \in Y$ with $\pi(\tilde{x}) = x$. An easy computation shows that $\mathrm{Ex}_\pi(\eta) = t \circ \zeta \circ t_\phi^{-1} = \mathrm{id}$, which is in particular smooth and thus proves that η is an isomorphism in $\mathfrak{Des}^1_\pi(i)^\infty$. □

5.2. Holonomy Maps

In this section, we show that important results of [**Bar91, CP94**] on holonomy maps of principal G-bundles with connection can be reproduced as particular cases.

Definition 5.5 ([**CP94**]). *A holonomy map on a smooth manifold M at a point $x \in M$ is a group homomorphism*

$$\mathcal{H}_x : \pi_1^1(M, x) \longrightarrow G,$$

which is smooth in the following sense: for every open subset $U \subset \mathbb{R}^k$ and every map $c : U \longrightarrow L_xM$ such that $\Gamma(u,t) := c(u)(t)$ is smooth on $U \times [0,1]$, also

$$U \xrightarrow{c} L_xM \xrightarrow{\mathrm{pr}} \pi_1^1(M,x) \xrightarrow{\mathcal{H}} G$$

is smooth.

Here, $L_xM \subset PM$ is the set of paths $\gamma : x \longrightarrow x$, whose image under the projection $\mathrm{pr} : PM \longrightarrow P^1M$ is, by definition, the thin homotopy group $\pi_1^1(M,x)$ of M at x. Also notice, that

- in the context of diffeological spaces reviewed in Appendix A.2, the definition of smoothness given here just means that $\mathcal{H}$ is a morphism between diffeological spaces, cf. Proposition A.6 ii).
- the notion of *intimate paths* from [**CP94**] and the notion of thin homotopy from [**MP02**] coincides with our notion of thin homotopy, while the notion of thin homotopy used in [**Bar91**] is different from ours.

In [**CP94**] it has been shown that parallel transport in a principal G-bundle over M around based loops defines a holonomy map. For connected manifolds M it was also shown how to reconstruct a principal G-bundle with connection from a given holonomy map $\mathcal{H}$ at x, such that the holonomy of this bundle around loops based at x equals $\mathcal{H}$. This establishes a bijection between holonomy maps and principal G-bundles with connection over connected manifolds. The same result has been proven (with the before mentioned different notion of thin homotopy) in [**Bar91**].

To relate these results to Theorem 5.4, we consider again transport functors $\mathrm{tra} : \mathcal{P}_1(M) \longrightarrow G\text{-Tor}$ with $\mathcal{B}G$-structure. Recall from Section 2.1 that for any point $x \in M$ and any identification $F(x) \cong G$ the functor tra produces a group homomorphism $F_{x,x} : \pi_1^1(M,x) \longrightarrow G$.

Proposition 5.6. *Let* $\mathrm{tra} : \mathcal{P}_1(M) \longrightarrow G\text{-Tor}$ *be a transport functor on M with $\mathcal{B}G$-structure. Then, for any point $x \in M$ and any identification $F(x) \cong G$, the group homomorphism*

$$\mathrm{tra}_{x,x} : \pi_1^1(M,x) \longrightarrow G$$

is a holonomy map.

Proof. The group homomorphism $\mathrm{tra}_{x,x}$ is a Wilson line of the transport functor tra, and hence smooth by Theorem 3.12. □

For illustration, let us combine Theorem 5.4 and Proposition 5.6 to the following diagram, which is evidently commutative:

$$\mathfrak{Bun}_G^{\nabla}(M) \xrightarrow{\quad [\mathbf{CP94}] \quad} \left\{ \begin{array}{c} \text{Holonomy maps} \\ \text{on } M \text{ at } x \end{array} \right\}$$

Theorem 5.4   Proposition 5.6

$$\mathrm{Trans}_{\mathrm{Gr}}^1(M, G\text{-Tor}).$$

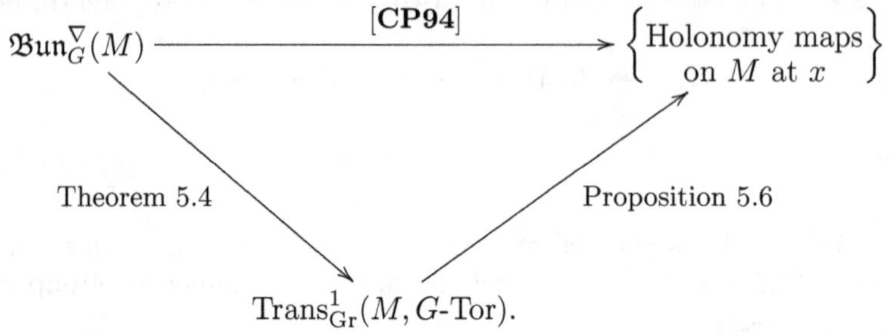

5.3. Associated Bundles and Vector Bundles with Connection

Recall that a principal G-bundle P together with a faithful representation $\rho : G \longrightarrow \mathrm{Gl}(V)$ of the Lie group G on a vector space V defines a vector bundle $P \times_\rho V$ with structure group G, called the vector bundle associated to P by the representation ρ. One can regard a (say, complex) representation of a group G conveniently as a functor $\rho : \mathcal{B}G \longrightarrow \mathrm{Vect}(\mathbb{C})$ from the one-point-category $\mathcal{B}G$ into the category of complex vector spaces: the object of $\mathcal{B}G$ is sent to the vector space V of the representation, and a group element $g \in G$ is sent to an isomorphism $g : V \longrightarrow V$ of this vector space. The axioms of a functor are precisely the axioms one demands for a representation. Furthermore, the representation is faithful, if and only if the functor is faithful.

Definition 5.7. *Let*

$$\rho : \mathcal{B}G \longrightarrow \mathrm{Vect}(\mathbb{C})$$

be any representation of the Lie group G. A transport functor

$$\mathrm{tra} : \mathcal{P}_1(M) \longrightarrow \mathrm{Vect}(\mathbb{C})$$

with $\mathcal{B}G$-structure is called <u>*associated transport functor*</u>.

As an example, we consider the defining representation of the Lie group $U(n)$ on the vector space $\mathbb{C}^n$, considered as a functor

$$\rho_n : \mathcal{B}U(n) \longrightarrow \mathrm{Vect}(\mathbb{C}_h^n) \tag{5.12}$$

to the category of n-dimensional hermitian vector spaces and isometries between those. Because we only include isometries in $\mathrm{Vect}(\mathbb{C}_h^n)$, the functor ρ_n is an equivalence of categories.

Similarly to Theorem 5.4, we find a geometric interpretation for associated transport functors on M with $\mathcal{B}U(n)$-structure, namely hermitian vector bundles of rank n with (unitary) connection over M. We denote the category of those vector bundles by $\mathrm{VB}(\mathbb{C}_h^n)_M^{\nabla}$. Let us just outline the very basics: given such a vector bundle E, we associate a functor

$$\mathrm{tra}_E : \mathcal{P}_1(M) \longrightarrow \mathrm{Vect}(\mathbb{C}_h^n),$$

which sends a point $x \in M$ to the vector space E_x, the fibre of E over x, and a path $\gamma : x \longrightarrow y$ to the parallel transport map $\tau : E_x \longrightarrow E_y$, which is linear and an isometry.

Theorem 5.8. *The functor* tra_E *obtained from a hermitian vector bundle* E *with connection over* M *is a transport functor on* M *with* $\mathcal{B}U(n)$*-structure; furthermore, the assignment* $E \longmapsto \mathrm{tra}_E$ *yields a functor*

$$\mathrm{VB}(\mathbb{C}_h^n)_M^\nabla \longrightarrow \mathrm{Trans}^1_{\mathcal{B}U(n)}(M, \mathrm{Vect}(\mathbb{C}_h^n)), \tag{5.13}$$

which is an equivalence of categories.

Proof. We proceed like in the first proof of Theorem 5.4. Here we use the correspondence between hermitian vector bundles with connection and their local data in $Z^1_\pi(U(n))^\infty$, for contractible surjective submersions π. Under this correspondence the functor (5.13) becomes naturally equivalent to the composite

$$Z^1_\pi(U(n))^\infty \xrightarrow{\;\Xi\;} \mathfrak{Des}^1_\pi(i)^\infty \xrightarrow{\mathrm{Rec}_\pi} \mathrm{Triv}^1_\pi(i)^\infty \xrightarrow{\;v^\infty\;} \mathrm{Trans}^1_{\mathcal{B}U(n)}(M, \mathrm{Vect}(\mathbb{C}_h^n))$$

which is, by Corollary 4.9, Theorem 2.9 and Proposition 3.14, an equivalence of categories. □

Let us also consider the Lie groupoid $\mathrm{Gr}_U := \bigsqcup_{n\in\mathbb{N}} \mathcal{B}U(n)$, whose set of objects is $\mathbb{N}$ (with the discrete smooth structure) and whose morphisms are

$$\mathrm{Mor}_{\mathrm{Gr}_U}(n,m) = \begin{cases} U(n) & \text{if } n = m \\ \emptyset & \text{if } n \neq m \end{cases}$$

so that $\mathrm{Mor}(\mathrm{Gr}_U)$ is a disjoint union of Lie groups. The functors ρ_n from (5.12) induce a functor

$$\rho_U : \mathrm{Gr}_U \longrightarrow \mathrm{Vect}(\mathbb{C}_h)$$

to the category of hermitian vector spaces (without a fixed dimension) and isometries between those.

The category $\mathrm{Vect}(\mathbb{C}_h)$ in fact a monoidal category, and its monoidal structure induces monoidal structures on the category $\mathrm{VB}(\mathbb{C}_h)_M^\nabla$ of hermitian vector bundles with connection over M as well as on the category of transport functors $\mathrm{Trans}^1_{\mathrm{Gr}_U}(M, \mathrm{Vect}(\mathbb{C}_h))$, as outlined in Section 2.1. Since parallel transport in vector bundles is compatible with tensor products, we have

Corollary 5.9. *The functor*

$$\mathrm{VB}(\mathbb{C}_h)_M^\nabla \longrightarrow \mathrm{Trans}^1_{\mathrm{Gr}_U}(M, \mathrm{Vect}(\mathbb{C}_h))$$

is a monoidal equivalence of monoidal categories.

In particular, we have the unit transport functor $\mathbf{I}_\mathbb{C}$ which sends every point to the complex numbers $\mathbb{C}$, and every path to the identity $\mathrm{id}_\mathbb{C}$. The following fact is easy to verify:

Lemma 5.10. *Let* $\mathrm{tra} : \mathcal{P}_1(M) \longrightarrow \mathrm{Vect}(\mathbb{C}_h)$ *be a transport functor with* Gr_U*-structure, corresponding to a hermitian vector bundle* E *with connection over* M*. Then, there is a canonical bijection between morphisms*

$$\eta : \mathbf{I}_\mathbb{C} \longrightarrow \mathrm{tra}$$

of transport functors with Gr_U*-structure and smooth flat section of* E*.*

5.4. Generalized Connections

In this section we consider functors

$$F : \mathcal{P}_1(M) \longrightarrow \mathcal{B}G.$$

By now, we can arrange such functors in three types:

1. We demand nothing of F: such functors are addressed as *generalized connections* [**AI92**].
2. We demand that F is a transport functor with $\mathcal{B}G$-structure: it corresponds to an ordinary principal G-bundle with connection.
3. We demand that F is smooth in the sense of Definition 3.1: by Proposition 4.7, one can replace such functors by 1-forms $A \in \Omega^1(M, \mathfrak{g})$, so that we can speak of a trivial G-bundle.

Note that for a functor $F : \mathcal{P}_1(M) \longrightarrow \mathcal{B}G$ and the identity functor $\mathrm{id}_{\mathcal{B}G}$ on $\mathcal{B}G$ the Wilson line

$$\mathcal{W}_{x_1,x_2}^{F,\mathrm{id}_{\mathcal{B}G}} : \mathrm{Mor}_{\mathcal{P}_1(M)}(x_1, x_2) \longrightarrow G$$

does not depend on choices of objects G_1, G_2 and morphisms $t_k : i(G_k) \longrightarrow F(x_k)$ as in the general setup described in Section 3.2, since $\mathcal{B}G$ has only one object and one can canonically choose $t_k = \mathrm{id}$. So, generalized connections have a particularly good Wilson lines. Theorem 3.12 provides a precise criterion to decide when a generalized connection is regular: if and only if all its Wilson lines are smooth.

6. Groupoid Bundles with Connection

In all examples we have discussed so far the Lie groupoid Gr is of the form $\mathcal{B}G$, or a union of those. In this section we discuss transport functors with Gr-structure for a general Lie groupoid Gr. We start with the local aspects of such transport functors, and then discuss two examples of target categories. Our main example is related to the notion of principal groupoid bundles [**MM03**]. In contrast to the examples in Section 5, transport functors with Gr-structure do not only reproduce the existing definition of a principal groupoid bundle, but also reveal precisely what a connection on such a bundle must be.

We start with the local aspects of transport functors with Gr-structure by considering smooth functors

$$F : \mathcal{P}_1(X) \longrightarrow \mathrm{Gr}. \tag{6.1}$$

Our aim is to obtain a correspondence between such functors and certain 1-forms, generalizing the one derived in Section 4. If we denote the objects of Gr by Gr_0 and the morphisms by Gr_1, F defines in the first place a smooth map $f : X \longrightarrow \mathrm{Gr}_0$. Using the technique introduced in Section 4, we obtain further a 1-form A on X with values in the vector bundle $f^*\mathrm{id}^*T\mathrm{Gr}_1$ over X. Only the fact that F respects targets and sources imposes two new conditions:

$$f^*\mathrm{d}s \circ A = 0 \quad \text{and} \quad f^*\mathrm{d}t \circ A + \mathrm{d}f = 0.$$

Here we regard $\mathrm{d}f$ as a 1-form on X with values in $f^*T\mathrm{Gr}_0$, and $\mathrm{d}s$ and $\mathrm{d}t$ are the differentials of the source and target maps.

Now we recall that the *Lie algebroid* E of Gr is the vector bundle

$$E := \mathrm{id}^*\ker(\mathrm{d}s)$$

over Gr_0 where $\mathrm{id} : \mathrm{Gr}_0 \longrightarrow \mathrm{Gr}_1$ is the identity embedding. The *anchor* is the morphism $a := \mathrm{d}t : E \longrightarrow T\mathrm{Gr}_0$ of vector bundles over Gr_0. Using this terminology, we see that the smooth functor (6.1) defines a smooth map $f : X \longrightarrow \mathrm{Gr}_0$ plus a 1-form $A \in \Omega^1(X, f^*E)$ such that $f^*a \circ A + \mathrm{d}f = 0$.

In order to deal with smooth natural transformations, we introduce the following notation. We denote by

$$c : \mathrm{Gr}_1\ {}_s\!\times_t \mathrm{Gr}_1 \longrightarrow \mathrm{Gr}_1 : (h, g) \longmapsto h \circ g$$

the composition in the Lie groupoid Gr, and for $g : x \longrightarrow y$ a morphism by

$$r_g : s^{-1}(y) \longrightarrow s^{-1}(x) : h \longmapsto h \circ g$$

the composition by g from the right. Notice that c and r_g are smooth maps. It is straightforward to check that one has a well-defined map

$$\mathrm{AD}_g : T_g\Gamma_1\ {}_{\mathrm{d}s}\!\times_a E_{s(g)} \longrightarrow E_{t(g)}$$

which is defined by

$$\mathrm{AD}_g(X, Y) := \mathrm{d}r_{g^{-1}}|_g(\mathrm{d}c|_{g,\mathrm{id}_{s(g)}}(X, Y)). \tag{6.2}$$

For example, if $\mathrm{Gr} = \mathcal{B}G$ for a Lie group G, the Lie algebroid is the trivial bundle $E = \Gamma_0 \times \mathfrak{g}$, the composition c is the multiplication of G, and (6.2) reduces to

$$\mathrm{AD}_g(X, Y) = \bar{\theta}_g(X) + \mathrm{Ad}_g(Y) \in \mathfrak{g}.$$

Suppose now that

$$\eta : F \Longrightarrow F'$$

is a smooth natural transformation between smooth functors F and F' which correspond to pairs (f, A) and (f', A'), respectively. It defines a smooth map $g : X \longrightarrow \mathrm{Gr}_1$ such that

$$s \circ g = f \quad \text{and} \quad t \circ g = f'. \tag{6.3}$$

Generalizing Lemma 4.2, the naturality of η implies additionally

$$A' + \mathrm{AD}_g(\mathrm{d}g, -A) = 0. \tag{6.4}$$

The structure obtained like this forms a category $Z^1_X(\mathrm{Gr})$ of Gr-*connections*: its objects are pairs (f, A) of smooth functions $f : X \longrightarrow \mathrm{Gr}_0$ and 1-forms $A \in \Omega^1(X, f^*E)$ satisfying $f^*\mathrm{d}t \circ A + \mathrm{d}f = 0$, and its morphisms are smooth maps $g : X \longrightarrow \mathrm{Gr}_1$ satisfying (6.3) and (6.4). The category $Z^1_X(\mathrm{Gr})$ generalizes the category of G-connections from Definition 4.6 in the sense that $Z^1_X(\mathcal{B}G) = Z^1_X(G)$ for G a Lie group. We obtain the following generalization of Proposition 4.7.

Proposition 6.1. *There is a canonical isomorphism of categories*

$$\mathrm{Funct}^\infty(X, \mathrm{Gr}) \cong Z^1_X(\mathrm{Gr}).$$

We remark that examples of smooth of functors with values in a Lie groupoid naturally appear in the discussion of transgression to loop spaces, see Section 4 of the forthcoming paper [**SW08b**].

Now we come to the global aspects of transport functors with Gr-structure. We introduce the category of Gr-torsors as an interesting target category of such transport functors. A *smooth* Gr-*manifold* [**MM03**] is a triple (P, λ, ρ) consisting a smooth manifold P, a surjective submersion $\lambda : P \longrightarrow \mathrm{Gr}_0$ and a smooth map

$$\rho : P \,{}_{\lambda}{\times}_{t}\, \mathrm{Gr}_1 \longrightarrow P$$

such that

1. ρ respects λ in the sense that $\lambda(\rho(p, \varphi)) = s(\varphi)$ for all $p \in P$ and $\varphi \in \mathrm{Gr}_1$ with $\lambda(p) = t(\varphi)$,
2. ρ respects the composition $\circ$ of morphisms of Gr.

A *morphism between* Gr-*manifolds* is a smooth map $f : P \longrightarrow P'$ which respects λ, λ' and ρ, ρ'. A Gr-*torsor* is a Gr-manifold for which ρ acts in a free and transitive way. Gr-torsors form a category denoted Gr-Tor. For a fixed object $X \in \mathrm{Obj}(\mathrm{Gr})$, $P_X := t^{-1}(X)$ is a Gr-torsor with $\lambda = s$ and $\rho = \circ$. Furthermore, a morphism $\varphi : X \longrightarrow Y$ in Gr defines a morphism $P_X \longrightarrow P_Y$ of Gr-torsors. Together, this defines a functor

$$i_{\mathrm{Gr}} : \mathrm{Gr} \longrightarrow \text{Gr-Tor}. \tag{6.5}$$

The functor (6.5) allows us to study transport functors

$$\mathrm{tra} : \mathcal{P}_1(M) \longrightarrow \text{Gr-Tor}$$

with Gr-structure. By a straightforward adaption of the Second Proof of Theorem 5.4 one can construct the total space P of a fibre bundle over M from the transition function $\tilde{g} : Y^{[2]} \longrightarrow \mathrm{Gr}_1$ of tra, in such a way that P is fibrewise a Gr-torsor. More precisely, we reproduce the following definition.

Definition 6.2 ([**MM03**]). *A principal Gr-bundle over M is a Gr-manifold (P, λ, ρ) together with a smooth map $p : P \longrightarrow M$ which is preserved by the action, such that there exists a surjective submersion $\pi : Y \longrightarrow M$ with a smooth map $f : Y \longrightarrow \mathrm{Gr}_1$ and a morphism*

$$\phi : P \times_M Y \longrightarrow Y \,{}_{f}{\times}_{t}\, \mathrm{Gr}_1$$

of Gr-manifolds that preserves the projections to Y.

Here we have used surjective submersions instead of open covers, like we already did for principal bundles (see Section 5.1). Principal Gr-bundles over M form a category denoted Gr-$\mathfrak{Bun}^{\nabla}(M)$, whose morphisms are morphisms of Gr-manifolds that preserve the projections to M.

The descent data of the transport functor tra not only consists of the transition function $\tilde{g}$ but also of a smooth functor $\mathrm{triv} : \mathcal{P}_1(Y) \longrightarrow \mathrm{Gr}$. Now, Proposition 6.1 *predicts* the notion of a connection 1-form on a principal Gr-bundle:

Definition 6.3. *Let* Gr *be a Lie groupoid and* E *be its Lie algebroid. A* <u>*connection on a principal* Gr*-bundle*</u> P *is a 1-form* $\omega \in \Omega^1(P, \lambda^*E)$ *such that*

$$\lambda^*\mathrm{d}t \circ \omega + \mathrm{d}\lambda = 0 \quad \text{and} \quad p_1^*\omega + \mathrm{AD}_g(\mathrm{d}g, -\rho^*\omega) = 0,$$

where $\lambda : P \longrightarrow \mathrm{Gr}_0$ *and* $\rho : P\ {}_\lambda\!\times_t \mathrm{Gr}_1 \longrightarrow P$ *are the structure of the* Gr*-manifold* P, p_1 *and* g *are the projections to* P *and* Gr_1*, respectively.*

By construction, we have

Theorem 6.4. *There is a canonical equivalence of categories*

$$\mathrm{Gr}\text{-}\mathfrak{Bun}^\nabla(M) \cong \mathrm{Trans}_{\mathrm{Gr}}(M, \mathrm{Gr}\text{-Tor}).$$

Indeed, choosing a local trivialization (Y, f, ϕ) of a principal Gr-bundle P, one obtains a section $s : Y \longrightarrow P : y \longmapsto p(\phi^{-1}(y, \mathrm{id}_{f(y)}))$. This section satisfies $\lambda \circ s = f$, so that the pullback of a connection 1-form $\omega \in \Omega^1(P, \lambda^*E)$ along s is a 1-form $A := s^*\omega \in \Omega^1(Y, f^*E)$. The first condition in Definition 6.3 implies that (A, f) is an object in $Z^1_Y(\mathrm{Gr})$, and thus by Proposition 6.1 a smooth functor $\mathrm{triv} : \mathcal{P}_1(Y) \longrightarrow \mathrm{Gr}$. The second condition implies that the transition function $\tilde{g} : Y^{[2]} \longrightarrow \mathrm{Gr}$ defined by $s(\pi_1(\alpha)) = \rho(s(\pi_2(\alpha)), \tilde{g}(\alpha))$ is a morphism in $Z^1_{Y^{[2]}}(\mathrm{Gr})$ from π_1^*F to π_2^*F. All together, this is descent data for a transport functor on M with Gr-structure.

We remark that this automatically induces a notion of parallel transport for a connection A on a principal Gr-bundle P: let $\mathrm{tra}_{P,A} : \mathcal{P}_1(M) \longrightarrow \mathrm{Gr}\text{-Tor}$ be the transport corresponding to (P, A) under the equivalence of Theorem 6.4. Then, the parallel transport of A along a path $\gamma : x \longrightarrow y$ is the Gr-torsor morphism

$$\mathrm{tra}_{P,A}(\gamma) : P_x \longrightarrow P_y.$$

In the remainder of this section we discuss a class of groupoid bundles with connection related to action groupoids. We recall that for V a complex vector space with an action of a Lie group G, the action groupoid $V/\!\!/G$ has V as its objects and $G \times V$ as its morphisms. The source map is the projection to V, and the target map is the action $\rho : G \times V \longrightarrow V$. Every action groupoid $V/\!\!/G$ comes with a canonical functor

$$i_{V/\!\!/G} : V/\!\!/G \longrightarrow \mathrm{Vect}_*(\mathbb{C})$$

to the category of pointed complex vector spaces, which sends an object $v \in V$ to the pointed vector space (V, v) and a morphism (v, g) to the linear map $\rho(g, -)$, which respects the base points.

Proposition 6.5. *A transport functor* $\mathrm{tra} : \mathcal{P}_1(M) \longrightarrow \mathrm{Vect}_*(\mathbb{C})$ *with* $V/\!\!/G$*-structure is a complex vector bundle over* M *with structure group* G *and a smooth flat section.*

Proof. We consider the strictly commutative diagram

$$\begin{array}{ccc} V/\!\!/G & \xrightarrow{i_{V/\!\!/G}} & \mathrm{Vect}_*(\mathbb{C}) \\ {\scriptstyle \mathrm{pr}}\downarrow & & \downarrow{\scriptstyle f} \\ \mathcal{B}G & \xrightarrow[\rho]{} & \mathrm{Vect}(\mathbb{C}) \end{array}$$

of functors, in which f is the functor that forgets the base point, pr is the functor which sends a morphism (g, v) in $V/\!\!/G$ to g, and ρ is the given representation. The diagram shows that the composition

$$f \circ \mathrm{tra} : \mathcal{P}_1(M) \longrightarrow \mathrm{Vect}(\mathbb{C})$$

is a transport functor with $\mathcal{B}G$-structure, and hence the claimed vector bundle E by (a slight generalization of) Theorem 5.8. Remembering the forgotten base point defines a natural transformation

$$\eta : \mathbf{I}_{\mathbb{C}} \longrightarrow f \circ \mathrm{tra}.$$

If we regard the identity transport functor $\mathbf{I}_{\mathbb{C}}$ as a transport functor with $\mathcal{B}G$-structure, the natural transformation η becomes a morphism of transport functors with $\mathcal{B}G$-structure, and thus defines by Lemma 5.10 a smooth flat section in E. □

7. Generalizations and further Topics

The concept of transport functors has generalizations in many aspects, some of which we want to outline in this section.

7.1. Transport n-Functors

The motivation to write this article was to find a formulation of parallel transport along curves, which can be generalized to higher dimensional parallel transport. Transport functors have a natural generalization to transport n-functors. In particular the case $n = 2$ promises relations between transport 2-functors and gerbes with connective structure [**BS07**], similar to the relation between transport 1-functors and bundles with connections presented in Section 5. We address these issues in a further publication [**SW08a**].

Let us briefly describe the generalization of the concept of transport functors to transport n-functors. The first generalization is that of the path groupoid $\mathcal{P}_1(M)$ to a path n-groupoid $\mathcal{P}_n(M)$. Here, n-groupoid means that every k-morphism is an equivalence, i.e. invertible up to $(k+1)$-isomorphisms. The set of objects is again the manifold M, the k-morphisms are smooth maps $[0,1]^k \longrightarrow M$ with sitting instants on each boundary of the k-cube, and the top-level morphisms $k = n$ are additionally taken up to thin homotopy in the appropriate sense.

We then consider n-functors

$$F : \mathcal{P}_n(M) \longrightarrow T \tag{7.1}$$

from the path n-groupoid $\mathcal{P}_n(M)$ to some target n-category T. Local trivializations of such n-functors are considered with respect to an n-functor $i : \mathrm{Gr} \longrightarrow T$, where

Gr is a Lie n-groupoid, and to a surjective submersions $\pi : Y \longrightarrow M$. A π-local i-trivialization then consists of an n-functor triv $: \mathcal{P}_n(Y) \longrightarrow$ Gr and an equivalence

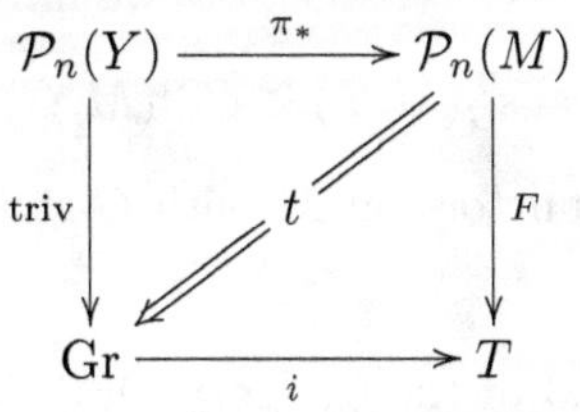

$$
\begin{array}{ccc}
\mathcal{P}_n(Y) & \xrightarrow{\;\pi_*\;} & \mathcal{P}_n(M) \\
\downarrow{\scriptstyle \text{triv}} & {\scriptstyle t} \nearrow & \downarrow{\scriptstyle F} \\
\text{Gr} & \xrightarrow{\;i\;} & T
\end{array}
\tag{7.2}
$$

of n-functors. Local trivializations lead to an n-category $\mathfrak{Des}_\pi^n(i)$ of descent data, which are descent n-categories in the sense of [**Str04**], similar to Remark 2.10 for $n = 1$.

The category $\mathfrak{Des}_\pi^n(i)$ has a natural notion of smooth objects and smooth k-morphisms. Then, n-functors (7.1) which allow local trivializations with smooth descent data will be called transport n-functors, and form an n-category $\mathrm{Trans}_{\mathrm{Gr}}^n(M, T)$.

In the case $n = 1$, the procedure described above reproduces the framework of transport functors described in this article. The case $n = 2$ will be considered in detail in two forthcoming papers. First we settle the local aspects: we derive a correspondence between smooth 2-functors and differential 2-forms (Theorem 2.20 in [**SW08b**]). Then we continue with the global aspects in [**SW08a**].

As a further example, we now describe the case $n = 0$. Note that a 0-category is a set, a Lie 0-groupoid is a smooth manifold, and a 0-functor is a map. To start with, we have the set $\mathcal{P}_0(M) = M$, a set T, a smooth manifold G and an injective map $i : G \longrightarrow T$. Now we consider maps $F : M \longrightarrow T$. Following the general concept, such a map is π-locally i-trivializable, if there exists a map triv $: Y \longrightarrow G$ such that the diagram

$$
\begin{array}{ccc}
Y & \xrightarrow{\;\pi\;} & M \\
\downarrow{\scriptstyle \text{triv}} & & \downarrow{\scriptstyle F} \\
G & \xrightarrow{\;i\;} & T
\end{array}
$$

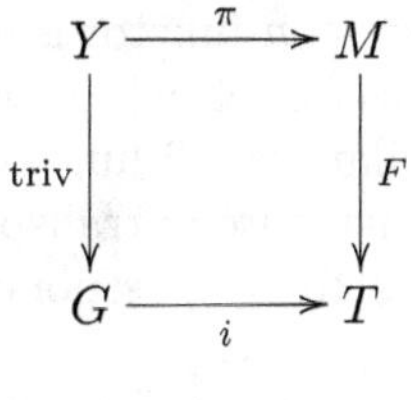

is commutative. Maps F together with π-local i-trivializations form the set $\mathrm{Triv}_\pi^0(i)$. The set $\mathfrak{Des}_\pi^0(i)$ of descent data is just the set of maps triv $: Y \longrightarrow G$ satisfying the equation

$$
\pi_1^* \mathrm{triv}_i = \pi_2^* \mathrm{triv}_i,
\tag{7.3}
$$

where we have used the notation $\pi_k^* \mathrm{triv}_i = i \circ \mathrm{triv} \circ \pi_k$ from Section 2. It is easy to see that every π-local i-trivialization triv of a map F satisfies this condition. This defines the map

$$
\mathrm{Ex}_\pi : \mathrm{Triv}_\pi^0(i) \longrightarrow \mathfrak{Des}_\pi^0(i).
$$

Similar to Theorem 2.9 in the case $n = 1$, this is indeed a bijection: every function triv $: Y \longrightarrow G$ satisfying (7.3) with i injective factors through π. Now it is easy to say when an element in $\mathfrak{Des}_\pi^0(i)$ is called smooth: if and only if the map triv :

$Y \longrightarrow G$ is smooth. Such maps form the set $\mathfrak{Des}^0_\pi(i)^\infty$, which in turn defines the set $\mathrm{Trans}^0_G(M,T)$ of transport 0-functors with G-structure. Due to (7.3), there is a canonical bijection $\mathfrak{Des}^0_\pi(i)^\infty \cong C^\infty(M,G)$. So, we have

$$\mathrm{Trans}^0_G(M,T) \cong C^\infty(M,G),$$

in other words: transport 0-functors on M with G-structure are smooth functions from M to G.

Let us revisit Definition 3.3 of the category $\mathfrak{Des}^1_\pi(i)^\infty$ of smooth descent data, which now can equivalently be reformulated as follows:

> *Let* Gr *be a Lie groupoid and let* $i : \mathrm{Gr} \longrightarrow T$ *be a functor. An object* (triv, g) *in* $\mathfrak{Des}^1_\pi(i)$ *is called* smooth, *if the functor* $\mathrm{triv} : \mathcal{P}_1(Y) \longrightarrow \mathrm{Gr}$ *is smooth the sense of Definition 3.1, and if the natural equivalence* $g : Y^{[2]} \longrightarrow \mathrm{Mor}(T)$ *is a transport* 0*-functor with* $\mathrm{Mor}(T)$*-structure. A morphism*
>
> $$h : (\mathrm{triv}, g) \longrightarrow (\mathrm{triv}', g')$$
>
> *between smooth objects is called* smooth, *if* $h : Y \longrightarrow \mathrm{Mor}(T)$ *is a transport 0-functor with* $\mathrm{Mor}(T)$*-structure.*

This gives an outlook how the definition of the n-category $\mathfrak{Des}^n_\pi(i)^\infty$ of smooth descent data will be for higher n: it will recursively use transport $(n-1)$-functors.

7.2. Curvature of Transport Functors

When we describe parallel transport in terms of functors, it is a natural question how related notions like curvature can be seen in this formulation. Interestingly, it turns out that the curvature of a transport functor is a transport 2-functor. More generally, the curvature of a transport n-functor is a transport $(n+1)$-functor. This becomes evident with a view to Section 4, where we have related smooth functors and differential 1-forms. In a similar way, 2-functors can be related to 2-forms. A comprehensive discussion of the curvature of transport functors is therefore beyond the scope of this article, and has to be postponed until after the discussion of transport 2-functors [**SW08a**].

We shall briefly indicate the basic ideas. We recall from Section 2.1 when a functor $F : \mathcal{P}_1(M) \longrightarrow T$ is flat: if it factors through the fundamental groupoid $\Pi_1(M)$, whose morphisms are smooth homotopy classes of paths in M. In general, one can associate to a transport functor tra a 2-functor

$$\mathrm{curv}(\mathrm{tra}) : \mathcal{P}_2(M) \longrightarrow \mathrm{Grpd}$$

into the 2-category of groupoids. This 2-functor is particularly trivial if tra is flat. Furthermore, the 2-functor curv(tra) is itself flat in the sense that it factors through the fundamental 2-groupoid of M: this is nothing but the Bianchi identity.

For smooth functors $F : \mathcal{P}_1(M) \longrightarrow \mathcal{B}G$, which corresponding by Proposition 4.7 to 1-forms $A \in \Omega^1(M, \mathfrak{g})$, it turns out that the 2-functor $\mathrm{curv}(F)$ corresponds to a 2-form $K \in \Omega^2(M, \mathfrak{g})$ which is related to A by the usual equality $K = \mathrm{d}A + A \wedge A$.

7.3. Alternatives to smooth Functors

The definition of transport functors concentrates on the smooth aspects of parallel transport. As we have outlined in Appendix A.2, our definition of smooth descent data $\mathfrak{Des}^1_\pi(i)^\infty$ can be regarded as the internalization of functors and natural transformations in the category D^∞ of diffeological spaces and diffeological maps.

Simply by choosing another ambient category C, we obtain possibly weaker notions of parallel transport. Of particular interest is the situation where the ambient category is the category Top of topological spaces and continuous maps. Indicated by results of [**Sta74**], one would expect that reconstruction theorems as discussed in Section 2.3 should also exist for Top, and also for transport n-functors for $n > 1$. Besides, parallel transport along topological paths of bounded variation can be defined, and is of interest for its own right, see, for example, [**Bau05**].

7.4. Anafunctors

The notion of smoothly locally trivializable functors is closely related to the concept of anafunctors. Following [**Mak96**], an anafunctor $F : A \longrightarrow B$ between categories A and B is a category $|F|$ together with a functor $\tilde{F} : |F| \longrightarrow B$ and a surjective equivalence $p : |F| \longrightarrow A$, denoted as a diagram

$$\begin{array}{ccc} |F| & \xrightarrow{\tilde{F}} & B \\ {\scriptstyle p}\big\downarrow & & \\ A & & \end{array} \qquad (7.4)$$

called a span. It has been shown in [**Bar04**] how to formulate the concept of an anafunctor internally to any category C.

Note that an anafunctor in C gives rise to an ordinary functor $A \longrightarrow B$ in C, if the epimorphism p has a section. In the category of sets, $C = \mathfrak{Set}$, every epimorphism has a section, if one assumes the axiom of choice (this is what we do). The original motivation for introducing anafunctors was, however, to deal with situations where one does not assume the axiom of choice [**Mak96**]. In the category $C = C^\infty$ of smooth manifolds, surjective submersions are particular epimorphisms, as they arise for example as projections of smooth fibre bundles. Since not every bundle has a global smooth section, an anafunctor in C^∞ does not produce a functor. The same applies to the category $C = D^\infty$ of diffeological spaces described in Appendix A.2.

Let us indicate how anafunctors arise from smoothly locally trivialized functors. Let $\mathrm{tra} : \mathcal{P}_1(M) \longrightarrow T$ be a transport functor with Gr-structure. We choose a π-local i-trivialization (triv, t), whose descent data (triv, g) is smooth. Consider the functor

$$R_{(\mathrm{triv},g)} : \mathcal{P}_1^\pi(M) \longrightarrow T$$

that we have defined in Section 2.3 from this descent data. By Definition 3.3 of smooth descent data, the functor $\mathrm{triv} : \mathcal{P}_1(Y) \longrightarrow \mathrm{Gr}$ is smooth and the natural equivalence g factors through a smooth natural equivalence $\tilde{g} : Y \longrightarrow \mathrm{Mor(Gr)}$. So, the functor $R_{(\mathrm{triv},g)}$ factors through Gr,

$$R_{(\mathrm{triv},g)} = i \circ A$$

for a functor $A : \mathcal{P}_1^{\pi}(M) \longrightarrow \mathrm{Gr}$. In fact, the category $\mathcal{P}_1^{\pi}(M)$ can be considered as a category internal to D^{∞}, so that the functor A is internal to D^{∞} as described in Appendix A.2, Proposition A.7 ii). Hence the reconstructed functor yields a span

$$\begin{array}{ccc} \mathcal{P}_1^{\pi}(M) & \xrightarrow{A} & \mathrm{Gr} \\ {\scriptstyle p^{\pi}}\downarrow & & \\ \mathcal{P}_1(M), & & \end{array}$$

internal to D^{∞}, i.e. an anafunctor $\mathcal{P}_1(M) \longrightarrow \mathrm{Gr}$. Because the epimorphism p^{π} is not invertible in D^{∞}, we do not get an ordinary functor $\mathcal{P}_1(M) \longrightarrow \mathrm{Gr}$ internal to D^{∞}: the weak inverse functor $s : \mathcal{P}_1(M) \longrightarrow \mathcal{P}_1^{\pi}(M)$ we have constructed in Section 2.3 is not internal to D^{∞}.

ACKNOWLEDGEMENTS We thank Bruce Bartlett, Uwe Semmelmann, Jim Stasheff and Danny Stevenson for helpful correspondences and Christoph Schweigert for helpful discussions. U.S. thanks John Baez for many valuable suggestions and discussions. We acknowledge support from the Sonderforschungsbereich "Particles, Strings and the Early Universe - the Structure of Matter and Space-Time".

A. More Background

A.1. The universal Path Pushout

Here we motivate Definition 2.11 of the groupoid $\mathcal{P}_1^{\pi}(M)$. Let $\pi : Y \longrightarrow M$ be a surjective submersion. A *path pushout of* π is a triple (A, b, ν) consisting of a groupoid A, a functor $b : \mathcal{P}_1(Y) \longrightarrow A$ and a natural equivalence $\nu : \pi_1^* b \longrightarrow \pi_2^* b$ with

$$\pi_{13}^* \nu = \pi_{23}^* \nu \circ \pi_{12}^* \nu.$$

A morphism

$$(R, \mu) : (A, b, \nu) \longrightarrow (A', b', \nu')$$

between path pushouts is a functor $R : A \longrightarrow A'$ and a natural equivalence $\mu : R \circ b \longrightarrow b'$ such that

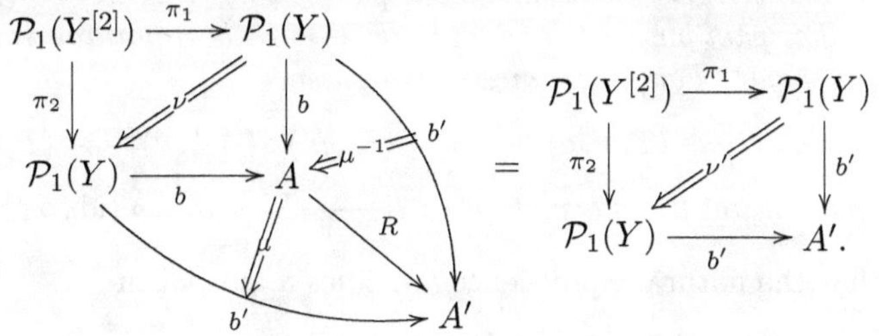

(A.1)

Among all path pushouts of π we distinguish some having a universal property.

Definition A.1. *A path pushout (A, b, ν) is* <u>*universal*</u>*, if, given any other path pushout (T, F, g), there exists a morphism $(R, \mu) : (A, b, \nu) \longrightarrow (T, F, g)$ such that, given any other such morphism (R', μ'), there is a unique natural equivalence $r : R \longrightarrow R'$ with*

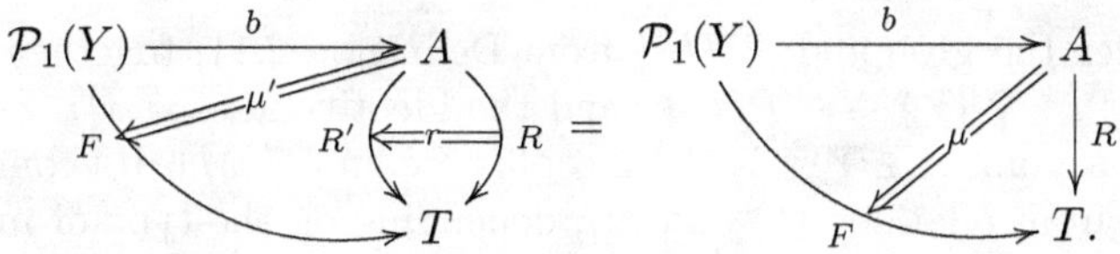

(A.2)

Now we show how two path pushouts having both the universal property, are related.

Lemma A.2. *Given two universal path pushouts (A, b, ν) and (A', b', ν') of the same surjective submersion $\pi : Y \longrightarrow M$, there is an equivalence of categories $a : A \longrightarrow A'$.*

Proof. We use the universal properties of both triples applied to each other. We obtain two choices of morphisms (R, μ) and $(\tilde{R}, \tilde{\mu})$, namely

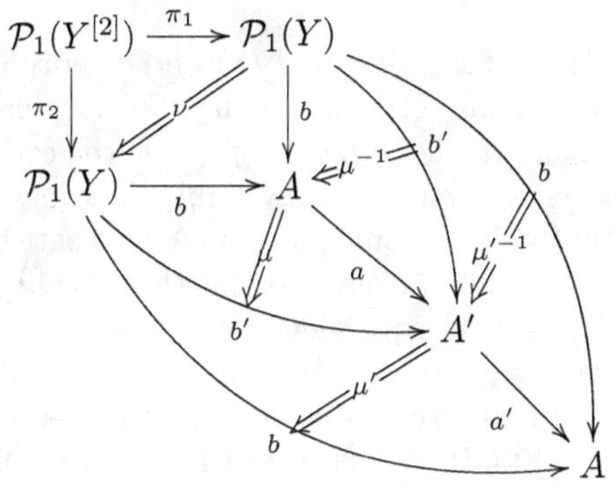

and

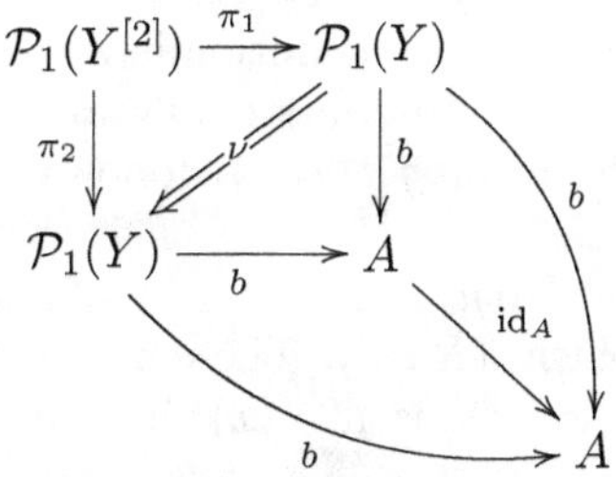

The unique natural transformation we get from the universal property is here $r : a' \circ a \longrightarrow \mathrm{id}_A$. Doing the same thing in the other order, we obtain a unique natural transformation $r' : a \circ a' \longrightarrow \mathrm{id}_{A'}$. Hence $a : A \longrightarrow A'$ is an equivalence of categories. □

We also need

Lemma A.3. *Let (A, b, ν) be a universal path pushout of π, let (T, F, g) and (T, F', g') two other path pushouts and let $h : F \longrightarrow F'$ be a natural transformation with $\pi_2^* h \circ g = \pi_1^* h \circ g'$. For any choice of morphisms*

$$(R, \mu) : (A, b, \nu) \longrightarrow (T, F, g) \quad \text{and} \quad (R', \mu') : (A, b, \nu) \longrightarrow (T, F', g')$$

there is a unique natural transformation $r : R \longrightarrow R'$ with $\mu \bullet (\mathrm{id}_b \circ r) = \mu'$.

Proof. Note that the natural equivalence h defines a morphism

$$(\mathrm{id}_T, h) : (T, F, g) \longrightarrow (T, F', g')$$

of path pushouts. The composition $(\mathrm{id}_T, h) \circ (R, \mu)$ gives a morphism

$$(R, h \circ \mu) : (A, b, \nu) \longrightarrow (T', F', g').$$

Since (R', μ') is universal, we obtain a unique natural transformation $r : R \longrightarrow R'$. □

Now consider the groupoid $\mathcal{P}_1^\pi(M)$ from Definition 2.11, together with the inclusion functor $\iota : \mathcal{P}_1(Y) \longrightarrow \mathcal{P}_1^\pi(M)$ and the identity $\mathrm{id}_{Y^{[2]}} : \pi_1^* \iota \longrightarrow \pi_2^* \iota$ whose component at a point $\alpha \in Y^{[2]}$ is the morphism α in $\mathcal{P}_1^\pi(M)$. Its commutative diagram follows from relations (1) and (2), depending on the type of morphism you apply it to. Its cocycle condition follows from (3). So, the triple $(\mathcal{P}_1^\pi(M), \iota, \mathrm{id}_{Y^{[2]}})$ is a path pushout.

Lemma A.4. *The triple $(\mathcal{P}_1^\pi(M), \iota, \mathrm{id}_{Y^{[2]}})$ is universal.*

Proof. Let (T, F, g) any path pushout. We construct the morphism

$$(R, \mu) : (\mathcal{P}_1^\pi(M), \iota, \mathrm{id}_{Y^{[2]}}) \longrightarrow (T, F, g)$$

as follows. The functor

$$R : \mathcal{P}_1^\pi(M) \longrightarrow T$$

sends an object $x \in Y$ to $F(x)$, a morphism $\gamma : x \longrightarrow y$ to $F(\gamma)$ and a morphism α to $g(\alpha)$. This definition is well-defined under the relations among the morphisms: (1) is the commutative diagram for the natural transformation g, (2) is the cocycle condition for g and (3) follows from the latter since g is invertible. The natural equivalence $\mu : R \circ \iota \longrightarrow F$ is the identity. By definition equation (A.1) is satisfied, so that (R, μ) is a morphism of path pushouts. Now we assume that there is another morphism (R', μ'). The component of the natural equivalence $r : R \longrightarrow R'$ at a point $x \in Y$ is $\mu'^{-1}(x)$, its naturality with respect to a morphisms $\gamma : x \longrightarrow y$ is then just the one of μ', and with respect to morphisms $\alpha \in Y^{[2]}$ comes from condition (A.1) on morphisms of path pushouts. It also satisfies the equality (A.2). Since this equation already determines r, it is unique. □

Notice that the construction of the functor R reproduces Definition 2.12. Let us finally apply Lemma A.3 to the universal path pushout $(\mathcal{P}_1^\pi(M), \iota, \mathrm{id}_{Y^{[2]}})$. Given the two functors $F, F' : \mathcal{P}_1(Y) \longrightarrow T$, the natural transformation $h : F \longrightarrow F'$, and the universal morphisms (R, μ) and (R', μ') as constructed in the proof of Lemma A.4, the natural transformation $r : R \longrightarrow R'$ has the component $h(x)$ at x. This reproduces Definition 2.13.

A.2. Diffeological Spaces and smooth Functors

This section puts Definition 3.1 of a smooth functor into the wider perspective of functors internal to some category C, here the category D^∞ of diffeological spaces [**Che77, Sou81**]. Diffeological spaces generalize the concept of a smooth manifold. While the set $C^\infty(X, Y)$ of smooth maps between smooth manifolds X and Y does not form, in general, a smooth manifold itself, the set $D^\infty(X, Y)$ of diffeological maps between diffeological spaces is again a diffeological space in a canonical way. In other words, the category D^∞ of diffeological spaces is closed.

Definition A.5. *A diffeological space is a set X together with a collection of plots: maps*

$$c : U \longrightarrow X,$$

each of them defined on an open subset $U \subset \mathbb{R}^k$ for some $k \in \mathbb{N}_0$, such that

a) for any plot $c : U \longrightarrow X$ and any smooth function $f : V \longrightarrow U$ also

$$c \circ f : V \longrightarrow X$$

is a plot.

b) every constant map $c : U \longrightarrow X$ is a plot.

c) if $f : U \longrightarrow X$ is a map defined on $U \subset \mathbb{R}^k$ and $\{U_i\}_{i \in I}$ is an open cover of U for which all restrictions $f|_{U_i}$ are plots of X, then also f is a plot.

A diffeological map between diffeological spaces X and Y is a map $f : X \longrightarrow Y$ such that for every plot $c : U \longrightarrow X$ of X the map $f \circ c : U \longrightarrow Y$ is a plot of Y. The set of all diffeological maps is denoted by $D^\infty(X, Y)$.

In fact Chen originally used convex subsets $U \subset \mathbb{R}^k$ instead of open ones, but this will not be of any importance for this article. For a comparison of various concepts see [**KM97**]. The following examples of diffeological spaces are important for us.

(1) First of all, every smooth manifold is a diffeological space, the plots being all smooth maps defined on all open subset of all $\mathbb{R}^n$. A map between two manifolds is smooth if and only if it is diffeological.

(2) For diffeological spaces X and Y the space $D^\infty(X, Y)$ of all diffeological maps from X to Y is a diffeological space in the following way: a map

$$c : U \longrightarrow D^\infty(X, Y)$$

is a plot if and only if for any plot $c' : V \longrightarrow X$ of X the composite

$$U \times V \xrightarrow{c \times c'} D^\infty(X, Y) \times X \xrightarrow{\text{ev}} Y$$

is a plot of Y. Here, ev denotes the evaluation map $\text{ev}(f, x) := f(x)$.

(3) Every subset Y of a diffeological space X is a diffeological space: its plots are those plots of X whose image is contained in Y.

(4) For a diffeological space X, a set Y and a map $p : X \longrightarrow Y$, the set Y is also a diffeological space: a map $c : U \longrightarrow Y$ is a plot if and only if there exists a cover of U by open sets U_α together with plots $c_\alpha : U_\alpha \longrightarrow X$ of X such that $c|_{U_\alpha} = p \circ c_\alpha$.

(5) Combining (1) and (2) we obtain the following important example: for smooth manifolds X and Y the space $C^\infty(X, Y)$ of smooth maps from X to Y is a diffeological space in the following way: a map

$$c : U \longrightarrow C^\infty(X, Y)$$

is a plot if and only if

$$U \times X \xrightarrow{c \times \mathrm{id}_X} C^\infty(X, Y) \times X \xrightarrow{\mathrm{ev}} Y$$

is a smooth map. This applies for instance to the free loop space $LM = C^\infty(S^1, M)$.

(6) Combining (3) and (5), the based loop space $L_x M$ and the path space PM of a smooth manifold are diffeological spaces.

(7) Combining (4) and (6) applied to the projection $\mathrm{pr} : PM \longrightarrow P^1 M$ to thin homotopy classes of paths, $P^1 M$ is a diffeological space. In the same way, the thin homotopy group $\pi_1^1(M, x)$ is a diffeological space.

From Example (7) we see that diffeological spaces arise naturally in the setup of transport functors introduced in this article.

Proposition A.6. *During this article, we encountered two examples of diffeological maps:*

i) A Wilson line

$$\mathcal{W}^{F,i}_{x_1,x_2} : \mathrm{Mor}_{\mathcal{P}_1(M)}(x_1, x_2) \longrightarrow \mathrm{Mor}_{\mathrm{Gr}}(G_1, G_2)$$

is smooth in the sense of Definition 3.10 if and only if it is diffeological.

ii) A group homomorphism $\mathcal{H} : \pi_1^1(M, x) \longrightarrow G$ is a holonomy map in the sense of Definition 5.5, if and only if it is diffeological.

Diffeological spaces and diffeological maps form a category D^∞ in which we can internalize categories and functors. Examples of such categories are:

- the path groupoid $\mathcal{P}_1(M)$: its set of objects is the smooth manifold M, which is by example (1) a diffeological space. Its set of morphisms $P^1 X$ is a diffeological space by example (7).
- the universal path pushout $\mathcal{P}_1^\pi(M)$ of a surjective submersion $\pi : Y \longrightarrow M$: its set of objects is the smooth manifold Y, and hence a diffeological space. A map

$$\phi : U \longrightarrow \mathrm{Mor}(\mathcal{P}_1^\pi(M))$$

is a plot if and only if there is a collection of plots $f_i : U \longrightarrow P^1Y$ and a collection of smooth maps $g_i : U \longrightarrow Y^{[2]}$ such that

$$g_N(x) \circ f_N(x) \cdots g_2(x) \circ f_2(x) \circ g_1(x) \circ f_1(x) = \phi(x).$$

We also have examples of functors internal to D^∞:

Proposition A.7. *During this article, we encountered two examples of functors internal to D^∞:*

i) A functor $F : \mathcal{P}_1(M) \longrightarrow \mathrm{Gr}$ is internal to D^∞ if and only if it is smooth in the sense of Definition 3.1.

ii) For a smooth object (triv, g) in $\mathfrak{Des}^1_\pi(i)$, the functor $R_{(\mathrm{triv},g)}$ factors smoothly through $i : \mathrm{Gr} \longrightarrow T$, i.e. there is a functor $A : \mathcal{P}^\pi_1(M) \longrightarrow \mathrm{Gr}$ internal to D^∞ such that $i \circ A = R_{(\mathrm{triv},g)}$.

B. Postponed Proofs

B.1. Proof of Theorem 2.9

Here we prove that the functor

$$\mathrm{Ex}_\pi : \mathrm{Triv}^1_\pi(i) \longrightarrow \mathfrak{Des}^1_\pi(i)$$

is an equivalence of categories. In Section 2.3 we have defined a reconstruction functor Rec_π going in the opposite direction. Now we show that Rec_π is a weak inverse of Ex_π and thus prove that both are equivalences of categories. For this purpose, we show (a) the equation $\mathrm{Ex}_\pi \circ \mathrm{Rec}_\pi = \mathrm{id}_{\mathfrak{Des}^1_\pi(i)}$ and (b) that there exists a natural equivalence

$$\zeta : \mathrm{id}_{\mathrm{Triv}^1_\pi(i)} \longrightarrow \mathrm{Rec}_\pi \circ \mathrm{Ex}_\pi.$$

To see (a), let (triv, g) be an object in $\mathfrak{Des}^1_\pi(i)$, and let $\mathrm{Rec}_\pi(\mathrm{triv}, g) = s^*R_{(\mathrm{triv},g)}$ be the reconstructed functor, coming with the π-local i-trivialization (triv, t) with $t := g \circ \iota^*\lambda$. Extracting descent data as described in Section 2.2, we find

$$(\pi_2^*t \circ \pi_1^*t^{-1})(\alpha) = g((\pi_2^*\lambda \circ \pi_1^*\lambda^{-1})(\alpha)) = g(\alpha)$$

so that $\mathrm{Ex}_\pi(\mathrm{Rec}_\pi(\mathrm{triv}, g)) = (\mathrm{triv}, g)$. Similar, if $h : (\mathrm{triv}, g) \longrightarrow (\mathrm{triv}', g')$ is a morphism in $\mathfrak{Des}^1_\pi(i)$, the reconstructed natural equivalence is $\mathrm{Rec}_\pi(h) := s^*R_h$. Extracting descent data, we obtain for the component at a point $x \in Y$

$$\begin{aligned}(t' \circ \pi^*s^*R_h \circ t^{-1})(x) &= g'(\lambda(x)) \circ R_h(s(\pi(x))) \circ g^{-1}(\lambda(x)) \\ &= g'(x, s(\pi(x))) \circ h(s(\pi(x))) \circ g^{-1}(x, s(\pi(x))) \\ &= h(x)\end{aligned}$$

where we have used Definition 2.13 and the commutativity of diagram (2.2). This shows that $\mathrm{Ex}_\pi(\mathrm{Rec}_\pi(h)) = h$.

To see (b), let $F : \mathcal{P}_1(M) \longrightarrow T$ be a functor with π-local i-trivialization (triv, t). Let us first describe the functor

$$\mathrm{Rec}_\pi(\mathrm{Ex}_\pi(F)) : \mathcal{P}_1(M) \longrightarrow T.$$

We extract descent data (triv, g) in $\mathfrak{Des}^1_\pi(i)$ as described in Section 2.2 by setting

$$g := \pi_2^* t \circ \pi_1^* t^{-1}. \tag{B.1}$$

Then, we have

$$\text{Rec}_\pi(\text{Ex}_\pi(F))(x) = \text{triv}_i(s(x))$$

for any point $x \in M$. A morphism $\overline{\gamma} : x \longrightarrow y$ is mapped by the functor s to some finite composition

$$s(\overline{\gamma}) = \alpha_n \circ \gamma_n \circ \alpha_{n-1} \circ ... \circ \gamma_2 \circ \alpha_1 \circ \gamma_1 \circ \alpha_0$$

of basic morphisms $\gamma_i : x_i \longrightarrow y_i$ and $\alpha_i \in Y^{[2]}$, so that we have

$$\text{Rec}_\pi(\text{Ex}_\pi(F))(\overline{\gamma}) = g(\alpha_n) \circ \text{triv}_i(\overline{\gamma_n}) \circ g(\alpha_{n-1}) \circ ... \circ \text{triv}_i(\overline{\gamma_1}) \circ g(\alpha_0). \tag{B.2}$$

Now we are ready define the component of the natural equivalence ζ at a functor F. This component is a morphism in $\text{Triv}^1_\pi(i)$ and thus itself a natural equivalence

$$\zeta(F) : F \longrightarrow \text{Rec}_\pi(\text{Ex}_\pi(F)).$$

We define the component of $\zeta(F)$ at a point $x \in M$ by

$$\zeta(F)(x) := t(s(x)) : F(x) \longrightarrow \text{triv}_i(s(x)). \tag{B.3}$$

Now we check that this is natural in x: let $\overline{\gamma} : x \longrightarrow y$ be a morphism like the above one. The diagram whose commutativity we have to show splits along the decomposition (B.2) into diagrams of two types:

$$\begin{array}{ccc} F(\pi(x_i)) & \xrightarrow{t(x_i)} & \text{triv}_i(x_i) \\ {\scriptstyle \pi_*\gamma_i}\downarrow & & \downarrow{\scriptstyle \text{triv}_i(\gamma_i)} \\ F(\pi(y_i)) & \xrightarrow[t(y_i)]{} & \text{triv}_i(y_i) \end{array} \quad \text{and} \quad \begin{array}{ccc} F(\pi(\pi_1(\alpha))) & \xrightarrow{t(\pi_1(\alpha))} & \text{triv}_i(\pi_1(\alpha)) \\ {\scriptstyle \text{id}}\downarrow & & \downarrow{\scriptstyle g(\gamma_i)} \\ F(\pi(\pi_2(\alpha))) & \xrightarrow[t(\pi_2(\alpha))]{} & \text{triv}_i(\pi_2(\alpha)). \end{array}$$

Both diagrams are indeed commutative, the one on the left because t is natural in $y \in Y$ and the one on the right because of (B.1).

It remains to show that ζ is natural in F, i.e. we have to prove the commutativity of the naturality diagram

$$\begin{array}{ccc} F & \xrightarrow{\zeta(F)} & \text{Rec}_\pi(\text{Ex}_\pi(F)) \\ {\scriptstyle \alpha}\downarrow & & \downarrow{\scriptstyle \text{Rec}_\pi(\text{Ex}_\pi(\alpha))} \\ F' & \xrightarrow[\zeta(F')]{} & \text{Rec}_\pi(\text{Ex}_\pi(F')) \end{array} \tag{B.4}$$

for any natural transformation $\alpha : F \longrightarrow F'$. Recall that $\text{Ex}_\pi(\alpha)$ is the natural transformation

$$h := t \circ \pi^*\alpha \circ t^{-1} : \text{triv}_i \longrightarrow \text{triv}_i'$$

and that $\text{Rec}_\pi(\text{Ex}_\pi(\alpha))$ is the natural transformation whose component at a point

$x \in M$ is the morphism

$$h(s(x)) : \mathrm{triv}_i(s(x)) \longrightarrow \mathrm{triv}'_i(s(x))$$

in T. Then, with definition (B.3), the commutativity of the naturality square (B.4) becomes obvious.

B.2. Proof of Theorem 3.12

We show that a Wilson line $\mathcal{W}^{\mathrm{tra},i}_{x_1,x_2}$ of a transport functor tra with Gr-structure is smooth. Let $c : U \longrightarrow PM$ be a map such that $\Gamma(u,t) := c(u)(t)$ is smooth, let $\pi : Y \longrightarrow M$ be a surjective submersion, and let (triv, t) be a π-local i-trivialization of the transport functor tra, for which $\mathrm{Ex}_\pi(\mathrm{triv}, t)$ is smooth. Consider the pullback diagram

$$\begin{array}{ccc} \Gamma^{-1}Y & \xrightarrow{\;a\;} & Y \\ {\scriptstyle p}\downarrow & & \downarrow{\scriptstyle \pi} \\ U \times [0,1] & \xrightarrow[\Gamma]{} & M \end{array}$$

with the surjective submersion $p : \Gamma^{-1}Y \longrightarrow U \times [0,1]$. We have to show that

$$\mathcal{W}^{\mathrm{tra},i}_{x_1,x_2} \circ \mathrm{pr} \circ c : U \longrightarrow G \tag{B.5}$$

is a smooth map. This can be checked locally in a neighbourhood of a point $u \in U$. Let $t_j \in I$ for $j = 0, ..., n$ be numbers with $t_{j-1} < t_j$ for $j = 1, ..., n$, and V_j open neighbourhoods of u chosen small enough to admit smooth local sections

$$s_j : V_j \times [t_{j-1}, t_j] \longrightarrow \Gamma^{-1}Y.$$

Then, we restrict all these sections to the intersection V of all the V_j. Let $\beta_j : t_{j-1} \longrightarrow t_j$ be paths through I defining smooth maps

$$\tilde{\Gamma}_j : V \times I \longrightarrow Y : (v,t) \longmapsto a(s_j(V, \beta_j(t))), \tag{B.6}$$

which can be considered as maps $\tilde{c}_j : V \longrightarrow PY$. Additionally, we define the smooth maps

$$\tilde{\alpha}_j : V \longrightarrow Y^{[2]} : v \longmapsto (\tilde{\Gamma}_{j-1}(v,1), \tilde{\Gamma}_j(v,0)).$$

Note that for any $v \in V$, both $\mathrm{pr}(\tilde{c}_j(v))$ and $\tilde{\alpha}_j(v)$ are morphisms in the universal path pushout $\mathcal{P}_1^\pi(M)$, namely

$$\mathrm{pr}(\tilde{c}_j(v)) : \tilde{\Gamma}_j(v,0) \longrightarrow \tilde{\Gamma}_j(v,1) \quad \text{and} \quad \tilde{\alpha}_j(v) : \tilde{\Gamma}_{j-1}(v,1) \longrightarrow \tilde{\Gamma}_j(v,0).$$

Taking their composition, we obtain a map

$$\phi : V \longrightarrow \mathrm{Mor}(\mathcal{P}_1^\pi(M)) : v \longrightarrow \tilde{c}_n(v) \circ \tilde{\alpha}_j(v) \circ ... \circ \tilde{\alpha}_1(v) \circ \tilde{c}_0(v).$$

Now we claim two assertions for the composite

$$i^{-1} \circ (R_{(\mathrm{triv},g)})_1 \circ \phi : V \longrightarrow \mathrm{Mor}(\mathrm{Gr}) \tag{B.7}$$

of ϕ with the functor $R_{(\mathrm{triv},g)} : \mathcal{P}_1^\pi(M) \longrightarrow \mathrm{Mor}(T)$ we have defined in Section 2.3: first, it is smooth, and second, it coincides with the restriction of $\mathcal{W}^{\mathrm{tra},i}_{x_1,x_2} \circ \mathrm{pr} \circ c$ to V,

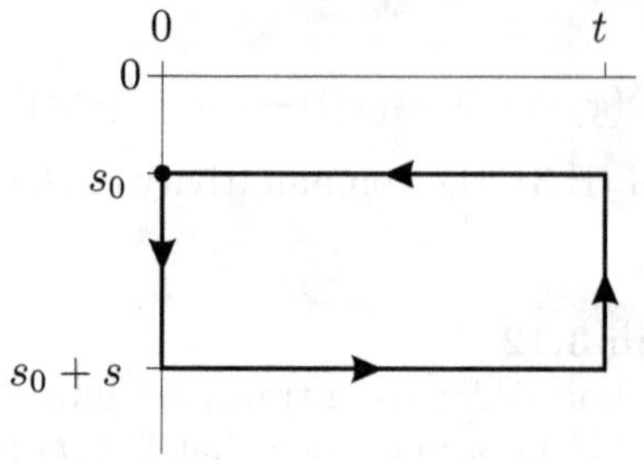

Figure 1: The path $\tau_{s_0}(s,t)$.

both assertions together prove the smoothness of (B.5). To show the first assertion, note that (B.7) is the following assignment:

$$v \longmapsto \mathrm{triv}(\tilde{c}_j(v)) \cdot \tilde{g}(\tilde{\alpha}_j(v)) \cdot \ldots \cdot \tilde{g}(\tilde{\alpha}_1(v)) \circ \mathrm{triv}(\tilde{c}_0(v)). \tag{B.8}$$

By definition, the descent data (triv, g) is smooth. Because triv is a smooth functor, and the maps $\tilde{c}_j$ satisfy the relevant condition (B.6), every factor $\mathrm{triv} \circ \tilde{c}_j : V \longrightarrow G$ is smooth. Furthermore, the maps $\tilde{g} : Y^{[2]} \longrightarrow G$ are smooth, so that also the remaining factors are smooth in v. To show the second assertion, consider a point $v \in V$. If we choose in the definition of the Wilson line $\mathcal{W}^{\mathrm{tra},i}_{x_1,x_2}$ the objects $G_k := \mathrm{triv}(\tilde{x}_k)$ and the isomorphisms $t_k := t(\tilde{x}_k)$ for some lifts $\pi(\tilde{x}_k) = x_k$, where t is the trivialization of tra from the beginning of this section, we find

$$(\mathcal{W}^{\mathrm{tra},i}_{x_1,x_2} \circ \mathrm{pr} \circ c)(v) = \mathrm{tra}(\overline{c(v)}).$$

The right hand side coincides with the right hand side of (B.8).

B.3. Proof of Proposition 4.3

We are going to prove that the map $k_A : PX \longrightarrow G$, defined by

$$k_A(\gamma) := f_{\gamma^* A}(0,1)$$

for a path $\gamma : [0,1] \longrightarrow X$ depends only on the thin homotopy class of γ. Due to the multiplicative property (4.5) of k_A, it is enough to show $k_A(\gamma_0^{-1} \circ \gamma_1) = 1$ for every thin homotopy equivalent paths γ_0 and γ_1. For this purpose we derive a relation to the pullback of the curvature $K := \mathrm{d}A + [A \wedge A]$ of the 1-form A along a homotopy between γ_0 and γ_1. If this homotopy is thin, the pullback vanishes.

Let us fix the following notation. $Q := [0,1] \times [0,1]$ is the unit square, $\gamma_{(a,b,c,d)} : (a,b) \longrightarrow (c,d)$ is the straight path in Q, and

$$\tau_{s_0} : Q \longrightarrow PQ$$

assigns for fixed $s_0 \in [0,1]$ to a point $(s,t) \in Q$ the closed path

$$\tau_{s_0}(s,t) := \gamma_{(s_0,t,s_0,0)} \circ \gamma_{(s_0+s,t,s_0,t)} \circ \gamma_{(s_0+s,0,s_0+s,t)} \circ \gamma_{(s_0,0,s_0+s,0)},$$

which goes counter-clockwise around the rectangle spanned by the points $(s_0, 0)$ and $(s_0 + s, t)$, see Figure 1. Now consider two paths $\gamma_0, \gamma_1 : x \longrightarrow y$ in X. Without loss of generality we can assume that the paths $\gamma_{(a,b,c,d)}$ used above have sitting instants, such that τ_{s_0} is smooth and

$$\gamma_0(\gamma_{(0,1,0,0)}(t)) = \gamma_0^{-1}(t) \quad \text{and} \quad \gamma_1(\gamma_{(0,1,1,1)}(t)) = \gamma_1(t). \tag{B.9}$$

Lemma B.1. *Let $h : Q \longrightarrow X$ be a smooth homotopy between the paths $\gamma_0, \gamma_1 : x \longrightarrow y$ with $h(0,t) = \gamma_0(t)$ and $h(1,t) = \gamma_1(t)$. Then, the map*

$$u_{A,s_0} := k_A \circ h_* \circ \tau_{s_0} : Q \longrightarrow G$$

is smooth and has the following properties

(a) $u_{A,0}(1,1) = k_A(\gamma_0^{-1} \circ \gamma_1)$

(b) $u_{A,s_0}(s,1) = u_{A,s_0}(s',1) \cdot u_{A,s_0+s'}(s-s',1)$

(c) with $\gamma_{s,t}$ the path defined by $\gamma_{s,t}(\tau) := h(s,\tau t)$ and $K := \mathrm{d}A + [A \wedge A]$ the curvature of A we have:

$$\frac{\partial}{\partial s}\frac{\partial}{\partial t} u_{A,s_0}\bigg|_{(0,t)} = -\mathrm{Ad}^{-1}_{k_A(\gamma_{s_0,t})}(h^*K)_{(s_0,t)}\left(\frac{\partial}{\partial s}, \frac{\partial}{\partial t}\right) \tag{B.10}$$

Proof. Since h is constant for $t = 0$ and $t = 1$, (a) follows from (B.9). The multiplicative property (4.5) of k_A implies (b). To prove (c), we define a further path $\gamma_{s_0,s,t}(\tau) := h(s_0 + s\tau, t)$ and write

$$u_{A,s_0}(s,t) = f_{\gamma^*_{s_0,t}A}(0,1)^{-1} \cdot f_{\gamma^*_{s_0,s,t}A}(0,1)^{-1} \cdot f_{\gamma^*_{s_0+s,t}A}(0,1) \tag{B.11}$$

where $f_\varphi : \mathbb{R} \times \mathbb{R} \longrightarrow G$ are the smooth functions that correspond to the a 1-form $\varphi \in \Omega^1(\mathbb{R}, \mathfrak{g})$ by Lemma 4.1 as the solution of initial value problems. By a uniqueness argument one can show that $f_{\gamma^*_{s,t}A}(0,1) = f_{\gamma^*_{s,1}A}(0,t)$. Then, we calculate with (B.11) and, for simplicity, in a faithful matrix representation of G,

$$\frac{\partial}{\partial t} u_{A,s_0}(s,t) = f^{-1}_{\gamma^*_{s_0,t}A}(0,1) \cdot \left((h^*A)_{(s_0,t)}\left(\frac{\partial}{\partial t}\right) \cdot f_{\gamma^*_{s_0,s,t}A}(0,1)^{-1} \right.$$
$$\left. + \frac{\partial}{\partial t} f_{\gamma^*_{s_0,s,t}A}(0,1)^{-1} - f_{\gamma^*_{s_0,s,t}A}(0,1)^{-1} \cdot (h^*A)_{(s_0+s,t)}\left(\frac{\partial}{\partial t}\right)\right) \cdot f_{\gamma^*_{s_0+s,t}A}(0,1).$$

To take the derivatives along s, we use $f_{\gamma^*_{s_0,s,t}A}(0,1) = f_{\gamma_{s_0,1,t}}(0,s)$ and $f_{\gamma_{s_0,0,t}}(0,1) = 1$, both together show

$$\frac{\partial}{\partial s}\bigg|_0 f_{\gamma^*_{s_0,s,t}A}(0,1)^{-1} = (h^*A)_{(s_0,t)}\left(\frac{\partial}{\partial s}\right).$$

Finally,

$$\frac{\partial}{\partial s}\frac{\partial}{\partial t} u_{A,s_0}\bigg|_{s=0} = f^{-1}_{\gamma^*_{s_0,t}A}(0,1) \cdot \left(\left((h^*A)_{(s_0,t)}\left(\frac{\partial}{\partial t}\right) \cdot (h^*A)_{(s_0,t)}\left(\frac{\partial}{\partial s}\right) \right.\right.$$
$$+ \frac{\partial}{\partial t}(h^*A)_{(s_0,t)}\left(\frac{\partial}{\partial s}\right) - (h^*A)_{(s_0,t)}\left(\frac{\partial}{\partial s}\right) \cdot (h^*A)_{(s_0,t)}\left(\frac{\partial}{\partial t}\right)$$
$$\left.\left. - \frac{\partial}{\partial s}\bigg|_0 (h^*A)_{(s_0+s,t)}\left(\frac{\partial}{\partial t}\right)\right)\right) \cdot f_{\gamma^*_{s_0,t}A}(0,1).$$

This yields the claimed equality. □

Notice that if h is a thin homotopy, $h^*K = 0$, so that the right hand side in (c) vanishes. Then we calculate at $(0,1)$

$$\frac{\partial}{\partial s}u_{A,s_0}\Big|_{(0,1)} = \int_0^1 \frac{\partial}{\partial s}\frac{\partial}{\partial t}u_{A,s_0}\Big|_{(0,t)} \mathrm{d}t = 0.$$

Using (b) we obtain the same result for all points $(s_0, 1)$,

$$\frac{\partial}{\partial s}u_{A,0}\Big|_{(s_0,1)} = u_{A,0}(s_0,1)\cdot\frac{\partial}{\partial s}u_{A,s_0}\Big|_{(0,1)} = u_{A,0}(s_0,1)\cdot 0 = 0.$$

This means that the function $u_{A,0}(s,1)$ is constant and thus determined by its value at $s=0$, namely

$$1 = u_{A,0}(0,1) = u_{A,0}(1,1) \overset{\text{(a)}}{=} k_A(\gamma_1^{-1}\circ\gamma_0) = k_A(\gamma_1)^{-1}\cdot k_A(\gamma_0).$$

This finishes the proof.

B.4. Proof of Proposition 4.7

We have to show that the functor $\mathcal{P} : Z^1_X(G)^\infty \longrightarrow \mathrm{Funct}^\infty(\mathcal{P}_1(X), \mathcal{B}G)$ is bijective on objects. For this purpose, we define an inverse map $\mathcal{D}$ that assigns to any smooth functor $F : \mathcal{P}_1(X) \longrightarrow \mathcal{B}G$ a $\mathfrak{g}$-valued 1-form $\mathcal{D}(F)$, such that $\mathcal{P}(\mathcal{D}(F)) = F$ and such that $\mathcal{D}(\mathcal{P}(A)) = A$ for any 1-form $A \in \Omega^1(X,\mathfrak{g})$.

Let $F : \mathcal{P}_1(X) \longrightarrow \mathcal{B}G$ be a smooth functor. We define the 1-form $A := \mathcal{D}(F)$ at a point $p \in X$ and for a tangent vector $v \in T_pX$ in the following way. Let $\gamma : \mathbb{R} \longrightarrow X$ be a smooth curve such that $\gamma(0) = p$ and $\dot\gamma(0) = v$. We consider the map

$$f_\gamma := F_1 \circ (\gamma_*)_1 : \mathbb{R}\times\mathbb{R} \longrightarrow G. \qquad \text{(B.12)}$$

The evaluation $\mathrm{ev}\circ((\gamma_*)_1\times\mathrm{id}) : U\times[0,1] \longrightarrow X$ with $U = \mathbb{R}\times\mathbb{R}$ is a smooth map because γ is smooth. Hence, by Definition 3.1, f_γ is smooth. The properties of the functor F further imply the cocycle condition

$$f_\gamma(y,z)\cdot f_\gamma(x,y) = f_\gamma(x,z). \qquad \text{(B.13)}$$

By Lemma 4.1, the smooth map $f_\gamma : \mathbb{R}\times\mathbb{R} \longrightarrow G$ corresponds to a $\mathfrak{g}$-valued 1-form A_γ on $\mathbb{R}$. We define

$$\alpha_{F,\gamma}(p,v) := A_\gamma|_0\left(\frac{\partial}{\partial t}\right) \in \mathfrak{g}. \qquad \text{(B.14)}$$

With a view to the definition (4.3) of A_γ, this is

$$\alpha_{F,\gamma}(p,v) = -\frac{\mathrm{d}}{\mathrm{d}t}f_\gamma(0,t)\Big|_{t=0}. \qquad \text{(B.15)}$$

Lemma B.2. *$\alpha_{F,\gamma}(p,v)$ is independent of the choice of the smooth curve γ.*

Proof. Let γ_0 and γ_1 be two choices, both with $\gamma_i(0) = p$ and $\dot\gamma_i(0) = v$, for $i = 0,1$. Let $h : [0,1]\times[0,1] \longrightarrow X$ be a smooth homotopy between γ_0 and γ_1, and $\epsilon > 0$ chosen so small that h restricted to $[0,1]\times(0,\epsilon]$ is injective onto its image U. Such an homotopy can always be constructed in a chart of a neighbourhood of x. We construct a map

$$p : [0,1] \times [0,1] \longrightarrow BX$$

by $p(i,\sigma,\tau)(s,t) := h(i,\sigma s,\tau t)$, choose an inverse $\bar{h} : U \longrightarrow [0,1] \times [0,1]$ of h and obtain a map $q := p \circ \bar{h} : U \longrightarrow PX$ which satisfies by construction $p = q \circ h$ on $[0,1] \times (0,\epsilon]$. In fact, the domain of q can be enlarged to $U_0 := U \cup \{x\}$ by $q(x) := \mathrm{id}_x$, so that $p = q \circ h$ on all of $[0,1] \times [0,\epsilon]$. The purpose of these constructions is that the map

$$F \circ p : [0,1] \times [0,\epsilon] \longrightarrow G$$

on the one hand satisfies $(F \circ p)(i,t) = f_{\gamma_i}(0,t)$ for $i = 0,1$, and on the other hand factors through two smooth maps $(F \circ q)$ and h, so that we can apply the chain rule:

$$\begin{aligned} -\alpha_{F,\gamma_i}(p,v) &= \frac{\mathrm{d}}{\mathrm{d}t} f_{\gamma_i}\bigg|_{t=0} = \frac{\mathrm{d}}{\mathrm{d}t}(F \circ p)\bigg|_{(i,0)} \\ &= \mathrm{d}(F \circ q)|_{h(i,0)}\left(\frac{\partial h}{\partial t}\bigg|_{(i,0)}\right) = \mathrm{d}(F \circ q)|_x(v) \end{aligned}$$

The right hand side is, in particular, independent of i. □

According to the result of Lemma B.2, we drop the index γ, and remain with a map $\alpha_F : TX \longrightarrow \mathfrak{g}$ defined canonically by the functor F. We show next that α_F is linear in v. For a multiple sv of v we can choose the curve γ_s with $\gamma_s(t) := \gamma(st)$. It is easy to see that then $f_{\gamma_s}(x,y) = f_\gamma(sx,sy)$. Again by the chain rule

$$\alpha(p,sv) = -\frac{\mathrm{d}}{\mathrm{d}t} f_{\gamma_s}(0,t)|_{t=0} = -\frac{\mathrm{d}}{\mathrm{d}t} f_\gamma(0,st)|_{t=0} = s\alpha_F(p,v).$$

In the same way one can show that $\alpha(p,v+w) = \alpha(p,v) + \alpha(p,w)$.

Lemma B.3. *The pointwise linear map $\alpha_F : TX \longrightarrow \mathfrak{g}$ is smooth, and thus defines a 1-form $A \in \Omega^1(X,\mathfrak{g})$ by $A|_p(v) := \alpha_F(p,v)$.*

Proof. If X is n-dimensional and $\phi : U \longrightarrow X$ is a coordinate chart with an open subset $U \subset \mathbb{R}^n$, the standard chart for the tangent bundle TX is

$$\phi_{TX} : U \times \mathbb{R}^n \longrightarrow TX : (u,v) \longmapsto \mathrm{d}\phi|_u(v).$$

We prove the smoothness of α_F in the chart ϕ_{TX}, i.e. we show that

$$A \circ \phi_{TX} : U \times \mathbb{R}^n \longrightarrow \mathfrak{g}$$

is smooth. For this purpose, we define the map

$$c : U \times \mathbb{R}^n \times \mathbb{R} \longrightarrow PX : (u,v,\tau) \longmapsto (t \longmapsto \phi(u + \beta(t\tau)v))$$

where β is some smoothing function, i.e. an orientation-preserving diffeomorphism of $[0,1]$ with sitting instants. Now, $\mathrm{ev} \circ (c \times \mathrm{id})$ is evidently smooth in all parameters, and since F is a smooth functor,

$$f_c := F_1 \circ \mathrm{pr} \circ c : U \times \mathbb{R}^n \times \mathbb{R} \longrightarrow G$$

is a smooth function. Note that $\gamma_{u,v}(t) := c(u,v,t)(1)$ defines a smooth curve in X with the properties

$$\gamma_{u,v}(0) = \phi(u) \quad \text{and} \quad \dot{\gamma}_{u,v} = \mathrm{d}\phi|_u(v), \tag{B.16}$$

and which is in turn related to c by

$$(\gamma_{u,v})_*(0,t) = c(u,v,t). \tag{B.17}$$

Using the path $\gamma_{u,v}$ in the definition of the 1-form A, we find

$$\begin{aligned}(A \circ \phi_{TX})(u,v) &= \alpha_F(\phi(u), \mathrm{d}\phi|_u(v)) \\ &\overset{(B.16)}{=} -\frac{\mathrm{d}}{\mathrm{d}t}(F_1 \circ (\gamma_{u,v})_*)(0,t)\Big|_{t=0} \\ &\overset{(B.17)}{=} -\frac{\mathrm{d}}{\mathrm{d}t} f_c(u,v,t)\Big|_{t=0}.\end{aligned}$$

The last expression is, in particular, smooth in u and v. □

Summarizing, we started with a given smooth functor $F : \mathcal{P}_1(X) \longrightarrow \mathcal{B}G$ and have derived a 1-form $\mathcal{D}(F) := A \in \Omega^1(X, \mathfrak{g})$. Next we show that this 1-form is the preimage of F under the functor

$$\mathcal{P} : Z^1_X(G)^\infty \longrightarrow \mathrm{Funct}^\infty(X, \mathcal{B}G)$$

from Proposition 4.7, i.e. we show

$$\mathcal{P}(A)(\overline{\gamma}) = F(\overline{\gamma})$$

for any path $\gamma \in PX$. We recall that the functor $\mathcal{P}(A)$ was defined by $\mathcal{P}(A)(\overline{\gamma}) := f_{\gamma^*A}(0,1)$, where $f_{\gamma^*A} : \mathbb{R} \times \mathbb{R} \longrightarrow G$ solves the differential equation

$$\frac{\mathrm{d}}{\mathrm{d}t} f_{\gamma^*A}|_{(0,t)} = -\mathrm{d}r_{f_{\gamma^*A}(0,t)}|_1 \left((\gamma^*A)_t \left(\frac{\partial}{\partial t} \right) \right) \tag{B.18}$$

with the initial value $f_{\gamma^*A}(0,0) = 1$. Now we use the construction of the 1-form A from the given functor F. For the smooth function $f_\gamma : \mathbb{R} \times \mathbb{R} \longrightarrow G$ from (B.12) we obtain using $\gamma_t(\tau) := \gamma(t+\tau)$ with $p := \gamma_t(0)$ and $v := \dot{\gamma}_t(0)$

$$\begin{aligned}\frac{\mathrm{d}}{\mathrm{d}\tau} f_\gamma(t, t+\tau)|_{\tau=0} &= \frac{\mathrm{d}}{\mathrm{d}\tau} f_{\gamma_t}(0,\tau)|_{\tau=0} \\ &\overset{(B.15)}{=} -\alpha_{F,\gamma_t}(p,v) = -A_p(v) = -(\gamma^*A)_t\left(\frac{\partial}{\partial t}\right).\end{aligned} \tag{B.19}$$

Then we have

$$\frac{\mathrm{d}}{\mathrm{d}t} f_\gamma(0,t) \overset{(B.13)}{=} \frac{\mathrm{d}}{\mathrm{d}\tau} f_\gamma(t,\tau)|_{\tau=t} \cdot f_\gamma(0,t) \overset{(B.19)}{=} -\mathrm{d}r_{f_\gamma(0,t)}|_1 \left((\gamma^*A)_t \left(\frac{\partial}{\partial t} \right) \right).$$

Hence, f_γ solves the initial value problem (B.18). By uniqueness, $f_{\gamma^*A} = f_\gamma$ and finally

$$F(\overline{\gamma}) = f_\gamma(0,1) = f_{\gamma^*A}(0,1) = \mathcal{P}(A)(\overline{\gamma}).$$

It remains to show that, conversely, for a given 1-form $A \in \Omega^1(X, \mathfrak{g})$,

$$\mathcal{D}(\mathcal{P}(A)) = A.$$

We test the 1-form $\mathcal{D}(\mathcal{P}(A))$ at a point $x \in X$ and a tangent vector $v \in T_xX$. Let $\Gamma : \mathbb{R} \longrightarrow X$ be a curve in X with $x = \Gamma(0)$ and $v = \dot{\Gamma}(0)$. If we further denote $\gamma_\tau := \Gamma_*(0, \tau)$ we have

$$-\mathcal{D}(\mathcal{P}(A))|_x(v) \overset{\text{(B.15)}}{=} \left.\frac{\partial f_{\gamma_\tau}}{\partial \tau}\right|_{(0,0)} \overset{\text{(B.12)}}{=} \left.\frac{\partial}{\partial \tau}\right|_0 \mathcal{P}(A)(\gamma_\tau) = \left.\frac{\partial}{\partial \tau}\right|_0 f_{\gamma_\tau^* A}(0,1)$$

Here, $f_{\gamma_\tau^* A}$ is the unique solution of the initial value problem (B.18) for the given 1-form A and the curve γ_τ. With a uniqueness argument $f_{A,\gamma_\tau}(t_0, t) = f_{A,\gamma_1}(\tau t_0, \tau t)$. Its derivative is

$$\left.\frac{\partial}{\partial \tau} f_{\gamma_\tau^* A}(0,t)\right|_{\tau=0,t=1} = \left.\frac{\partial}{\partial t} f_{\gamma_1^* A}(0,t)\right|_{t=0} = -A_p(v),$$

this yields $\mathcal{D}(\mathcal{P}(A)) = A$.

Table of Notations

$\mathcal{P}_1^\pi(M)$	the universal path pushout of a surjective submersion $\pi : Y \longrightarrow M$.	Page 197
Ex_π	the functor $\mathrm{Ex}_\pi : \mathrm{Triv}_\pi^1(i) \longrightarrow \mathfrak{Des}_\pi^1(i)$ which extracts descent data.	Page 197
f_*	the functor $f_* : \mathcal{P}_1(M) \longrightarrow \mathcal{P}_1(N)$ of path groupoids induced by a smooth map $f : M \longrightarrow N$.	Page 193
$\mathrm{Funct}(S, T)$	the category of functors between categories $S$and T.	Page 198
$\mathrm{Funct}^\infty(\mathcal{P}_1(X), \mathrm{Gr})$	the category of smooth functors from $\mathcal{P}_1(X)$ to a Lie groupoid Gr.	Page 210
G-Tor	the category of smooth principal G-spaces and smooth equivariant maps.	Page 188
Gr	a Lie groupoid	Page 207
i	a functor $i : \mathrm{Gr} \longrightarrow T$, which relates the typical fibre Gr of a functor to the target category T.	Page 207
C^∞	the category of smooth manifolds and smooth maps between those.	Page 231
$\pi_1^1(M, x)$	the thin homotopy group of the smooth manifold M at the point $x \in M$.	Page 193
p^π	the projection functor $p^\pi : \mathcal{P}_1^\pi(M) \longrightarrow \mathcal{P}_1(M)$.	Page 200

PM	the set of paths in M	Page 192
P^1M	the set of thin homotopy classes of paths in M	Page 192
$\mathcal{P}_1(M)$	the path groupoid of the smooth manifold M.	Page 193
$\Pi_1(M)$	the fundamental groupoid of a smooth manifold M	Page 194
Rec_π	the functor $\mathrm{Rec}_\pi : \mathfrak{Des}^1_\pi(i) \longrightarrow \mathrm{Triv}^1_\pi(i)$ which reconstructs a functor from descent data.	Page 197
s	the section functor $s : \mathcal{P}_1(M) \longrightarrow \mathcal{P}^\pi_1(M)$ associated to a surjective submersion $\pi : Y \longrightarrow M$.	Page 197
$\mathcal{B}G$	the category with one object whose set of morphisms is the Lie group G.	Page 210
D^∞	the category of smooth spaces	Page 231
T	the target category of transport functors – the fibres of a bundle are objects in T, and the parallel transport maps are morphisms in T.	Page 207
$\mathfrak{Des}^1_\pi(i)$	the category of descent data of π-locally i-trivialized functors.	Page 196
$\mathfrak{Des}^1_\pi(i)^\infty$	the category of smooth descent data of π-locally i-trivialized functors	Page 202
$\mathrm{Trans}^1_{\mathrm{Gr}}(M,T)$	the category of transport functors with Gr- structure.	Page 204
$\mathrm{Triv}^1_\pi(i)$	the category of functors $F : \mathcal{P}_1(M) \longrightarrow T$ together with π-local i-trivializations	Page 195
$\mathrm{Triv}^1_\pi(i)^\infty$	the category of transport functors on M with Gr-structure together with π-local i-trivializations.	Page 203
$\mathrm{Vect}(\mathbb{C}^n_h)$	the category of n-dimensional hermitian vector spaces and isometries between those.	Page 218
$\mathrm{VB}(\mathbb{C}^n_h)^\nabla_M$	the category of hermitian vector bundles of rank n with unitary connection over M.	Page 218
$Z^1_X(G)^\infty$	the category of differential cocycles on X with gauge group G.	Page 210
$Z^1_\pi(G)^\infty$	the category of differential cocycles of a surjective submersion π with gauge group G.	Page 211

References

[AI92] A. Ashtekar and C. J. Isham, *Representations of the Holonomy Algebras of Gravity and nonabelian Gauge Theories*, Class. Quant. Grav. **9**, 1433–1468 (1992), `arxiv:hep-th/9202053`

[Bae07] J. C. Baez, *Quantization and Cohomology*, (2007), Lecture Notes, UC Riverside.

[Bar91] J. W. Barrett, *Holonomy and Path Structures in General Relativity and Yang-Mills Theory*, Int. J. Theor. Phys. **30**(9), 1171–1215 (1991).

[Bar04] T. Bartels, *2-Bundles and Higher Gauge Theory*, PhD thesis, University of California, Riverside, 2004, `arxiv:math/0410328`

[Bau05] F. Baudoin, *An Introduction to the Geometry of stochastic Flows*, World Scientific, 2005.

[BS04] J. Baez and U. Schreiber, *Higher Gauge Theory: 2-Connections on 2-Bundles*, preprint, `arxiv:hep-th/0412325`

[BS07] J. C. Baez and U. Schreiber, Higher Gauge Theory, in *Categories in Algebra, Geometry and Mathematical Physics*, edited by A. Davydov, Proc. Contemp. Math, AMS, Providence, Rhode Island, 2007, `arxiv:math/0511710`

[Che77] K.-T. Chen, *Iterated Path Integrals*, Bull. Amer. Math. Soc. **83**, 831–879 (1977).

[CJM02] A. L. Carey, S. Johnson and M. K. Murray, *Holonomy on D-Branes*, J. Geom. Phys. **52**(2), 186–216 (2002), `arxiv:hep-th/0204199`

[CP94] A. Caetano and R. F. Picken, *An axiomatic Definition of Holonomy*, Int. J. Math. **5**(6), 835–848 (1994).

[Del91] P. Deligne, *Le Symbole modéré*, Publ. Math. IHES **73**, 147–181 (1991).

[Gaw88] K. Gawędzki, Topological Actions in two-dimensional Quantum Field Theories, in *Non-perturbative Quantum Field Theory*, edited by G. Hooft, A. Jaffe, G. Mack, K. Mitter and R. Stora, pages 101–142, Plenum Press, 1988.

[GR02] K. Gawędzki and N. Reis, *WZW Branes and Gerbes*, Rev. Math. Phys. **14**(12), 1281–1334 (2002), `arxiv:hep-th/0205233`

[KM97] A. Kriegl and P. W. Michor, *The convenient Setting of global Analysis*, AMS, 1997.

[Mac87] K. C. H. Mackenzie, *Lie Groupoids and Lie Algebroids in Differential Geometry*, volume 124 of *London Math. Soc. Lecture Note Ser.*, Cambridge Univ. Press, 1987.

[Mak96] M. Makkai, *Avoiding the Axiom of Choice in general Category Theory*, Journal of Pure and Applied Algebra **108**(2), 109–174 (1996).

[MM03] I. Moerdijk and J. Mrčun, *Introduction to Foliations and Lie Groupoids*, volume 91 of *Cambridge Studies in Adv. Math.*, Cambridge Univ. Press, 2003.

[Moe02] I. Moerdijk, *Introduction to the Language of Stacks and Gerbes*, Summer school lecture notes, `arxiv:math.AT/0212266`

[MP02] M. Mackaay and R. Picken, *Holonomy and parallel Transport for abelian Gerbes*, Adv. Math. **170**(2), 287–339 (2002), `arxiv:math/0007053`

[Mur96] M. K. Murray, *Bundle Gerbes*, J. Lond. Math. Soc. **54**, 403–416 (1996), `arxiv:dg-ga/9407015`

[Sou81] J.-M. Souriau, Groupes différentiels, in *Lecture Notes in Mathematics*, volume 836, pages 91–128, Springer, 1981.

[SSW07] U. Schreiber, C. Schweigert and K. Waldorf, *Unoriented WZW Models and Holonomy of Bundle Gerbes*, Commun. Math. Phys. **274**(1), 31–64 (2007), `arxiv:hep-th/0512283`

[Sta74] J. D. Stasheff, *Parallel Transport and Classification of Fibrations*, volume 428 of *Lecture Notes in Mathematics*, Springer, 1974.

[Str04] R. Street, *Categorical and combinatorial Aspects of Descent Theory*, Applied Categorical Structures **12**(5-6) (2004), `arxiv:math/0303175v2`

[SW08a] U. Schreiber and K. Waldorf, *Connections on non-abelian Gerbes and their Holonomy*, preprint, `arxiv:0808.1923`

[SW08b] U. Schreiber and K. Waldorf, *Smooth Functors vs. Differential Forms*, preprint, `arxiv:0802.0663`

This article may be accessed via WWW at `http://www.rmi.acnet.ge/jhrs/`

Urs Schreiber
`urs.schreiber@math.uni-hamburg.de`
Konrad Waldorf
`konrad.waldorf@math.uni-hamburg.de`

Organisationseinheit Mathematik
Schwerpunkt Algebra und Zahlentheorie
Universität Hamburg
Bundesstraße 55
D–20146 Hamburg

Journal of Homotopy and Related Structures, vol. 4(1), 2009, pp.245–253

ON THE COFIBRANT GENERATION OF MODEL CATEGORIES

GEORGE RAPTIS

(*communicated by Jiri Rosicky*)

Abstract

The paper studies the problem of the cofibrant generation of a model category. We prove that, assuming Vopěnka's principle, every cofibrantly generated model category is Quillen equivalent to a combinatorial model category. We discuss cases where this result implies that the class of weak equivalences in a cofibrantly generated model category is accessibly embedded. We also prove a necessary condition for a model category to be cofibrantly generated by a set of generating cofibrations between cofibrant objects.

1. Introduction and Statement of Results

The purpose of the paper is to study the problem of the cofibrant generation of a model category and relate it to ideas from the theory of combinatorial model categories. A combinatorial model category is a cofibrantly generated model category whose underlying category is locally presentable. They were first introduced by J. Smith and have been studied fruitfully ever since [**2**] [**5, 6**], [**11**]. Locally presentable categories allow empowered uses of Quillen's small-object argument that greatly facilitate the construction of combinatorial model structures. In fact, the problem of the existence of a model category structure on a locally presentable category often reduces to the problem of the cofibrant generation for the candidate class of cofibrations, as long as the class of weak equivalences is known to satisfy some closure and smallness conditions. Moreover, combinatorial model categories share remarkable categorical and homotopical properties and include many important examples. Simplicial sets and, more generally Grothendieck topoi, are locally presentable categories that carry combinatorial model structures [**4**].

Whereas not every model category is combinatorial, most of the important examples are at least Quillen equivalent to one. Assuming some input from set theory, one of our main results formalises this claim as follows.

I would like to thank my supervisor Ulrike Tillmann for her support. I would also like to gratefully acknowledge the support of a partial EPSRC Studentship and a scholarship from the Onassis' Public Benefit Foundation.
Received January 30, 2009, revised June 6, 2009; published on July 20, 2009.
2000 Mathematics Subject Classification: 18G55, 18C35.
Key words and phrases: cofibrantly generated model category, combinatorial model category, Vopěnka's principle.

Theorem 1.1. *Assuming Vopěnka's principle, every cofibrantly generated model category is Quillen equivalent to a combinatorial model category.*

Vopěnka's principle is a set-theoretical axiom that characteristically appears in the study of locally presentable categories because it gives a simple characterisation of them. Under the assumption that Vopěnka's principle holds, a category is locally presentable if and only if it is cocomplete and has a dense subcategory [**1**, Theorem 6.14]. We will discuss cases where the assumption of Vopěnka's principle in the theorem is not needed.

In the most general case, the problem of the cofibrant generation for a cofibrantly closed class of morphisms $\mathcal{S}$ in a cocomplete category $\mathcal{C}$ asks whether there exists a set of morphisms $\mathcal{I}$ whose cofibrant closure in $\mathcal{C}$ is $\mathcal{S}$. This is difficult to decide in this generality. The problem is especially interesting in the following two closely related cases:

A When is a model category $\mathcal{M}$ cofibrantly generated?

B Let $\mathcal{C}$ be a locally presentable category with a class of weak equivalences $\mathcal{W}$ which satisfies the 2-out-of-3 property, it is closed under retracts and it is accessible and accessibly embedded in $\mathcal{C}^{\to}$. Let $\mathcal{C}of$ be a cofibrantly closed class of maps in $\mathcal{C}$ such that $\mathcal{C}of \cap \mathcal{W}$ is closed under pushouts and transfinite compositions. When is $\mathcal{C}of$ cofibrantly generated?

Regarding the first question, $\mathcal{M}$ is cofibrantly generated only if there is a set of "test"-cofibrations (resp. trivial "test"-cofibrations) such that every morphism is a trivial fibration (resp. fibration) if and only if it has the right lifting property with respect to this set. Cofibrantly generated model categories include most examples of interest in applications and they allow certain constructions to be possible directly (e.g. model structures on diagram categories, constructions of homotopy colimits, etc. see [**9**]) by essentially giving a grip on the cofibrations the same way that CW complexes do in the homotopy theory of topological spaces.

A set S of objects in a category $\mathcal{C}$ is called left adequate if, for every map $f : X \to Y$, f is an isomorphism if and only if $\mathcal{C}(K, X) \to \mathcal{C}(K, Y)$ is a bijection for all $K \in S$. This concept is due to Heller [**8**]. A set S of objects in a category $\mathcal{C}$ with a terminal object is called left weakly adequate if, for every object X in $\mathcal{C}$, X is isomorphic to the terminal object if and only if $\mathcal{C}(K, X) = \{\star\}$ for every $K \in S$.

We prove the following necessary smallness condition. This was previously known for cofibrantly generated *pointed* model categories [**10**, Theorem 7.3.1].

Theorem 1.2. *Let $\mathcal{M}$ be a cofibrantly generated model category. Suppose that there is a set I of generating cofibrations between cofibrant objects. Then the homotopy category $Ho\mathcal{M}$ of $\mathcal{M}$ admits a left weakly adequate set of objects.*

Not every model category is cofibrantly generated and examples are known in the literature, e.g. [**3**]. In [**12**], Strøm discovered a model structure on the category of topological spaces whose weak equivalences are the homotopy equivalences. We conjecture that Theorem 1.2 applies to show that Strøm's model category is not cofibrantly generated.

Regarding the second question above, it is, in practice, a crucial step in order to deduce that $\mathcal{C}of$ and $\mathcal{W}$ determine a combinatorial model category structure on

$\mathcal{C}$. This observation rests on J. Smith's main theorem, a version of which will be recalled below.

The organisation of the paper is as follows. In Section 2, we recall the definitions of cofibrantly generated and combinatorial model categories. Its purpose is mainly to establish some notation and terminology. In this we follow the conventions of [**9**],[**10**] that we recommend. For background in the theory of locally presentable categories and accessibility, the reader should consult [**1**]. A nice exposition of the theory of combinatorial model categories can be found in [**11**].

In Section 3, we prove Theorem 1.1 and we discuss cases where the class of weak equivalences in a cofibrantly generated model category is accessibly embedded. Finally, in Section 4, we prove Theorem 1.2.

2. Recollections

Let $\mathcal{C}$ be a cocomplete category and I a set of morphisms. The cofibrant closure $Cof(I)$ of I (in $\mathcal{C}$) is the smallest class of morphisms that contains I and is closed under retracts, pushouts and transfinite compositions. A class of morphisms $\mathcal{S}$ in $\mathcal{C}$ is cofibrantly generated if there exists a set I of morphisms in $\mathcal{C}$ such that $\mathcal{S} = Cof(I)$.

A *model* category is a category $\mathcal{M}$ together with three classes of morphisms called weak equivalences, fibrations and cofibrations, each of which is closed under composition and contains the identities and satisfies the following axioms:

M1 (Limits) All small limits and colimits exist in $\mathcal{M}$.

M2 (2-out-of-3) If two out of three morphisms f,g and $f \circ g$ are weak equivalences, then so is the third.

M3 (Retracts) If f is a retract of g and g is a fibration, cofibration or weak equivalence, then so is f.

M4 (Lifting) Given a commutative diagram in $\mathcal{M}$

$$\begin{array}{ccc} A & \xrightarrow{f} & X \\ {\scriptstyle i}\downarrow & & \downarrow{\scriptstyle p} \\ B & \xrightarrow{g} & Y \end{array}$$

then there exists a lift $h : B \to X$ if either:

- i is a cofibration and p is a trivial fibration (i.e. a fibration and a weak equivalence), or
- i is a trivial cofibration (i.e. a cofibration and a weak equivalence) and p is a fibration.

M5 (Factorisation) Every morphism f in $\mathcal{M}$ can be factorised functorially in the following two ways:

- $f = qi$, where i is a cofibration and q is a trivial fibration
- $f = pj$, where j is a trivial cofibration and p is a fibration.

A model category $\mathcal{M}$ is *cofibrantly generated* if there exist sets of morphisms I and J such that the following hold:

1. the domains of I are small relative to $I-cellular$ morphisms; the domains of J are small relative to $J-cellular$ morphisms.
2. the fibrations are the $J-injective$ morphisms; the trivial fibrations are the $I-injective$ morphisms.

I is called a set of generating cofibrations and J a set of generating trivial cofibrations. Equivalently, a model category is cofibrantly generated if the classes of cofibrations and trivial cofibrations are cofibrantly generated and the required smallness condition of the definition is satisfied. Note that the smallness condition is automatically satisfied when the underlying category consists solely of presentable objects.

A model category $\mathcal{M}$ is *combinatorial* if it is cofibrantly generated and its underlying category is locally presentable. Recall that a cocomplete category $\mathcal{C}$ is locally presentable if, for some regular cardinal λ, it has a set S of λ-presentable objects such that every object is a λ-directed colimit of objects in S [**1**].

The tools for comparison between model categories are called Quillen functors; these are functors which preserve part of the structure and they have good derivability properties. Let $\mathcal{M}$ and $\mathcal{N}$ be model categories. A functor $F : \mathcal{M} \to \mathcal{N}$ is a left Quillen functor if F is a left adjoint and it preserves cofibrations and trivial cofibrations. Equivalently, if its right adjoint $G : \mathcal{N} \to \mathcal{M}$ is a right Quillen functor, i.e. it preserves fibrations and trivial fibrations. The adjunction $F : \mathcal{M} \leftrightarrows \mathcal{N} : G$ is called a Quillen adjunction.

It can be shown that left Quillen functors preserve weak equivalences between cofibrant objects. The subcategory of cofibrant objects is a (left) "deformation retract" of the whole category because, by the factorisation axiom, every object has a functorial cofibrant replacement. It follows that every left Quillen functor $F : \mathcal{M} \to \mathcal{N}$ admits a total left derived functor $LF : Ho\mathcal{M} \to Ho\mathcal{N}$ (and dually, every right Quillen functor G admits a total right derived functor RG). Moreover, every Quillen adjunction $F : \mathcal{M} \leftrightarrows \mathcal{N} : G$ induces a derived adjunction $LF : Ho\mathcal{M} \leftrightarrows Ho\mathcal{N} : RG$ [**10**, Chapter 1].

A Quillen adjunction $F : \mathcal{M} \leftrightarrows \mathcal{N} : G$ is called a Quillen equivalence if $LF : Ho\mathcal{M} \leftrightarrows Ho\mathcal{N} : RG$ is an adjoint equivalence of categories.

3. The Proof of Theorem 1.1

Let S be a set of objects in a cocomplete category $\mathcal{C}$. Denote by $\mathcal{S}$ the full subcategory of $\mathcal{C}$ with objects in S. Every object X in $\mathcal{C}$ determines a comma category $\mathcal{S} \downarrow X$ and a canonical forgetful diagram $\underline{X} : \mathcal{S} \downarrow X \to \mathcal{C}$ with respect to $\mathcal{S}$. Let $\kappa_{\mathcal{S}}(X)$ denote the colimit of this diagram in $\mathcal{C}$. There is an induced canonical morphism $\eta_X : \kappa_{\mathcal{S}}(X) \to X$ in $\mathcal{C}$ which is natural in X. We say that X is S-generated if η_X is an isomorphism, cf. [**7**]. Let $\mathcal{C}_{\mathcal{S}}$ denote the full subcategory of S-generated objects.

Proposition 3.1. *Let $\mathcal{C}$ be a cocomplete category and S a set of objects.*

(i) The functor $\kappa_{\mathcal{S}} : \mathcal{C} \to \mathcal{C}_{\mathcal{S}}$ is right adjoint to the inclusion functor.

(ii) *$\mathcal{C}_S$ is locally presentable if and only if every object $A \in S$ is λ-presentable in $\mathcal{C}_S$ for some regular ordinal λ.*

(iii) *Assuming Vopěnka's principle, $\mathcal{C}_S$ is locally presentable.*

Proof. (i) The adjunction isomorphism follows from the universal property of the morphism η_X. (ii) The "only if" is obvious. For the "if" part, note the S is a strong generator of λ-presentable objects in $\mathcal{C}_S$. (iii) $\mathcal{C}_S$ is cocomplete and it has a dense subcategory. □

Proof. (of Theorem 1.1) Let $\mathcal{M}$ be a cofibrantly generated model category with a generating set of cofibrations I and trivial cofibrations J. Let S be the set of objects that appear as domains or codomains of $I \cup J$. Assuming Vopěnka's principle, $\mathcal{M}_S$ is a locally presentable category. We claim that the model structure on $\mathcal{M}$ restricts to a model structure on $\mathcal{M}_S$ and the adjunction $i : \mathcal{M}_S \leftrightarrows \mathcal{M} : \kappa_S$ is a Quillen equivalence. The only non-trivial part of the first claim is to show that factorisations exist in $\mathcal{M}_S$. This is true in virtue of the fact that the factorisations given by the small-object argument can be performed in $\mathcal{M}_S$ because i preserves colimits and therefore the pushout and the directed colimit of S-generated objects is S-generated. The unit transformation $1 \to \kappa_S \circ i$ is a natural isomorphism and therefore also a natural weak equivalence of functors. Each component of the counit $i \circ \kappa_S(X) \to X$ has the right lifting property with respect to I by the universal property of the arrow $\kappa_S(X) \to X$. Hence it is a trivial fibration and so, in particular, a weak equivalence. It follows that the Quillen adjunction $i : \mathcal{M}_S \leftrightarrows \mathcal{M} : \kappa_S$ is a Quillen equivalence. □

Remark. Note that property (1) in the definition of cofibrantly generated model categories from Section 2 is not needed in the last proof. This observation is due to Jiří Rosický.

The assumption of Vopěnka's principle can be dropped as long as the local presentability of $\mathcal{M}_S$ can be asserted otherwise. For example, if the objects in S are presentable, then $\mathcal{M}_S$ is locally presentable by Proposition 3.1. For many purposes, the condition that the objects in S are presentable seems to be almost as good as knowing that $\mathcal{M}$ is a combinatorial model category. An example of this will be shown in Proposition 3.3 below.

Moreover, $\mathcal{M}_S$ is always locally presentable when $\mathcal{M}$ is a fibre-small topological category by [**7**, Theorem 3.6]. The category of topological spaces clearly has this property for example.

The following remarkable result, due to J. Smith, is very useful for generating model category structures on locally presentable categories.

Theorem 3.2. *Let $\mathcal{C}$ be a locally presentable category, I be a set of morphisms and $\mathcal{W}$ a class of morphisms. Suppose that the following are satisfied:*

(i) *$\mathcal{W}$ satisfies the 2-out-of-3 property and it is closed under retracts in $\mathcal{C}^{\to}$,*

(ii) *$I -$ injective $\subseteq \mathcal{W}$,*

(iii) *$Cof(I) \cap \mathcal{W}$ is closed under pushouts and transfinite compositions,*

(iv) *$\mathcal{W}$ is accessible and accessibly embedded in $\mathcal{C}^{\to}$.*

Then the classes $\mathcal{W}$, $Cof(I)$ and $(Cof(I) \cap \mathcal{W}) - injective$ define a combinatorial model category structure on $\mathcal{C}$.

Proof. This is a version of J. Smith's theorem [**2**, Theorem 1.7]. It reduces to it from the fact that an accessible and accessibly embedded subcategory of a locally presentable category is cone-reflective (see [**1**, Theorem 2.53]) and therefore it satisfies the solution-set condition at every object. □

Let $\mathcal{C}$ be a locally presentable category and $\mathcal{W}$ a class of weak equivalences that satisfies (i) and (iv) of the Theorem. The class of formal cofibrations Cof with respect to the pair $(\mathcal{C}, \mathcal{W})$ is the largest cofibrantly closed class of morphisms such that $Cof \cap \mathcal{W}$ is closed under pushouts and transfinite compositions. Clearly, every cofibration of a model category structure on $\mathcal{C}$ with weak equivalences $\mathcal{W}$ is a formal cofibration. Conversely, given a set I of formal cofibrations that is big enough in the sense that $I - injective \subseteq \mathcal{W}$, then $Cof(I)$ is a class of cofibrations for such a (combinatorial) model structure.

Proposition 3.3. *(i) Let $\mathcal{M}$ be a cofibrantly generated model category with I and J sets of generating cofibrations and trivial cofibrations respectively. Suppose that the domains and codomains of the maps in $I \cup J$ are presentable objects. Then the class of weak equivalences $\mathcal{W}$ is accessibly embedded in $\mathcal{M}^{\rightarrow}$.*

(ii) Let $\mathcal{M}$ be a cofibrantly generated model category with I and J sets of generating cofibrations and trivial cofibrations respectively. Suppose that the codomains of the maps in I are presentable objects. Assuming Vopěnka's principle, the class of weak equivalences $\mathcal{W}$ is accessibly embedded in $\mathcal{M}^{\rightarrow}$.

Proof. (i) By Proposition 3.1 (ii) and Theorem 1.1, there is a set of objects S in $\mathcal{M}$ and a Quillen equivalence $\mathcal{M}_{\mathcal{S}} \leftrightarrows \mathcal{M}$ where $\mathcal{M}_{\mathcal{S}}$ is a cominatorial model category. By [**6**, Proposition 7.3], the class of weak equivalences $\mathcal{W}_{\mathcal{S}}$ of $\mathcal{M}_{\mathcal{S}}$ is accessibly embedded in $\mathcal{M}_{\mathcal{S}}^{\rightarrow}$, so they are closed under μ-directed colimits for some μ. Let λ be a regular ordinal larger than μ such that every object that appears as the domain or codomain of I is λ-presentable. We show that the weak equivalences $\mathcal{W}$ of $\mathcal{M}$ are closed under λ-directed colimits. Let $F : J \to \mathcal{M}^{\rightarrow}$ be a λ-directed diagram such that $F(j) \in \mathcal{W}$ for all j. Then $F_{\mathcal{S}} := \kappa_{\mathcal{S}} \circ F : J \to \mathcal{M}_{\mathcal{S}}^{\rightarrow}$ satisfies $F_{\mathcal{S}}(j) \in \mathcal{W}_{\mathcal{S}}$ and therefore $colim_J F_{\mathcal{S}} \in \mathcal{W}_{\mathcal{S}}$. The arrows $F_{\mathcal{S}}(j) \to F(j)$ in $\mathcal{M}^{\rightarrow}$ are objectwise trivial fibrations by the proof of Theorem 1.1. Then the induced arrow $colim_J F_{\mathcal{S}} \to colim_J F$ in $\mathcal{M}^{\rightarrow}$ is also a trivial fibration objectwise because every member of I is λ-presentable in $\mathcal{M}^{\rightarrow}$. It follows, by the 2-out-of-3 property, that $colim_J F \in \mathcal{W}$. Thus $\mathcal{W}$ is accessibly embedded in $\mathcal{M}^{\rightarrow}$.

(ii) The same argument essentially as in (i) applies to show that $\mathcal{W}$ is accessibly embedded in $\mathcal{M}^{\rightarrow}$. The assertion that $\mathcal{M}_{\mathcal{S}}$ is locally presentable requires Vopěnka's principle in this case. Also, the members of I are not necessarily presentable in $\mathcal{M}^{\rightarrow}$ in this case, but they are presentable in the full subcategory of arrows in $\mathcal{M}^{\rightarrow}$ whose domains are in $\mathcal{M}_{\mathcal{S}}$, which suffices for the argument. □

4. The proof of Theorem 1.2

Proof. (of Theorem 1.2) Let $\mathcal{M}$ be a cofibrantly generated model category with generating set $I = \{A_i \to X_i\}_i$ of cofibrations such that A_i is cofibrant for all i. Let 1 denote the terminal object in $\mathcal{M}$. Suppose that X is an object in $Ho\mathcal{M}$ which we can assume to be fibrant in $\mathcal{M}$ and write $e : X \to 1$ for the unique morphism. Suppose that $e_* : Ho\mathcal{M}(K, X) \to Ho\mathcal{M}(K, 1) = \{\star\}$ is a bijection for every $K \in \{A_i, X_i\}_i$. Then every diagram

$$\begin{array}{ccc} A_i & \xrightarrow{f} & X \\ \downarrow{\scriptstyle i} & & \downarrow \\ X_i & \longrightarrow & 1 \end{array}$$

admits a lift h up to homotopy. Since A_i is cofibrant by assumption, there is a homotopy $H : Cyl(A_i) \cup_{A_i} X_i \to X$ that restricts to f and h respectively. Note that $Cyl : \mathcal{M} \to \mathcal{M}$ denotes a functorial choice of cylinder objects in $\mathcal{M}$. We will show that f extends to X_i. There is a (cofibration, trivial fibration)-factorisation $j = p \circ t : Cyl(A_i) \cup_{A_i} X_i \to C'X_i \to Cyl(X_i)$. j is weak equivalence, therefore so is t. The diagram

$$\begin{array}{ccc} Cyl(A_i) \cup_{A_i} X_i & \longrightarrow & X \\ \downarrow{\scriptstyle t} & \overset{g}{\nearrow} & \downarrow \\ C'X_i & \longrightarrow & 1 \end{array}$$

admits a lift $g : C'X_i \to X$. Also let $s : X_i \to C'X_i$ be a lift of

$$\begin{array}{ccc} A_i & \xrightarrow{i_1} & C'X_i \\ \downarrow{\scriptstyle i} & \overset{s}{\nearrow} & \downarrow{\scriptstyle p} \\ X_i & \xrightarrow{i_i} & Cyl(X_i) \end{array}$$

Then $g \circ s : X_i \to X$ makes the first diagram commute. It follows that e is a weak equivalence and $\{A_i, X_i\}$ a left weakly adequate set of $Ho\mathcal{M}$. □

Remark. Note that the proof only requires that the class of cofibrations is cofibrantly generated by a set of cofibrations between cofibrant objects.

Strøm's model category [**12**] is a model category structure on the category *Top* of topological spaces which is different from the usual cofibrantly generated model category of spaces [**10**]. The cofibrations are the closed (Hurewicz) cofibrations (or closed NDR-pairs), the weak equivalences are the homotopy equivalences and the fibrations are the Hurewicz fibrations. With respect to this model category structure, every object is both cofibrant and fibrant. It follows that the homotopy category is the quotient of *Top* with respect to the relation of homotopy on the morphism sets. It seems possible that it does not have a left weakly adequate set, but we do not know if this is true.

Note that the stronger assertion that the homotopy category of a cofibrantly generated model category has a left adequate set is not true. For example, the usual cofibrantly generated model category of topological spaces does not have this property [**8**].

References

[1] J. ADÁMEK AND J. ROSICKÝ, *Locally presentable and accessible categories.* London Mathematical Society Lecture Note Series, 189. Cambridge University Press, Cambridge, 1994.

[2] T. BEKE, *Sheafifiable homotopy model categories*, Math. Proc. Cambridge Philos. Soc. 129 (2000), no. 3, 447–475.

[3] B. CHORNY, *The model category of maps of spaces is not cofibrantly generated*, Proc. Amer. Math. Soc. 131 (2003), no. 7, 2255–2259.

[4] D. CISINSKI, *Théories homotopiques dans les topos*, J. Pure Appl. Algebra 174 (2002), no. 1, 43–82.

[5] D. DUGGER, *Universal homotopy theories*, Adv. Math. 164 (2001), no. 1, 144–176.

[6] D. DUGGER, *Combinatorial model categories have presentations*, Adv. Math. 164 (2001), no. 1, 177–201.

[7] L. FAJSTRUP AND J. ROSICKÝ, *A convenient category for directed homotopy*, Theory Appl. Categ. 21 (2008), No. 1, 7–20.

[8] A. HELLER, *On the representability of homotopy functors*, J. London Math. Soc. (2) 23 (1981), no. 3, 551–562.

[9] P. S. HIRSCHHORN, *Model categories and their localizations.* Mathematical Surveys and Monographs, Vol. 99. American Mathematical Society, Providence, RI, 2003.

[10] M. HOVEY, *Model categories.* Mathematical Surveys and Monographs, Vol. 63. American Mathematical Society, Providence, RI, 1999.

[11] J. ROSICKÝ, *On Combinatorial Model Categories*, Appl. Categ. Struct. 17 (2009), 303-316.

[12] A. STRØM, *The homotopy category is a homotopy category*, Arch. Math. (Basel) 23 (1972), 435–441.

This article may be accessed via WWW at http://www.rmi.acnet.ge/jhrs/

George Raptis
raptis@maths.ox.ac.uk

Mathematical Institute
24-29 St Giles'
Oxford
OX1 3LB
England

Journal of Homotopy and Related Structures, vol. 4(1), 2009, pp.255–264

ENSEMBLES SIMPLICIAUX RÉGULIERS

MICHEL ZISMAN

(*communicated by Lionel Schwartz*)

Résumé

On définit une sous-catégorie intéressante $\mathcal{S}_{\text{reg}}$ de la catégorie des ensembles simpliciaux, dont les objets sont appelés *réguliers.* Cette catégorie, ainsi que sa sous-catégorie $\mathcal{S}_{f-\text{reg}}$ dont les objets sont les ensembles simpliciaux réguliers et finis, ont de bonnes propriétés de stabilité par limite et union. La catégorie $\mathcal{S}_{f-\text{reg}}$ est cartésienne fermée, en contraste de celle des enembles simpliciaux finis qui n'est pas cartésienne fermée.

Le but de cette note est de répondre à une question d'André Joyal, qui m'a été posée par Georges Maltsiniotis. Désignons par *Hom* le Hom interne de la catégorie $\mathcal{S}$ des ensembles simpliciaux, et soient U et X deux ensembles simpliciaux finis (i.e. qui n'ont qu'un nombre fini de simplexes non dégénérés) : dans ces conditions $Hom(U, X)$ est-il fini ? Curieusement, la réponse est négative en général. Le premier contre-exemple, dû à Jacob Lurie [1], consiste à prendre $U = \Delta[1]$ et $X = \Delta[3]/\dot{\Delta}[3]$. Je n'ai pas réussi jusqu'à présent à caractériser les ensembles simpliciaux finis X pour lesquels la réponse est oui pour tous les ensembles simpliciaux finis U, mais la réponse est oui pour une large famille d'ensembles simpliciaux, que l'on appellera ensembles simpliciaux *réguliers* (confer 1.3.), en vertu du théorème suivant :

Théorème 1. Si X est un ensemble simplicial régulier de dimension finie, alors, pour tout ensemble simplicial fini U, l'ensemble simplicial $Hom(U, X)$ est aussi de dimension finie.

(On dit qu'un ensemble simplicial est de *dimension finie* si le degré de ses simplexes non dégénérés est borné).

En particulier si X est régulier et fini, alors, pour tout ensemble simplicial fini U, l'ensemble simplicial $Hom(U, X)$ est fini.

Je remercie Georges Maltsiniotis pour ses encouragements, le soin avec lequel il a relu de précédents états du manuscrit, et ses suggestions pour faciliter la lisibilité du texte. Je remercie aussi le referee dont les questions judicieuses m'ont permis de compléter utilement certains énoncés et de rendre plus agréable la lecture de quelques démonstrations.
Received September 10, 2008, revised June 15, 2009; published on September 21, 2009.
2000 Mathematics Subject Classification: 18G30.
Key words and phrases: Finite simplicial sets, cartesian closed categories, degeneracy operators.

1. au cours d'une conversation avec A. Joyal et G.Maltsiniotis

Nous verrons aussi que les sous-catégories pleines $\mathcal{S}_{\text{reg}}$ de $\mathcal{S}$ formées par ces ensemble simpliciaux et $\mathcal{S}_{f-\text{reg}}$ formées par ceux qui sont en plus finis, possèdent de bonnes propriétés de stabilité.

1. Préliminaires.

1.1. Rappelons que les p-simplexes de $Hom(U, X)$ sont les morphismes $f : \Delta[p] \times U \to X$ de $\mathcal{S}$. Il en résulte immédiatement que l'on a

$$Hom(\text{colim}_\alpha U_\alpha, X) = \lim_\alpha Hom(U_\alpha, X).$$

Comme tout ensemble simplicial U est colimite d'une famille de $\Delta[n]_x$ (indicée par les simplexes non dégénérés x de U) on voit que pour démontrer que quel que soit U fini, $Hom(U, X)$ est fini pour un certain X, il suffit de le vérifier pour les $U = \Delta[n]$, et on se limitera à ce cas dans toute la suite.

1.2. Voici quelques notations utiles dans la suite. Les objects de la catégorie simpliciale Δ sont désignés comme d'habitude par des entiers entre crochets. Les morphismes faces et dégénérescences sont notés respectivement ∂_i et σ_i. Soient X un ensemble simplicial et $\varphi : [p] \to [q]$ un morphisme de Δ. On note $X(\varphi) : X_q \to X_p$ l'application associée par la structure simpliciale de X. Soit $\varphi(i, i+r) : [1] \to [p]$ le morphisme défini par $0 \mapsto i$ et $1 \mapsto i+r$.

1.3. Introduisons maintenant quelques définitions.

On dit qu'un ensemble simplicial X est *fortement régulier* si pour tout simplexe x non dégénéré de X, les faces $d_i x$ sont aussi non dégénérées.

Une *arête élémentaire* d'un n-simplexe x d'un ensemble simplicial X est un 1-simplexe y de X égal à $X(\varphi(i, i+1))x$ pour un certain i, $0 \leqslant i \leqslant n-1$.

On dit qu'un ensemble simplicial vérifie la propriété P_r si, étant donné un simplexe x de X tel que $X(\varphi(i, i+r))x$ soit dégénéré, alors il existe $y \in X$ tel que $x = s_{i+r-1} \dots s_i y$.

On dit qu'un ensemble simplicial est *régulier* s'il vérifie la propriété P_1. Comme le laisse entendre la terminologie proposée, nous verrons qu'un ensemble simplicial fortement régulier est régulier.

1.4. Quelques propriétés élémentaires

1.4.1. On vérifie facilement le résultat suivant :

(*) Soit $x = s_{i+r-1} s_{i+r-2} \dots s_i y$ un simplexe de X. Alors l'arête $X(\varphi(i, i+r))x$ est dégénérée.

On en déduit immédiatement le

Lemme 1. Soit $y \in X$ un q-simplexe, soient $\alpha : [p] \to [q]$ un morphisme surjectif, et $x = X(\alpha)y$. Alors $X(\varphi(i, i+1))x$ est non dégénérée pour au plus q valeurs de i.

En effet $X(\alpha) = s_{i_1} s_{i_2} \cdots s_{i_{p-q}}$ avec $i_1 > i_2 > \cdots > i_{p-q}$.

Lorsque X est fortement régulier, le résultat $(*)$ possède une réciproque :

Lemme 2. Un ensemble simplicial fortement régulier vérifie la propriété P_r pour tout $r > 0$ (en particulier, il est régulier). Réciproquement si un ensemble simplicial satisfait aux conditions P_1 et P_2, il est fortement régulier.

Démonstration : Pour la première partie, on procède par récurrence sur le degré p des simplexes, le cas $p = 1$ étant tautologique. Soit $x \in X_p$ un p-simplexe tel que $X(\varphi(i, i+r))x$ soit dégénéré. Puisque X est fortement régulier, x est donc dégénéré, disons $x = s_k y$ pour un certain $k \leqslant p-1$. Il vient donc $X(\varphi(i, i+r))x = X(\sigma_k \circ \varphi(i, i+r))y$. Les relations de commutation

$$\sigma_k \circ \varphi(i, i+r) = \begin{cases} \varphi(i-1, i+r-1) & \text{si} \quad k < i \\ \varphi(i, i+r-1) & \text{si} \quad i \leqslant k < i+r \\ \varphi(i, i+r) & \text{si} \quad k \geqslant i+r \end{cases} .$$

permettent d'écrire suivant les cas

$$X(\varphi(i, i+r))x = \begin{cases} X(\varphi(i-1, i+r-1))y \\ X(\varphi(i, i+r-1))y \\ X(\varphi(i, i+r))y \end{cases} .$$

Puisque cette arête est dégénérée, l'hypothèse de récurrence montre qu'il existe un z de degré $p-2$ tel que, selon les valeurs de k, on a

$$y = \begin{cases} s_{i+r-2} \cdots s_{i-1} z \\ s_{i+r-2} \cdots s_i z \\ s_{i+r-1} \cdots s_i z \end{cases} .$$

Mais alors il vient

$$x = s_k y = \begin{cases} s_{i+r-1} \cdots s_i s_k z \\ s_{i+r-1} \cdots s_i z \\ s_{i+r-1} \cdots s_i s_{k-r} z. \end{cases}$$

Supposons maintenant que X vérifie P_1 et P_2. Soit $x \in X$ et supposons que $d_k x$ est dégénéré, disons $d_k x = s_l y$. Alors (confer $(*)$) $X(\varphi(l, l+1))d_k x$ est dégénéré. Or on a

$$X(\varphi(l, l+1))d_k = X(\partial_k \circ \varphi(l, l+1))$$

et on vérifie les égalités suivantes :

$$\partial_k \circ \varphi(l, l+1) = \begin{cases} \varphi(l+1, l+2) & \text{si} \quad k \leqslant l \\ \varphi(l, l+2) & \text{si} \quad k = l+1 \\ \varphi(l, l+1) & \text{si} \quad k > l+1. \end{cases}$$

L'hypothèse implique donc que, dans le premier cas, on a $x = s_{l+1} z$, dans le second on a $x = s_{l+1} s_l z$ et dans le troisième $x = s_l z$ pour un certain z. Dans tous les cas, x est dégénéré.

Lemme 3. Soit X un ensemble simplicial. Les trois propriétés suivantes sont équivalentes :

(i) X est régulier.

(ii) Les arêtes élémentaires d'un simplexe non dégénéré de X sont toutes non dégénérées.

(iii) Un simplexe x de X est non dégénéré si et seulement si toutes ses arêtes élémentaires sont non dégénérées.

Démonstration : (i) $\Rightarrow$ (ii) par définition même de régulier. (iii) n'est autre que (ii) à laquelle on ajoute la propriété vraie sans restriction sur X à savoir qu'un dégénéré possède toujours une arête élémentaire dégénérée (confer $(*)$). Reste à montrer (ii) $\Rightarrow$ (i) et pour cela que si X ne satisfait pas à (i), il existe un simplexe non dégénéré de X dont une arête élémentaire est dégénérée. L'hypothèse dit qu'il existe un $z \in X_p$ et un i tel que $X(\varphi(i, i+1))z$ est dégénéré, mais qu'il n'existe aucun y tel que $z = s_i y$. Écrivons $z = X(\phi)x$ avec x non dégénéré et ϕ surjective : $\phi(i)$ et $\phi(i+1)$ sont donc soit égaux, soit deux entiers successifs. Le premier cas ne peut se présenter car il impliquerait l'existence d'un ψ tel que $\phi = \psi \circ \sigma_i$, et donc on aurait $z = s_i X(\psi)x$ en contradiction avec l'hypothèse. Reste donc le second. Posons $\phi(i) = j$. Comme on a $X(\varphi(i, i+1))z = X(\varphi(j, j+1))x$ et que x est non dégénéré, la démonstration est achevée.

La propriété (ii) du lemme précédent est très commode pour reconnaître un ensemble simplicial régulier. Par exemple, si $n \geqslant 2$, le quotient de $\Delta[n]$ par son arête $0n$, qui n'est évidemment pas fortement régulier, est régulier. De même on démontre sans peine la proposition suivante :

Proposition 1. La catégorie $\mathcal{S}_{\text{reg}}$ est stable par limites et sommes. Un sous-objet (dans $\mathcal{S}$) d'un objet de $\mathcal{S}_{\text{reg}}$ est dans $\mathcal{S}_{\text{reg}}$. Si X_a est une famille de sous-ensembles simpliciaux de X, et si chaque X_a est régulier, la réunion $\bigcup X_a$ l'est aussi. Un nerf est toujours régulier.

(Pour montrer par exemple la stabilité de la catégorie $\mathcal{S}_{\text{reg}}$ par produits, il suffit d'utiliser la définition ; pour montrer qu'un sous ensemble simplicial d'un ensemble simplicial régulier est régulier, on utilise la propriété (ii) et le fait qu'un simplexe d'un sous-ensemble simplicial est dégénéré si et seulement si il est dégénéré dans l'ensemble simplicial tout entier ; la stabilité par limites résulte de ces deux résultats.)

2. Un critère de dégénérescence.

2.1. Rappelons que les w-simplexes de $\Delta[p] \times \Delta[n]$ sont les fonctions croissantes $[w] \to [p] \times [n]$. Si la deuxième coordonnée reste constante, on dira que le simplexe est *horizontal*, et si la première coordonnée reste constante, on dira qu'il est *vertical.*

Appelons *chemin* a du réseau $[p] \times [n]$ un simplexe non dégénéré, de longueur maximale pour origine et extrémité fixées. Géométriquement, le chemin explicite les valeurs successives de $a : [w] \to [p] \times [n]$, deux valeurs successives ayant toujours soit même abscisse soit même ordonnée. Notons $\mathcal{C}$ l'ensemble des $\binom{n+p}{p}$ chemins maximaux i.e. ceux qui relient $(0,0)$ à (p,n), et faisons la remarque, triviale mais utile, que si c est un chemin maximal passant par le point de coordonnées (i,j), alors on a $c(i+j)=(i,j)$.

Pour se donner un p-simplexe f de $Hom(\Delta[n],X)$, donc un morphisme $f:\Delta[p]\times\Delta[n]\to X$, il suffit de se donner les $(p+n)$-simplexes $z_a=f(a)$ de X lorsque a parcourt $\mathcal{C}$, ces données étant soumises aux relations naturelles qui expriment que "sur l'intersection $a\cap b$ de deux chemins maximaux, z_a et z_b coïncident" [2]. On se propose dans ce paragraphe de donner un critère qui exprime que f est en fait un dégénéré d'un $(p-1)$-simplexe g. Commençons par le diagramme commutatif suivant.

2.2. Étant donné un chemin maximal a de $[p]\times[n]$ et un entier positif ou nul $k<p$, il existe un unique entier t tel que l'on a

$$a(k+t)=(k,t)$$
$$a(k+t+1)=(k+1,t)$$

Ayant ainsi fixé t, on définit un chemin maximal m de $[p-1]\times[n]$ en posant

$$m(j)=\begin{cases} a(j) & \text{si} \quad j\leqslant k+t \\ a(j+1)-(1,0) & \text{si} \quad j\geqslant k+t \end{cases}$$

et on vérifie sans peine que le diagramme

$$\begin{array}{ccc} [p+n] & \xrightarrow{\;a\;} & [p]\times[n] \\ {\scriptstyle \sigma_{k+t}}\downarrow & & \downarrow{\scriptstyle \sigma_k\times \mathrm{id}} \\ [p+n-1] & \xrightarrow[m]{} & [p-1]\times[n] \end{array}$$

commute. Mais alors si $g:\Delta[p-1]\times\Delta[n]\to X$ est un $(p-1)$-simplexe de $Hom(\Delta[n],X)$ et si h désigne le p-simplexe dégénéré $s_k g$, il vient :

$$h(a)=s_{k+t}g(m).$$

D'après l'assertion $(*)$ de 1.4.1., l'image par h de l'arête $((k,t),(k+1,t))$ est dégénérée. Introduisons, pour désigner ce phénomène, les définitions suivantes :

Soit k un entier, $0\leqslant k\leqslant p-1$. On dit qu'un p-simplexe f de $Hom(\Delta[n],X)$ est *k-presque dégénéré* si pour tout $0\leqslant j\leqslant n$, le 1-simplexe de X égal au composé $[1]\longrightarrow[p]\times[n]\xrightarrow{\;f\;}X$ (où la première flèche est défini par $0\mapsto(k,j)$, $1\mapsto(k+1,j)$) est dégénéré. On dira qu'il est *presque-dégénéré* s'il existe un k tel qu'il est k-presque dégénéré.

2. Confer par exemple P. Gabriel and M. Zisman *Calculus of fractions and homotopy theory*, Ergebnisse der Mathematik, Band 35, Chapter II, 5.5.

Nous avons donc constaté qu'un p-simplexe dégénéré de $Hom(\Delta[n], X)$ est presque dégénéré. Nous verrons que moyennant des conditions sur X, cet énoncé possède une réciproque.

2.2.1. Nous allons dans ce but préciser la forme des chemins maximaux dans la tranche verticale limitée par les points d'abscisse k et $k+1$. Pour tout triplet α, t, β d'entiers avec $0 \leqslant \alpha \leqslant t \leqslant \beta \leqslant n$, soit $b_{\alpha t \beta}$ l'unique chemin du réseau $[p] \times [n]$ passant par les points $(k, \alpha), (k, t), (k+1, t)$ et $(k+1, \beta)$ et soit $l_{\alpha\beta}$ l'unique chemin du réseau $[p-1] \times [n]$ passant par les points (k, α) et (k, β). Si $k = 0$, on prend $\alpha = 0$, si $k = p-1$, on prend $\beta = n$.

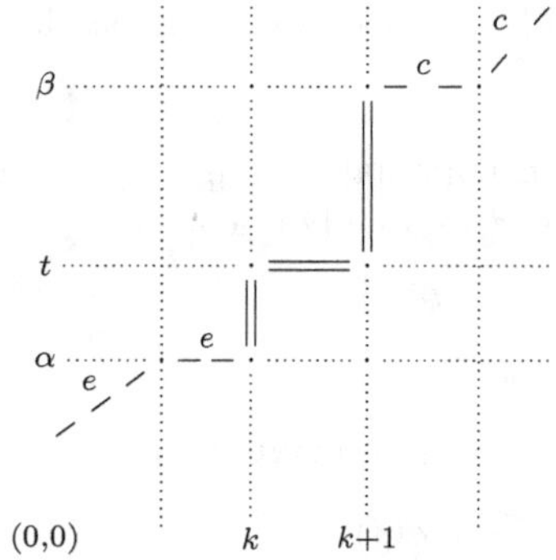

en pointillé : le réseau $[p] \times [n]$, en trait double : $b_{\alpha t \beta}$, en tirets : e et c.

On introduit aussi les ensembles de chemins E_α qui joignent les points (0,0) à (k, α) et qui *arrivent horizontalement* en (k, α) (i.e. qui passent par $(k-1, \alpha)$; ensemble vide si $k = 0$), et C_β qui joignent $(k+1, \beta)$ à (p, n) et qui *partent horizontalement* de $(k+1, \beta)$ (i.e. qui passent par $(k+2, \beta)$; ensemble vide si $k = p-1$). Enfin à $c \in C_\beta$, on associe le chemin c^- de $[p-1] \times [n]$, défini par $c^-(i) = c(i) - (1, 0)$.

Ces notations étant fixées, il est clair que tout chemin maximal $a \in \mathcal{C}$ s'écrit d'une et d'une seule manière comme une somme $a = e + b_{\alpha t \beta} + c$, avec $e \in E_\alpha$ et $c \in C_\beta$, l'addition $+$ désignant la concaténation : $a(i) = e(i)$ pour $i \leqslant k + \alpha$, $a(i) = b_{\alpha t \beta}(i - k - \alpha)$ pour $k + \alpha \leqslant i \leqslant k + 1 + \beta$ et $a(i) = c(i - k - 1 - \beta)$ pour $k + 1 + \beta \leqslant i$. Par ailleurs, la relation de 2.2. s'écrit maintenant

$$h(e + b_{\alpha t \beta} + c) = s_{k+t}\, g(e + l_{\alpha\beta} + c^-)$$

pour $h = s_k g$.

2.2.2. Le lemme principal.

Lemme 4. Soient X un ensemble simplicial régulier, et f un simplexe k-presque dégénéré de $Hom(\Delta[n], X)$. Alors il existe un simplexe g tel que $f = s_k g$. Les simplexes presque dégénérés de $Hom(\Delta[n], X)$ sont donc dégénérés.

Démonstration : Soit f un p-simplexe de $Hom(\Delta[n], X)$ presque dégeneré. Il existe donc un entier k, $0 \leqslant k \leqslant p-1$, tel que pour tout $0 \leqslant j \leqslant n$, le 1-simplexe de X égal au composé $\Delta[1] \longrightarrow \Delta[p] \times \Delta[n] \xrightarrow{f} X$ (où la première flèche est définie par $0 \mapsto (k,j)$, $1 \mapsto (k+1,j)$) est dégénéré. Soit a un chemin maximal que l'on écrit $a = e + b_{\alpha t \beta} + c$. L'arête $X(\varphi(k+t, k+t+1))f(a)$ est donc dégénérée. Puisque X est régulier, cela signifie que l'on a

$$f(a) = s_{k+t} y_t$$

pour un certain $y_t \in X$. Si $\alpha < t \leqslant \beta$, choisissons maintenant $a' = e + b_{\alpha(t-1)\beta} + c$. Il vient de même $f(a') = s_{k+t-1} y_{t-1}$. Comme a et a' ne diffèrent que sur le carré formé par les deux verticales du réseau, d'abscisse k et $k+1$ d'une part et les deux horizontales d'ordonnée $t-1$ et t d'autre part, les relations de compatibilité indiquent que l'on a $d_{k+t} f(a) = d_{k+1+t-1} f(a')$ ce qui impose

$$y_t = y_{t-1}.$$

Les y_t sont donc indépendants de t, et tous égaux à $d_{k+\beta+1} f(e + b_{\alpha\beta\beta} + c)$, ne dépendant que de e, c, α, β. On les notera $g(e + l_{\alpha\beta} + c^-)$. Remarquons que si on a $\alpha = \beta$, alors le chemin $b_{\alpha\beta\beta}$ est réduit à une arête et le chemin $l_{\alpha\beta}$ à un point.

Ainsi, à tout chemin maximal $e + l_{\alpha\beta} + c^-$ du réseau $[p-1] \times [n]$, nous avons associé le $(p+n-1)$-simplexe $g(e + l_{\alpha\beta} + c^-)$ de X. Par construction, on a $f(e + b_{\alpha t\beta} + c) = s_{k+t} g(e + l_{\alpha\beta} + c^-)$ et il reste à vérifier que les relations de compatibilité sont satisfaites par les $g(e + l_{\alpha\beta} + c^-)$. Soient m et m' deux chemins maximaux du réseau $[p-1] \times [n]$ qui ne diffèrent que sur le carré passant par les deux points (i,j) et $(i+1, j+1)$. On doit avoir $d_{i+j+1} g(m) = d_{i+j+1} g(m')$. Il y a quatre cas, selon que $i < k-1, i = k-1, i = k$ et $i > k$. Les deux cas extrêmes sont évidents. Traitons par exemple le cas $i = k$. On a $m = e + l_{\alpha\beta} + c^-$ et $m' = e + l_{\alpha(\beta-1)} + c'^-$, avec $c'(0) = (k+1, \beta-1), c'(1) = (k+2, \beta-1)$ et $c'(r+1) = c(r)$ pour $r > 1$. Nous devons vérifier l'égalité

$$d_{k+\beta} d_{k+\beta+1} f(e + b_{\alpha\beta\beta} + c) = d_{k+\beta} d_{k+\beta} f(e + b_{\alpha(\beta-1(\beta-1)} + c')$$

ce qui ne pose aucune difficulté : c'est exactement la compatibilité des deux $(p+n)$-simplexes qui figurent de part et d'autre de l'égalité.

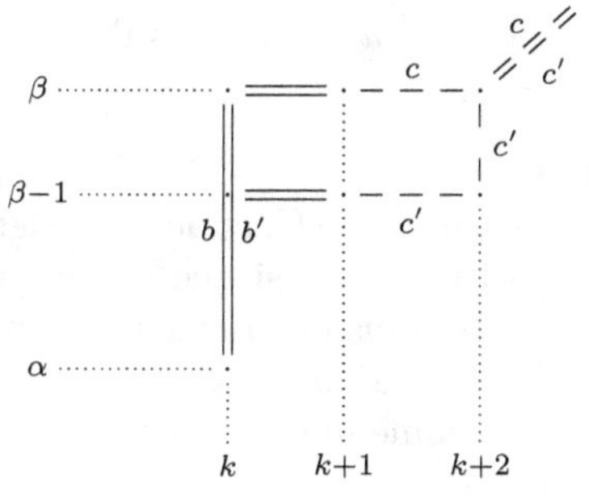

on a écrit b au lieu de $b_{\alpha\beta\beta}$ et b' au lieu de $b_{\alpha(\beta-1)(\beta-1)}$.

3. Les théorèmes.

3.1. La proposition suivante est un cas particulier du théorème 1 :

Proposition 2. Si X est un ensemble simplicial régulier de dimension q, alors $Hom(\Delta[n], X)$ est de dimension au plus $(n+1)q$.

Démonstration : Soit f un p-simplexe de $Hom(\Delta[n], X)$, i.e. un morphisme $\Delta[p] \times \Delta[n] \to X$. Pour tout $0 \leqslant j \leqslant n$, soit $x_j : [p] \to [p] \times [n]$ le p-simplexe de $\Delta[p] \times \Delta[n]$ défini par $i \mapsto (i, j)$. Son image par f est un p-simplexe de X. On dira que l'arête $((i, j), (i+1, j))$ du réseau $[p] \times [n]$ est *efficace* si l'arête élémentaire $X(\varphi(i, i+1))f(x_j)$ est non dégénérée. Supposons maintenant que X est de dimension finie q, et prenons $p > q$: sous ces conditions le simplexe $f(x_j)$ est dégénéré; le lemme 1 nous dit que ce simplexe possède au plus q arêtes efficaces. Comme j prend $n+1$ valeurs distinctes, le réseau possède au plus $(n+1)q$ arêtes efficaces distinctes. Choisissons donc $p > (n+1)q$, ce qui nous assure de l'existence d'un entier k tel que, pour tout $j = 0, \ldots n$, l'arête $((k, j), (k+1, j))$ est non efficace. En d'autres termes, le simplexe f est presque dégénéré. D'après le lemme 4, il est donc dégénéré.

3.1.1. La borne $(n+1)q$ est la meilleure possible puisque $Hom(\Delta[n], \Delta[q])$ est de dimension $(n+1)q$. En effet les p-simplexes de $Hom(\Delta[n], \Delta[q])$ sont les applications croissantes $f : [p] \times [n] \to [q]$ et ce simplexe est dégénéré si et seulement si f prend les mêmes valeurs sur deux colonnes voisines du réseau $[p] \times [n]$. Soit alors $f : [(n+1)q] \times [n] \to [q]$ l'application définie par les égalités suivantes, où j parcourt les valeurs $0 \leqslant j \leqslant n$, et où l'on a écrit $i = kq + a$ pour $i > 0$, $k \in \{0, \ldots, n\}$ et $1 \leqslant a \leqslant q$: on pose $f(0, j) = 0$ et, pour $i > 0$,

$$f(i,j) = \begin{cases} 0 & \text{si} \quad j < n-k \\ a & \text{si} \quad j = n-k \\ q & \text{si} \quad j > n-k. \end{cases}$$

On définit ainsi un $(n+1)q$-simplexe non dégénéré de $Hom(\Delta[n], \Delta[q])$ qui est donc de dimension au moins q. Montrons que la dimension est exactement $(n+1)q$. Soit f un simplexe non dégénéré de degré p. Pour tout $i \in [p]$, il existe au moins un $j \in [n]$ tel que $f(i, j) < f(i+1, j)$. Donc, si $\sigma(i)$ désigne la somme $f(i, 0) + f(i, 1) + \cdots + f(i, n)$ des valeurs prises par la fonction sur la i-ème colonne du réseau, il vient $\sigma(i) < \sigma(i+1)$ et finalement $\sigma(p) \geqslant p$. Mais par ailleurs il est clair que l'on a, pour tout i, $\sigma(i) \leqslant (n+1)q$, puisqu'il y a $n+1$ termes dans chaque colonne du réseau. Donc il vient $p \leqslant \sigma(p) \leqslant (n+1)q$.

3.1.2. *Démonstration du théorème 1.* On écrit $U = \operatorname{colim}_u \Delta[|u|]_x$ où u parcourt l'ensemble des simplexes non dégénérés de U, et où $|u|$ désigne le degré de u. L'égalité donnée dans (1.1) et la proposition 2, ainsi que les propriétés élémentaires de la dimension permettent de conclure. Plus précisément, soient A et B deux ensembles simpliciaux de dimension finie. On a dim $(A \times B) =$ dim $A +$ dim B, et si $A \subset B$ alors il vient dim $A <$ dim B. Comme une limite finie n'est autre qu'un sous objet d'un produit fini, nous obtenons :

Théorème 1bis. Soient X un ensemble simplicial de dimension finie et U un ensemble simplicial fini. On a :

$$\dim Hom(U, X) \leqslant \sum_u (|u| + 1).\dim X$$

où, dans la somme, u parcourt l'ensemble des simplexes non dégénérés de U.

3.2. Partant d'un ensemble simplicial régulier X qu'en est-il de $Hom(U, X)$? Le théorème 2 répond à la question :

Théorème 2. Soient X un ensemble simplicial régulier et U un ensemble simplicial quelconque. Alors $Hom(U, X)$ est régulier.

Démonstration : Utilisant la remarque 1.1. et la proposition 1, nous voyons qu'il suffit de démontrer le théorème dans le cas où $U = \Delta[n]$. Posons $Y = Hom(\Delta[n], X)$. Soit $f \in Y_p$, et supposons que $Y(\varphi(k, k+1))f$ est dégénéré. Cela signifie qu'il existe un h qui fait commuter le diagramme suivant :

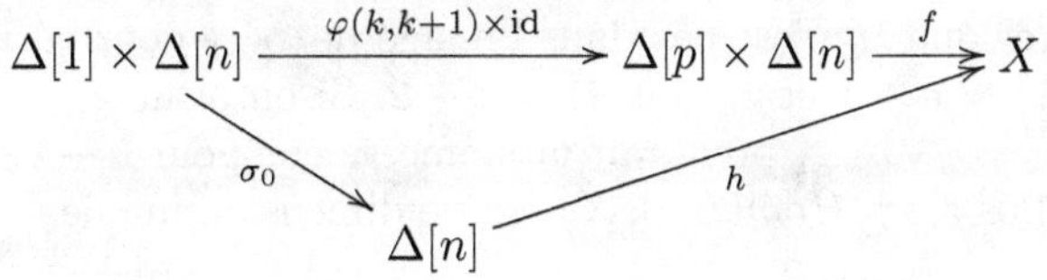

Mais alors pour tout $j \in \{0, \cdots, n\}$, les 1-simplexes $((k, j), (k+1, j))$ ne sont pas efficaces. Comme dans la démonstration du théorème 1, cela signifie que f est k-presque dégénéré. Le lemme 4 implique qu'il existe g tel que $f = s_k g$. Donc Y vérifie P_1.

Lorsque l'on se restreint aux ensembles simpliciaux finis, on obtient (confer proposition 1) :

Théorème 3. La catégorie $\mathcal{S}_{f-\text{reg}}$ est stable par limites finies et cartésienne fermée. Elle est aussi stable par sous-objets et sommes finies (et même réunions finies).

4. Quelques compléments. Disons qu'un ensemble simplicial X est fortement fini si $Hom(U, X)$ est fini pour tout ensemble simplicial fini U.

On peut montrer que tous les quotients de $\Delta[2]$ sont fortement finis. Il est probable qu'on peut en déduire que les ensembles simpliciaux finis de dimension 2 sont fortement finis.

Notons F_i la i-ème face de $\Delta[q]$.Voici un résultat qui généralise légèrement celui de Lurie annoncé au début :

Proposition 3. Soit $F \subset \dot{\Delta}[q]$ $(q \geqslant 3)$ une réunion de faces qui contient $F_a \cup F_{a+1}$ pour un certain a, $0 < a < q-1$). Alors $X = \Delta[q]/F$ n'est pas fortement fini.

Démonstration (Lurie) : Introduisons la fonction coupe définie sur $\mathbb{Z}$ et à valeurs dans $\mathbb{N}$ par

$$\text{coupe}(i) = \begin{cases} 0 & \text{si} \quad i \leqslant 0 \\ i & \text{si} \quad 0 \leqslant i \leqslant q \\ q & \text{si} \quad q \leqslant i \end{cases} .$$

Soit $p > q$ un entier, soit $0 \leqslant u \leqslant p$ et soit z_u le $(p+1)$-simplexe de $\Delta[q]$, i.e. le morphisme $[p+1] \to [q]$ de la catégorie Δ, défini par $z_u(i) = \text{coupe}(i-u+a)$. Il est clair que $z_u \circ \partial_u$ ne prend pas la valeur a et que $z_u \circ \partial_{u+1}$ ne prend pas la valeur $a+1$. Ainsi $d_u z_u$ et $d_{u+1} z_u$ sont des simplexes de F. Les $p+1$ simplexes $\bar{z}_u$, images des précédents dans le quotient X vérifient donc les relations de compatibilité $d_{u+1}\bar{z}_u = d_{u+1}\bar{z}_{u+1}$ et définissent ainsi un p-simplexe de $Hom(\Delta[1], X)$. Ce simplexe n'est jamais dégénéré. En effet (d'après 2.2., ou un raisonnement direct), la condition pour qu'un p-simplexe f de $Hom(\Delta[1], Z)$, donné par $p+1$ simplexes f_u, $u \in \{0, \dots, p\}$, de degré $p+1$ de Z, soit le dégénéré $s_k g$ d'un simplexe g donné par des $g_v \in Z_p$, est

$$f_u = s_{k+1} g_u \quad \text{si} \quad u \leqslant k \quad \text{et} \quad f_u = s_k g_{u-1} \quad \text{si} \quad u > k.$$

Il suffit de montrer que $\bar{z}_k$ n'est pas dans l'image de s_{k+1} pour vérifier l'assertion. Or on a $z_u(u+1) = a+1$ et $z_u(u+2) = a+2$. Si on avait $\bar{z}_k = s_{k+1}\bar{y}$ pour un certain y, alors $z_k = y \circ \sigma_{k+1}$ prendrait la même valeur pour $k+1$ et $k+2$, ce qui est impossible. Mais alors $Hom(\Delta[1], X)$ est de dimension infinie.

Remarque. Qu'en est-il de $\Delta[q]/F_i$? La question reste ouverte. Je ne sais pas non plus si l'on peut remplacer dans le théorème 2, régulier par fortement régulier. C'est vrai pour $X = \Delta[q]$, d'après un raisonnement analogue à celui de (3.1.1.) utilisant la croissance de σ.

This article may be accessed via WWW at `http ://www.rmi.acnet.ge/jhrs/`

Michel Zisman
zisman@math.jussieu.fr

Université Paris 7
Paris, France

Journal of Homotopy and Related Structures, vol. 4(1), 2009, pp.265–273

POPATHS AND HOLINKS

DAVID A. MILLER

(*communicated by Pascal Lambrechts*)

Abstract

In the study of stratified spaces it is useful to examine spaces of popaths (paths which travel from lower strata to higher strata) and holinks (those spaces of popaths which immediately leave a lower stratum for their final stratum destination). It is not immediately clear that for adjacent strata these two path spaces are homotopically equivalent, and even less clear that this equivalence can be constructed in a useful way (with a deformation of the space of popaths which fixes start and end points and where popaths instantly become members of the holink). The advantage of such an equivalence is that it allows a stratified space to be viewed categorically because popaths, unlike holink paths (which are easier to study), can be composed. This paper proves the aforementioned equivalence in the case of Quinn's homotopically stratified spaces [**1**].

1. Introduction

There are many different notions of a stratified space. One such notion is that of a homotopically stratified space. These were introduced by Frank Quinn. Here the strata are related by "homotopy rather than geometric conditions" [**1**]. This makes them ideal for studying the topology of stratified spaces. Two such tools for studying that topology are holinks and popath spaces.

The popaths between two strata are any paths which travel from one stratum to the other passing only from "lower strata to higher strata". On the other hand the holink between two strata consists only of popaths which instantly leave one stratum for the other.

Popaths are very useful in obtaining a categorical view of stratified spaces. However for a space with many strata the space of popaths between two strata could be difficult to compute or visualize. The holink between two strata only depends on the two strata involved and with this in mind is easier to deal with, but problems

---

I would like to take this opportunity to acknowledge the continuing guidance and assistance of my PhD supervisor Michael Weiss for which I am very grateful. I would also like to thank Jon Woolf for many useful comments and correspondences.

Received August 30, 2008, revised November 10, 2008; published on September 24, 2009.

2000 Mathematics Subject Classification: 54E20 (Primary), 55R65 (Secondary).

Key words and phrases: stratified spaces, homotopy link.

may arises because holink paths can't be composed. Therefore a result connecting these concepts becomes desirable.

The result obtained here is that for the space of popaths and the holink between two fixed strata there exists a homotopy h : popaths $\times I \to$ popaths which fixes the start and end points of paths, image$\{h_s\} \subset$ holink when $s \in (0,1]$ and $h_0 =$ identity. It may seem strange to require a result which is stronger than just the inclusion being a homotopy equivalence, but exactly this result has already found a relevance in the work of Jon Woolf [**2**] and seems to be required to construct a particular map to prove that a stratified space can be reconstructed from its popath category (the author hopes to show this in the near future).

2. Homotopically Stratified Spaces

Definition 2.1. A topological space X is **filtered** if there are designated closed subspaces X^i indexed by a finite poset S_X such that $X = \bigcup_{i \in S_X} X^i$ and $X^j \subseteq X^i \Leftrightarrow j \leqslant i$. The **strata** are defined as path connected components of $X_i = X^i - \bigcup_{j<i} X^j$. If $j < i$ we may say X_i is a higher stratum than X_j or equivalently X_j is a lower stratum than X_i.

Definition 2.2. A closed subspace K of a filtered space is said to be **pure** if it is a union of strata.

Definition 2.3. Assume W is a subspace of a filtered space and W contains the distinct strata X_b and X_a. Let the space of **popaths** from X_a through W to X_b be denoted by $\text{pop}(X_b, W, X_a)$ and defined as the space (with the compact open topology) of all order preserving paths $\omega : [0, T_\omega] \to W$ such that $\omega(0) \in X_a$, $\omega(T_\omega) \in X_b$. Here order preserving means if $t_1 \leqslant t_2$ and $\omega(t_1) \in X_j$ while $\omega(t_2) \in X_i$ then $j \leqslant i$ (meaning ω cannot flow from higher strata into lower strata).

Definition 2.4. For a space S and subspace $Y \subset S$ the **holink** (also called **homotopy link**) between Y and $S - Y$ denoted $\text{hol}(S, Y)$ is defined as the space (with the compact open topology) of paths $\omega : [0, T_\omega] \to S$ where $\omega(0) \in Y$ and $\omega(t) \in S - Y$ for $t \in (0, T_\omega]$.

Remark 2.5. It should be clear that $\text{hol}(X_b \cup X_a, X_a)$ is the subspace of $\text{pop}(X_b, X, X_a)$ consisting of popaths which immediately leave X_a and travel straight into X_b.

Definition 2.6. Suppose K, L are unions of strata in X and $L \subseteq K$. Then L is said to be **tame** in K if there is a neighborhood N of L in K and a nearly stratum preserving strong deformation retraction r of N onto L. Here **nearly stratum preserving** means points of $K - L$ remain in the same stratum until the last moment when they get pushed into L.

Definition 2.7. Let $R : N - L \to \text{hol}(N, L)$ be defined by $R(x)(t) = r(x, 1-t)$ for all $x \in N - L$.

Definition 2.8. A filtered metric space X is a **homotopically stratified space** if for every $j < i$, X_j is tame in $X_j \cup X_i$ and the map from $\text{hol}(X_i \cup X_j, X_j)$ to X_j given by evaluation at the start point is a fibration.

3. Popaths and Holinks in the 2 Strata Case

Definition 3.1. Let $\kappa : \mathrm{pop}(X_b, X_b \cup X_a, X_a) \to \mathbb{R}$ be the map which sends $\omega \in \mathrm{pop}(X_b, X_b \cup X_a, X_a)$ to the unique point, $\kappa(\omega)$, in $\mathbb{R}$ such that $\omega(t) \in X_a$ for $t \leqslant \kappa(\omega)$ and $\omega(t) \in X_b$ for $t > \kappa(\omega)$. Note κ is not a continuous map, but it is upper-continuous. We define upper-continuity as meaning for any $\omega \in \mathrm{pop}(X_b, X_b \cup X_a, X_a)$ and any neighborhood V of $\kappa(\omega)$ having the form $[0, r)$ there exists a neighborhood U of ω such that $f(U) \subset V$.

Lemma 3.2. Let N be a neighborhood of tameness for X_a in $X_a \cup X_b$ where X_a, X_b are distinct. There exists a continuous map $\lambda : \mathrm{pop}(X_b, X_b \cup X_a, X_a) \to \mathbb{R}$ that for any popath $\omega : [0, T_\omega] \to X$ satisfies:

1. $\lambda(\omega) \in (\kappa(\omega), T_\omega)$
2. $\omega(t) \in N - X_a$ for $t \in (\kappa(\omega), \lambda(\omega)]$

Remark 3.3. This will be useful when we wish to use the κ map but cannot because it is not continuous.

Proof. Since X is metric, $\mathrm{pop}(X_b, X_b \cup X_a, X_a)$ is also metric and so paracompact. Therefore by a partition of unity type argument it suffices to show it is true locally. Fix $\sigma \in \mathrm{pop}(X_b, X_b \cup X_a, X_a)$, clearly we can choose a point Γ in $[0, T_\sigma]$ which satisfies the conditions of $\lambda(\sigma)$. Now since κ is upper continuous the same value Γ satisfies the conditions to be $\lambda(\omega)$ for all ω within a small enough neighborhood of σ. Hence the lemma holds locally and therefore holds globally. □

Definition 3.4. Let E denote the space

$$\{(\omega, t) \in \mathrm{pop}(X_b, X_b \cup X_a, X_a) \times \mathbb{R} : t \in (\kappa(\omega), \lambda(\omega)]\}$$

and p denote the canonical map $p : E \to \mathrm{pop}(X_b, X_b \cup X_a, X_a)$. In fact p is a fiber bundle homeomorphic to the trivial bundle with total space $\mathrm{pop}(X_b, X_b \cup X_a, X_a) \times (0, 1]$. A trivialisation is given by $(\omega, t) \mapsto (\omega, \beta_\omega(t))$, where $\beta_\omega(t)$ is the unique member of $(0, 1]$ such that

$$\int_{\kappa(\omega)}^{t} \mathrm{dist}_{X_a}(\omega(t'))dt' = \beta_\omega(t) \int_{\kappa(\omega)}^{\lambda(\omega)} \mathrm{dist}_{X_a}(\omega(t'))dt'.$$

Lemma 3.5. The inclusion map i induces a homotopy equivalence

$$\mathrm{hol}(X_b \cup X_a, X_a) \simeq \mathrm{pop}(X_b, X_b \cup X_a, X_a).$$

Furthermore it has a homotopy inverse $\varphi : \mathrm{pop}(X_b, X_b \cup X_a, X_a) \to \mathrm{hol}(X_b \cup X_a, X_a)$ where there exists a homotopy h from the identity map to $i \circ \varphi$ which fixes the start and end points of paths and $h_s(\omega) \in \mathrm{hol}(X_b \cup X_a, X_a)$ for all popaths ω when $s \in (0, 1]$.

Remark 3.6. This lemma means in the two strata case we have a continuous way of deforming the space of popaths into the holink so that popaths instantly become holink paths.

Proof. We will in fact directly construct the maps $h_s : \mathrm{pop}(X_b, X_b \cup X_a, X_a) \to \mathrm{pop}(X_b, X_b \cup X_a, X_a)$ which fix start and end points, have image in $\mathrm{hol}(X_b \cup X_a, X_a)$ when $s \in (0,1]$ and where h_0 is the identity map. Then by setting $\varphi = h_1$ we have proved the lemma.

Let N be a neighborhood of tameness for X_a in $X_a \cup X_b$ and r be the corresponding strong deformation retraction. Define a map $F : \mathrm{pop}(X_b, X_b \cup X_a, X_a) \times (0,1] \times I \to X_a$ by for all $\omega \in \mathrm{pop}(X_b, X_b \cup X_a, X_a)$ sending (ω, s, t) to $r\left(\omega(t \cdot \beta_\omega^{-1}(s)), 1\right)$. Define another map $G : \mathrm{pop}(X_b, X_b \cup X_a, X_a) \times (0,1] \to \mathrm{hol}(X_b \cup X_a, X_a)$ by $\omega \mapsto R\left(\omega(\beta_\omega^{-1}(s))\right)$ (see Definition 1.7). The definition of Quinn stratified spaces tells us the map E_0, evaluating a holink path at its start point, is a fibration. So there is a lift $\widetilde{F}$,

$$\begin{array}{ccc} \mathrm{pop}(X_b, X_b \cup X_a, X_a) \times (0,1] \times \{1\} & \xrightarrow{G} & \mathrm{hol}(X_b \cup X_a, X_a) \\ \downarrow & \overset{\widetilde{F}}{\nearrow} & \downarrow E_0 \\ \mathrm{pop}(X_b, X_b \cup X_a, X_a) \times (0,1] \times I & \xrightarrow{F} & X_a \end{array}$$

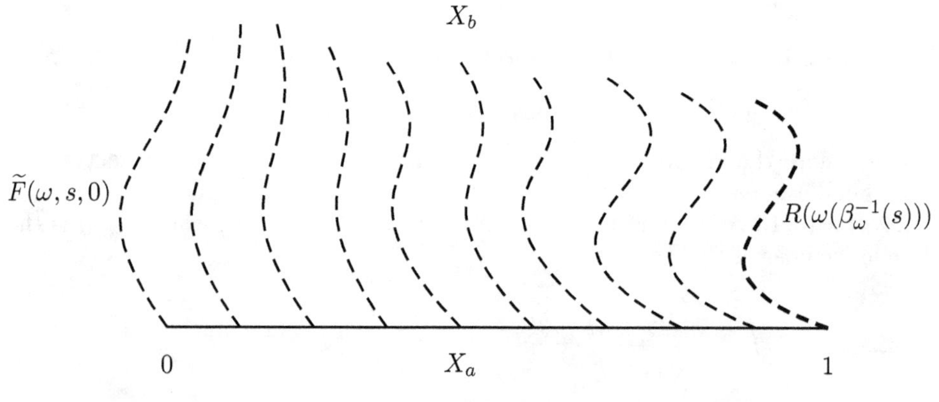

Now we can use $\widetilde{F}$ to construct to h. When $s = 0$, h_s is the identity and when $s \in (0,1]$

$$h_s(\omega)(t) = \begin{cases} \widetilde{F}\left(\omega, s, \frac{t + s\beta_\omega^{-1}(s)}{\beta_\omega^{-1}(s)}(1-s)\right)\left(\frac{t}{\beta_\omega^{-1}(s)} s\right) & 0 \leqslant t \leqslant (1-s)\beta_\omega^{-1}(s) \\ \widetilde{F}\left(\omega, s, \frac{t}{\beta_\omega^{-1}(s)}\right)\left(\frac{s(s-1)(t-\beta_\omega^{-1}(s)) + (t-(1-s)\beta_\omega^{-1}(s))}{s\beta_\omega^{-1}(s)}\right) & (1-s)\beta_\omega^{-1}(s) < t \leqslant \beta_\omega^{-1}(s) \\ \omega(t) & \beta_\omega^{-1}(s) \leqslant t \leqslant T_\omega \end{cases}$$

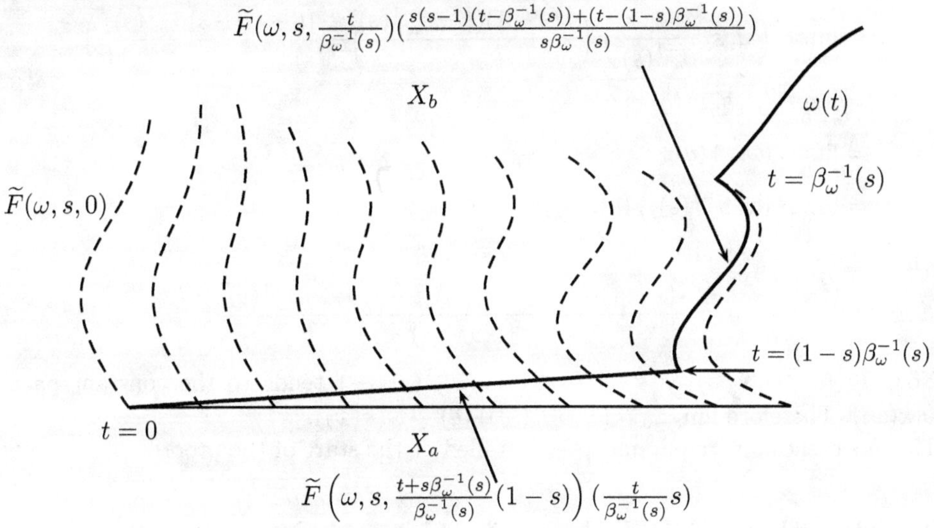

Clearly this is continuous within the three intervals for $s \in (0,1]$. To see it is continuous at $t = (1-s)\beta_\omega^{-1}(s)$ just substitute for t and get $\widetilde{F}(\omega, s, 1-s)(s(1-s))$ for both expressions. To see it is continuous at $t = \beta_\omega^{-1}(s)$ substitute into the second expression to get $\widetilde{F}(\omega, s, 1)(1)$, now

$$\widetilde{F}(\omega, s, 1)(1) = G(\omega, s)(1) = R\left(\omega(\beta_\omega^{-1}(s))\right)(1) = \omega(\beta_\omega^{-1}(s))$$

so it is continuous at $t = \beta_\omega^{-1}(s)$.

To prove $h_s \to Id$ as $s \to 0$ we will prove it is true in each of the three intervals in the definition of h_s (in the last interval there is nothing to prove).

In the first interval

$$\begin{aligned}
&\lim_{s\to 0} \widetilde{F}\left(\omega, s, \frac{t+s\beta_\omega^{-1}(s)}{\beta_\omega^{-1}(s)}(1-s)\right)(\frac{t}{\beta_\omega^{-1}(s)}s) \\
&= \lim_{s\to 0} \widetilde{F}\left(\omega, s, \frac{t}{\beta_\omega^{-1}(s)}\right)(0) \\
&= \lim_{s\to 0} F\left(\omega, s, \frac{t}{\beta_\omega^{-1}(s)}\right) \\
&= \lim_{s\to 0} r\left(\omega(\frac{t}{\beta_\omega^{-1}(s)}\beta_\omega^{-1}(s))\right), 1) \\
&= \lim_{s\to 0} r(\omega(t)), 1),
\end{aligned}$$

and since we are within $0 \leqslant t \leqslant (1-s)\beta_\omega^{-1}(s)$ which tends towards $0 \leqslant t \leqslant \kappa(\omega)$ as s tends to 0 then $r(\omega(t)), 1)$ tends to $\omega(t)$ as s tends to 0 for $0 \leqslant t \leqslant (1-s)\beta_\omega^{-1}(s)$.

The second interval tends to $t = \lim_{s\to 0} \beta_\omega^{-1}(s) = \kappa(\omega)$ as s tends to 0 and

$$\begin{aligned}
&\lim_{s\to 0} \widetilde{F}\left(\omega, s, \frac{t}{\beta_\omega^{-1}(s)}\right)\left(\frac{s(s-1)(t-\beta_\omega^{-1}(s)) + (t-(1-s)\beta_\omega^{-1}(s))}{s\beta_\omega^{-1}(s)}\right) \\
&= \lim_{s\to 0} \widetilde{F}(\omega, s, 1)(q) \\
&= \lim_{s\to 0} G(\omega, s)(q) \\
&= \lim_{s\to 0} R\left(\omega(\beta_\omega^{-1}(s))\right)(q),
\end{aligned}$$

where

$$q = \frac{s(s-1)(t-\beta_\omega^{-1}(s)) + (t-(1-s)\beta_\omega^{-1}(s))}{s\beta_\omega^{-1}(s)}.$$

Now $\beta_\omega^{-1}(s) \to \kappa(\omega)$ as $s \to 0$ so $r\left(\omega(\beta_\omega^{-1}(s)), -\right)$ tends to the constant path $\omega(\kappa(\omega))$. Therefore $\lim_{s\to 0} R\left(\omega(\beta_\omega^{-1}(s))\right)(q) = \omega(\kappa(\omega))$.
This h satisfies our requirements as detailed at the start of the proof. □

4. Popaths and Holinks in the General Case

In a slight abuse of notation we will throughout this section redefine and use symbols like κ, N, λ, E and β in a more general context.

Definition 4.1. Consider X_a, X_b to be distinct strata in a homotopically stratified space X. Let $\kappa : \mathrm{pop}(X_b, X, X_a) \to \mathbb{R}$ be the map which sends $\omega \in \mathrm{pop}(X_b, X, X_a)$ to the unique point, $\kappa(\omega)$, in $\mathbb{R}$ such that $\omega(t) \in X_a$ for $t \leqslant \kappa(\omega)$ and $\omega(t) \notin X_a$ for $t > \kappa(\omega)$. Note κ is not a continuous map, but as in Definition 2.1 it is upper-continuous.

Lemma 4.2. Suppose X is a homotopically stratified space and $K \subset X$ is pure. Then there is a nearly stratum preserving strong deformation retract r of a neighborhood N of K in X to K. The neighborhood may be referred to as a neighborhood of tameness.

Proof. This is the first part of Proposition 3.2 of "Homotopically Stratified Sets" by Frank Quinn [**1**]. □

Lemma 4.3. Consider X_a as a lowest possible stratum in X by if necessary discarding any lower strata (which are of no consequence when considering $\mathrm{pop}(X_b, X, X_a)$). Let N be a neighborhood of tameness for X_a in X. There exists a continuous map $\lambda : \mathrm{pop}(X_b, X, X_a) \to \mathbb{R}$ that for all popaths $\omega : [0, T_\omega] \to X$ satisfies:

1. $\lambda(\omega) \in (\kappa(\omega), T_\omega)$
2. $\omega(t) \in N - X_a$ for $t \in (\kappa(\omega), \lambda(\omega)]$

Remark 4.4. Again this will be useful when we wish to use the κ map but cannot because it is not continuous.

Proof. This is proved in exactly the same way as Lemma 2.2. □

Definition 4.5. Let E denote the space $\{(\omega, t) \in \text{pop}(X_b, X, X_a) \times \mathbb{R} : t \in (\kappa(\omega), \lambda(\omega)]\}$ and p denote the canonical map $p : E \to \text{pop}(X_b, X, X_a)$. In fact p is a fiber bundle homeomorphic to the trivial bundle with total space $\text{pop}(X_b, X, X_a) \times (0, 1]$. Let β_ω denote a reparametrization of $(\kappa(\omega), \lambda(\omega)]$ to $(0, 1]$ giving a trivialization. The trivialisation can be obtained in the same way as in Definition 2.4.

Definition 4.6. Let B be a path space and A a subspace of B. We will say the inclusion $A \subset B$ is a special path inclusion if there exists a homotopy $h : B \times I \to B$ which fixes the start and end points of paths, $\text{image}\{h_s\} \subset A$ when $s \in (0, 1]$ and $h_0 = \text{identity}$. Note this implies the inclusion map is a homotopy equivalence. Also note the composition (not concatenation) of two special path inclusions is again a special path inclusion.

Remark 4.7. Lemma 2.5 proves that $\text{hol}(X_b \cup X_a, X_a) \subset \text{pop}(X_b, X_b \cup X_a, X_a)$ is a special path inclusion. The aim of this paper is to show that $\text{hol}(X_b \cup X_a, X_a) \subset \text{pop}(X_b, X, X_a)$ is a special path inclusion for any two strata X_a and X_b of a homotopically stratified space X.

Lemma 4.8. Consider X_b as any stratum in X and X_a as a lowest possible stratum in X by if necessary discarding any lower strata. If $\text{hol}(X_b \cup X_a, X_a) \subset \text{pop}(X_b, X, X_a)$ is a special path inclusion then

$$\{\sigma \in \text{pop}(X_b, X, X) : \sigma(\delta) \notin X_a \text{ for all } \delta > 0\} \subset \text{pop}(X_b, X, X)$$

is also a special path inclusion.

Proof. First let us show $\{\sigma \in \text{pop}(X_b, X, N) : \sigma(\delta) \notin X_a \text{ for all } \delta > 0\} \subset \text{pop}(X_b, X, N)$ is a special path inclusion for N a neighborhood of tameness of X_a in X.
Given a path $\sigma \in \text{pop}(X_b, X, N)$ define $\sigma^+ \in \text{pop}(X_b, X, X_a)$ by

$$\sigma^+(t) = \begin{cases} r(\sigma(0), 1-t) & 0 \leqslant t \leqslant 1 \\ \sigma(t-1) & 1 \leqslant t \leqslant T_\omega + 1 \end{cases}.$$

Let f be the map from $\text{pop}(X_b, X, X_a) \times I$ to Moore paths onto I defined by

$$f(\omega, s)(t) = \begin{cases} 0 & 0 \leqslant t \leqslant 1 \\ (\min\{1, (t-1)(\lambda(\omega) - t)\})\, s & 1 \leqslant t \leqslant \lambda(\omega) \\ 0 & \lambda(\omega) \leqslant t \leqslant T_\omega \end{cases}$$

for all $\omega \in \text{pop}(X_b, X, X_a)$ where $1 < \lambda(\omega)$ and defined by $f(\omega, s)(t) = 0$ for all $t \in [0, T_\omega]$ if $\lambda(\omega) \leqslant 1$.
Now define a suitable homotopy for special path inclusion $g : \text{pop}(X_b, X, N) \times I \to \text{pop}(X_b, X, N)$ by $g_s(\sigma)(t) = \left(h_{f(\sigma^+, s)(t+1)}\sigma^+\right)(t+1)$ where h is a homotopy for the special path inclusion $\text{hol}(X_b \cup X_a, X_a) \subset \text{pop}(X_b, X, X_a)$. Intuitively this can be thought of as concatenating a path $[0, 1] \to X$ to the start of σ then manipulating the $(1, \lambda(\sigma))$ part of the new path away from X_a using h and finally removing the $[0, 1)$ part that was added.
Now since for all s, $g_s(\sigma)(t) = \sigma(t)$ when $t \geqslant t_0$ for any t_0 where $\sigma(t_0) \notin N$ we can extend g to give a special path inclusion $\{\sigma \in \text{pop}(X_b, X, X) : \sigma(\delta) \notin X_a \text{ for all } \delta > 0\} \subset \text{pop}(X_b, X, X)$ by setting $g_s(\sigma)$ as σ when $\sigma \in \text{pop}(X_b, X, X - N)$. □

Theorem 4.9. $\mathrm{hol}(X_b \cup X_a, X_a) \subset \mathrm{pop}(X_b, X, X_a)$ *is a special path inclusion for any two distinct strata* X_a *and* X_b *of a homotopically stratified space* X.

Proof. In the case when X only has two strata then $X = X_a \cup X_b$ and the proposition is proved as Lemma 2.5. Likewise if there are no other strata which popaths from X_a to X_b can pass through then $\mathrm{pop}(X_b, X, X_a) = \mathrm{pop}(X_b, X_a \cup X_b, X_a)$ and the proposition is again proved by Lemma 2.5. Therefore we can assume X_a is a lowest stratum in X and there is at least one other stratum, X_c, which popaths from X_a to X_b may pass through, let X_c denote a lowest stratum of this type.
We will assume inductively that the lemma holds for $\mathrm{hol}(X_b \cup X_c, X_c) \subset \mathrm{pop}(X_b, X, X_c)$ and using this show $\mathrm{pop}(X_b, X - X_c, X_a) \subset \mathrm{pop}(X_b, X, X_a)$ is a special path inclusion. Then the proposition can be proven by removing strata like X_c until we are in the two strata case which has been proved already as Lemma 2.5.
We will define the homotopy we require as a composition (not concatenation) of two homotopies.
Define the first homotopy j by $j_0 =$ identity and for $s \in (0,1]$ and all $\omega \in \mathrm{pop}(X_b, X, X_a)$

$$j_s(\omega)(t) = \begin{cases} r\left(\omega(t \cdot \frac{\beta_\omega^{-1}(s)}{\beta_\omega^{-1}(\frac{1}{2}s)}), 1\right) & 0 \leqslant t \leqslant \beta_\omega^{-1}(\frac{1}{2}s) \\ R\left(\omega(\beta_\omega^{-1}(s))\right)\left(\frac{t - \beta_\omega^{-1}(\frac{1}{2}s)}{\beta_\omega^{-1}(s) - \beta_\omega^{-1}(\frac{1}{2}s)}\right) & \beta_\omega^{-1}(\frac{1}{2}s) \leqslant t \leqslant \beta_\omega^{-1}(s) \\ \omega(t) & \beta_\omega^{-1}(s) \leqslant t \leqslant T_\omega \end{cases}.$$

Intuitively this homotopy retracts the start of the path back into X_a and then travels along the reverse of the tameness retraction of $\omega(\beta_\omega^{-1}(s))$ before continuing along the original path from then onwards.
For the second homotopy k again define $k_0 =$ identity and for $s \in (0,1]$ and all $\omega \in \mathrm{pop}(X_b, X, X_a)$

$$k_s(\omega)(t) = \begin{cases} \omega(t) & 0 \leqslant t \leqslant \beta_\omega^{-1}(\frac{1}{2}s) \\ g_s\left(\omega|_{[\beta_\omega^{-1}(\frac{1}{2}s), T_\omega]}\right)(t) & \beta_\omega^{-1}(\frac{1}{2}s) \leqslant t \leqslant T_\omega \end{cases},$$

where g is the homotopy constructed in the previous lemma with respect to $\{\sigma \in \mathrm{pop}(X_b, X - X_a, X - X_a) | \sigma(\delta) \notin X_c \text{ for all } \delta > 0\} \subset \mathrm{pop}(X_b, X - X_a, X - X_a)$ which can used because of the inductive hypothesis. This homotopy ensures that for $s > 0$, $k_s(\omega)(t) \notin X_c$ for $t \in (\beta_\omega^{-1}(\frac{1}{2}s), T_\omega]$.
Now if we define a homotopy $l : \mathrm{pop}(X_b, X, X_a) \times I \to \mathrm{pop}(X_b, X, X_a)$ by $l_s = j_s \circ k_s$ we get $\mathrm{image}\{l_s\} \subset \mathrm{pop}(X_b, X - X_c, X_a)$ for $s > 0$ because the start of the path is the retraction to X_a by a nearly stratum preserving retraction of the point $g_s\left(\omega|_{[\beta_\omega^{-1}(\frac{1}{2}s), T_\omega]}\right)(\beta_\omega^{-1}(s))$ which is not in X_c or a lower strata. Clearly $l_0 =$ identity because $j_0 = k_0 =$ identity and l fixes start and end points of paths because both j and k fix start and end points of paths.
Thus we have satisfied the requirements for $\mathrm{pop}(X_b, X - X_c, X_a) \subset \mathrm{pop}(X_b, X, X_a)$ to be a special path inclusion and so using induction to remove strata like X_c we reach the two strata case and so conclude $\mathrm{hol}(X_b \cup X_a, X_a) \subset \mathrm{pop}(X_b, X, X_a)$ is a special path inclusion by Lemma 2.5. □

References

[**1**] Frank Quinn, *Homotopically Stratified Sets*, Journal of the American Mathematical Society, Vol. 1, No. 2. (Apr., 1988), pp441-499.

[**2**] Jon Woolf, *The fundamental category of a stratified space*, arXiv: 0811.2580v1, 2008.

This article may be accessed via WWW at `http://www.rmi.acnet.ge/jhrs/`

David A. Miller
`d.miller@maths.abdn.ac.uk`

Mathematics Department
University of Aberdeen
Aberdeen
United Kingdom
AB24 5UE

Journal of Homotopy and Related Structures, vol. 4(1), 2009, pp.275–315

ON HOMOTOPY INVARIANCE FOR ALGEBRAS OVER COLORED PROPS

MARK W. JOHNSON AND DONALD YAU

(*communicated by James Stasheff*)

Abstract

Over a monoidal model category, under some mild assumptions, we equip the categories of colored PROPs and their algebras with projective model category structures. A Boardman-Vogt style homotopy invariance result about algebras over cofibrant colored PROPs is proved. As an example, we define *homotopy* topological conformal field theories and observe that such structures are homotopy invariant.

1. Introduction

A PROP (short for PROduct and Permutation) is a very general algebraic machinery that can encode algebraic structures with multiple inputs and multiple outputs. It was invented by Mac Lane [**Mac67**] and was used prominently by Boardman and Vogt [**BV73**] in their seminal paper on homotopy invariant algebraic structures in algebraic topology. By forgetting structures, every PROP P has an underlying operad $U\mathsf{P}$. Operads are "smaller" algebraic machines that encode algebraic structures with multiple inputs and one output. PROPs are also widely used in string topology and mathematical physics. Indeed, several types of topological field theories, such as topological conformal field theories (TCFT), are described by the so-called Segal PROP Se [**Seg88, Seg01, Seg04**]. This is a topological PROP consisting of moduli spaces of Riemann surfaces with boundary holes. Closely related is the PROP $\mathcal{RCF}(g)$ [**Cha05, CG04**] in string topology that is built from spaces of reduced metric Sullivan chord diagrams with genus g.

There is an important generalization of PROPs, called *colored PROPs*, that can encode even more general types of algebraic structures, including morphisms and more general diagrams of algebras over a (colored) PROP. The PROP $\mathcal{RCF}(g)$ mentioned in the previous paragraph is actually a colored PROP. Also, the Segal PROP Se has an obvious colored analogue in which the Riemann surfaces are allowed boundary holes with varying circumferences. Multi-categories (a.k.a. colored

---

The authors thank Benoit Fresse for his helpful comments about earlier drafts of this paper and the anonymous referee for his/her helpful suggestions. M. Johnson was partially supported by U.S.-Israel Binational Science Foundation grant 2006039.
Received February 11, 2009, revised July 23, 2009; published on September 25, 2009.
2000 Mathematics Subject Classification: 18D50, 55U35, 81T45
Key words and phrases: Colored PROP, Quillen model category, homotopy algebra, topological conformal field theory

operads) are to operads what colored PROPs are to PROPs. In the simplest case, the set of colors $\mathfrak{C}$ is the one-element set, and in this case we have a 1-colored PROP. For example, if P is a 1-colored PROP, then diagrams of the form $A \to B$, consisting of two P-algebras and a P-algebra morphism between them, are encoded as algebras over a 2-colored PROP $\mathsf{P}_{\bullet\to\bullet}$ [**FMY08**, Example 2.10].

The purpose of this paper is to study homotopy theory of colored PROPs and homotopy invariance properties of algebras over colored PROPs. To ensure that our results are applicable in a variety of categories, we work over a symmetric monoidal closed category $\mathcal{E}$ with a compatible model category structure [**Qui67**]. To "do homotopy theory" of colored PROPs over $\mathcal{E}$ means that we lift the model category structure on $\mathcal{E}$ to the category of colored PROPs. Since colored PROPs are important mainly because of their algebras, under suitable conditions, we will also lift the model category structure on $\mathcal{E}$ to the category of P-algebras for a colored PROP P over $\mathcal{E}$.

Several of the results here are generalizations of those in Berger and Moerdijk [**BM03, BM07**], in which the main focus are operads and colored operads. The preprint [**Har07**] by Harper has results that are similar to those in [**BM03**] with slightly different hypotheses and were obtained by different methods. An earlier precedent was the work of Hinich [**Hin97**], who obtained such model category structures in the setting of chain complexes and operads over them. In [**Mar06**, p.373 (6)], Markl stated the problem of developing a theory of homotopy invariant algebraic structures over PROPs in the setting of chain complexes. In a recent preprint, Fresse [**Fre08**] independently studied homotopy theory of 1-colored PROPs and homotopy invariance properties of algebras over 1-colored PROPs in an arbitrary symmetric monoidal model category.

A description of some of our main results follows. To study homotopy invariance properties of algebras, first we need a suitable model category structure on the category of colored PROPs. The following result will be proved in Section 3, where the notion of strongly cofibrantly generated is defined in Definition 3.1.

Theorem 1.1 (Model Category of Colored PROPs). *Let $\mathfrak{C}$ be a non-empty set. Suppose that $\mathcal{E}$ is a strongly cofibrantly generated symmetric monoidal model category with a symmetric monoidal fibrant replacement functor, and either:*

1. *a cofibrant unit and a cocommutative interval, or*
2. *functorial path data.*

Then the category $\mathbf{PROP}^{\mathfrak{C}}_{\mathcal{E}}$ of $\mathfrak{C}$-colored PROPs over $\mathcal{E}$ is a strongly cofibrantly generated model category with fibrations and weak equivalences defined entrywise in $\mathcal{E}$.

We will define *cocommutative interval* (Definition 3.8) and *functorial path data* (Definition 3.9) precisely in Section 3. Both of these assumptions are used to construct path objects for fibrant $\mathfrak{C}$-colored PROPs, which are needed to use the Lifting Lemma 3.3 to lift the model category structure. For example, the category of simplicial sets has a cofibrant unit and a cocommutative interval [**BM07**, Section 2]. On the other hand, the categories of chain complexes over a characteristic 0 ring and of simplicial modules over a commutative ring admit functorial path data [**Fre08**, Section 5]. In particular, Theorem 1.1 applies to each of these categories. In fact, there is a modification of this projective structure on $\mathbf{PROP}^{\mathfrak{C}}_{\mathcal{E}}$, having more weak equivalences, which will be shown (Proposition 9.4) to be strongly Quillen equivalent to the projective structure on operads considered in [**BM07**]. As a consequence, one can view the homotopy theory of $\mathbf{PROP}^{\mathfrak{C}}_{\mathcal{E}}$ in the projective structure as a refinement of the homotopy theory of operads.

Next we describe our homotopy invariance result. An algebraic structure is considered to be homotopy invariant if it can be transferred back and forth through a weak equivalence, at least under mild assumptions. Historically, homotopy invariant structures in algebraic topology were first studied in the case of associative topological monoids. Stasheff [**Sta63**] found that a homotopy invariant substitute for an associative topological monoid is an A_∞-space. In [**BV73**] Boardman and Vogt vastly generalized this with the so-called W-construction. Given a topological (colored) operad O, the algebras over $W\mathsf{O}$ are homotopy invariants, where $W\mathsf{O}$ is a cofibrant resolution (i.e., cofibrant replacement) of O. Analogues of the results of Boardman and Vogt in the setting of chain complexes over a characteristic 0 field were obtained by Markl [**Mar04**].

One cannot expect that algebras over an arbitrary colored PROP P be homotopy invariants. As suggested by the Boardman-Vogt W-construction, to obtain homotopy invariance of algebras, one should consider *cofibrant* colored PROPs. In fact, following Markl [**Mar04**, Principle 1], one can define *homotopy algebras* as algebras over a cofibrant colored PROP. The following result, which will be proved in Section 4, says that homotopy algebras are homotopy invariants.

Theorem 1.2 (Homotopy Invariance of Homotopy Algebras). *Let $\mathfrak{C}$ and $\mathcal{E}$ be as in Theorem 1.1, and let P be a cofibrant $\mathfrak{C}$-colored PROP in $\mathcal{E}$. Let $f = \{f_c \colon X_c \to Y_c\}_{c\in\mathfrak{C}}$ be a collection of maps in $\mathcal{E}$. Then the following statements hold.*

1. *Suppose that all of the $Y_{\underline{c}}$ are fibrant and that all of the $f_{\underline{c}} \colon X_{\underline{c}} \to Y_{\underline{c}}$ are acyclic cofibrations. Then any P-algebra structure on $X = \{X_c\}$ induces one on $Y = \{Y_c\}$ such that f is a morphism of P-algebras.*
2. *Suppose that all of the $X_{\underline{c}}$ are cofibrant and that all of the $f_{\underline{c}} \colon X_{\underline{c}} \to Y_{\underline{c}}$ are acyclic fibrations. Then any P-algebra structure on $Y = \{Y_c\}$ induces one on $X = \{X_c\}$ such that f is a morphism of P-algebras.*

In the Theorem above, $X_{\underline{c}} = X_{c_1} \otimes \cdots \otimes X_{c_n}$ if $\underline{c} = (c_1, \ldots, c_n)$ with each $c_i \in \mathfrak{C}$, and similarly for $f_{\underline{c}}$. In the hypotheses of the two assertions, $\underline{c}$ runs through all of the finite non-empty sequences of elements in $\mathfrak{C}$. Theorem 1.2 may be considered as a colored PROP, model category theoretic version of the Boardman-Vogt philosophy of homotopy invariant structures [**BV73**]. It should be compared with [**Mar04**, (M1)-(M3), (M1')] and [**Mar02**, Theorem 15 and Corollary 16]. Moreover, under suitable additional assumptions (see Theorem 1.4), there is a model category structure on the category of P-algebras with entrywise fibrations and weak equivalences. With this model category structure, we can therefore strengthen the conclusions in Theorem 1.2: $f \colon X \to Y$ is a weak equivalence of P-algebras in the first case and an acyclic fibration of P-algebras in the second case.

As an illustration of Theorems 1.1 and 1.2, consider the category $\mathbf{Ch}(R)$ of non-negatively graded chain complexes over a field R of characteristic 0 with its usual model category structure [**DS95, Hov99, Qui67**]. In Section 5, we will recall the details of the Segal PROP Se mentioned earlier. An algebra over the chain Segal PROP $\mathbf{Se} = C_*(\mathsf{Se})$ is called a *topological conformal field theory* (TCFT) [**Seg88, Seg01, Seg04**]. Denote by $\mathbf{Se}_\infty$ the functorial cofibrant replacement of $\mathbf{Se}$ given by Quillen's small object argument, and call an algebra over $\mathbf{Se}_\infty$ a *homotopy topological conformal field theory* (HTCFT). A TCFT is also an HTCFT via an extended action associated to the acyclic fibration $\mathbf{Se}_\infty \to \mathbf{Se}$.

The following result, which will be discussed in Section 5, is a special case of Theorem 1.2. Keep in mind that all objects in $\mathbf{Ch}(R)$ are both fibrant and cofibrant, which simplifies the statement.

Corollary 1.3 (Homotopy Invariance of HTCFTs). *Let $f\colon A \to B$ be a map in $\mathbf{Ch}(R)$, where R is a field of characteristic* 0.

1. *Suppose that f is an injective quasi-isomorphism. Then any HTCFT structure on A induces one on B such that f becomes a morphism of HTCFTs.*
2. *Suppose that f is a surjective quasi-isomorphism. Then any HTCFT structure on B induces one on A such that f becomes a morphism of HTCFTs.*

In other words, HTCFT is a homotopy invariant analogue of TCFT.

Next we consider the category $\mathbf{Alg}(\mathsf{P})$ of algebras over a colored PROP P. Let $\mathcal{E}^{\mathfrak{C}}$ be the product category $\prod_{c\in\mathfrak{C}} \mathcal{E}$. Given a $\mathfrak{C}$-colored PROP P in $\mathcal{E}$, each P-algebra X has an underlying object in $\mathcal{E}^{\mathfrak{C}}$. Let $U\colon \mathbf{Alg}(\mathsf{P}) \to \mathcal{E}^{\mathfrak{C}}$ be the forgetful functor.

Theorem 1.4 (Model Category of Algebras). *Let $\mathfrak{C}$ and $\mathcal{E}$ be as in Theorem 1.1, and let P be a cofibrant $\mathfrak{C}$-colored PROP in $\mathcal{E}$. Suppose in addition that*

1. *finite tensor products of fibrations with fibrant targets remain fibrations in $\mathcal{E}$, and*
2. *the forgetful functor U admits a left adjoint.*

Then there is a lifted model category structure on $\mathbf{Alg}(\mathsf{P})$ with fibrations and weak equivalences defined entrywise in $\mathcal{E}^{\mathfrak{C}}$.

As B. Fresse explained to the authors, for a general (colored) PROP P, the forgetful functor U may not admit a left adjoint, which is why we need this assumption. On the other hand, if the monoidal category $\mathcal{E}$ is Cartesian (i.e., the monoidal product is the Cartesian product), then U always admits a left adjoint. For example, any category of presheaves of sets is Cartesian, which includes simplicial sets. The assumption about finite tensor products preserving fibrations is used in a technical result (Lemma 6.1). Many symmetric monoidal model categories $\mathcal{E}$ have this property, including chain complexes over a characteristic 0 ring and all Cartesian categories.

Below we describe the organization of the rest of this paper.

In Section 2, we define colored PROPs using two monoidal products $\boxtimes_v$ and $\boxtimes_h$. Colored PROPs can be defined as $\boxtimes_v$-monoidal $\boxtimes_h$-monoids. In Section 3, we lift the model category structure from the base category $\mathcal{E}$ to the category of colored PROPs over $\mathcal{E}$ (Theorem 3.11), which is Theorem 1.1 above. A key ingredient is a Lifting Lemma 3.3. Different forms of this Lifting Lemma have been employed by various authors, e.g., [**Hin97, SS00**], to lift model category structures. In Section 4, we consider homotopy invariant versions of algebras over a colored PROP P and prove Theorem 1.2 above. Section 5 provides an illustration of the Homotopy Invariance Theorem 1.2 and of the necessity of the (colored) PROP setting. We discuss the Segal PROP Se, its chain level algebras (TCFTs), and observe the validity of Corollary 1.3. In Section 6, we consider the category of algebras over a colored PROP P and prove Theorem 1.4 above. In Section 7, we consider Quillen pairs on the categories of colored Σ-bimodules, colored PROPs, and algebras induced by various adjoint pairs.

In the last two sections, we present some further results about colored PROPs for future references. In Section 8, it is shown that colored PROPs can also be regarded as $\boxtimes_h$-monoidal $\boxtimes_v$-monoids. In Section 9, we discuss the free-forgetful (Quillen) adjunction between colored operads and colored PROPs. Using this adjunction we observe that, for a colored operad O, the categories of O-algebras and O_{prop}-algebras are equivalent, where O_{prop} is the colored PROP generated by O.

2. Colored PROPs

Our setting throughout this paper assumes $\mathcal{E} = (\mathcal{E}, \otimes, I)$ is an underlying symmetric monoidal model category that is strongly cofibrantly generated with a zero object 0. The reader is referred to [**SS00**] for the definition of a monoidal model category (which subsumes the closed symmetric monoidal condition) and the pushout-product axiom. It is assumed that $\mathcal{E}$ in particular has all small colimits and limits. We will say more about strong cofibrant generation in the next section.

To facilitate our discussion of model category structure on colored PROPs, in this section we give a formal definition of colored PROPs in $\mathcal{E}$. We work with colored PROPs without units. As in the classical case of operads, we will build our colored PROPs from a form of Σ-objects. Analogous to the description of operads as monoids with respect to the circle product, one can think of colored PROPs as *monoidal monoids* (Proposition 2.17 and Definition 2.16). This description of colored PROPs involves *two* monoidal products $\boxtimes_v$ (2.10.1) and $\boxtimes_h$ (2.14.1), the first one for the vertical composition and the other for the horizontal composition. Thus, colored PROPs are monoids in a category of monoids.

We begin with a discussion of colors and colored Σ-bimodules.

2.1. Colored Σ-bimodules

Let $\mathfrak{C}$ be a non-empty set, whose elements are called **colors**. Our PROPs have a base set of colors $\mathfrak{C}$. The simplest case is when $\mathfrak{C} = \{*\}$, which gives 1-colored PROPs.

Let $\mathcal{P}(\mathfrak{C})$ denote the category whose objects, called **profiles** or **$\mathfrak{C}$-profiles**, are finite non-empty sequences of colors. If $\underline{d} = (d_1, \ldots, d_m) \in \mathcal{P}(\mathfrak{C})$, then we write $|\underline{d}| = m$. Our convention is to use a normal alphabet, possibly with a subscript (e.g., d_1) to denote a color and to use an underlined alphabet (e.g., $\underline{d}$) to denote an object in $\mathcal{P}(\mathfrak{C})$.

Permutations $\sigma \in \Sigma_{|\underline{d}|}$ act on such a profile $\underline{d}$ from the left by permuting the $|\underline{d}|$ colors. Given two profiles $\underline{c} = (c_1, \ldots, c_n)$ and $\underline{d} = (d_1, \ldots, d_m)$, a **morphism** in $\mathcal{P}(\mathfrak{C})$, $\underline{c} \to \underline{d}$ is a permutation σ such that $\sigma(\underline{c}) = \underline{d}$. Such a morphism exists if and only if $\underline{d}$ is in the orbit of $\underline{c}$. Of course, if such a morphism exists, then $|\underline{c}| = |\underline{d}|$. Clearly, this defines a groupoid and the **orbit type** of a $\mathfrak{C}$-profile $\underline{c}$, or equivalently the set of objects of the same connected component of the groupoid, is denoted by $[\underline{c}]$.

To emphasize that the permutations act on the profiles from the left, we will also write $\mathcal{P}(\mathfrak{C})$ as $\mathcal{P}_l(\mathfrak{C})$. If we let the permutations act on the profiles from the right instead, then we get an equivalent category $\mathcal{P}_r(\mathfrak{C}) = \mathcal{P}_l(\mathfrak{C})^{op}$.

Given profiles as above, we define the concatenation of $\underline{c}$ and $\underline{d}$,

$$(\underline{c}, \underline{d}) = (c_1, \ldots, c_n, d_1, \ldots, d_m) \in \mathcal{P}(\mathfrak{C}). \tag{2.1.1}$$

The category of **$\mathfrak{C}$-colored Σ-bimodules** over $\mathcal{E}$ is defined to be the diagram category $\mathcal{E}^{\mathcal{P}_l(\mathfrak{C}) \times \mathcal{P}_r(\mathfrak{C})}$. In other words, a $\mathfrak{C}$-colored Σ-bimodule is a functor $\mathsf{P}\colon \mathcal{P}_l(\mathfrak{C}) \times \mathcal{P}_r(\mathfrak{C}) \to \mathcal{E}$, and a morphism of $\mathfrak{C}$-colored Σ-bimodules is a natural transformation of such functors. Unpacking this definition, a $\mathfrak{C}$-colored Σ-bimodule P consists of the following data:

1. For any profiles $\underline{d} \in \mathcal{P}_l(\mathfrak{C})$ and $\underline{c} \in \mathcal{P}_r(\mathfrak{C})$, it has an object

$$\mathsf{P}\binom{\underline{d}}{\underline{c}} = \mathsf{P}\binom{d_1, \ldots, d_m}{c_1, \ldots, c_n} \in \mathcal{E}.$$

This object should be thought of as the space of operations with $|\underline{c}| = n$ inputs and $|\underline{d}| = m$ outputs. The n inputs have colors $c_1, \ldots, c_n$, and the m outputs have colors $d_1, \ldots, d_m$.

2. Given permutations $\sigma \in \Sigma_{|\underline{d}|}$ and $\tau \in \Sigma_{|\underline{c}|}$, there is a morphism

$$(\sigma; \tau)\colon \mathsf{P}\binom{\underline{d}}{\underline{c}} \to \mathsf{P}\binom{\sigma\underline{d}}{\underline{c}\tau} \tag{2.1.2}$$

in $\mathcal{E}$ such that:

(a) $(1; 1)$ is the identity morphism,
(b) $(\sigma'\sigma; \tau\tau') = (\sigma'; \tau') \circ (\sigma; \tau)$, and
(c) $(1; \tau) \circ (\sigma; 1) = (\sigma; \tau) = (\sigma; 1) \circ (1; \tau)$.

Assembling these morphisms, there are commuting left Σ_m action (acting on the d_j) and right Σ_n action (acting on the c_i) on the object

$$\mathsf{P}(m, n) = \operatorname{colim} \mathsf{P}\binom{d_1, \ldots, d_m}{c_1, \ldots, c_n} = \operatorname{colim} \mathsf{P}\binom{\underline{d}}{\underline{c}}, \tag{2.1.3}$$

where the colimit is taken over all $\underline{d}$ and $\underline{c}$ with $|\underline{d}| = m$ and $|\underline{c}| = n$. The object $\mathsf{P}(m, n)$ is said to have **biarity** (m, n), and $\mathsf{P}\binom{\underline{d}}{\underline{c}}$ is called a **component** of $\mathsf{P}(m, n)$. If there is only one color (i.e., $\mathfrak{C} = \{*\}$), then $\mathsf{P}(m, n)$ has only one component, which has a left Σ_m action and a right Σ_n action that commute with each other. We will sometimes abuse notation and refer to the left or right action of σ on $\mathsf{P}\binom{\underline{d}}{\underline{c}}$ to describe the structure maps from it to $\mathsf{P}\binom{\sigma\underline{d}}{\underline{c}}$ or $\mathsf{P}\binom{\underline{d}}{\underline{c}\sigma}$.

Likewise, one observes that a morphism $f\colon \mathsf{P} \to \mathsf{Q}$ of $\mathfrak{C}$-colored Σ-bimodules consists of color-preserving morphisms

$$\left\{\mathsf{P}\binom{\underline{d}}{\underline{c}} \xrightarrow{f} \mathsf{Q}\binom{\underline{d}}{\underline{c}} : (\underline{d};\underline{c}) \in \mathcal{P}_l(\mathfrak{C}) \times \mathcal{P}_r(\mathfrak{C})\right\}$$

that respect the Σ_m-Σ_n action. In other words, each square

$$\begin{array}{ccc} \mathsf{P}\binom{\underline{d}}{\underline{c}} & \xrightarrow{f} & \mathsf{Q}\binom{\underline{d}}{\underline{c}} \\ \downarrow{\scriptstyle(\sigma;\tau)} & & \downarrow{\scriptstyle(\sigma;\tau)} \\ \mathsf{P}\binom{\sigma\underline{d}}{\underline{c}\tau} & \xrightarrow{f} & \mathsf{Q}\binom{\sigma\underline{d}}{\underline{c}\tau} \end{array}$$

is commutative.

To simplify the notation, from now on the category of $\mathfrak{C}$-colored Σ-bimodules over $\mathcal{E}$ is denoted by $\mathbf{\Sigma}^{\mathfrak{C}}_{\mathcal{E}}$ rather than $\mathcal{E}^{\mathcal{P}_l(\mathfrak{C}) \times \mathcal{P}_r(\mathfrak{C})}$.

2.2. Colored PROPs as monoidal monoids

Recall that an operad can be equivalently defined as a monoid in the category of Σ-objects [**May97**, Lemma 9]. We now describe a colored PROP analogue of this conceptual description of operads as monoids. Since a colored PROP has both a vertical composition and a compatible horizontal composition, it makes sense that a colored PROP is described by *two* monoidal products instead of just one. First we build the vertical composition; then we build the horizontal composition on top of it.

To build the vertical composition, first we want to decompose $\mathbf{\Sigma}^{\mathfrak{C}}_{\mathcal{E}}$ into smaller pieces with each piece corresponding to a pair of orbit types of colors. To do this we need to use smaller indexing categories than all of the $\mathfrak{C}$-profiles.

Definition 2.3. Let $\underline{b} = (b_1, \ldots, b_k)$ be a $\mathfrak{C}$-profile. Define the category $\Sigma_{\underline{b}}$ to be the maximal connected sub-groupoid of $\mathcal{P}_l(\mathfrak{C})$ containing $\underline{b}$.

To introduce notation, this is the category whose objects are the $\mathfrak{C}$-profiles $\tau\underline{b} = \left(b_{\tau(1)}, \ldots, b_{\tau(k)}\right) \in \mathcal{P}(\mathfrak{C})$ obtained from $\underline{b}$ by permutations $\tau \in \Sigma_k$. Given two (possibly equal) objects $\tau\underline{b}$ and $\tau'\underline{b}$ in $\Sigma_{\underline{b}}$, a morphism $\tau''\colon \tau\underline{b} \to \tau'\underline{b}$ is a permutation in Σ_k such that $\tau''\tau\underline{b} = \tau'\underline{b}$ as $\mathfrak{C}$-profiles.

Notice that when we write $\tau\underline{b}$ as an object in $\Sigma_{\underline{b}}$, the permutation τ is not necessarily unique when $\underline{b}$ contains repeated colors. Indeed, $\tau'\underline{b}$ is the same object as $\tau\underline{b}$ if and only if they are equal as ordered sequences of colors. It is easy to see that there is an isomorphism $\Sigma_{\underline{b}} \cong \Sigma_{\tau\underline{b}}$ of groupoids for any $\tau \in \Sigma_{|\underline{b}|}$.

Example 2.4. In the one-colored case, i.e., $\mathfrak{C} = \{*\}$, a $\mathfrak{C}$-profile $\underline{b}$ is uniquely determined by its length $|\underline{b}| = k$. In this case, there is precisely one object $\underline{b} = (*, \ldots, *)$ (k entries) in the category $\Sigma_{\underline{b}}$, since $\underline{b}$ is unchanged by any permutation in Σ_k. In other words, in the one-colored case, $\Sigma_{\underline{b}}$ is the permutation group $\Sigma_{|\underline{b}|}$, regarded as a category with one object, whose classifying space is $B\Sigma_{|\underline{b}|}$. □

Example 2.5. In the other extreme, suppose that $\underline{b} = (b_1, \ldots, b_k)$ consists of distinct colors, i.e., $b_i \neq b_j$ if $i \neq j$. There are now $k!$ different permutations of $\underline{b}$, one for each $\tau \in \Sigma_k$. So there are $k!$ objects in $\Sigma_{\underline{b}}$. Given two objects $\tau\underline{b}$ and $\tau'\underline{b}$ in $\Sigma_{\underline{b}}$, there is a unique morphism $\tau'\tau^{-1}\colon \tau\underline{b} \to \tau'\underline{b}$. Thus, the classifying space of $\Sigma_{\underline{b}}$ in this case is a model for $E\Sigma_{|\underline{b}|}$ as in [**Dwy01**, 5.9]. □

To decompose $\mathfrak{C}$-colored Σ-bimodules, we actually need a pair of $\mathfrak{C}$-profiles at a time. So we introduce the following groupoid.

Definition 2.6. Given any pair of $\mathfrak{C}$-profiles $\underline{d}$ and $\underline{c}$, define $\Sigma_{\underline{d};\underline{c}} = \Sigma_{\underline{d}} \times \Sigma_{\underline{c}}^{op}$, where $\Sigma_{\underline{d}}$ and $\Sigma_{\underline{c}}$ are introduced in Definition 2.3.

Of course, one could equivalently describe $\Sigma_{\underline{d};\underline{c}}$ as the maximal connected subgroupoid of $\mathcal{P}_l(\mathfrak{C}) \times \mathcal{P}_r(\mathfrak{C})$ containing the ordered pair $(\underline{d};\underline{c})$. If $\underline{d} = (d_1, \ldots, d_m)$ and $\underline{c} = (c_1, \ldots, c_n)$, then we write the objects in $\Sigma_{\underline{d};\underline{c}}$ as pairs

$$\begin{pmatrix} \sigma\underline{d} \\ \underline{c}\tau \end{pmatrix} = \begin{pmatrix} d_{\sigma(1)}, \ldots, d_{\sigma(m)} \\ c_{\tau^{-1}(1)}, \ldots, c_{\tau^{-1}(n)} \end{pmatrix}$$

for $\sigma \in \Sigma_m$ and $\tau \in \Sigma_n$.

Given any $\mathfrak{C}$-profile $\underline{d}$, we denote by $[\underline{d}]$ the orbit type of $\underline{d}$ under permutations in $\Sigma_{|\underline{d}|}$. The following result is the decomposition of colored Σ-bimodules that we have been referring to.

Lemma 2.7. *There is a canonical isomorphism*

$$\mathbf{\Sigma}_{\mathcal{E}}^{\mathfrak{C}} \cong \prod_{[\underline{d}],[\underline{c}]} \mathcal{E}^{\Sigma_{\underline{d};\underline{c}}}, \quad \mathsf{P} \mapsto \left\{ \mathsf{P}\begin{pmatrix} [\underline{d}] \\ [\underline{c}] \end{pmatrix} \right\} \tag{2.7.1}$$

of categories, in which the product runs over all the pairs of orbit types of $\mathfrak{C}$-profiles.

Proof. First recall that a (small) groupoid is canonically isomorphic to the coproduct of its connected components $\mathcal{P}_l(\mathfrak{C}) \times \mathcal{P}_r(\mathfrak{C}) \approx \coprod \Sigma_{\underline{d};\underline{c}}$. Now notice that the universal properties imply the category of functors $Fun(\coprod \Sigma_{\underline{d};\underline{c}}, \mathcal{E})$ is canonically isomorphic to the product category $\prod Fun(\Sigma_{\underline{d};\underline{c}}, \mathcal{E})$. □

Example 2.8. If $\mathfrak{C} = \{*\}$, then the decomposition (2.7.1) becomes

$$\mathbf{\Sigma}_{\mathcal{E}}^{\mathfrak{C}} \cong \prod_{m,n \geqslant 1} \mathcal{E}^{\Sigma_m \times \Sigma_n^{op}}.$$

An object in the diagram category $\mathcal{E}^{\Sigma_m \times \Sigma_n^{op}}$ is simply an object $\mathsf{P}(m,n)$ in $\mathcal{E}$ with a left Σ_m-action and a right Σ_n-action that commute with each other. □

An important ingredient in building the vertical composition in a colored PROP is a functor

$$\otimes_{\Sigma_{\underline{b}}} \colon \mathcal{E}^{\Sigma_{\underline{d};\underline{b}}} \times \mathcal{E}^{\Sigma_{\underline{b};\underline{c}}} \to \mathcal{E}^{\Sigma_{\underline{d};\underline{c}}},$$

which in the one-colored case reduces to the usual $\otimes_{\Sigma_k}$. So let $X \in \mathcal{E}^{\Sigma_{\underline{d};\underline{b}}}$ and $Y \in \mathcal{E}^{\Sigma_{\underline{b};\underline{c}}}$. First we specify what $X \otimes_{\Sigma_{\underline{b}}} Y$ does to objects in $\Sigma_{\underline{d};\underline{c}}$.

Fix an object $\binom{\sigma \underline{d}}{\underline{c}\mu} \in \Sigma_{\underline{d};\underline{c}}$. Consider the diagram $D = D(X, Y; \sigma\underline{d}, \underline{c}\mu) \colon \Sigma_{\underline{b}} \to \mathcal{E}$ defined as

$$D(\tau\underline{b}) = X\begin{pmatrix}\sigma\underline{d}\\ \underline{b}\tau^{-1}\end{pmatrix} \otimes Y\begin{pmatrix}\tau\underline{b}\\ \underline{c}\mu\end{pmatrix} \tag{2.8.1}$$

for each object $\tau\underline{b} \in \Sigma_{\underline{b}}$. (Note that $\underline{b}\tau^{-1} = \tau\underline{b}$ as $\mathfrak{C}$-profiles.) The image of a morphism $\tau'' \in \Sigma_{\underline{b}}(\tau\underline{b}, \tau'\underline{b})$ under D is the map

$$X\begin{pmatrix}\sigma\underline{d}\\ \underline{b}\tau^{-1}\end{pmatrix} \otimes Y\begin{pmatrix}\tau\underline{b}\\ \underline{c}\mu\end{pmatrix} \xrightarrow{\binom{1}{\tau''^{-1}} \otimes \binom{\tau''}{1}} X\begin{pmatrix}\sigma\underline{d}\\ \underline{b}\tau'^{-1}\end{pmatrix} \otimes Y\begin{pmatrix}\tau'\underline{b}\\ \underline{c}\mu\end{pmatrix}. \tag{2.8.2}$$

Now we define the object

$$(X \otimes_{\Sigma_{\underline{b}}} Y)\begin{pmatrix}\sigma\underline{d}\\ \underline{c}\mu\end{pmatrix} = \operatorname{colim} D(X, Y; \sigma\underline{d}, \underline{c}\mu) \in \mathcal{E}, \tag{2.8.3}$$

the colimit of the diagram $D = D(X, Y; \sigma\underline{d}, \underline{c}\mu)$.

Next we define the image under $X \otimes_{\Sigma_{\underline{b}}} Y$ of a morphism in $\Sigma_{\underline{d};\underline{c}}$. So consider the morphism

$$(\sigma''; \mu'') \in \Sigma_{\underline{d};\underline{c}}\left(\begin{pmatrix}\sigma\underline{d}\\ \underline{c}\mu\end{pmatrix}, \begin{pmatrix}\sigma'\underline{d}\\ \underline{c}\mu'\end{pmatrix}\right).$$

For each morphism $\tau'' \in \Sigma_{\underline{b}}(\tau\underline{b}, \tau'\underline{b})$, the square

$$\begin{array}{ccc}
X\begin{pmatrix}\sigma\underline{d}\\ \underline{b}\tau^{-1}\end{pmatrix} \otimes Y\begin{pmatrix}\tau\underline{b}\\ \underline{c}\mu\end{pmatrix} & \xrightarrow{\binom{\sigma''}{1} \otimes \binom{1}{\mu''}} & X\begin{pmatrix}\sigma'\underline{d}\\ \underline{b}\tau^{-1}\end{pmatrix} \otimes Y\begin{pmatrix}\tau\underline{b}\\ \underline{c}\mu'\end{pmatrix} \\
\Big\downarrow {\scriptstyle \binom{1}{\tau''^{-1}} \otimes \binom{\tau''}{1}} & & \Big\downarrow {\scriptstyle \binom{1}{\tau''^{-1}} \otimes \binom{\tau''}{1}} \\
X\begin{pmatrix}\sigma\underline{d}\\ \underline{b}\tau'^{-1}\end{pmatrix} \otimes Y\begin{pmatrix}\tau'\underline{b}\\ \underline{c}\mu\end{pmatrix} & \xrightarrow{\binom{\sigma''}{1} \otimes \binom{1}{\mu''}} & X\begin{pmatrix}\sigma'\underline{d}\\ \underline{b}\tau'^{-1}\end{pmatrix} \otimes Y\begin{pmatrix}\tau'\underline{b}\\ \underline{c}\mu'\end{pmatrix}
\end{array}$$

is commutative because both composites are equal to $\binom{\sigma''}{\tau''^{-1}} \otimes \binom{\tau''}{\mu''}$. These commutative squares give us a map

$$\begin{pmatrix}\sigma''\\ 1\end{pmatrix} \otimes \begin{pmatrix}1\\ \mu''\end{pmatrix} \colon D(X, Y; \sigma\underline{d}, \underline{c}\mu) \to D(X, Y; \sigma'\underline{d}, \underline{c}\mu')$$

in the diagram category $\mathcal{E}^{\Sigma_{\underline{b}}}$. Taking colimits we obtain a map

$$(X \otimes_{\Sigma_{\underline{b}}} Y)(\sigma''; \mu'') \colon (X \otimes_{\Sigma_{\underline{b}}} Y)\begin{pmatrix}\sigma\underline{d}\\ \underline{c}\mu\end{pmatrix} \to (X \otimes_{\Sigma_{\underline{b}}} Y)\begin{pmatrix}\sigma'\underline{d}\\ \underline{c}\mu'\end{pmatrix} \in \mathcal{E}. \tag{2.8.4}$$

The naturality of the constructions above is clear. So we have the following result.

Lemma 2.9. *There is a functor*

$$\otimes_{\Sigma_{\underline{b}}} \colon \mathcal{E}^{\Sigma_{\underline{d};\underline{b}}} \times \mathcal{E}^{\Sigma_{\underline{b};\underline{c}}} \to \mathcal{E}^{\Sigma_{\underline{d};\underline{c}}}, \quad (X, Y) \mapsto (X \otimes_{\Sigma_{\underline{b}}} Y) \tag{2.9.1}$$

which restricts to the usual $\otimes_{\Sigma_{|\underline{b}|}}$ *in the* 1*-colored case.*

The functor (2.9.1) should be compared with the tensoring over a category construction in [**MM92**, VII Section 2], but the details of this expanded version are exploited below.

Example 2.10. If $\mathfrak{C} = \{*\}$, then the functor (2.9.1) takes the form

$$\otimes_{\Sigma_{\underline{b}}} : \mathcal{E}^{\Sigma_m \times \Sigma_k^{op}} \times \mathcal{E}^{\Sigma_k \times \Sigma_n^{op}} \to \mathcal{E}^{\Sigma_m \times \Sigma_n^{op}}$$

if $|\underline{b}| = k$. An object $X \in \mathcal{E}^{\Sigma_m \times \Sigma_k^{op}}$ is an object $X \in \mathcal{E}$ with a left Σ_m-action and a right Σ_k-action that commute with each other. Likewise, an object $Y \in \mathcal{E}^{\Sigma_k \times \Sigma_n^{op}}$ is an object in $\mathcal{E}$ equipped with commuting Σ_k-Σ_n actions. Since $\Sigma_{\underline{b}} = \Sigma_k$ (as a category with one object), the only object in the diagram $D(X, Y; \sigma\underline{d}, \underline{c}\mu)$ above is $X \otimes Y \in \mathcal{E}$. The map (2.8.2) now takes the form

$$X \otimes Y \xrightarrow{g^{-1} \otimes g} X \otimes Y$$

for $g \in \Sigma_k$. Therefore, we have

$$X \otimes_{\Sigma_{\underline{b}}} Y = \operatorname*{colim}_{g \in \Sigma_k} \left(X \otimes Y \xrightarrow{g^{-1} \otimes g} X \otimes Y \right) = X \otimes_{\Sigma_k} Y.$$

The maps (2.8.4)

$$(X \otimes_{\Sigma_k} Y)(\sigma; \mu) \colon X \otimes_{\Sigma_k} Y \to X \otimes_{\Sigma_k} Y$$

for $(\sigma; \mu) \in \Sigma_m \times \Sigma_n^{op}$ give the Σ_m-Σ_n actions on $X \otimes_{\Sigma_k} Y$, which are induced by those on X and Y. □

Using the construction $\otimes_{\Sigma_{\underline{b}}}$, now we want to describe colored Σ-bimodules equipped with a vertical composition. So let $\mathsf{P} = \left\{ \mathsf{P}\binom{[\underline{d}]}{[\underline{c}]} \right\}$ and $\mathsf{Q} = \left\{ \mathsf{Q}\binom{[\underline{d}]}{[\underline{c}]} \right\}$ be $\mathfrak{C}$-colored Σ-bimodules over $\mathcal{E}$. Recall the decomposition $\mathbf{\Sigma}_{\mathcal{E}}^{\mathfrak{C}} \cong \prod \mathcal{E}^{\Sigma_{\underline{d};\underline{c}}}$ (Lemma 2.7), in which $\mathsf{P}\binom{[\underline{d}]}{[\underline{c}]} \in \mathcal{E}^{\Sigma_{\underline{d};\underline{c}}}$. We define a functor

$$\boxtimes_v \colon \mathbf{\Sigma}_{\mathcal{E}}^{\mathfrak{C}} \times \mathbf{\Sigma}_{\mathcal{E}}^{\mathfrak{C}} \to \mathbf{\Sigma}_{\mathcal{E}}^{\mathfrak{C}}$$

by setting

$$(\mathsf{P} \boxtimes_v \mathsf{Q}) \binom{[\underline{d}]}{[\underline{c}]} = \coprod_{[\underline{b}]} \mathsf{P}\binom{[\underline{d}]}{[\underline{b}]} \otimes_{\Sigma_{\underline{b}}} \mathsf{Q}\binom{[\underline{b}]}{[\underline{c}]} \in \mathcal{E}^{\Sigma_{\underline{d};\underline{c}}}, \tag{2.10.1}$$

where the coproduct is taken over all the orbit types of $\mathfrak{C}$-profiles.

Lemma 2.11. *The functor $\boxtimes_v$ gives $\mathbf{\Sigma}_{\mathcal{E}}^{\mathfrak{C}}$ the structure of a monoidal category.*

Proof. The unit of $\boxtimes_v$ is the object $\mathbf{1} \in \mathbf{\Sigma}_{\mathcal{E}}^{\mathfrak{C}}$ defined as

$$\mathbf{1}\binom{\underline{d}}{\underline{c}} = \begin{cases} 0 & \text{if } [\underline{d}] \neq [\underline{c}], \\ I & \text{if } [\underline{d}] = [\underline{c}], \end{cases}$$

where I is the unit of $\mathcal{E}$. The morphism $(\sigma; \tau) \colon \mathbf{1}\binom{\underline{d}}{\underline{c}} \to \mathbf{1}\binom{\sigma\underline{d}}{\underline{c}\tau}$ is the identity map of either 0 or I, depending on whether $[\underline{d}]$ and $[\underline{c}]$ are equal or not.

The required associativity of $\boxtimes_v$ boils down to the associativity of the construction $\otimes_{\Sigma_{\underline{b}}}$. In other words, we need to show that the diagram

$$\begin{array}{ccc} \mathcal{E}^{\Sigma_{\underline{d};\underline{b}}} \times \mathcal{E}^{\Sigma_{\underline{b};\underline{c}}} \times \mathcal{E}^{\Sigma_{\underline{c};\underline{a}}} & \xrightarrow{\otimes_{\Sigma_{\underline{b}}} \times Id} & \mathcal{E}^{\Sigma_{\underline{d};\underline{c}}} \times \mathcal{E}^{\Sigma_{\underline{c};\underline{a}}} \\ {\scriptstyle Id \times \otimes_{\Sigma_{\underline{c}}}} \downarrow & & \downarrow {\scriptstyle \otimes_{\Sigma_{\underline{c}}}} \\ \mathcal{E}^{\Sigma_{\underline{d};\underline{b}}} \times \mathcal{E}^{\Sigma_{\underline{b};\underline{a}}} & \xrightarrow[\otimes_{\underline{b}}]{} & \mathcal{E}^{\Sigma_{\underline{d};\underline{a}}} \end{array} \tag{2.11.1}$$

is commutative. Suppose that $X \in \mathcal{E}^{\Sigma_{\underline{d};\underline{b}}}$, $Y \in \mathcal{E}^{\Sigma_{\underline{b};\underline{c}}}$, $Z \in \mathcal{E}^{\Sigma_{\underline{c};\underline{a}}}$, and $\binom{\sigma\underline{d}}{\underline{a}\nu} \in \Sigma_{\underline{d};\underline{a}}$. Note that $\otimes$ (being a left adjoint) commutes with colimits and is associative. Therefore, either one of the two composites in (2.11.1), when applied to (X, Y, Z) and then to $\binom{\sigma\underline{d}}{\underline{a}\nu}$, gives the object

$$\operatorname{colim}\left[X\begin{pmatrix}\sigma\underline{d}\\ \underline{b}\tau^{-1}\end{pmatrix} \otimes Y\begin{pmatrix}\tau\underline{b}\\ \underline{c}\mu^{-1}\end{pmatrix} \otimes Z\begin{pmatrix}\mu\underline{c}\\ \underline{a}\nu\end{pmatrix} \xrightarrow{\binom{1}{\tau''^{-1}}\otimes\binom{\tau''}{\mu''^{-1}}\otimes\binom{\mu''}{1}} X\begin{pmatrix}\sigma\underline{d}\\ \underline{b}\tau'^{-1}\end{pmatrix} \otimes Y\begin{pmatrix}\tau'\underline{b}\\ \underline{c}\mu'^{-1}\end{pmatrix} \otimes Z\begin{pmatrix}\mu'\underline{c}\\ \underline{a}\nu\end{pmatrix}\right]$$

in $\mathcal{E}$. This is the colimit of a diagram D in $\mathcal{E}$ indexed by the category $\Sigma_{\underline{b}} \times \Sigma_{\underline{c}}$. For each pair of morphisms $(\tau'' \in \Sigma_{\underline{b}}(\tau\underline{b}; \tau'\underline{b}); \mu'' \in \Sigma_{\underline{c}}(\mu\underline{c}; \mu'\underline{c})) \in \Sigma_{\underline{b}} \times \Sigma_{\underline{c}}$, the diagram D has a morphism as indicated. A similar observation shows that the two composites in (2.11.1) agree on morphisms in $\Sigma_{\underline{d};\underline{a}}$ as well. □

We will not use the unit **1** of $\boxtimes_v$ below and will consider $(\mathbf{\Sigma}^{\mathfrak{C}}_{\mathcal{E}}, \boxtimes_v)$ as a monoidal category without unit.

Definition 2.12. Denote by $\mathbf{vPROP}^{\mathfrak{C}}_{\mathcal{E}}$ the category of monoids in the monoidal category $(\mathbf{\Sigma}^{\mathfrak{C}}_{\mathcal{E}}, \boxtimes_v)$ (without unit), whose objects are called **vPROPs**, with v standing for *vertical.*

Unwrapping the definitions of $\boxtimes_v$ and an associative map $\mathsf{P} \boxtimes_v \mathsf{P} \to \mathsf{P}$, we have the following description of a vPROP.

Proposition 2.13. *A vPROP consists of precisely the following data:*

1. *An object* $\mathsf{P} \in \mathbf{\Sigma}^{\mathfrak{C}}_{\mathcal{E}}$.
2. *A* ***vertical composition***

$$\mathsf{P}\begin{pmatrix}[\underline{d}]\\ [\underline{b}]\end{pmatrix} \otimes_{\Sigma_{\underline{b}}} \mathsf{P}\begin{pmatrix}[\underline{b}]\\ [\underline{c}]\end{pmatrix} \xrightarrow{\circ} \mathsf{P}\begin{pmatrix}[\underline{d}]\\ [\underline{c}]\end{pmatrix}$$

in $\mathcal{E}^{\Sigma_{\underline{d};\underline{c}}}$ *that is associative in the obvious sense.*

Moreover, a morphism of vPROPs is precisely a morphism in $\mathbf{\Sigma}^{\mathfrak{C}}_{\mathcal{E}}$ *that preserves the vertical compositions.*

By considering the definition of $\otimes_{\Sigma_{\underline{b}}}$, one can unwrap the vertical composition one step further and describe it as an associative map

$$\mathsf{P}\binom{\underline{d}}{\underline{b}} \otimes \mathsf{P}\binom{\underline{b}}{\underline{c}} \xrightarrow{\circ} \mathsf{P}\binom{\underline{d}}{\underline{c}} \tag{2.13.1}$$

in $\mathcal{E}$ for any $\mathfrak{C}$-profiles $\underline{b}$, $\underline{c}$, and $\underline{d}$ (not just representatives of orbit types). It is equivariant, in the sense that the diagram

$$\begin{array}{ccc}
\mathsf{P}\binom{\underline{d}}{\underline{b}\tau^{-1}} \otimes \mathsf{P}\binom{\tau\underline{b}}{\underline{c}} & = & \mathsf{P}\binom{\underline{d}}{\underline{b}\tau^{-1}} \otimes \mathsf{P}\binom{\tau\underline{b}}{\underline{c}} \\
{\scriptstyle (1;\tau^{-1})\otimes(\tau;1)}\uparrow & & \downarrow{\scriptstyle \circ} \\
\mathsf{P}\binom{\underline{d}}{\underline{b}} \otimes \mathsf{P}\binom{\underline{b}}{\underline{c}} & \xrightarrow{\circ} & \mathsf{P}\binom{\underline{d}}{\underline{c}} \\
{\scriptstyle (\sigma;1)\otimes(1;\mu)}\downarrow & & \downarrow{\scriptstyle (\sigma;\mu)} \\
\mathsf{P}\binom{\sigma\underline{d}}{\underline{b}} \otimes \mathsf{P}\binom{\underline{b}}{\underline{c}\mu} & \xrightarrow{\circ} & \mathsf{P}\binom{\sigma\underline{d}}{\underline{c}\mu}
\end{array} \tag{2.13.2}$$

is commutative for $\sigma \in \Sigma_{|\underline{d}|}$, $\mu \in \Sigma_{|\underline{c}|}$, and $\tau \in \Sigma_{|\underline{b}|}$.

Next we build the horizontal composition in a colored PROP. To do this, we need to construct a functor

$$\boxdot \colon \mathcal{E}^{\Sigma_{\underline{d};\underline{c}}} \times \mathcal{E}^{\Sigma_{\underline{b};\underline{a}}} \to \mathcal{E}^{\Sigma_{(\underline{d},\underline{b});(\underline{c},\underline{a})}}.$$

This functor is used to construct a monoidal product $\boxtimes_h$ on $\mathbf{vPROP}^{\mathfrak{C}}_{\mathcal{E}}$. We then use $\boxtimes_h$ to describe PROPs as monoids in $(\mathbf{vPROP}^{\mathfrak{C}}_{\mathcal{E}}, \boxtimes_h)$. Remembering that $\mathbf{vPROP}^{\mathfrak{C}}_{\mathcal{E}}$ is the category of $\boxtimes_v$-monoids in $\mathbf{\Sigma}^{\mathfrak{C}}_{\mathcal{E}}$, this says that PROPs are *$\boxtimes_v$-monoidal $\boxtimes_h$-monoids*, or *monoidal monoids* for short.

The functor $\boxdot$ is constructed as an inclusion functor followed by a left Kan extension. Indeed, there is a(n external product) functor

$$\iota \colon \mathcal{E}^{\Sigma_{\underline{d};\underline{c}}} \times \mathcal{E}^{\Sigma_{\underline{b};\underline{a}}} \to \mathcal{E}^{\Sigma_{\underline{d}} \times \Sigma_{\underline{b}} \times \Sigma^{op}_{\underline{c}} \times \Sigma^{op}_{\underline{a}}}, \quad (X, Y) \mapsto X \otimes Y$$

that sends $(X, Y) \in \mathcal{E}^{\Sigma_{\underline{d};\underline{c}}} \times \mathcal{E}^{\Sigma_{\underline{b};\underline{a}}}$ to the diagram $X \otimes Y$ with

$$(X \otimes Y)\left(\sigma\underline{d}; \mu\underline{b}; \underline{c}\tau^{-1}; \underline{a}\nu^{-1}\right) = X\binom{\sigma\underline{d}}{\underline{c}\tau^{-1}} \otimes Y\binom{\mu\underline{b}}{\underline{a}\nu^{-1}}, \tag{2.13.3}$$

and similarly for maps in $\Sigma_{\underline{d}} \times \Sigma_{\underline{b}} \times \Sigma^{op}_{\underline{c}} \times \Sigma^{op}_{\underline{a}}$. On the other hand, the subcategory inclusion

$$\left(\Sigma_{\underline{d}} \times \Sigma_{\underline{b}}\right) \times \left(\Sigma^{op}_{\underline{c}} \times \Sigma^{op}_{\underline{a}}\right) \xrightarrow{i} \Sigma_{(\underline{d},\underline{b});(\underline{c},\underline{a})} = \Sigma_{(\underline{d};\underline{b})} \times \Sigma^{op}_{(\underline{c},\underline{a})}$$

induces a functor on the diagram categories

$$\mathcal{E}^i \colon \mathcal{E}^{\Sigma_{(\underline{d},\underline{b});(\underline{c},\underline{a})}} \to \mathcal{E}^{\Sigma_{\underline{d}} \times \Sigma_{\underline{b}} \times \Sigma^{op}_{\underline{c}} \times \Sigma^{op}_{\underline{a}}}. \tag{2.13.4}$$

This last functor has a left adjoint given by left Kan extension

$$K \colon \mathcal{E}^{\Sigma_{\underline{d}} \times \Sigma_{\underline{b}} \times \Sigma^{op}_{\underline{c}} \times \Sigma^{op}_{\underline{a}}} \to \mathcal{E}^{\Sigma_{(\underline{d},\underline{b});(\underline{c},\underline{a})}}, \tag{2.13.5}$$

which is left adjoint to the functor $\mathcal{E}^i$ ([**Mac98**, pp.236-240]). Then we define the functor $\boxdot$ as the composite $\boxdot = K\iota$.

Lemma 2.14. *The functor*

$$\boxdot = K\iota \colon \mathcal{E}^{\Sigma_{\underline{d};\underline{c}}} \times \mathcal{E}^{\Sigma_{\underline{b};\underline{a}}} \to \mathcal{E}^{\Sigma_{(\underline{d},\underline{b});(\underline{c},\underline{a})}}$$

is associative in the obvious sense.

Proof. The associativity of $\boxdot$ is a consequence of the associativity of $\otimes$ in $\mathcal{E}$ (2.13.3) and the universal properties of left Kan extensions. $\square$

Now we want to use $\boxdot$ to define a monoidal product $\boxtimes_h$ on $\mathbf{vPROP}^{\mathfrak{C}}_{\mathcal{E}}$. Let P and Q be vPROPs. First define the object $\mathsf{P} \boxtimes_h \mathsf{Q} \in \mathbf{\Sigma}^{\mathfrak{C}}_{\mathcal{E}} \cong \prod \mathcal{E}^{\Sigma_{\underline{d};\underline{c}}}$ (Lemma 2.7) by setting

$$(\mathsf{P} \boxtimes_h \mathsf{Q})\binom{[\underline{d}]}{[\underline{c}]} = \coprod_{\substack{\underline{d}=(\underline{d}_1,\underline{d}_2)\\ \underline{c}=(\underline{c}_1,\underline{c}_2)}} \mathsf{P}\binom{[\underline{d}_1]}{[\underline{c}_1]} \boxdot \mathsf{Q}\binom{[\underline{d}_2]}{[\underline{c}_2]} \in \mathcal{E}^{\Sigma_{\underline{d};\underline{c}}} \tag{2.14.1}$$

for any pair of orbit types $[\underline{d}]$ and $[\underline{c}]$.

Lemma 2.15. *The definition* (2.14.1) *gives a monoidal product*

$$\boxtimes_h \colon \mathbf{vPROP}^{\mathfrak{C}}_{\mathcal{E}} \times \mathbf{vPROP}^{\mathfrak{C}}_{\mathcal{E}} \to \mathbf{vPROP}^{\mathfrak{C}}_{\mathcal{E}}$$

on the category $\mathbf{vPROP}^{\mathfrak{C}}_{\mathcal{E}}$.

Proof. In (2.14.1) we already defined $\mathsf{P} \boxtimes_h \mathsf{Q}$ as an object in $\mathbf{\Sigma}^{\mathfrak{C}}_{\mathcal{E}}$. To make it into a vPROP, we need an associative vertical composition (Proposition 2.13)

$$(\mathsf{P} \boxtimes_h \mathsf{Q})\binom{[\underline{d}]}{[\underline{b}]} \otimes_{\Sigma_{\underline{b}}} (\mathsf{P} \boxtimes_h \mathsf{Q})\binom{[\underline{b}]}{[\underline{c}]} \xrightarrow{\circ} (\mathsf{P} \boxtimes_h \mathsf{Q})\binom{[\underline{d}]}{[\underline{c}]} \in \mathcal{E}^{\Sigma_{\underline{d};\underline{c}}}.$$

Since $\boxtimes_h$ is defined as a coproduct, we only need to define $\circ$ when restricted to a typical summand of the source:

$$\left[\mathsf{P}\binom{[\underline{d}_1]}{[\underline{b}_1]} \boxdot \mathsf{Q}\binom{[\underline{d}_2]}{[\underline{b}_2]}\right] \otimes_{\Sigma_{\underline{b}}} \left[\mathsf{P}\binom{[\underline{b}'_1]}{[\underline{c}_1]} \boxdot \mathsf{Q}\binom{[\underline{b}'_2]}{[\underline{c}_2]}\right]$$
$$\xrightarrow{\circ} \mathsf{P}\binom{[\underline{d}_1]}{[\underline{c}_1]} \boxdot \mathsf{Q}\binom{[\underline{d}_2]}{[\underline{c}_2]} \hookrightarrow (\mathsf{P} \boxtimes_h \mathsf{Q})\binom{[\underline{d}]}{[\underline{c}]}.$$

This restriction of $\circ$ is defined as the 0 map, unless $\underline{b}_1 = \underline{b}'_1$ (which implies $\underline{b}_2 = \underline{b}'_2$), in which case this $\circ$ is induced by those on P and Q (in the form (2.13.1)), using the left Kan extension description of $\boxdot = K\iota$. The associativity of $\circ$ follows from those on P and Q, the associativity of $\otimes_{\Sigma_{\underline{b}}}$, and the naturality of the construction $\boxdot$. So $\mathsf{P} \boxtimes_h \mathsf{Q}$ is indeed a vPROP. The associativity of $\boxtimes_h$ follows from that of $\boxdot$ (Lemma 2.14) and the definition (2.14.1). $\square$

Definition 2.16. Denote by $\mathbf{PROP}^{\mathfrak{C}}_{\mathcal{E}}$ the category of monoids in the monoidal category (without unit) $(\mathbf{vPROP}^{\mathfrak{C}}_{\mathcal{E}}, \boxtimes_h)$, whose objects are called **$\mathfrak{C}$-colored PROPs**.

Unwrapping the definition of an associative map $\mathsf{P} \boxtimes_h \mathsf{P} \to \mathsf{P}$ of vPROPs and using Proposition 2.13, we have the following description of $\mathfrak{C}$-colored PROPs.

Proposition 2.17. *A $\mathfrak{C}$-colored PROP consists of exactly the following data:*

1. *An object* $\mathsf{P} \in \mathbf{\Sigma}^{\mathfrak{C}}_{\mathcal{E}}$.

2. *An associative* ***vertical composition***

$$\mathsf{P}\binom{[\underline{d}]}{[\underline{b}]} \otimes_{\Sigma_{\underline{b}}} \mathsf{P}\binom{[\underline{b}]}{[\underline{c}]} \xrightarrow{\circ} \mathsf{P}\binom{[\underline{d}]}{[\underline{c}]}$$

in $\mathcal{E}^{\Sigma_{\underline{d};\underline{c}}}$.

3. *An associative* ***horizontal composition***

$$\mathsf{P}\binom{[\underline{d}_1]}{[\underline{c}_1]} \boxdot \mathsf{P}\binom{[\underline{d}_2]}{[\underline{c}_2]} \xrightarrow{\otimes} \mathsf{P}\binom{[\underline{d}_1, \underline{d}_2]}{[\underline{c}_1, \underline{c}_2]} \tag{2.17.1}$$

in $\mathcal{E}^{\Sigma_{(\underline{d}_1,\underline{d}_2);(\underline{c}_1,\underline{c}_2)}}$.

The assembled map

$$\otimes \colon \mathsf{P} \boxtimes_h \mathsf{P} \to \mathsf{P}$$

in $\mathbf{\Sigma}_{\mathcal{E}}^{\mathfrak{C}}$ *is required to be a map of vPROPs, a condition called the* ***interchange rule***. *Moreover, a map of* $\mathfrak{C}$*-colored PROPs is exactly a map in* $\mathbf{\Sigma}_{\mathcal{E}}^{\mathfrak{C}}$ *that preserves both the vertical and the horizontal compositions.*

As in the case of the vertical composition (2.13.1), we may go one step further in unwrapping the horizontal composition. At the level of $\mathcal{E}$, the horizontal composition consists of associative maps

$$\mathsf{P}\binom{\underline{d}_1}{\underline{c}_1} \otimes \mathsf{P}\binom{\underline{d}_2}{\underline{c}_2} \xrightarrow{\otimes} \mathsf{P}\binom{\underline{d}_1, \underline{d}_2}{\underline{c}_1, \underline{c}_2} \tag{2.17.2}$$

for any $\mathfrak{C}$-profiles $\underline{d}_1$, $\underline{d}_2$, $\underline{c}_1$, and $\underline{c}_2$. These maps are bi-equivariant, in the sense that the square

$$\begin{array}{ccc} \mathsf{P}\binom{\underline{d}_1}{\underline{c}_1} \otimes \mathsf{P}\binom{\underline{d}_2}{\underline{c}_2} & \xrightarrow{\quad\otimes\quad} & \mathsf{P}\binom{\underline{d}_1, \underline{d}_2}{\underline{c}_1, \underline{c}_2} \\ {\scriptstyle (\sigma_1;\tau_1)\otimes(\sigma_2;\tau_2)}\downarrow & & \downarrow{\scriptstyle (\sigma_1\times\sigma_2;\tau_1\times\tau_2)} \\ \mathsf{P}\binom{\sigma_1\underline{d}_1}{\underline{c}_1\tau_1} \otimes \mathsf{P}\binom{\sigma_2\underline{d}_2}{\underline{c}_2\tau_2} & \xrightarrow{\quad\otimes\quad} & \mathsf{P}\binom{\sigma_1\underline{d}_1, \sigma_2\underline{d}_2}{\underline{c}_1\tau_1, \underline{c}_2\tau_2} \end{array} \tag{2.17.3}$$

is commutative for all $\sigma_i \in \Sigma_{|\underline{d}_i|}$ and $\tau_i \in \Sigma_{|\underline{c}_i|}$. In totally unwrapped form, the interchange rule says that the diagram

$$\begin{array}{ccc} \left[\mathsf{P}\binom{\underline{d}_1}{\underline{b}_1} \otimes \mathsf{P}\binom{\underline{d}_2}{\underline{b}_2}\right] \otimes \left[\mathsf{P}\binom{\underline{b}_1}{\underline{c}_1} \otimes \mathsf{P}\binom{\underline{b}_2}{\underline{c}_2}\right] & \xrightarrow[\cong]{\text{switch}} & \left[\mathsf{P}\binom{\underline{d}_1}{\underline{b}_1} \otimes \mathsf{P}\binom{\underline{b}_1}{\underline{c}_1}\right] \otimes \left[\mathsf{P}\binom{\underline{d}_2}{\underline{b}_2} \otimes \mathsf{P}\binom{\underline{b}_2}{\underline{c}_2}\right] \\ & & \downarrow{\scriptstyle (\circ,\circ)} \\ {\scriptstyle (\otimes,\otimes)}\downarrow & & \mathsf{P}\binom{\underline{d}_1}{\underline{c}_1} \otimes \mathsf{P}\binom{\underline{d}_2}{\underline{c}_2} \\ & & \downarrow{\scriptstyle \otimes} \\ \mathsf{P}\binom{\underline{d}_1, \underline{d}_2}{\underline{b}_1, \underline{b}_2} \otimes \mathsf{P}\binom{\underline{b}_1, \underline{b}_2}{\underline{c}_1, \underline{c}_2} & \xrightarrow{\qquad\circ\qquad} & \mathsf{P}\binom{\underline{d}_1, \underline{d}_2}{\underline{c}_1, \underline{c}_2} \end{array} \tag{2.17.4}$$

is commutative.

Note that the interchange rule (2.17.4) is symmetric with respect to the vertical and the horizontal compositions. In fact, it is possible to describe $\mathfrak{C}$-colored PROPs as monoidal monoids in the other order, i.e., as $\boxtimes_h$-monoidal $\boxtimes_v$-monoids. We will prove this in Section 8.

The following Lemma, which will be needed in the following sections, follows easily from the construction (see [**FMY08**, 2.5-2.8] and comments above Theorem 3.11) of (free) $\mathfrak{C}$-colored PROPs and Proposition 2.17. A slight modification of the discussion in [**Fre08**, 4.3-4.5] gives another proof.

Lemma 2.18. *The category of $\mathfrak{C}$-colored PROPs over $\mathcal{E}$ has all small limits and colimits. Filtered colimits, reflexive coequalizers, and all limits are constructed entrywise as in the underlying category $\mathbf{\Sigma}_{\mathcal{E}}^{\mathfrak{C}}$ of $\mathfrak{C}$-colored Σ-bimodules.*

2.19. Colored endomorphism PROP

Before we talk about P-algebras, let us first spell out the colored endomorphism PROP construction through which a P-algebra is defined. Given objects X, Y, and Z in $\mathcal{E}$, there is a natural map

$$\eta\colon Z^Y \otimes Y^X \to Z^X, \tag{2.19.1}$$

which is adjoint to the composition of the maps $Z^Y \otimes Y^X \otimes X \to Z^Y \otimes Y \to Z$. Here $Y^X \otimes X \to Y$ is the adjoint of the identity map on Y^X and similarly for the right-most map. The map η is associative in the obvious sense.

Definition 2.20. A $\mathfrak{C}$-colored *endomorphism PROP* E_X is associated to a $\mathfrak{C}$-graded object $X = \{X_c\}_{c\in\mathfrak{C}}$ in $\mathcal{E}$. Given $m, n \geqslant 1$ and colors $c_1, \ldots, c_n, d_1, \ldots, d_m$, it has the component

$$E_X\binom{\underline{d}}{\underline{c}} = (X_{d_1} \otimes \cdots \otimes X_{d_m})^{(X_{c_1} \otimes \cdots \otimes X_{c_n})} = X_{\underline{d}}^{X_{\underline{c}}}.$$

The $\Sigma_{\underline{d}}$ and $\Sigma_{\underline{c}}$ acts as expected, with $\Sigma_{\underline{d}}$ permuting the m factors $X_{\underline{d}} = X_{d_1} \otimes \cdots \otimes X_{d_m}$ and $\Sigma_{\underline{c}}$ permuting the n factors in the exponent. The horizontal composition in E_X is given by the naturality of exponentiation. The vertical composition is induced by the natural map $\eta\colon X_{\underline{d}}^{X_{\underline{c}}} \otimes X_{\underline{c}}^{X_{\underline{a}}} \to X_{\underline{d}}^{X_{\underline{a}}}$ discussed above (2.19.1).

Thinking of the case of pointed sets or modules, $E_X\binom{\underline{d}}{\underline{c}} = \mathrm{Hom}(X_{\underline{c}}, X_{\underline{d}})$, and the horizontal composition is simply tensoring of functions. The vertical composition is composition of functions with matching colors.

Definition 2.21. For a $\mathfrak{C}$-colored PROP P, a P-*algebra* structure on X is a morphism $\lambda\colon \mathsf{P} \to E_X$ of $\mathfrak{C}$-colored PROPs. In this case, we say that X is a P-algebra with structure map λ.

As usual one can unpack this definition and, via adjunction, express the structure map as a collection of maps

$$\lambda\colon \mathsf{P}\binom{\underline{d}}{\underline{c}} \otimes X_{\underline{c}} \to X_{\underline{d}}$$

with $(\underline{d};\underline{c}) \in \mathcal{P}_l(\mathfrak{C}) \times \mathcal{P}_r(\mathfrak{C})$ that are associative (with respect to both the horizontal and the vertical compositions) and bi-equivariant.

A *morphism* $f\colon X \to Y$ of P-algebras is a collection of maps $f = \{f_c\colon X_c \to Y_c\}_{c\in\mathfrak{C}}$ such that the diagram

$$\begin{array}{ccc} \mathsf{P}\binom{\underline{d}}{\underline{c}} \otimes X_{\underline{c}} & \xrightarrow{\lambda_X} & X_{\underline{d}} \\ {\scriptstyle Id\otimes f_{\underline{c}}}\downarrow & & \downarrow{\scriptstyle f_{\underline{d}}} \\ \mathsf{P}\binom{\underline{d}}{\underline{c}} \otimes Y_{\underline{c}} & \xrightarrow{\lambda_Y} & Y_{\underline{d}} \end{array} \tag{2.21.1}$$

commutes for all $m, n \geqslant 1$ and colors $c_1, \ldots, c_n$ and $d_1, \ldots, d_m$. Here $f_{\underline{c}} = f_{c_1} \otimes \cdots \otimes f_{c_n}$.

3. Model structure on colored PROPs

The purpose of this section is to prove Theorem 1.1, i.e., to lift the model category structure on $\mathcal{E}$ to the category $\mathbf{PROP}_{\mathcal{E}}^{\mathfrak{C}}$ of $\mathfrak{C}$-colored PROPs over $\mathcal{E}$ (Theorem 3.11). This is achieved by first lifting the model category structure on $\mathcal{E}$ to the category $\mathbf{\Sigma}_{\mathcal{E}}^{\mathfrak{C}}$ of $\mathfrak{C}$-colored Σ-bimodules (Proposition 3.5). The resulting model category structure on $\mathbf{\Sigma}_{\mathcal{E}}^{\mathfrak{C}}$ is then lifted to $\mathbf{PROP}_{\mathcal{E}}^{\mathfrak{C}}$ via a standard Lifting Lemma 3.3. A variation defining the model structure on $\mathbf{\Sigma}_{\mathcal{E}}^{\mathfrak{C}}$ using only a subset of the orbits of profiles is also discussed briefly at the end of the section, since it could be useful for change of colors operations, as considered in [**CGMV08**].

Definition 3.1. Suppose K is a set of morphisms in a model category $\mathcal{E}$. Then:

- the *relative K-cell complexes* will denote those morphisms which can be written as a transfinite composition of cobase changes of morphisms in K. (See [**Hir02**, 10.5.8].)
- we say *sources in K are small* if for each $f : A \to B$ in K, there exists a cardinal κ_A such that for every regular cardinal $\lambda \geqslant \kappa$ and every λ-sequence $X : \lambda \to \mathcal{E}$ the natural map
$$\operatorname*{colim}_{\beta<\lambda} \mathcal{E}(A, X_\beta) \to \mathcal{E}(A, \operatorname*{colim}_{\beta<\lambda} X_\beta)$$
is a bijection. (See [**Hir02**, 10.4.1].)
- $\mathcal{E}$ is a *strongly cofibrantly generated* model category if there exists sets of maps I and J with sources in both sets small, and a map p is a fibration (resp. acyclic fibration) if and only if p has the right lifting property with respect to every morphism in J (resp. I). (See [**Hir02**, 11.1.1].)
- given a functor $R : \mathcal{D} \to \mathcal{E}$, an *$R$-fibration (resp. R-weak equivalence)* will denote a morphism p in $\mathcal{D}$ with $R(p)$ a fibration (resp. weak equivalence). An *R-fibrant replacement* for an object $Y \in \mathcal{D}$ will indicate an R-weak equivalence $g : Y \to Z$ with Z an R-fibrant object. The R-lifted structure on $\mathcal{D}$ will refer to these classes together with the R-cofibrations, defined as those maps with the left lifting property with respect to each R-fibration which is also an R-weak equivalence.

- an *R-path object* (often just called a path object below) in $\mathcal{E}$ for an object $X \in \mathcal{E}$ will denote the intermediate object in a factorization of the diagonal map $X \to Path(X) \to X \times X$ as an R-weak equivalence followed by an R-fibration. (See [**Hir02**, 7.3.2(3)].)

Remark 3.2. We emphasize by the adjective 'strongly' that our sources are assumed to be small, rather than the weaker condition of assuming that our sources are small with respect to the class of relative I-cell complexes. Important examples excluded by this stronger assumption include almost all topological examples, while simplicial examples, chain complexes and so forth satisfy this strengthened assumption. This assumption greatly simplifies the exposition, and our later results on Quillen equivalences (see section 7) help to justify this restriction.

We will often use a stronger notion of path object, where the first morphism is assumed to be an acyclic cofibration, rather than just a weak equivalence.

Lemma 3.3. *Suppose $\mathcal{E}$ is a strongly cofibrantly generated model category and $R : \mathcal{D} \to \mathcal{E}$ has a left adjoint L. Then $\mathcal{D}$ becomes a cofibrantly generated model category (and (L, R) form a strong Quillen pair) under the R-lifted model structure provided:*

- *$\mathcal{D}$ has all small limits and colimits, while R creates/preserves filtered colimits,*
- *there is a functorial R-fibrant replacement Q in $\mathcal{D}$, and*
- *for every R-fibrant object in $\mathcal{D}$, there is a path object construction which is preserved by R.*

For clarity of presentation, we separate the key portion of the proof, adapted from [**Sch99**, Lemma B.2].

Sublemma 3.4. *Under the last two assumptions of Lemma 3.3, (retracts of) relative $L(J)$-cell complexes are R-weak equivalences.*

Proof. Suppose j is a relative $L(J)$-cell complex. By R-fibrancy of $Q(X)$ and the adjunction argument for the RLP against $L(J)$, there is a lift r in the following diagram.

$$\begin{array}{ccc} X & \xrightarrow{\ \eta\ } & Q(X) \\ {\scriptstyle j}\big\downarrow & \overset{r}{\nearrow} & \big\downarrow \\ Y & \longrightarrow & * \end{array} \qquad (3.4.1)$$

This implies $R(rj) = R(r)R(j)$ is a weak equivalence in $\mathcal{E}$ by the assumption of η an R-weak equivalence. Since the weak equivalences in $\mathcal{E}$ are precisely the maps whose image in the homotopy category are isomorphisms, it then suffices to verify there is a right inverse for $R(j) \approx R(Q(j))$ in the homotopy category of $\mathcal{E}$. Our candidate will be $R(r)$.

Define a map $Y \to QY \times QY$ by taking η on the first factor and $Q(j)r$ on the second. Then for any path object $P(QY)$ for QY one has the following diagram, and the indicated dotted lift

$$\begin{array}{ccccc} X & \longrightarrow & QX & \longrightarrow & P(QY) \\ {\scriptstyle j}\downarrow & & {\scriptstyle H} \nearrow & & \downarrow \\ Y & & \longrightarrow & & QY \times QY, \end{array} \tag{3.4.2}$$

where the map $QX \to QY \to P(QY)$ comes from functoriality of Q and the construction of the path object $P(QY)$. Now apply R to this diagram to yield

$$\begin{array}{ccccc} R(X) & \longrightarrow & R(P(QY)) & \xrightarrow{\approx} & P(R(QY)) \\ {\scriptstyle R(j)}\downarrow & {\scriptstyle R(H)} \nearrow & & & \downarrow \\ R(Y) & \longrightarrow & R(QY \times QY) & \xrightarrow[\approx]{} & R(QY) \times R(QY), \end{array} \tag{3.4.3}$$

which exhibits a right homotopy between $R(Q(j)r)$ and $R(\eta)$. Since $R(QY)$ is fibrant by assumption, it follows as usual for a fibrant target that $R(Q(j)r)$ and $R(\eta)$ are identified in the homotopy category. This suffices to imply $R(Q(j))R(r)$ is an isomorphism in the homotopy category of $\mathcal{E}$, since $R(\eta)$ is a weak equivalence by assumption. □

Proof of Lemma 3.3. We can appeal to [**Hir02**, 11.3.2], with the assumption that R preserves filtered colimits sufficient to verify that sources in $L(I)$ and $L(J)$ are small, since sources in I and J are assumed to be small. This implies that $L(I)$ and $L(J)$ permit the small object argument, satisfying condition 11.3.2(1), while condition 11.3.2(2) is verified by the Sublemma. □

Proposition 3.5. *The category $\Sigma_{\mathcal{E}}^{\mathfrak{C}}$ of $\mathfrak{C}$-colored Σ-bimodules carries a projective strongly cofibrantly generated model structure, where fibrations and weak equivalences are defined entrywise (in $\mathcal{E}^{\mathcal{P}_l(\mathfrak{C})\times\mathcal{P}_r(\mathfrak{C})}$), while the entries of any cofibration are cofibrations in $\mathcal{E}$. In addition, the properties of being simplicial, or proper are inherited from $\mathcal{E}$ in this structure.*

Proof. Since this is a category of diagrams, Hirschhorn's [**Hir02**, 11.6.1] applies. For the simplicial condition use [**Hir02**, 11.7.3], for the entries of cofibrations use [**Hir02**, 11.6.3], and for the proper condition use [**Hir02**, 13.1.14] (or the definitions and the claim on cofibrations). □

Remark 3.6. Keeping in mind our decomposition of the category of bimodules as a product, we point out that each piece of the decomposition also carries a similar model structure by the same arguments. As a consequence, a morphism is a cofibration precisely when each projection in the decomposition (2.7.1) is a cofibration.

When the classes of fibrations and weak equivalences in a model category are defined by considering all entries in an underlying structure, it has become common to call it a projective model structure. We propose to refer to modified projective structures (relative to a subset of the entries) when we define fibrations as those maps having only a specified set of entries fibrations, and similarly for weak equivalences. We will see later, in considering both changes of colors (Corollary 7.7) and the comparison with colored operads (Proposition 9.4) that modified structures can be useful. However, unless otherwise noted, we will focus on projective structures throughout the remainder of this article. Let Σ denote the groupoid $\mathcal{P}_l(\mathfrak{C}) \times \mathcal{P}_r(\mathfrak{C})$ defined in Section 2.1.

Corollary 3.7. *Given a subset of the components of Σ, we have a modified projective structure on $\Sigma_{\mathcal{E}}^{\mathfrak{C}}$ where fibrations and weak equivalences are defined only by considering entries in the chosen components. Once again, any entry of a cofibration will remain a cofibration in $\mathcal{E}$, and this structure also inherits the properties of being simplicial or proper from $\mathcal{E}$.*

Proof. One approach follows from a slight modification of Hirschhorn's [**Hir02**, 11.6.1], where only the generating cofibrations from the chosen components are used to define I and J. Hence, the class of cofibrations is contained in those for the projective structure, which suffices to imply the entries are cofibrations in $\mathcal{E}$ as above.

Alternatively, one exploits the decomposition (2.7.1) and considers each factor as one of only two types. One observes that the components not chosen have strongly cofibrantly generated model structures where all maps are acyclic fibrations, the cofibrations are precisely the isomorphisms, and the sets I and J both consist of only the identity map on the initial object, by [**Hir02**, 11.3.1]. Now apply the arguments in the previous proof to give the components chosen projective structures. The required modified projective structure is then the product of these structures, as in [**Hir02**, 11.1.10]. The claim about cofibrations then follows from the previous remark. □

Let $U \colon \mathbf{PROP}_{\mathcal{E}}^{\mathfrak{C}} \to \Sigma_{\mathcal{E}}^{\mathfrak{C}}$ denote the underlying $\mathfrak{C}$-colored Σ-bimodule functor. In order to lift the above model category structure on $\Sigma_{\mathcal{E}}^{\mathfrak{C}}$ to $\mathbf{PROP}_{\mathcal{E}}^{\mathfrak{C}}$ using the Lifting Lemma 3.3, we need path objects for U-fibrant $\mathfrak{C}$-colored PROPs. One way to obtain a path object construction for U-fibrant $\mathfrak{C}$-colored PROPs is by using a *cocommutative interval* in $\mathcal{E}$, which we now discuss.

Definition 3.8. We say that $\mathcal{E}$ admits a *cocommutative interval* (called a *cocommutative coalgebra interval* in [**BM07**]) if the fold map $\nabla \colon I \sqcup I \to I$ can be factored as

$$I \sqcup I \xrightarrow{\alpha} J \xrightarrow{\beta} I, \tag{3.8.1}$$

in which α is a cofibration and β is a weak equivalence, $J = (J, \Delta)$ is a coassociative cocommutative comonoid, and α and β are both maps of comonoids.

For example, the categories of (pointed) simplicial sets and of symmetric spectra [**HSS00**] both admit cocommutative intervals [**BM07**, Section 2].

Definition 3.9. We say that $\mathcal{E}$ admits *functorial path data* if there exist a symmetric monoidal functor $Path$ on $\mathcal{E}$ and monoidal natural transformations $s \colon Id \to Path$, $d_0, d_1 \colon Path \to Id$ such that $X \xrightarrow{s} Path(X) \xrightarrow{(d_0,d_1)} X \times X$ is a path object for X whenever X is fibrant.

This definition is adapted from Fresse [**Fre08**, Fact 5.3]. Fresse showed that functorial path data exists when $\mathcal{E}$ is either the category of chain complexes over a characteristic 0 ring or of simplicial modules, among others. We would like to consider other examples by using the following technical result.

Lemma 3.10. *If $\mathcal{E}$ admits a cocommutative interval and I is cofibrant, then $\mathcal{E}$ admits functorial path data.*

Proof. Let J be a cocommutative interval (3.8.1) in $\mathcal{E}$. Then define $Path(X) = X^J$ with the required transformations coming from the diagram

$$X \cong X^I \xrightarrow{\beta^*} X^J \xrightarrow{\alpha^*} X^{I \sqcup I} \cong X \times X$$

(so d_0 and d_1 are the projections of α^*). Notice the transformations are monoidal since α and β are assumed to be comonoidal, while X^J is symmetric monoidal via

$$X^J \otimes Y^J \to (X \otimes Y)^{J \otimes J} \xrightarrow{\Delta^*} (X \otimes Y)^J$$

since J is a coassociative, cocommutative comonoid.

When X is fibrant, by the pushout-product axiom it follows from α a cofibration that α^* is a fibration. Similarly, from β a weak equivalence between cofibrant objects it follows that β^* is a weak equivalence. Hence, X^J serves as a path object when X is fibrant. □

Next we want to lift the model category structure on $\mathbf{\Sigma}^{\mathfrak{C}}_{\mathcal{E}}$ (Proposition 3.5) to $\mathbf{PROP}^{\mathfrak{C}}_{\mathcal{E}}$ using the Lifting Lemma 3.3. As mentioned above, the forgetful functor $U\colon \mathbf{PROP}^{\mathfrak{C}}_{\mathcal{E}} \to \mathbf{\Sigma}^{\mathfrak{C}}_{\mathcal{E}}$ has a left adjoint, the *free $\mathfrak{C}$-colored PROP* functor $F\colon \mathbf{\Sigma}^{\mathfrak{C}}_{\mathcal{E}} \to \mathbf{PROP}^{\mathfrak{C}}_{\mathcal{E}}$. The existence of this left adjoint under set-theoretic assumptions on $\mathcal{E}$ can be determined using Freyd's adjoint functor theorem. However, an explicit description of this functor in the 1-colored case can be found in [**Mar08**, Proposition 57], [**MV07**, 1.2], or [**Fre08**, Appendix A]. The functor F is defined as a colimit over a certain groupoid of directed graphs. A straightforward extension of this construction, in which the edges of the directed graphs are $\mathfrak{C}$-colored, works for the $\mathfrak{C}$-colored case [**FMY08**, 2.5-2.8].

Theorem 3.11. *Let $\mathcal{E}$ be a strongly cofibrantly generated symmetric monoidal model category with:*

1. *a symmetric monoidal fibrant replacement functor, and*
2. *functorial path data (e.g. with cofibrant unit and admitting a cocommutative interval by Lemma 3.10).*

Then the category $\mathbf{PROP}^{\mathfrak{C}}_{\mathcal{E}}$ of $\mathfrak{C}$-colored PROPs over $\mathcal{E}$ is a strongly cofibrantly generated model category with fibrations and weak equivalences defined entrywise.

Proof. We would like to apply Lemma 3.3 to the free-forgetful adjoint pair (F, U) above. Recall the model structure on $\mathbf{\Sigma}^{\mathfrak{C}}_{\mathcal{E}}$ comes from Proposition 3.5 and Lemma 2.18 deals with the first condition. For the second and third conditions, simply apply the assumed symmetric monoidal fibrant replacement functor and functorial path data entrywise. The result remains a $\mathfrak{C}$-colored PROP by the symmetric monoidal assumption, and gives a U-fibrant replacement or U-path object for U-fibrant objects by the entrywise definitions in $\mathbf{\Sigma}_{\mathcal{E}}^{\mathfrak{C}}$. □

In fact, the same proof applies to our other categories of monoids as well, since the constructions are all symmetric monoidal and the left adjoint is the free monoid functor in both cases.

Theorem 3.12. *Under the assumptions of Theorem 3.11, the category of hPROPs (Definition 8.1) and the category of vPROPs are similarly strongly cofibrantly generated model categories with fibrations and weak equivalences defined entrywise.*

Remark 3.13. Rather than working with all entries, in Theorems 3.11 and 3.12 one could instead choose a subset of the orbits of pairs of $\mathfrak{C}$-profiles and define fibrations and weak equivalences by considering only those entries. For example, one could use the 'intersection' (see [**IJ02**, Definition 8.5 and Proposition 8.7]) of the model structures given by lifting over each chosen evaluation functor and its left adjoint. Since there is such a modified projective structure on $\mathbf{\Sigma}_{\mathcal{E}}^{\mathfrak{C}}$ (Corollary 3.7), and all other structures are lifted from there, such an approach is equally valid for $\mathbf{PROP}_{\mathcal{E}}^{\mathfrak{C}}$, vPROPs, and hPROPs. This is particularly useful in applications dealing with change of colors, as discussed near the end of Section 7, and a comparison between colored operads and colored PROPs (Proposition 9.4).

4. Homotopy invariance of homotopy algebras

The purpose of this section is to prove Theorem 1.2, which says that, under some reasonable conditions, an algebra structure over a cofibrant colored PROP is a homotopy invariant. The proof is adapted from [**BM03**, Theorem 3.5]. Throughout this short section, $f : X \to Y$ will denote a morphism in $\prod_{c\in\mathfrak{C}} \mathcal{E}$, and $\mathsf{P} \in \mathbf{PROP}_{\mathcal{E}}^{\mathfrak{C}}$ a cofibrant $\mathfrak{C}$-colored PROP.

Definition 4.1. First, we define the $\mathfrak{C}$-colored Σ-bimodule $E_{X,Y}$ as a mixed endomorphism construction, having components

$$E_{X,Y}\binom{\underline{d}}{\underline{c}} = (Y_{d_1} \otimes \cdots \otimes Y_{d_m})^{X_{c_1}\otimes\cdots\otimes X_{c_n}} = Y_{\underline{d}}^{X_{\underline{c}}}$$

for $(\underline{d};\underline{c}) \in \mathcal{P}_l(\mathfrak{C}) \times \mathcal{P}_r(\mathfrak{C})$.

Now define the relative endomorphism construction E_f via the pullback square in $\mathbf{\Sigma}_{\mathcal{E}}^{\mathfrak{C}}$ (recall pullbacks are defined entrywise):

$$\begin{array}{ccc} E_f & \xrightarrow{\overline{f}^*} & E_X \\ {\scriptstyle \overline{f}_*}\downarrow & & \downarrow{\scriptstyle f_*} \\ E_Y & \xrightarrow{f^*} & E_{X,Y} \end{array} \tag{4.1.1}$$

The map f^* is given on entries by

$$f^* = Y_{\underline{d}}^{f_{\underline{c}}} \colon E_Y\binom{\underline{d}}{\underline{c}} = Y_{\underline{d}}^{Y_{\underline{c}}} \to Y_{\underline{d}}^{X_{\underline{c}}} = E_{X,Y}\binom{\underline{d}}{\underline{c}}.$$

The map f_* is given on entries by

$$f_* = f_{\underline{d}}^{X_{\underline{c}}} \colon E_X\binom{\underline{d}}{\underline{c}} = X_{\underline{d}}^{X_{\underline{c}}} \to Y_{\underline{d}}^{X_{\underline{c}}} = E_{X,Y}\binom{\underline{d}}{\underline{c}}.$$

Lemma 4.2. *The relative endomorphism construction $E_f \in \mathbf{PROP}^{\mathfrak{C}}_{\mathcal{E}}$, and both morphisms $\overline{f}^*$ and $\overline{f}_*$ lie in $\mathbf{PROP}^{\mathfrak{C}}_{\mathcal{E}}$. Furthermore, $f : X \to Y$ is a morphism of P-algebras if and only if the P-algebra structures on X and Y both descend from the same morphism $\mathsf{P} \to E_f$ in $\mathbf{PROP}^{\mathfrak{C}}_{\mathcal{E}}$.*

Proof. To obtain the horizontal composition for E_f, consider the diagram

$$\begin{array}{ccccc}
E_f\binom{\underline{d}}{\underline{c}} \otimes E_f\binom{\underline{b}}{\underline{a}} & \xrightarrow{\overline{f}^*\otimes\overline{f}^*} & E_X\binom{\underline{d}}{\underline{c}} \otimes E_X\binom{\underline{b}}{\underline{a}} & \xrightarrow{\text{horizontal comp}} & E_X\binom{\underline{d},\underline{b}}{\underline{c},\underline{a}} \\
\Big\downarrow{\scriptstyle \overline{f}_*\otimes\overline{f}_*} & & & & \Big\downarrow{\scriptstyle f_*} \\
E_Y\binom{\underline{d}}{\underline{c}} \otimes E_Y\binom{\underline{b}}{\underline{a}} & \xrightarrow{\text{horizontal comp}} & E_Y\binom{\underline{d},\underline{b}}{\underline{c},\underline{a}} & \xrightarrow{f^*} & E_{X,Y}\binom{\underline{d},\underline{b}}{\underline{c},\underline{a}}.
\end{array}$$

This diagram is commutative, so by the universal property of pullbacks there is a unique induced map

$$\otimes \colon E_f\binom{\underline{d}}{\underline{c}} \otimes E_f\binom{\underline{b}}{\underline{a}} \to E_f\binom{\underline{d}, \underline{b}}{\underline{c}, \underline{a}},$$

which is the horizontal composition in E_f. The vertical composition in E_f is defined similarly using the vertical compositions in E_X and E_Y and the universal property of pullbacks. As a consequence, the morphisms $\overline{f}^*$ and $\overline{f}_*$ also lie in $\mathbf{PROP}^{\mathfrak{C}}_{\mathcal{E}}$ by inspection.

In fact, E_f enjoys a stronger than usual universal property, saying it is as close to a pullback in $\mathbf{PROP}^{\mathfrak{C}}_{\mathcal{E}}$ as possible, given that $E_{X,Y}$ is only in $\mathbf{\Sigma}^{\mathfrak{C}}_{\mathcal{E}}$. In more detail, suppose $\theta : \mathsf{P} \to E_f$ is a morphism of bimodules equivalent by the universal property of pullbacks to a pair of maps in $\mathbf{\Sigma}^{\mathfrak{C}}_{\mathcal{E}}$, $\theta_X : \mathsf{P} \to E_X$ and $\theta_Y : \mathsf{P} \to E_Y$ which agree when composed into $E_{X,Y}$. The strengthened pullback condition is that θ underlies a morphism in $\mathbf{PROP}^{\mathfrak{C}}_{\mathcal{E}}$ precisely when both θ_X and θ_Y underly morphisms in $\mathbf{PROP}^{\mathfrak{C}}_{\mathcal{E}}$. One direction of this implication follows from $\overline{f}^*$ and $\overline{f}_*$ morphisms in $\mathbf{PROP}^{\mathfrak{C}}_{\mathcal{E}}$, so now suppose both θ_X and θ_Y are morphisms in $\mathbf{PROP}^{\mathfrak{C}}_{\mathcal{E}}$ and combine commutativity of the diagram above with the commutative diagram for θ_X (and similarly for θ_Y)

$$\begin{array}{ccc}
\mathsf{P}\binom{\underline{d}}{\underline{c}} \otimes \mathsf{P}\binom{\underline{b}}{\underline{a}} & \longrightarrow & E_X\binom{\underline{d}}{\underline{c}} \otimes E_X\binom{\underline{b}}{\underline{a}} \\
\Big\downarrow & & \Big\downarrow \\
\mathsf{P}\binom{\underline{d},\underline{b}}{\underline{c},\underline{a}} & \longrightarrow & E_X\binom{\underline{d},\underline{b}}{\underline{c},\underline{a}}.
\end{array} \tag{4.2.1}$$

Uniqueness of the induced map $\mathsf{P}\binom{\underline{d}}{\underline{c}} \otimes \mathsf{P}\binom{\underline{b}}{\underline{a}} \to E_f\binom{\underline{d},\underline{b}}{\underline{c},\underline{a}}$ then suffices to show θ is compatible with horizontal composition, and a similar argument verifies compatibility with vertical composition.

The second claim follows from the strengthened universal property above, since an inspection of the definition shows $f : X \to Y$ is a morphism of P-algebras precisely when

$$\begin{array}{ccc} \mathsf{P} & \longrightarrow & E_X \\ \big\downarrow & & \big\downarrow {\scriptstyle f_*} \\ E_Y & \xrightarrow{f^*} & E_{X,Y} \end{array}$$

commutes in $\mathbf{\Sigma}_{\mathcal{E}}^{\mathfrak{C}}$. □

Proposition 4.3. *Suppose the morphism* $\overline{f}^*$ *(respectively,* $\overline{f}_*$*) has the right lifting property with respect to the initial morphism into* P *(in* $\mathbf{PROP}_{\mathcal{E}}^{\mathfrak{C}}$*). Then any* P*-algebra structure on* X *(respectively, on* Y*) extends to make* f *a morphism of* P*-algebras.*

Proof. If $\overline{f}^*$ has the right lifting property with respect to the initial morphism, then any morphism $\eta : \mathsf{P} \to E_X$ lifts to a morphism $\overline{\eta} : \mathsf{P} \to E_f$. (In fact, this lift is unique up to homotopy over E_X.) Now apply Lemma 4.2. □

Remark 4.4. In fact, given a morphism $f : X \to Y$, one could use this lifting property approach to define the class of PROPs for which algebra structures always transfer along f. Note that in Theorem 1.2, the hypotheses are made so that the maps $Y_{\underline{d}}^{f_{\underline{c}}}$ and $f_{\underline{d}}^{X_{\underline{c}}}$ are acyclic fibrations, so it suffices to require P to be cofibrant. However, the same argument implies any cofibrant contractible PROP P has the property that algebra structures extend for any f with f^* (or f_*) fibrations.

Proof of Theorem 1.2. Under the hypotheses of (1) and an equivalent form of the pushout-product axiom, the map $f^* = Y_{\underline{d}}^{f_{\underline{c}}}$ (obtained by exponentiating a fibrant object to an acyclic cofibration) in (4.1.1) is an acyclic fibration in $\mathcal{E}$. Since acyclic fibrations in $\mathcal{E}$ are closed under pullback, and acyclic fibrations in $\mathbf{PROP}_{\mathcal{E}}^{\mathfrak{C}}$ are defined entrywise, this implies $\overline{f}^*$ is an acyclic fibration in $\mathbf{PROP}_{\mathcal{E}}^{\mathfrak{C}}$. Similarly, $\overline{f}_*$ is an acyclic fibration under the hypotheses of (2). As $\mathsf{P} \in \mathbf{PROP}_{\mathcal{E}}^{\mathfrak{C}}$ is assumed to be cofibrant, the claim follows in either case from Proposition 4.3. □

5. Homotopy topological conformal field theory

The purpose of this section is to provide an example to illustrate (i) the Homotopy Invariance Theorem 1.2 and (ii) the necessity of our (colored) PROP approach. We will define a homotopy version of topological conformal field theory, and observe that it is, in fact, a homotopy invariant.

Our discussion will mostly focus on the 1-colored *Segal PROP* (§5.1) and topological conformal field theory (§5.2) following [**Seg88, Seg01, Seg04**] for clarity. Some other sources that discuss TCFT in a similar context are [**CV06, Get94**]. The Segal PROP comes from considering moduli spaces of Riemann surfaces with boundary holes, and it is natural to consider varying (positive) circumferences of these boundary holes as the colors in this setting. The vertical composition for the Segal PROP comes from holomorphically sewing along boundary holes, so the colored generalization would only allow sewing when the circumferences match.

Considering varying circumferences in the boundary holes is not unprecedented. For example, in the setting of string topology, there is a combinatorially defined colored PROP $\mathcal{RCF}(g)$ [**Cha05, CG04**] that is built from spaces of reduced metric Sullivan chord diagrams with genus g. Such a Sullivan chord diagram is a marked *fat graph* (also known as ribbon graph) that represents a surface with genus g that has a certain number of input and output circles in its boundary. These boundary circles are allowed to have different circumferences and these form the set of colors for the colored PROP $\mathcal{RCF}(g)$. However, for the remainder of this section, we will take $\mathfrak{C} = \{*\}$.

5.1. Segal PROP

For integers $m, n \geqslant 1$, let $\mathsf{Se}(m,n)$ be the moduli space of (isomorphism classes of) complex Riemann surfaces whose boundaries consist of $m+n$ labeled holomorphic holes that are mutually non-overlapping. In the literature, $\mathsf{Se}(m,n)$ is sometimes denoted by $\widehat{\mathcal{M}}(m,n)$. The holomorphic holes are actually bi-holomorphic maps from $m+n$ copies of the closed unit disk to the Riemann surface. The first m labeled holomorphic holes are called the *outputs* and the last n are called the *inputs*. Note that these Riemann surfaces M can have arbitrary genera and are *not* required to be connected.

One can visualize a Riemann surface $M \in \mathsf{Se}(m,n)$ as a pair of "alien pants" in which there are n legs (the inputs) and m waists (the outputs). See Figure 5.1. With this picture in mind, such a Riemann surface is also known as a *worldsheet* in the physics literature. In this interpretation, a worldsheet is an embedding of closed strings in space-time. We think of such a Riemann surface M as a machine that provides an operation with n inputs and m outputs.

The collection of moduli spaces $\{\mathsf{Se}(m,n)\colon m,n \geqslant 1\}$ forms a (1-colored) topological PROP Se, called the *Segal PROP*, also known as the *Segal category*. This Segal PROP Se is an honest PROP, in the sense that it is not generated by an operad, hence the need for additional machinery.

Using the characterization of PROPs from Proposition 2.17, its horizontal composition

$$\mathsf{Se}(m_1,n_1) \times \mathsf{Se}(m_2,n_2) \xrightarrow{\otimes = \sqcup} \mathsf{Se}(m_1+m_2, n_1+n_2)$$

is given by disjoint union $M_1 \sqcup M_2$. In other words, put two pairs of alien pants side-by-side. Its vertical composition

$$\mathsf{Se}(m,n) \times \mathsf{Se}(n,k) \xrightarrow{\circ} \mathsf{Se}(m,k), \quad (M,N) \mapsto M \circ N$$

is given by holomorphically sewing the n output holes (the waists) of N with the n input holes (the legs) of M. The Σ_m-Σ_n action on $\mathsf{Se}(m,n)$ is given by permuting the labels of the m output and the n input holomorphic holes.

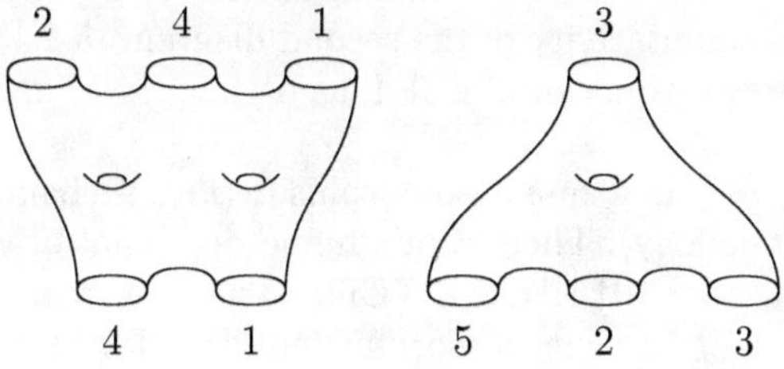

Figure 1: An element in Se(4, 5) with two connected components.

5.2. Topological conformal field theory

We are interested in the chain level algebras of the Segal PROP Se. So we first pass to a suitable category of chain complexes. Let $\mathbf{Ch}(R)$ be the category of non-negatively graded chain complexes of modules over a fixed commutative ring R, in applications a field of characteristic 0. Let C_* be the singular chain functor with coefficients in R from topological spaces to $\mathbf{Ch}(R)$. Applying C_* to the Segal PROP Se, we obtain the *chain Segal PROP* $\mathbf{Se} = C_*(\mathsf{Se})$, which is a (1-colored) PROP over $\mathbf{Ch}(R)$.

A *topological conformal field theory* (TCFT) is defined as an $\mathbf{Se}$-algebra. It is also known as a *string background* in the literature. In other words, a TCFT has an underlying chain complex $A = \{A_n\}$ (the *states space*) together with chain maps

$$\lambda_{m,n} \colon \mathbf{Se}(m,n) \otimes A^{\otimes n} \to A^{\otimes m}$$

for $m, n \geqslant 1$ such that:

1. The map $\lambda_{m,n}$ is Σ_m-Σ_n equivariant.
2. The following two diagrams are commutative, where $B(m,n) = \mathbf{Se}(m,n) \otimes A^{\otimes n}$.

$$\begin{array}{ccc} \mathbf{Se}(m_1,n_1) \otimes \mathbf{Se}(m_2,n_2) \otimes A^{\otimes(n_1+n_2)} & \xrightarrow{\text{switch}} & B(m_1,n_1) \otimes B(m_2,n_2) \\ \downarrow{\scriptstyle \sqcup\otimes Id} & & \downarrow{\scriptstyle \lambda\otimes\lambda} \\ \mathbf{Se}(m_1+m_2,n_1+n_2) \otimes A^{\otimes(n_1+n_2)} & \xrightarrow{\lambda} & A^{\otimes m_1} \otimes A^{\otimes m_2}, \end{array}$$

$$\begin{array}{ccc} \mathbf{Se}(m,n) \otimes \mathbf{Se}(n,k) \otimes A^{\otimes k} & \xrightarrow{Id\otimes\lambda} & \mathbf{Se}(m,n) \otimes A^{\otimes n} \\ \downarrow{\scriptstyle \circ\otimes Id} & & \downarrow{\scriptstyle \lambda} \\ \mathbf{Se}(m,k) \otimes A^{\otimes k} & \xrightarrow{\lambda} & A^{\otimes m}. \end{array} \qquad (5.2.1)$$

A *morphism* of *TCFTs* is a chain map $f \colon A \to B$ that is compatible with the structure maps λ (see (2.21.1)).

To simplify typography, we have abused notations and used $\sqcup$ and $\circ$ here to denote the images of the horizontal and vertical compositions in Se under the singular chain functor C_*. The commutativity of the second diagram (5.2.1) is usually called the *sewing axiom*. One can think of a TCFT as a chain level manifestation of topological string theory.

Now suppose that $f\colon A \to B$ is a quasi-isomorphism (i.e., a chain map that induces an isomorphism on homology). Then from a homotopy point-of-view A and B should be considered "the same." If A has a TCFT structure, then morally so should B, and vice versa, where f becomes a morphism of TCFTs. In other words:

Topological conformal field theory, or a close variant of it, should be homotopy invariant.

For the rest of this section, we will describe a suitable setting in which this is true.

5.3. Homotopy TCFT

We will fit TCFTs into the setting of the Homotopy Invariance Theorem 1.2 by enlarging the category to a homotopy version of TCFT. To have a good notion of homotopy, let us first describe a model category structure on $\mathbf{Ch}(R)$.

In $\mathbf{Ch}(R)$, where R is a field of characteristic 0, a map $f\colon A \to B$ is a:

1. *weak equivalence* if and only if it is a quasi-isomorphism;
2. *fibration* if and only if $f_k\colon A_k \to B_k$ is a surjection for $k > 0$;
3. *cofibration* if and only if it is a degree-wise injection.

Full details of the existence of this model category structure can be found in [**DS95**, **Qui67**]. Moreover, in [**Hov99**] it is shown that this model category is (strongly) cofibrantly generated. Since the 0 object in $\mathbf{Ch}(R)$ is the 0 chain complex, every object $A \in \mathbf{Ch}(R)$ is both fibrant and cofibrant. Note that by the 5-Lemma, an acyclic fibration f is surjective in dimension 0 as well. So the acyclic fibrations in $\mathbf{Ch}(R)$ are exactly the surjective quasi-isomorphisms.

As discussed in the Introduction, the model category $\mathbf{Ch}(R)$ satisfies the hypotheses of Theorem 1.1. Therefore, the category $\mathbf{PROP}(\mathbf{Ch}(R))$ of $\mathfrak{C}$-colored PROPs over $\mathbf{Ch}(R)$ inherits a cofibrantly generated model category structure in which the fibrations and weak equivalences are defined entrywise.

Now consider the chain Segal PROP $\mathbf{Se}$. To use the Homotopy Invariance Theorem 1.2, we need a cofibrant version of $\mathbf{Se}$. There is a cofibrant replacement functor [**Hir02**, 10.5.16] in the model category $\mathbf{PROP}(\mathbf{Ch}(R))$, provided by Quillen's small object argument. Applying this cofibrant replacement functor to $\mathbf{Se}$, we have an acyclic fibration

$$r\colon \mathbf{Se}_\infty \overset{\sim}{\twoheadrightarrow} \mathbf{Se} \tag{5.3.1}$$

in which $\mathbf{Se}_\infty$ is a cofibrant PROP over $\mathbf{Ch}(R)$. We call $\mathbf{Se}_\infty$ *the cofibrant chain Segal PROP*. Since weak equivalences and fibrations in $\mathbf{PROP}(\mathbf{Ch}(R))$ are defined entrywise, each chain map

$$r(m,n)\colon \mathbf{Se}_\infty(m,n) \overset{\sim}{\twoheadrightarrow} \mathbf{Se}(m,n) = C_*(\mathsf{Se}(m,n))$$

is a surjective quasi-isomorphism.

Definition 5.4. The category of *homotopy topological conformal field theories* (HTCFTs), or *homotopy string backgrounds*, is defined to be the category of $\mathbf{Se}_\infty$-algebras.

One can unpack this definition and interpret an HTCFT in terms of structure maps $\lambda_{m,n}$ as in section 5.2. Notice that TCFTs are HTCFTs. Indeed, a TCFT is given by a map $\lambda\colon \mathbf{Se} \to E_A$ of PROPs over $\mathbf{Ch}(R)$, where E_A is the endomorphism PROP of a chain complex A. Composing with r of (5.3.1), we have a map $\lambda r\colon \mathbf{Se}_\infty \to E_A$ of PROPs, which gives an HTCFT structure on A.

We now obtain Corollary 1.3, which is a restatement of Theorem 1.2 in the current setting: $\mathsf{P} = \mathbf{Se}_\infty$ and $\mathcal{E} = \mathbf{Ch}(R)$. Since TCFTs are, in particular, HTCFTs, we also have the following alternative version of Corollary 1.3. Again, keep in ming that all objects in $\mathbf{Ch}(R)$ being both fibrant and cofibrant simplifies the statement.

Corollary 5.5. *Let $f\colon A \to B$ be a map in $\mathbf{Ch}(R)$, where R is a field of characteristic 0.*

1. *Suppose that f is an injective quasi-isomorphism. If A is a TCFT, then there exists an HTCFT structure on B such that f becomes a morphism of HTCFTs.*
2. *Suppose that f is a surjective quasi-isomorphism. If B is a TCFT, then there exists an HTCFT structure on A such that f becomes a morphism of HTCFTs.*

6. Model structure on algebras

The purpose of this section is to prove Theorem 1.4, which says that, under some suitable assumptions on $\mathcal{E}$, there is a lifted model category structure on the category of P-algebras.

We begin with the following key technical result, keeping in mind that $\mathcal{E}^{\mathfrak{C}}$ has a projective model structure by [**Hir02**, 11.6.1]. This result can be considered as a colored PROP version of the discussion in [**Rez96**, §4.1.14], which focuses on operads.

Lemma 6.1. *Suppose finite tensor products of fibrations with fibrant targets remain fibrations in $\mathcal{E}$ and that P is a cofibrant object of $\mathbf{PROP}^{\mathfrak{C}}_{\mathcal{E}}$. Then for any factorization of a morphism of P-algebras $g : A \to C$ with (entrywise) fibrant target as $A \xrightarrow{f} B \xrightarrow{h} C$ with f an acyclic cofibration in $\mathcal{E}^{\mathfrak{C}}$ and h a fibration in $\mathcal{E}^{\mathfrak{C}}$, one may equip B with a P-algebra structure making both f and h morphisms of P-algebras.*

Proof. First, by Lemma 4.2, f is a morphism of P-algebras precisely when the structure maps $\mathsf{P} \to E_A$ and $\mathsf{P} \to E_B$ both descend from a common morphism $\mathsf{P} \to E_f$ in $\mathbf{PROP}^{\mathfrak{C}}_{\mathcal{E}}$ and similarly for h. If we form the following pullback

$$\begin{array}{ccc} E_1 & \longrightarrow & E_h \\ \downarrow & & \downarrow \\ E_f & \longrightarrow & E_B \end{array} \tag{6.1.1}$$

in $\mathbf{PROP}_{\mathcal{E}}^{\mathfrak{C}}$, we can extend it to form the commutative diagram

$$\begin{array}{ccccc} E_1 & \longrightarrow & E_h & \longrightarrow & E_C \\ \downarrow & & \downarrow & & \downarrow \\ E_f & \longrightarrow & E_B & \longrightarrow & E_{B,C} \\ \downarrow & & \downarrow & & \downarrow \\ E_A & \longrightarrow & E_{A,B} & \longrightarrow & E_{A,C} \end{array} \tag{6.1.2}$$

in $\mathbf{\Sigma}_{\mathcal{E}}^{\mathfrak{C}}$. Notice E_g is the pullback of the largest square here, so this induces a morphism in $\mathbf{PROP}_{\mathcal{E}}^{\mathfrak{C}}$, $\theta : E_1 \to E_g$ by the strengthened universal property discussed in the proof of Lemma 4.2 and the fact that the composites $E_1 \to E_C$ and $E_1 \to E_A$ are morphisms in $\mathbf{PROP}_{\mathcal{E}}^{\mathfrak{C}}$ by construction. We will show θ is an acyclic fibration in $\mathbf{PROP}_{\mathcal{E}}^{\mathfrak{C}}$, hence that there must be a lift in the diagram

$$\begin{array}{ccc} & & E_1 \\ & \nearrow & \downarrow \theta \\ \mathsf{P} & \longrightarrow & E_g \end{array} \tag{6.1.3}$$

so the induced composite $\mathsf{P} \to E_1 \to E_B$ gives B a P-algebra structure in which both f and h become morphisms of P-algebras.

In fact, we need only show θ is an acyclic fibration in $\mathbf{\Sigma}_{\mathcal{E}}^{\mathfrak{C}}$, so for the remainder of the proof we work solely in $\mathbf{\Sigma}_{\mathcal{E}}^{\mathfrak{C}}$. Let the following pullback diagrams define their upper left corners

$$\begin{array}{ccc} E_2 & \longrightarrow & E_{B,C} \\ \downarrow & & \downarrow \\ E_{A,B} & \longrightarrow & E_{A,C} \end{array} \qquad \begin{array}{ccc} E_3 & \longrightarrow & E_C \\ \downarrow & & \downarrow \\ E_2 & \longrightarrow & E_{B,C} \end{array} \qquad \begin{array}{ccc} E_4 & \longrightarrow & E_2 \\ \downarrow & & \downarrow \\ E_A & \longrightarrow & E_{A,B} \end{array} \qquad \begin{array}{ccc} E_5 & \longrightarrow & E_3 \\ \downarrow & & \downarrow \\ E_4 & \longrightarrow & E_2 \end{array} \tag{6.1.4}$$

and notice E_5 is then canonically isomorphic to E_g, so we now consider $E_1 \to E_5$. Also notice there is an induced map $E_B \to E_2$, whose entries are precisely the pushout-product maps

$$\begin{tikzcd} B_{\underline{d}}^{B_{\underline{c}}} \arrow[dr] \arrow[drr, "(h_{\underline{d}})_*"] \arrow[ddr, "(f_{\underline{c}})^*"'] & & \\ & E_2\binom{\underline{d}}{\underline{c}} \arrow[r] \arrow[d] & C_{\underline{d}}^{B_{\underline{c}}} \arrow[d, "(f_{\underline{c}})^*"] \\ & B_{\underline{d}}^{A_{\underline{c}}} \arrow[r, "(h_{\underline{d}})_*"'] & C_{\underline{d}}^{A_{\underline{c}}}. \end{tikzcd} \tag{6.1.5}$$

The assumption about tensors preserving fibrations is needed to conclude $h_{\underline{d}}$ is a fibration in $\mathcal{E}$, while $f_{\underline{c}}$ is an acyclic cofibration in $\mathcal{E}$, so the pushout-product axiom implies $E_B \to E_2$ is an (entrywise) acyclic fibration.

Using the universal property of pullbacks, we now have a commutative cube:

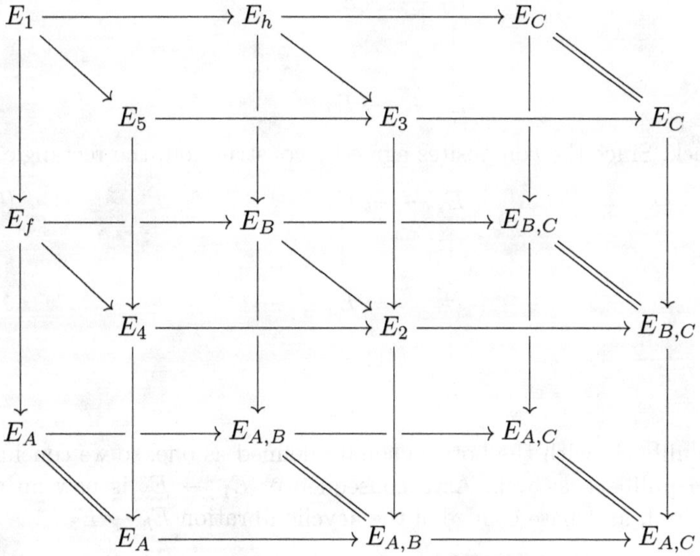

(6.1.6)

Now the diagram

$$\begin{array}{ccccc} E_h & \longrightarrow & E_3 & \longrightarrow & E_C \\ \downarrow & & \downarrow & & \downarrow \\ E_B & \longrightarrow & E_2 & \longrightarrow & E_{B,C} \end{array} \tag{6.1.7}$$

in which the right square and the large rectangle are pullbacks, allows us to conclude the left square is a pullback, hence that $E_h \to E_3$ is an acyclic fibration as a base change of $E_B \to E_2$. Similarly,

$$\begin{array}{ccc} E_f & \longrightarrow & E_B \\ \downarrow & & \downarrow \\ E_4 & \longrightarrow & E_2 \\ \downarrow & & \downarrow \\ E_A & \longrightarrow & E_{A,B} \end{array} \tag{6.1.8}$$

with the rectangle and the bottom square pullbacks, allows us to conclude the top square is a pullback, and so the rectangle in

$$\begin{array}{ccc} E_1 & \longrightarrow & E_h \\ \downarrow & & \downarrow \\ E_f & \longrightarrow & E_B \\ \downarrow & & \downarrow \\ E_4 & \longrightarrow & E_2 \end{array} \tag{6.1.9}$$

is also a pullback. Since the composites agree by construction, the rectangle in

$$\begin{array}{ccc} E_1 & \longrightarrow & E_h \\ \downarrow & & \downarrow \\ E_5 & \longrightarrow & E_3 \\ \downarrow & & \downarrow \\ E_4 & \longrightarrow & E_2 \end{array} \tag{6.1.10}$$

is then also a pullback, with the bottom square defined as one, so we conclude the top square is a pullback as well. As a consequence, $E_1 \to E_5$ is now an acyclic fibration as desired, as a base change of the acyclic fibration $E_h \to E_3$. □

Proof of Theorem 1.4. The forgetful functor U preserves filtered colimits, so we can use Lifting Lemma 3.3. For a fibrant replacement functor, we notice $A \in \mathbf{Alg}(\mathsf{P})$ implies $\tilde{A} \in \mathbf{Alg}(\tilde{\mathsf{P}})$ by functoriality of the symmetric monoidal fibrant replacement functor $(\tilde{-})$. Here $\tilde{\mathsf{P}}$ indicates the fibrant replacement in $\mathbf{PROP}^{\mathfrak{C}}_{\mathcal{E}}$, obtained from P by applying the symmetric monoidal fibrant replacement functor to each component and map of P. Now $\tilde{A}$ equipped with the P-algebra structure associated to the composite $\mathsf{P} \to \tilde{\mathsf{P}} \to E_{\tilde{A}}$ serves as a fibrant replacement for A in $\mathbf{Alg}(\mathsf{P})$.

For the required path object construction, let A be an entrywise fibrant P-algebra, and form a strong path object in $\mathcal{E}^{\mathfrak{C}}$, by factoring the diagonal map $A \xrightarrow{f} B \xrightarrow{h} A \times A$ with f an acyclic cofibration and h a fibration. Since products are formed entrywise, $A \times A$ is also entrywise fibrant, so Lemma 6.1 implies B may be given a P-algebra structure such that both f and h are maps of P-algebras. By construction, f is a weak equivalence (although we no longer claim it has the requisite lifting property of a cofibration of P-algebras) and h is a fibration, providing the requisite path object for fibrant objects in $\mathbf{Alg}(\mathsf{P})$. □

7. Quillen pairs and Quillen equivalences

We would like to understand how various adjoint pairs interact with our model structures. Throughout this section, we will simply assume the relevant "projective" model structures exist, and focus on comparing them. Keep in mind that the model structures may be constructed via methods technically different from those introduced in this article, but the results of this section will still apply. We will also speak about algebras over PROPs, but similar definitions for algebras over *vPROPs* or *hPROPs* (Definition 8.1) would work equally well in what follows.

Lemma 7.1. *Suppose (L, R) is an adjoint pair, $L : \mathcal{E} \to \mathcal{E}'$, where both functors are symmetric monoidal and R preserves arbitrary coproducts and colimits indexed on connected groupoids. Then they induce adjoint pairs at the level of $\mathfrak{C}$-colored Σ-bimodules, $\mathfrak{C}$-colored PROPs, vPROPs, hPROPs, and algebras.*

Proof. First, (L, R) extend entrywise to bimodules, as a diagram category. In the three PROP cases, we exploit the fact that L and R must preserve the relevant monoidal operations, which are constructed from the symmetric monoidal operation $\otimes$ in a natural way. Notice here we require the assumption on R, since the constructions $\Sigma_{\underline{b}}$ (for $\boxtimes_v$) and $\Sigma_{\underline{d};\underline{c}} \times \Sigma_{\underline{b};\underline{a}}$ (for $\boxtimes_h$) involve coproducts and colimits indexed on connected groupoids.

Finally, for the case of algebras, given Y an $L\mathsf{P}$-algebra, there is an induced P-algebra structure on RY with structure maps

$$\begin{array}{ccccc} \mathsf{P}\binom{\underline{d}}{\underline{c}} \otimes RY_{\underline{c}} & \xrightarrow{\eta\otimes 1} & RL\mathsf{P}\binom{\underline{d}}{\underline{c}} \otimes RY_{\underline{c}} & \longrightarrow & R\left(L\mathsf{P}\binom{\underline{d}}{\underline{c}} \otimes Y_{\underline{c}}\right) \\ & & & & \downarrow \\ & & & & RY_{\underline{d}} \end{array} \tag{7.1.1}$$

with the last map given by the structure map of Y as an $L\mathsf{P}$-algebra. Once again, we have extended L and R entrywise here in the product category underlying algebras, where they remain an adjoint pair. Then an exercise involving the triangular identities for an adjunction and the assumption that both L and R are symmetric monoidal implies

$$\begin{array}{ccc} \mathsf{P}\binom{\underline{d}}{\underline{c}} \otimes X_{\underline{c}} & \longrightarrow & X_{\underline{d}} \\ \downarrow & & \downarrow \\ \mathsf{P}\binom{\underline{d}}{\underline{c}} \otimes RY_{\underline{c}} & \longrightarrow & RY_{\underline{d}} \end{array} \tag{7.1.2}$$

commutes in $\mathcal{E}$ precisely when the corresponding diagram

$$\begin{array}{ccc} L\mathsf{P}\binom{\underline{d}}{\underline{c}} \otimes LX_{\underline{c}} & \longrightarrow & LX_{\underline{d}} \\ \downarrow & & \downarrow \\ L\mathsf{P}\binom{\underline{d}}{\underline{c}} \otimes Y_{\underline{c}} & \longrightarrow & Y_{\underline{d}} \end{array} \tag{7.1.3}$$

commutes in $\mathcal{E}'$, which completes the proof. □

Proposition 7.2. *Suppose (L, R) is a strong Quillen pair, $L : \mathcal{E} \to \mathcal{E}'$, where the component functors are symmetric monoidal and R preserves arbitrary coproducts and colimits indexed on connected groupoids. Then they induce strong Quillen pairs at the level of $\mathfrak{C}$-colored Σ-bimodules, $\mathfrak{C}$-colored PROPs, vPROPs, hPROPs, and* *algebras. If, in addition, (L, R) form a Quillen equivalence, then the same is true of each induced adjunction, whenever the entries of cofibrant objects are cofibrant in $\mathcal{E}$.*

Proof. To verify that each adjunction forms a strong Quillen pair, it is simplest to observe that (acyclic) fibrations are defined in terms of entries in $\mathcal{E}$ or $\mathcal{E}'$, hence preserved by the entrywise right adjoints of Lemma 7.1. For verifying the induced Quillen equivalence condition, it is key to know that the entries of a cofibrant object are cofibrant objects in $\mathcal{E}$ in each case. □

Remark 7.3. The ability to change our underlying model category up to Quillen equivalence is particularly gratifying since our technical assumptions have incidentally excluded important examples like (pointed) topological spaces. Now we can conveniently say working instead with (pointed) simplicial sets is equally valid, as the cofibrant entries condition is vacuously satisfied over simplicial sets. This cofibrant entries condition is also automatically satisfied for any category of $\mathfrak{C}$-colored Σ-bimodules, as in Proposition 3.5.

There are also adjoint pairs induced by changing colors; thus we assume we have a map $\alpha : \mathfrak{C} \to \mathfrak{C}'$ of sets.

Lemma 7.4. *There are adjoint pairs given by changing colors, denoted $(\alpha_!, \alpha^*)$, at the level of colored Σ-bimodules, colored PROPs, vPROPs, hPROPs, and algebras.*

Proof. Here α^* is just a precomposition functor, where we again use α to denote the induced functor at the level of pairs of profiles $\mathcal{P}_l(\mathfrak{C}) \times \mathcal{P}_r(\mathfrak{C}) \to \mathcal{P}_l(\mathfrak{C}') \times \mathcal{P}_r(\mathfrak{C}')$. Hence, the usual left Kan extension formula for diagram categories gives the left adjoint at the level of colored Σ-bimodules. Alternatively, we notice defining

$$\alpha_! \mathsf{P}\binom{\underline{d}'}{\underline{c}'} = \coprod_{\alpha(\underline{c})=\underline{c}', \alpha(\underline{d})=\underline{d}'} \mathsf{P}\binom{\underline{d}}{\underline{c}}$$

has the requisite universal property to define this left adjoint. With this formulation, it is not particularly difficult to verify that the operations from P induce those of $\alpha_! \mathsf{P}$ for the three PROP cases.

For the case of algebras, we notice a similar coproduct formula will define the left adjoint to the precomposition functor we will again call α^* at the level of product categories underlying algebras. Then we exploit the fact that $\otimes$ must distribute over coproducts by the existence of exponential objects, and the universal properties of coproducts to verify directly that the squares defining $\alpha_! X \to Y$ as a morphism of $\alpha_! \mathsf{P}$-algebras commute precisely when those defining $X \to \alpha^* Y$ as a morphism of P-algebras commute. □

Proposition 7.5. *There is a strong Quillen pair $(\alpha_!, \alpha^*)$ at the level of colored Σ-bimodules, colored PROPs, vPROPs, hPROPs, and algebras associated to any $\alpha : \mathfrak{C} \to \mathfrak{C}'$.*

Proof. It again follows directly from the definitions that α^* preserves the class of (acyclic) fibrations in each case. □

Remark 7.6. Notice in this change of colors adjunction, there is no claim of a Quillen equivalence. However, the following more general result does imply a Quillen equivalence of projective structures as expected in the case of a bijection of colors.

Corollary 7.7. *If $\alpha : \mathfrak{C} \to \mathfrak{C}'$ is injective, there is a strong Quillen equivalence $(\alpha_!, \alpha^*)$ at the level of colored Σ-bimodules, colored PROPs, vPROPs, hPROPs, and algebras, only after using the modified projective structures (as in Corollary 3.7 and Remark 3.13), choosing only orbits consisting of colors in the image of α.*

Proof. The point of using the modified structure on the $\mathfrak{C}'$-colored structures is that weak equivalences are then created by the adjunction, while it remains a strong Quillen pair. Hence, the pair is a Quillen equivalence precisely when the unit of the derived adjunction $X \to \alpha^* Q(\alpha_! X)$ is a weak equivalence for every cofibrant $\mathfrak{C}$-colored object X, where Q indicates a fibrant replacement in the $\mathfrak{C}'$-colored structures [**Hov99**, 1.3.16]. However, when α is injective, the formula for $\alpha_!$ given in the proof of Lemma 7.4 implies the unit of the adjunction $X \to \alpha^*\alpha_! X$ is the identity. As noted above, α^* creates weak equivalences with this modified projective structure on the $\mathfrak{C}'$-colored structures, so $\alpha_! X \to Q(\alpha_! X)$ a weak equivalence in this structure means precisely that $\alpha^*\alpha_! X \to \alpha^* Q(\alpha_! X)$ is a weak equivalence, and so $X \to \alpha^* Q(\alpha_! X)$ is a weak equivalence. □

8. Another description of colored PROPs

In Definition 2.16, we defined $\mathfrak{C}$-colored PROPs as $\boxtimes_v$-monoidal $\boxtimes_h$-monoids. In this section, we observe that the roles of $\boxtimes_v$ and $\boxtimes_h$ can be interchanged. In other words, $\mathfrak{C}$-colored PROPs can be equivalently defined as $\boxtimes_h$-monoidal $\boxtimes_v$-monoids. We can first build $\mathfrak{C}$-colored Σ-bimodules that have a horizontal composition $\otimes$ using a monoidal structure $\boxtimes_h$ on $\mathbf{\Sigma}^{\mathfrak{C}}_{\mathcal{E}}$. Then we build the vertical composition on top of the horizontal composition by considering a monoidal structure $\boxtimes_v$ on $\mathbf{Mon}(\mathbf{\Sigma}^{\mathfrak{C}}_{\mathcal{E}}, \boxtimes_h)$. The monoids in this last monoidal category are exactly the $\mathfrak{C}$-colored PROPs.

In more details, the functor $\boxtimes_h$ defined in (2.14.1) gives a monoidal product $\boxtimes_h \colon \mathbf{\Sigma}^{\mathfrak{C}}_{\mathcal{E}} \times \mathbf{\Sigma}^{\mathfrak{C}}_{\mathcal{E}} \to \mathbf{\Sigma}^{\mathfrak{C}}_{\mathcal{E}}$ on $\mathbf{\Sigma}^{\mathfrak{C}}_{\mathcal{E}}$, since its definition only involves the $\mathfrak{C}$-colored Σ-bimodule structures on the two arguments.

Definition 8.1. Denote by $\mathbf{hPROP}^{\mathfrak{C}}_{\mathcal{E}}$ the category of monoids in the monoidal category $(\mathbf{\Sigma}^{\mathfrak{C}}_{\mathcal{E}}, \boxtimes_h)$ (without unit), whose objects are called **hPROPs**.

From the definitions of $\boxtimes_h$ and an associative map $\mathsf{P} \boxtimes_h \mathsf{P} \to \mathsf{P}$, an hPROP consists of exactly the data:

1. An object $\mathsf{P} \in \mathbf{\Sigma}^{\mathfrak{C}}_{\mathcal{E}}$.
2. An associative horizontal composition $\otimes$ as in (2.17.1).

Equivalently, the horizontal composition can be described in the level of $\mathcal{E}$ as an associative operation (2.17.2) that is bi-equivariant (2.17.3).

To build the vertical compositions on top of the horizontal compositions, we consider the monoidal product $\boxtimes_v$ on $\mathbf{\Sigma}^{\mathfrak{C}}_{\mathcal{E}}$ (Lemma 2.11). We need to upgrade $\boxtimes_v$

to a monoidal product on $\mathbf{hPROP}^{\mathfrak{C}}_{\mathcal{E}}$. So suppose that P and Q are hPROPs. Since $\mathsf{P} \boxtimes_v \mathsf{Q} \in \mathbf{\Sigma}^{\mathfrak{C}}_{\mathcal{E}}$, to make $\mathsf{P} \boxtimes_v \mathsf{Q}$ into an hPROP, we need a horizontal composition

$$(\mathsf{P} \boxtimes_v \mathsf{Q})\binom{[\underline{d}_1]}{[\underline{c}_1]} \boxdot (\mathsf{P} \boxtimes_v \mathsf{Q})\binom{[\underline{d}_2]}{[\underline{c}_2]} \xrightarrow{\otimes} (\mathsf{P} \boxtimes_v \mathsf{Q})\binom{[\underline{d}_1, \underline{d}_2]}{[\underline{c}_1, \underline{c}_2]} \tag{8.1.1}$$

that is induced by those on P and Q. Since $\boxtimes_v$ is defined as a coproduct, to define (8.1.1) it suffices to define the operation

$$X \boxdot Y \xrightarrow{\otimes} Z \tag{8.1.2}$$

in $\mathcal{E}^{\Sigma_{\underline{d};\underline{c}}}$, where

$$X = \mathsf{P}\binom{[\underline{d}_1]}{[\underline{b}_1]} \otimes_{\Sigma_{\underline{b}_1}} \mathsf{Q}\binom{[\underline{b}_1]}{[\underline{c}_1]},\; Y = \mathsf{P}\binom{[\underline{d}_2]}{[\underline{b}_2]} \otimes_{\Sigma_{\underline{b}_2}} \mathsf{Q}\binom{[\underline{b}_2]}{[\underline{c}_2]},\; Z = \mathsf{P}\binom{[\underline{d}]}{[\underline{b}]} \otimes_{\Sigma_{\underline{b}}} \mathsf{Q}\binom{[\underline{b}]}{[\underline{c}]},$$

$\underline{d} = (\underline{d}_1, \underline{d}_2)$, $\underline{c} = (\underline{c}_1, \underline{c}_2)$, and $\underline{b} = (\underline{b}_1, \underline{b}_2)$. Recall that $\boxdot = K\iota$, where K is a left Kan extension (2.13.5). Using the universal properties of left Kan extensions, to define (8.1.2) it suffices to define the operation

$$X \otimes Y \xrightarrow{\otimes} \mathcal{E}^i(Z) \in \mathcal{E}^{\Sigma_{\underline{d}_1} \times \Sigma_{\underline{d}_2} \times \Sigma^{op}_{\underline{c}_1} \times \Sigma^{op}_{\underline{c}_2}}, \tag{8.1.3}$$

where $\mathcal{E}^i$ and $X \otimes Y$ are defined in (2.13.4) and (2.13.3), respectively. To define the operation (8.1.3), we go back to the definition of the functor $\otimes_{\Sigma_{\underline{b}}}$ ((2.8.3) and (2.8.4)).

Consider an object $(\sigma_1 \underline{d}_1; \sigma_2 \underline{d}_2; \underline{c}_1 \tau_1^{-1}; \underline{c}_2 \tau_2^{-1}) \in \Sigma_{\underline{d}_1} \times \Sigma_{\underline{d}_2} \times \Sigma^{op}_{\underline{c}_1} \times \Sigma^{op}_{\underline{c}_2}$. Then

$$\begin{aligned}(X \otimes Y)\left(\sigma_1 \underline{d}_1; \sigma_2 \underline{d}_2; \underline{c}_1 \tau_1^{-1}; \underline{c}_2 \tau_2^{-1}\right) &= X\binom{\sigma_1 \underline{d}_1}{\underline{c}_1 \tau_1^{-1}} \otimes Y\binom{\sigma_2 \underline{d}_2}{\underline{c}_2 \tau_2^{-1}} \\ &= (\operatorname{colim} D_X) \otimes (\operatorname{colim} D_Y),\end{aligned}$$

where

$$\begin{aligned} D_X &= D_X\left(\mathsf{P}\binom{[\underline{d}_1]}{[\underline{b}_1]}, \mathsf{Q}\binom{[\underline{b}_1]}{[\underline{c}_1]}; \sigma_1 \underline{d}_1, \underline{c}_1 \tau_1^{-1}\right) \in \mathcal{E}^{\Sigma_{\underline{b}_1}}, \\ D_Y &= D_Y\left(\mathsf{P}\binom{[\underline{d}_2]}{[\underline{b}_2]}, \mathsf{Q}\binom{[\underline{b}_2]}{[\underline{c}_2]}; \sigma_2 \underline{d}_2, \underline{c}_2 \tau_2^{-1}\right) \in \mathcal{E}^{\Sigma_{\underline{b}_2}}\end{aligned}$$

are the diagrams defined in (2.8.1) and (2.8.2). Likewise, we have

$$\mathcal{E}^i(Z)(\sigma_1 \underline{d}_1; \sigma_2 \underline{d}_2; \underline{c}_1 \tau_1^{-1}; \underline{c}_2 \tau_2^{-1}) = \operatorname{colim} D_Z,$$

where

$$D_Z = D_Z\left(\mathsf{P}\binom{[\underline{d}]}{[\underline{b}]}, \mathsf{Q}\binom{[\underline{b}]}{[\underline{c}]}; (\sigma_1 \underline{d}_1, \sigma_2 \underline{d}_2), (\underline{c}_1 \tau_1^{-1}, \underline{c}_2 \tau_2^{-1})\right) \in \mathcal{E}^{\Sigma_{\underline{b}}}.$$

Let $\mu_1 \underline{b}_1 \in \Sigma_{\underline{b}_1}$ and $\mu_2 \underline{b}_2 \in \Sigma_{\underline{b}_2}$. Then we have a natural map

$$\varphi(\mu_1 \underline{b}_1, \mu_2 \underline{b}_2) \colon D_X(\mu_1 \underline{b}_1) \otimes D_Y(\mu_2 \underline{b}_2) \to \operatorname{colim} D_Z \tag{8.1.4}$$

in $\mathcal{E}$ that is defined as the composite:

$$\begin{array}{ccc}
D_{X,Y} & = & \left[\mathsf{P}\begin{pmatrix}\sigma_1\underline{d}_1\\ \underline{b}_1\mu_1^{-1}\end{pmatrix} \otimes \mathsf{Q}\begin{pmatrix}\mu_1\underline{b}_1\\ \underline{c}_1\tau_1^{-1}\end{pmatrix}\right] \otimes \left[\mathsf{P}\begin{pmatrix}\sigma_2\underline{d}_2\\ \underline{b}_2\mu_2^{-1}\end{pmatrix} \otimes \mathsf{Q}\begin{pmatrix}\mu_2\underline{b}_2\\ \underline{c}_2\tau_2^{-1}\end{pmatrix}\right] \\
 & & \downarrow \cong\ \text{switch} \\
 & & \left[\mathsf{P}\begin{pmatrix}\sigma_1\underline{d}_1\\ \underline{b}_1\mu_1^{-1}\end{pmatrix} \otimes \mathsf{P}\begin{pmatrix}\sigma_2\underline{d}_2\\ \underline{b}_2\mu_2^{-1}\end{pmatrix}\right] \otimes \left[\mathsf{Q}\begin{pmatrix}\mu_1\underline{b}_1\\ \underline{c}_1\tau_1^{-1}\end{pmatrix} \otimes \mathsf{Q}\begin{pmatrix}\mu_2\underline{b}_2\\ \underline{c}_2\tau_2^{-1}\end{pmatrix}\right] \\
\downarrow \varphi(\mu_1\underline{b}_1,\mu_2\underline{b}_2) & & \downarrow (\otimes,\otimes) \\
 & & \mathsf{P}\begin{pmatrix}\sigma_1\underline{d}_1, \sigma_2\underline{d}_2\\ \underline{b}_1\mu_1^{-1}, \underline{b}_2\mu_2^{-1}\end{pmatrix} \otimes \mathsf{Q}\begin{pmatrix}\mu_1\underline{b}_1, \mu_2\underline{b}_2\\ \underline{c}_1\tau_1^{-1}, \underline{c}_2\tau_2^{-1}\end{pmatrix} \\
 & & \Vert \\
\operatorname{colim} D_Z & \xleftarrow{\ \eta\ } & D_Z(\mu_1\underline{b}_1, \mu_2\underline{b}_2).
\end{array}$$

Here $D_{X,Y} = D_X(\mu_1\underline{b}_1) \otimes D_Y(\mu_2\underline{b}_2)$, and the map $(\otimes,\otimes)$ has components the horizontal compositions (in the form (2.17.2)) in P and Q, respectively. The map η is the natural map that comes with a colimit.

The maps $\varphi(\mu_1\underline{b}_1, \mu_2\underline{b}_2)$ are natural with respect to $\mu_1\underline{b}_1$ and $\mu_2\underline{b}_2$. Now we fix $\mu_1\underline{b}_1 \in \Sigma_{\underline{b}_1}$ and let $\mu_2\underline{b}_2$ vary through the category $\Sigma_{\underline{b}_2}$. Using the commutativity of $\otimes$ in $\mathcal{E}$ with colimits, we obtain an induced map

$$\varphi(\mu_1\underline{b}_1)\colon D_X(\mu_1\underline{b}_1) \otimes (\operatorname{colim} D_Y) \to \operatorname{colim} D_Z.$$

Now, letting $\mu_1\underline{b}_1$ vary through the category $\Sigma_{\underline{b}_1}$, we obtain an induced map

$$\varphi\colon\ (\operatorname{colim} D_X) \otimes (\operatorname{colim} D_Y) \to \operatorname{colim} D_Z.$$

This map φ is the required map (8.1.3) when applied to a typical object

$$(\sigma_1\underline{d}_1; \sigma_2\underline{d}_2; \underline{c}_1\tau_1^{-1}; \underline{c}_2\tau_2^{-1}) \in \Sigma_{\underline{d}_1} \times \Sigma_{\underline{d}_2} \times \Sigma_{\underline{c}_1}^{op} \times \Sigma_{\underline{c}_2}^{op}.$$

The naturality of the construction φ with respect to maps in $\Sigma_{\underline{d}_1} \times \Sigma_{\underline{d}_2} \times \Sigma_{\underline{c}_1}^{op} \times \Sigma_{\underline{c}_2}^{op}$ is easy to check. So we have constructed the operation (8.1.3), and hence the operation (8.1.1). The associativity of (8.1.1) follows from that of (8.1.3), which in turn follows from the naturality of the construction φ and the associativity of the horizontal compositions in P and Q. Thus, we have shown that $\boxtimes_v$ (Lemma 2.11) extends to a monoidal product on the category $\mathbf{hPROP}_{\mathcal{E}}^{\mathfrak{C}}$.

As before, we consider the category of monoids $\mathbf{Mon}\left(\mathbf{hPROP}_{\mathcal{E}}^{\mathfrak{C}}, \boxtimes_v\right)$ in the monoidal category $\left(\mathbf{hPROP}_{\mathcal{E}}^{\mathfrak{C}}, \boxtimes_v\right)$. Unwrapping the meaning of a monoid, it is straightforward to check that $\mathbf{Mon}\left(\mathbf{hPROP}_{\mathcal{E}}^{\mathfrak{C}}, \boxtimes_v\right)$ is canonically isomorphic to the category $\mathbf{PROP}_{\mathcal{E}}^{\mathfrak{C}}$ (Proposition 2.17). In this case, the interchange rule says that the monoid map $\circ : \mathsf{P} \boxtimes_v \mathsf{P} \to \mathsf{P}$ is a map of hPROPs. This is, in fact, equivalent to the original interchange rule (2.17.4) due to its symmetry. In summary, $\mathfrak{C}$-colored PROPs can be equivalently described as:

1. $\boxtimes_v$-monoidal $\boxtimes_h$-monoids, as in Proposition 2.17, or
2. $\boxtimes_h$-monoidal $\boxtimes_v$-monoids.

Symbolically, we have

$$\begin{aligned}\mathbf{PROP}^{\mathfrak{C}}_{\mathcal{E}} &= \mathbf{Mon}\left(\mathbf{vPROP}^{\mathfrak{C}}_{\mathcal{E}}, \boxtimes_h\right) = \mathbf{Mon}\left(\mathbf{Mon}\left(\mathbf{\Sigma}^{\mathfrak{C}}_{\mathcal{E}}, \boxtimes_v\right), \boxtimes_h\right) \\ &\cong \mathbf{Mon}\left(\mathbf{hPROP}^{\mathfrak{C}}_{\mathcal{E}}, \boxtimes_v\right) = \mathbf{Mon}\left(\mathbf{Mon}\left(\mathbf{\Sigma}^{\mathfrak{C}}_{\mathcal{E}}, \boxtimes_h\right), \boxtimes_v\right).\end{aligned}$$

In the first description, we consider a $\mathfrak{C}$-colored PROP P as a $\mathfrak{C}$-colored Σ-bimodule with a vertical composition $\circ$ (i.e., a vPROP) together with a horizontal composition $\otimes$ that is built on top of $\circ$. In the second description, we consider P as a $\mathfrak{C}$-colored Σ-bimodule with a horizontal composition $\otimes$ (i.e., an hPROP) together with a vertical composition $\circ$ that is built on top of $\otimes$.

9. Colored operads and colored PROPs

In this section, we construct the (Quillen) adjunction between colored operads and colored PROPs (Proposition 9.2 and Corollary 9.3), generalizing what is often used in the 1-colored chain case [**Mar08**, Example 60]. In one direction, it associates to a colored operad the free colored PROP generated by it. This free functor involves a functor $\boxdot$ (Lemma 2.14) that is the main ingredient of the monoidal product $\boxtimes_h$ (2.14.1). In the other direction, it associates to a colored PROP its underlying colored operad. In fact, we will show that a modified projective structure on $\mathbf{PROP}^{\mathfrak{C}}_{\mathcal{E}}$ can be chosen to turn this adjunction into a strong Quillen equivalence (Proposition 9.4). As a consequence, the homotopy theory of $\mathbf{PROP}^{\mathfrak{C}}_{\mathcal{E}}$ in its projective structure, having fewer weak equivalences than this modified projective structure, is a refinement of the projective homotopy theory for colored operads. We will also see that, for a colored operad O, its free colored PROP O_{prop} has an equivalent category of algebras (Corollary 9.5). In other words, going from a colored operad O to the colored PROP O_{prop} does not change the algebras.

9.1. Colored operads and colored PROPs

We refer the reader to [**May72, May97**] for the well-known definitions of operads, endomorphism operads, and algebras over an operad. The definitions in the colored case can be found in, e.g., [**Mar04**, Section 2]. The category of $\mathfrak{C}$-colored operads over $\mathcal{E}$ is denoted by $\mathbf{Operad}^{\mathfrak{C}}_{\mathcal{E}}$.

Proposition 9.2. *There is a pair of adjoint functors*

$$(-)_{prop} \colon \mathbf{Operad}^{\mathfrak{C}}_{\mathcal{E}} \rightleftarrows \mathbf{PROP}^{\mathfrak{C}}_{\mathcal{E}} \colon U \tag{9.2.1}$$

between the categories of $\mathfrak{C}$-colored PROPs and $\mathfrak{C}$-colored operads, where the right adjoint U is the forgetful functor.

Proof. First we construct the forgetful functor U. Suppose that $d, c_i, b^i_j \in \mathfrak{C}$ are colors, where $1 \leqslant i \leqslant n$ and, for each i, $1 \leqslant j \leqslant k_i$. Write $\underline{c} = (c_1, \ldots, c_n)$, $\underline{b}^i = (b^i_1, \ldots, b^i_{k_i})$, and $\underline{b} = (\underline{b}^1, \ldots, \underline{b}^n)$. If P is a $\mathfrak{C}$-colored PROP, then the components in the $\mathfrak{C}$-colored operad $U\mathsf{P}$ are

$$(U\mathsf{P})\binom{d}{c_1,\ldots,c_n} = \mathsf{P}\binom{d}{c_1,\ldots,c_n} = \mathsf{P}\binom{d}{\underline{c}}. \tag{9.2.2}$$

The structure map ρ of the $\mathfrak{C}$-colored operad $U\mathsf{P}$ is the composition

$$\begin{array}{ccc} \mathsf{P}\binom{d}{\underline{c}} \otimes \mathsf{P}\binom{c_1}{\underline{b}^1} \otimes \cdots \otimes \mathsf{P}\binom{c_n}{\underline{b}^n} & & \\ \Big\downarrow {\scriptstyle Id\otimes(\text{horizontal})} & \searrow^{\rho} & \\ \mathsf{P}\binom{d}{\underline{c}} \otimes \mathsf{P}\binom{\underline{c}}{\underline{b}} & \xrightarrow{\circ} & \mathsf{P}\binom{d}{\underline{b}}. \end{array} \tag{9.2.3}$$

The associativity of the horizontal and the vertical compositions in P together with the interchange rule (2.17.4) imply that ρ is associative. The equivariance of ρ follows from those of $\otimes$ and $\circ$.

Now we construct the unique colored PROP O_{prop} generated by a colored operad O. Let O be a $\mathfrak{C}$-colored operad with components

$$\mathsf{O}\binom{d}{c_1,\ldots,c_n} = \mathsf{O}\binom{d}{\underline{c}}$$

for $d, c_i \in \mathfrak{C}$. First we define the underlying $\mathfrak{C}$-colored Σ-bimodule of O_{prop}. We have to specify the diagrams

$$\mathsf{O}_{prop}\binom{[\underline{d}]}{[\underline{c}]} \in \mathcal{E}^{\Sigma_{\underline{d};\underline{c}}} = \mathcal{E}^{\Sigma_{\underline{d}} \times \Sigma^{op}_{\underline{c}}},$$

where $\underline{d} = (d_1, \ldots, d_m)$ and $\underline{c} = (c_1, \ldots, c_n)$ are $\mathfrak{C}$-profiles. To each partition $r_1 + \cdots + r_m = n$ of n with each $r_i \geqslant 1$, we can associate to the $\mathfrak{C}$-colored operad O the diagrams

$$\mathsf{O}\binom{[d_i]}{[\underline{c}_i]} \in \mathcal{E}^{\Sigma_{d_i;\underline{c}_i}} = \mathcal{E}^{\Sigma_{d_i} \times \Sigma^{op}_{\underline{c}_i}} = \mathcal{E}^{\{*\} \times \Sigma^{op}_{\underline{c}_i}}$$

for $1 \leqslant i \leqslant m$, where

$$\underline{c}_i = \left(c_{r_1+\cdots+r_{i-1}+1}, \ldots, c_{r_1+\cdots+r_i}\right).$$

Recall the associative functor

$$\boxdot = K\iota \colon \mathcal{E}^{\Sigma_{\underline{d}_1;\underline{c}_1}} \times \mathcal{E}^{\Sigma_{\underline{d}_2;\underline{c}_2}} \to \mathcal{E}^{\Sigma_{(\underline{d}_1,\underline{d}_2);(\underline{c}_1,\underline{c}_2)}}$$

from Lemma 2.14, where ι is an inclusion functor and K is a left Kan extension. Using the associativity of $\boxdot$, we define the object

$$\mathsf{O}_{prop}\binom{[\underline{d}]}{[\underline{c}]} = \coprod_{r_1+\cdots+r_m=n} \mathsf{O}\binom{[d_1]}{[\underline{c}_1]} \boxdot \cdots \boxdot \mathsf{O}\binom{[d_m]}{[\underline{c}_m]} \tag{9.2.4}$$

in $\mathcal{E}^{\Sigma_{\underline{d};\underline{c}}}$, where the coproduct is taken over all the partitions $r_1 + \cdots + r_m = n$ with each $r_i \geqslant 1$. This defines O_{prop} as an object in $\mathbf{\Sigma}^{\mathfrak{C}}_{\mathcal{E}}$.

The horizontal composition in O_{prop} is given by concatenation of $\boxdot$ products and inclusion of summands. Using the universal properties of left Kan extensions, the vertical composition in O_{prop} is uniquely determined by the operad composition in O. It is straightforward to check that $(-)_{prop}$ is left adjoint to the forgetful functor U from $\mathfrak{C}$-colored PROPs to $\mathfrak{C}$-colored operads. □

Note that the left adjoint $(-)_{prop}$ is an embedding. In fact, for a $\mathfrak{C}$-color operad O, it follows from the definitions of $(-)_{prop}$ and U that $\mathsf{O} = U(\mathsf{O}_{prop})$.

Under the hypotheses of Theorem 1.1, the category $\mathbf{Operad}^{\mathfrak{C}}_{\mathcal{E}}$ of $\mathfrak{C}$-colored operads in $\mathcal{E}$ also has a cofibrantly generated model category structure in which the fibrations and weak equivalences are defined entrywise in $\mathcal{E}$ (see [**BM03**, Theorem 3.2] and [**BM07**, Theorem 2.1 and Example 1.5.7]).

Corollary 9.3. *Under the hypotheses of Theorem 1.1, the adjoint pair* $((-)_{prop}, U)$ *in Proposition 9.2 is a Quillen pair.*

Proof. From the definition (9.2.2), U preserves (acyclic) fibrations. □

Proposition 9.4. *Under the hypotheses of Theorem 1.1, the adjoint pair* $((-)_{prop}, U)$ *in Proposition 9.2 is a Quillen equivalence, if* $\mathbf{PROP}^{\mathfrak{C}}_{\mathcal{E}}$ *is given the modified projective structure (as discussed in Remark 3.13) determined by components with a single object in the target.*

Proof. This proof is quite similar to that for Corollary 7.7. Again, the choice of modified projective structure implies the right adjoint creates weak equivalences, with the unit of the adjunction an isomorphism. As before, this implies the right adjoint preserves the weak equivalence built into the unit of the derived adjunction, so [**Hov99**, 1.3.16] suffices. □

We now observe that passing from a colored operad O to the colored PROP O_{prop} does not alter the category of algebras.

Corollary 9.5. *Let* O *be a* $\mathfrak{C}$*-colored operad. Then there are functors*

$$\Phi\colon \mathbf{Alg}(\mathsf{O}) \rightleftarrows \mathbf{Alg}(\mathsf{O}_{prop})\colon \Psi$$

that give an equivalence between the categories $\mathbf{Alg}(\mathsf{O})$ *of* O*-algebras and* $\mathbf{Alg}(\mathsf{O}_{prop})$ *of* O_{prop}*-algebras.*

Proof. First observe that in each of the two categories, an algebra has an underlying $\mathfrak{C}$-graded object $\{A_c\}$. Given an O-algebra A, the formula (9.2.4) for O_{prop} together with the universal properties of left Kan extensions extend A to an O_{prop}-algebra. This is the functor Φ.

Conversely, an O_{prop}-algebra is a map $\lambda\colon \mathsf{O}_{prop} \to E_X$ of $\mathfrak{C}$-colored PROPs, where E_X is the $\mathfrak{C}$-colored endomorphism PROP of a $\mathfrak{C}$-graded object $X = \{X_c\}$. Using the free-forgetful adjunction from Proposition 9.2, this O_{prop}-algebra is equivalent to a map $\lambda'\colon \mathsf{O} \to U(E_X)$ of $\mathfrak{C}$-colored operads. From the definition ((9.2.2) and (9.2.3)) of the forgetful functor U, one observes that $U(E_X)$ is the $\mathfrak{C}$-colored endomorphism operad of X. Therefore, the map λ' is actually giving an O-algebra structure on X. This is the functor Ψ. One can check that the functor Φ and Ψ give an equivalence of categories. □

References

[BM03] C. Berger and I. Moerdijk, Axiomatic homotopy theory for operads, Comm. Math. Helv. 78 (2003) 805–831.

[BM07] C. Berger and I. Moerdijk, Resolution of colured operads and rectification of homotopy algebras, Contemp. Math. 431 (2007), 31–58.

[BV73] J.M. Boardman and R.M. Vogt, Homotopy invariant algebraic structures on topological spaces, Lecture Notes in Math. 347, 1973.

[CGMV08] C. Casacuberta, J.J. Gutierrez, I. Moerdijk, and R.M. Vogt, Localization of algebras over coloured operads, `arXiv:0806.3983`.

[Cha05] D. Chataur, A bordism approach to string topology, Int. Math. Res. Not. (2005), 2829–2875.

[CG04] R.L. Cohen and V. Godin, A polarized view of string topology, London Math. Soc. Lecture Note Ser. 308, 127–154, Cambridge Univ. Press, Cambridge, 2004.

[CV06] R.L. Cohen and A.A. Voronov, Notes on string topology, in: String topology and cyclic homology, 1–95, Adv. Courses Math. CRM Barcelona, Birkhäuser, Basel, 2006.

[Dwy01] W.G. Dwyer, Classifying spaces and homology decompositions, in: Homotopy theoretical methods in group cohomology, Adv. Courses Math. CRM Barcelona, Birkhäuser, Basel, 2001.

[DS95] W.G. Dwyer and J. Spalinski, Homotopy theories and model categories, in: Handbook of algebraic topology, 73–126, North-Holland, Amsterdam, 1995.

[FMY08] Y. Frégier, M. Markl, and D. Yau, The L_∞-deformation complex of diagrams of algebras, `arXiv:0812.2981v2`.

[Fre08] B. Fresse, PROPs in model categories and homotopy invariance of structures, `arXiv:0812.2738v1`.

[Get94] E. Getzler, Batalin-Vilkovisky algebras and two-dimensional topological field theories, Comm. Math. Phys. 159 (1994), no. 2, 265–285.

[Har07] J.E. Harper, Homotopy theory of modules over operads and non-Σ operads in monoidal model categories, `arXiv:0801.0191v1`.

[Hin97] V. Hinich, Homological algebra of homotopy algebras, Comm. Algebra 25 (1997), 3291–3323.

[Hir02] P. Hirschhorn, Model categories and their localizations, Mathematical Surveys and Monographs 99, Amer. Math. Soc., Providence, RI, 2003.

[Hov99] M. Hovey, Model categories, Mathematical Surveys and Monographs 63, Amer. Math. Soc., Providence, RI, 1999.

[HSS00] M. Hovey, B. Shipley, and J. Smith, Symmetric spectra, J. Amer. Math. Soc. 13 (2000), 149–208.

[IJ02] M. Intermont and M.W. Johnson, Model structures on the category of ex-spaces, Topology and its Applications, 119 (2002) 325–353.

[Mac67] S. Mac Lane, Natural associativity and commutativity, Rice Univ. Stud. 49(1) (1963), 28-46.

[Mac98] S. Mac Lane, Categories for the working mathematician, Grad. Texts in Math. 5, 2nd ed., Springer, New York, 1998.

[MM92] S. Mac Lane and I. Moerdijk, Sheaves in geometry and logic: A first introduction to topos theory, Universitext, Springer, New York, 1992.

[Mar02] M. Markl, Homotopy diagrams of algebras, Rend. del Circ. Mat. di Palermo, Series II, Suppl. 69 (2002), 161-180.

[Mar04] M. Markl, Homotopy algebras are homotopy algebras, Forum Math. 16 (2004) 129–160.

[Mar06] M. Markl, A resolution (minimal model) of the PROP for bialgebras, J. Pure Appl. Alg. 205 (2006), 341-374.

[Mar08] M. Markl, Operads and PROPs, Handbook of algebra, Elsevier, 2008. `arXiv:math.AT/0601129`.

[MV07] M. Markl and A. A. Voronov, PROPped up graph cohomology, `arXiv:math/0307081`, to appear in Maninfest.

[May72] J.P. May, The geometry of iterated loop spaces, Springer Lecture Notes in Math. 271, Springer-Verlag, New York, 1972.

[May97] J.P. May, Definitions: operads, algebras and modules, Contemp. Math. 202 (1997), 1–7.

[Qui67] D.G. Quillen, Homotopical algebra, Lecture Notes in Math. 43, Springer-Verlag, Berlin-New York, 1967.

[Rez96] C. Rezk, Spaces of algebra structures and cohomology of operads, Ph.D. Thesis, Massachusetts Institute of Technology, 1996. Available at `http://www.math.uiuc.edu/~rezk/rezk-thesis.dvi`.

[Sch99] S. Schwede, Stable homotopical algebra and Γ-spaces, Math. Proc. Cambridge Philos. Soc. 126 (1999), 329–356.

[SS00] S. Schwede and B. E. Shipley, Algebras and modules in monoidal model categories, Proc. London Math. Soc. (3) 80 (2000), no. 2, 491–511.

[Seg88] G. Segal, Elliptic cohomology (after Landweber-Stong, Ochanine, Witten, and others), Séminaire Bourbaki, Vol. 1987/88, Astérisque No. 161-162 (1988), Exp. No. 695, 4, 187–201 (1989).

[Seg01] G. Segal, Topological structures in string theory, R. Soc. Lond. Philos. Trans. Ser. A Math. Phys. Eng. Sci. 359 (2001), no. 1784, 1389–1398.

[Seg04] G. Segal, The definition of conformal field theory, London Math. Soc. Lecture Note Ser. 308 (2004), 421–577.

[Sta63] J.D. Stasheff, Homotopy associativity of H-spaces I, II, Trans. Amer. Math. Soc. 108 (1963), 275–312.

This article may be accessed via WWW at `http://www.rmi.acnet.ge/jhrs/`

Mark W. Johnson
`mwj3@psu.edu`

Department of Mathematics
Penn State Altoona
Altoona, PA 16601-3760

Donald Yau
`dyau@math.ohio-state.edu`

Department of Mathematics
The Ohio State University at Newark
1179 University Drive
Newark, OH 43055

Journal of Homotopy and Related Structures, vol. 4(1), 2009, pp.317–330

EMBEDDING THE FLAG REPRESENTATION IN DIVIDED POWERS

GEOFFREY M.L. POWELL

(*communicated by Lionel Schwartz*)

Abstract

A generalization of a theorem of Crabb and Hubbuck concerning the embedding of flag representations in divided powers is given, working over an arbitrary finite field $\mathbb{F}$, using the category of functors from finite-dimensional $\mathbb{F}$-vector spaces to $\mathbb{F}$-vector spaces.

1. Introduction

Let V be a finite-dimensional $\mathbb{F}$-vector space over a finite field $\mathbb{F}$; the flag variety of complete flags of length r in V induces a permutation representation $\mathbb{F}[\mathfrak{Flag}_r](V)$ of the general linear group $GL(V)$, which is of interest in representation theory. The notation is derived from the fact that the flag representation arises as the evaluation on the space V of a functor $\mathbb{F}[\mathfrak{Flag}_r]$ in the category $\mathscr{F}$ of functors from finite-dimensional $\mathbb{F}$-vector spaces to $\mathbb{F}$-vector spaces; similarly, the divided power functors Γ^n induce $GL(V)$-representations $\Gamma^n(V)$. Motivated by questions from algebraic topology (and working over the prime field $\mathbb{F}_2$), Crabb and Hubbuck [**1**, **3**] associated to a sequence $\underline{s}$ of integers, $s_1 \geqslant \ldots \geqslant s_r > s_{r+1} = 0$, a morphism

$$\mathbb{F}_2[\mathfrak{Flag}_r](V) \to \Gamma^{[\underline{s}]_2}(V) \tag{1}$$

or $GL(V)$-modules, where $[\underline{s}]_q$ is the integer $\Sigma_{i=1}^{r}(q^{s_i} - 1)$. The motivating observation of this paper is that this is defined globally as a natural transformation

$$\phi_{\underline{s}} : \mathbb{F}[\mathfrak{Flag}_r] \to \Gamma^{[\underline{s}]_q} \tag{2}$$

in the functor category $\mathscr{F}$ and for each finite field $\mathbb{F}$, where $q = |\mathbb{F}|$.

The Crabb-Hubbuck morphism $\phi_{\underline{s}}$ arises in the construction of the ring of lines (developed independently in the dual situation by Repka and Selick [**8**]). Namely, the divided power functors form a commutative graded algebra in $\mathscr{F}$ and the ring of lines is the graded sub-functor generated by the images of the morphisms $\phi_{\underline{s}}$, which forms a sub-algebra of Γ^*.

The ring of lines is of interest in relation to the study of the primitives under the action of the Steenrod algebra on the singular homology $H_*(BV; \mathbb{F}_2)$ of the

---

The author would like to thank Grant Walker for conversations which led to the current approach to the result of Crabb and Hubbuck and for his interest.
Received December 19, 2008, revised July 26, 2009; published on September 25, 2009.
2000 Mathematics Subject Classification: Primary 55S10; Secondary 18E10.
Key words and phrases: Flag representation, divided power, functor category

classifying space of V; the primitives arise in a number of questions in algebraic topology. This relation can be explained from the point of view of the functor category $\mathscr{F}$ as follows, by identifying $H_*(BV;\mathbb{F}_2)$ with the graded vector space $\Gamma^*(V)$. Steenrod reduced power operations correspond to natural transformations of the form $\Gamma^a \to \Gamma^b$, $a \geqslant b$ (see [**6**]). Define the Steenrod kernel functors K^a by:

$$K^a := \mathrm{Ker}\{\Gamma^a \to \bigoplus_{f\in \mathrm{Hom}(\Gamma^a,\Gamma^b), a>b} \Gamma^b\}.$$

These functors form a commutative graded algebra in $\mathscr{F}$. The primitives in $H_*(BV;\mathbb{F}_2)$ are obtained by evaluating on V. The analysis of the primitives is dual to the study of the indecomposables for the action of the Steenrod reduced powers on $H^*(BV;\mathbb{F}_2)$; this is a difficult problem which has attracted much interest. For the field $\mathbb{F}_2$, the complete structure is known only for spaces of dimension at most four; the case of dimension three is due to Kameko and a published account is available in Boardman [**5, 2**]; Kameko also announced the case of dimension four, which has since been calculated by Sum, a student of Nguyen Hung.

The morphism $\phi_{\underline{s}}$ of equation (2) maps to $K^{[\underline{s}]_2}$ for elementary reasons and the ring of lines is a graded sub-algebra in $\mathscr{F}$ of K^*. A fundamental question is to determine in which degrees the ring of lines coincides with the Steenrod kernel. Motivated by this question, Crabb and Hubbuck [**3**, Proposition 3.10] gave an explicit criterion upon the sequence (s_i) with respect to the dimension of V for the morphism $\phi_{\underline{s}}$ to be a monomorphism.

The purpose of this note is two-fold; to present a proof exploiting the category $\mathscr{F}$ and to generalize the result to an arbitrary finite field $\mathbb{F}$, with $q = |\mathbb{F}|$. The main result of the paper is the following:

Theorem 1. *Let r be a natural number and $\underline{s} = (s_1 > \ldots > s_r > s_{r+1} = 0)$ be a sequence of integers which satisfies the condition $[s_i - s_{i+1}]_q \geqslant (q-1)(\dim V - i + 1)$, for $1 \leqslant i \leqslant r$; then the morphism $\phi_{\underline{s}}$ induces a monomorphism*

$$\mathbb{F}[\mathfrak{Flag}_r](V) \hookrightarrow \Gamma^{[\underline{s}]_q}(V).$$

The proof sheds light upon the method proposed by Crabb and Hubbuck; namely, the proof establishes the stronger result that the composite with a morphism induced by the iterated diagonal on divided power algebras and the Verschiebung morphism is a monomorphism. This is of interest in the light of recent work over the field $\mathbb{F}_2$ by Grant Walker, Reg Wood [**9**] and Tran Ngoc Nam generalizing the result of Crabb and Hubbuck, relaxing the required hypothesis on the sequence (s_i).

These techniques can be used to provide further information on the nature of the embedding results; for instance:

Proposition 2. *Let $\underline{s}$ be a sequence of integers $(s_1 > \ldots > s_r > s_{r+1} = 0)$ and V be a finite-dimensional vector space for which the morphism $\phi_{\underline{s}}(V) : \mathbb{F}[\mathfrak{Flag}_r](V) \to \Gamma^{[\underline{s}]_q}(V)$ is a monomorphism.*

Let $\underline{s^+}$ denote the sequence given by $s_i^+ = s_i + 1$, for $1 \leqslant i \leqslant r$, and $s_{r+1}^+ = 0$. Then the morphism

$$\phi_{\underline{s^+}}(V) : \mathbb{F}[\mathfrak{Flag}_r](V) \to \Gamma^{[\underline{s^+}]_q}(V)$$

is a monomorphism.

Contents

2. Preliminaries

Fix a finite field $\mathbb{F} = \mathbb{F}_q$, and write $q = p^m$, where p is the characteristic of $\mathbb{F}$. Let $\mathscr{F}$ be the category of functors from finite-dimensional $\mathbb{F}$-vector spaces to $\mathbb{F}$-vector spaces. The group of units $\mathbb{F}^\times$ is isomorphic to the cyclic group $\mathbb{Z}/(q-1)\mathbb{Z}$ via $i \mapsto \lambda^i$, for a generator λ. In particular, the group has order prime to p, hence the category of $\mathbb{F}[\mathbb{F}^\times]$-modules is semi-simple. This gives rise to the weight splitting of $\mathscr{F}$ (as in [**6**]):

$$\mathscr{F} \cong \prod_{i \in \mathbb{Z}/(q-1)\mathbb{Z}} \mathscr{F}^i,$$

where $\mathscr{F}^i$ is the full subcategory of functors such that $F(\lambda 1_V) = \lambda^i F(1_V)$ for all finite-dimensional spaces V and $\lambda \in \mathbb{F}^\times$. By reduction mod $q-1$, the weight category $\mathscr{F}^k$ can be taken to be defined for k an integer.

The duality functor $D : \mathscr{F}^{\mathrm{op}} \to \mathscr{F}$ is defined by $DF(V) := F(V^*)^*$ and D restricts to a functor $D : (\mathscr{F}^k)^{\mathrm{op}} \to \mathscr{F}^k$

2.1. Divided powers and the Verschiebung

Recall that Γ^k denotes the kth divided power functor, defined as the invariants $\Gamma^k := (T^k)^{\mathfrak{S}_k}$ under the action of the symmetric group permuting the factors of T^k, the kth tensor power functor. (By convention, the divided power Γ^0 functor is the constant functor $\mathbb{F}$ and $\Gamma^i = 0$ for $i < 0$). The divided power functor Γ^k is dual to the kth symmetric power functor S^k and the functors Γ^k, T^k, S^k all belong to the weight category $\mathscr{F}^k$.

The divided power functors Γ^* form a graded exponential functor; namely for finite-dimensional vector spaces U, V and a natural number n, there is a binatural isomorphism

$$\Gamma^n(U \oplus V) \cong \bigoplus_{i+j=n} \Gamma^i(U) \otimes \Gamma^j(V).$$

This has important consequences (see [**4**], for example); in particular, for pairs of natural numbers (a, b), there are cocommutative coproduct morphisms $\Gamma^{a+b} \xrightarrow{\Delta} \Gamma^a \otimes$

Γ^b and commutative product morphisms $\Gamma^a \otimes \Gamma^b \xrightarrow{\mu} \Gamma^{a+b}$, which are coassociative (respectively associative) in the appropriate graded sense.

The Verschiebung is a natural surjection $\mathcal{V} : \Gamma^{qn} \twoheadrightarrow \Gamma^n$, for integers $n \geqslant 0$, dual to the Frobenius qth power map on symmetric powers. More generally there is a Verschiebung morphism $\mathcal{V}_p : \Gamma^{np} \twoheadrightarrow \Gamma^{(1)}$, dual to the Frobenius pth power map, where $(-)^{(1)}$ denotes the Frobenius twist functor (see [**4**]). The Verschiebung $\mathcal{V}$ is obtained by iterating $\mathcal{V}_p$ m times, where $q = p^m$.

The pth truncated symmetric power functor $\overline{S}^n$ is defined by imposing the relation $v^p = 0$; similarly the qth truncated symmetric power functor $\tilde{S}^n$ is given by forming the quotient by the relation $v^q = 0$. There is a natural surjection $\tilde{S}^n \twoheadrightarrow \overline{S}^n$; over a prime field the functors coincide.

Dualizing gives the following:

Definition 2.1. For n a natural number,

1. let $\tilde{\Gamma}^n$ denote the kernel of the composite

$$\Gamma^n \xrightarrow{\Delta} \Gamma^{n-q} \otimes \Gamma^q \xrightarrow{1 \otimes \mathcal{V}} \Gamma^{n-q} \otimes \Gamma^1;$$

2. let $\overline{\Gamma}^n$ be the kernel of the composite

$$\Gamma^n \xrightarrow{\Delta} \Gamma^{n-p} \otimes \Gamma^p \xrightarrow{1 \otimes \mathcal{V}_p} \Gamma^{n-p} \otimes (\Gamma^1)^{(1)}.$$

Lemma 2.2. *Let n be a natural number.*

1. *The functor $\tilde{\Gamma}^n$ is dual to $\tilde{S}^n$.*
2. *The functor $\overline{\Gamma}^n$ is isomorphic to $\overline{S}^n$ and is simple.*
3. *$\tilde{\Gamma}^n(\mathbb{F}^d) = 0$ if $d \leqslant (n-1)/(q-1)$.*

Proof. The first statement follows from the definitions and the identification $DS^n = \Gamma^n$. The simplicity of $\overline{S}^n$ is a standard fact [**6**]; the isomorphism follows from the fact that the simple functors of $\mathscr{F}$ are self-dual [**7**].

The final statement is an elementary verification, which is a consequence of the observation that the maximal degree of a free q-truncated symmetric algebra on d variables of degree one is $d(q-1)$. □

2.2. Further properties of divided powers

Notation 2.3. For s a natural number, let $[s]_q$ denote the integer $q^s - 1$ and, for $\underline{s}$ a sequence of integers, $s_1 \geqslant \ldots \geqslant s_r > s_{r+1} = 0$, let $[\underline{s}]_q$ denote $\Sigma_i [s_i]_q$.

Notation 2.4. The element of $\Gamma^k(V)$ corresponding to the symmetric tensor $x^{\otimes k} \in T^k(V)$, for x an element of V, will be denoted simply by $x^{\otimes k}$.

Lemma 2.5. *Let s be a positive integer. For any $x \in V$, the class $x^{\otimes [s]_q} \in \Gamma^{[s]_q}(V)$ is equal to the product*

$$\prod_{j=0}^{sm-1} x^{\otimes (p-1)p^j},$$

where $|\mathbb{F}| = q = p^m$.

This Lemma is a consequence of the following well-known general property of the divided power functors.

Lemma 2.6. *Let $a_0, \ldots, a_t$ and $0 = r_0 < r_1 < \ldots < r_t$ be sequences of natural numbers such that, for $0 \leqslant i < t$, $a_i p^{r_i} < p^{r_{i+1}}$. Then the composite*

$$\Gamma^{\Sigma_{i=0}^t a_i p^{r_i}} \to \bigotimes_{i=0}^{t} \Gamma^{a_i p^{r_i}} \to \Gamma^{\Sigma_{i=0}^t a_i p^{r_i}}$$

is an isomorphism, where the first morphism is the coproduct and the second the product.

Proof. Using the (co)associativity of the product (respectively the coproduct), it suffices to prove the result for $t = 1$. The composite morphism in this case is multiplication by the scalar $\binom{a_0 + a_1 p^{r_1}}{a_0}$, which is equal to one modulo p, since $a_0 < p^{r_1}$, by hypothesis. □

The following Lemma is the key to the construction of the Crabb-Hubbuck morphism, $\phi_{\underline{s}}$, in Section 4.

Lemma 2.7. *Let x be an element of V, s be a natural number and $0 < i \leqslant [s]_q$ be an integer. The product $x^{\otimes [s]_q} x^{\otimes i}$ is zero in $\Gamma^{[s]_q + i}(V)$.*

Proof. The element $x^{\otimes [s]_q}$ is equal to the product $\prod_{j=0}^{sm-1} x^{\otimes (p-1)p^j}$, by Lemma 2.5. Similarly, considering the p-adic expansion $i = \Sigma_{j=0}^{sm-1} i_j p^j$, where $0 \leqslant i_j < p$ and at least one i_j is non-zero, there is an equality $x^{\otimes i} = \prod_{j=0}^{sm-1} x^{\otimes i_j p^j}$. The product is associative and commutative, hence it suffices to show that, if $i_j \neq 0$, then $x^{\otimes (p-1)p^j} x^{\otimes i_j p^j}$ is zero. Up to non-zero scalar in $\mathbb{F}_p^\times$, the element $x^{\otimes i_j p^j}$ is the i_j-fold product of $x^{\otimes p^j}$ (since $0 < i_j < p$, by hypothesis), hence it suffices to show that $x^{\otimes (p-1)p^j} x^{\otimes p^j}$ is zero. This element identifies with $\binom{p^{j+1}}{p^j} x^{\otimes p^{j+1}}$ and the scalar is zero in $\mathbb{F}$. □

The behaviour of the Verschiebung morphism with respect to products is important.

Lemma 2.8. *Let $\beta_1, \ldots, \beta_k$ be positive integers such that $\Sigma_{i=1}^k \beta_i = pN$ for some integer N. The composite*

$$\bigotimes_{i=1}^{k} \Gamma^{\beta_i} \xrightarrow{\mu} \Gamma^{pN} \xrightarrow{\mathcal{V}_p} (\Gamma^N)^{(1)},$$

where $\mathcal{V}_p$ is the Verschiebung and μ is the product, is trivial unless for each i, $\beta_i = p\beta_i'$, $\beta_i' \in \mathbb{N}$. In this case, there is a commutative diagram

$$\begin{array}{ccc}
\bigotimes_{i=1}^k \Gamma^{p\beta_i'} & \xrightarrow{\mu} & \Gamma^{pN} \\
\downarrow{\scriptstyle \otimes \mathcal{V}_p} & & \downarrow{\scriptstyle \mathcal{V}_p} \\
\bigotimes_{i=1}^k (\Gamma^{\beta_i'})^{(1)} & \xrightarrow[\mu^{(1)}]{} & (\Gamma^N)^{(1)}.
\end{array}$$

Proof. The statement is more familiar in the dual situation, where it corresponds to the fact that the diagonal of the symmetric power algebra commutes with the Frobenius. □

Remark 2.9. An analogous statement holds for iterates of $\mathcal{V}_p$ and for the Verschiebung $\mathcal{V} : \Gamma^{qN} \to \Gamma^N$.

3. Projectives and flags

This section introduces the flag functors and relates them to the standard projective generators of the category $\mathscr{F}$.

3.1. The projective and flag functors

For r a natural number, the standard projective object $P_{\mathbb{F}^r}$ in $\mathscr{F}$ is given by $P_{\mathbb{F}^r}(V) = \mathbb{F}[\mathrm{Hom}(\mathbb{F}^r, V)]$ and is determined up to isomorphism by $\mathrm{Hom}_{\mathscr{F}}(P_{\mathbb{F}^r}, G) = G(\mathbb{F}^r)$. In particular, the projective $P_{\mathbb{F}}$ is the functor $V \mapsto \mathbb{F}[V]$.

The weight splitting determines a direct sum decomposition

$$P_{\mathbb{F}} \cong \bigoplus_{i \in \mathbb{Z}/(q-1)\mathbb{Z}} P^i_{\mathbb{F}}$$

in which $P^i_{\mathbb{F}}$ is indecomposable for $i \neq 0$ and $P^0_{\mathbb{F}}$ admits a decomposition $P^0_{\mathbb{F}} = \mathbb{F} \oplus \overline{P^0_{\mathbb{F}}}$ (Cf [**6**, Lemma 5.3], noting that Kuhn uses a splitting associated to the multiplicative semigroup $\mathbb{F}$).

There is a Knneth isomorphism for projectives so that, for r a positive integer, $P_{\mathbb{F}}^{\otimes r}$ is projective and identifies with the projective functor.

Definition 3.1. For r a positive integer, let $\mathbb{F}[\mathfrak{Flag}_r]$ be the functor which is defined in terms of complete flags of length r as follows.

As a vector space, $\mathbb{F}[\mathfrak{Flag}_r](V)$ has basis the set of complete flags of length r. A morphism $V \to W$ sends a complete flag to its image, if this is a complete flag, and to zero otherwise.

Recall that a functor F of $\mathscr{F}$ is said to be constant-free if $F(0) = 0$.

Lemma 3.2. *Let $r \geqslant s > 0$ be integers.*

1. *The functor $\mathbb{F}[\mathfrak{Flag}_r]$ is constant-free and belongs to $\mathscr{F}^0$.*
2. *There is a diagonal morphism in $\mathscr{F}$:*

$$\mathbb{F}[\mathfrak{Flag}_r] \to \mathbb{F}[\mathfrak{Flag}_r] \otimes \mathbb{F}[\mathfrak{Flag}_r].$$

3. *There is a surjection*

$$\pi_{r,s} : \mathbb{F}[\mathfrak{Flag}_r] \twoheadrightarrow \mathbb{F}[\mathfrak{Flag}_s]$$

which forgets the subspaces of dimension greater than s.

3.2. Structure of the projectives

Lemma 3.3. *Let $n \geqslant 1$ be an integer, then* $\mathrm{Hom}_{\mathscr{F}}(\overline{P^0_{\mathbb{F}}}, \Gamma^{n(q-1)}) = \mathbb{F}$.

Proof. The result follows from the Yoneda lemma, the weight splitting and the fact that $n \geqslant 1$ allows passage to the constant-free part, $\overline{P^0_{\mathbb{F}}}$. □

Recall that a functor is said to be finite if it has a finite composition series.

Proposition 3.4. *[6, 7]*

1. *The surjection $P_{\mathbb{F}} \twoheadrightarrow \mathbb{F}[\mathfrak{Flag}_1]$ induces an isomorphism $\overline{P^0_{\mathbb{F}}} \cong \mathbb{F}[\mathfrak{Flag}_1]$.*
2. *There is an inverse system of finite functors $\ldots \to \mathfrak{q}_k\overline{P^0_{\mathbb{F}}} \to \mathfrak{q}_{k-1}\overline{P^0_{\mathbb{F}}} \to \ldots$ such that*
   (a) $\overline{P^0_{\mathbb{F}}} \cong \lim_{\leftarrow} \mathfrak{q}_k\overline{P^0_{\mathbb{F}}}$;
   (b) *for $k \geqslant 1$, $\mathfrak{q}_k\overline{P^0_{\mathbb{F}}}$ is isomorphic to the image of any non-trivial morphism $\overline{P^0_{\mathbb{F}}} \to \Gamma^{k(q-1)}$;*
   (c) *for $k \geqslant 2$, there is a non-split short exact sequence*
   $$0 \to \tilde{\Gamma}^{k(q-1)} \to \mathfrak{q}_k\overline{P^0_{\mathbb{F}}} \to \mathfrak{q}_{k-1}\overline{P^0_{\mathbb{F}}} \to 0.$$
3. *The functor $\overline{P^0_{\mathbb{F}}}$ is dual to a locally-finite functor.*
4. *If $\mathbb{F}$ is the prime field $\mathbb{F}_p$, then the functor $\overline{P^0_{\mathbb{F}}}$ is uniserial with composition factors $\{\overline{\Gamma}^{k(p-1)} | k \geqslant 1\}$, each occurring with multiplicity one.*

Proof. It is more straightforward to deduce the result from the description of the dual $D\overline{P^0_{\mathbb{F}}}$; this is isomorphic to the functor $(\bigoplus_{k=0}^{\infty} S^{k(q-1)})/\langle v^q - v\rangle$, where the relation is induced by the weight zero part of the ideal $\langle v^q - v\rangle$ (this is deduced from [**6**, Lemma 4.12] by applying the evident weight splitting). □

It is important to have a measure of how good an approximation $\mathfrak{q}_k\overline{P^0_{\mathbb{F}}}$ is to $\overline{P^0_{\mathbb{F}}}$.

Lemma 3.5. *Let $k \geqslant 1$ be an integer. Up to scalar in $\mathbb{F}^\times$, there is a unique non-trivial morphism $\overline{P^0_{\mathbb{F}}} \to \Gamma^{k(q-1)}$. Any non-trivial morphism $\overline{P^0_{\mathbb{F}}} \to \Gamma^{k(q-1)}$*

1. *factors as*
   $$\overline{P^0_{\mathbb{F}}} \twoheadrightarrow \mathfrak{q}_k\overline{P^0_{\mathbb{F}}} \hookrightarrow \Gamma^{k(q-1)}$$
   and
2. *induces a monomorphism*
   $$\overline{P^0_{\mathbb{F}}}(V) \hookrightarrow \Gamma^{k(q-1)}(V)$$
   if $\dim V \leqslant k$.

Proof. The unicity follows from Lemma 3.3 and the factorization follows from this unicity together with the identification of $\mathfrak{q}_k\overline{P^0_{\mathbb{F}}}$ which is given in Proposition 3.4.

The kernel of the surjection $\overline{P^0_{\mathbb{F}}} \twoheadrightarrow \mathfrak{q}_k\overline{P^0_{\mathbb{F}}}$ has a filtration with subquotients of the form $\tilde{\Gamma}^{l(q-1)}$ with $l > k$. The functor $\tilde{\Gamma}^{l(q-1)}$ is zero when evaluated on spaces with $\dim V \leqslant k$, by Lemma 2.2 (3). It follows that the kernel is zero when evaluated on such spaces. Thus $\overline{P^0_{\mathbb{F}}}(V) \to \mathfrak{q}_k\overline{P^0_{\mathbb{F}}}(V)$ is an isomorphism when $\dim V \leqslant k$ and the result follows from the first part of the Lemma. □

4. The flag morphism of Crabb and Hubbuck

Throughout this section, let r denote a fixed positive integer and $\underline{s}$ denote a fixed decreasing sequence of positive integers, $s_1 \geqslant s_2 \geqslant \ldots \geqslant s_r > s_{r+1} = 0$.

Notation 4.1. For each positive integer s, let ϕ_s be the element of $\mathrm{Hom}_{\mathscr{F}}(\overline{P^0_{\mathbb{F}}}, \Gamma^{[s]_q})$ which sends the canonical generator of $\overline{P^0_{\mathbb{F}}}(\mathbb{F}) \cong \mathbb{F}[\mathfrak{Flag}_1](\mathbb{F})$ to $\iota^{\otimes [s]_q}$, where ι is any generator of $\mathbb{F}$. (The morphism is independent of the choice of ι).

Definition 4.2. For $\underline{s}$ a sequence of positive integers, let $\tilde{\phi}_{\underline{s}}$ denote the morphism

$$\tilde{\phi}_{\underline{s}} : P_{\mathbb{F}^r} \cong P_{\mathbb{F}}^{\otimes r} \stackrel{\otimes \phi_{s_i}}{\longrightarrow} \bigotimes_{i=1}^{r} \Gamma^{[s_i]_q} \rightarrow \Gamma^{[\underline{s}]_q}$$

in which the second morphism is induced by the product.

The following Proposition is proved in the case $q = 2$ in [**3**].

Proposition 4.3. *The morphism $\tilde{\phi}_{\underline{s}}$ factorizes as*

$$P_{\mathbb{F}}^{\otimes r} \twoheadrightarrow \mathbb{F}[\mathfrak{Flag}_r] \stackrel{\phi_{\underline{s}}}{\rightarrow} \Gamma^{[\underline{s}]_q}.$$

Proof. Fixing a basis of $\mathbb{F}^r$, a canonical basis element of $P_{\mathbb{F}^r}(V)$ is an ordered sequence (v_i) of r elements of V. The morphism $\tilde{\phi}_{\underline{s}}$ sends this generator to $\prod_{i=1}^{r} v_i^{\otimes [s_i]_q} = \prod_{i=1}^{r} \prod_{j=0}^{s_i - 1} v_i^{\otimes (q-1) q^j}$.

Using this notation, define a natural surjection $P_{\mathbb{F}^r} \twoheadrightarrow \mathbb{F}[\mathfrak{Flag}_r]$ by sending (v_i) to the flag $\langle v_1 \rangle < \langle v_1, v_2 \rangle < \ldots < \langle v_1, \ldots, v_r \rangle$ if the elements are linearly independent and zero otherwise. The proposition asserts that $\tilde{\phi}_{\underline{s}}$ factorizes across this surjection.

The result follows as in the proof of [**3**, Lemma 3.1], by applying Lemma 2.7. □

5. The embedding theorem

The purpose of this section is to prove the main result of the paper, stated here as Theorem 5.17, which gives a criterion for

$$\phi_{\underline{s}}(V) : \mathbb{F}[\mathfrak{Flag}_r](V) \rightarrow \Gamma^{[\underline{s}]_q}(V)$$

to be a monomorphism, where r is a positive integer and $\underline{s} = s_1 > s_2 > \ldots > s_r > s_{r+1} = 0$ is a strictly decreasing sequence of integers.

Remark 5.1. The morphism $\phi_{\underline{s}}$ is clearly not a monomorphism of functors, since the functor $\Gamma^{[\underline{s}]_q}$ is finite whereas $\mathbb{F}[\mathfrak{Flag}_r]$ is highly infinite.

However, Lemma 3.5 (2) provides the key calculational input, which is restated as the following:

Lemma 5.2. *Let s be a natural number. The morphism $\phi_s : \mathbb{F}[\mathfrak{Flag}_1] \cong \overline{P^0_{\mathbb{F}}} \rightarrow \Gamma^{[s]_q}$ induces a monomorphism $\phi_s(V)$ if $[s]_q \geqslant (q-1) \dim V$.*

The theorem is proved by an induction using Lemma 5.2 to provide the inductive step. The strategy involves composing $\phi_{\underline{s}}$ with a morphism $\delta_{\underline{s}}$ (defined below) to give a morphism $\psi_{\underline{s}}$ which is amenable to induction. The key to setting up the induction is Lemma 5.6.

Write $[\underline{s}]_q = r[s_r]_q + \Sigma_{i=1}^{r-1}([s_i]_q - [s_r]_q)$; thus the coproduct gives a morphism $\Delta : \Gamma^{[\underline{s}]_q} \rightarrow \Gamma^{r[s_r]_q} \otimes \Gamma^{\Sigma([s_i]_q - [s_r]_q)}$. For each i, $[s_i]_q - [s_r]_q = q^{s_r}[s_i - s_r]_q$, hence there is an iterated Verschiebung morphism $\mathcal{V}^{s_r} : \Gamma^{\Sigma([s_i]_q - [s_r]_q)} \twoheadrightarrow \Gamma^{\Sigma[s_i - s_r]_q}$.

Definition 5.3.

1. Let $\delta_{\underline{s}} : \Gamma^{[\underline{s}]_q} \rightarrow \Gamma^{r[s_r]_q} \otimes \Gamma^{\Sigma_{i=1}^{r-1}[s_i - s_r]_q}$ be the composite morphism
$$\Gamma^{[\underline{s}]_q} \xrightarrow{\Delta} \Gamma^{r[s_r]_q} \otimes \Gamma^{\Sigma([s_i]_q - [s_r]_q)} \xrightarrow{1 \otimes \mathcal{V}^{s_r}} \Gamma^{r[s_r]_q} \otimes \Gamma^{\Sigma_{i=1}^{r-1}[s_i - s_r]_q}.$$
2. Let $\psi_{\underline{s}} : \mathbb{F}[\mathfrak{Flag}_r] \rightarrow \Gamma^{r[s_r]_q} \otimes \Gamma^{\Sigma_{j=1}^{r-1}[s_j - s_r]_q}$ be the composite morphism $\delta_{\underline{s}} \circ \phi_{\underline{s}}$.

The following elementary observation is recorded as a Lemma.

Lemma 5.4. *Let V be a finite-dimensional vector space. If the morphism $\psi_{\underline{s}}(V)$ is a monomorphism, then $\phi_{\underline{s}}(V)$ is a monomorphism.*

Notation 5.5. Let $\underline{s}'$ denote the sequence (of length $r-1$) of positive integers $(s_1 - s_r > \ldots > s_{r-1} - s_r > 0)$.

The following Lemma is the key to the inductive proof, and relies upon the fact that the iterated Verschiebung is used in the definition of $\psi_{\underline{s}}$. Observe that the Crabb-Hubbuck morphism associated to the sequence of integers $(s_r, \ldots, s_r)$ of length r induces a morphism
$$\phi_{(s_r,\ldots,s_r)} : \mathbb{F}[\mathfrak{Flag}_r] \rightarrow \Gamma^{r[s_r]_q}.$$

Lemma 5.6. *The morphism $\psi_{\underline{s}}$ identifies with the composite morphism*
$$\begin{array}{ccccc} \mathbb{F}[\mathfrak{Flag}_r] & \xrightarrow{\text{diag}} & \mathbb{F}[\mathfrak{Flag}_r] \otimes \mathbb{F}[\mathfrak{Flag}_r] & \xrightarrow{1 \otimes \pi_{r,r-1}} & \mathbb{F}[\mathfrak{Flag}_r] \otimes \mathbb{F}[\mathfrak{Flag}_{r-1}] \\ & & & & \downarrow{\scriptstyle \phi_{(s_r,\ldots,s_r)} \otimes \phi_{\underline{s}'}} \\ & & & & \Gamma^{r[s_r]_q} \otimes \Gamma^{\Sigma_{j=1}^{r-1}[s'_j]_q} \end{array}$$

Proof. We are required to prove that the following diagram is commutative.
$$\begin{array}{ccccc} \mathbb{F}[\mathfrak{Flag}_r] & \xrightarrow{\text{diag}} & \mathbb{F}[\mathfrak{Flag}_r] \otimes \mathbb{F}[\mathfrak{Flag}_r] & \xrightarrow{1 \otimes \pi_{r,r-1}} & \mathbb{F}[\mathfrak{Flag}_r] \otimes \mathbb{F}[\mathfrak{Flag}_{r-1}] \\ {\scriptstyle \phi_{\underline{s}}}\downarrow & & & & \downarrow{\scriptstyle \phi_{(s_r,\ldots,s_r)} \otimes \phi_{\underline{s}'}} \\ \Gamma^{[\underline{s}]_q} & & \xrightarrow[\delta_{\underline{s}}]{\qquad\qquad\qquad} & & \Gamma^{r[s_r]_q} \otimes \Gamma^{\Sigma_{j=1}^{r-1}[s'_j]_q}. \end{array}$$

Choose a surjection $P_{\mathbb{F}^r} \twoheadrightarrow \mathbb{F}[\mathfrak{Flag}_r]$ as in the proof of Proposition 4.3; it is equivalent to prove the commutativity of the diagram obtained by composition, replacing the top left entry by $P_{\mathbb{F}^r}$. The analysis of the composite morphism
$$P_{\mathbb{F}^r} \rightarrow \Gamma^{[\underline{s}]_q} \rightarrow \Gamma^{r[s_r]_q} \otimes \Gamma^{\Sigma_{j=1}^{r-1}[s'_j]_q}. \qquad (3)$$
around the bottom of the diagram can then be carried out as follows.

The definition of the morphism $\delta_{\underline{s}}$ and of the morphism $\phi_{\underline{s}}$ implies that this composite factors across

$$\bigotimes_{i=1}^{r}\Gamma^{[s_i]_q}\xrightarrow{\mu}\Gamma^{[\underline{s}]_q}\xrightarrow{\Delta}\Gamma^{r[s_r]_q}\otimes\Gamma^{q^{s_r}\Sigma_{j=1}^{r-1}[s'_j]_q}$$

where μ denotes the product on divided powers and Δ the diagonal.

The exponential algebra structure of Γ^* (essentially the fact that these functors take values in bicommutative Hopf algebras) implies that there is a commutative diagram

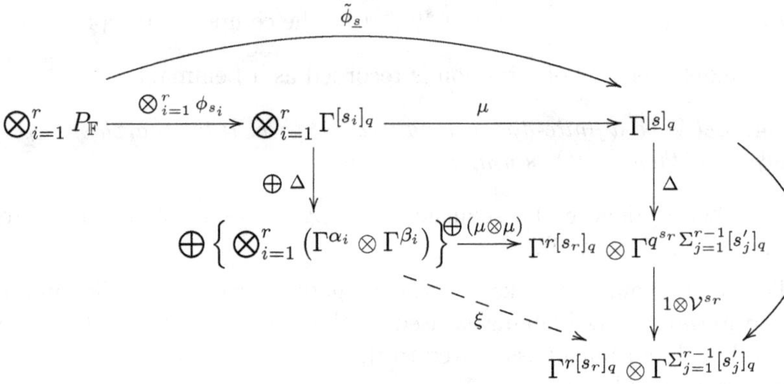

where the sum is labelled over sequences of pairs of natural numbers (α_i,β_i) satisfying $\alpha_i+\beta_i=[s_i]_q$ for each i and $\Sigma\alpha_i=r[s_r]_q$.

Consider the composite morphism ξ in the diagram; Lemma 2.8 implies that the only components of this morphism which are non-trivial are those corresponding to sequences (α_i,β_i) for which $\beta_i=q^{s_r}\beta'_i$ for natural numbers β'_i. The condition $\alpha_i+q^{s_r}\beta'_i=[s_i]_q$ implies that α_i is non-zero; it follows that $\alpha_i\geqslant[s_r]_q$, for each i, since $\alpha_i\equiv[s_r]_q \mod (q^{s_r})$. The condition $\Sigma_i\alpha_i=r[s_r]_q$ therefore implies that $\alpha_i=[s_r]_q$ for each i. It follows that ξ has only one non-zero component and a straightforward verification shows that the composite corresponds to the composite around the top of the diagram given in the statement of the Lemma. □

The inductive argument is simplified using the following:

Lemma 5.7. *Let V be a finite-dimensional vector space for which the morphism $\phi_{\underline{s}'}(V)$ is a monomorphism. Then the morphism $\psi_{\underline{s}}(V)$ is a monomorphism if and only if the composite morphism*

$$\mathbb{F}[\mathfrak{Flag}_r]\xrightarrow{\text{diag}}\mathbb{F}[\mathfrak{Flag}_r]\otimes\mathbb{F}[\mathfrak{Flag}_r]\xrightarrow{\phi_{s_r,\ldots,s_r}\otimes\pi_{r,r-1}}\Gamma^{r[s_r]_q}\otimes\mathbb{F}[\mathfrak{Flag}_{r-1}]$$

induces a monomorphism when evaluated upon V.

Proof. This follows from the identification of $\psi_{\underline{s}}$ which is given in Lemma 5.6. □

Using the fact that $\mathbb{F}[\mathfrak{Flag}_{r-1}](V)$ is generated by complete flags of length $r-1$, this allows the decomposition into components.

Notation 5.8. For Φ a complete flag of length $r-1$ in V, let

1. $\langle\Phi\rangle \leqslant V$ denote the $(r-1)$-dimensional subspace of V defined by Φ;
2. $\mathbb{F}[\mathfrak{Flag}_r]_\Phi(V)$ denote the subspace generated by flags containing Φ;
3. γ_Φ denote the image of $[\Phi] \in \mathbb{F}[\mathfrak{Flag}_{r-1}](V)$ under the Crabb-Hubbuck morphism $\phi_{s_r,\ldots,s_r}(V) : \mathbb{F}[\mathfrak{Flag}_{r-1}](V) \to \Gamma^{(r-1)[s_r]_q}(V)$.

Remark 5.9. The space $\mathbb{F}[\mathfrak{Flag}_r]_\Phi(V)$ is isomorphic to $\mathbb{F}[\mathfrak{Flag}_1](V/\langle\Phi\rangle)$.

Lemma 5.10. *Let V be a finite-dimensional vector space for which the morphism $\phi_{\underline{s}'}(V)$ is a monomorphism. The morphism $\psi_{\underline{s}}(V)$ is a monomorphism if and only if, for each complete flag Φ in V of length $r-1$, the restriction of*

$$\mathbb{F}[\mathfrak{Flag}_r](V) \overset{\phi_{s_r,\ldots,s_r}}{\longrightarrow} \Gamma^{r[s_r]_q}(V)$$

to $\mathbb{F}[\mathfrak{Flag}_r]_\Phi(V)$ is a monomorphism.

Proof. A straightforward consequence of Lemma 5.7. □

Notation 5.11. Let V be a finite-dimensional vector space and Φ be a complete flag in V of length $r-1$. Let ρ_Φ denote the composite linear map

$$\mathbb{F}[\mathfrak{Flag}_r]_\Phi(V) \overset{\cong}{\to} \mathbb{F}[\mathfrak{Flag}_1](V/\langle\Phi\rangle) \overset{\phi_{s_r}}{\to} \Gamma^{[s_r]_q}(V/\langle\Phi\rangle)$$

induced by the projection $V \twoheadrightarrow V/\langle\Phi\rangle$ and the morphism ϕ_{s_r}.

Notation 5.12. For V a finite-dimensional vector space, Φ a complete flag in V of length $r-1$ and σ a section of the projection $V \twoheadrightarrow V/\langle\Phi\rangle$, let

$$\gamma_\Phi \cap_\sigma : \Gamma^{[s_r]_q}(V/\langle\Phi\rangle) \to \Gamma^{r[s_r]_q}(V)$$

denote the linear morphism induced by the section σ followed by the product with γ_Φ with respect to the algebra structure of $\Gamma^*(V)$.

Lemma 5.13. *Let V, Φ and σ be as above. The linear morphism*

$$\gamma_\Phi \cap_\sigma : \Gamma^{[s_r]_q}(V/\langle\Phi\rangle) \to \Gamma^{r[s_r]_q}(V)$$

is a monomorphism.

Proof. The result follows from the exponential structure of the divided power functors, since the element γ_Φ is the image of an element of $\Gamma^{(r-1)[s_r]_q}(\langle\Phi\rangle)$ under the morphism induced by the natural inclusion. □

Lemma 5.14. *Let V, Φ, σ be as above. The restriction of $\phi_{s_r,\ldots,s_r}(V)$ to $\mathbb{F}[\mathfrak{Flag}_r]_\Phi(V)$ identifies with the linear morphism $(\gamma_\Phi \cap_\sigma) \circ \rho_\Phi$.*

Proof. The result follows from the definition of the morphism $\phi_{s_r,\ldots,s_r}(V)$. □

Lemmas 5.13 and 5.14 together imply the following result:

Lemma 5.15. *Let V be a finite-dimensional vector space and Φ be a complete flag in V of length $r-1$. The restriction of $\phi_{s_r,\ldots,s_r}(V)$ to $\mathbb{F}[\mathfrak{Flag}_r]_\Phi(V)$ is a monomorphism if and only if*

$$\phi_{s_r} : \mathbb{F}[\mathfrak{Flag}_1](V/\langle\Phi\rangle) \to \Gamma^{[s_r]_q}(V/\langle\Phi\rangle)$$

is a monomorphism.

Remark 5.16. By lemma 5.2, a sufficient condition is

$$[s_r]_q \geqslant (q-1)\dim(V/\langle\Phi\rangle) = (q-1)(\dim V - r + 1).$$

When $q = 2$, this is an equivalent condition.

Putting these results together, one obtains the following generalization of [**3**, Proposition 3.10].

Theorem 5.17. *Suppose that the sequence* $\underline{s}$ *satisfies the condition* $[s_i - s_{i+1}]_q \geqslant (q-1)(\dim V - i + 1)$, *for* $1 \leqslant i \leqslant r$. *Then the morphism* $\phi_{\underline{s}}$ *induces a monomorphism*

$$\mathbb{F}[\mathfrak{Flag}_r](V) \hookrightarrow \Gamma^{[\underline{s}]_q}(V).$$

Proof. The result is proved by induction upon r, starting with the initial case, $r = 1$, which is provided by Lemma 5.2. For the inductive step, by Lemma 5.4, it is sufficient to show that $\psi_{\underline{s}}(V)$ is a monomorphism, under the given hypotheses.

Observe that the hypotheses upon $\underline{s}$ imply that $\underline{s}'$ also satisfy the hypotheses with respect to V, so that the morphism $\phi_{\underline{s}'}(V)$ is injective, by induction. Hence Lemma 5.10 reduces the proof to showing that the restriction of $\phi_{s_r,\ldots,s_r}$ to $\mathbb{F}[\mathfrak{Flag}_r]_\Phi(V)$ is a monomorphism, for each complete flag Φ of length $r-1$ in V. The inductive step is completed by combining Lemma 5.15 with Lemma 5.2. □

6. A stabilization result

The techniques employed in the proof of Theorem 5.17 can be used to provide further information on the nature of the embedding results. For instance, one has a direct proof of the following stabilization result.

Proposition 6.1. *Let* $\underline{s}$ *be a sequence of integers* $(s_1 > \ldots > s_r > s_{r+1} = 0)$ *and* V *be a finite-dimensional vector space for which the morphism* $\phi_{\underline{s}}(V) : \mathbb{F}[\mathfrak{Flag}_r](V) \to \Gamma^{[\underline{s}]_q}(V)$ *is a monomorphism.*

Let $\underline{s^+}$ *denote the sequence given by* $s_i^+ = s_i + 1$, *for* $1 \leqslant i \leqslant r$, *and* $s_{r+1}^+ = 0$. *Then the morphism*

$$\phi_{\underline{s^+}}(V) : \mathbb{F}[\mathfrak{Flag}_r](V) \to \Gamma^{[\underline{s^+}]_q}(V)$$

is a monomorphism.

Proof. The diagonal induces a morphism $\Gamma^{[\underline{s^+}]_q} \to \Gamma^{q[\underline{s}]_q} \otimes \Gamma^{(q-1)r}$. Hence, composing with the Verschiebung on the first morphism gives $\eta : \Gamma^{[\underline{s^+}]_q} \to \Gamma^{[\underline{s}]_q} \otimes \Gamma^{(q-1)r}$, as in Definition 5.3.

There is a commutative diagram

$$\begin{array}{ccc}
\mathbb{F}[\mathfrak{Flag}_r] & \xrightarrow{\phi_{\underline{s^+}}} & \Gamma^{[\underline{s^+}]_q} \\
\downarrow{\scriptstyle \mathrm{diag}} & & \downarrow{\scriptstyle \eta} \\
\mathbb{F}[\mathfrak{Flag}_r] \otimes \mathbb{F}[\mathfrak{Flag}_r] & \xrightarrow[\phi_{\underline{s}} \otimes \phi_{1,\ldots,1}]{} & \Gamma^{[\underline{s}]_q} \otimes \Gamma^{(q-1)r},
\end{array}$$

the commutativity of which is established by an argument similar to that employed in the proof of Lemma 5.6.

It suffices to show that the composite

$$\mathbb{F}[\mathfrak{Flag}_r](V) \xrightarrow{\phi_{\underline{s}^+}} \Gamma^{[\underline{s}^+]_q}(V) \xrightarrow{\eta} \Gamma^{[\underline{s}]_q} \otimes \Gamma^{(q-1)r}(V)$$

is a monomorphism. By hypothesis, the morphism $\phi_{\underline{s}}(V)$ is a monomorphism. As in the inductive step of the proof of Theorem 5.17, the result then follows from the fact that the morphism $\mathbb{F}[\mathfrak{Flag}_r](V) \rightarrow \Gamma^{(q-1)r}(V)$ is non-trivial. The latter follows from the fact that the hypothesis upon $\phi_{\underline{s}}(V)$ implies that V has dimension at least r. □

Remark 6.2. This argument is related to standard techniques using the Kameko Sq^0 operation [**5**], which is based in an essential way upon the analysis of the Verschiebung morphism.

References

[1] Mohamed Ali Alghamdi, M. C. Crabb, and J. R. Hubbuck, *Representations of the homology of BV and the Steenrod algebra. I*, Adams Memorial Symposium on Algebraic Topology, 2 (Manchester, 1990), London Math. Soc. Lecture Note Ser., vol. 176, Cambridge Univ. Press, Cambridge, 1992, pp. 217–234. MR MR1232208 (94i:55022)

[2] J. Michael Boardman, *Modular representations on the homology of powers of real projective space*, Algebraic topology (Oaxtepec, 1991), Contemp. Math., vol. 146, Amer. Math. Soc., Providence, RI, 1993, pp. 49–70. MR MR1224907 (95a:55041)

[3] M. C. Crabb and J. R. Hubbuck, *Representations of the homology of BV and the Steenrod algebra. II*, Algebraic topology: new trends in localization and periodicity (Sant Feliu de Guíxols, 1994), Progr. Math., vol. 136, Birkhäuser, Basel, 1996, pp. 143–154. MR MR1397726 (97h:55018)

[4] Vincent Franjou, Eric M. Friedlander, Alexander Scorichenko, and Andrei Suslin, *General linear and functor cohomology over finite fields*, Ann. of Math. (2) **150** (1999), no. 2, 663–728. MR MR1726705 (2001b:14076)

[5] Masaki Kamako, *Products of projective spaces as Steenrod modules*, Ph.D. thesis, Johns Hopkins University, 1990.

[6] Nicholas J. Kuhn, *Generic representations of the finite general linear groups and the Steenrod algebra. I*, Amer. J. Math. **116** (1994), no. 2, 327–360. MR MR1269607 (95c:55022)

[7] ———, *Generic representations of the finite general linear groups and the Steenrod algebra. II*, *K*-Theory **8** (1994), no. 4, 395–428. MR MR1300547 (95k:55038)

[**8**] J. Repka and P. Selick, *On the subalgebra of $H_*((\mathbf{RP}^\infty)^n; F_2)$ annihilated by Steenrod operations*, J. Pure Appl. Algebra **127** (1998), no. 3, 273–288. MR MR1617199 (99d:55012)

[**9**] Grant Walker and Reg Wood, *Embedding the Steinberg and the flag representations of $GL(n, \mathbb{F}_q)$ in the divided power algebra*, Notes from talk at Larry Smith Emertierungsfeier, September 2007.

This article may be accessed via WWW at `http://www.rmi.acnet.ge/jhrs/`

Geoffrey M.L. Powell
`powell@math.univ-paris13.fr`

Laboratoire Analyse, Géométrie et Applications, UMR 7539
Institut Galilée, Université Paris 13, 93430 Villetaneuse, France

Journal of Homotopy and Related Structures, vol. 4(1), 2009, pp.331–346

HOMOTOPY CLASSIFICATION OF MAPS INTO HOMOGENEOUS SPACES

SERGIY KOSHKIN

(*communicated by James Stasheff*)

Abstract

We give an alternative to Postnikov's homotopy classification of maps from 3-dimensional CW-complexes to homogeneous spaces G/H of Lie groups. It describes homotopy classes in terms of lifts to the group G and is suitable for extending the notion of homotopy to Sobolev maps. This is required in applications to variational problems of mathematical physics.

Introduction

A classical theorem of Postnikov [**Ps, WJ**] gives a homotopy classification of continuous maps from a 3–dimensional CW–complex M to a connected simply connected complex X of any dimension. First it gives the primary invariant that characterizes when $M \xrightarrow{\psi,\varphi} X$ are 2–homotopic, i.e. their restrictions to a 2–skeleton of M are homotopic. When this happens a secondary invariant is defined along with a condition that makes ψ and φ homotopic. Unfortunately, the secondary invariant is hard to compute because one has to homotop one of the maps into the other on the 2–skeleton. While the primary invariant can be characterized in terms of deRham cohomology and thus extended to discontinuous (Sobolev) maps, the secondary one of Postnikov is tied too closely to restrictions and homotopy that do not make sense without continuity.

In physical applications one is often forced to extend topological notions to Sobolev maps. This requires rethinking characterizations of homotopy classes in a way that makes such extension possible. One way is to interpret homotopy classes as connected components in the space of continuous maps. A natural generalization is to study components in spaces of Sobolev maps [**HL, Wh**]. However, in applications to mathematical physics a more hands on approach is usually taken. One identifies invariants that characterize homotopy classes and then extends them to Sobolev maps. This approach is taken in recent works on the Faddeev model [**AK2, LY**] with maps into $S^2 = SU(2)/U(1)$. The present work was motivated by considering its generalization, the Faddeev-Niemi model [**FN**] with targets $SU(N)/T$, T the maximal torus. In this paper we give an alternative description only for homotopy

Received August 1, 2008, revised July 21, 2009; published on September 30, 2009.
2000 Mathematics Subject Classification: 57T15, 55Q25, 46M20, 58D30.
Key words and phrases: Homotopy class, homogeneous space, Hopf invariant, Faddeev model

classes of continuous maps. The extension to Sobolev maps and applications will be published elsewhere [**K**].

We consider continuous maps from 3–dimensional CW–complexes to compact simply connected homogeneous spaces of Lie groups. In exchange for loss in generality one gets a much more explicit description of the secondary invariant and its deRham presentation. It sheds new light on the primary invariant as well.

A smooth manifold is called a homogenous space under an action of a Lie group G if the action is transitive. Any homogenous space X can be identified with the coset space G/H, where H is the isotropy subgroup of a point. Up to a diffeomorphism, different pairs G, H may produce the same space X. Throughout this paper X *will always denote* G/H *where* G *is compact, connected and simply connected and* $H \subset G$ *is connected.* This can be done without loss of generality by switching to a maximal compact subgroup, universal cover and/or identity component of G as appropriate [**BtD, Mg**]. Under this assumtion, we prove that ψ and φ are 2-homotopic if and only if there exists a continuous lift $M \xrightarrow{u} G$ such that $\psi = u\varphi$ (Theorem 2), where $u\varphi$ refers to the action of G on X. This easily generalizes if we allow u to be a Sobolev map but the real advantage is that the secondary invariant for ψ, φ becomes the primary one for u.

It is convenient to introduce the basic class of a space F. Suppose $\pi_0(F) = \ldots = \pi_{n-1}(F) = 0$ then by the Hurewicz theorem $H_n(F,\mathbb{Z}) \simeq \pi_n(F)$. The *basic class* $\mathbf{b}_F \in H^n(F, \pi_n(F))$ is the cohomology class that maps every homology class in $H_n(F,\mathbb{Z})$ into its image in $\pi_n(F)$ under the Hurewicz isomorphism ($\mathbf{b}_F$ is called the fundamental class of F by Steenrod). It follows essentially from the Eilenberg classification theorem [**St**] that $\psi^*\mathbf{b}_F$ is the primary invariant for homotopy. Namely, two maps $M \xrightarrow{\psi,\varphi} F$ are n-homotopic, i.e. their restrictions to the n-skeleton are homotopic iff $\psi^*\mathbf{b}_F = \varphi^*\mathbf{b}_F$.

Let $\mathbf{b}_G \in H^3(F, \pi_3(G))$ be the basic class of G then the secondary invariant is $u^*\mathbf{b}_G$. However, ψ, φ being homotopic is not quite equivalent to its vanishing. The problem is that u in $\psi = u\varphi$ is not unique. Potentially, there are maps $M \xrightarrow{w} G$ with $w\varphi = \varphi$ but $w^*\mathbf{b}_G \neq 0$. One has to factor out the subgroup generated by such maps

$$\mathcal{O}_\varphi := \{w^*\mathbf{b}_G \mid w\varphi = \varphi\} \subset H^3(M, \pi_3(G)).$$

Despite its appearence, this subgroup depends only on the 2-homotopy class of φ (Lemma 5). Our main results (Theorems 2 and 3) can be summarized as follows.

Theorem. *Let* $X = G/H$ *be a compact simply connected homogeneous space and* M *a 3-dimensional CW–complex. Then two maps* $M \xrightarrow{\psi,\varphi} X$ *are homotopic if and only if there exists a map* $M \xrightarrow{u} G$ *such that* $\psi = u\varphi$ *and* $u^*\mathbf{b}_G \in \mathcal{O}_\varphi$*. Within the 2-homotopy class of* φ*, the homotopy classes are in one-to-one correspondence with* $H^3(M, \pi_3(G))/\mathcal{O}_\varphi$.

When $H^2(M,\mathbb{Z}) = 0$ any two maps $M \xrightarrow{\psi,\varphi} X$ are 2-homotopic and therefore related by a lift u. We can always choose φ to be the constant map and define the secondary invariant for a single map ψ. The subgroup $\mathcal{O}_{\mathrm{const}}$ is trivial and homotopy classes are in one-to-one correspondence with $H^3(M, \pi_3(G))$. When $M = S^3$ and

$X = S^2 = SU_2/U_1$ one has $\pi_3(G) \simeq \mathbb{Z}$ and our $u^*\mathbf{b}_G$ is essentially the Hopf invariant, cf. [**LY**]. The class $\mathbf{b}_G$ has a particularly nice deRham presentation when the group G is simple. Then always $\pi_3(G) \simeq \mathbb{Z}$ and one can identify it with the class of the integral form $\Theta := c_G \operatorname{tr}(g^{-1}dg \wedge g^{-1}dg \wedge g^{-1}dg)$. Here c_G are normalizing constants computed in [**AK1**], for example $c_{SU_N} = -\frac{1}{96\pi^2 N}$. The pullback is

$$u^*\Theta = c_G \operatorname{tr}(u^{-1}du \wedge u^{-1}du \wedge u^{-1}du)$$

and $\int_M u^*\Theta$ can be made sense of for Sobolev u.

It is instructive to compare our secondary invariant with Postnikov's [**Ps, WJ**]. His definition requires homotoping ψ into a function $\widetilde{\psi}$ equal to φ on the 2-skeleton of M and computing the primary difference $\overline{d}(\varphi, \psi) \in H^3(M, \pi_3(X))$. Homotopy occurs when this difference takes value in a subgroup with a complicated description that involves the Whitehead product [**WG**] and the Postnikov square [**Nk1, Ps, WJ**]. Note that by the homotopy exact sequence

$$0 = \pi_2(H) \xleftarrow{\partial} \pi_3(G/H) \xleftarrow{\pi_*} \pi_3(G) \xleftarrow{\imath_*} \pi_3(H) \longleftarrow \cdots \tag{1}$$

and $\pi_3(X) \simeq \pi_3(G)/\imath_*\pi_3(H)$. If $\imath_*\pi_3(H) = 0$ as in the case of U_1 in SU_2 our invariant can be identified with Postnikov's.

In Section 1 we solve the relative lifting problem for two maps as a problem in obstruction theory. A key role is played by the *bundle of shifts* that helps characterize existence of the lift in terms of the *primary characteristic class* of the quotient bundle $G \to G/H$. In Section 2 we show that the primary characteristic class is essentially the basic class $\mathbf{b}_X$ and prove our characterization of 2-homotopy classes. Finally, in Section 3 the secondary invariant is introduced and the homotopy classification is completed. We also give a deRham interpretation of the secondary invariant.

1. Primary characteristic class and lifting

In this section we will define a cohomology class on G/H that regulates existence of a relative lift u such that $\psi = u\varphi$ for two maps $M \xrightarrow{\psi,\varphi} G/H$. This class is the primary characteristic class [**MS, St**] of the bundle $G \to G/H$ denoted $\varkappa(G)$. Of course, $\varkappa(G)$ also depends on $H \subset G$ but we follow the usual abuse of notation. The lift exists iff pullbacks of $\varkappa(G)$ by both maps are the same (Theorem 1). We will prove this by reducing the lifting problem to a problem in the obstruction theory [**MS, St**]. In the next section we will identify $\varkappa(G)$ with the primary obstruction to homotopy.

Here is the basic idea of the proof. Given two maps $M \xrightarrow{\varphi,\psi} X$ define $M \xrightarrow{(\varphi,\psi)} X \times X$ and consider the *ratio bundle* over M:

$$Q_{\varphi,\psi} := \{(m, g) \in M \times G | \psi(m) = g\varphi(m)\} \tag{2}$$

Obviously, sections of this bundle $M \xrightarrow{\sigma} Q_{\varphi,\psi} \subset M \times G$ have the form $\sigma(m) = (m, u(m))$, where $\psi = u\varphi$. In other words, they play the role of non-existent ratios ψ/φ. Hence the problem of finding a lift u is equivalent to constructing a section of the bundle $Q_{\varphi,\psi}$, which is a standard problem in the obstruction theory.

First, we have to establish that $Q_{\varphi,\psi}$ are indeed fiber bundles. Note that

$$Q_{\varphi,\psi} \simeq \{(m,x,g) \in M \times X \times G | (\varphi(m), \psi(m) = (x, gx)\} \tag{3}$$

and by definition of pullback $Q_{\varphi,\psi} \simeq (\varphi,\psi)^*Q$, where Q is the bundle of shifts defined next. A particular case of this bundle was used in [**AS**] for similar purposes.

Definition 1 (The bundle of shifts). *The bundle of shifts of a homogeneous space $G/H = X$ is the fiber bundle Q over $X \times X$ given by:*

$$\begin{aligned} X \times G &\xrightarrow{\alpha} X \times X. \\ (x, g) &\longmapsto (x, gx) \end{aligned} \tag{4}$$

Thus, it suffices to prove that Q itself is a fiber bundle. We will do more and identify the principal bundle it is associated to. It turns out to be the Cartesian double $G \times G \xrightarrow{\pi\times\pi} X \times X$ of the quotient bundle $G \xrightarrow{\pi} X = G/H$. This is a principal bundle with the structure group $H \times H$.

Lemma 1. *Let G be a compact Lie group, $H \subset G$ a closed subgroup and $G \xrightarrow{\pi} X = G/H$ the corresponding quotient bundle. Then the bundle of shifts $Q \xrightarrow{\alpha} X \times X$ is a fiber bundle associated to $G \times G \xrightarrow{\pi\times\pi} X \times X$.*

Proof. Recall that given a principal bundle P over X and a space F where the structure group T acts on the left by μ, one can form a set of equivalence classes

$$P \times_\mu F := \{[p, f] \in P \times F | (p, f) \sim (pt, \mu(t^{-1})f)\}. \tag{5}$$

Then $[p, f] \mapsto \pi(p)$ is a bundle projection that turns $P \times_\mu F$ into a fiber bundle over X associated to P by μ [**Hus, MS**].

We will construct an explicit isomorphism between Q and the following associated bundle. The group $T := H \times H$ acts on $F := H$ on the left by

$$\begin{aligned} (H \times H) \times H &\xrightarrow{\mu} H \\ ((\lambda_1, \lambda_2), h) &\longmapsto \lambda_2 h \lambda_1^{-1} \end{aligned}$$

Set $E_1 := ((G \times G) \times_\mu H \xrightarrow{\pi} X)$, $E_2 := Q$ and consider the following map

$$\begin{aligned} E_1 &\xrightarrow{\mathcal{F}} E_2 \\ [g_1, g_2, h] &\longmapsto (g_1 H, g_2 h g_1^{-1}) \end{aligned}$$

To begin with $\mathcal{F}$ is well defined:

$$g_1\lambda_1 H, g_2\lambda_2, \lambda_2^{-1} h \lambda_1) \longmapsto (g_1, \lambda_1 H, g_2 h g_1^{-1}) = (g_1 H, g_2 h g_1^{-1}).$$

The inverse is given by $(x, g) \xrightarrow{\mathcal{F}^{-1}} [g_1, gg_1, 1]$, where $g_1 H = x$. If $g_1\lambda$ is chosen instead with $\lambda \in H$ then $[g_1\lambda, gg_1\lambda, \lambda^{-1}1\lambda] = [g_1, gg_1, 1]$ so $\mathcal{F}^{-1}$ is well-defined. It is easy to see that it is indeed the inverse to $\mathcal{F}$.

We claim that both diagrams commute

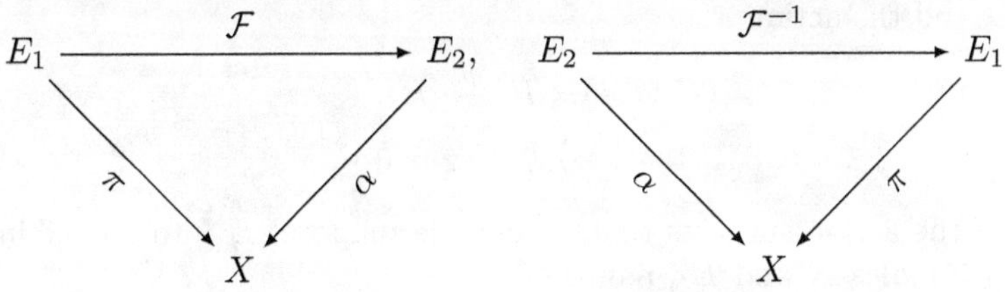

(6)

For instance,

$$(\alpha \circ \mathcal{F})([g_1, g_2, h]) = \alpha(g_1H, g_2Hg_1^{-1}) = (g_1H, g_2hH) = (g_1H, g_2H) = \pi([g_1, g_2, h]).$$

Therefore the bundle of shifts $Q = E_2$ is indeed a fiber bundle and $\mathcal{F}$ is a bundle isomorphism. □

For obstruction theory we follow terminology and notation of Steenrod [**St**]. We say that n is the *lowest homotopy non-trivial dimension* of F if $\pi_k(F) = 0$ for $1 \leqslant k \leqslant n-1$ but $\pi_n(F) \neq 0$. Assume that in a fiber bundle $F \overset{\imath}{\hookrightarrow} E \overset{\pi}{\longrightarrow} B$ the base B is a CW–complex and the fiber F is *homotopy simple* up to this dimension, i.e. $\pi_1(F)$ acts trivially on $\pi_k(F)$ for $1 \leqslant k \leqslant n$. This means that there is no obstruction to constructing a section up to dimension n and we may assume that $B^{(n)} \overset{\sigma}{\longrightarrow} E$ is already constructed, here $B^{(n)}$ is the n-skeleton of B. Let $\Delta \subset B$ be an $(n+1)$ cell of B which we may assume to be contractible or even a simplex. Choosing a local trivialization we get a map $f_\sigma : \partial\Delta \longrightarrow F$ that defines an element of $\pi_n(F)$. It turns out that this element does not depend on a choice of trivialization and $c_\sigma(\Delta) := [f_\sigma] \in \pi_n(F)$ is a $\pi_n(F)$-valued cochain and in fact a cocycle. Its cohomology class $\overline{c}_\sigma \in H^{n+1}(B, \pi_n(F))$ is called the primary obstruction to extending σ. This cohomology class does not even depend on a choice of σ on the n-skeleton of B and is an invariant of the bundle $E \overset{\pi}{\longrightarrow} B$ itself.

Definition 2 (Primary characteristic class). *The invariant* $\varkappa(E) := \overline{c}_\sigma$ *is called the primary characteristic class of* E.

The characteristic class is natural with respect to the pullback of bundles:

$$\varkappa(\varphi^* E) = \varphi^* \varkappa(E)$$

and the Eilenberg extension theorem claims that a section σ can be altered on $B^{(n)}$ so as to be extendable to $B^{(n+1)}$ if and only if $\overline{c}_\sigma = 0$. This completely solves the sectioning problem when $\pi_k(F) = 0$ for $n+1 \leqslant k < \dim B$, i.e. there are no further obstructions: a section exists if and only if $\varkappa(E) = 0$.

In our case the bundle in question is $H \overset{\imath}{\hookrightarrow} Q_{\varphi,\psi} \overset{\pi}{\longrightarrow} M$. The fiber is a Lie group so it is homotopy simple in all dimensions. The first non-trivial dimension is $n = 1$ as H is connected and $\varkappa(Q_{\varphi,\psi}) \in H^2(M, \pi_1(H))$. Since $\pi_2(H) = 0$ for all finite-dimensional Lie groups and $\dim M = 3$ there are no further obstructions and a section exists if and only if $\varkappa(Q_{\varphi,\psi}) = 0$. Thus, we want to compute this characteristic class. By naturality $\varkappa(Q_{\varphi,\psi}) = \varkappa((\varphi,\psi)^* Q) = (\varphi,\psi)^* \varkappa(Q)$ and we need to compute $\varkappa$ for the bundle of shifts.

By Lemma 1 Q is isomorphic to the associated bundle $\widehat{E} := \widehat{P} \times_{\widehat{\mu}} H$ with $\widehat{P} = G \times G$ and the action

$$(H \times H) \times H \xrightarrow{\widehat{\mu}} H$$
$$((\lambda_1, \lambda_2), h) \longmapsto \lambda_2 h \lambda_1^{-1}$$

The form of the action suggests that we can 'decompose' $\widehat{E}$ into a combination of two simpler bundles E and E', namely

$$E := P \times_\mu H \quad \text{with} \quad \mu(\lambda)h := \lambda h$$

and its dual

$$E' := P \times_{\mu'} H \quad \text{with} \quad \mu'(\lambda)h := h\lambda^{-1}$$

(in our case $P = G$ and one can multiply on both sides). We will not explain precisely what the decomposition means in this case but it should be clear from the proof of Lemma 2(ii). Note that E is bundle isomorphic to P itself by $p \mapsto [p, 1]$ so we write $\varkappa(P)$ for $\varkappa(E)$.

Lemma 2. *Let $P \xrightarrow{\pi} X$ be a principal bundle with the structure group H. Define $\widehat{P} := (P \times P \longrightarrow X \times X)$, E, E', $\widehat{E}$ as above and let π_1, π_2 denote the projections from $X \times X$ to the first and the second components. Then*

(i) $\varkappa(P) = \varkappa(E) = -\varkappa(E')$.

(ii) $\varkappa(\widehat{E}) = \pi_2^* \varkappa(P) - \pi_1^* \varkappa(P)$, if in addition $H^k(X, \mathbb{Z}) = 0$ for $0 \leqslant k \leqslant n$, where n is the lowest homotopy non-trivial dimension of H.

Proof. (i) Note that if $\sigma(x) = [p, h]$ gives a section of E then $\sigma'(x) = [p, h^{-1}]$ gives a section of E'. Also if $\Delta \xrightarrow{S_\Delta} P|_\Delta$ is a local section of P then

$$\Delta \times F \xrightarrow{\Phi|_\Delta} (P \times_\mu F)|_\Delta$$
$$(x, f) \longmapsto [S_\Delta(x), f]$$
$$(\pi(p), \mu(\lambda^{-1})f) \longleftarrow [p, f], \quad \text{with } S_\Delta(\pi(p)) = p\lambda,$$

is a local trivialization of the associated bundle.

We choose a section S_Δ of P and denote Φ_Δ, Φ'_Δ the corresponding trivializations of E, E'. Also if σ is the chosen section of E on $B^{(n)}$ then the σ' is the one we choose for E'. By definition,

$$\begin{aligned} f_{\sigma'}(x) = \pi_2 \circ \Phi_\Delta^{-1} \circ \sigma'(x) &= \pi_2 \circ \Phi_\Delta^{-1}([p, h^{-1}]), & \pi(p) = x \\ &= (\pi(p), \mu'(\lambda^{-1})h^{-1}), & S_\Delta(\pi(p)) = S_\Delta(x) = p\lambda \\ &= h^{-1}(\lambda^{-1})^{-1} = (\lambda^{-1}h)^{-1} = (\mu(\lambda^{-1})h)^{-1} \\ &= (\pi_2 \circ \Phi_\Delta^{-1}([p, h])^{-1}) = (\pi_2 \circ \Phi_\Delta^{-1} \circ \sigma(x))^{-1} = f_\sigma(x)^{-1}. \end{aligned}$$

In other words, $c_{\sigma'}(\Delta) = [f_\sigma^{-1}]$. But in $\pi_n(H)$ one has $[o^{-1}] = -[o]$ [**Dy**] for any o and $\varkappa(E') = \overline{c}_{\sigma'} = -\overline{c}_\sigma = -\varkappa(E)$.

(ii) Under our assumptions the Künneth formula and the universal coefficients theorem imply that

$$H^{n+1}(X \times X, \pi_n(H)) \simeq H^{n+1}(X, \pi_n(H)) \oplus H^{n+1}(X, \pi_n(H)),$$
$$\omega \longmapsto (\imath_1^*\omega, \imath_2^*\omega)$$
$$\pi_1^*\omega^* + \pi_2^*\omega_2 \longleftarrow (\omega_1, \omega_2),$$

where $x \overset{\imath_1}{\longmapsto} (x, x_0)$, $x \overset{\imath_2}{\longmapsto} (x_0, x)$ for some fixed point $x_0 \in X$. Let $p_0 \in P$ be any point with $\pi(p_0) = x_0$, then

$$\begin{aligned} \imath_1^*\widehat{E} &= \{(x, [p, p_0, h]) \in X \times \widehat{E}|\ (x, x_0) = (\pi(p), \pi(p_0))\} \\ &\simeq \{(x, [p, h]) \in X \times E|\ \pi(p) = x\} \simeq E' \end{aligned}$$

since p_0 is fixed and $\widehat{\mu}$ reduces to μ' on the first component. Analogously, $\imath_2^*\widehat{E} \simeq E$. Therefore from naturality and (i)

$$\begin{aligned} \varkappa(\widehat{E}) &= \pi_1^*\imath_1^*\varkappa(\widehat{E}) + \pi_2^*\imath_2^*\varkappa(\widehat{E}) = \pi_1^*\varkappa(\imath_1^*\widehat{E}) + \pi_2^*\varkappa(\imath_2^*\widehat{E}) \\ &= \pi_1^*\varkappa(E') + \pi_2^*\varkappa(E) = \pi_2^*\varkappa(P) - \pi_1^*\varkappa(P) \end{aligned}$$

□

Now we are ready for the main result of this section.

Theorem 1. *Let $X = G/H$ be a simply connected homogeneous space, M be a 3-dimensional CW-complex and $M \xrightarrow{\psi,\varphi} X$ be continuous maps. Then a continuous $M \xrightarrow{u} G$ with $\psi = u\varphi$ exists if and only if*

$$\psi^*\varkappa(G) = \varphi^*\varkappa(G),$$

where $\varkappa(G)$ is the primary characteristic class of the quotient bundle $G \to X$.

Proof. In our case P is the quotient bundle $G \longrightarrow X$ and we write $\varkappa(G)$ for its primary characteristic class. It is easy to compute $\varkappa(Q_{\varphi,\psi})$ now since $Q_{\varphi,\psi} = (\varphi, \psi)^*Q$ and $Q = \widehat{E}$ for the quotient bundle $G \longrightarrow X$:

$$\begin{aligned} \varkappa(Q_{\varphi,\psi}) &= \varkappa((\varphi, \psi)^*Q) = (\varphi, \psi)^*\varkappa(Q)) && \text{by naturality} \\ &= (\varphi, \psi)^*(\pi_2^*\varkappa(G) - \pi_1^*\varkappa(G)) && \text{by Lemma 2} \\ &= (\pi_2 \circ (\varphi, \psi))^*\varkappa(G) - (\pi_1 \circ (\varphi, \psi))^*\varkappa(G) \\ &= \psi^*\varkappa(G) - \varphi^*\varkappa(G). \end{aligned}$$

□

In fact the conditions of Lemma 2 are satisfied with $n = 1$ if H is connected and X is simply connected (simple connectedness of G is not necessary). Hence Theorem 1 can be applied directly to U_n homogeneous spaces without reducing them to SU_n ones as long as the subgroup $H \subset U_n$ is already connected.

2. Characterization of 2-homotopy classes

In the previous section we reduced the lifting problem to equality of pullbacks of the primary characteristic class. By obstruction theory the primary obstruction to homotopy is described in the same fashion. It turns out that this is not a coincidence and the primary characteristic class of $G \to G/H$ is essentially the same as the basic class of G/H. As a consequence, a lift u in $\psi = u\varphi$ exists iff ψ and φ are 2-homotopic (Theorem 2).

As before we follow terminology and notation of Steenrod [**St**]. Let B be a $CW-$ complex and $B \overset{\psi,\varphi}{\longrightarrow} F$ be two maps homotopic on $B^{(n-1)}$ by Φ. If $\Delta \subset B^{(n)}$ is an n-cell then $\partial(\Delta \times I) \simeq S^n$ and we can set

$$d_\Phi(\varphi, \psi)(\Delta) := [\Phi(\partial(\Delta \times I))] \in \pi_n(F)$$

This defines a $\pi_n(F)$–valued cochain on B called *the difference cochain.* It turns out to be a cocycle and its cohomology class

$$\overline{d}(\varphi, \psi) := \overline{d_\Phi(\varphi, \psi)}$$

does not depend on a choice of homotopy on $B^{(n-1)}$. Obviously $\overline{d}(\varphi, \psi) \in H^n(B, \pi_n(F))$. The homotopy Φ can be extended from $B^{(n-2)}$ to $B^{(n)}$ (it may have to be altered on $B^{(n-1)}$) if and only if $\overline{d}(\varphi, \psi) = 0$. The difference is natural $\overline{d}(\varphi \circ f, \psi \circ f) = f^*\overline{d}(\varphi, \psi)$ and additive $\overline{d}(\varphi, \chi) = \overline{d}(\varphi, \psi) + \overline{d}(\psi, \chi)$. Since φ is always homotopic to itself $\overline{d}(\varphi, \varphi) = 0$ and additivity implies $\overline{d}(\psi, \varphi) = -\overline{d}(\varphi, \psi)$.

Now let n be the lowest homotopy non-trivial dimension of F and F be homotopy simple up to this dimension. Then any two maps into F are homotopic on $B^{(n-1)}$ and $\overline{d}(\varphi, \psi)$ is defined for any pair. It is called *the primary difference* between φ and ψ [**St**].

Theorem (Eilenberg classification theorem). *If the primary difference is the only obstruction to homotopy i.e.* $\pi_k(F) = 0$ *for* $n + 1 \leqslant k \leqslant \dim B$, *then* φ, ψ *are homotopic if and only if* $\overline{d}(\varphi, \psi) = 0$. *Moreover, for any* $\omega \in H^n(B, \pi_n(F))$ *and a given* $B \overset{\varphi}{\longrightarrow} F$ *there is* $B \overset{\psi}{\longrightarrow} F$ *such that* $\overline{d}(\varphi, \psi) = \omega$.

In other words, under the conditions of the theorem, maps are classified up to homotopy by their primary differences with a fixed map φ, and there is a one-to-one correspondence between homotopy classes and $H^n(B, \pi_n(F))$. In our case $B = M$, $F = X$, $n = 2$ since X is simply connected and $q = 1$ since generally speaking $\pi_3(X) \neq 0$. So $M \overset{\psi,\varphi}{\longrightarrow} X$ are 2-homotopic if and only if $\overline{d}(\varphi, \psi) = 0$. We will reexpress this condition first in terms of the basic class and then of the primary characteristic class.

For any connected space F there are two special self-maps, the identity id_F and the constant map $\mathrm{pt}_F(x) = x_0 \in F$. The primary difference $\overline{d}(\mathrm{id}_F, pt_F)$ only depends on F itself since all constant maps into a connected space are homotopic to each other. Note that $\overline{d}(\mathrm{id}_F, \mathrm{pt}_F) \in H^n(F, \pi_n(F))$ and one can show [**St**] that

$$\overline{d}(\mathrm{id}_F, \mathrm{pt}_F) = \mathbf{b}_F$$

Now let $M \overset{\psi,\varphi}{\longrightarrow} X$ be any continuous maps and $M \overset{\mathrm{pt}_{M,X}}{\longrightarrow} X$ be a constant map. Then by naturality and additivity

$$\overline{d}(\varphi, \psi) = \varphi^*\overline{d}(\mathrm{id}_X, \mathrm{pt}_X) - \psi^*\overline{d}(\mathrm{id}_X, \mathrm{pt}_X) = \varphi^*\mathbf{b}_X - \psi^*\mathbf{b}_X. \qquad (7)$$

In general, $\varkappa(G) \in H^2(X, \pi_1(H))$ and $\mathbf{b}_X \in H^2(X, \pi_2(X))$ but from (11) we have $\pi_1(H) \simeq \pi_2(X)$ under the connecting homomorphism ∂. This suggests that $\varkappa(G) = \pm\partial \circ \mathbf{b}_X$. To prove the equality we need to use the transgression [**MT, St**].

Definition 3 (Transgression). *Let $F \overset{\imath}{\hookrightarrow} E \overset{\pi}{\longrightarrow} B$ be a fiber bundle and $\mathbb{A}$ an Abelian group. An element $\alpha \in H^n(F, \mathbb{A})$ is called transgressive if there are cochains $\xi \in C^n(E, \mathbb{A})$ and $\eta \in C^{n+1}(B, \mathbb{A})$ such that $\overline{\imath^*\xi} = \alpha$ and $\delta\xi = \pi^*\eta$, where the bar denotes the corresponding cohomology class and δ is the cohomology differential. When α is transgressive, classes $\tau^\#\alpha := \overline{\eta} \in H^{n+1}(B, \mathbb{A})$ are called its (cohomology) transgressions.*

Our definition follows Steenrod, but the reader is cautioned that there is another tradition in differential geometry, where the transgression goes the opposite way. There is also an analogous notion of transgression $\tau_\#$ in homology and the two are dual to each other, i.e. when α and a are transgressive $\tau^\#\alpha(a) = \alpha(\tau_\# a)$. Unlike the connecting homomorphism ∂ which is everywhere defined and unambiguous, both transgressions $\tau^\#, \tau_\#$ in general map from a subgroup to a quotient of the corresponding (co)homology groups. The homology transgression in a sense imitates the non-existent connecting homomorphism in homology. In particular, spherical classes in $H_{n+1}(B, \mathbb{Z})$ are always transgressive and the diagram

$$\begin{array}{ccc} \pi_{n+1}(B) & \xrightarrow{\partial} & \pi_n(F) \\ \Big\downarrow \mathcal{H}_B & & \Big\downarrow \mathcal{H}_F \\ H_{n+1}(B, \mathbb{Z}) & \xrightarrow{\tau_\#} & H_n(F, \mathbb{Z}) \end{array} \tag{8}$$

commutes. Here $\mathcal{H}_B$, $\mathcal{H}_F$ are Hurewicz homomorphisms and it is understood that $\mathcal{H}_F(\partial(z))$ is just one of the transgressions of $\mathcal{H}_B(z)$. Commutativity can be established by inspecting the definitions of $\tau_\#$ and ∂ (see [**Hu**]).

There is a case when the transgression is unambiguous. When $H^i(B, \mathbb{A}) = 0$ for $0 < i < k$ and $H^j(F, \mathbb{A}) = 0$ for $0 < j < l$ a result of J.-P. Serre says that $H^m(F, \mathbb{A}) \xrightarrow{\tau^\#} H^{m+1}(B, \mathbb{A})$ is well-defined and one has the *Serre exact sequence* [**MT**]:

$$0 \longrightarrow H^1(B, \mathbb{A}) \xrightarrow{\pi^*} H^1(E, \mathbb{A}) \xrightarrow{\imath^*} H^1(F, \mathbb{A}) \xrightarrow{\tau^\#} H^2(B, \mathbb{A}) \xrightarrow{\pi^*} \ldots \xrightarrow{\imath^*} H^{k+l-1}(F, \mathbb{A}). \tag{9}$$

An analogous statement is also true for the homology transgression. Conditions of the Serre exact sequence are satisfied in particular if n, $n+1$ are the lowest homotopy non-trivial dimensions for F and B respectively and $k = n+1$, $l = n$. Under these assumtions the primary characteristic class is related straightforwardly to the basic class of the base.

Lemma 3. *Let $F \overset{\imath}{\hookrightarrow} E \overset{\pi}{\longrightarrow} B$ be a fiber bundle with the fiber F being homotopy simple up to dimension n and let n, $n+1$ be the lowest homotopy non-trivial* *dimensions of F and B respectively. Then*

$$\varkappa(E) = -\partial \circ \mathbf{b}_B, \tag{10}$$

where $\pi_{n+1}(B) \xrightarrow{\partial} \pi_n(F)$ *is the connecting homomorphism (cf. [**Nk2**]).*

Proof. By the universal coefficients theorem:

$$0 \longrightarrow \mathrm{Ext}(H_n(B,\mathbb{Z}),\pi_n(F)) \longrightarrow H^{n+1}(B,\pi_n(F)) \longrightarrow \mathrm{Hom}(H_{n+1}(B,\mathbb{Z}),\pi_n(F)) \longrightarrow 0$$

is exact and since $n+1$ is the lowest homotopy non-trivial dimension of B the group $H_n(B,\mathbb{Z}) = 0$ and the Ext term vanishes. Hence the elements of $H^{n+1}(B,\pi_n(F))$ are completely determined by their pairing with integral homology classes. By the Serre exact sequences both transgressions $H^n(F,\pi_n(F)) \xrightarrow{\tau^{\#}} H^{n+1}(B,\pi_n(F))$ and $H_{n+1}(B,\mathbb{Z}) \xrightarrow{\tau_{\#}} H_n(F,\mathbb{Z})$ are unambiguous and the Whitehead transgression theorem [**St**] (see also [**BH**], Appendix 1) gives $\varkappa(E) = -\tau^{\#}\mathbf{b}_F$. Using also the duality of transgressions and (8)

$$\begin{aligned}\varkappa(E)(a) &= -\tau^{\#}\mathbf{b}_F(a) = -\mathbf{b}_F(\tau_{\#}a) = -\mathcal{H}_F^{-1}(\tau_{\#}a) = -\partial(\mathcal{H}_B^{-1}(a)) \\ &= -\partial(\mathbf{b}_B(a)) = -\partial \circ \mathbf{b}_B(a).\end{aligned}$$

Since $a \in H_{n+1}(B,\mathbb{Z})$ is arbitrary (all elements are spherical by the Hurewicz theorem and hence transgressive) we get (10). □

Recall that the basic class regulates 2-homotopy and the primary characteristic class regulates existence of a lift between two maps into G/H. We now establish the desired equivalence.

Theorem 2. *Let $X = G/H$ be a compact simply connected homogeneous space and M a 3-dimensional CW-complex. Then three conditions are equivalent for continuous* $M \xrightarrow{\psi,\varphi} X$*:*

(i) φ, ψ *are 2-homotopic, i.e. homotopic on the 2-skeleton of M;*

(ii) $\psi^*\mathbf{b}_X = \varphi^*\mathbf{b}_X \in H^2(M,\pi_2(X))$, *where $\mathbf{b}_X$ is the basic class of X;*

(iii) *There exists a continuous* $M \xrightarrow{u} G$ *such that $\psi = u\varphi$, where $u\varphi$ refers to the action of G on X.*

Proof. Equivalence of the first two conditions is just a particular case of the Eilenberg classification theorem. To prove (iii) we apply Lemma 3 to the bundle $H \hookrightarrow G \longrightarrow X = G/H$ with $n = 1$ since H is connected and get $\varkappa(G) = -\partial \circ \mathbf{b}_X \in H^2(X,\pi_1(H))$. Since $\pi_2(G) = 0$ for any finite-dimensional Lie group we have from the homotopy exact sequence

$$0 = \pi_1(G) \longleftarrow \pi_1(H) \xleftarrow{\partial} \pi_2(G/H) \longleftarrow \pi_2(G) = 0 \tag{11}$$

that the connecting homomorphism is an isomorphism. Since it also commutes with pullbacks $\psi^*\mathbf{b}_X = \varphi^*\mathbf{b}_X$ if and only if $\psi^*\varkappa(G) = \varphi^*\varkappa(G)$. Application of Theorem 1 now concludes the proof. □

3. Secondary invariant and homotopy classes

By the Eilenberg classification theorem, maps $M \to X$ are 2–homotopic if and only if they have the same pullbacks of the basic class $\mathbf{b}_X$. This pullback $\varphi^*\mathbf{b}_X$ is the primary invariant of a map φ. If $\pi_3(X) = 0$ then the 2-homotopy class is already the homotopy class (recall that we only consider a 3–dimensional M). Otherwise, secondary invariants have to be specified. Unlike the primary invariants, classical secondary invariants are not defined constructively [**MT**]. From Theorem 2 we know that even 2-homotopy of ψ and φ implies that $\psi = u\varphi$. In this section we derive an explicit characterization for such u in terms of $u^*\mathbf{b}_G$, where $\mathbf{b}_G$ is the basic class of G. In other words, we are using $u^*\mathbf{b}_G$ as a secondary invariant of a pair ψ, φ while for the lift u it is a primary invariant and is defined straightforwardly.

We start with a simple observation that follows directly from the homotopy lifting property in the bundle of shifts.

Lemma 4. *Let G be a compact connected Lie group, $H \subset G$ a closed subgroup, $X = G/H$ and M a CW–complex. Then two continuous maps $M \xrightarrow{\varphi, \psi} X$ are homotopic if and only if there exists a nullhomotopic $M \xrightarrow{u_0} G$ such that $\psi = u_0\varphi$. Given an arbitrary map $M \xrightarrow{u} G$ maps φ, $u\varphi$ are homotopic if and only if $u = u_0 w$, where u_0 is nullhomotopic and $w\varphi = \varphi$.*

Proof. Let $M \xrightarrow{1} G$ denote the constant map that maps every point into the identity of G. If u_0^t is a homotopy that translates u_0 into 1 then $\psi_t := u_0^t\varphi$ translates $u_0\varphi$ into φ and $\Phi(m, t) := (\varphi(m), \psi_t(m))$ translates (φ, φ) into (φ, ψ). The former admits a lift $(\varphi, 1)$ into Q, indeed $\alpha \circ (\varphi, 1) = (\varphi, \varphi)$. Since Q is a fiber bundle by Lemma 1 the homotopy lifting property implies that the following diagram can be completed as indicated:

$$
\begin{array}{ccc}
M \times \{0\} & \xrightarrow{(\varphi, 1)} & X \times G \\
\Big\uparrow & \nearrow^{\widetilde{\Phi}} \quad \alpha \Big\downarrow & \\
M \times I & \xrightarrow{\Phi} & X \times X
\end{array}
$$

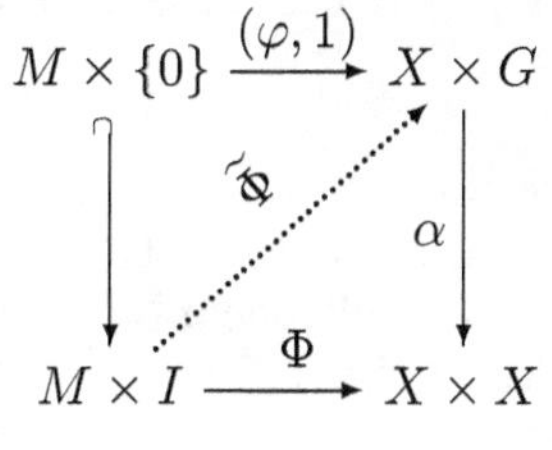

By the upper triangle $\widetilde{\Phi}_2(m, 0) = 1$ and by the lower one $\widetilde{\Phi}_1(m, t) = \Phi_1(m, t) = \varphi(m)$, $\widetilde{\Phi}_2(m, t)\widetilde{\Phi}_1(m, t) = \widetilde{\Phi}_2(m, t)\varphi(m) = \psi_t(m)$. Set $u_0(m) := \widetilde{\Phi}_2(m, 1)$ then $u_0\varphi = \psi$ and $\widetilde{\Phi}_2(\cdot, t)$ is a homotopy that translates the constant map 1 into u_0 as required.

For the second claim note that $u = u_0 w$ implies $u\varphi = u_0 w\varphi = u_0\varphi$ and is homotopic to φ. Conversely, if $u\varphi$ is homotopic to φ then by the first claim there is also a second nullhomotopic u_0 such that $u_\varphi = u_0\varphi$. It suffices to set $w := u_0^{-1}u$. □

Let (M, G) denote the space of continuous maps $M \to G$ and $(M, G)\varphi$ the space of maps $M \to X$ that have the form $u\varphi$ for $u \in (M, G)$. Lemma 4 suggests that the following maps play a special role.

Definition 4 (Stabilizer). *Given a map $M \xrightarrow{\varphi} X$ we call*

$$\mathrm{Stab}_\varphi := \{w \in (M,G) | w\varphi = \varphi\}$$

the stabilizer of φ.

We have a natural inclusion $\mathrm{Stab}_\varphi \overset{\imath}{\hookrightarrow} (M,G)$. Denote

$$[M,G] := \pi_0((M,G)),$$
$$[(M,G)\varphi] := \pi_0((M,G)\varphi).$$

Note that $[M,G]$ is the set of homotopy classes of continuous maps $M \longrightarrow G$ and by Theorem 2 $[(M,G)\varphi]$ is the set of homotopy classes of continuous maps into $X = G/H$, which are 2–homotopic to φ.

If G is compact simply connected $\pi_1(G) = \pi_2(G) = 0$ and it follows from the Eilenberg classification theorem that

$$\begin{aligned} [M,G] &\simeq H^3(M,\pi_3(G)) \\ [u] &\longmapsto u^*\mathbf{b}_G \end{aligned} \tag{12}$$

is a group isomorphism. Under this isomorphism the subgroup $\imath_*\pi_0(\mathrm{Stab}_\varphi) = \pi_0(\imath(\mathrm{Stab}_\varphi))$ is mapped into a subgroup of $H^3(M,\pi_3(G))$ that we denote $\mathcal{O}_\varphi$, i.e

$$\mathcal{O}_\varphi := \{w^*\mathbf{b}_G \mid w \in \mathrm{Stab}_\varphi\} \subset H^3(M,\pi_3(G)). \tag{13}$$

Although the definition (13) uses the map φ explicitly, we will show

Lemma 5. *$\mathcal{O}_\varphi$ only depends on the 2-homotopy class of φ or equivalently on $\varphi^*\mathbf{b}_X$ and not on φ itself.*

Proof. If ψ is 2–homotopic to φ, there is $M \xrightarrow{u} G$ with $\psi = u\varphi$ by Theorem 2. Hence

$$\mathrm{Stab}_\psi = \{w|w\psi = \psi\} = \{w|wu\varphi = u\varphi\} = \{w|u^{-1}wu \in \mathrm{Stab}_\varphi\} = u(\mathrm{Stab}_\varphi)u^{-1}$$

Let π_1, π_2 be the natural projections from $G \times G$ to the first and the second factor and $G \times G \xrightarrow{m} G$ be the multiplication map. Then it follows from the Hopf-Samelson theorem [**Dy, WG**] that

$$m^*\mathbf{b}_G = \pi_1^*\mathbf{b}_G + \pi_2^*\mathbf{b}_G. \tag{14}$$

This implies that given two maps $M \xrightarrow{u,v} G$ we have

$$(u \cdot v)^*\mathbf{b}_G = (m \circ (u,v))^*\mathbf{b}_G = (u,v)^*(\pi_1^*\mathbf{b}_G + \pi_2^*\mathbf{b}_G) = u^*\mathbf{b}_G + v^*\mathbf{b}_G$$

Using this formula we derive from definition (13)

$$\begin{aligned} \mathcal{O}_\psi = \{w^*\mathbf{b}_G | w \in \mathrm{Stab}_\psi\} &= \{(uw'u^{-1})^*\mathbf{b}_G | w' \in \mathrm{Stab}_\varphi\} \\ &= \{u^*\mathbf{b}_G + (w')^*\mathbf{b}_G - u^*\mathbf{b}_G | w' \in \mathrm{Stab}_\varphi\} = \mathcal{O}_\varphi \end{aligned}$$

□

Hence $\mathcal{O}_\varphi = \mathcal{O}_{\varphi^*\mathbf{b}_X}$ and since every $\varkappa \in H^2(M,\pi_2(X))$ is representable by a φ one can talk about $\mathcal{O}_\varkappa$.

Theorem 3. *Let $X = G/H$ be a compact simply connected homogeneous space and M a 3-dimensional CW-complex. Two continuous maps $M \xrightarrow{\psi,\varphi} X$ are homotopic if and only if $\psi = u\varphi$ and $u^*\mathbf{b}_G \in \mathcal{O}_\varphi$ for some $M \xrightarrow{u} G$. Moreover*

$$[(M,G)\varphi] \simeq H^3(M, \pi_3(G))/\mathcal{O}_\varphi \qquad (\simeq \textit{ means bijection}). \tag{15}$$

Proof. By definition of the isomorphism (12) having $u^*\mathbf{b}_G \in \mathcal{O}_\varphi$ is equivalent to $[u] \in \imath_*\pi_0(\mathrm{Stab}_\varphi)$ or u homotopic to $w \in \mathrm{Stab}_\varphi$. But then uw^{-1} is nullhomotopic and $\psi = (uw^{-1})w\varphi$ is homotopic to φ by Lemma 4.

To prove the bijection consider the map

$$\begin{gathered}(M,G) \xrightarrow{\Pi} (M,G)\varphi \\ u \mapsto u\varphi.\end{gathered}$$

We will show that this is a fibration following an idea from [**AS**]. By definition we need to complete the diagram as indicated for A arbitrary and $I := [0,1]$

$$\begin{array}{ccc} A \times \{0\} & \xrightarrow{F_0} & (M,G) \\ \downarrow & \nearrow & \downarrow \Pi \\ A \times I & \xrightarrow{f} & (M,G)\varphi \end{array} \tag{16}$$

Set $\overline{F}_0(m,a) := F_0(a)(m)$ and $\overline{f}(m,a,t) := f(a,t)(m)$. Recall from Lemma 1 that the bundle of shifts (4) is a fiber bundle and therefore a fibration so the following diagram can be completed as indicated:

$$\begin{array}{ccc} (M \times A) \times 0 & \xrightarrow{(\overline{F}_0,\varphi)} & G \times X \\ \downarrow & \overline{\Phi} \nearrow & \downarrow \alpha \\ (M \times A) \times I & \xrightarrow{(\overline{f},\varphi)} & X \times X \end{array}$$

Inspecting the definitions of $\overline{F}_0$, $\overline{f}$ one concludes that the original diagram can be completed as well using $\overline{\Phi}$.

If $v\varphi = u\varphi$ then $w := u^{-1}v \in \mathrm{Stab}_\varphi$ and the fiber of this fibration is exactly Stab_φ. Using the homotopy exact sequence of the fibration

$$\pi_0(\mathrm{Stab}_\varphi) \xrightarrow{\imath_*} \pi_0((M,G)) \xrightarrow{\pi_*} \pi_0((M,G)\varphi) \longrightarrow 0.$$

one gets

$$[(M,G)\varphi] \simeq \frac{[M,G]}{\imath_*\pi_0(\mathrm{Stab}_\varphi)} \qquad (\simeq \text{ means bijection}).$$

Under the isomorphism (12) this becomes (15). □

For applications it is convenient to reinterpret the secondary invariant in terms of the deRham cohomology. Let us start with the group $H^3(G, \pi_3(G))$. Recall that we assume that G is compact connected and simply connected. By the universal coefficients theorem the following sequence is exact:

$$0 \longrightarrow \mathrm{Tor}(H^2(G,\mathbb{Z}),\pi_3(G)) \longrightarrow H^3(G,\pi_3(G)) \longrightarrow H^3(G,\mathbb{Z})\otimes\pi_3(G) \longrightarrow 0.$$

Since G is a simply connected Lie group $H^2(G,\mathbb{Z}) = 0$ and the torsion term vanishes so

$$H^3(G,\pi_3(G)) \simeq H^3(G,\mathbb{Z})\otimes\pi_3(G).$$

Since G is also compact it is a direct product of simple components $G = G_1\times\cdots\times G_N$ and therefore

$$\pi_3(G) \simeq \pi_3(G_1)\oplus\cdots\oplus\pi_3(G_N).$$

The sum on the right $\simeq \mathbb{Z}^N$ because $\pi_3(\Gamma)\simeq\mathbb{Z}$ for any simple Lie group Γ [**BtD**]. Thus

$$H^3(G,\pi_3(G)) \simeq H^3(G,\mathbb{Z})\otimes\mathbb{Z}^N$$

Assume additionally that M is a closed connected 3–manifold. Both third cohomology groups $H^3(G,\mathbb{Z})$, $H^3(M,\mathbb{Z})$ are free Abelian, the first one by the Hurewicz theorem and the second by Poincare duality. This means that not only are elements of $H^3(G,\mathbb{Z})\otimes\mathbb{Z}^N$ completely represented by integral classes in $H^3(G,\mathbb{R})\otimes\mathbb{R}^N$ but also that their pullbacks are completely characterized as integral classes in $H^3(M,\mathbb{R})\otimes\mathbb{R}^N$. But real cohomology classes from $H^3(G,\mathbb{R})\otimes\mathbb{R}^N$ are represented by $\mathbb{R}^N$–valued differential 3–forms by the deRham theorem [**GHV, MS**].

Let Θ be a differential form that represents $\mathbf{b}_G$. Being $\mathbb{R}^N$–valued it is a collection $\Theta = (\Theta_1,\ldots,\Theta_N)$ of N scalar 3–forms and the pullback

$$u^*\Theta := (u^*\Theta_1,\ldots,u^*\Theta_N)$$

is defined as a vector–valued 3–form. We can go one step further. Assuming M is orientable $H^3(M,\mathbb{Z})\simeq\mathbb{Z}$ and again by the universal coefficients:

$$H^3(M,\pi_3(G)) \simeq H^3(M,\mathbb{Z})\otimes\pi_3(G) \simeq H^3(M,\mathbb{Z})\otimes\mathbb{Z}^N \simeq \mathbb{Z}^N.$$

The last isomorphism is given by evaluation of cohomology classes on the fundamental class of M or in terms of differential forms, by integration over M [**GHV, MS**]. Thus we get a combined isomorphism

$$\begin{aligned} H^3(M,\pi_3(G)) &\xrightarrow{\sim} \mathbb{Z}^N \\ u^*\mathbf{b}_G &\longmapsto \int_M u^*\Theta := \Big(\int_M u^*\Theta_1,\ldots,\int_M u^*\Theta_N\Big). \end{aligned} \tag{17}$$

Under this isomorphism the subgroup $\mathcal{O}_\varphi \subset H^3(M,\pi_3(G))$ is transformed into a subgroup of $\mathbb{Z}^N$ and we denote its image by the same symbol, explicitly

$$\mathcal{O}_\varphi := \{\int_M w^*\Theta \mid w\in\mathrm{Stab}_\varphi\} \subset \mathbb{Z}^N. \tag{18}$$

If G is a simple group then $H^3(M,\pi_3(G))\simeq\mathbb{Z}$ and we have explicitly

$$\Theta = c_G\,\mathrm{tr}(g^{-1}dg\wedge g^{-1}dg\wedge g^{-1}dg),$$

where c_G are numerical coefficients computed in [**AK1**] for every simple group. Thus

$$u^*\Theta = c_G \operatorname{tr}(u^{-1}du \wedge u^{-1}du \wedge u^{-1}du). \tag{19}$$

In general,

$$\Theta_k = c_{G_k} \operatorname{tr}(\operatorname{pr}_{\mathfrak{g}_k}(g^{-1}dg) \wedge \operatorname{pr}_{\mathfrak{g}_k}(g^{-1}dg) \wedge \operatorname{pr}_{\mathfrak{g}_k}(g^{-1}dg)),$$

where $\mathfrak{g}_k$ are the Lie algebras of G_k. Theorem 3 can be restated as

Corollary 1. *In conditions of Theorem 3 let M be a closed connected 3-manifold. Then two continuous maps $M \xrightarrow{\psi,\varphi} X$ are homotopic if and only if $\psi = u\varphi$ and $\int_M u^*\Theta \in \mathcal{O}_\varphi$ for some $M \xrightarrow{u} G$.*

If M is not orientable then $H^3(M,\mathbb{Z}) = 0$ and the secondary invariant is always 0.

References

[**AK1**] D. Auckly, L. Kapitanski, *Holonomy and Skyrme's model.* Comm. Math. Phys. **240**(2003), no. 1-2, 97–122.

[**AK2**] D. Auckly, L. Kapitanski, *Analysis of S^2-valued maps and Faddeev's model.* Comm. Math. Phys. **256**(2005), no. 3, 611–620.

[**AS**] D. Auckly, M. Speight, *Fermionic quantization and configuration spaces for the Skyrme and Faddeev-Hopf models.* Comm. Math. Phys. **263**(2006), no. 1, 173–216.

[**BH**] A. Borel, F. Hirzebruch, *Characteristic classes and homogeneous spaces II.* Amer. J. Math. **81**(1959), no. 2, 315–382.

[**BtD**] T. Bröker, T. Dieck, *Representations of compact Lie groups.* Graduate Texts in Mathematics **98**, Springer-Verlag, New York, 1985.

[**Ch**] C. Chevalley, *Theory of Lie groups, I.* Princeton Mathematical Series, **8**, Princeton Landmarks in Mathematics, Princeton University Press, Princeton, NJ, 1999.

[**Dy**] E. Dynkin, *Homologies of compact Lie groups.* Amer. Math. Soc. Transl. **12**(2)(1959), 251–300.

[**FN**] L. Faddeev, A. Niemi, *Partial duality in $SU(N)$ Yang-Mills theory.* Phys. Lett. B **387**(1999), 214–222.

[**GHV**] W. Greub, S. Halperin, R. Vanstone, *Connections, curvature, and cohomology, vol. I, II.* Pure and Applied Mathematics, **47**, Academic Press, New York-London, 1972-73.

[**HL**] F. Hang, F. Lin, *Topology of Sobolev mappings.* Math. Res. Lett. **8**(2001), no. 3, 321–330; *Topology of Sobolev mappings, II.* Acta Math. **191**(2003), no. 1, 55–107.

[**Hu**] S. Hu, *Homotopy theory.* Pure and Applied Mathematics, vol. VIII, Academic Press, New York-London, 1959.

[Hus] D. Husemoller, *Fibre bundles.* Graduate Texts in Mathematics **20**, Springer-Verlag, New York, 1994.

[K] S. Koshkin, *Gauge theory of Faddeev-Skyrme functionals.* arxiv: `math-ph/0907.0899` (submitted to *Commun. Contemp. Math.*)

[LY] F. Lin, Y. Yang, *Existence of energy minimizers as stable knotted solitons in the Faddeev model.* Comm. Math. Phys. **249**(2004), no. 2, 273–303.

[MS] J. Milnor, J. Stasheff, *Characteristic classes.* Annals of Mathematics Studies, No. 76. Princeton University Press, Princeton, N. J, 1974.

[Mg] D. Montgomery, *Simply connected homogeneous spaces.* Proc. Amer. Math. Soc. **1**(1950),no. 4, 467–469.

[MT] R. Mosher, M. Tangora, *Cohomology operations and applications in homotopy theory.* Harper & Row, Publishers, New York-London, 1968.

[Ms] G. Mostow, *The extensibility of local Lie groups of transformations and groups on surfaces.* Ann. of Math. (2) **52**(1950), no.3, 606–636.

[Nk1] N. Nakaoka, *Classification of mappings of a complex into a special kind of complex.* J. Inst. Polytech. Osaka City Univ. Ser. A. Math. **3**(1952), 101–143.

[Nk2] M. Nakaoka, *Transgression and the invariant* k_n^{q+1}. Proc. Japan Acad. **30**(1954), 363–368.

[Ps] M. Postnikov, *The classification of continuous mappings of a three-dimensional polyhedron into a simply connected polyhedron of arbitrary dimension.* Dokl. Akad. Nauk SSSR (N.S.) **64**(1949), 461–462. (Russian)

[St] N. Steenrod, *The topology of fibre bundles.* Princeton Landmarks in Math., Princeton Paperbacks, Princeton Univ. Press, Princeton, NJ, 1999.

[Wh] B. White, *Homotopy classes in Sobolev spaces and the existence of energy minimizing maps.* Acta Math. **160**(1988), no. 1-2, 1–17.

[WG] G.W. Whitehead, *Elements of homotopy theory.* Graduate Texts in Mathematics **61**, Springer-Verlag, New York-Berlin, 1978.

[WJ] J.H.C. Whitehead, *On the theory of obstructions.* Ann. of Math. (2) **54**(1951), no. 1, 68–84.

This article may be accessed via WWW at `http://www.rmi.acnet.ge/jhrs/`

Sergiy Koshkin
`koshkinS@uhd.edu`

Department of Computer and Mathematical Sciences
University of Houston-Downtown
1 Main Street #S705
Houston, TX 77002

Journal of Homotopy and Related Structures, vol. 4(1), 2009, pp.347–357

ON THE HOMOTOPY CLASSIFICATION OF MAPS

SAMSON SANEBLIDZE

(*communicated by James Stasheff*)

To Nodar Berikashvili

Abstract

We establish certain conditions which imply that a map $f : X \to Y$ of topological spaces is null homotopic when the induced integral cohomology homomorphism is trivial; one of them is: $H^*(X)$ and $\pi_*(Y)$ have no torsion and $H^*(Y)$ is polynomial.

1. Introduction

We give certain classification theorems for maps via induced cohomology homomorphism. Such a classification is based on a new aspects of obstruction theory to the section problem in a fibration beginning in [**4**], [**5**] and developed in some directions in [**24**], [**25**]. Given a fibration $F \to E \xrightarrow{\xi} X$, the obstructions to the section problem of ξ naturally lay in the groups $H^{i+1}(X; \pi_i(F)), i \geqslant 0$. A basic method here is to use the Hurewicz homomorphism $u_i : \pi_i(F) \to H_i(F)$ for passing the above obstructions into the groups $H^{i+1}(X; H_i(F)), i \geqslant 0$. In particular, this suggests the following condition on a fibration: The induced homomorphism

$$(1.1)_m \qquad u^* : H^{i+1}(X; \pi_i(F)) \to H^{i+1}(X; H_i(F)),\ 1 \leqslant i < m,$$

is an inclusion (assuming $u_1 : \pi_1(F) \to H_1(F)$ is an isomorphism). Note also that the idea of using the Hurewicz map in the obstruction theory goes back to the paper [**23**]. (Though its main result was erroneous, it became one crucial point for applications of characteristic classes (see [**7**]).)

For the homotopy classification of maps $X \to Y$, the space F in $(1.1)_m$ is replaced by ΩY and we establish the following statements. Below all topological spaces are assumed to be path connected (hence, Y is also simply connected) and the ground coefficient ring is the integers $\mathbb{Z}$. Given a commutative graded algebra (cga) H^* and an integer $m \geqslant 1$, we say that H^* is *m-relation free* if H^i is torsion free for $i \leqslant m$ and also there is no multiplicative relation in H^i for $i \leqslant m+1$; in particular, $H^{2i-1} = 0$ for $1 \leqslant i \leqslant [\frac{m+2}{2}]$. We also allow $m = \infty$ for H to be polynomial on even degree generators.

This research described in this publication was made possible in part by the grant GNF/ST06/3-007 of the Georgian National Science Foundation. I am grateful to Jesper Grodal for helpful comments. I thank to Jim Stasheff for helpful comments and suggestions.
Received October 28, 2008, revised June 8, 2009; published on October 14, 2009.
2000 Mathematics Subject Classification: Primary 55S37, 55R35; Secondary 55S05, 55P35.
Key words and phrases: cohomology homomorphism, functor D, polynomial algebra, section.

Theorem 1. *Let $f : X \to Y$ be a map such that the pair $(X, \Omega Y)$ satisfies $(1.1)_m$, X is an m-dimensional polyhedron and $H^*(Y)$ is m-relation free. Then f is null homotopic if and only if*

$$0 = H^*(f) : H^*(Y) \to H^*(X).$$

Theorem 2. *Let X and Y be spaces such that the Hurewicz map $u_i : \pi_i(\Omega Y) \to H_i(\Omega Y)$ is an inclusion for $1 \leqslant i < m$, and $\operatorname{Tor}\left(H^{i+1}(X), H_i(\Omega Y)/\pi_i(\Omega Y)\right) = 0$ when $\pi_i(\Omega Y) \neq 0$, X is an m-dimensional polyhedron and $H^*(Y)$ is m-relation free. Then a map $f : X \to Y$ is null homotopic if and only if*

$$0 = H^*(f) : H^*(Y) \to H^*(X).$$

Theorem 3. *Let X be an m-dimensional polyhedron and G a topological group such that $\pi_i(G)$ is torsion free for $1 \leqslant i < m$, and $\operatorname{Tor}\left(H^{i+1}(X), \operatorname{Coker} u_i\right) = 0$, $u_i : \pi_i(G) \to H_i(G)$ when $\pi_i(G) \neq 0$. Suppose that the cohomology algebra $H^*(BG)$ of the classifying space BG is m-relation free. Then a map $f : X \to BG$ is null homotopic if and only if*

$$0 = H^*(f) : H^*(BG) \to H^*(X).$$

In fact the two last Theorems follow from the first one, since their hypotheses imply $(1.1)_m$, too. A main example of G in Theorem 3 is the unitary group $U(n)$ with $m = 2n$, since u_{2i} is a trivial inclusion and u_{2i-1} is an inclusion given by multiplication by the integer $(i-1)!$ for $1 \leqslant i \leqslant n$. A $U(n)$-principal fibre bundle over X is classified by a map $X \to BU(n)$. Suppose that all its Chern classes are trivial, then $H^*(f) = 0$ and by Theorem 3, f is null homotopic. Therefore the $U(n)$-principal fibre bundle is trivial. Thus, we have in fact deduced the following statement, the main result of [**22**] (compare also [**29**]).

Corollary 1. *Let ξ be a $U(n)$-principal fibre bundle over X with $\dim X \leqslant 2n$ and the only torsion in $H^{2i}(X)$ is relatively prime to $(i-1)!$. Then ξ is trivial if and only if the Chern classes $c_k(\xi) = 0$ for $1 \leqslant k \leqslant n$.*

While the proof of this statement in [**22**] does not admit an immediate generalization for an infinite dimensional X, Theorem 3 does by taking $m = \infty$. Furthermore, for $G = U$ and $X = BU$ recall that $[BU, BU]$ is an abelian group, so we get that two maps $f, g : BU \to BU$ are homotopic if and only if $H^*(f) = H^*(g) : H^*(BU; \mathbb{Q}) \to H^*(BU; \mathbb{Q})$ (compare [**14**], [**21**]). Note also that when $m = \infty$ in Theorem 3, $H^*(Y)$ must have infinitely many polynomial generators (e.g. $Y = BU, BSp$) as it follows from the solution of the Steenrod problem for finitely generated polynomial rings [**1**] (the underlying spaces do not have torsion free homotopy groups in all degrees).

Finally, note that beside obstruction theory we apply a main ingredient of the proof of Theorem 1 is an explicit form of minimal multiplicative (non-commutative) resolution of an m-relation free cga (of a polynomial algebra when $m = \infty$) in total degrees $\leqslant m$ (compare [**24**], [**26**]). Namely, the generator set of the resolution in the above range only consists of monomials formed by $\smile_1$ products. Remark that the idea of using $\smile_1$ product when dealing with polynomial cohomology, especially in the context of homogeneous spaces, has been realized by several authors [**17**], [**9**], [**20**], [**13**] (see also [**18**] for further references).

In sections 2 and 3 we recall certain basic definitions and constructions, including the functor $D(X; H_*)$ [**2**], [**3**], for the aforementioned obstruction theory, and in section 4 prove Theorems 1-3.

2. Functor D(X;H)

Given a bigraded differential algebra $A = \{A^{i,j}\}$ with $d : A^{i,j} \to A^{i+1,j}$ and total degree $n = i + j$, let $D(A)$ be the set [**3**] defined by $D(A) = M(A)/G(A)$ where

$$\begin{aligned} M(A) &= \{a \in A^1 \,|\, da = -aa,\ a = a^{2,-1} + a^{3,-2} + \cdots\}, \\ G(A) &= \{p \in A^0 \,|\, p = 1 + p^{1,-1} + p^{2,-2} + \cdots\}, \end{aligned}$$

and the action $M(A) \times G(A) \to M(A)$ is given by the formula

$$a * p = p^{-1}ap + p^{-1}dp. \tag{2.1}$$

In other words, two elements $a, b \in M(A)$ are on the same orbit if there is $p \in G(A)$, $p = 1 + p'$, with

$$b - a = ap' - p'b + dp'. \tag{2.2}$$

Note that an element $a = \{a^{*,*}\}$ from $M(A)$ is of total degree 1 and referred to as *twisting*; we usually suppress the second degree below. There is a distinguished element in the set $D(A)$, the class of $0 \in A$, and denoted by the same symbol.

There is simple but useful (cf. [**24**])

Proposition 1. *Let $f, g : A^{*,*} \to B^{*,*}$ be two dga maps that preserve the bigrading. If they are (f, g)-derivation homotopic via $s : A^{i,j} \to B^{i-1,j}$, i.e., $f - g = sd + ds$ and $s(ab) = (-1)^{|a|} fasb + sagb$, then $D(f) = D(g) : D(A) \to D(B)$.*

Proof. Given $a \in M(A)$, apply the (f, g)-derivation homotopy s to get $fa - ga = dsa + sda = dsa + s(-aa) = dsa + fasa - saga$. From this we deduce that fa and ga are equivalent by (2.2) for $p' = -sa$. □

Another useful property of D is fixed by the following comparison theorem [**2**], [**3**]:

Theorem 4. *If $f : A \to B$ is a cohomology isomorphism, then $D(f) : D(A) \to D(B)$ is a bijection.*

For our purposes the main example of $D(A)$ is the following (cf. [**2**], [**3**])

Example 1. *Fix a graded (abelian) group H_*. Let*

$$\rho : (R_{\geqslant 0}H_q, \partial^R) \to H_q,\ \ \partial^R : R_iH_q \to R_{i-1}H_q,$$

be its free group resolution. Form the bigraded Hom *complex*

$$(\mathcal{R}^{*,*}, d^R) = \left(Hom(RH_*, RH_*), d^R\right), \quad d^R : \mathcal{R}^{s,t} \to \mathcal{R}^{s+1,t};$$

an element $f \in \mathcal{R}^{*,*}$ has bidegree (s, t) if $f : R_jH_q \to R_{j-s}H_{q-t}$. Note also that $\mathcal{R}^{*,*}$ becomes a dga with respect to the composition product.

Given a topological space X, consider the dga

$$(\mathcal{H}, \nabla) = (C^*(X; \mathcal{R}), \nabla = d^C + d^R)$$

which is bigraded via $\mathcal{H}^{r,t} = \prod_{r=i+j} C^i(X;\mathcal{R}^{j,t})$. *Thus we get*

$$\mathcal{H} = \{\mathcal{H}^n\}, \qquad \mathcal{H}^n = \prod_{n=r+t} \mathcal{H}^{r,t}, \qquad \nabla : \mathcal{H}^{r,t} \to \mathcal{H}^{r+1,t}.$$

We refer to r *as the* perturbation *degree which is mainly exploited by inductive arguments below. For example, for a twisting cochain* $h \in M(\mathcal{H})$, *we have*

$$h = h^2 + \cdots + h^r + \cdots, \qquad h^r \in \mathcal{H}^{r,1-r},$$

satifying the following sequence of equalities:

$$\nabla(h^2) = 0, \;\; \nabla(h^3) = -h^2h^2, \;\; \nabla(h^4) = -h^2h^3 - h^3h^2, \dots. \tag{2.3}$$

Define

$$D(X;H_*) = D(\mathcal{H},\nabla).$$

Then $D(X;H_*)$ *becomes a functor on the category of topological spaces and continuous maps to the category of pointed sets.*

Example 2. *Given two dga's* B^* *and* $C^{*,*}$ *with* $d^B : B^i \to B^{i+1}$ *and* $d_1^C : C^{j,t} \to C^{j+1,t}$, $d_2^C = 0$, *let* $A = B\hat{\otimes}C$. *View* (A,d) *as bigraded via* $A = \{A^{r,t}, d\}$, $A^{r,t} = \prod_{r=i+j} B^i \otimes C^{j,t}$, $d = d^B \otimes 1 + 1 \otimes d_1^C$. *Note also that the dga* $(\mathcal{H},\nabla)$ *in the previous example can also be viewed as a special case of the above tensor product algebra by setting* $B^* = C^*(X)$ *and* $C^{*,*} = \mathcal{R}^{*,*}$.

3. Predifferential $d(\xi)$ of a fibration

Let $F \to E \xrightarrow{\xi} X$ be a fibration. In [**2**] a unique element of $D(X;H_*(F))$ is naturally assigned to ξ; this element is denoted by $d(\xi)$ and referred to as the *predifferential* of ξ. The naturalness of $d(\xi)$ means that for a map $f : Y \to X$,

$$d(f(\xi)) = D(f)(d(\xi)), \tag{3.1}$$

where $f(\xi)$ denotes the induced fibration on Y.

Originally $d(\xi)$ appeared in homological perturbation theory for measuring the non-freeness of the Brown-Hirsch model: First, in [**11**] G. Hirsch modified E. Brown's twisting tensor product model $(C_*(X) \otimes C_*(F), d_\phi) \to (C_*(E), d_E)$ [**6**], [**8**] by replacing the chains $C_*(F)$ by its homology $H_*(F)$ provided the homology is a free module. In [**2**] the Hirsch model was extended for arbitrary $H_*(F)$ by replacing it by a free module resolution $RH_*(F)$ to obtain $(C_*(X) \otimes RH_*(F), d_h)$ in which $d_h = d_X \otimes 1 + 1 \otimes d_F + -\cap h$ and h is just an element of $M(\mathcal{H})$ in Example 1 with $H_* = H_*(F)$. Furthermore, to an isomorphism $p : (C_*(X) \otimes RH_*(F), d_h) \to (C_*(X) \otimes RH_*(F), d_{h'})$ between two such models answers an equivalence relation $h \sim_p h'$ in $M(\mathcal{H})$, and the class of h in $D(X;H_*(F))$ is identified as $d(\xi)$. More precisely, we recall some basic constructions for the definition of $d(\xi)$ we need for the obstruction theory in question.

For convenience, assume that X is a polyhedron and that $\pi_1(X)$ acts trivially on $H_*(F)$. Then ξ defines the following colocal system of chain complexes over X :

To each simplex $\sigma \in X$ is assigned the singular chain complex $(C_*(F_\sigma), \gamma_\sigma)$ of the space $F_\sigma = \xi^{-1}(\sigma)$:

$$X \ni \sigma \longrightarrow (C_*(F_\sigma), \gamma_\sigma) \subset (C_*(E), d_E),$$

and to a pair $\tau \subset \sigma$ of simplices an induced chain map

$$C_*(F_\tau) \to C_*(F_\sigma).$$

Set $\mathcal{C}_\sigma = \{\mathcal{C}_\sigma^{s,t}\}$, $\mathcal{C}_\sigma^{s,t} = \mathrm{Hom}^{s,t}(R_*H_*(F), C_*(F_\sigma))$ where C_* is regarded as bigraded via $C_{0,*} = C_*, C_{i,*} = 0$, $i \neq 0$, and $f : R_jH_q(F) \to C_{j-s,q-t}(F_\sigma)$ is of bidegree (s,t). Then we obtain a colocal system of cochain complexes $\mathcal{C} = \{\mathcal{C}_\sigma^{*,*}\}$ on X. Define $\mathcal{F}$ as the simplicial cochain complex $C^*(X;\mathcal{C})$ of X with coefficients in the colocal system $\mathcal{C}$. Then

$$\mathcal{F} = \{\mathcal{F}^{i,j,t}\}, \quad \mathcal{F}^{i,j,t} = C^i(X;\mathcal{C}^{j,t}).$$

Furthermore, obtain the bicomplex $\mathcal{F} = \{\mathcal{F}^{r,t}\}$ via

$$\mathcal{F}^{r,t} = \prod_{r=i+j} \mathcal{F}^{i,j,t}, \ \delta : \mathcal{F}^{r,t} \to \mathcal{F}^{r+1,t}, \ \gamma : \mathcal{F}^{r,t} \to \mathcal{F}^{r,t+1}, \ \delta = d^C + \partial^R, \ \gamma = \{\gamma_\sigma\},$$

and finally set

$$\mathcal{F} = \{\mathcal{F}^m\}, \quad \mathcal{F}^m = \prod_{m=r+t} \mathcal{F}^{r,t}.$$

We have a natural dg pairing

$$(\mathcal{F}, \delta + \gamma) \otimes (\mathcal{H}, \nabla) \to (\mathcal{F}, \delta + \gamma)$$

defined by $\smile$ product on $C^*(X;-)$ and the obvious pairing $\mathcal{C}_\sigma \otimes \mathcal{R} \to \mathcal{C}_\sigma$ in coefficients; in particular we have $\gamma(fh) = \gamma(f)h$ for $f \otimes h \in \mathcal{F} \otimes \mathcal{H}$. Denote $\mathcal{R}_\# = Hom(RH_*(F), H_*(F))$ and define

$$(\mathcal{F}_\#, \delta_\#) := (H(\mathcal{F}, \gamma), \delta_\#) = (C^*(X;\mathcal{R}_\#), \delta_\#).$$

Clearly, the above pairing induces the following dg pairing

$$(\mathcal{F}_\#, \delta_\#) \otimes (\mathcal{H}, \nabla) \to (\mathcal{F}_\#, \delta_\#).$$

In other words, this pairing is also defined by $\smile$ product on $C^*(X;-)$ and the pairing $\mathcal{R}_\# \otimes \mathcal{R} \to \mathcal{R}_\#$ in coefficients. Note that ρ induces an epimorphism of chain complexes

$$\rho^* : (\mathcal{H}, \nabla) \to (\mathcal{F}_\#, \delta_\#).$$

In turn, ρ^* induces an isomorphism in cohomology.

Consider the following equation

$$(\delta + \gamma)(f) = fh \tag{3.2}$$

with respect to a pair $(h, f) \in \mathcal{H}^1 \times \mathcal{F}^0$,

$$\begin{aligned} h &= h^2 + \cdots + h^r + \cdots, \quad & h^r \in \mathcal{H}^{r,1-r}, \\ f &= f^0 + \cdots + f^r + \cdots, \quad & f^r \in \mathcal{F}^{r,-r}, \end{aligned}$$

satisfying the initial conditions:

$$\nabla(h) = -hh$$
$$\gamma(f^0) = 0, \qquad [f^0]_\gamma = \rho^*(1) \in \mathcal{F}_{\#}^{0,0}, \ \ 1 \in \mathcal{H}.$$

Let (h, f) be a solution of the above equation. Then $d(\xi) \in D(X; H_*(F))$ is defined as the class of h. Moreover, the transformation of h by (2.1) is extended to pairs (h, f) by the map

$$(M(\mathcal{H}) \times \mathcal{F}^0) \times (G(\mathcal{H}) \times \mathcal{F}^{-1}) \to M(\mathcal{H}) \times \mathcal{F}^0$$

given for $((h, f), (p, s)) \in (M(\mathcal{H}) \times \mathcal{F}^0) \times (G(\mathcal{H}) \times \mathcal{F}^{-1})$ by the formula

$$(h, f) * (p, s) = (h * p\,,\ fp + s(h * p) + (\delta + \gamma)(s))\,. \tag{3.3}$$

We have that a solution (h, f) of the equation exists and is unique up to the above action. Therefore, $d(\xi)$ is well defined.

Note that action (3.3) in particular has a property that if $(\bar{h}, \bar{f}) = (h, f) * (p, s)$ and $h^r = 0$ for $2 \leqslant r \leqslant n$, then in view of (2.2) one gets the equalities

$$\bar{h}^{n+1} = h * (1 + p^n) = h^{n+1} + \nabla(p^n). \tag{3.4}$$

3.1. Fibrations with $d(\xi) = 0$

The main fact of this subsection is the following theorem from [**4**]:

Theorem 5. *Let $F \to E \xrightarrow{\xi} X$ be a fibration such that (X, F) satisfies $(1.1)_m$. If the restriction of $d(\xi) \in D(X; H_*(F))$ to $d(\xi)|_{X^m} \in D(X^m; H_*(F))$ is zero, then ξ has a section on the m-skeleton of X. The case of $m = \infty$, i.e., $d(\xi) = 0$, implies the existence of a section on X.*

Proof. Given a pair $(h, f) \in \mathcal{H} \times \mathcal{F}$, let (h_{tr}, f_{tr}) denote its component that lies in

$$C^*(X; Hom(H_0(F), RH_*(F))) \times C^*(X; Hom(H_0(F), C_*(F_\sigma))).$$

Below (h_{tr}, f_{tr}) is referred to as the *transgressive* component of (h, f). Observe that since $RH_0(F) = H_0(F) = \mathbb{Z}$, we can view (h_{tr}^{r+1}, f_{tr}^r) as a pair of cochains laying in $C^{>r}(X; RH_r(F)) \times C^r(X; C_r(F_\sigma))$. Such an interpretation allows us to identify a section $\chi^r : X^r \to E$ on the r-skeleton $X^r \subset X$ with a cochain, denoted by c_χ^r, in $C^r(X; C_r(F_\sigma))$ via $c_\chi^r(\sigma) = \chi^r|_\sigma : \Delta^r \to F_\sigma \subset E$, $\sigma \subset X^r$ is an r-simplex, $r \geqslant 0$.

The proof of the theorem just consists of choosing a solution (h, f) of (3.2) so that the transgressive component $f_{tr} = \{f_{tr}^r\}_{r \geqslant 0}$ is specified by $f_{tr}^r = c_\chi^r$ with χ a section of ξ. Indeed, since F is path connected, there is a section χ^1 on X^1; consequently, we get the pairs $(0, f_{tr}^0) := (0, c_\chi^0)$ and $(0, f_{tr}^1) := (0, c_\chi^1)$ with $\gamma(f_{tr}^1) = \delta(f_{tr}^0)$. Then $\delta(f_{tr}^1) \in C^2(X; C_1(F))$ is a γ-cocycle and $[\delta(f_{tr}^1)]_\gamma \in C^2(X; H_1(F))$ becomes the obstruction cocycle $c(\chi^1) \in C^2(X; \pi_1(F))$ for extending of χ^1 on X^2; moreover, one can choose h_{tr}^2 to be satisfying $\rho^*(h_{tr}^2) = [\delta(f_{tr}^1)]_\gamma$ (since ρ^* is an epimorphism and a weak equivalence).

Suppose by induction that we have constructed a solution (h, f) of (3.2) and a section χ^n on X^n such that $h^r = 0$ for $2 \leqslant r \leqslant n$, $f_{tr}^n = c_\chi^n$ and

$$\rho^*(h_{tr}^{n+1}) = [\delta(f_{tr}^n)]_\gamma \in C^{n+1}(X; H_n(F)).$$

In view of (2.3) we have $\nabla(h^{n+1}) = 0$ and from the above equality immediately follows that

$$u^{\#}(c(\chi^n)) = \rho^*(h^{n+1}_{tr})$$

in which $c(\chi^n) \in C^{n+1}(X;\pi_n(F))$ is the obstruction cocycle for extending of χ^n on X^{n+1} and $u^{\#} : C^{n+1}(X;\pi_n(F)) \to C^{n+1}(X;H_n(F))$.

Since $d(\xi)|_{X^m} = 0$, there is $p \in G(\mathcal{H})$ such that $(h * p)|_{X^m} = 0$; in particular, $(h*p)^{n+1} = 0 \in \mathcal{H}^{n+1,-n}$ and in view of (3.4) we establish the equality $h^{n+1} = -\nabla(p^n)$, i.e., $[h^{n+1}] = 0 \in H^*(\mathcal{H},\nabla)$; in particular, $[h^{n+1}_{tr}] = 0 \in H^{n+1}(X;H_n(F))$. Consequently, $[u^{\#}(c(\chi^n))] = 0 \in H^{n+1}(X;H_n(F))$. Since $(1.1)_n$ is an inclusion induced by $u^{\#}$, $[c(\chi^n)] = 0 \in H^{n+1}(X;\pi_n(F))$. Therefore, we can extend χ^n on X^{n+1} without changing it on X^{n-1} in a standard way. Finally, put $f^{n+1}_{tr} = c^{n+1}_{\chi}$ and choose a ∇-cocycle h^{n+2}_{tr} satisfying $\rho^*(h^{n+2}_{tr}) = [\delta(f^{n+1}_{tr})]_{\gamma}$. The induction step is completed. □

4. Proof of Theorems 1, 2 and 3

First we recall the following application of Theorem 5 ([**4**])

Theorem 6. *Let $f : X \to Y$ be a map such that X is an m-polyhedron and the pair $(X, \Omega Y)$ satisfies $(1.1)_m$. If $0 = D(f) : D(Y; H_*(\Omega Y)) \to D(X; H_*(\Omega Y))$, then f is null homotopic.*

Proof. Let $\Omega Y \to PY \xrightarrow{\pi} Y$ be the path fibration and $f(\pi)$ the induced fibration. It suffices to show that $f(\pi)$ has a section. Indeed, (3.1) together with $D(f) = 0$ implies $d(f(\pi)) = 0$, so Theorem 5 guaranties the existence of the section. □

Now we are ready to prove the theorems stated in the introduction. Note that just below we shall heavily use multiplicative, non-commutative resolutions of cga's that are enriched with $\smile_1$ products. Namely, given a space Z, recall its filtered model $f_Z : (RH(Z), d_h) \to C^*(Z)$ [**24**], [**26**] in which the underlying differential (bi)graded algebra $(RH(Z), d)$ is a non-commutative version of Tate-Jozefiak resolution of the cohomology algebra $H^*(Z)$ ([**28**], [**15**]), while h denotes a perturbation of d similar to [**10**]. Moreover, given a map $X \to Y$, there is a dga map $RH(f) : (RH(Y), d_h) \to (RH(X), d_h)$ (not uniquely defined!) such that the following diagram

$$\begin{array}{ccc} (RH(Y), d_h) & \xrightarrow{RH(f)} & (RH(X), d_h) \\ {\scriptstyle f_Y}\downarrow & & \downarrow{\scriptstyle f_X} \\ C^*(Y) & \xrightarrow{C(f)} & C^*(X) \end{array} \tag{4.1}$$

commutes up to (α,β)-derivation homotopy with $\alpha = C(f)\circ f_Y$ and $\beta = f_X \circ RH(f)$ (see, [**12**], [**24**]).

Proof of Theorem 1. The non-trivial part of the proof is to show that $H(f) = 0$ implies f is null homotopic. In view of Theorem 6 it suffices to show that $D(f) = 0$.

By (4.1) and Proposition 1 we get the commutative diagram of pointed sets

$$\begin{array}{ccc} D(\mathcal{H}_Y) & \overset{D(\mathcal{H}(f))}{\longrightarrow} & D(\mathcal{H}_X) \\ {\scriptstyle D(f_Y)}\downarrow & & \downarrow{\scriptstyle D(f_X)} \\ D(Y;H_*(\Omega Y)) & \overset{D(f)}{\longrightarrow} & D(X;H_*(\Omega Y)) \end{array}$$

in which

$$\mathcal{H}_X = RH^*(X)\hat{\otimes}Hom(RH_*(\Omega Y), RH_*(\Omega Y)),$$

$$\mathcal{H}_Y = RH^*(Y)\hat{\otimes}Hom(RH_*(\Omega Y), RH_*(\Omega Y))$$

(see Example 2) and the vertical maps are induced by $f_X \otimes 1$ and $f_Y \otimes 1$; these maps are bijections by Theorem 4. Below we need an explicit form of $RH(f)$ to see that $H(f) = 0$ necessarily implies $RH(f)|_{V^{(m)}} = 0$ with $V^{(m)} = \bigoplus_{1\leqslant i+j\leqslant m} V^{i,j}$; hence, the restriction of the map $\mathcal{H}(f) := RH(f)\otimes 1$ to $RH^{(m)}\otimes 1$, $RH^{(m)} = \bigoplus_{1\leqslant i+j\leqslant m} R^iH^j(Y)$, is zero, and, consequently,

$$D(f_X)\circ D(\mathcal{H}(f)) = 0. \tag{4.2}$$

First observe that any multiplicative resolution $(RH,d) = (T(V^{*,*}),d)$, $V = \langle \mathcal{V}\rangle$, of a cga H admits a sequence of multiplicative generators, denoted by

$$a_1 \smile_1 \cdots \smile_1 a_{n+1} \in \mathcal{V}^{-n,*}, \quad a_i \in \mathcal{V}^{0,*}, \quad n \geqslant 1, \tag{4.3}$$

where $a_i \smile_1 a_j = (-1)^{(|a_i|+1)(|a_j|+1)} a_j \smile_1 a_i$ and $a_i \neq a_j$ for $i \neq j$. Furthermore, the expression $ab \smile_1 uv$ also has a sense by means of formally (successively) applying the Hirsch formula

$$c\smile_1 (ab) = (c\smile_1 a)b + (-1)^{|a|(|c|+1)} a(c\smile_1 b). \tag{4.4}$$

The resolution differential d acts on (4.3) by iterative application of the formula

$$d(a\smile_1 b) = da\smile_1 b - (-1)^{|a|} a\smile_1 db + (-1)^{|a|}ab - (-1)^{|a|(|b|+1)}ba.$$

Consequently, we get

$$d(a_1\smile_1\cdots\smile_1 a_n) = \sum_{(\mathbf{i};\mathbf{j})} (-1)^{\epsilon}(a_{i_1}\smile_1\cdots\smile_1 a_{i_k})\cdot(a_{j_1}\smile_1\cdots\smile_1 a_{j_\ell})$$

where the summation is over unshuffles $(\mathbf{i};\mathbf{j}) = (i_1 < \cdots < i_k\,; j_1 < \cdots < j_\ell)$ of $\underline{n}$.

In the case of H to be m-relation free with a basis $U^i \subset H^i$, $i \leqslant m$, we have that the minimal multiplicative resolution RH of H can be built by taking $\mathcal{V}$ with $\mathcal{V}^{0,i} \approx \mathcal{U}^i, i \leqslant m$, and $\mathcal{V}^{-n,i}, n > 0$, to be the set consisting of monomials (4.3) for $1 \leqslant i - n \leqslant m$ (compare [**26**]). The verification of the acyclicity in the negative resolution degrees of RH restricted to the range $RH^{(m)}$ is straightforward (see also Remark 1). Regarding the map $RH(f)$, we can choose it on $RH^{(m)}$ as follows. Let $R_0H(f) : R_0H(Y) \to R_0H(X)$ be determined by $H(f)$ in an obvious way and then define $RH(f)$ for $a \in \mathcal{V}^{(m)}$ by

$$RH(f)(a) = \begin{cases} R_0H(f)(a), & a \in \mathcal{V}^{0,*}, \\ R_0H(f)(a_1)\smile_1\cdots\smile_1 R_0H(f)(a_n), & a = a_1\smile_1\cdots\smile_1 a_{n+1}, \\ & a\in\mathcal{V}^{-n,*}, a_i \in \mathcal{V}^{0,*}, n\geqslant 1, \end{cases}$$

and extend to $RH^{(m)}$ multiplicatively. Furthermore, f_X and f_Y are assumed to be preserving the generators of the form (4.3) with respect to the right most association of $\smile_1$ products in question. Since h annihilates monomials (4.3) and the existence of formula (4.4) in a simplicial cochain complex, f_X and f_Y are automatically compatible with the differentials involved. Then the maps α and β in (4.1) also preserve $\smile_1$ products, and become homotopic by an (α,β)-derivation homotopy $s : RH(Y) \to C^*(X)$ defined as follows: choose s on $\mathcal{V}^{0,*}$ by $ds = \alpha - \beta$ and extend on $\mathcal{V}^{-n,*}$ inductively by

$$s(a_0 \smile_1 z_n) = -\alpha(a_0) \smile_1 s(z_n) + s(a_0) \smile_1 \beta(z_n) + s(z_n)s(a_0), \quad n \geqslant 1,$$

in which $z_1 = a_1$ and $z_k = a_1 \smile_1 \cdots \smile_1 a_k$ for $k \geqslant 2$, $a_i \in \mathcal{V}^{0,*}$. Clearly, $H(f) = 0$ implies $RH(f)|_{V^{(m)}} = 0$. Since (4.2), $D(f) = 0$ and so f is null homotopic by Theorem 6. Theorem is proved.

Remark 1. *Let $\mathcal{V}_n^{(m)}$ be a subset of $\mathcal{V}^{(m)}$ consisting of all monomials formed by the $\cdot$ and $\smile_1$ products evaluated on a string of variables $a_1, ..., a_n$. Then there is a bijection of $\mathcal{V}_n^{(m)}$ with the set of all faces of the permutahedron P_n ([**19**], [**27**]) such that the resolution differential d is compatible with the cellular differential of P_n (compare [**16**]). In particular, the monomial $a_1 \smile_1 \cdots \smile_1 a_n$ is assigned to the top cell of P_n, while the monomials $a_{\sigma(1)} \cdots a_{\sigma(n)}, \sigma \in S_n$, to the vertices of P_n (see Fig. 1 for $n = 3$). Thus, the acyclicity of P_n immediately implies the acyclicity of $RH^{(m)}$ in the negative resolution degrees as desired.*

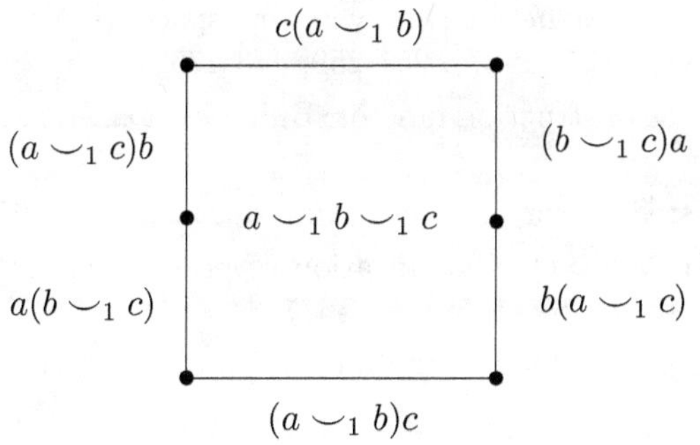

Figure 1. Geometrical interpretation of some syzygies involving $\smile_1$ product as homotopy for commutativity in the resolution RH.

Remark 2. *An example provided by the Hopf map $f : S^3 \to S^2$ shows that the implication $H(f) = 0 \Rightarrow RH(f)|_{V^{(k)}} = 0$, $k < m$ for $RH(f)$ making (4.1) commutative up to (α,β)-derivation homotopy is not true in general. More precisely, let $x \in R^0H^2(S^2)$ and $y \in R^0H^3(S^3)$ with $\rho x \in H^2(S^2)$ and $\rho y \in H^3(S^3)$ to be the generators, and let $x_1 \in R^{-1}H^4(S^2)$ with $dx_1 = x^2$. Then $s(x^2) = \alpha(x)s(x)$ is a cocycle in $C^3(S^3)$ with $d_{S^3}s(x) = \alpha(x)$ (since $\beta = 0$) and $[\alpha(x)s(x)] = \rho y$. Consequently, while $H(f) = 0 = R^0H(f)$, a map $RH(f) : RH(S^2) \to RH(S^3)$ required in (4.1) has a non-trivial component increasing the resolution degree: Namely, $R^{-1}H^4(S^2) \to R^0H^3(S^3)$, $x_1 \to y$.*

Proof of Theorem 2. The conditions that $u_i : \pi_i(\Omega Y) \to H_i(\Omega Y)$ is an inclusion and $\mathrm{Tor}\left(H^{i+1}(X), H_i(\Omega Y)/\pi_i(\Omega Y)\right) = 0$ for $1 \leqslant i < m$, immediately implies $(1.1)_m$. So the theorem follows from Theorem 1.

Proof of Theorem 3. Since the homotopy equivalence $\Omega BG \simeq G$, the conditions of Theorem 2 are satisfied: Indeed, there is the following commutative diagram

$$\begin{array}{ccc} \pi_k(G) & \xrightarrow{u_k} & H_k(G) \\ i_\pi \downarrow & & \downarrow i_H \\ \pi_k(G) \otimes \mathbb{Q} & \xrightarrow{u_k \otimes 1} & H_k(G) \otimes \mathbb{Q} \end{array}$$

where i_π, i_H and $u_k \otimes 1$ are the standard inclusions (the last one is a consequence of a theorem of Milnor-Moore). Consequently, $u_k : \pi_k(\Omega BG) \to H_k(\Omega BG), k < m$, is an inclusion, too. Theorem is proved. □

References

[1] K.K.S. Andersen and J. Grodal, The Steenrod problem of realizing polynomial cohomology rings, J. Topology, 1 (2008), 747-460.

[2] N. Berikashvili, On the differentials of spectral sequences (Russian), Proc. Tbilisi Mat. Inst., 51 (1976), 1-105.

[3] ———-, Zur Homologietheorie der Faserungen I, Proc. A. Razmadze Math. Inst. 116 (1998), 1–29.

[4] ———-, On the obstruction theory in fibre spaces (in Russian), Bull. Acad. Sci. Georgian SSR, 125 (1987), 257-259, 473-475.

[5] ———-, On the obstruction functor, Bull. Georgian Acad. Sci., 153 (1996), 25-30.

[6] E. Brown, Twisted tensor products, Ann. of Math., 69 (1959), 223-246.

[7] A. Dold and H. Whitney, Classification of oriented sphere bundles over 4-complex, Ann. Math., 69 (1959), 667-677.

[8] V.K.A.M. Gugenheim, On the chain complex of a fibration, Ill. J. Math., 16 (1972), 398-414.

[9] V.K.A.M. Gugenheim and J.P. May, On the theory and applications of differential torsion products, Memoirs of AMS, 142 (1974), 1–93.

[10] S. Halperin and J. D. Stasheff, Obstructions to homotopy equivalences, Adv. in Math., 32 (1979), 233-279.

[11] G. Hirsch, Sur les groups d'homologies des espaces fibres, Bull. Soc. Math. de Belg., 6 (1953), 76-96.

[12] J. Huebschmann, Minimal free multi-models for chain algebras, Georgian Math. J., 11 (2004), 733-752.

[13] D. Husemoller, J.C. Moore and J. Stasheff, Differential homological algebra and homogeneous spaces, J. Pure and Applied Algebra, 5 (1974), 113–185.

[14] S. Jackowski, J. McClure and R. Oliver, Homotopy classification of self-maps of *BG* via *G*-actions, I,II, Ann. Math., 135 (1992), 183–226, 227–270.

[15] J.T. Jozefiak, Tate resolutions for commutative graded algebras over a local ring, Fund. Math., 74 (1972), 209-231.

[16] S. MacLane, Natural associativity and commutativity, Rice University Studies, 49 (1963), 28-46.

[17] J.P. May, The cohomology of principal bundles, homogeneous spaces, and two-stage Postnikov systems, Bull. AMS, 74 (1968), 334-339.

[18] J. McCleary, "Users' guide to spectral sequences "(Publish or Perish. Inc., Wilmington, 1985).

[19] R.J. Milgram, Iterated loop spaces, Ann. of Math. 84 (1966), 386-403.

[20] H. J. Munkholm, The Eilenberg-Moore spectral sequence and strongly homotopy multiplicative maps, J. Pure and Applied Algebra, 5 (1974), 1–50.

[21] D. Notbohm, "Classifying spaces of compact Lie groups and finite loop spaces,"Handbook of algebraic topology (Ed. I.M. James), Chapter **21**, North-Holland, 1995.

[22] F.P. Peterson, Some remarks on Chern classes, Ann. Math., 69 (1959), 414-420.

[23] L. Pontrjagin, Classification of some skew products, Dokl. Acad. Nauk. SSSR, 47 (1945), 322-325.

[24] S. Saneblidze, Perturbation and obstruction theories in fibre spaces, Proc. A. Razmadze Math. Inst., 111 (1994), 1-106.

[25] ———, Obstructions to the section problem in a fibration with a weak formal base, Georgian Math. J., 4 (1997), 149-162.

[26] ———, Filtered Hirsch algebras, preprint math.AT/0707.2165.

[27] S. Saneblidze and R. Umble, Diagonals on the Permutahedra, Multiplihedra and Associahedra, J. Homology, Homotopy and Appl., 6 (2004), 363-411.

[28] J. Tate, Homology of noetherian rings and local rings, Illinois J. Math., 1 (1957), 14-27.

[29] E. Thomas, Homotopy classification of maps by cohomology homomorphisms, Trans. AMS, 111 (1964), 138-151.

This article may be accessed via WWW at `http://www.rmi.acnet.ge/jhrs/`

Samson Saneblidze
`sane@rmi.acnet.ge`

A. Razmadze Mathematical Institute
Department of Geometry and Topology
M. Aleksidze st., 1
0193 Tbilisi, Georgia

Journal of Homotopy and Related Structures, vol. 4(1), 2009, pp.359–387

THE FUNDAMENTAL CATEGORY OF A STRATIFIED SPACE

JON WOOLF

(*communicated by Michael Weiss*)

Abstract

The fundamental groupoid of a locally 0 and 1-connected space classifies covering spaces, or equivalently local systems. When the space is topologically stratified, Treumann, based on unpublished ideas of MacPherson, constructed an 'exit category' (in the terminology of this paper, the 'fundamental category') which classifies constructible sheaves, equivalently stratified etale covers. This paper generalises this construction to homotopically stratified sets, in addition showing that the fundamental category dually classifies constructible cosheaves, equivalently stratified branched covers.

The more general setting has several advantages. It allows us to remove a technical 'tameness' condition which appears in Treumann's work; to show that the fundamental groupoid can be recovered by inverting all morphisms and, perhaps most importantly, to reduce computations to the two-stratum case. This provides an approach to computing the fundamental category in terms of homotopy groups of strata and homotopy links. We apply these techniques to compute the fundamental category of symmetric products of $\mathbb{C}$, stratified by collisions.

Two appendices explain the close relations respectively between filtered and pre-ordered spaces and between cosheaves and branched covers (technically locally-connected uniquely-complete spreads).

1. Introduction

Covers of a (nice) topological space X are classified by the fundamental group $\pi_1 X$ or, if we wish to avoid assuming that X is connected and has a basepoint, by the fundamental groupoid $\Pi_1 X$. If X is a stratified space then it is natural to allow the covers to be stratified too. MacPherson observed (unpublished) that local homeomorphisms onto X which are covers when restricted to each stratum are classified by a modified version of the fundamental groupoid. The objects of this are the points of X and the morphisms are homotopy classes of paths which 'wind

Received November 18, 2008, revised July 01, 2009; published on November 28, 2009.
2000 Mathematics Subject Classification: Primary 55P99, Secondary 14H30.
Key words and phrases: Stratified spaces, homotopy theory, covering spaces.

outwards from the deeper strata', i.e. once they leave a stratum they do not re-enter it. The notion of homotopy here is that of a family of such paths. MacPherson phrased his result in terms of constructible sheaves rather than covering spaces, but the notions are equivalent under the well-known correspondence of sheaves and their étale spaces.

MacPherson's ideas were published and extended by Treumann [**Tre07**] in which he refers to paths which wind outwards as exit paths. For any topologically stratified space he defines the 'exit 1-category' — objects are points, morphisms are homotopy classes of exit paths (with a tameness assumption on the homotopy) — and the 'exit 2-category' — objects are points, 1-morphisms are exit paths and 2-morphisms 'tame' homotopies through exit paths. He shows that the set-valued functors on the exit 1-category are equivalent to constructible sheaves and that category-valued functors on the exit 2-category are equivalent to constructible stacks. Interesting examples of constructible stacks are given by categories of perverse sheaves on X. However, there does not yet seem to be any way of identifying the particular representations which correspond to these.

This paper develops these ideas by defining a 'fundamental category' for any 'pre-ordered space'. Stratified spaces are particular examples, and in this case our definition reduces to Treumann's exit 1-category (but without the tameness condition on homotopies). The fundamental category has good properties for a very wide class of stratified spaces, namely homotopically stratified sets with locally 0 and 1-connected strata. For these:

- The fundamental category can be computed in terms of homotopy groups of the strata and of the homotopy links — see §3.1.
- The fundamental groupoid can be recovered by localising the fundamental category at all morphisms — see Corollary 4.3.
- Covariant set-valued functors from the fundamental category classify constructible sheaves and, dually, contravariant functors classify constructible cosheaves — see Theorem 4.6. Geometrically these can be interpreted respectively as classifications of stratified étale and branched covers.

This class of spaces includes Whitney stratified spaces, Thom-Mather stratified spaces, topologically stratified spaces and Siebenmann's locally cone-like spaces. In particular we recover MacPherson's result. We do not consider the analogue of Treumann's exit 2-category but it seems likely that his results would generalise to homotopically stratified sets.

1.1. Structure of the paper

A pre-ordered space, or po-space, has a distinguished subset of paths: the po-paths are those paths which are also order-preserving maps, where $[0,1]$ is ordered by $\leqslant$. When X is a stratified space po-paths are precisely Treumann's exit paths. The fundamental category $\Pi_1^{po}X$ of a po-space X, defined in §2, is an ordered analogue of the fundamental groupoid in which morphisms are given by homotopy classes of po-paths.

In §3 we restrict the discussion to a homotopically nice class of filtered spaces, Quinn's homotopically stratified sets. These spaces have two advantages for our purposes. Firstly they are very general, subsuming almost any other notion of stratified space. Secondly, the language used in their definition, particularly that of the homotopy link, is well-suited for talking about po-paths. In particular it allows us to reduce to the simpler case of a space with only two strata. In §3.1 we explain an approach to computing the fundamental category based on this reduction.

The fundamental groupoid of a locally 0 and 1-connected space classifies covers, equivalently local systems. In §4 we show that the fundamental category plays a similar rôle for homotopically stratified spaces with locally 0 and 1-connected strata. However, because po-paths are not necessarily reversible, we obtain two different classification results. Covariant set-valued representations of the fundamental category of a homotopically stratified set correspond to constructible sheaves, or equivalently to stratified étale covers. Dually, contravariant representations correspond to constructible cosheaves, or equivalently to stratified branched covers (which we define in terms of Fox's notion of a complete spread [**Fox57**]).

As a corollary of this classification result we deduce that the fundamental groupoid of a homotopically stratified set with locally 0 and 1-connected strata can be recovered by localising the fundamental category at the set of all morphisms: $\Pi_1 X$ is the 'groupoidification' of $\Pi_1^{po} X$.

In §5 we consider an example, the symmetric product $SP^n\mathbb{C}$ with the natural stratification indexed by partitions of n. Morphisms in the fundamental category can be expressed in terms of various subgroups of the braid group B_n and symmetric group S_n associated to partitions.

Appendix A explains how filtered spaces arise as po-spaces with a certain natural compatibility between the pre-order and the topology. This is included to make the case that filtered and stratified spaces are not exotic examples in the world of ordered topology but the bread and butter of the subject. Finally, since cosheaves and complete spreads are less familiar than sheaves and étale maps we give a brief review of the relevant theory in Appendix B. This includes what seems to be a new result: the correspondence between cosheaves and uniquely-complete spreads on a topological space. This is a small extension of known results in the special case in which X is a complete metric space.

1.2. Related work

We have already discussed the relation to MacPherson and Treumann's ideas. In a less geometric vein, there has been work on homotopy theory for ordered or directed spaces in category theory and theoretical computer science. Unfortunately there is a plethora of slightly different definitions. In [**Kah06**] Kahl shows that the category of spaces equipped with partial orders and maps between them has a closed model structure with *unordered* homotopies as the notion of homotopy. Bubenik and Worytkiewicz [**BW06**] follow a similar programme for *locally ordered* spaces — spaces X with a partial order which need only be transitive locally, but which is closed as a subset of $X \times X$. However, here it is only possible to show that such spaces are *contained* in a closed model category. Finally, Grandis [**Gra03**] considers a more general notion of *directed* spaces — spaces equipped with a suitable class of 'directed' paths. He shows that the directed category has good properties (existence of limits and colimits, exponentiable directed interval etc) and develops directed homotopy theory within it using *ordered* homotopies (so that directed homotopy is not an equivalence relation). The intended application in all three cases is in theoretical computer science, to the theory of concurrent systems. Grandis also uses directed homotopy to produce interesting examples of higher-dimensional categories, see [**Gra06**].

1.3. Acknowledgments

I am grateful to Michael Loenne who explained the ideas behind the example in §5 and to Beverley O'Neill (generously funded by a Nuffield Undergraduate Bursary) who worked out the details. I am indebted to David Miller for sparing my blushes by pointing out a serious error in an earlier draft. I would like to thank Ivan Smith and particularly Tom Leinster for helpful discussions and suggestions. And finally, I would like to thank the referee for his meticulous reading and helpful corrections and comments.

2. The fundamental category

There are various notions of 'ordered space'. The one we use, the notion of a po-space, is the simplest. It is a topological space with a pre-order on the set of points. Recall that a *pre-order* is a set equipped with a reflexive and transitive relation $\leqslant$ and that a map is *increasing* if $p \leqslant q \Rightarrow f(p) \leqslant f(q)$. Po-spaces form a category with maps between them being both continuous and increasing. We will refer to these as po-maps for brevity.

Notice that there need be no compatibility between the topology and the pre-order. However, if we impose a natural compatibility condition then the resulting po-spaces are filtered spaces (see Appendix A.1). These are the po-spaces which arise most often in geometry and topology; all the examples we consider will be of this kind, indeed they will be stratified spaces (see §3). The relation between po-spaces and filtered spaces is explained more fully in Appendix A.

From now on we will assume that all spaces are compactly generated. Spaces of maps are topologised with the k-ification of the compact-open topology (so that they too are compactly generated).

For po-spaces X and Y let $\mathrm{Map}^{\leqslant}(X,Y)$ be the set of po-maps between them. We topologise this as a subspace of $\mathrm{Map}\,(X,Y)$. Let I be the ordered interval, i.e. $[0,1]$ equipped with the standard order. An element of $\mathrm{Map}^{\leqslant}(I,X)$ is a continuous path $\gamma : [0,1] \to X$ such that $\gamma(s) \leqslant \gamma(t)$ whenever $s \leqslant t$; we call it a *po-path* in X. The start and end of a po-path determine a continuous map

$$E_0 \times E_1 : \mathrm{Map}^{\leqslant}(I,X) \to X^2 : \gamma \mapsto (\gamma(0), \gamma(1))\,.$$

Definition 2.1. *The fundamental category $\Pi_1^{po}X$ of a po-space X is the category whose objects are the points of X and whose morphisms from x to x' are the (path) connected components $\pi_0(E_0 \times E_1)^{-1}(x,x')$ of the space of po-paths from x to x'. That is, a morphism from x to x' is a homotopy class of po-paths from x to x' where the homotopy is through po-paths. Composition is defined by concatenation of po-paths. The fundamental category is functorial: a po-map $f : X \to Y$ induces a functor $\Pi_1^{po}(f) : \Pi_1^{po}X \to \Pi_1^{po}Y$.*

Example 2.2. *Let P be a poset with the descending chain condition, i.e. any descending chain of elements of P is eventually constant. Consider P as a po-space by giving it the Alexandrov topology (see Appendix A). Any po-path in P is of the form*

$$\gamma(t) = p_i \text{ for } t \in (t_{i-1}, t_i] \cap [0,1] \qquad i = 0, \ldots, n$$

where $p_0 \leqslant \cdots \leqslant p_n$ is a chain in P and $t_{-1} < 0 < t_0 < \cdots < t_n = 1$ an increasing sequence in $\mathbb{R}$. There is a homotopy through po-paths

$$\eta(s,t) = \begin{cases} \gamma(t) & s = 0 \\ p_0 & t = 0 \\ p_n & \text{otherwise,} \end{cases}$$

from γ to the po-path starting at p_0 and moving instantly to p_n. Hence $\Pi_1^{po} P \cong P$ thought of as a category with a single morphism from p to p' when $p \leqslant p'$.

3. Homotopically stratified sets

A stratified space is a filtered space together with some information on how the strata glue together. There are many ways in which we can specify glueing data and hence many types of stratified space. In a homotopy-theoretic context the most flexible is Quinn's notion of a homotopically stratified set (defined below). This is the notion of stratified space we will use. It is very general, in particular Siebenmann's locally cone-like stratified spaces, Thom-Mather stratified spaces and Whitney stratified spaces are all homotopically stratified sets. The main example in this paper is the symmetric product $SP^n\mathbb{C}$. We stratify this with one stratum for each partition of n; the corresponding stratum is the subset of configurations in which the n points coalesce according to the partition (see §5).

The fundamental category of a homotopically stratified set is equivalent to a similar category in which morphisms are homotopy classes of po-paths which only pass through one or two strata. This allows us to describe morphisms in terms of the homotopy groups of strata and of homotopy links of pairs of strata.

In order to give the definition of a homotopically stratified set we introduce some terminology. Suppose B is a filtered space. A subspace $A \subset B$ is *tame* if it is a nearly stratum-preserving deformation retract of a neighbourhood N of A, i.e. there is a deformation retraction of N onto A such that points remain in the same stratum under the retraction until the last possible moment, when they must flow into A. The *homotopy link* holink (B, A) of a subset A of B is the space of paths (equipped with the compact-open topology) in B with $\gamma(0) \in A$ and $\gamma(0,1] \subset B - A$, i.e. the space of paths which start in A but leave it immediately. Evaluation at t defines a map $E_t : \text{holink}(A \cup B, A) \to A \cup B$.

Definition 3.1. *A* homotopically stratified set *is a filtered space X with finitely many connected strata X^i such that for any pair $i \leqslant j$*

1. *the inclusion $X^i \hookrightarrow X^i \cup X^j$ is tame and*
2. *the evaluation map $E_0 : \text{holink}\left(X^i \cup X^j, X^i\right) \to X^i$ is a fibration.*

It is not immediately apparent why this is a good definition. One reason is that, if we assume that X is a metric space, then

$$\begin{array}{ccc} \text{holink}\left(X^i \cup X^j, X^i\right) & \xrightarrow{E_1} & X^j \\ {\scriptstyle E_0}\downarrow & & \downarrow \\ X^i & \hookrightarrow & X^i \cup X^j \end{array}$$

is a homotopy push-out [**Qui88**, Lemma 2.4]. Intuitively the homotopy link plays the rôle of the boundary of a regular neighbourhood of X^i in $X^i \cup X^j$.

A homotopically stratified set is naturally a po-space, where we give it the preorder corresponding to the underlying filtration. Homotopy links can then be described in terms of po-maps. Let I_1 be the interval $[0,1]$ made into a po-space via the filtration $\{0\} \subset [0,1]$ — see Appendix A. Then

$$\text{holink}\left(X^i \cup X^j, X^i\right) = (E_0 \times E_1)^{-1}\left(X^i \times X^j\right) \subset \text{Map}^{\leqslant}(I_1, X).$$

In fact it is a good heuristic when working with homotopically stratified sets that their homotopy theory can be understood in terms of maps from I_1. The next lemma, which shows that po-paths are homotopic to elements of the holink, is an illustration of this principle.

Lemma 3.2. *Let X be a homotopically stratified set and $\gamma : [0,1] \to X$ a po-path in X from $x_i \in X^i$ to $x_j \in X^j$. Then there is a homotopy of γ, relative to its end points, to a po-path $\tilde{\gamma} \in$ holink $\left(X^i \cup X^j, X^i\right)$. Moreover, $\tilde{\gamma}$ is unique up to homotopy through po-paths in the homotopy link.*

Proof. In general the po-path $\gamma : [0,t] \to X$ will pass through several strata. Let $t_{-1} < 0 \leqslant t_0 < \cdots < t_{n-1} < t_n = 1$ be such that $\gamma(t)$ is in one stratum for $t \in (t_{k-1}, t_k] \cap [0,1]$ for $k = 0, 1, \ldots, n$. In particular the trace of γ is in X^i for $t \in [0, t_1]$ and in X^j for $t \in (t_n, 1]$. There is a homotopy of γ which moves it off the intermediate strata one by one starting from the highest. The segment $\gamma : (t_{n-1}, t_n] \to X$ is a path in some stratum X^k. The final segment $\gamma : [t_n, 1] \to X$ is a lift (inverse image) of $\gamma(t_n)$ along the start point map $E_0 :$ holink $\left(X^k \cup X^j, X^k\right) \to X^k$. By definition E_0 is a fibration so we can extend this lift along $\gamma : (t_{n-1}, t_n] \to X^k$. The result is a map

$$\eta : (t_{n-1}, t_n] \times [t_n, 1] \to X$$

with $\eta(-, t_n) = \gamma(-)$, $\eta(t_n, -) = \gamma(-)$ and $\eta(s,t) \in X^j$ for $t \neq t_n$. This provides a homotopy, through po-paths, between γ and

$$\gamma_1(t) = \begin{cases} \gamma(t) & 0 \leqslant t \leqslant t_{n-1} \\ \eta(t_{n-1}, t_n - t_{n-1} + t) & t_{n-1} < t \leqslant t_{n-1} + 1 - t_n \\ \eta(t_n - 1 + t, 1) & t_{n-1} + 1 - t_n < t \leqslant 1. \end{cases}$$

The po-path γ_1 now passes through one fewer intermediate strata (since it avoids X^k). Continuing inductively we obtain the desired $\tilde{\gamma}$. Uniqueness up to homotopy follows from multiple applications of the uniqueness up to homotopy of the extension of a lift along a fibration. The situation is most easily apprehended by looking at the following diagram.

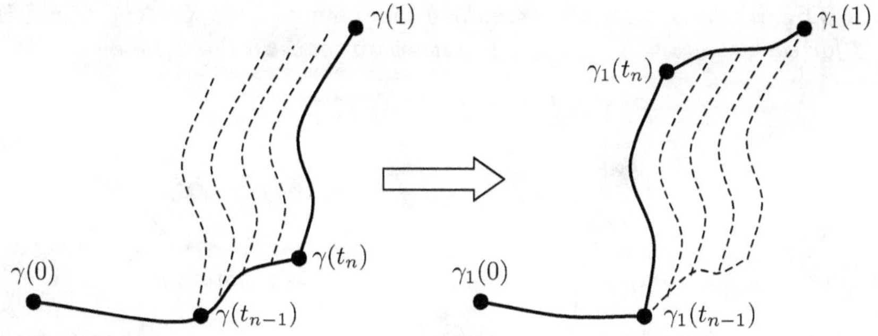

□

Of course we would really like a relative version of this result, allowing us to find homotopies of families of po-paths to families in the holink. This does not follow for free from the above lemma; the main difficulty is that the point at which po-paths in a family leave a given stratum is not a continuous function of the parameters of the family. Nevertheless Miller has recently announced

Theorem 3.3 (See [**Mil09**, Theorem 3.9].). *Let X be a homotopically stratified set and suppose X is a metric space. Then the space of po-paths beginning in X^i and ending in X^j is homotopy equivalent to* holink $\left(X^i \cup X^j, X^i\right)$ *by a homotopy which fixes start and end points and takes po-paths 'instantly' into the homotopy link.*

This theorem allows us to give an equivalent, but simpler, definition of the fundamental category. Let $\Pi_1^{ho}X$ be the category with objects the points of X and with morphisms from $x \in X^i$ to $x' \in X^j$ given by the (path) connected components of

$$(E_0 \times E_1)^{-1}(x, x') \subset \text{holink}\left(X^i \cup X^j, X^i\right).$$

In order to define composition we need to concatenate paths and then choose a homotopic (through po-paths) path in the holink. The above theorem guarantees the result is well-defined, up to homotopy through po-paths.

Corollary 3.4. *The functor $\Pi_1^{ho}X \to \Pi_1^{po}X$ induced by the inclusion of the homotopy link in the space of po-paths is an equivalence.*

The proof is immediate from Theorem 3.3. Later, in Corollary 4.5, we will give an independent proof of this in the case when the strata of X are locally 0 and 1-connected. This result allows us to reduce to the two-stratum case: morphisms in the fundamental category of a homotopically stratified set can be described in terms of the homotopy links, they do not depend on any intermediate strata.

3.1. Computing the fundamental category

In this section we give several fibration sequences for computing morphisms in the fundamental category of a homotopically stratified set X.

Lemma 3.5. *Write H^{ij} for the homotopy link* $\operatorname{holink}(X^i \cup X^j, X^i)$. *Fix a basepoint x_i for each stratum X^i. The four downward maps in the diagram*

$$\begin{array}{ccccc} E_1^{-1}(x_j) & \hookrightarrow & H^{ij} & \hookleftarrow & E_0^{-1}(x_i) \\ \downarrow{\scriptstyle E_0} & \swarrow{\scriptstyle E_0} & & {\scriptstyle E_1}\searrow & \downarrow{\scriptstyle E_1} \\ X^i & & & & X^j \end{array}$$

are fibrations. In addition $E_0 \times E_1 : H^{ij} \to X^i \times X^j$ is a fibration.

Proof. Since X is homotopically stratified $E_0 : H^{ij} \to X^i$ is a fibration. Now consider the restriction $E_1^{-1}(x') \to X^i$ of E_0 to paths ending at x'. Given a family $F : A \times [0,1] \to X^i$ of paths in X^i and a lift $G : A \times \{0\} \to E_1^{-1}(x')$ of the start points we have a lift $G : A \times [0,1] \to H^{ij}$. Then the 'diagonal' family $\widetilde{F} : A \times [0,1] \to E_1^{-1}(x')$ given by

$$\widetilde{F}(a,s)(t) = G\left(a, s(1-t)\right)(t)$$

is a lift of F to $E_1^{-1}(x')$ along E_0 starting at $G(-,0)$. Hence the restriction of E_0 to $E_1^{-1}(x')$ is a fibration.

The end point map $E_1 : H^{ij} \to X^j$ is a fibration because, given a family $F : A \times [0,1] \to X^j$ of paths in X^j and a lift $G : A \times \{0\} \to H^{ij}$ of the starting points, the family given by the composition of paths

$$\widetilde{F}(a,t) = G(a)(-) \circ F(a,-)|_{[0,t]}$$

is a lift of F to H^{ij}. It is easy to see that this construction of a lift also shows that the restriction of E_1 to $E_0^{-1}(x_i)$ is a fibration.

Finally, we must show that $E_0 \times E_1 : H^{ij} \to X^i \times X^j$ is a fibration. If $F = (F_1, F_2) : A \times [0,1] \to X^i \times X^j$ is a family of paths and $G : A \times \{0\} \to H^{ij}$ a lift of the starting points then, using essentially the same argument which showed that $E_0 : E_1^{-1}(x') \to X^i$ is a fibration, we can find a lift $G : A \times [0,1] \to H^{ij}$ of F_1 along E_0 such that $G(a,s)(1) = G(a,0)(1) = F_2(a,0)$. Then the composition of paths

$$\widetilde{F}(a,s) = G(a,s) \circ F_2(a,-)|_{[0,s]}$$

is a lift of F along $E_0 \times E_1$. □

A compatible choice of basepoints for the spaces appearing in Lemma 3.5 is given by choosing a basepoint $\gamma^{ij} \in H^{ij}$ with $E_0(\gamma^{ij}) = x_i$ and $E_1(\gamma^{ij}) = x_j$ for each $i < j$, i.e. a path with $\gamma^{ij}(0) = x_i, \gamma^{ij}(1) = x_j$ and $\gamma^{ij}(0,1] \subset X^j$. With this choice we obtain five long exact sequences corresponding to the five fibrations. These are displayed in a single commutative 'braided' diagram in Figure 1.

The long exact sequences in Figure 1 give us several methods for computing sets of morphisms in $\Pi_1^{ho}X$. We would also like to be able to compute compositions of morphisms. In general this is fiddly; the complications arise in keeping track of the combinatorics when the homotopy links and their fibres are disconnected. Here we will treat only the simpler case when they are connected. In §5 we treat a more difficult example where this fails.

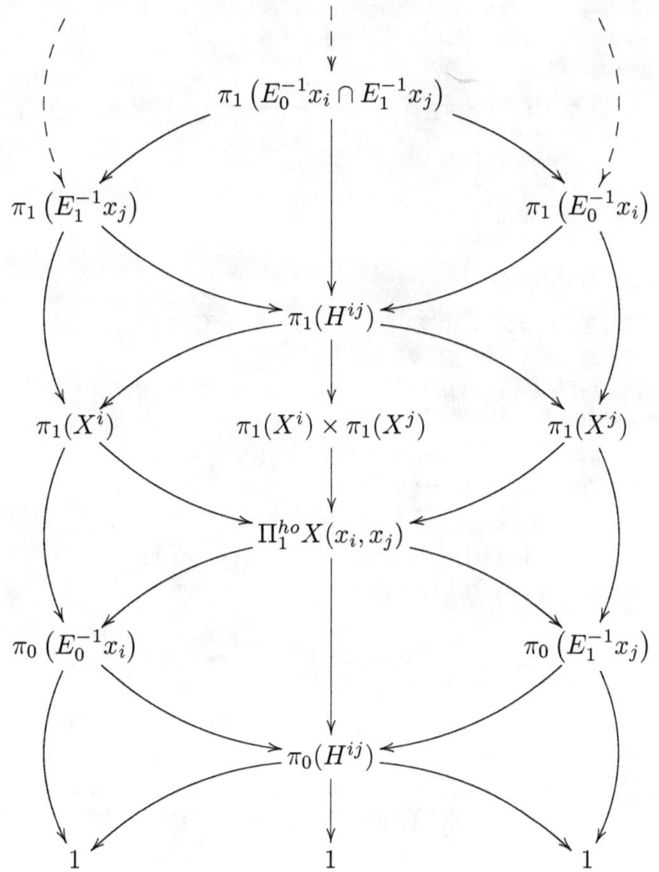

Figure 1: A commutative braided diagram showing the (lower parts of the) long exact sequences arising from the fibrations in Lemma 3.5. To aid reading we have suppressed all basepoints; they are the points $x_i \in X^i, x_j \in X^j$ or the path $\gamma^{ij} \in \operatorname{holink}\left(X^i \cup X^j, X^i\right)$ from x_i to x_j as appropriate. Recall that $\Pi_1^{ho}X(x_i, x_j) \cong \pi_0\left(E_0^{-1}x_i \cap E_1^{-1}x_j\right)$ and that we assume strata are connected, hence the row of 1s at the bottom. See §5 for an example of a computation using these sequences.

Lemma 3.6. *Let X be a homotopically stratified set. Assume that the fibres $E_0^{-1}(x_i) \subset H^{ij}$ are connected for each $i \leqslant j$ so that we have an exact sequence*

$$\cdots \longrightarrow \pi_1 E_0^{-1}(x_i) \xrightarrow{\pi_1 E_1} \pi_1(X^j) \xrightarrow{\Gamma^{ij}} \Pi_1^{ho} X(x_i, x_j) \longrightarrow 1$$

where Γ^{ij} is the map induced by pre-composition by γ^{ij}. (For the sake of readability we have suppressed the basepoints.) Then for $i \leqslant j \leqslant k$

$$\begin{array}{ccc}
\pi_1(H^{jk}) \times \pi_1(X^k) & \xrightarrow{\pi_1 E_1 \times 1} & \pi_1(X^k) \times \pi_1(X^k) \\
\downarrow {\scriptstyle \pi_1 E_0 \times 1} & & \downarrow {\scriptstyle \text{compose}} \\
\pi_1(X^j) \times \pi_1(X^k) & & \pi_1(X^k) \\
\downarrow {\scriptstyle \Gamma^{ij} \times \Gamma^{jk}} & & \downarrow {\scriptstyle \Gamma^{ik}} \\
\Pi_1^{ho} X(x_i, x_j) \times \Pi_1^{ho} X(x_j, x_k) & \longrightarrow & \Pi_1^{ho} X(x_i, x_k)
\end{array}$$

commutes, where the bottom map is composition in $\Pi_1^{ho} X$ and the bottom right map Γ^{ik} is pre-composition by $\gamma^{ij}\gamma^{jk}$. Hence, at least in principle, we can compose a pair of morphisms by choosing a lift to $\pi_1(H^{jk}) \times \pi_1(X^k)$ along the left hand surjection and then applying the clockwise sequence of maps.

Proof. In order to see that the diagram commutes, given $[g] \in \pi_1(X^j)$ let $[\tilde{g}] \in \pi_1(X^k)$ be any choice of 'lift' of $[g]$ via

$$\pi_1(X^j, x_j) \xleftarrow{\pi_1 E_0} \pi_1(H^{jk}, \gamma^{jk}) \xrightarrow{\pi_1 E_1} \pi_1(X^k, x_k).$$

The assumption that $E_0^{-1}(x_j) \subset H^{jk}$ is connected ensures that the left hand map is surjective so that this is always possible. Then, possibly after replacing g by another representative of its class in $\pi_1(X^j, x_j)$, there is a homotopy $\gamma^{jk}\tilde{g} \simeq g\gamma^{jk}$ (see Figure 2 below) so that for any $[h] \in \pi_1(X^k)$ we have

$$\Gamma^{ik}[\tilde{g}h] = [\gamma^{ij}\gamma^{jk}\tilde{g}h] = [\gamma^{ij}g\gamma^{jk}h] = \Gamma^{ij}[g] \circ \Gamma^{jk}[h]$$

as required. □

4. Stratified covers

Suppose X is locally path-connected and locally simply-connected. Then the categories of covers of X, of spaces over X with the unique lifting property for homotopies, of local systems on X and of set-valued representations of the fundamental groupoid of X, i.e. functors from the fundamental groupoid to sets, are equivalent. (Here we say that a map $p : Y \to X$ has the unique lifting property for

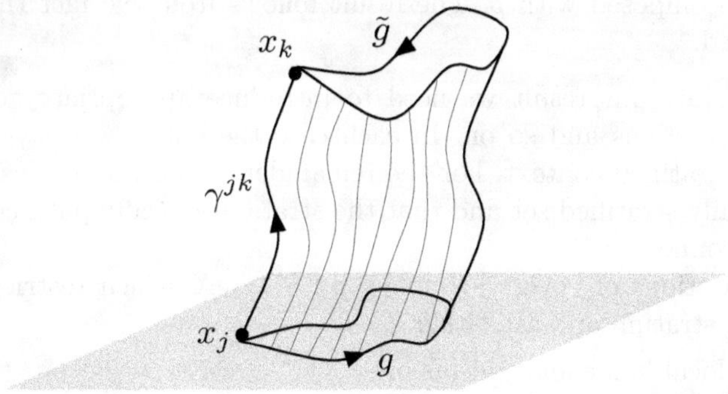

Figure 2: Picture of the homotopy $\gamma^{jk}\tilde{g} \simeq g\gamma^{jk}$ from Lemma 3.6.

homotopies if there is a unique solution to the lifting problem

$$\begin{array}{ccc} A \times \{0\} & \longrightarrow & Y \\ \downarrow & \nearrow & \downarrow p \\ A \times [0,1] & \longrightarrow & X \end{array}$$

where A is any CW-complex.)

Now suppose X is homotopically stratified. We will generalise these equivalences to the case when the strata of X are locally path-connected and locally simply-connected. This condition is equivalent to asking that X be 'locally po-path-connected and locally po-simply-connected' in the following sense.

Lemma 4.1. *Suppose X is homotopically stratified. Then the strata of X are locally path-connected and locally simply-connected if and only if each point x of X has a neighbourhood U such that, up to homotopy through po-paths, there is a unique po-path from x to any $x' \in U$, equivalently, if and only if x is an initial object in $\Pi_1^{po}U$.*

Proof. If each x has a neighbourhood U such that x is initial in $\Pi_1^{po}U$ then it is clear that the strata must be locally path-connected and simply-connected. Suppose then that the strata have these properties. Each stratum X^i is an almost stratum-preserving deformation retract of a neighbourhood N of X^i in X [**Qui88**, Proposition 3.2]. Let U be the inverse image under this retraction of a connected and simply-connected neighbourhood of x in X^i. Let $x' \in U$. We show that $\Pi_1^{po}U(x, x')$ has a unique element. Let ρ be the path from a point of $X^i \cap U$ to x' given by the retraction. Then there is a po-path from x to x': compose a path in $X^i \cap U$ from x to $\rho(0)$ with ρ. To see that this po-path is unique up to homotopy, note that any po-path γ from x to x' is homotopic through po-paths to the composition of a path in $X^i \cap U$ with ρ. Specifically γ is homotopic to the result of applying the retraction to γ composed with ρ. The result follows from the fact that $U \cap X^i$ is simply-connected. □

In order to state our result we need to introduce appropriate generalisations of covers, local systems and so on. In each case there are two ways to relax the definition in a stratified context. For the remainder of this section, assume that X is a homotopically stratified set and that the strata are locally path-connected and locally simply-connected.

The generalisations of covers are maps $p : Y \to X$ which restrict to covering maps over each stratum and are either

1. étale, i.e. local homeomorphisms or
2. locally-connected uniquely-complete spreads (see Appendix B).

The second class are branched covers in the topological sense, see Appendix B for the definition. We refer to such maps as stratified étale covers and stratified branched covers respectively. Figure 3 illustrates motivating examples. Let $\mathbf{Et}\downarrow X$ and $\mathbf{Br}\downarrow X$ be the categories with respective objects the stratified étale and branched covers over X and with maps the continuous maps over X.

Figure 3: The 'line with two origins' is naturally a stratified étale cover of the real line stratified by $\{0\}$ and $\mathbb{R} - \{0\}$; two transversely intersecting lines a stratified branched cover.

More evidently, the two ways to relax the homotopy-theoretic notion of cover as a space with the unique homotopy lifting property are to consider maps $p : Y \to X$ which have the unique lifting property either

1. for families of po-paths or
2. for families of op-paths.

Here by an op-path we mean the reverse $\overline{\gamma} : [0,1] \to X : t \mapsto \gamma(1-t)$ of a po-path. A *family* of po-paths is a continuous map

$$F : A \times [0,1] \to X$$

where A is a CW-complex and for each $a \in A$ the restriction $F(a,-)$ is a po-path. Families of op-paths are defined analogously. Let $\mathbf{UL^{po}}\downarrow X$ and $\mathbf{UL^{op}}\downarrow X$ respectively be the categories of spaces over X with these properties. In both cases morphisms are continuous maps over X.

Proposition 4.2. *A stratified étale cover has the unique lifting property for families of po-paths. A stratified branched cover has the unique lifting property for families of op-paths.*

Proof. We begin with the cases of a single po-path or op-path and then generalise to families. Note that if $\gamma : [0,1] \to X$ is a po-path and $\gamma(t)$ is in a stratum X^i then there is some $\epsilon > 0$ such that $\gamma(s) \in X^i$ for $s \in (t-\epsilon, t]$. Obviously for an op-path there is a similar statement but with $\gamma(s) \in X^i$ for $s \in [t, t+\epsilon)$.

Suppose that $p : Y \to X$ is a stratified étale cover and that $\gamma : [0,1] \to X$ is a po-path. Let $\widetilde{\gamma}(0)$ be a lift of $\gamma(0)$ to Y and let $L \subset [0,1]$ be the set of s for which $\gamma|_{[0,s]}$ has a unique lift starting at $\widetilde{\gamma}(0)$. Clearly $0 \in L$. Suppose $[0,t) \subset L$. Then by the first observation $\gamma(t)$ lies in the same stratum as $\gamma(s)$ for $s \in (t-\epsilon, t]$. Since p is a cover over each stratum the unique lifting property of covers shows that $t \in L$. Hence L is closed. On the other hand, it $t \in L$ then $\widetilde{\gamma}(t)$ has a neighbourhood U such that $p|_U : U \to p(U)$ is a homeomorphism onto an open neighbourhood of $\gamma(t)$. It follows that L is open. Since $[0,1]$ is connected we have $L = [0,1]$ and we can lift po-paths uniquely.

Suppose that $p : Y \to X$ is a stratified branched cover and that $\gamma : [0,1] \to X$ is an op-path. Let $\widetilde{\gamma}(0)$ be a lift of $\gamma(0)$ to Y and let $L \subset [0,1]$ be the set of s for which $\gamma|_{[0,s]}$ has a unique lift $\widetilde{\gamma} : [0,s] \to Y$ starting at $\widetilde{\gamma}(0)$. Clearly $0 \in L$. Suppose $[0,t) \subset L$. Then the trace of $\widetilde{\gamma}$ determines an element of

$$\lim_{U \ni \gamma(t)} \pi_0(p^{-1}U) \cong p^{-1}\gamma(t).$$

This is the unique continuous extension $\widetilde{\gamma}(t)$. It follows that L is closed. Now suppose $[0,t] \subset L$. Then by the observation at the beginning of the proof, $\gamma(s)$ is in the same stratum for $s \in [t, t+\epsilon)$ for some $\epsilon > 0$. Since p restricts to a covering of each stratum we can extend the lift uniquely and L is open. Therefore $L = [0,1]$ and we can uniquely lift op-paths.

To deal with the family case it remains only to show that these unique lifts fit into continuous families. The continuity of the lift of a family of po-paths to a stratified étale cover follows easily from the local homeomorphism property of the cover. Thus we will focus on the case of lifting a family $F : A \times [0,1] \to X$ of op-paths, parameterised by a CW-complex A, to a stratified branched cover $p : Y \to X$. Lifting each op-path individually uniquely defines a lift $\widetilde{F} : A \times [0,1] \to Y$ which is certainly continuous on $\{a\} \times [0,1]$ for each $a \in A$. To show that it is continuous on $A \times [0,1]$ we use the fact that $f : Z \to Y$ is continuous at z

$\iff$ for each open neighbourhood $V \ni f(z)$ the inverse image $f^{-1}V$ contains an open neighbourhood of z,

$\iff$ for each open neighbourhood $U \ni pf(z)$ there is an open neighbourhood of z mapping to the component V of $f(z)$ in $p^{-1}U$.

If $\widetilde{F}$ is continuous on $A \times [0,t)$ then we can use this together with the continuity of $\widetilde{F}$ on each $\{a\} \times [0,1]$ and the local-connectivity of A to show that $\widetilde{F}$ is continuous on $A \times [0,t]$.

Now suppose $\widetilde{F}$ is continuous on $A \times [0,t]$. Fix $a \in A$. By the initial observation $F(a,s)$ is in the same stratum for $s \in [t, t+\delta]$ for some $\delta > 0$. As Y is a stratified branched cover the cosheaf of components of Y (see appendix B) is locally-constant on strata. This means that we can cover

$$\{F(a,s) \mid s \in [t, t+\delta]\}$$

by a finite sequence of open neighbourhoods $U_i \ni F(a,t_i)$ for $t = t_0 < t_1 < \cdots <$

$t_n = t + \delta$ each having the property that the evident map

$$p^{-1}x \to \pi_0\left(p^{-1}U_i\right)$$

is an isomorphism for each $x \in U_i$. It follows that if $\widetilde{F}(a,t)$ and $\widetilde{F}(a',t)$ are in the same component of $p^{-1}U_0$ and

$$F(a,s), F(a',s) \in U_0 \cup \cdots \cup U_n \qquad \forall s \in [t, t+\delta]$$

then $\widetilde{F}(a,t+\delta)$ and $\widetilde{F}(a',t+\delta)$ are in the same component of $p^{-1}U_n$. Thus continuity propagates from (a,t) to $(a,t+\delta)$ and $\widetilde{F}$ is continuous on the whole of $A \times [0,1]$. □

It follows immediately from the definition that a space over X with the unique lifting property for families of po-paths defines a functor $\Pi_1^{po}X \to \mathbf{Set}$. Similarly, a space over X with the unique lifting property for families of op-paths defines a functor $\Pi_1^{op}X \to \mathbf{Set}$, where $\Pi_1^{op}X$ is the analogue of $\Pi_1^{po}X$ but with po-paths replaced by op-paths. We now explain how to construct generalisations of local systems from such functors.

A local system is a locally-constant sheaf, but it is also a locally-constant cosheaf (see Appendix B for the definition). The two appropriate generalisations of a local system to the stratified context are constructible sheaves and constructible cosheaves. These are, respectively, sheaves and cosheaves which are locally-constant when restricted to each stratum of X. Denote the resulting full subcategories of sheaves and cosheaves by $\mathbf{Sh^c}\downarrow X$ and $\mathbf{Cosh^c}\downarrow X$.

Given a functor $F : \Pi_1^{po}X \to \mathbf{Set}$ we define a presheaf $\mathcal{F}$ on X with $\mathcal{F}(U)$ being the set of functions $f : U \to \prod_{x\in U} F(x)$ such that $f(x) \in F(x)$ and $f(x') = F(\gamma)(f(x))$ whenever γ is a po-path in U from x to x'. It follows from Lemma 4.1 that the stalk $\mathcal{F}_x = F(x)$. Furthermore, the restriction of $\mathcal{F}$ to a stratum is a locally-constant presheaf. Hence the sheafification is a constructible sheaf with stalk $F(x)$ at $x \in X$.

Similarly given a functor $G : \Pi_1^{op}X \to \mathbf{Set}$ we let $\mathcal{G}$ be the precosheaf with cosections $\mathcal{G}(U) = \sum_{x\in U} G(x)/\sim$ where $\alpha \sim \alpha'$ if there is an op-path γ in U from x to x' with $\alpha' = G(\gamma)(\alpha)$. This is locally-constant on strata and, using Lemma 4.1, we see that the costalk $\mathcal{G}_x = G(x)$. Thus the cosheafification (see Appendix B) is a constructible cosheaf with costalk $G(x)$ at $x \in X$.

Finally, we can construct an étale space over X from any sheaf on X. If the sheaf is constructible then the étale space will be a cover over each stratum of X, i.e. it will be a stratified étale cover. Less well-known is the fact (see Appendix B) that we can construct a locally-connected, uniquely-complete spread from a cosheaf. When the cosheaf is constructible the corresponding spread is a stratified branched cover.

Theorem 4.3. *There are equivalences of categories*

$$\mathbf{Et}\downarrow X \simeq \mathbf{UL^{po}}\downarrow X \simeq [\Pi_1^{po}(X), \mathbf{Set}] \simeq \mathbf{Sh^c}\downarrow X$$

and $\mathbf{Br}\downarrow X \simeq \mathbf{UL^{op}}\downarrow X \simeq [\Pi_1^{op}(X), \mathbf{Set}] \simeq \mathbf{Cosh^c}\downarrow X$.

Proof. We have shown that there are maps of the objects

$$\begin{array}{ccc} \mathbf{Et}\downarrow X & \longrightarrow & \mathbf{UL^{po}}\downarrow X \\ \uparrow & & \downarrow \\ \mathbf{Sh^c}\downarrow X & \longleftarrow & [\Pi_1^{po}(X), \mathbf{Set}] \end{array} \quad \text{and} \quad \begin{array}{ccc} \mathbf{Br}\downarrow X & \longrightarrow & \mathbf{UL^{op}}\downarrow X \\ \uparrow & & \downarrow \\ \mathbf{Cosh^c}\downarrow X & \longleftarrow & [\Pi_1^{op}(X), \mathbf{Set}]. \end{array}$$

It is tedious but easy to check that each construction is functorial and that starting at any category and cycling around the diagram is an auto-equivalence. We leave this as an exercise! □

Remark 4.4. *If a functor $F : \mathbf{C} \to \mathbf{D}$ induces an equivalence $F^* : [\mathbf{D}, \mathbf{Set}] \to [\mathbf{C}, \mathbf{Set}]$ then F is fully faithful. Therefore F is an equivalence if, in addition, it is essentially surjective. To see that F is fully faithful, note that F^* has a left adjoint L (which is also fully faithful). It follows from the Yoneda lemma that L takes the representable functor $\hom_C(c, -)$ to the representable $\hom_D(Fc, -)$, and that it takes composition with $f : c' \to c$ to composition with $Ff : Fc' \to Fc$, i.e. that the diagram*

$$\begin{array}{ccc} \mathbf{C}^{op} & \xrightarrow{F^{op}} & \mathbf{D}^{op} \\ \downarrow & & \downarrow \\ [\mathbf{C}, \mathbf{Set}] & \xrightarrow[L]{} & [\mathbf{D}, \mathbf{Set}], \end{array}$$

in which the vertical functors are the Yoneda embeddings, commutes. The claim follows.

Corollary 4.5. *If X is homotopically stratified and each stratum is locally path-connected and locally simply-connected then the functor $\Pi_1^{ho}X \to \Pi_1^{po}X$ arising from the inclusion of the homotopy link in the space of po-paths is an equivalence.*

Proof. Theroem 4.3 and the preceding discussion can be repeated with $\Pi_1^{ho}X$ in place of $\Pi_1^{po}X$. We conclude that $\Pi_1^{ho}X \to \Pi_1^{po}X$ induces an equivalence

$$[\Pi_1^{po}X, \mathbf{Set}] \to [\Pi_1^{ho}X, \mathbf{Set}],$$

and so is an equivalence by Remark 4.4. □

Corollary 4.6. *Suppose X is a homotopically stratified space with locally connected and locally simply-connected strata. Let $\Pi_1^{po}X_{loc}$ be the category obtained by localising the fundamental category at the set of all morphisms — thus the objects of $\Pi_1^{po}X_{loc}$ are the points of X and the morphisms are equivalence classes of words in the morphisms of $\Pi_1^{po}X$ and formal inverses thereof under obvious relations arising from composition and cancellation. The functor $\Pi_1^{po}X_{loc} \to \Pi_1X$ given by composing the terms in these words is an equivalence, i.e. Π_1X is the 'groupoidification' of $\Pi_1^{po}X$.*

Proof. The obvious functors $\Pi_1^{po}X \to \Pi_1^{po}X_{loc} \to \Pi_1X$ induce functors

$$[\Pi_1X, \mathbf{Set}] \to [\Pi_1^{po}X_{loc}, \mathbf{Set}] \to [\Pi_1^{po}X, \mathbf{Set}].$$

It follows from the connectivity assumptions on the strata that X is locally connected and locally simply-connected. Hence $[\Pi_1 X, \mathbf{Set}]$ is equivalent to the category of covers of X. On the other hand $[\Pi_1^{po} X_{loc}, \mathbf{Set}] \to [\Pi_1^{po} X, \mathbf{Set}]$ is the inclusion of the full subcategory of functors which take all maps to isomorphisms. A stratified étale cover is a genuine cover precisely when the monodromy induced by any po-path is an isomorphism. Thus

$$[\Pi_1^{po} X_{loc}, \mathbf{Set}] \to [\Pi_1^{po} X, \mathbf{Set}]$$

corresponds to the inclusion of the category of covers in the category of stratified étale covers. Hence $[\Pi_1 X, \mathbf{Set}] \to [\Pi_1^{po} X_{loc}, \mathbf{Set}]$ is an equivalence. It follows from Remark 4.4 that $\Pi_1^{po} X_{loc} \to \Pi_1 X$ is an equivalence. □

Thus the fundamental category of a homotopically stratified set (with locally connected and simply connected strata) contains at least as much information as the fundamental groupoid. In some cases this fact can be used to compute the fundamental group by computing a skeleton of $\Pi_1^{po} X$, localising this to obtain a groupoid — which is equivalent to the fundamental groupoid by the above theorem — and then reading off the fundamental group as the automorphisms of an object. The following example is offered as a 'proof of concept', not because it is an elegant way to compute $\pi_1 \mathbb{RP}^n$!

Example 4.7. *Real projective space $\mathbb{RP}^n$ has a filtration $\mathbb{RP}^0 \subset \mathbb{RP}^1 \subset \cdots \subset \mathbb{RP}^n$. Let $f : S^n \to \mathbb{RP}^n$ be the standard 2 to 1 covering, and $g : \mathbb{R}^{n+1} - \{0\} \to \mathbb{RP}^n$ the standard quotient map. Let*

$$y_i = (0, \ldots, 0, 1, 0, \ldots, 0)$$

where the non-zero entry is in the ith place and choose the basepoint $x_i = f(y_i)$ for the stratum $X^i := \mathbb{RP}^i - \mathbb{RP}^{i-1}$. The full subcategory on the objects x_i is a skeleton of $\Pi_1^{po} \mathbb{RP}^n$ because every point is connected to precisely one of the x_i by a reversible po-path. In order to characterise po-paths it is helpful to define the function

$$L : \mathbb{RP}^n \to \{0, \ldots, n\} : [p_0 : \ldots : p_n] \mapsto \max\{j \mid p_j \neq 0\}.$$

Then $p \in X^i \iff L(p) = i$, and a path $\gamma : [0,1] \to \mathbb{RP}^n$ is a po-path if and only if the composite $L \circ \gamma : [0,1] \to \{0, \ldots, n\}$ is increasing.

Let $\gamma : [0,1] \to \mathbb{RP}^n$ be a po-path from x_i to x_j where $i < j$. (The case $i = j$ is uninteresting because the strata are simply-connected.) Let $\tilde{\gamma} : [0,1] \to S^n$ be the unique lift of γ along the covering map f starting at y_i. The end point of $\tilde{\gamma}$ is then at $\epsilon_\gamma y_j$ where $\epsilon_\gamma \in \{\pm 1\}$. Furthermore, by homotopy lifting, the end point is the same for any po-path homotopic to γ through po-paths so there is a well-defined map

$$\epsilon : \Pi_1^{po} \mathbb{RP}^n(x_i, x_j) \to \{\pm 1\}.$$

In fact this is a bijection. To see this consider the map

$$\tilde{\eta} : [0,1]^2 \to \mathbb{R}^{n+1} - \{0\} : (s,t) \mapsto (1-s)\tilde{\gamma}(t) + s(0, \ldots, 1-t, \ldots, \epsilon_\gamma t, \ldots, 0)$$

where the only non-zero entries of the right hand term are in the ith and jth places. The composite $\eta = g \circ \tilde{\eta}$ is a homotopy relative to end points from γ to an element of holink $\left(X^i \cup X^j, X^i\right)$. *To see that it is a homotopy through po-paths note that,* *because γ is a po-path, we have $\epsilon_\gamma \tilde{\gamma}_j(t) \geqslant 0$ for $t \in [0,1]$. It follows that $L(\eta(s,t))$ is an increasing function of t for each $s \in [0,1]$ as required.*

Hence $\Pi_1^{po}\mathbb{RP}^n(x_i, x_j) = \{\alpha, \beta\}$ is a two element set. (Properly we should write α_{ij} and β_{ij} but we omit the subscripts for ease of reading.) It is easy to check that $\epsilon_{\gamma.\gamma'} = \epsilon_\gamma \epsilon_{\gamma'}$ so that composing paths we have $\alpha^2 = \beta^2$ and $\alpha\beta = \beta\alpha$. Localising introduces inverses α^{-1} and β^{-1} satisfying

$$\alpha\beta^{-1} = \alpha^{-1}\beta = \beta\alpha^{-1} = \beta^{-1}\alpha.$$

The automorphisms of x_0 in the resulting groupoid are

$$\langle \alpha\beta^{-1} \mid (\alpha\beta^{-1})^2 = 1 \rangle \cong \mathbb{Z}/2\mathbb{Z} \cong \pi_1(\mathbb{RP}^n, x_0)$$

as expected from Corollary 4.6.

5. Configuration spaces of points in the plane

In this section we compute (a skeleton of) $\Pi_1^{po}X$ when X is the configuration space of n indistinguishable but not necessarily distinct points in $\mathbb{C}$, i.e. X is the symmetric product $SP^n\mathbb{C} := \mathbb{C}^n/S_n$ where the symmetric group S_n acts on $\mathbb{C}^n$ by permuting coordinates. A point x of $SP^n\mathbb{C}$ determines a configuration of n indistinguishable points in $\mathbb{C}$ given by the set of coordinates of any pre-image of x in $\mathbb{C}^n$.

The symmetric product has a natural stratification by orbit types indexed by partitions of n. The strata are the subsets of points corresponding to configurations where the n points coalesce according to the indexing partition. More precisely, let C be a concrete partition of $\{1,\ldots,n\}$, i.e. C is a set $\{C_1,\ldots,C_k\}$ of disjoint subsets of $\{1,\ldots,n\}$ whose union is the entire set. Define

$$Y^C = \{(z_1,\ldots,z_n\} \in \mathbb{C}^n \mid z_a = z_b \iff a,b \in C_i \text{ for some } 1 \leqslant i \leqslant k\}.$$

Then Y^C is a closed complex submanifold of an open subset of $\mathbb{C}^n$ and has codimension

$$\sum_{i=1}^{k} (|C_i| - 1).$$

It is obtained by deleting complex linear subspaces from a complex linear subspace and is therefore connected. For a concrete partition C let $\mathcal{P}(C)$ be the corresponding (abstract) partition of n with cardinalities $|C_1|,\ldots,|C_k|$. For a partition P set

$$Y^P = \bigcup_{\mathcal{P}(C)=P} Y^C.$$

Note that there is an element $\pi \in S_n$ inducing a complex-analytic isomorphism $Y^C \cong Y^{C'}$ exactly when $\mathcal{P}(C) = \mathcal{P}(C')$. Hence the Y^P are invariant under the action of S_n on $\mathbb{C}^n$. The quotients $X^P = Y^P/S_n$ form a stratification of X by connected complex analytic strata.

The poset of strata is the set of partitions with the relation $P \leqslant Q$ if Q is a refinement of P, that is if Q is obtained by further partitioning the parts of

P. We denote the number of parts in P by $|P|$ and write the partition with k parts of cardinalities $p_1, \ldots, p_k$ as $(p_1|\cdots|p_k)$. In this notation, the top element, corresponding to the open stratum, is $(1|1|\cdots|1)$ and the bottom, corresponding to the stratum where all n points coalesce, is (n).

Fix a basepoint x_P in each stratum. Elements of $\Pi_1^{po}X(x_P, x_Q)$ are represented by po-paths from x_P to x_Q in the homotopy link. Thinking of points of the symmetric product as configurations in $\mathbb{C}$, the graph of such a po-path corresponds to a set of strings in $\mathbb{C} \times [0,1]$ joining a configuration of type P to one of type Q. The condition that it is a po-path in the homotopy link means that the set of strings is a braid on $|Q|$ strings where the starts of the strings are glued together according to the partition P. Furthermore, each string is labelled by a natural number which records the cardinality of the corresponding part of Q. The sum of these for the set of strings emanating from a point, which corresponds to a part of P, is the cardinality of that part. Two representations give the same morphism if they are isotopic relative to the end points. Thus there are the usual braid relations on $|Q|$ strings, but also relations coming from 'internal' braiding within the parts of P which becomes 'external' braiding of the $|Q|$ strings. Figure 4 illustrates this in a simple example.

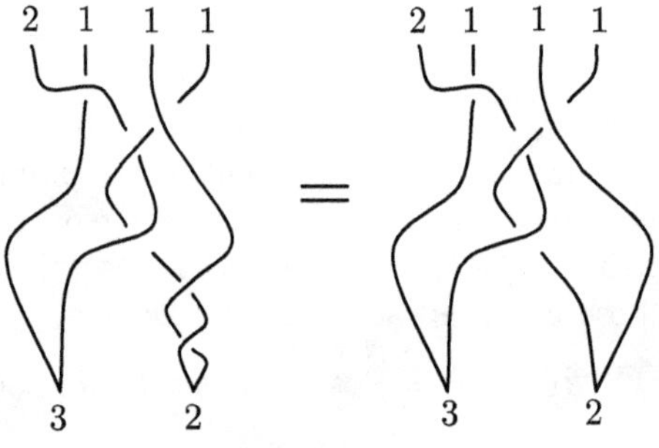

Figure 4: Two representatives of a morphism in the fundamental category from x_P to x_Q where $P = (3|2)$ and $Q = (2|1|1|1)$.

Let's make this precise. In order that we may work with subgroups of S_n and of the braid group B_n we choose a po-path $\gamma^{P,Q}$ from x_P to x_Q whenever $P < Q$. We do this in such a way that $\gamma^{P,R}$ is the composite path $\gamma^{P,Q}.\gamma^{Q,R}$ whenever $P < Q < R$. Label the points in the configuration corresponding to the basepoint in the open stratum by $1, \ldots, n$. The chosen po-paths then identify a concrete partition C of $\{1, \ldots, n\}$ with $\mathcal{P}(C) = P$ corresponding to each basepoint x_P: namely i and j are in the same part at x_P if the points i and j coalesce along the reverse of the po-path from x_P to $x_{(1|\cdots|1)}$. Henceforth we will abuse notation by denoting this choice of concrete partition by the same letter as the corresponding abstract partition.

To each concrete partition P of $\{1, \ldots, n\}$ into k subsets of cardinality $p_1, \ldots, p_k$ we associate three subgroups of S_n and three subgroups of B_n. Let

$$S_P = \{\sigma \in S_n : i \sim_P j \iff \sigma(i) \sim_P \sigma(j)\},$$

where $\sim_P$ is the equivalence relation on $\{1, \ldots, n\}$ corresponding to P. This group of symmetries of P is a product:

$$S_P \cong IS_P \times ES_P$$

where the internal symmetries $IS_P \cong S_{p_1} \times \cdots \times S_{p_k}$ permute the elements within the parts and the external symmetries $ES_P \leqslant S_k$ permute the parts amongst themselves, preserving the cardinality. Similarly we define B_P to be the product $IB_P \times EB_P$ of the internal braids $IB_P \cong B_{p_1} \times \cdots \times B_{p_k}$, which braid the strings within the parts, and the external braids EB_P. The latter are defined by the pull-back square

$$\begin{array}{ccc} EB_P & \longrightarrow & B_k \\ \downarrow & & \downarrow \\ ES_P & \longrightarrow & S_k. \end{array}$$

The partition P defines embeddings of IB_P and EB_P into B_n as commuting sub-groups. This determines an embedding of B_P into B_n.

It is well-known — and visually obvious — that the fundamental group of the top stratum $X_{(1|\cdots|1)}$ is the braid group B_n on n strings. More generally, for a partition P into subsets of cardinality $p_1, \ldots, p_k$, the fundamental group of the stratum X^P is isomorphic to EB_P. We do not get the full braid group on k strings because points of the stratum correspond to sets of k points *labelled by the cardinalities* $p_1, \ldots, p_k$ and this labelling must be preserved by the braiding. There are also no 'internal' braids.

Pick partitions $P \leqslant Q$. Let $E_0 : \operatorname{holink}\left(X^P \cup X^Q, X^P\right) \to X^P$ be the evaluation at 0 map. From Figure 1 there is an exact sequence

$$\pi_1\left(E_0^{-1}x_P, \gamma^{P,Q}\right) \to \pi_1\left(X^Q, x_Q\right) \to \Pi_1^{po}(x_P, x_Q) \to \pi_0\left(E_0^{-1}x_P\right) \to 1. \qquad (1)$$

The fibre $E_0^{-1}x_P$ is the space of 'geometric braids' embedded in $\mathbb{C} \times [0,1]$ starting at the configuration x_P and immediately splitting into $|Q|$ strings. These strings are labelled by the cardinalities of the parts of Q in such a way that the sum of the labels on the set of strings emanating from the same point of x_P is the cardinality of the corresponding part of P.

Since the ends are free to move the set of components $\pi_0\left(E_0^{-1}x_P\right)$ is the set of ways in which P can be refined to Q. We can describe this in terms of concrete partitions of $\{1, \ldots, n\}$ as follows. Define a sub*set*

$$S_{P,Q} = \{\sigma \in S_n : \sigma(i) \sim_Q \sigma(j) \Rightarrow i \sim_P j\}.$$

The subgroup IS_P acts on the left and S_Q acts on the right. The set of ways of refining P to Q is the double orbit space:

$$\pi_0\left(E_0^{-1}x_P\right) \cong IS_P \backslash S_{P,Q} / S_Q.$$

The image of $\pi_1 E_1 : \pi_1\left(E_0^{-1}x_P, \gamma^{P,Q}\right) \to \pi_1\left(X^Q, x_Q\right) \cong EB_Q$ consists of braids on $|Q|$ strings (respecting labels) which can be continuously deformed by gathering together strings in the same part of P so that they become the trivial braid on $|P|$

strings. Thus the image is precisely the subgroup $IB_P \cap EB_Q$ (where the intersection is taken by embedding both in B_n).

Therefore (1) gives an exact sequence

$$1 \to IB_P \cap EB_Q \backslash EB_Q \to \Pi_1^{po}(x_P, x_Q) \to IS_P \backslash S_{P,Q}/S_Q \to 1. \tag{2}$$

Here is an explicit way to construct the middle term. Let $EB_{P,Q}$ be the sub*set* of $B_{|Q|}$ defined by the pullback square

$$\begin{array}{ccc} EB_{P,Q} & \longrightarrow & B_{|Q|} \\ \downarrow & & \downarrow \\ S_{P,Q} & \longrightarrow & S_n \end{array}$$

where the map $B_{|Q|} \to S_n$ is obtained by permuting the parts of the partition at x_Q whilst leaving the ordering within parts fixed. Note that $S_{Q,Q} = S_Q$, and also $EB_{Q,Q} = EB_Q$ because EB_Q fits into the pullback square

$$\begin{array}{ccc} EB_Q & \longrightarrow & B_{|Q|} \\ \downarrow & & \downarrow \\ S_Q & \longrightarrow & S_n. \end{array}$$

It follows from this and the fact that $S_Q \subset S_{P,Q}$ that there is an inclusion $EB_Q \to EB_{P,Q}$. Furthermore, the group S_Q acts on the right on the set $S_{P,Q}$ and it follows that EB_Q acts on $EB_{P,Q}$ with orbit space $S_{P,Q}/S_Q$. I.e., appropriately interpreted, there is an exact sequence

$$1 \to EB_Q \to EB_{P,Q} \to S_{P,Q}/S_Q \to 1.$$

The partition Q determines an embedding of $EB_{P,Q}$ into B_n, and henceforth we identify $EB_{P,Q}$ with its image in B_n. Denote the image of $EB_{P,Q}$ in the orbit space $IB_P \backslash B_n$ by $IB_P \backslash EB_{P,Q}$. Comparing with (2) it follows that

$$\Pi_1^{po}(x_P, x_Q) \cong IB_P \backslash EB_{P,Q}. \tag{3}$$

The composition in $\Pi_1^{po} X$ is simple to describe in terms of the composition in B_n. Suppose $P \leqslant Q \leqslant R$. Composition within S_n determines a map $S_{P,Q} \times S_{Q,R} \to S_{P,R}$ and it follows that composition in B_n gives a map

$$EB_{P,Q} \times EB_{Q,R} \to EB_{P,R}.$$

Noting that the internal braids IB_Q commute with $EB_{P,Q}$ and are a subgroup of the internal braids IB_P we see that the above map descends to

$$IB_P \backslash EB_{P,Q} \times IB_Q \backslash EB_{Q,R} \to IB_R \backslash EB_{P,R} : ([\alpha], [\beta]) \mapsto [\alpha\beta].$$

This is the composition in $\Pi_1^{po} X$.

The quotient map $\mathbb{C}^n \to \mathbb{C}^n/S_n = X$ is a stratified branched cover. It corresponds to the functor

$$x_P \mapsto IS_P \backslash S_n \quad \text{and} \quad \Pi_1^{po}(x_P, x_Q) \cong IB_P \backslash EB_{P,Q} \mapsto IS_P \backslash S_{P,Q}$$

where the map $IS_Q\backslash S_n \to IS_P\backslash S_n$ corresponding to an element of $IS_P\backslash S_{P,Q}$ is induced by composition $S_{P,Q} \times S_n \to S_n$ in S_n.

A. Pre-orders and spaces

A pre-order P is a set equipped with a reflexive and transitive relation $\leqslant$. An increasing map $f : P \to Q$ is one such that $f(p) \leqslant f(p')$ whenever $p \leqslant p'$. Let **Preorder** be the category of pre-orders and increasing maps.

If $\leqslant$ is also antisymmetric, i.e. $p \leqslant q$ and $q \leqslant p$ implies $p = q$, then P is a poset. Each pre-order P has an associated poset given by quotienting by the equivalence relation

$$p \sim p' \iff p \leqslant p' \text{ and } p' \leqslant p.$$

The quotient map $P \to P/\!\sim$ is increasing. This construction is right adjoint to the natural inclusion **Poset** $\hookrightarrow$ **Preorder**.

Pre-orders arise naturally from topology: if X is a topological space there is a pre-order on the points of X given by

$$x \leqslant y \iff (x \in U \Rightarrow y \in U) \iff x \in \overline{y}$$

where $\overline{y}$ is the closure of y, and U is an open set. This is called the *specialisation* pre-order because lower points are more 'special' and higher ones more 'generic' (in the sense familiar to algebraic geometers). It defines a functor S : **Top** $\to$ **Preorder**.

Conversely, given a pre-order P we can topologise it (not necessarily uniquely) so that the resulting specialisation pre-order is P. The coarsest topology with this property has closed sets generated (under finite union and arbitrary intersection) by the $D_p = \{q \mid q \leqslant p\}$ for $p \in P$. The finest topology with this property is the *Alexandrov topology*, in which the sets $U_p = \{q \mid p \leqslant q\}$ form a basis. Write $D(P)$ for P with the coarsest topology and $U(P)$ for P with the Alexandrov topology. Here D stands for 'downward-closed' and U for 'upward-open'. The identity maps

$$US(X) \to X \to DS(X)$$

are continuous, so U is a candidate left adjoint and D a candidate right adjoint for S. It turns out that D is not a right adjoint, indeed it is not even a functor; an increasing map $P \to Q$ need not induce a continuous map $D(P) \to D(Q)$. For example if we order $[0, 1]$ in the usual way then

$$t \mapsto \begin{cases} t/2 & t \in [0,1) \\ 1 & t = 1 \end{cases}$$

is increasing but not continuous as a map $D[0, 1] \to D[0, 1]$. However, a map of pre-orders is increasing if, and only if, it is continuous in the Alexandrov topologies. In particular

$$U : \textbf{Preorder} \to \textbf{Top}$$

is a functor and is left adjoint to specialisation. The identity maps on the underlying sets give the unit and counit of the adjunction

$$P \to SU(P) \quad \text{and} \quad US(X) \to X.$$

It follows from the equivalences $p \leqslant q \iff q \in U_p \iff U_q \subset U_p$ that $P \cong SU(P)$. However, the topology of $US(X)$ can be much finer than that of X. For instance if X is a metric space then $US(X)$ has the discrete topology. In fact, $US(X) \cong X$ if and only if $X \cong U(P)$ for some pre-order P, in which case we say X is an *Alexandrov space.*

Alexandrov spaces can be alternatively characterised as those spaces for which each point x has a unique minimal open neighbourhood U_x (by minimal we mean that for any open U we have $x \in U \iff U_x \subset U$). A simple consequence is that any space with a finite topology is Alexandrov. Finally note that, if P is a pre-order whose associated poset is finite then $U(P) \cong D(P)$ and there is a unique topological space whose specialisation pre-order is P.

A.1. Pre-ordered and filtered spaces

A pre-ordered space, or *po-space* for short, is a pair $(X, \leqslant)$ consisting of a topological space X and a pre-order $\leqslant$ on the points of X. A map of po-spaces, or *po-map*, $f : X \to Y$ is a continuous and increasing map. Let **PoSpace** be the resulting category.

We can think of po-spaces in two other ways. The first is as a space equipped with two topologies, a 'spatial' topology and an Alexandrov topology which defines the pre-order. Po-maps correspond to maps which are continuous with respect to the spatial and Alexandrov topologies. The second is as a space over a poset, i.e. as a space X together with a surjective map $\sigma_X : X \to P_X$ where P_X is the poset associated to the pre-order on the points of X and σ_X the quotient map. We call P_X the *poset of strata* of X, for reasons which will become apparent in a moment. In this picture po-maps are commutative squares

$$\begin{array}{ccc} X & \xrightarrow{f} & Y \\ {\scriptstyle \sigma_X}\downarrow & & \downarrow{\scriptstyle \sigma_Y} \\ P_X & \xrightarrow[g]{} & P_Y \end{array}$$

in which f is continuous and g increasing.

The definition of po-space assumes no compatibility between the topology and the pre-order. We now consider two compatibility conditions

C1 : the down-sets $D_x = \{x' \mid x' \leqslant x\}$ are closed for all $x \in X$;

C2 : the up-sets $U_x = \{x' \mid x' \geqslant x\}$ are open for all $x \in X$.

The first of these is perhaps the most natural, we expect $\leqslant$ to be a closed condition. The second condition implies the first.

Po-spaces satisfying C1 are better known as filtered spaces: a topological space X is *filtered* if there are non-empty closed subspaces X_i indexed by a poset P_X such that

1. $X_i \subset X_j \iff i \leqslant j$;
2. the subsets $X^i = X_i - \bigcup_{j<i} X_j$ partition X.

The X^i are known as the *strata* of the filtration and P_X is referred to as the *poset of strata.* There is an induced pre-order on the points of a filtered space X coming from the map $\sigma_X : X \to P_X$ taking points to the stratum in which they lie. I.e. we define $x \leqslant x' \iff \sigma_X(x) \leqslant \sigma_X(x')$. By definition, for $x \in X_i$

$$D_x = \{x' \mid x' \leqslant x\} = X_i$$

is closed, so that a filtered space is a po-space satisfying C1. Conversely, if $(X, \leqslant)$ is a po-space satisfying C1 then we can filter it by the the non-empty closed subsets $X_i = \{x' \mid x' \leqslant x\}$.

The second compatibility condition C2 is equivalent, from the filtered perspective, to asking that upward unions of strata

$$\bigcup_{j \geqslant i} X^i = X - \bigcup_{k \not\geqslant i} X_k$$

are open for each i. Equivalently, the surjection $\sigma_X : X \to P_X$ is continuous with respect to the Alexandrov topology on P_X. We will say X is *well-filtered* when this holds — it is automatic for a space with a finite filtration. The strata of a well-filtered space are locally-closed, i.e. each is a closed subset of an open subset of the space.

Example A.1. *We define a well-filtered space I_n whose underlying space is the interval $[0,1]$ with the standard topology and whose filtration is*

$$\{0\} \subset \left[0, \frac{1}{n}\right] \subset \cdots \subset \left[0, \frac{n-1}{n}\right] \subset [0,1].$$

The strata are the subsets $\{0\}, (0, \frac{1}{n}], \ldots, (\frac{n-1}{n}, 1]$ and the poset of strata is $0 \leqslant 1 \leqslant \cdots \leqslant n$. The Alexandrov topology on this has open sets

$$\{0, 1, \ldots, n\}, \{1, 2, \ldots, n\}, \ldots, \{n\} \text{ and } \emptyset.$$

The continuous surjection corresponding to the filtration is given by $t \mapsto \lceil nt \rceil$.

The ordered interval I is the space $[0,1]$ with the standard topology and order. It is filtered because $D_t = \{s \mid s \leqslant t\} = [0,t]$ is closed, but not well-filtered because $U_t = [t,1]$ is not open for $t \neq 0$.

A continuous map $f : X \to Y$ of filtered spaces is *filtered* if there is an increasing map of the indexing posets $g : P_X \to P_Y$ such that $f(X_i) \subset Y_{g(i)}$. This is a rather weak condition and we will instead consider the stronger notion of a *stratified* map, i.e. a map for which

$$f(X^i) \subset Y^{g(i)}.$$

Every stratified map is filtered. The map g can be recovered from f (but the requirement that g be increasing is a restriction on f). Note that $f : X \to Y$ is stratified if and only if

$$\begin{array}{ccc} X & \xrightarrow{f} & Y \\ {\scriptstyle \sigma_X}\downarrow & & \downarrow{\scriptstyle \sigma_Y} \\ P_X & \xrightarrow[g]{} & P_Y \end{array}$$

commutes so that stratified maps are po-maps and vice versa. Hence the two compatibility conditions C1 and C2 between the topology and the pre-order cut out two full subcategories

$$\mathbf{PoSpace} \supset \mathbf{Filt} \supset \mathbf{WellFilt}$$

consisting respectively of the filtered and the well-filtered spaces, and the stratified maps between them. Most interesting examples of po-spaces arising 'in nature' seem to be filtered or even well-filtered, certainly we will focus on these. However, for the purposes of theory it is convenient to work with po-spaces.

B. Cosheaves and complete spreads

The notion of a sheaf and the correspondence of sheaves and étale spaces are well-known. In contrast cosheaves are rarely discussed and there are few references. A further complication is that, unlike the case of sheaves, a cosheaf of abelian groups is not simply a cosheaf of sets whose cosections have compatible abelian group structures. Thus the theories of cosheaves of sets and of abelian groups are different. (The reason is that, whilst the underlying set of a product of abelian groups is just the product of the underlying sets of the abelian groups, the same is not true for coproducts — the coproduct of sets is the disjoint union whereas the coproduct of abelian groups is the direct sum.)

This appendix provides the necessary background for the use of cosheaves in this paper, and in particular explains the correspondence between cosheaves on a space X and locally-connected, uniquely complete spreads over X. Surprisingly, this result seems to be new — it is a minor generalisation of the results of [**Fun95**, §5,6] which treats the special case in which X is a complete metric space.

A *precosheaf* of sets on a topological space X is a functor $\mathcal{F} : \mathbf{U}(X) \to \mathbf{Set}$ from the category of open subsets of X and inclusions to the category of sets. Elements of $\mathcal{F}(U)$ are called *cosections* over U, and the maps $\mathcal{F}(U) \to \mathcal{F}(V)$ for $U \subset V$ are called *extensions*. A *cosheaf* of sets on X is a precosheaf which preserves colimits, i.e. for any collection $\{U_i\}$ of open sets $\mathcal{F}\left(\bigcup_i U_i\right)$ is the colimit of

$$\sum_{i,j} \mathcal{F}\left(U_i \cap U_j\right) \to \sum_i \mathcal{F}\left(U_i\right).$$

The displayed map is induced from the inclusions of $U_i \cap U_j$ into U_i and U_j in the obvious way. To be concrete, this means that

$$\mathcal{F}\left(\bigcup_i U_i\right) = \sum_i \mathcal{F}\left(U_i\right) / \sim$$

is the quotient of the disjoint union by the equivalence relation generated by $\alpha_i \sim \alpha_j$ if there is $\beta \in \mathcal{F}\left(U_i \cap U_j\right)$ which extends to both $\alpha_i \in \mathcal{F}\left(U_i\right)$ and $\alpha_j \in \mathcal{F}\left(U_j\right)$. Maps of precosheaves and cosheaves are natural transformations. We denote the categories of precosheaves and cosheaves on X by $\mathbf{Precosh}{\downarrow}\mathbf{X}$ and $\mathbf{Cosh}{\downarrow}\mathbf{X}$ respectively.

Example B.1. *Given a continuous map $p_Y : Y \to X$ the assignment*

$$U \mapsto \pi_0\left(p_Y^{-1} U\right)$$

is a precosheaf CY. *(Here* π_0 *denotes components, not path components.) This defines a functor* $C : \mathbf{Top}\downarrow\mathbf{X} \to \mathbf{Precosh}\downarrow\mathbf{X}$.

If Y *is locally-connected then* CY *is in fact a cosheaf. To see this, recall that a space is locally-connected if and only if the connected components of open subsets are open. It follows that*

$$\pi_0 : \mathbf{LCTop} \to \mathbf{Set},$$

where **LCTop** *is the full subcategory of locally-connected spaces, is left adjoint to the discrete space functor. Since left adjoints preserve colimits* CY *is a cosheaf when* Y *is locally-connected; we call it the* cosheaf of components *of* Y.

This example shows that we can naturally turn spaces over a given space into precosheaves on that space. Conversely, every precosheaf $\mathcal{F}$ has an associated *display space* $D\mathcal{F} \in \mathbf{Top}\downarrow\mathbf{X}$. As a set the display space is the disjoint union

$$\sum_{x\in X} \mathcal{F}_x$$

where $\mathcal{F}_x = \lim_{U\ni x} \mathcal{F}(U)$ is the *costalk* of $\mathcal{F}$ at x. (An element β of the costalk $\mathcal{F}_x$ is simply a set of consistent choices β_U of cosections over each open neighbourhood U of x.) We topologise the display space by declaring

$$V_\alpha = \{\beta \in \mathcal{F}_x \mid x \in U, \beta_U = \alpha\}$$

for each open $U \subset X$ and $\alpha \in \mathcal{F}(U)$ to be a basis of opens. The obvious projection

$$p_\mathcal{F} : D\mathcal{F} \to X$$

with fibres the costalks is then continuous because $p_\mathcal{F}^{-1}U = \sum_{\alpha\in\mathcal{F}(U)} V_\alpha$ is open. It is easy to check that D determines a functor $\mathbf{Precosh}\downarrow\mathbf{X} \to \mathbf{Top}\downarrow\mathbf{X}$.

One might imagine that C and D were adjoint, but this is not quite so. There is a natural map $CD\mathcal{F} \to \mathcal{F}$ for any precosheaf $\mathcal{F}$ given by

$$CD\mathcal{F}(U) \to \mathcal{F}(U) : [\beta] \mapsto \beta_U$$

where $[\beta]$ is the component of $p_\mathcal{F}^{-1}U$ containing $\beta \in \mathcal{F}_x$. This is well-defined since $p_\mathcal{F}^{-1}U = \sum_{\alpha\in\mathcal{F}(U)} V_\alpha$ is a disjoint union of opens so each component is contained within a unique V_α. However, the natural map

$$Y \to DCY : y \mapsto \{[y] \in \pi_0(p_Y^{-1}U)\}_{U\ni x}$$

need not be continuous. The inverse image of a basic open subset is a connected component of $p_Y^{-1}U$ for some U. This need not be open unless Y is locally-connected. Fortunately, the solution to this difficulty is simple. The inclusion $\mathbf{LCTop}\downarrow\mathbf{X} \hookrightarrow \mathbf{Top}\downarrow\mathbf{X}$ has a right adjoint $Y \mapsto \hat{Y}$ given by the unique minimal refinement of the topology on Y which is locally-connected, see e.g. [**Fun95**, §5]. Write $\hat{D}$ for the functor $\mathcal{F} \mapsto \widehat{D\mathcal{F}}$.

Proposition B.2. *The functor* $C : \mathbf{LCTop}\downarrow\mathbf{X} \to \mathbf{Precosh}\downarrow\mathbf{X}$ *is left adjoint to* $\hat{D} : \mathbf{Precosh}\downarrow\mathbf{X} \to \mathbf{LCTop}\downarrow\mathbf{X}$.

Proof. The unit $Y \to DCY$ and counit $CD\mathcal{F} \to \mathcal{F}$ were constructed above *when Y was locally-connected.* We can easily check that in this case these determine a natural isomorphism

$$\mathbf{Precosh}\downarrow\mathbf{X}\,(CY, \mathcal{F}) \cong \mathbf{Top}\downarrow\mathbf{X}\,(Y, D\mathcal{F})\,.$$

We obtain the result by composing with the natural isomorphism

$$\mathbf{Top}\downarrow\mathbf{X}\,(Y, D\mathcal{F}) \cong \mathbf{LCTop}\downarrow\mathbf{X}\left(Y, \hat{D\mathcal{F}}\right).$$

□

We have already seen that CY is a cosheaf when Y is locally-connected. Conversely, it follows from the lemma below that $D\mathcal{F}$ is locally-connected when $\mathcal{F}$ is a cosheaf.

Lemma B.3. *If $\mathcal{F}$ is a cosheaf, $U \subset X$ is open and $\alpha \in \mathcal{F}(U)$ then the basic open set $V_\alpha = \{\beta \in \mathcal{F}_x \mid x \in U, \beta_U = \alpha\} \subset D\mathcal{F}$ is non-empty and connected.*

Proof. Suppose $V_\alpha = \emptyset$. Then each $x \in U$ has an open neighbourhood $V_x \subset U$ such that α is not in the image of the extension $\mathcal{F}(V_x) \to \mathcal{F}(U)$. These V_x form a cover of U and so the cosheaf condition exhibits $\mathcal{F}(U)$ as the quotient of

$$\sum_{x \in X} \mathcal{F}(V_x)$$

by an equivalence relation. Since the extensions are the composites $\mathcal{F}(V_x) \to \sum_{x\in X} \mathcal{F}(V_x) \to \mathcal{F}(U)$ this contradicts the fact that α is not in the image of any of these. Hence V_α is non-empty.

It remains to show that V_α is connected. Consider an arbitrary cover of V_α by basic open subsets V_{α_i} for $\alpha_i \in \mathcal{F}(U_i)$ where $U_i \subset U$. For ease of reading we write V_i for V_{α_i}.

Let $U' = \bigcup_i U_i$. First we show that $V_\alpha = V_{\alpha'}$ for some unique $\alpha' \in \mathcal{F}(U')$. Note that if $x \in U - U'$ then x has a neighbourhood $W_x \subset U$ such that α is not in the image of the extension $\mathcal{F}(W_x) \to \mathcal{F}(U)$. Otherwise there is an element $\beta \in \mathcal{F}_x$ with $\beta_U = \alpha$. This implies that $\beta \in V_\alpha$ which contradicts the fact that the V_i cover V_α. Cover U by U' and the W_x for $x \in U - U'$. The cosheaf condition exhibits $\mathcal{F}(U)$ as a quotient of

$$\mathcal{F}(U') + \sum_{x\in U-U'} \mathcal{F}(W_x)$$

by the equivalence relation generated by equating extensions from the intersections. Since $V_\alpha \neq \emptyset$, and there are no elements in any of the $\mathcal{F}(W_x)$ whose extensions are α, there is a unique $\alpha' \in \mathcal{F}(U')$ whose extension to $\mathcal{F}(U)$ is α.

Without loss of generality we may now assume that $U' = U$. Considering the cosheaf condition for $U = \bigcup_i U_i$ we have

$$\mathcal{F}(U) = \operatorname{colim}\left(\sum_{i,j} \mathcal{F}(U_i \cap U_j) \to \sum_i \mathcal{F}(U_i)\right).$$

By construction the extension of each α_i to $\mathcal{F}(U)$ is α. Since the α_i are identified in the colimit we must, for any pair of indices i and j, be able to find a finite sequence of indices $i = i_1, \ldots, i_n = j$ and elements

$$\beta_k \in \mathcal{F}(U_{i_k} \cap U_{i_{k+1}})$$

which extend to $\alpha_{i_k} \in \mathcal{F}(U_{i_k})$ and $\alpha_{i_{k+1}} \in \mathcal{F}(U_{i_{k+1}})$. Thus for any V_i and V_j we have found a finite sequence V_{i_k} with $V_i = V_{i_1}, V_j = V_{i_n}$ and

$$V_{i_k} \cap V_{i_{k+1}} \neq \emptyset$$

for $i = 1, \ldots, n-1$. It follows that V_α is connected, for if $V_\alpha = V + V'$ is disconnected we could cover V and V' by basic opens to obtain a cover with a pair i and j of indices for which there was no such sequence. □

Remark B.4. *Note that we did not need to assume that the subset U of the previous lemma was connected. The cosheaf condition implies that the open subset V_α is contained within the inverse image via $p : D\mathcal{F} \to X$ of a single connected component of U.*

It follows immediately that $D\mathcal{F}$ is locally-connected when $\mathcal{F}$ is a cosheaf, but in fact $D\mathcal{F}$ has even better properties. The idea of a *complete spread* was introduced by Fox in [**Fox57**] to give a purely topological notion of a branched cover. A map $p_Y : Y \to X$ is a *spread* if the set of connected components of $p_Y^{-1}U$ for open U in X forms a basis for the topology of Y. It is a *complete spread* if whenever we make a consistent choice of component $\alpha_U \subset p_Y^{-1}U$ for each neighbourhood U of some fixed $x \in X$ — consistent meaning that $\alpha_U \subset \alpha_V$ whenever $U \subset V$ — then the intersection $\cap_{U \ni x} \alpha_U \neq \emptyset$. If, in addition, there is a unique point in this intersection we say Y is *uniquely-complete*. More succinctly, a spread Y is uniquely-complete if and only if

$$p_Y^{-1}x \to \lim_{U \ni x} \pi_0 \left(p_Y^{-1}U\right) : y \mapsto \left\{[y] \in \pi_0(p_Y^{-1}U)\right\}_{U \ni x}$$

is a bijection for each $x \in X$. Write $\mathbf{UCS} \downarrow \mathbf{X}$ for the category of locally-connected, uniquely-complete spreads over X, maps are continuous maps over X.

Corollary B.5. *If $\mathcal{F}$ is a cosheaf then $D\mathcal{F}$ is a locally-connected, uniquely-complete spread.*

Proof. Assume $\mathcal{F}$ is a cosheaf on X. Let $U \subset X$ be open. Then $p_{D\mathcal{F}}^{-1}U = \sum_{\alpha \in \mathcal{F}(U)} V_\alpha$. The above lemma shows that the right hand side is the decomposition into connected components. By definition these form a basis of the topology of $D\mathcal{F}$, which is thus a spread. As remarked above, $D\mathcal{F}$ is locally-connected, and

$$\lim_{U \ni x} \pi_0 \left(p_{D\mathcal{F}}^{-1}U\right) = \lim_{U \ni x} \pi_0 \left(\sum_{\alpha \in \mathcal{F}(U)} V_\alpha\right) \cong \lim_{U \ni x} \mathcal{F}(U) = \mathcal{F}_x$$

so that $D\mathcal{F}$ is uniquely-complete. □

Proposition B.6. *The unit $Y \to \hat{D}CY$ is a homeomorphism if and only if Y is a locally-connected, uniquely-complete spread over X. The counit $C\hat{D}\mathcal{F} \to \mathcal{F}$ is an isomorphism if and only if $\mathcal{F}$ is a cosheaf.*

Proof. First consider the unit. If Y is locally-connected, CY is a cosheaf and so DCY is locally-connected by the above lemma. Hence $\hat{D}CY = DCY$. Consider the map $Y \to DCY$. The restriction to the fibre over x is

$$p_Y^{-1}x \to \varinjlim_{U \ni x} \pi_0\left(p_Y^{-1}U\right) = CY_x = p_{DCY}^{-1}x$$

which is a bijection precisely when Y is uniquely-complete. When this holds, the map is a homeomorphism if and only if the inverse images of the basis $\{V_\alpha\}$ for the topology of DCY are a basis for the topology of Y. The inverse image of V_α, where $\alpha \in CY(U) = \pi_0(p_Y^{-1}U)$, is simply the connected component α of $p_Y^{-1}U$. So the map is a homeomorphism precisely when these components form a basis, i.e. precisely when Y is a spread.

Now consider the counit. If $C\hat{D}\mathcal{F} \cong \mathcal{F}$ then $\mathcal{F}$ is a cosheaf because $\hat{D}\mathcal{F}$ is locally-connected. On the other hand if $\mathcal{F}$ is a cosheaf then $D\mathcal{F}$ is locally-connected so $\hat{D}\mathcal{F} = D\mathcal{F}$. Furthermore we have seen, in the proof of the above corollary, that

$$CD\mathcal{F}(U) = \pi_0\left(p_{D\mathcal{F}}^{-1}U\right) \cong \mathcal{F}(U),$$

i.e. that the counit is an isomorphism. $\qquad\square$

We have shown that the adjoint functors C and $\hat{D}$ restrict to an equivalence between the full subcategories of locally-connected, uniquely-complete spreads and cosheaves. The situation is summarised in the following commutative diagram (the vertical arrows are the inclusions).

$$
\begin{array}{ccc}
\mathbf{LCTop}\!\downarrow\!\mathbf{X} & \xleftarrow{\ \hat{D}\ } & \mathbf{Precosh}\!\downarrow\!\mathbf{X} \\
\uparrow & \underset{C}{\searrow} & \uparrow \\
\mathbf{UCS}\!\downarrow\!\mathbf{X} & \xleftarrow{\ \sim\ } & \mathbf{Cosh}\!\downarrow\!\mathbf{X}
\end{array}
$$

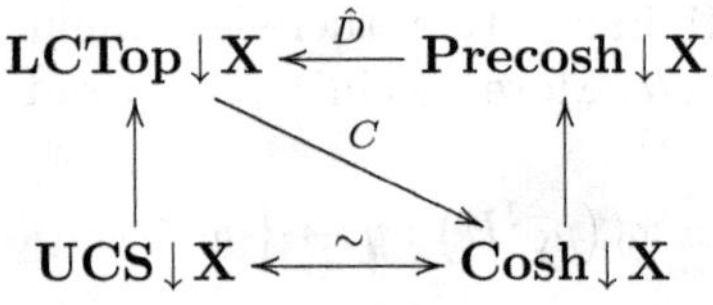

Remark B.7. *The composite $C\hat{D} :$ **Precosh**$\downarrow$**X** $\to$ **Cosh**$\downarrow$**X** is a cosheafification functor. That is, for any precosheaf $\mathcal{F}$ there is a map $C\hat{D}\mathcal{F} \to \mathcal{F}$ with the universal property that for any cosheaf $\mathcal{E}$ and map $\varphi : \mathcal{E} \to \mathcal{F}$ of precosheaves there is a unique factorisation:*

$$
\begin{array}{ccc}
 & & C\hat{D}\mathcal{F} \\
 & \overset{\exists!}{\nearrow} & \downarrow \\
\mathcal{E} & \xrightarrow{\ \varphi\ } & \mathcal{F}
\end{array}
$$

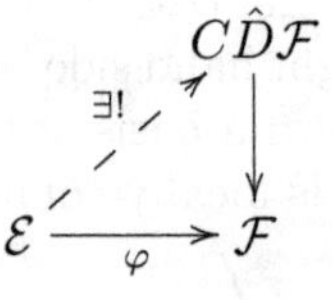

We will not go into a full discussion of the functoriality of cosheaves here. However note that a cosheaf can be restricted to a subspace $\imath : Y \to X$ by defining

$$\imath^*\mathcal{F}(V) = \varinjlim_{U \supset V} \mathcal{F}(U).$$

This corresponds to restricting the corresponding locally-connected uniquely-complete spread $D\mathcal{F}$ to Y. The restriction of a cosheaf to a point is simply the costalk.

References

[BW06] P. Bubenik and K. Worytkiewicz. A model category for local po-spaces. *Homology, Homotopy Appl.*, 8(1):263–292 (electronic), 2006. math.AT/0506352.

[Fox57] R. H. Fox. Covering spaces with singularities. In *A symposium in honor of S. Lefschetz*, pages 243–257. Princeton University Press, Princeton, N.J., 1957.

[Fun95] J. Funk. The display locale of a cosheaf. *Cahiers Top. Géom. Diff. Catég.*, 36(1):53–93, 1995.

[Gra03] M. Grandis. Directed homotopy theory, I. The fundamental category. *Cahiers Top. Géom. Diff. Catég*, 44:281–316, 2003.

[Gra06] M. Grandis. Modelling fundamental 2-categories for directed homotopy. *Homology Homotopy Appl.*, 8:31–70, 2006.

[Kah06] T. Kahl. Relative directed homotopy theory of partially ordered spaces. *J. Homotopy Relat. Struct.*, 1(1):79–100 (electronic), 2006. math.AT/0601079.

[Mil09] D. Miller. Popaths and holinks. arXiv:0909.1201, September 2009. To appear in Journal of Homotopy and Related Structures.

[Qui88] F. Quinn. Homotopically stratified sets. *J. Amer. Math. Soc.*, 1(2):441–499, 1988.

[Tre07] D. Treumann. Exit paths and constructible stacks. arXiv:0708.0659v1, August 2007.

This article may be accessed via WWW at `http://www.rmi.acnet.ge/jhrs/`

Jon Woolf
`jonathan.woolf@liv.ac.uk`

Department of Mathematical Sciences
University of Liverpool
Liverpool
L69 7ZL
U.K.

Journal of Homotopy and Related Structures, vol. 4(1), 2009, pp.389–428

DOUBLE BICATEGORIES AND DOUBLE COSPANS

JEFFREY C. MORTON

(*communicated by Ronnie Brown*)

Abstract

Interest in weak cubical n-categories arises in various contexts, in particular in topological field theories. In this paper, we describe a concept of *double bicategory* in terms of bicategories internal to **Bicat**. We show that in a special case one can reduce this to what we call a *Verity double bicategory*, after Domenic Verity. This is a weakened version of a double category, in the sense that composition in both horizontal and vertical directions satisfy associativity and unit laws only up to (coherent) isomorphisms. We describe examples in the form of double bicategories of "double cospans" (or "double spans") in any category with pushouts (pullbacks, respectively). We also give a construction from this which involves taking isomorphism classes of objects, and gives a Verity double bicategory of double cospans. Finally, we describe how to use a minor variation on this to describe cobordism of manifolds with boundary.

1. Introduction

The need to generalize the concept of a category was implicit from the beginning of the subject. Saunders Mac Lane stated that the concept of category was introduced to study not categories themselves, nor even functors from one category to another, but natural transformations between functors, which are naturally seen as 2-morphisms in a 2-category of all categories. This was an early seed of the notion of higher categories. Once explicitly recognized, however, the concept proved to be ambiguous.

There has been considerable work toward a general definition of a (weak) n-category. This has $(n+1)$ layers of structure, including objects, morphisms between

---

The author would like to acknowledge the help of John Baez in forming this project. Thanks and recognition are due to Dominic Verity for providing, in his Ph.D. thesis, the crucial definition; to Marco Grandis for significantly developing of the theory of cubical cospans since the original version of this paper was written; to Aaron Lauda for references on manifolds with corners; and to Tom Fiore for helpful discussion on weak double categories. Acknowledgment is also due to Dan Christensen, James Dolan, Derek Wise, Toby Bartels, Mike Stay, Alex Hoffnung, and John Huerta for discussions of the work in progress.
Received April 22, 2009, revised July 06, 2009; published on December 01, 2009.
2000 Mathematics Subject Classification: Primary 18D05; Secondary 57Q20.
Key words and phrases: double categories, bicategories, double bicategories, cospans, cobordism, manifolds with corners.

objects, 2-morphisms between morphisms, and so on up to n-morphisms. Several possible alternative definitions exist, as discussed by Cheng and Lauda [**CL**], and by Leinster [**Le2**]. One of the features which varies among such definitions is the *shape* of higher-dimensional morphisms, with different choices suitable to different applications. Our aim in this paper is to develop one particular notion of higher category, in particular a *double bicategory*, which we shall define. We also show that there is a broad class of examples of this type in the form of *double cospans*.

The author's original motivation here was to describe rigorously a bicategory of *cobordisms with corners*. The most natural development of this idea turned out to be a special case of such double cospans. This in turn made it clear that the most natural structure for such things is not a bicategory, but the double bicategories discussed here. However, as we will prove in Theorem 3.4.1, given a double bicategory satisfying some simple conditions, one can get a bicategory, which is a better-understood and simpler structure. Our class of double span examples can be made to be of this type. The development of the topological material involved in cobordisms with corners will be given in a companion paper, but here we aim to be accessible to readers of that paper seeking background, and thus will give a relatively expository description of double bicategories and their double cospan examples.

The related concept of a "weak double category", or "pseudo double category" has also been defined (for further discussion, see e.g Marco Grandis and Robert Paré [**GP1**], Thomas Fiore [**Fi**], or Richard Garner [**Ga**]). In this setting, the weakening only occurs in only one direction, say the horizontal. That is, the associativity of composition, and unit laws, in the vertical direction apply only up to certain higher *associator* and *unitor* isomorphisms. In the horizontal direction, category axioms hold strictly. In fact, this must be so when weakening uses just the square 2-cells of the double category. For the composition in a double bicategory to be weak in both directions, it must be that the associator isomorphisms are (globular) 2-morphisms, rather than (square) 2-cells.

We note here that the feature that one direction is strict also appears in the *weak n-cubical categories* discussed by Grandis ([**Gr1**], [**Gr3**]), but that these are well defined for any dimension n. In particular, they are defined so as to have one direction in which composition is strict, while all others are weak. However, a Verity double bicategory can be taken to be a weak 3-cubical category with no nontrivial morphisms in the strict direction; on the other hand, a weak 2-cubical category can be seen as a Verity double bicategory in which the composition in one direction happens to be strict. For some purposes, this asymmetry is useful, but for our motivating application to cobordisms with corners, we want to define a "fully" weak cubical 2-category. We shall comment on the structures described by Grandis again when we consider double cospans, which give examples of both double bicategories and weak n-cubical categories.

In Section 2 we briefly describe some of the necessary category-theoretic background for readers who may be unfamiliar with it. This includes the concepts of *enrichment* and *internalization* in category theory, which give rise to bicategories and double categories respectively. The concept of a double bicategory is a mutual generalization of that of *bicategory* and of *double category*, and is best understood in this light.

Bicategories gave the first precise, explicit notion of weak higher categories. They were described by Bénabou [**Be**] in 1967, introducing the concept of 2-morphisms between morphisms:

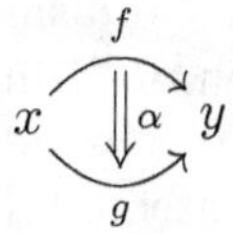

$$x \overset{f}{\underset{g}{\Downarrow \alpha}} y \tag{1}$$

Equations in the axioms for a category are replaced by 2-isomorphisms, which themselves satisfy coherence laws given in equations. This is known as *weakening*.

The original example used to illustrate this concept was the bicategory of *spans* in a suitable category $\mathbf{C}$, namely diagrams of the form:

$$X \to S \leftarrow Y \tag{2}$$

There is a natural concept of a map of spans, and an operation of composition for spans which is not strictly associative—rather, it is only associative up to isomorphism. Weakening thus appeared naturally in the setting of spans. The double cospans defined here lead to weakening in just the same way, but the concept being weakened is that of a double category.

Double categories, introduced by Ehresmann [**Eh1, Eh2**], have objects, horizontal and vertical morphisms which can be represented diagrammatically as edges, and squares:

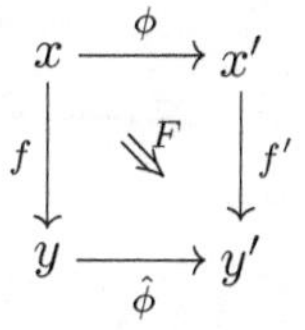

$$\begin{array}{ccc} x & \xrightarrow{\phi} & x' \\ {\scriptstyle f}\downarrow & \Swarrow{\scriptstyle F} & \downarrow{\scriptstyle f'} \\ y & \xrightarrow{\hat{\phi}} & y' \end{array} \tag{3}$$

These can be composed in geometrically obvious ways.

Double categories distinguish between horizontal and vertical 1-morphisms, which in general can only be composed with other morphisms of the same type. On the other hand, 2-morphisms are "squares", with both horizontal and vertical source and target, which can be composed in either direction with other squares having a common boundary.

Moskaliuk and Vlassov [**MV**] discuss the application of double categories to mathematical physics, and particularly to topological quantum field theories (TQFT's), and to dynamical systems with changing boundary conditions—that is, with inputs and outputs. Kerler and Lyubashenko [**KL**] describe "extended" TQFT's as "double pseudofunctors" between double categories. This formulation involves, among other things, a double category of cobordisms with corners. This sort of topological category has manifolds for objects, and manifolds with boundary or with corners as higher morphisms. This makes it possible to describe systems with changing boundary conditions, and the most natural way to do this is by allowing both initial and final states, and changing boundary conditions, as part of the boundary in a more general sense. This is one of the main motivations for the concepts we describe here, and we shall return to it in a subsequent paper. Double categories are too strict to be really natural for our purpose, however. Composition in a double category must be strictly associative, and in order to achieve this, one considers only *equivalence classes* of cobordisms as morphisms.

Thus, the principle here is to weaken the definition of a double category. This had previously been done in the definition of a *pseudocategory*, as described, for instance, by Fiore [**Fi**]. However, in a pseudocategory, just one direction of composition is weak: that is, the associative and unit laws satisfied by composition are replaced by associator and unitor isomorphisms. In a double bicategory, composition is weak in both directions.

In Section 3 we introduce double bicategories using a form of internalization, analogous to that which gives double categories as categories internal to **Cat**. In Section 3.2, we describe a somewhat different concept of double bicategory, due to Dominic Verity (which we denote a *Verity double bicategory* for clarity), a structure which also has both horizontal and vertical bicategories, and square 2-cells, with weak composition in both directions. In Section 3.3, we explain how a special case of our double bicategories can be reduced to Verity's definition. In turn, we show in Section 3.4 how a Verity double bicategory satisfying certain conditions, in turn yields a bicategory in the usual sense. Thus, this presents a series of increasingly manageable simplifications.

We finish in Section 4 by describing a rather broad general class of examples of double bicategories, which arise in rather the same way as the fact that Span(**C**) is a bicategory. A "double cospan" in a category with pushouts is a diagram of the following form:

$$\begin{array}{ccccc} X_1 & \longrightarrow & S & \longleftarrow & X_2 \\ \downarrow & & \downarrow & & \downarrow \\ T_1 & \longrightarrow & M & \longleftarrow & T_2 \\ \uparrow & & \uparrow & & \uparrow \\ X_1' & \longrightarrow & S' & \longleftarrow & X_2' \end{array} \tag{4}$$

These diagrams can be composed horizontally and vertically, and in either case composition is by pushout, just as with ordinary spans or cospans. In Section 4.1 we describe double cospans and their composition in detail, and show that they naturally form a double bicategory in our original sense. In Section 4.2 we show how reducing to certain natural equivalence classes of double cospans yields a Verity double bicategory, following the procedure in Section 3.3.

2. Bicategories and Double Categories

Since this paper is intended as a companion to another of a more topological nature, we will recall for the reader with less category-theoretic background some relevant ideas about bicategories and double categories. Other readers may wish to skip to Section 3 when we introduce double bicategories.

We want to weaken the notion of a double category. Weakening a concept X in category theory generally involves creating a new concept in which defining equations in the original concept (such as associativity) are replaced by specified isomorphisms. Thus, one says that the defining equations hold with equality in a *strict* X and hold only "up to" isomorphism in a *weak* X.

Before describing our weakened concept of double category, we recall how this process works, and examine the strict form of the concept we want to weaken. So we begin by recalling some facts about bicategories, to illustrate weakening, and double categories, to provide a starting point.

2.1. Bicategories

A **bicategory** is a "weak globular 2-category". That is, if $\mathbf{B}$ is a bicategory, and $x, y \in \mathbf{B}$, then $\hom(x, y) \in \mathbf{Cat}$, allowing isomorphisms between morphisms where formerly we had equations. The morphisms in $\hom(x, y)$ are thought of as "2-morphisms" in $\mathbf{B}$. Moreover, the strict version of a bicategory, usually called a "2-category", has the same unit and associativity axioms as a category. However, the weak form replaces these with 2-isomorphisms satisfying some coherence properties. So in particular, we have the following definition, due to Bénabou [**Be**], and discussed in more detail, for instance, in [**Le**].

Definition 2.1.1. *A **bicategory** $\mathcal{B}$ consists of the following data:*

- *A collection of **objects** $\mathbf{Obj}$*
- *For each pair $x, y \in \mathbf{Obj}$, a category $\hom(x, y)$ whose objects are called **morphisms** of $\mathcal{B}$ and whose morphisms are called **2-morphisms** of $\mathcal{B}$*
- *For each object $x \in \mathbf{Obj}$, an identity $1_x \in \hom(x, x)$*
- *For each triple x, y, z of objects, a **composition** functor $\circ : \hom(x, y) \times \hom(y, z) \to \hom(x, z)$*
- *For each composable triple f, g, h of morphisms, a 2-isomorphism (i.e. invertible 2-morphism) $\alpha_{f,g,h} : h \circ (g \circ f) \to (h \circ g) \circ f$ called the **associator***
- *For each morphism $f : x \to y$, left and right **unitor** 2-isomorphisms $l_f : 1_y \circ f \to f$ and $r_f : f \circ 1_x \to f$*

The associator is subject to the Pentagon identity, namely that the following *diagram commutes for any 4-tuple of composable morphisms (f, g, h, j):*

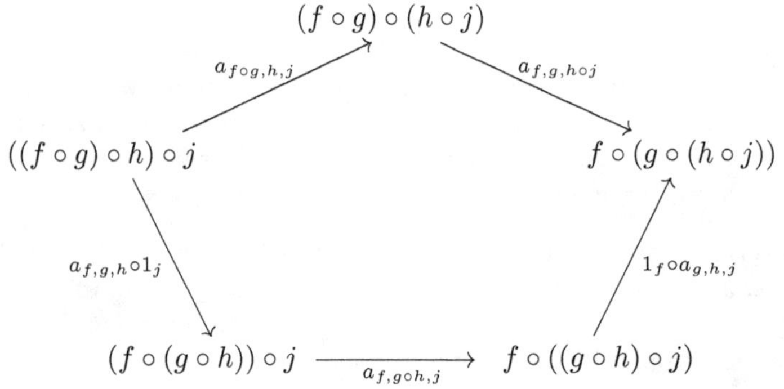

(5)

Also, the unitors and associator make the following commute for all composable g, f:

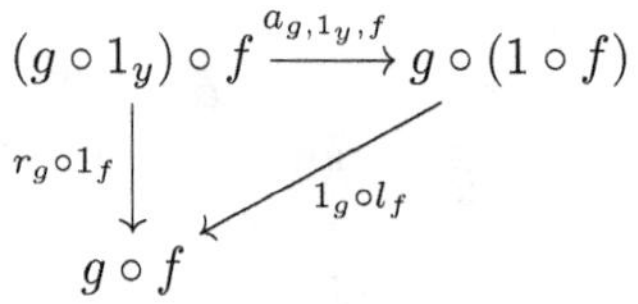

(6)

(where $y = t(f) = s(g)$).

Remark 2.1.2. This is a compact definition of a bicategory, but it is possible to describe the same data in different ways, which will be more directly relevant to subsequent discussion of double bicategories. In particular, this definition is related to the definition of a *strict* bicategory (a *2-category*) as a *category enriched in* **Cat**. That is, for any objects x and y, there is a category $\hom(x, y)$. However, there is a more elementary, although perhaps less elegant, way of describing the data of a bicategory.

One can form the collection $\mathbf{Mor} = \coprod \mathrm{ob}(\hom(x, y))$ of all morphisms of $\mathcal{B}$, and $\mathbf{2Mor} = \coprod \mathrm{mor}(\hom(x, y))$ of all 2-morphisms of $\mathcal{B}$. Then the existence of composition functors imply that there is, just as in categories, a partially defined composition function $\circ : \mathbf{Mor} \times \mathbf{Mor} \to \mathbf{Mor}$, which is defined for pairs (f, g) for which $t(f) = s(g)$ (and two such functions giving composition of 2-morphisms). These functions will have properties determined by the fact that they must give composition functors as defined above. The existence of identity morphisms means that there is an **identity** map $i : \mathbf{Obj} \to \mathbf{Mor}$, and this satisfies the usual relations with the source and target maps, and the composition.

As well as the source and target maps

$$s, t : \mathbf{Mor} \to \mathbf{Obj}$$

given in this definition, there are the source and target maps in each $\hom(x, y)$. These imply the existence of $s, t : \mathbf{2Mor} \to \mathbf{Mor}$, with the property that for any 2-morphism α, $s(s(\alpha)) = s(t(\alpha))$, and $t(s(\alpha)) = t(t(\alpha))$. A similar condition will apply to double categories, and indeed double bicategories, as we shall see. Together these describe the picture summarized in the diagram (1), depicting 2-morphisms as 2-dimensional cells between arrow-shaped 1-morphisms.

Jean Bénabou [**Be**] introduced bicategories in a 1967 paper, and one broad class of examples introduced there comes from the notion of a *span*. Since we will want to use a similar construction later, we remark on this here:

Definition 2.1.3. *(**Bénabou**) Given any category* **C**, *a **span** (S, π_1, π_2) between objects $X_1, X_2 \in$ **C** *is a diagram in* **C** *of the form*

$$P_1 \xleftarrow{\;\pi_1\;} S \xrightarrow{\;\pi_2\;} P_2 \tag{7}$$

*Given two spans (S, s, t) and (S', s', t') between X_1 and X_2 a **morphism of spans** is a morphism $g : S \to S'$ making the following diagram commute:*

$$
\begin{array}{ccc}
 & S & \\
 \pi_1 \swarrow & \downarrow g & \searrow \pi_2 \\
 X_1 \xleftarrow{\;\pi'_1\;} & S' & \xrightarrow{\;\pi'_2\;} X_2
\end{array}
\tag{8}
$$

Composition of spans S from X_1 to X_2 and S' from X_2 to X_3 is given by pullback: that is, an object R with maps f_1 and f_2 making the following diagram commute:

$$
\begin{array}{ccccc}
 & & R & & \\
 & f_1 \swarrow & & \searrow f_2 & \\
 & S & & S' & \\
 \pi_1 \swarrow & & \searrow \pi_2 \quad \pi'_2 \swarrow & & \searrow \pi'_3 \\
 X_1 & & X_2 & & X_3
\end{array}
\tag{9}
$$

which is terminal among all such objects. That is, given any other Q with maps g_1 and g_2 which make the analogous diagram commute, these maps factor through a unique map $Q \to R$. R becomes a span from X_1 to X_3 with the maps $\pi_1 \circ f_1$ and $\pi_2 \circ f_2$.

The span construction has a dual concept:

Definition 2.1.4. *A **cospan** in* **C** *is a span in* **C**^{op}, *morphisms of cospans are morphisms of spans in* **C**^{op}, *and composition of cospans is given by pullback in* **C**^{op}—*that is, by pushout in* **C**.

Remark 2.1.5. ([**Be, ex. 2.6**]) Given any category **C** with all limits, there is a bi-category Span(**C**), whose objects are the objects of **C**, whose *hom*-sets of morphisms Span(**C**)(X_1, X_2) consist of all spans between X_1 and X_2 with composition as defined above, and whose 2-morphisms are morphisms of spans. Span(**C**) as defined above forms a bicategory (dually, there is a bicategory Cosp(**C**) of cospans).

One should note that there is some choice in the precise definition of this bi-category since pushout (or pullback) is only defined up to isomorphism. However,

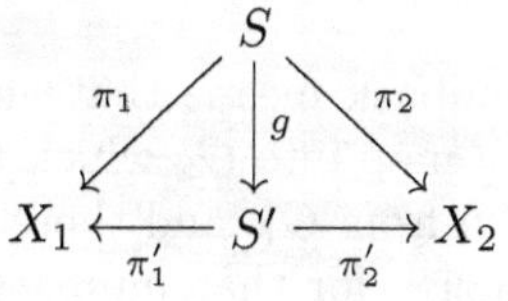

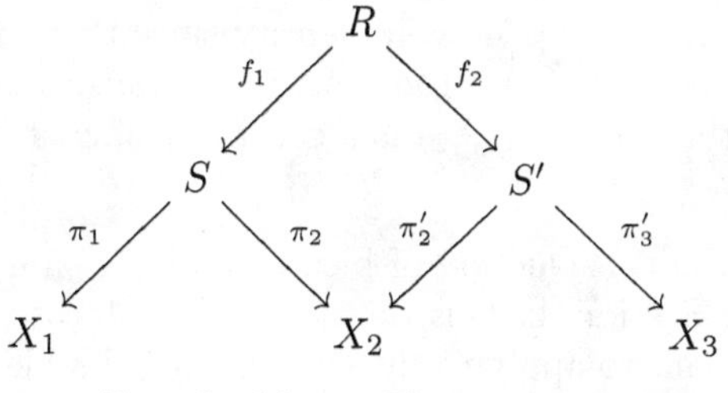

one can make a particular choice of pushout (or pullback) as a given composite, and given this choice get corresponding associators and unitors. Different choices of composite will of course give different such maps. However, all such choices are equivalent. This is due in part to the fact that the pullback is a universal construction (universal properties of Span(**C**) are discussed by Dawson, Paré and Pronk [**DPP**]).

We briefly describe the proof of Bénabou that Span(**C**) is a bicategory:

The identity for X is $X \stackrel{id}{\leftarrow} X \stackrel{id}{\rightarrow} X$, which has an obvious unitor whose properties are easy to check.

The associator arises from the fact that the pullback is a *universal* construction. Given morphisms $f : X \to Y$, $g : Y \to Z$, $h : Z \to W$ in Span(**C**), the composites $((f \circ g) \circ h)$ and $(f \circ (g \circ h))$ are pullbacks consisting of objects O_1 and O_2 with maps into X and W. The universal property of pullbacks gives an isomorphism between O_1 and O_2 as follows.

The universal property of pullback means that any object with maps into the objects f and $(g \circ h)$ will have a map into O_2 which they factor through. We have maps into the objects f, g, and h from O_1, and therefore a unique compatible map into $g \circ h$ by the universal property for that pullback. Therefore, there is again a unique compatible map into O_2. This we take to be the associator. We notice that in particular, the same argument works in reverse, and so the two maps we get are inverses, hence isomorphisms.

These associators satisfy the pentagon identity since they are unique (in particular, both sides of the pentagon give the same isomorphism).

It is easy to check that $\hom(X_1, X_2)$ is a category, since it inherits all the usual properties from **C**.

We will generalize the construction of bicategories of spans to give examples of double bicategories in Section 4. This development of bicategories illustrates the sort of weakening we want to apply to the concept of a double category. So we will first describe the strict notion in Section 2.2, before considering how to weaken it, in Section 3.2.

2.2. Double Categories

The concept of a double category extends that of a category in a different way than the concept of bicategory. Both, however, can be visualized as having both "arrow-like" morphisms, and also two-dimensional cells thought of as higher morphisms.

Just as with bicategories, we recall the definition here first by giving an abstract definition, then showing an equivalent, more concrete, version. The first definition of a bicategory highlighted its relation to the idea of an enriched category. Here we begin by illustrating how double categories illustrate *internalization*, which we will return to in Section 3 when describing double bicategories.

Definition 2.2.1. *A double category is a category internal to* **Cat**.

This is a generalization of the more broadly familiar terminology in which, for instance, a group internal to **Top** is called a *group object* in **Top**, or topological group. However, not all structures we might want to internalize are determined by a single object. In particular, a category (by default, internal to **Set**) consists of not one but two sets, namely the sets of *objects* and *morphisms*. A category internal to **C** (or "in **C**") has two objects of **C** playing the same roles.

So a category in **Cat** is a structure having a category **Ob** of objects and a category **Mor** of morphisms, with functors such as s and t satisfying the usual category axioms. Note that these axioms give conditions at both the object and morphism level, in addition to those which follow from the fact that they are functors.

We thus have sets of objects and morphisms in **Ob**, which satisfy the usual axioms for a category. The same is true for **Mor**. In addition, the category axioms for the double category are imposed on the composition and identity functors, and these must be compatible with the category axioms in the other direction. Thus we can think of both the objects in **Mor** and the morphisms in **Obj** as morphisms between the objects in **Obj**. A double category is often thought of as including the morphisms of two (potentially) different categories on the same collection of objects. These are the *horizontal* and *vertical* morphisms.

Here, the objects in the diagram can be thought of as objects in **Obj**, the vertical morphisms f and f' can be thought of as morphisms in **Obj** and the horizontal morphisms ϕ and $\hat{\phi}$ as objects in **Mor**. Vertical composition is given by composition in **Mor**, and horizontal composition by the morphism map of the composition functor $\circ$. (In fact, we can adopt either convention for distinguishing horizontal and vertical morphisms). However, we also have morphisms in **Mor**. We represent these as 2-cells, or *squares*, like the 2-cell S represented in (11). The fact that the composition map $\circ$ is a functor means that horizontal and vertical composition of square 2-cells commutes.

A double category can therefore be seen more directly. It consists of:

- a set O of objects
- *horizontal* and *vertical* categories, whose sets of objects are both O
- for any diagram of the form

$$\begin{array}{ccc} x & \xrightarrow{\phi} & x' \\ {\scriptstyle f}\downarrow & & \downarrow{\scriptstyle f'} \\ y & \xrightarrow[\phi']{} & y' \end{array} \qquad (10)$$

  a collection of *2-cells*, having horizontal source and target f and f', and vertical source and target ϕ and ϕ'

along with additional data such as the source and target maps, identities, and so forth, all satisfying category-like axioms in both horizontal and vertical directions. In particular, the 2-cells can be composed either horizontally or vertically in the obvious way. We denote a 2-cell filling the above diagram like this:

$$\begin{array}{ccc} x & \xrightarrow{\phi} & x' \\ {\scriptstyle f}\downarrow & \Swarrow S & \downarrow{\scriptstyle f'} \\ y & \xrightarrow[\phi']{} & y' \end{array} \tag{11}$$

and think of the composition of 2-cells as pasting them along an edge. The resulting 2-cell fills a square whose boundaries are the corresponding composites of the morphisms along its edges.

Next, in Section 3 we take our descriptions of double categories and bicategories, and see how to find some common generalizations of both.

3. Double Bicategories

We wish to describe a structure which is sufficient to reproduce the various types of composition found in a double category, but in such a way that all are weakened. This means we should have horizontal composition for horizontal morphisms and vertical composition for vertical morphisms. Square 2-cells should be composable in both directions. Composition of morphisms in each direction is to be "weak", in the sense of having associator and unitor isomorphisms rather than associativity and unit laws. This means there will also be 2-morphisms of some appropriate shape to act as unitors and associators (and, of course, there will in general be other 2-morphisms as well). In particular, in place of the mere categories found in a double category, we have horizontal and vertical *bicategories*, with their (globular) 2-morphisms, as well as (square) 2-cells.

The natural choice of name for such a structure is a *double bicategory.* This term seems to have been originally introduced by Dominic Verity [**Ve**]. There is some ambiguity here. By analogy with "double category", the term "double bicategory" might be expected to describe is an internal bicategory in **Bicat**, the category of all bicategories . Indeed, it is what we will mean by a double bicategory here. However, this is not quite the concept given by Verity. The two turn out to be closely related, and both will be important, so we will refer to double bicategories in the sense of Verity by the term *Verity double bicategories*, while reserving *double bicategory* for internal bicategories in **Bicat**. For more discussion of the relation between these, see Section 3.2.

3.1. Double Bicategories and Internalization

Here we present a more precise definition of the concept of a double bicategory as a bicategory internal to **Bicat**. This will be somewhat more complex than the analogous process for a double category, but runs along similar lines.

Thus, we will have bicategories Obj, Mor and 2Mor. Then one can consider a bicategory internal to **Bicat**. It is straightforward to treat $F(\text{Obj})$ as a horizontal bicategory, and the objects of Obj, Mor and 2Mor as forming a vertical bicategory. Note, however, that a diagrammatic representation of, for instance, 2-morphisms in

2Mor would require a 4-dimensional diagram element. The comparison can be seen by contrasting tables 1 and 2 in Section 3.3.

Definition 3.1.1. *A double bicategory consists of:*

- *bicategories* **Obj** *of* ***objects***, **Mor** *of* ***morphisms***, **2Mor** *of* ***2-morphisms***
- ***source*** *and* ***target*** *2-functors*
  - $s, t : \mathbf{Mor} \to \mathbf{Obj}$
  - $s, t : \mathbf{2Mor} \to \mathbf{Obj}$
  - $s, t : \mathbf{2Mor} \to \mathbf{Mor}$
- *Composition 2-functors:*
  - $\circ : \mathrm{MPairs} \to \mathbf{Mor}$
  - $\circ : \mathrm{HPairs} \to \mathbf{2Mor}$
  - $\cdot : \mathrm{VPairs} \to \mathbf{2Mor}$

  satisfying the interchange law, where
  - $\mathrm{MPairs} = \mathbf{Mor} \times_{\mathbf{Obj}} \mathbf{Mor}$
  - $\mathrm{HPairs} = \mathbf{2Mor} \times_{\mathbf{Mor}} \mathbf{2Mor}$
  - $\mathrm{VPairs} = \mathbf{2Mor} \times_{\mathbf{Obj}} \mathbf{2Mor}$

  are (strict) pullbacks
- *an* ***associator*** *2-functor*
  - $a : \mathrm{Triples} \to \mathrm{2Mor}$

  where
  - $\mathrm{Triples} = \mathbf{Mor} \times_{\mathbf{Obj}} \mathbf{Mor} \times_{\mathbf{Obj}} \mathbf{Mor}$
- ***unitors***
  - $l, r : \mathbf{Obj} \to \mathbf{Mor}$

such that a makes the following diagram commute:

$$\begin{array}{ccccc} \mathrm{Pairs} & \xleftarrow{\circ\times 1} & \mathrm{Triples} & \xrightarrow{1\times\circ} & \mathrm{Pairs} \\ \downarrow{\scriptstyle\circ} & & \downarrow{\scriptstyle a} & & \downarrow{\scriptstyle\circ} \\ \mathrm{Mor} & \xleftarrow[s]{} & \mathrm{2Mor} & \xrightarrow[t]{} & \mathrm{Mor} \end{array}$$

and additional diagrams with the interpretation that a gives invertible 2-morphisms. The unitors must satisfy $s(l(x)) = t(l(x)) = x$ *and* $s(r(x)) = t(r(x)) = x$*, and the associator should satisfy the pentagon identity (5), and the unitors should satisfy the unitor laws (6).*

We interpret these morphisms involving pullbacks (the fibred products) as giving *partially defined* composition 2-functors $\circ : \mathbf{Mor}^2 \to \mathbf{Mor}$, $\circ : \mathbf{2Mor}^2 \to \mathbf{2Mor}$ and $\cdot : \mathbf{2Mor}^2 \to \mathbf{2Mor}$, and associator 2-functor $a : \mathbf{Mor}^3 \to \mathbf{2Mor}$.

Remark 3.1.2. The Pentagon identity is shown in (5) for a bicategory (i.e. a bicategory in **Sets**). In **Bicat**, this holds for objects, morphisms, and 2-morphisms. We can express this condition formally, in any category (with pullbacks), building from composable quadruples, so that the pentagon identity is expressed in a commuting diagram which includes the one built by pasting the two following diagrams together along the outside edges:

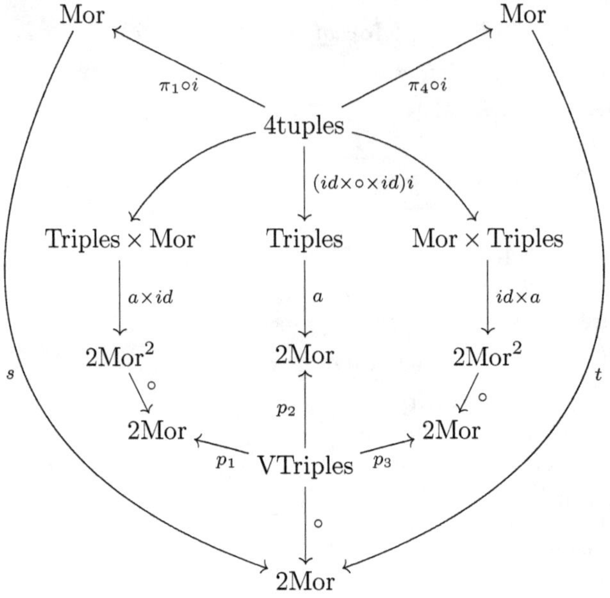

(12)

and

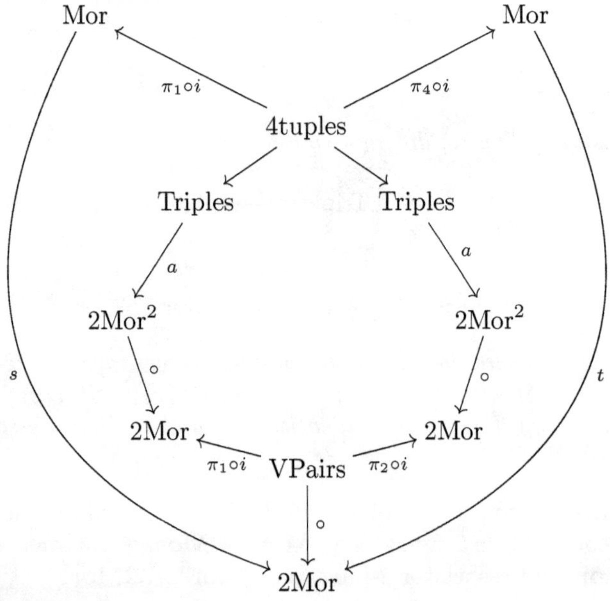

(13)

Note that diagram (12) denotes the three 2-morphism sequence in the pentagon, and (13) the sequence of two 2-morphisms.

Similar remarks apply to give “element-free” versions of the interchange laws for composition of 2-morphisms and the unitor laws shown in (6).

To fully expand this definition without assuming the concept of a bicategory would be much longer than the form given here. One would have to specify nine types of data - objects, morphisms, and 2-morphisms in each of **Obj**, **Mor**, and **2Mor**, and describe all the axioms in detail, such as the conditions implied by the fact that $\circ$ and $\cdot$ are functors. This is a rather complicated structure, as we see in more detail when we return to it in Section 3.3 (and in particular we show the types of data implied by this definition in Table 2).

In particular, the most natural geometric representation of a 2-morphism in **2Mor** is as a 4-dimensional object. We had hoped for a common generalization of double categories and bicategories, each of which is represented graphically with morphisms being cells having dimension at most 2. One could hope that such a structure would also have at most 2-dimensional morphisms. The definition of a Verity double bicategory satisfies this, as we describe in Section 3.2, and in Section 3.3 we show how it is related to the definition we have given here.

First, we briefly remark that one can cast this description of internal bicategories in terms of models of the *finite limit theory* of bicategories, **Th(Bicat)**. This is a category with finite limits, which can be described in terms of its generators and relations. It is generated by objects O, M and B, together with morphisms, and subject relations, as given in the definition (where in that case these are in **Bicat**. A model of such a theory in a category **C** with finite limits is a functor $F : \mathbf{Th(Bicat)} \to \mathbf{C}$.

Here we are considering *strict* models of the theory of categories in **Cat**, and bicategories in **Bicat**, rather than a weak model, which one might also consider. In particular, **Bicat** is a tricategory (defined by Gordon, Power and Street [**GPS**]): it has objects which are bicategories, morphisms which are (weak) 2-functors between bicategories, 2-morphisms which are natural transformations between bifunctors, and 3-morphisms which are modifications of such transformations. However, for both double categories and double bicategories, we ignore the higher morphisms in this setting and think of **Bicat** as a mere category, taking equivalence classes of morphisms between bicategories (that is, 2-functors) where needed. Thus, equations in the theory are mapped to equations (not isomorphisms) in **Bicat**.

Even such strict models, however, are fairly complex structures, so we now consider one way to simplify them.

3.2. Verity Double Bicategories

The following definition of a "double bicategory" is due to Dominic Verity [**Ve**], and will henceforth be referred to as a Verity double bicategory. It is readily seen as a natural weakening of the definition of a double category. Just as the concept of *bicategory* is weaker than that of *2-category* by weakening the associative and unit laws, Verity double bicategories will be weaker than double categories.

Definition 3.2.1. *(Verity) A* ***Verity double bicategory*** **C** *is a structure* $\mathcal{V}$ *consisting of the following data:*

- *a class of* ***objects*** Obj,

- ***horizontal*** *and* ***vertical bicategories*** **Hor** *and* **Ver** *having* Obj *as their objects*
- *for every square of horizontal and vertical morphisms of the form*

$$\begin{array}{ccc} a & \xrightarrow{h} & b \\ {\scriptstyle v}\big\downarrow & & \big\downarrow{\scriptstyle v'} \\ c & \xrightarrow{h'} & d \end{array} \tag{14}$$

a class of ***squares*** **Squ**, *with maps* s_h, t_h : **Squ** $\to$ Mor(**Hor**) *and* s_v, t_v : **Squ** $\to$ Mor(**Ver**), *satisfying an equation for each corner, namely:*

$$\begin{aligned} s(s_h) &= s(s_v) \\ t(s_h) &= s(t_v) \\ s(t_h) &= t(s_v) \\ t(t_h) &= t(t_v) \end{aligned} \tag{15}$$

The squares should have horizontal and vertical composition operations, defining the vertical composite $F \otimes_V G$

$$\begin{array}{ccc} x & \longrightarrow & x' \\ \big\downarrow & \Downarrow F & \big\downarrow \\ y & \longrightarrow & y' \\ \big\downarrow & \Downarrow G & \big\downarrow \\ z & \longrightarrow & z' \end{array} \quad = \quad \begin{array}{ccc} x & \longrightarrow & x' \\ \big\downarrow & \Downarrow F \otimes_V G & \big\downarrow \\ z & \longrightarrow & z' \end{array} \tag{16}$$

and horizontal composite $F \otimes_H G$:

$$\begin{array}{ccccc} x & \longrightarrow & y & \longrightarrow & z \\ \big\downarrow & \Downarrow F & \big\downarrow & \Downarrow G & \big\downarrow \\ x' & \longrightarrow & y' & \longrightarrow & z' \end{array} \quad = \quad \begin{array}{ccc} x & \longrightarrow & z \\ \big\downarrow & \Downarrow F \otimes_H G & \big\downarrow \\ x' & \longrightarrow & z' \end{array} \tag{17}$$

The composites have the usual relation to source and target maps, satisfy the interchange law

$$(F \otimes_V F') \otimes_H (G \otimes_V G') = (F \otimes_H G) \otimes_V (F' \otimes_H G') \tag{18}$$

and there is a unit for composition of squares:

$$\begin{array}{ccc} x & \xrightarrow{1_x} & x \\ {\scriptstyle f}\big\downarrow & \Downarrow 1_f & \big\downarrow{\scriptstyle f} \\ y & \xrightarrow{1_y} & y \end{array} \tag{19}$$

(and similarly for vertical composition).

There is a left and right action by the horizontal and vertical 2-morphisms on

Squ, *giving* $F \star_V \alpha$,

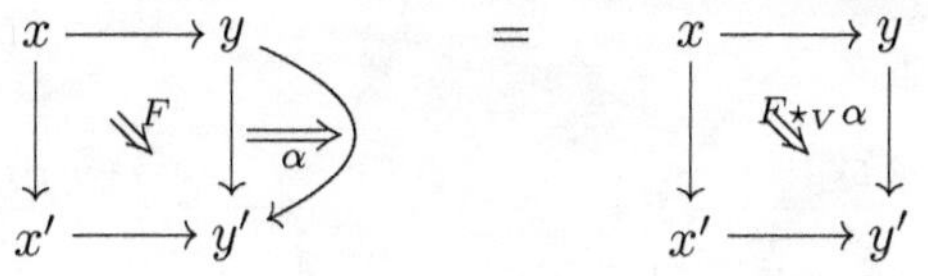

(20)

(and similarly on the left) and $F \star_H \alpha$,

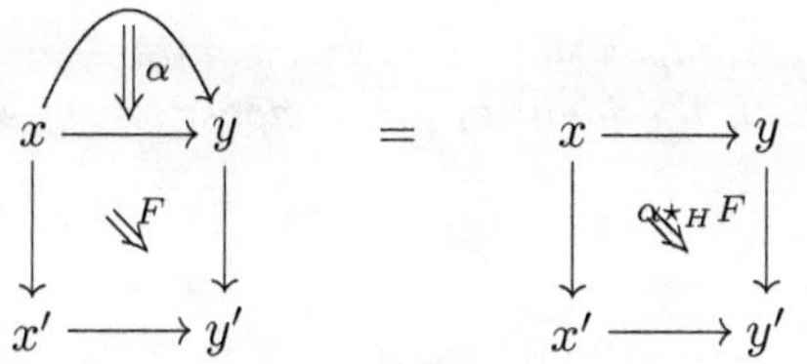

(21)

The actions also satisfy interchange laws:

$$(F \otimes_H F') \star_H (\alpha \otimes_V \alpha') = (F \star_H \alpha) \otimes_h (F' \star_H \alpha') \tag{22}$$

(and similarly for the vertical case) and are compatible with composition:

$$(F \otimes_H G) \star_V \alpha = F \otimes_H (G \star_V \alpha) \tag{23}$$

(and analogously for vertical composition). They also satisfy additional compatibility conditions: the left and right actions of both vertical and horizontal 2-morphisms satisfy the "associativity" properties,

$$\alpha \star (S \star \beta) = (\alpha \star S) \star \beta \tag{24}$$

for both $\star_H$ *and* $\star_V$. *Moreover, horizontal and vertical actions are independent:*

$$\alpha \star_H (\beta \star_V S) = \beta \star_V (\alpha \star_H S) \tag{25}$$

and similarly for the right action.

Finally, the composition of squares agrees with the associators for composition *by the action in the sense that given three composable squares F, G, and H:*

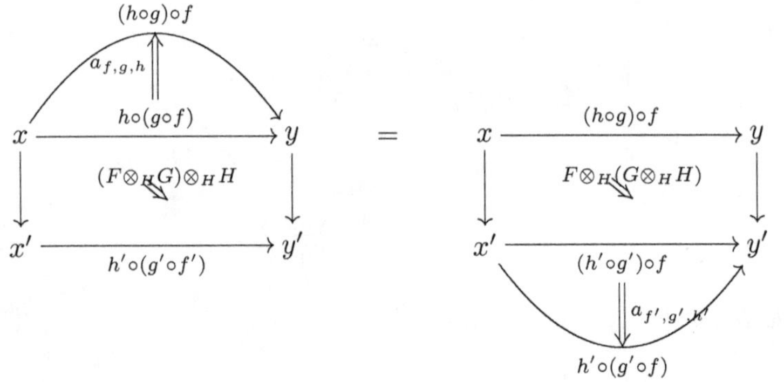

(26)

and similarly for vertical composition. Likewise, unitors in the horizontal and vertical bicategories agree with the identity for composition of squares:

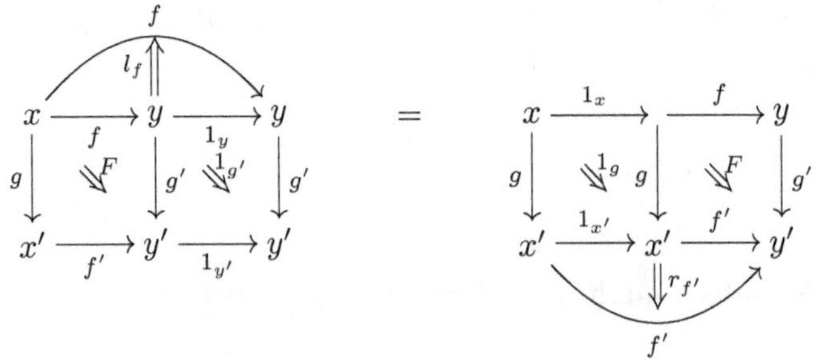

(27)

and similarly for vertical unitors.

Remark 3.2.2. This is rather more unwieldy than the definition of either a bicategory or a double category, but is simpler than a similarly elementary description of a double bicategory (in the sense of Section 3.1) would be. In particular, where there are compatibility conditions involving equations in this definition, a double bicategory would have only higher isomorphisms, themselves satisfying additional coherence laws. In particular, in Verity double bicategories, the action of 2-morphisms on squares is described by strict equations, rather than by a specified isomorphism satisfying coherence laws.

To help make sense of this definition, we note that it is possible (following [**Ve**, sec. 1.4]) to define categories $\mathbf{Cyl_H}$ (respectively, $\mathbf{Cyl_V}$) of *cylinders*. The objects of these categories are squares, and maps are pairs of vertical (respectively, horizontal) 2-morphisms joining the vertical (respectively, horizontal) source and targets of pairs of squares which share the other two sides (this is shown in Table 2, in Section 3.3: the cylinders are "thin" versions of higher morphisms appearing there). These are categories in the usual sense, with strict associativity and unit laws. These conditions would be weakened in a double bicategory (in which maps would include not just pairs of 2-morphisms, but also a 3-dimensional interior of the cylinder, which is a morphism in 2 Mor, or 2-morphism in Mor, satisfying properties only up to a 4-dimensional 2-morphism in 2 Mor).

3.3. Decategorification

The main idea we are pursuing in this section is that of "decategorification". This rather vague term refers to a process in which category-theoretic information is discarded from a structure. Typically, it refers to replacing isomorphisms with equations—for example, a decategorification of the category of finite sets is the set of their cardinalities, $\mathbb{N}$. Similarly, one can turn a bicategory into a category by taking new morphisms to be 2-isomorphism classes of old morphisms, and discarding all 2-morphisms.

To establish a relationship between the apparently conflicting terms for the two types of "double bicategory", we will now show how a Verity double bicategory can arise as a decategorification of a double bicategory satisfying some conditions. We will show later that this can be done with the double cospan examples of Section 4. The conditions which are needed allow us to speak of the "action of 2-cells upon squares". To see what these are, we first consider a "lower dimensional" example of a similar process. What we want to do to obtain Verity double bicategories has an analog in the case of double categories.

	Obj	**Mor**
Objects	$\bullet^{x}$	$\bullet \xrightarrow{f} \bullet$
Morphisms	$\bullet \overset{g}{\downarrow} \bullet$	$\begin{array}{ccc} \bullet & \longrightarrow & \bullet \\ \downarrow & \Downarrow F & \downarrow \\ \bullet & \longrightarrow & \bullet \end{array}$

Table 1: Data of a Double Category

In a double category, thought of as an internal category in **Cat**, we have data of four sorts, as shown in Table 1. That is, a double category **DC** has categories **Obj** of objects and **Mor** of morphisms. The first column of the table shows the data of **Obj**: its objects are the objects of **DC**; its morphisms are the *vertical* morphisms. The second column shows the data of **Mor**: its objects are the *horizontal* morphisms of **DC**; its morphisms are the squares of **DC**.

There is a condition we can impose which effectively turns the double category into a category, where the horizontal and vertical morphisms are composable, and the squares can be ignored. The sort of condition involved is similar to the *horn-filling conditions* introduced by Ross Street [**St**] in his first introduction of the idea of weak ω-categories, or *quasicategories*, in which all morphisms are n-simplexes for some n. A horn filling condition says that, given some hollow simplex with just one face (morphism) missing from the boundary, there will be a morphism to fill that face, and a compatible "filler" for the inside of the simplex. In a double category, there is an analogous "niche-filler" condition.

Definition 3.3.1. *A double category $\mathcal{DC}$ satisfies the* **composability condition** *if the following holds. For any pair (f,g) of a horizontal and vertical morphism where* *the target object of f is the source object of g, there is a unique pair $(h, \star)$ consisting of a unique vertical morphism h and unique invertible square $\star$ making the following diagram commute:*

$$\begin{array}{ccc} x & \xrightarrow{f} & y \\ {\scriptstyle h}\,\vert & \Searrow^{\star} & \downarrow{\scriptstyle g} \\ z & \xrightarrow[1_z]{} & z \end{array} \tag{28}$$

and similarly when the source of f is the target of g.

Notice that taking f to be the identity in this condition implies $\star$ is the identity square. This defines a composition:

Theorem 3.3.2. *If $\mathcal{DC}$ satisfies the composability condition and there are no other squares in $\mathcal{DC}$, there is a category $\mathcal{DC}_0$ with the same objects as $\mathcal{DC}$, and all horizontal and vertical morphisms as its morphisms.*

Proof. Begin by defining composition from $\star$, so that if f is horizontal and g is vertical, then $g \circ f = h$. If f and g are both horizontal (or both vertical), then define $g \circ f$ to be the usual composite. Then this composition is associative and has identities. We only need to check this for the composition using $\star$. For example, given morphisms as in the diagram:

$$\begin{array}{ccccc} w & \xrightarrow{f} & x & \xrightarrow{f'} & y \\ & & & & \downarrow{\scriptstyle g} \\ z & \xrightarrow[1_z]{} & z & \xrightarrow[1_z]{} & z \end{array} \tag{29}$$

there are two ways to use the unique-filler principle to fill this rectangle. One way is to first compose the pairs of horizontal morphisms on the top and bottom, then fill the resulting square. The square we get is unique, and the morphism is denoted $g \circ (f' \circ f)$. The second way is to first fill the right-hand square, and then using the unique morphism we call $g \circ f'$, we get another square on the left hand side, which our principle allows us to fill as well. The square is unique, and the resulting morphism is called $(g \circ f') \circ f$. Composing the two squares obtained this way must give the square obtained the other way, since both make the diagram commute, and both are unique. So we have:

$$\begin{array}{ccccc} w & \xrightarrow{f} & x & \xrightarrow{f'} & y \\ {\scriptstyle (g\circ f')\circ f}\,\vert & \Searrow^{\star} & \vert & \Searrow^{\star} & \downarrow{\scriptstyle g} \\ z & \xrightarrow[1_z]{} & z & \xrightarrow[1_z]{} & z \end{array} \quad = \quad \begin{array}{ccc} w & \xrightarrow{f'\circ f} & y \\ {\scriptstyle g\circ(f'\circ f)}\,\vert & \Searrow^{\star} & \downarrow{\scriptstyle g} \\ z & \xrightarrow[1_z]{} & z \end{array} \tag{30}$$

□

Remark 3.3.3. Note that the composability condition does not require a square for every possible combination of source and target morphisms. In particular, there must be an identity morphism on the boundary of the square—on the bottom in (28). If instead of the identity 1_z, one could have any morphism h, then by choosing f and g to be identities, this would imply that every morphism must be invertible (at least weakly), since there must then be an h^{-1} with $h^{-1} \circ h$ isomorphic to the identity, but of course we do not insist that all morphisms should have inverses. When a filler square does exist, it indicates there is a commuting square in $\mathcal{DC}_0$: the square $\star$ becomes an equation between the composites along the upper right and lower left.

The decategorification of a double bicategory to give a Verity double bicategory is similar, except that with a double category we were removing only the squares (the lower-right quadrant of Table 1). There will be a similar condition to satisfy, but we need to do more with a double bicategory, since there are more sorts of data (and, therefore, a more complex condition). These fall into a similar arrangement, as shown in Table 2.

	Obj	**Mor**	**2Mor**
Objects		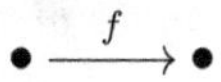	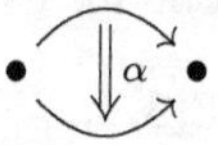
Morphisms	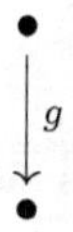	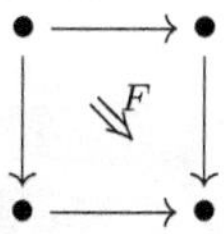	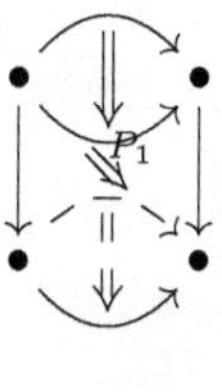
2-Cells	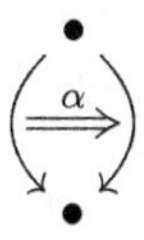	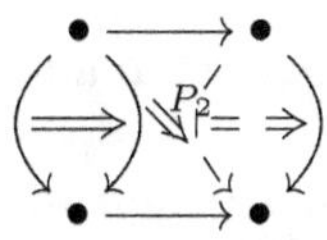	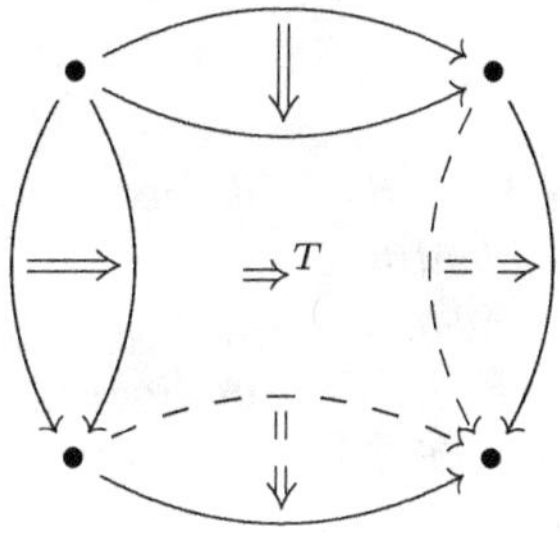

Table 2: The data of a double bicategory

This table shows the data of the bicategories **Obj**, **Mor**, and **2Mor**, each of which has objects, morphisms, and 2-cells. Note that the morphisms in the three entries in the lower right hand corner—2-cells in **Mor**, and morphisms and 2-cells in **2Mor**—are not 2-dimensional. The 2-cells in **Mor** and morphisms in **2Mor** are the three-dimensional "filling" inside the illustrated cylinders, which each have two square faces and two bigonal faces. The 2-cells in **2Mor** should be drawn as 4-dimensional. The picture illustrated can be thought of as taking both square faces of one cylinder P_1 to those of another, P_2, by means of two other cylinders (S_1 and S_2, say), in such a way that P_1 and P_2 share their bigonal faces. This description works whether we consider the P_i to be horizontal and the S_j vertical, or vice versa. These describe the "frame" of this sort of morphism: the filling is the 4-dimensional "track" taking P_1 to P_2, or equivalently, S_1 to S_2, just as a square in a double category can be read horizontally or vertically. (Not all relevant parts of the diagrams have been labeled here, for clarity.)

Next we want to describe a condition similar to the composability condition for a double category. In that case, we got a condition which effectively allowed us to treat any square as an identity, so that we only had objects and morphisms. Here, we want a condition which lets us throw away the three entries of dimension greater than two in Table 2 in the bottom right. This condition, when satisfied, should allow us to treat a double bicategory as a Verity double bicategory. It comes in three parts, one for each type of data we want to discard:

Definition 3.3.4. *We say that a double bicategory satisfies the* ***vertical action condition*** *if, for any morphism $F_1 \in$ **Mor** and 2-morphism $\alpha \in$ **Obj** such that $s(F_1) = t(\alpha)$, there is a unique morphism $F_2 \in$ **Mor** and unique invertible 2-morphism $P \in$ **Mor** such that P fills the "pillow diagram":*

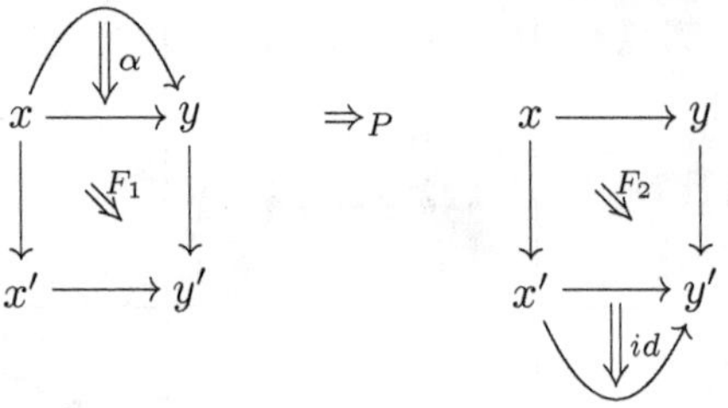

(31)

*where F_2 is the back face of this diagram, and the 2-morphism in **Obj** at the bottom is the identity.*

A double bicategory satisfies the ***horizontal action condition*** *if for any morphism $F_1 \in$ **Mor** and object α in **2Mor** with $s(F_1) = t(\alpha)$ there is a unique morphism $F_2 \in$ **Mor** and unique invertible morphism $P \in$ **2Mor** such that P fill the pillow diagram:*

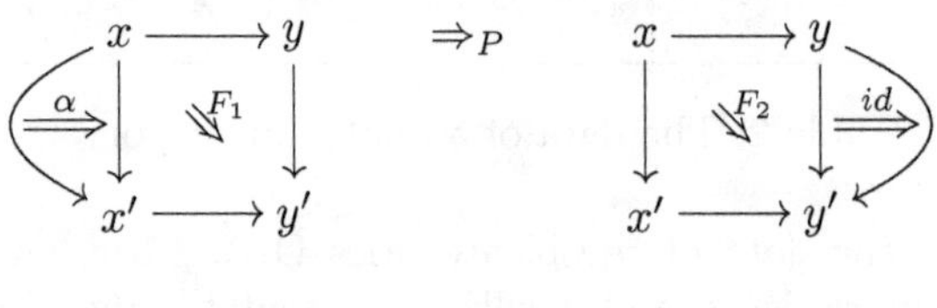

(32)

In (31), F_2 is the square which will eventually be named $F_1 \star_H \alpha$ when we define an action of 2-cells on squares, and in (32), F_2 is the square will eventually be named $F_1 \star_V \alpha$.

Remark 3.3.5. One can see that this condition is analogous to the filler condition (28) in a double category by imagining the diagram (31) viewed obliquely. The diagram says that given a square with two bigons—the top one arbitrary and the bottom one the identity—there is another square F_2 (the back face of a pillow diagram) and a filler 2-morphism $P \in$ **2Mor** which fills the diagram. If one imagines turning this diagram on its side and viewing it obliquely, one sees precisely (28), as a dimension has been suppressed. The role played by cylinders (2-morphism in **2Mor**) in (31) and (32 is played by a square in (28); the roles of both squares and bigons in (31) and (32) are played by arrows in (28); the role of arrows in (31) and (32) is filled by point-like objects in (28).

This gives horizontal and vertical actions, but to get the compatibility between them, we need a further condition. In particular, since these conditions involve both horizontal and vertical cylinders, the compatibility condition must correspond to the 4-dimensional 2-cells in **2Mor**, shown in the lower right corner of Table 2.

To draw the necessary condition is difficult, since the necessary diagram is four-dimensional, but we can describe it as follows:

Definition 3.3.6. *We say a double bicategory satisfies the* ***action compatibility condition*** *if the following holds. Suppose we are given*

- *a morphism $F \in$* **Mor**
- *an object $\alpha \in$* **2Mor** *whose target in* **Mor** *is a source of F*
- *a 2-cell $\beta \in$* **Obj** *whose target morphism is a source of F*
- *an invertible morphism $P_1 \in$* **2Mor** *with F as source, and the objects α and* id *in* **2Mor** *as source and target*
- *an invertible 2-cell $P_2 \in$* **Mor** *with F as source, and the 2-cells β and* id *in* **Mor** *as source and target*

where P_1 and P_2 have, as targets, morphisms in **Mor** *we call $\alpha \star F$ and $\beta \star F$ respectively. Then there is a unique morphism $\hat{F}$ in* **Mor** *and unique invertible 2-cell T in* **2Mor** *having all of the above as sources and targets.*

Geometrically, the unique 2-cell in **2Mor** looks like the structure in the bottom right corner of Table 2. This can be seen as taking one horizontal cylinder to another in a way that fixes the (vertical) bigons on its sides. It does this by means of a translation which acts on the front and back faces with a pair of vertical cylinders (have the same top and bottom bigonal faces). Alternatively, it can be seen as taking one vertical cylinder to another, acting on the faces with a pair of horizontal cylinders. In either case, the cylinders involved in the translation act on the faces, but the four-dimensional interior, T, acts on the original cylinder to give another. The simplest interpretation of this condition is that it is precisely the condition needed to give the compatibility condition (25).

Remark 3.3.7. Notice that the two conditions given imply the existence of unique data of three different sorts in our double bicategory. If these are the only data of these kinds, we can effectively omit them (since it suffices to know information about their sources and targets). This omission is part of a decategorification of the same kind we saw for a double category **DC**.

In particular, we show how a double bicategory **D** satisfying the above conditions gives a Verity double bicategory. We know that **D** consists of bicategories $(\mathbf{Obj}, \mathbf{Mor}, \mathbf{2Mor})$ together with all required maps (three kinds of source and target maps, two kinds of identity, three partially-defined compositions, left and right unitors, and the associator), satisfying the usual properties. To begin with, we describe how the elements of a Verity double bicategory **V** (Definition 3.2.1) arise from this:

Definition 3.3.8. *If* **D** *is a double bicategory satisfying the horizontal and vertical action conditions and the action compatibility condition, then* $\mathbf{V(D)}$ *is the Verity double bicategory with:*

- *The objects* Obj *are the objects of* **Obj**.
- *The horizontal bicategory* **Hor** *of* $\mathbf{V(D)}$ *is* **Obj**
- *The vertical bicategory* **Ver** *of* $\mathbf{V(D)}$ *has:*
  - *Objects: Objects of* **Obj**
  - *Morphisms: Objects of* **Mor**
  - *2-morphisms: Objects of* **2Mor**

  The source, target and composition maps for **Ver** *are the object maps from the source, target, and composition 2-functors for* **D**.
- *The squares* **Squ** *of* $\mathbf{V(D)}$ *are isomorphism classes of morphisms of* **Mor**. *These are equipped with:*
  - *Vertical source and target maps: the morphism maps from the functors* $s, t : \mathbf{Mor} \to \mathbf{Obj}$.
  - *Horizontal source and target maps: the internal ones in* **Mor**.
  - *Horizontal composition (17): the composition of morphisms in* **Mor**.
  - *Vertical composition (16): the morphism maps for the partially defined functor* $\circ$ *for* **Mor**,
  - *Horizontal Identity: The identity square for a morphism* g *in* **Ver** *(i.e.* g *an object in* **Mor** *is* $1_f \in \mathbf{Mor}$.
  - *Vertical Identity: The identity square for a morphism* f *in* **Hor** *(i.e. a morphism* f *in* **Obj***) is given by* $\mathrm{id}(f)$ *for the unit functor* $\mathrm{id} : \mathbf{Obj} \to$ Mor.
- *The horizontal action defines* $F \star_H \alpha$ *to be (the isomorphism class of) the unique morphism in* **Mor** *whose existence is required by the horizontal action condition.*
- *The vertical action defines* $F \star_V \alpha$ *to be (the isomorphism class of) the unique morphism in* **Mor** *whose existence is required by the vertical action condition.*

Of course, we must check this is really a Verity double bicategory:

Theorem 3.3.9. *Suppose* **D** *is a double bicategory satisfying the horizontal and vertical action conditions and the action compatibility condition. Then* **V(D)** *is a Verity double bicategory.*

Proof. We check all the properties in the definition of a Verity double bicategory:

- By assumption, **Hor** is a bicategory.
- **Ver** is a bicategory since the source and target functors in **D** for **Ver** satisfy all the usual axioms for a bicategory, hence their object maps do also. Similarly, the composition maps have natural isomorphisms giving associators and unitors: they are just object maps of functors which satisfy the same conditions: in **D**, the associator a satisfies the pentagon identity. The object maps for a give the associator in **Ver**. Since the associator 2-natural transformation satisfies the pentagon identity, so do these object maps. The other properties are shown similarly, so that **Ver** is a bicategory.
- The source and target maps for **Squ** satisfies equations (15) because the source and target maps of **D** are functors.
- The composition lawsfor squares have the usual relation to source and target maps because, by assumption, **Mor** is a bicategory, but taking **Squ** to be 2-isomorphism classes of morphisms in **Mor**, and disregarding all other 2-morphisms, we get that horizontal composition in **Squ** is exactly associative and has exact identities, so the squares are the morphisms of a category with respect to horizontal composition.
  Vertical composition for squares in **D** satisfies the axioms for a bicategory by the same argument as given above for **Ver**, since it is the morphism map for the functor $\circ$. In particular, it has an associator and a unitor: but these must be morphisms in **2Mor** since we take the morphism maps from the associator and unitor functors for $\circ$. These are 2-isomorphisms, but since we defined squares to be 2-isomorphism classes (any isomorphism in **2Mor** becomes an equation), this composition is exactly associative and has a unit. Also, we disregard any morphisms in **2Mor**, so the squares are the morphisms of a category under vertical composition.
- The interchange rule (18) follows from functoriality of the composition functors.
- The actions $\star_H$ and $\star_V$ defined by the horizontal and vertical action conditions is well defined. In particular, by composition of in **Mor** or **2Mor**, we guarantee the existence of the categories of horizontal and vertical cylinders $\mathbf{Cyl_H}$ and $\mathbf{Cyl_V}$, respectively. These come from the 2-morphisms in **Mor** or morphisms in **2Mor** respectively which those conditions demand must exist. Taking these to be identities, the cylinders consist of commuting cylindrical diagrams with two bigons and two squares.
  In the case where one bigon is the identity, and the other is any bigon α, the conditions guarantee the existence of an invertible cylinder, which is now the identity because we have taken squares to be isomorphism classes. This defines the effect of the action of α on the square whose source is the target of α. If this square is F, we denote the other square $\alpha \star_H F$ or $\alpha \star_V F$ as appropriate.

- The horizontal action condition gives a well-defined action satisfies (22) and (23) by an argument exactly analogous to that in the proof of Proposition 3.3.2. That is, the horizontal action condition means that certain fillers are unique. When they can be obtained in two ways, these are equal.
- The vertical action satisfies the vertical equivalent of (22) and (23) for the same reason.
- The condition (25) guaranteeing independence of the horizontal and vertical actions follows from the action compatibility condition. For suppose we have a square F whose horizontal and vertical source arrows are the targets of 2-cells α and β, and attach to its opposite faces two identity 2-cells. Then the horizontal and vertical action conditions mean that there will be a square $\alpha \star_H F$ and a square $\beta \star_V F$). Then the action compatibility condition applies (the P_i are the identities we get from the action condition), and there is a morphism in **Mor**, namely a square in **V** and a 2-cell $T \in$ **2Mor**. Consider the remaining face, which the action condition suggests we call $\alpha \star_H (\beta \star_V F)$ or $\beta \star_V (\alpha \star_H F)$, depending on the order in which we apply them. The compatibility condition says that there is a unique square which fills this spot so the two must be equal.
- We next check that composition for squares agrees with composition as in (26). Suppose we have three composable squares—that is, morphisms F, G, and H in **Mor**, which are composable along shared source and target objects in **Mor**. The associator functor has an object map, giving objects in **2Mor** at the "top" and "bottom" of the squares. It also has a morphism map, giving morphisms in **2Mor**. But by assumption there is only a unique such map between , these associators must be the unique morphism in **2Mor** with source $(H \circ G) \circ F$ and target $H \circ (G \circ F)$. Then by the vertical action condition, we have a filler 2-morphism in **Mor** for the action on the composite square by the top associator, and then, taking the result and composing with the bottom associator, we get another filler. This must be the unique map between the two composites, which is the identity since they have the same sources and targets. So we get a commuting cylinder. Composing squares along source and target morphisms in **Obj** works the same way by a symmetric argument.
- The condition (27) is similar. The unitor functor will give the unique morphism in **2Mor**, and the action compatibility condition gives the commuting cylinder for unitors on the composite of squares.

So indeed the construction of **V(D)** defines a Verity double bicategory. □

Next, in Section 3.4, we continue the process of reducing the complexity of these structures. In particular, we see how Verity double bicategories can give rise to ordinary bicategories, which are frequently easier to use.

3.4. Bicategories from Double Bicategories

It is well known that double categories can yield 2-categories in three different ways. Two obvious cases are when there are only identity horizontal morphisms, or only identity vertical morphisms, so that squares simply collapse into bigons with the two nontrivial sides. Notice that it is also true that a Verity double bicategory in which **Hor** is trivial (equivalently, if **Ver** is trivial) is again a bicategory. The squares become 2-morphisms in the obvious way, the action of 2-morphisms on squares is then just composition, and the composition rules for squares in the double category become the rules for composing 2-morphisms, and the result is clearly a bicategory.

The other, less obvious, case, is when the horizontal and vertical categories on the objects are the same: this is the case of *path-symmetric* double categories, and the recovery of a bicategory was shown by Brown and Spencer [**BS**]. Fiore [**Fi**] shows how their demonstration of this fact is equivalent to one involving *folding structures.*

In this case we can interpret squares as bigons by composing the top and right edges, and the left and bottom edges. Introducing identity bigons completes the structure. These new bigons have a natural composition inherited from that for squares. It turns out that this yields a bicategory. Here, our goal will be to show half of an analogous result, that a Verity double bicategory similarly gives rise to a bicategory when the horizontal and vertical bicategories are equal. We will also show that a double bicategory for which the horizontal (or vertical) bicategory is trivial can be seen as a bicategory. The condition that $\mathbf{Hor} = \mathbf{Ver}$ will hold in our general example of double cospans.

Theorem 3.4.1. *Any Verity double bicategory*

$$\mathbf{V} = (\mathrm{Obj}, \mathbf{Hor}, \mathbf{Ver}, \mathbf{Squ}, \otimes_H, \otimes_V, \star_H, \star_V)$$

for which $\mathbf{Hor} = \mathbf{Ver}$ *produces a bicategory* $\mathbf{B}$ *by taking the 2-morphisms to be 2-morphisms in* **Hor** *and squares in* **Squ**.

Proof. We begin by defining the data of **B**. Its objects and morphisms are the same as those of **Hor** (equivalently, **Ver**). We describe the 2-morphisms by observing that **B** must contain all those in **Hor** (equivalently, **Ver**), but also some others, which correspond to the squares in **Squ**.

In particular, given a square

$$\begin{array}{ccc} a & \xrightarrow{f} & b \\ {\scriptstyle g}\downarrow & \Downarrow S & \downarrow{\scriptstyle g'} \\ c & \xrightarrow[f']{} & d \end{array} \tag{33}$$

there should be a 2-morphism

$$a \underset{f' \circ g}{\overset{g' \circ f}{\Downarrow S}} d \tag{34}$$

The composition of squares corresponds to either horizontal or vertical composition of 2-morphisms in **B**, and the relation between these two is given in terms of the interchange law in a bicategory:

Given a composite of squares,

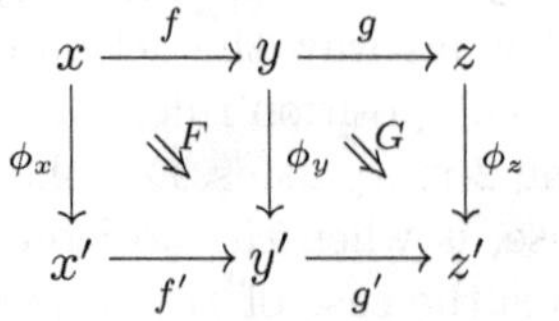

(35)

there will be a corresponding diagram in **B**:

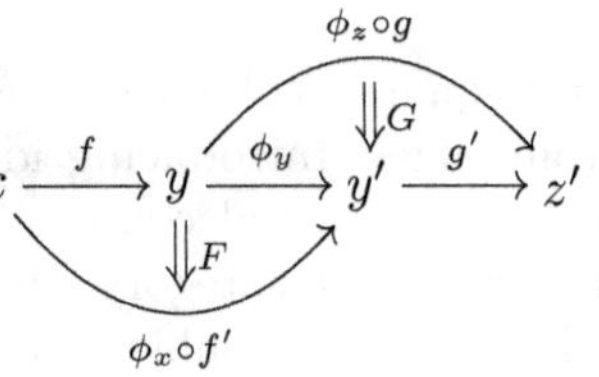

(36)

Using horizontal composition with identity 2-morphisms ("whiskering"), we can write this as a vertical composition:

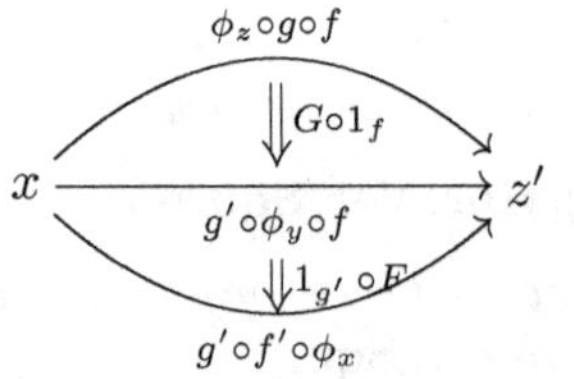

(37)

So the square $F \otimes_H G$ corresponds to $(1 \circ G) \cdot (F \circ 1)$ for appropriate identities 1. Similarly, the vertical composite of $F' \otimes_V G'$ must be the same as $(1 \circ F) \cdot (G \circ 1)$. Thus, every composite of squares which can be built from horizontal and vertical composition, gives a corresponding composite of 2-morphisms in **B**, which are generated by those corresponding to squares in **Squ**, subject to the relations imposed by the composition rules in a bicategory.

Now we want to show that Verity double bicategory **V** gives the entire bicategory **B**. That is, that **B** has no other 2-morphisms than those which arise by the above process. It suffices to show that all such 2-morphisms not already in **Hor** arise as squares (that is, the structure is closed under composition). So suppose we have any composable pair of 2-morphisms which arise from squares F and G. If F and G have an edge in common, then we have the situation depicted above (or possibly the corresponding form in the vertical direction). In this case, the composite 2-morphism corresponds exactly to the composite of squares, and the axioms for composition of squares ensure that all 2-morphisms generated this way are already in our bicategory. In particular, the unit squares become unit 2-morphisms when composed with left and right unitors.

Now, if there is no edge in common to two squares, the 2-morphisms in **B** must be made composable by whiskering with identities. In this case, all the identities can be derived from 2-morphisms in **Hor**, or from identity squares in **Squ** (inside commuting diagrams). Clearly, any identity 2-morphism can be factored this way. Then, again, the composite 2-morphisms in **B** will correspond exactly to the composite of all such squares in **Squ** and 2-morphisms **Hor**.

Finally, the associativity condition (26) for the action of 2-morphisms on squares ensures that composition of squares agrees with that for 2-morphisms, so there are no extra squares from composites of more than two squares. □

Remark 3.4.2. When producing the bicategory **B** from **V**, we made a particular choice of orientation for the 2-morphisms obtained from squares. The square $S \in$ **Squ** shown in (33) has vertical source f and target f', and horizontal source g and target g'. However, the corresponding 2-morphism $S \in \mathbf{B}$ has source $g' \circ f$, which combines vertical source and horizontal target; on the other hand, the target of $S \in \mathbf{B}$ is $f' \circ g$, combining vertical target and horizontal source. We could equally well have chosen the opposite convention. This would give $\mathbf{B}^{\mathrm{co}}$, which is **B** with the orientation of its 2-morphisms reversed. (See, e.g. [**Le**]).

It is also worth considering here the situation of a double bicategory in which all horizontal morphisms and 2-morphisms are identities. In this case, one can define a 2-morphism from a square with and bottom edges being identities, whose source is the object whose identity is the corresponding edge, and similarly for the target. The composition rules for squares in the vertical direction, then, are just the same as those for a bicategory. Likewise, the axioms for action of a 2-morphism on a square reduce to the composition laws for a bicategory if one replaces the square by a 2-cell.

Next we describe a broad class of examples of double bicategories, in the spirit of the use of spans to give examples of bicategories.

4. Double Cospans

In Remark 2.1.5 we described Bénabou's demonstration that Span(**C**) is a bicategory for any category **C** with pullbacks. Similarly, there is a bicategory of cospans in a category **C** with pushouts. There will be an analogous fact giving a double bicategory of double spans. In fact, we describe this in terms of double *cospans*, since our aim in a subsequent paper will be to use these to describe cobordisms, which have a natural description as cospans. Since cospans in **C** are the same as spans in the opposite category, $\mathbf{C}^{\mathrm{op}}$, this distinction is a matter of taste.

We remark here that similar constructions are described by Grandis [**Gr3**], and related "profunctor-based examples" of pseudo-double categories are described by Grandis and Paré [**GP2**].

4.1. The Double Cospan Example

We begin by defining a double bicategory of double cospans:

Definition 4.1.1. 2Cosp(**C**) *is a double bicategory of* ***double cospans*** *in* **C**, *consisting of the following:*

- *the bicategory of objects is* **Obj** $=$ Cosp(**C**)
- *the bicategory of morphisms* **Mor** *has:*

– *as objects, cospans in* **C**;
– *as morphisms, commuting diagrams of the form*

$$\begin{array}{ccccc} X_1 & \longrightarrow & S & \longleftarrow & X_2 \\ \downarrow & & \downarrow & & \downarrow \\ T_1 & \longrightarrow & M & \longleftarrow & T_2 \\ \uparrow & & \uparrow & & \uparrow \\ X'_1 & \longrightarrow & S' & \longleftarrow & X'_2 \end{array} \qquad (38)$$

(in subsequent diagrams we suppress the labels for clarity);
– *as 2-morphisms, cospans of cospan maps, namely commuting diagrams of the following shape:*

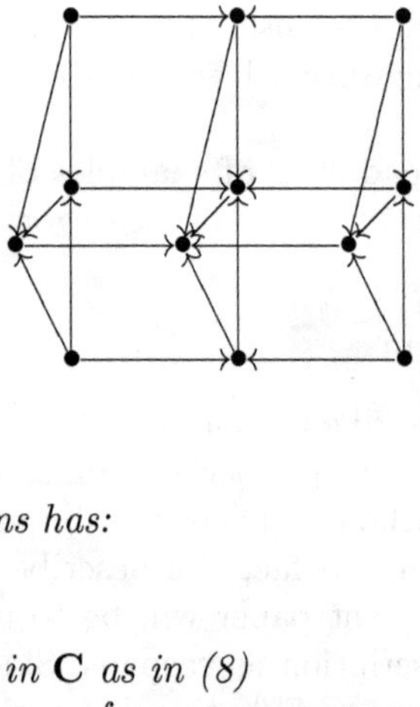

(39)

- *the bicategory of 2-morphisms has:*

– *as objects, cospan maps in* **C** *as in (8)*
– *as morphisms, cospan maps of cospans:*

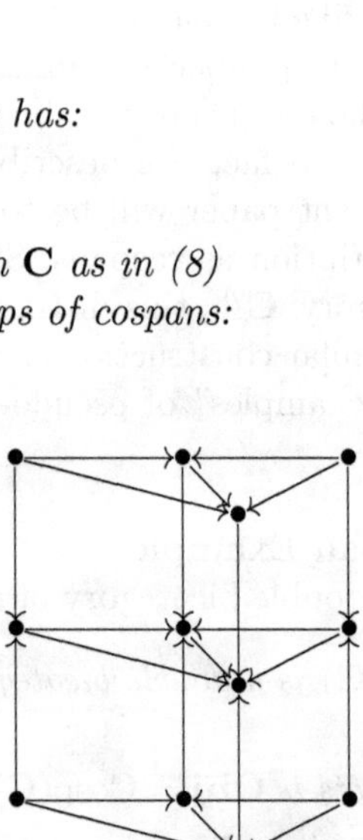

(40)

– *as 2-morphisms, cospan maps of cospan maps:*

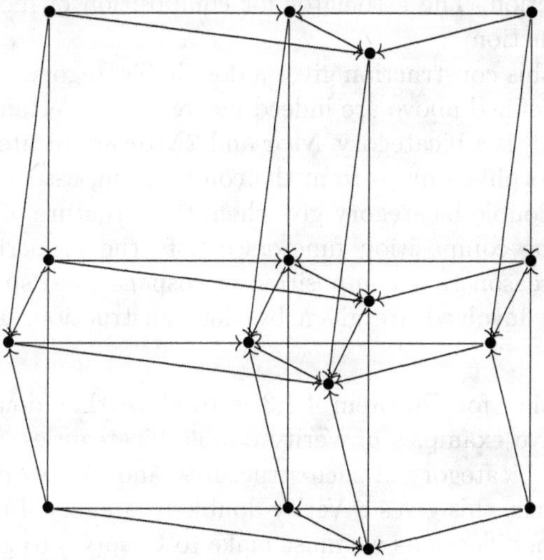

(41)

All composition operations are by pushout; source and target operations are the same as those for cospans. The associators and unitors in the horizontal and vertical bicategories are the maps which come from the universal property of pushouts.

Remark 4.1.2. Just as 2-morphisms in **Mor** and morphisms in **2Mor** can be seen as diagrams which are "products" of a cospan with a map of cospans, 2-morphisms in **2Mor** are given by diagrams which are products (as diagrams) of horizontal and vertical cospan maps. These have, in either direction, four maps of cospans, with objects joined by maps of cospans. Composition again is by pushout in composable pairs of diagrams.

Note that all these diagrams are products of smaller diagrams, each of which is either a cospan, or a cospan map. This suggests that the horizontal and vertical directions should in some way behave like a bicategory of cospans. The next theorem shows this is indeed the case:

Theorem 4.1.3. *For any category* **C** *with pushouts,* 2Cosp(**C**) *forms a double bicategory.*

Proof. **Mor** and **2Mor** are bicategories since the composition functors act just like composition in Cosp(**C**), the bicategory of cospans in **C**, in each column, and therefore satisfies the same axioms.

Now, the horizontal and vertical directions have composition operations defined in the same way. Thus we can construct functors between **Obj**, **Mor**, and **2Mor** with the properties of a bicategory simply by using the same constructions that turn each into a bicategory in its own right. In particular, the source and target maps $s, t : \mathbf{Mor} \to \mathbf{Obj}$ and $s, t : \mathbf{2Mor} \to \mathbf{Mor}$ are the obvious maps giving the domains of the maps in (38). The partially defined (horizontal) composition maps $\circ : \mathbf{Mor}^2 \to \mathbf{Mor}$ and $\otimes_H : \mathbf{2Mor}^2 \to \mathbf{2Mor}$ are defined by taking pushouts of diagrams in **C**, which exist for any composable pairs of diagrams because **C** has pushouts. They are functorial since they are independent of composition in the horizontal direction. The associator for composition of morphisms is given in the pushout construction.

To see that this construction gives a double bicategory, we note that **Obj**, **Mor**, and **2Mor** as defined above are indeed bicategories. Certainly, **Obj** is a bicategory because Cosp(**C**) is a bicategory. **Mor** and **2Mor** are bicategories because the morphism and 2-morphism maps from the composition, associator, and other functors required for a double bicategory give them the structure of bicategories as well.

Moreover, the composition functors satisfy the properties of a bicategory for just the same reason that composition of cospans (and spans) does, since each of the three maps involved are given by this construction. Thus, we have a double bicategory. □

Our motivation for Theorem 4.1.3 is to show that double cospans in suitable categories **C** give examples of Verity double bicategories. We have described how to get a double bicategory of such structures, and we saw in Section 3.3 that given certain conditions, this gives a Verity double bicategory. In Section 4.2 we describe explicitly the modifications we must make to Cosp(**C**) to get these conditions.

4.2. A Verity Double Bicategory of Double Cospans

As we saw in Section 3.3, double bicategories have higher morphisms of dimension up to 4, but given certain conditions, these can be omitted to give a Verity double bicategory.

Definition 4.2.1. *For a category* **C** *with pushouts, the Verity double bicategory* $2\mathrm{Cosp}(\mathbf{C})_0$*, has:*

- *the objects are objects of* **C**
- *the horizontal and vertical bicategories* **Hor** = **Ver** *are both equal to a sub-bicategory of* Cosp(**C**)*, which includes only invertible cospan maps*
- *the squares are isomorphism classes of commuting diagrams of the form (38)*

where two diagrams of the form (38) are isomorphic if they differ only in the middle objects, say M and M', and the maps into these objects, and if there is an isomorphism $f : M \to M'$ making the combined diagram commute.

The action of 2-morphisms α in **Hor** *and* **Ver** *on squares is by composition in diagrams of the form:*

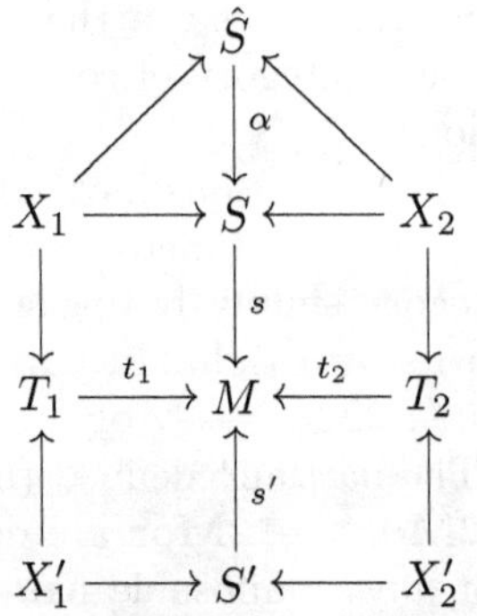

(42)

(where the resulting square is as in 38, with $\hat{S}$ in place of S and $s \circ \alpha$ in place of s).

Composition (horizontal or vertical) of squares of cospans is, as in 2Cosp(**C**), *given by composition (by pushout) of the three cospans of which the square is composed. The composition for diagrams of cospan maps are as usual in* Cosp(**C**).

Remark 4.2.2. Notice that **Hor** and **Ver** as defined are indeed bicategories: eliminating all but the invertible 2-morphisms leaves a collection which is closed under composition by pushouts.

We will show more fully that this is a Verity double bicategory in Theorem 4.2.4. First one must show that horizontal and vertical composition of squares is well defined is defined on equivalence classes. We will get this result indirectly as a result of Theorems 4.1.3 and 3.3.9, but it is instructive to see directly how this works in Cosp(**C**).

Theorem 4.2.3. *In any category with pushouts, composition of squares in Definition 4.2.1 is well-defined.*

Proof. Suppose we have two representatives of a square, bounded by horizontal cospans $X_1 \to S \leftarrow X_2$ and $X_1' \to S' \leftarrow X_2'$, and vertical cospans $X_1 \to T_1 \leftarrow X_1'$ and $X_2 \to T_2 \leftarrow X_2'$. Suppose the middle objects are M and $\hat{M}$ as in the diagram (38). Given a composable diagram which coincides along an edge (morphism in **Hor** or **Ver**) with the first, we need to know that the two pushouts of the different representatives are also isomorphic (that is, represent the same composite square).

In the horizontal and vertical composition of these squares, the maps to the middle object M of the new square from the middle objects of the new sides (given by composition of cospans) arise from the universal property of the pushouts on the sides being composed (and the induced maps from M to the corners, via the maps in the cospans on the other sides). Since the middle objects are defined only up to isomorphism class, so is the pushout: so the composition is well defined, since the result is again a square of the form (38). □

We use this, together with Theorems 3.3.9 and 4.1.3, (proved in Section 3), to show the following:

Theorem 4.2.4. *If* **C** *is a category with pushouts, then* $2\mathrm{Cosp}(\mathbf{C})_0$ *is a Verity double bicategory.*

Proof. In the construction of $2\mathrm{Cosp}(\mathbf{C})_0$, we take isomorphism classes of double cospans as the squares. We also restrict to invertible cospan maps in the horizontal and vertical bicategories.

That is, take 2-isomorphism classes of morphisms in **Mor** in the double bicategory, where the 2-isomorphisms are invertible cospan maps, in both horizontal and vertical directions. We are then effectively discarding all non-invertible morphisms and 2-morphisms in **2Mor**, and all non-invertible 2-morphisms in **Mor**. In particular, there may be "squares" of the form (38) in 2Cosp(**C**) with non-invertible maps joining their middle objects M, but we have ignored these, and also ignore non-invertible cospan maps in the horizontal and vertical bicategories. Thus, we consider no diagrams of the form (39) except those in which the span maps are invertible, in which case the middle objects are representatives of the same isomorphism class. Similar reasoning applies to the 2-morphisms in **2Mor**.

The structure we get from discarding these will again be a double bicategory. In particular, the new **Mor** and **2Mor** will be bicategories, since they are, respectively, just a category and a set made into a discrete bicategory by adding identity morphisms or 2-morphisms as needed. On the other hand, for the composition, source and target maps to be bifunctors, the structures built from the objects, morphisms, and 2-cells respectively must be bicategories. This is since the composition, source, and target maps are the object, morphism, and 2-morphism maps of these bifunctors, which satisfy the usual category axioms. But the same argument applies to those built from the morphisms and 2-cells as within **Mor** and **2Mor**. So we have a double bicategory.

Next we show that the horizontal and vertical action conditions (Definition 3.3.4 of Section 3.3) hold in 2Cosp(**C**). A square in 2Cosp(**C**) is a diagram of the form (38), and a 2-cell is a map of cospans. Given a square M and 2-cell α with compatible source and targets as in the action conditions, we have a diagram of the form shown in (42). Here, M is the square diagram at the bottom, whose top row is the cospan containing S. The 2-cell α is the cospan map including the arrow $\alpha : \hat{S} \to S$. There is a unique square built using the same objects as M, but using the cospan containing $\hat{S}$ as the top row. The map from $\hat{S}$ to M is then $s \circ \alpha$.

To satisfy the action condition, we want this square $\hat{M}$, which is the candidate for $M_1 \star_V \alpha$, to be unique. But suppose there were another $\hat{M}_2$ with a map from $\hat{S}$. Since we are in 2Cosp$(\mathbf{C})_0$, α must be invertible, which would give a map to $\hat{M}_2$ from S. We then find that $\hat{M}_2$ and $\hat{M}$ are representatives of the same isomorphism class, so in fact this is the same square. That is, there is a unique morphism in **2Mor** taking M to $\hat{M}$ (a diagram of the form 40) with invertible cospan maps in the middle and bottom rows. This is the unique filler for the pillow diagram required by definition 3.3.4.

The argument that 2Cosp$(\mathbf{C})_0$ satisfies the action compatibility condition is similar.

So 2Cosp$(\mathbf{C})_0$ is a double bicategory in which, there there is at most one unique morphism in **Mor**, and at most unique morphisms and 2-morphisms in **2Mor**, for any specified source and target, and the horizontal and vertical action conditions hold. So 2Cosp$(\mathbf{C})_0$ can be interpreted as a Verity double bicategory (Theorem 3.3.9). □

Remark 4.2.5. We observe here that the compatibility condition (26) relating the associator in the horizontal and vertical bicategories to composition for squares is due to the fact that the associators are maps which come from the universal property of pushouts. This is by the parallel argument to that we gave for spans in Remark 2.1.5. The same argument applies to the middle objects of the squares, and gives associator isomorphisms for that composition. When we reduce to isomorphism classes, these isomorphisms become identity maps, so we get a commuting pillow as in (26). A similar argument shows the compatibility condition for the unitor, (27).

It is interesting to note how the arguments in the proof of Theorem 3.3.9 apply to the case of 2Cosp(**C**).

In particular, the interchange rules hold because the middle objects in the four squares being composed form the vertices of a new square. The pushouts in the vertical and horizontal direction form the middle objects of vertical and horizontal cospans over these. The interchange law means that the pushout (in the horizontal direction) of the objects from the vertical cospans is in the same isomorphism class as the pushout (in the vertical direction) of the objects from the horizontal cospans. This follows from the universal property of the pushout.

4.3. Example: Cobordisms with Corners

One important example of a category of cospans involves cobordism of manifolds, although to realize this example requires some additional structure. In particular, the category $\mathbf{nCob}_2$ of cobordisms with corners is not 2Cosp(**C**) for a category **C** with pushouts, since the objects of this category are manifolds, and **Man** does not have pushouts.

Recall that two manifolds S_1, S_2 are *cobordant* if there is a compact manifold with boundary, M, such that ∂M is isomorphic to the disjoint union of S_1 and S_2. This M is called a *cobordism* between S_1 and S_2. So in particular, a cobordism is a cospan $S_1 \rightarrow M \leftarrow S_2$, where both maps are inclusions of the boundary components. A cobordism with corners is then a manifold with corners, where the boundary components and corners are included in a double cospan.

In particular, for topological cobordisms (i.e. cobordisms which are topological manifolds with boundary), all the pushouts required to compose such double cospans will still be topological manifolds. For smooth manifolds, to ensure that the result of gluing is smooth we need to specify an additional condition, using "collars" on the boundaries and corners.

4.3.1. Cobordisms and Collars

To begin with, recall that a smooth manifold with corners is a topological manifold with boundary, together with a maximal compatible set of coordinate charts $\phi : \Omega \rightarrow [0, \infty)^n$ - into the positive sector of *latex*$\mathbb{R}^n$. (where ϕ_1, ϕ_2 are compatible if $\phi_2 \circ \phi_1^{-1}$ is a diffeomorphism).

Jänich [**Ja**] introduces the notion of $\langle n \rangle$-manifold, reviewed by Laures [**Lau**]. This is build on a manifold with corners, using the notion of a *face*:

Definition 4.3.1. *(**Jänich**)A **face** of a manifold with corners is the closure of some connected component of the set of points with exactly one zero component in any coordinate chart. An $\langle n \rangle$-manifold is a manifold with faces together with an n-tuple $(\partial_0 M, \ldots, \partial_{n-1} M)$ of faces of M, such that*

- $\partial_0 M \cup \ldots \partial_{n-1} M = \partial M$
- $\partial_i M \cap \partial_j M$ *is a face of* $\partial_i M$ *and* $\partial_j M$

The case we will be interested in here is the case of $\langle 2 \rangle$-manifolds. In this notation, a $\langle 0 \rangle$-manifold is just a manifold without boundary, a $\langle 1 \rangle$-manifold is a manifold with boundary, and a $\langle 2 \rangle$-manifold is a manifold with corners whose boundary decomposes into two components (of codimension 1), whose intersections form the corners (of codimension 2). We can think of $\partial_0 M$ and $\partial_1 M$ as the "horizontal" and "vertical" part of the boundary of M.

Now, for a point $x \in S$, there will be a neighborhood U of x which restricts to $U_1 \subset M_1$ and $U_2 \subset M_2$ with smooth maps $\phi_i : U_i \to [0,\infty)^n$ with $\phi_i(x)$ on the boundary of $[0,\infty)^n$ with exactly one coordinate equal to 0. One can easily combine these to give a homeomorphism $\phi : U \to \mathbb{R}^n$, but this will not necessarily be a diffeomorphism along the boundary S. While *topological* cobordisms can be composed along their boundaries, *smooth* cobordisms M_1 and M_2 should be composed differently, to ensure that every point—including points on the boundary of M_i—will have a neighborhood with a smooth coordinate chart. To solve this problem, we use *collars*, which is also done in the category **nCob**.

The *collaring theorem* says that for any smooth manifold with boundary M, ∂M has a *collar*: an embedding $f : \partial M \times [0,\infty) \to M$, with $(x,0) \mapsto x$ for $x \in \partial M$. This is a well-known result (for a proof, see e.g. [**Hi**], sec. 4.6). It is an easy corollary that we can choose to use the interval $[0,1]$ in place of $[0,\infty)$ here.

Laures ([**Lau**], Lemma 2.1.6) describes a generalization of this theorem to $\langle n\rangle$-manifolds, so that for any $\langle n\rangle$-manifold M, there is an n-dimensional cubical diagram (*$\langle n\rangle$-diagram*) of embeddings of cornered neighborhoods of the faces. It is then standard that one can compose two smooth cobordisms with corners, equipped with such smooth collars, by gluing along S. The composite is then the topological pushout of the two inclusions. Along the collars of S in M_1 and M_2, charts $\phi_i : U_i \to [0,\infty)^n$ are equivalent to charts mapping into $\mathbb{R}^{n-1} \times [0,\infty)$, and since the the composite has a smooth structure defined up to a diffeomorphism which is the identity along S. The precise smooth structure on this cobordism depends on the collar chosen, but there is always such a choice, and the resulting composites are all equivalent up to diffeomorphism.

Now, for each n, one can define:

Definition 4.3.2. *The bicategory* $\mathbf{nCob_2}$ *is given by the following data:*

- *The objects of* $\mathbf{nCob_2}$ *are of the form* $P = \hat{P} \times I^2$ *where* $\hat{P}$ *may be any* $(n-2)$ *manifolds without boundary and* $I = [0,1]$.
- *The morphisms of* $\mathbf{nCob_2}$ *are cobordisms* $P_1 \xrightarrow{i_1} S \xleftarrow{i_2} P_2$ *where* $S = \hat{S} \times I$ *and* $\hat{S}$ *is an* $(n-1)$*-dimensional collared cobordism with corners such that: the* $\hat{P}_i \times I$ *are objects, the maps are injections into* S*, a manifold with boundary, such that* $i_1(P_1) \cup i_2(P_2) = \partial S \times I$, $i_1(P_1) \cap i_2(P_2) = \emptyset$,
- *The 2-morphisms of* $\mathbf{nCob_2}$ *are generated by:*
  - *diffeomorphisms of the form* $f \times \mathrm{id} : T \times [0,1] \to T' \times [0,1]$ *where* T *and* T' *have a common boundary, and* f *is a diffeomorphism*$T \to T'$ *compatible with the source and target maps, i.e. fixing the collar.*
  - *2-cells: diffeomorphism classes of* n*-dimensional manifolds* M *with corners satisfying the properties of* M *in the diagram of equation (38), where isomorphisms are diffeomorphisms preserving the boundary*

  where the composite of the diffeomorphisms with the 2-cells (classes of manifolds M*) is given by composition of diffeomorphisms of the boundary cobordisms with the injection maps of the boundary* M

The source and target objects of any cobordism S are specified by saying that the source of S is the collection of components of $\partial S \times I$ for which the image of $(x, 0)$ lies on the boundary for $x \in \partial S$, and the target has the image of $(x, 1)$. The source and target objects are the collars, embedded in the cobordism in such a way that the source object $P = \hat{P} \times I^2$ is embedded in the cobordism $S = \hat{S} \times I$ by a map which is the identity on I taking the first interval in the object to the interval for a horizontal morphism, and the second to the interval for a vertical morphism. The same condition distinguishing source and target applies as above.

Composition of 2-cells works by gluing along common boundaries.

Lemma 4.3.3. *Composition of morphisms and 2-morphisms in* $\mathbf{nCob}_2$ *is well-defined and* $\mathbf{nCob}_2$ *is closed under composition.*

Proof. The horizontal and vertical morphisms are products of the interval I with $\langle 1 \rangle$-manifolds, whose boundary is $\partial_0 S$), equipped with collars. Suppose we are given two such cobordisms S_1 and S_2, and an identification of the source of S_2 with the target of S_1 (say this is $P = \hat{P} \times I$). Then the composite $S_2 \circ S_1$ is topologically the pushout of S_1 and S_2 over P. Now, P is smoothly embedded in S_1 and S_2, and any point in the pushout will be in the interior of either S_1 or S_2 since for any point on $\hat{P}$ each end of the interval I occurs as the boundary of only one of the two cobordisms. So the result is smooth. Thus, **2Cob** is closed under such composition of morphisms.

The same argument holds for 2-cells, since it holds for any representative of the equivalence class of some manifold with corners, M, and the differentiable structure will be the same, since we consider equivalence up to diffeomorphisms which preserve the collar exactly. □

Examples of such cobordisms with corners in 2 and 3 dimensions, as illustrations of (38), are shown in Figures 1 and 2, respectively.

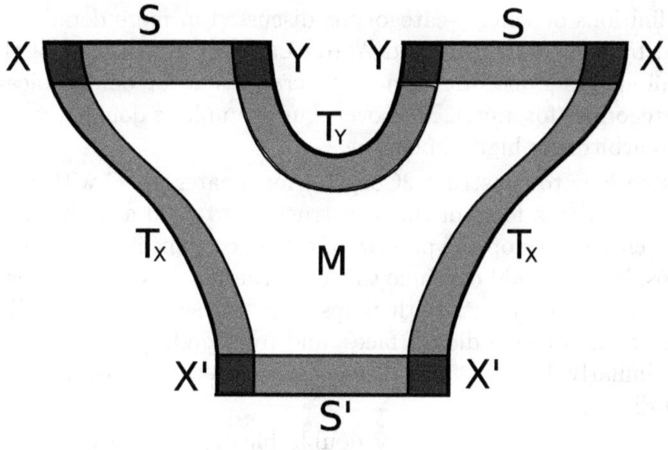

Figure 1: A 2-Dimensional Cobordism with Corners

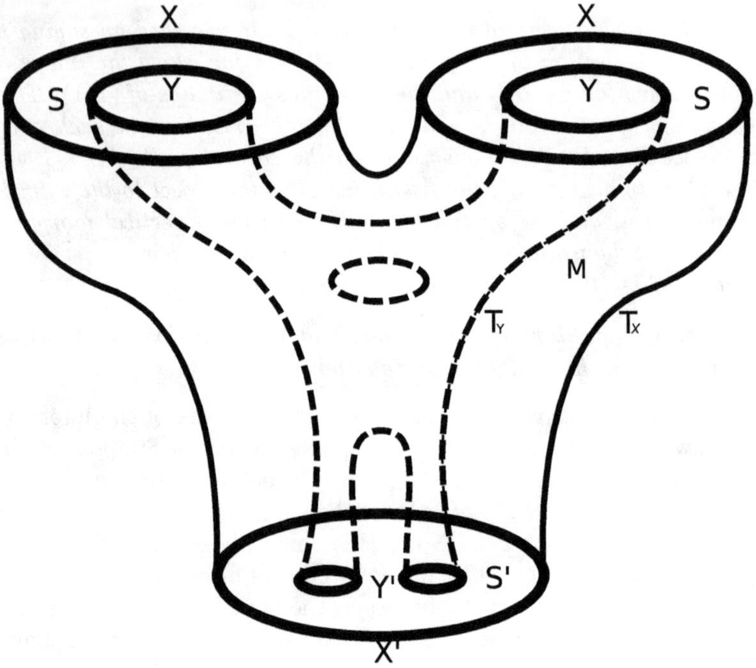

Figure 2: A 3-Dimensional Cobordism with Corners

4.4. Prospects for n-tuple Bicategories

We have discussed both double bicategories and Verity double bicategories, both of which can be seen as forms of *weak cubical* n-category. The broader question of various definitions of weak n-category is discussed in more detail by Tom Leinster [**Le**], and by Eugenia Cheng and Aaron Lauda [**CL**]. In light of this context, Theorem 4.1.3 suggests one direction of generalization for double bicategories, to "n-tuple bicategories" for any n. Moreover, our example of double (co)spans can be generalized to arbitrarily high dimension.

We have seen how to construct 2Cosp($\mathbf{C}$) for a category $\mathbf{C}$ with pushouts, and how we take a restricted form of this construction to yield a Verity double bicategory. We have chosen to stop the process of taking cospans in a category of cospans after two steps, but we could continue this construction. Taking cospans in this new category gives cubes of objects with maps from corners to the middles of edges, from middles of edges to middles of faces, and from middles of faces to the middle of the cube. Similarly, for any finite n, we can iterate the process of taking cospans to yield an n-dimensional cube.

In particular, we note that "Verity double bicategories" arise from special examples of bicategories internal to **Bicat**. There is a category of all such structures, namely the functor category of all maps $F : Th(\mathbf{Bicat}) \to \mathbf{Bicat}$, denoted $[Th(\mathbf{Bicat}), \mathbf{Bicat}]$. There will be an analogous concept of "triple bicategories", namely bicategories internal to $[Th(\mathbf{Bicat}), \mathbf{Bicat}]$. In general, a "$k$-tuple bicategory" will be a bicategory internal to the category of weak $(k-1)$-tuple categories.

We expect that for all k, a k-tuply iterated process of taking cospans of cospans (or similarly for spans) will yield examples of these structures. These k-dimensional (co)spans will naturally form a weak k-tuple category. Marco Grandis [**Gr3**] describes this in terms of a somewhat different description of weak n-cubical categories.

A further direction of generalization would be to substitute tricategories, tetracategories, and so forth in place of bicategories in the preceding construction, perhaps making different choices each stage. The question then arises what sort of structures it would be possible to define by selectively decategorifying, and what sorts of "filler" conditions this would need. Another potentially interesting question is whether the examples based on cospans also generalize—perhaps by taking cospans, not in a category, but in an n-category.

References

[**Ab**] Lowell Abrams. *Two-dimensional topological quantum field theories and Frobenius algebras.* J. Knot Theory Ramifications 5 (1996), 569-587. Available at http://home.gwu.edu/labrams/docs/tqft.ps.

[**Ati**] Michael Atiyah. *The Impact of Thom's Cobordism Theory.* Bulletin of the AMS, Vol 21, No. 3, pp337-340. 2004.

[**At1, At2**] Michael Atiyah. *Topological Quantum Field Theory.* Cambridge University Press, 1990.
Michael Atiyah. *Topological quantum field theory.* Publications Mathmatiques de l'IHS, 68 (1988), p. 175-186

[**BD**] John Baez, James Dolan. *Higher-Dimensional Algebra and Topological Quantum Field Theory.* Available as preprint http://arxiv.org/abs/q-alg/9503002. 1995.

[**BS**] John Baez, Urs Schreiber. *2-Connections on 2-Bundles.* Preprint.

[**BN**] Dror Bar-Natan. *Khovanov's Categorification of the Jones Polynomial.* Algebr. Geom. Topol. 2. 2002, pp337-370. Available as http://arXiv.org/abs/math.QA/0201043.

[**Ba**] Bruce Bartlett. *Categorical Aspects of Topological Quantum Field Theories.* Masters Thesis, Utrecht University. September 2005.

[**Be**] Jean Bénabou. *Introduction to Bicategories, pp1-77, Reports of the Midwest Category Seminar - Springer Lecture Notes in Mathematics 47.* Springer Verlag, New York/Berlin, 1967.

[BS] Ronald Brown, Christopher Spencer. *Double groupoids and crossed modules.* Cahiers Topologie Gom. Différentielle, **17(4)**, 343-362. 1976.

[CL] Eugenia Cheng, Aaron Lauda. *Higher-Dimensional Categories: An Illustrated Guidebook.* Available as . 2004.

[CY] Louis Crane, David Yetter. *On Algebraic Structures Implicit in Topological Quantum Field Theories.* J.Knot Theor.Ramifications 8 (1999) pp 125-163. Available as preprint http://arxiv.org/abs/hep-th/9412025. 1994.

[DPP] R. J. MacG. Dawson, R. Paré, D. A. Pronk. *Universal Properties of Span. Theory and Applications of Categories, Vol. 13, no. 4, pp61-85.* 2004.

[Eh1, Eh2] C. Ehresmann. *Catégories Structurées.* Ann. Sci. Ecole Norm. Sup **80**, pp 349-425. 1963.
C. Ehresmann. *Catégories et Structures*, Dunod, Paris, 1965.

[Fi] Thomas M. Fiore. *Pseudo Algebras and Pseudo Double Categories.* Journal of Homotopy and Related Structures Vol. 2 No. 2, pp 119-170, 2007.

[Fr] Daniel S. Freed. *Higher Algebraic Structures and Quantization.* Commun.Math.Phys. **159** 343-398. 1994.

[Ga] Richard Garner. *Polycategories.* Ph.D. Thesis, University of Cambridge. 2005.

[Gr1] Marco Grandis. *Higher Cospans and Weak Cubical Categories (Cospans in Algebraic Topology I).* Theory and Applications of Categories Vol. 18 No. 12, pp321-347. 2007.

[Gr3] Marco Grandis. *Cubical Cospans and Higher Cobordisms (Cospans in Algebraic Topology III).* Journal of Homotopy and Related Structures, Vol 3, No. 1. 2008.

[GP1] Marco Grandis, Robert Paré. *Limits in Double Categories.* Cahiers Topologie Géom. Différentielle Catég. **40**, pp162-220. 1999.

[GP2] Marco Grandis, Robert Paré. *Adjoints for Double Categories. Cah. Topologie Géom. Différentielle Catég. **45**, pp193-240.* 2004.

[GPS] R. Gordon, A.J. Power, Ross Street. *Coherence for Tricategories. Memoirs of the AMS* Vol. 117, No. 558. 1995.

[Ja] K. Jänich. *On the classification of $O(n)$-manifolds.* Math. Annalen **176**, pp 53-76. 1968.

[Jo] P.T. Johnstone. *Sketches of an Elephant: A Topos Theory Compendium: Volume 1.* Oxford University Press, 2002.

[Hi] Morris W. Hirsch. *Differential Topology.* Springer-Verlag, 1976.

[KS] G.M. Kelly, R. Street. *Proceedings Sydney Category Theory Seminar 1972/1973.* Lecture Notes in Mathematics **420**. Springer Verlag, 1974.

[KL] Thomas Kerler, Vlyodymyr V. Lyubashenko. *Non-semisimple Topological Quantum Field Theories for 3-Manifolds with Corners.* Lecture Notes in Mathematics **1765**. Springer Verlag, 2001.

[Ko] Joachim Kock. *Frobenius Algebras and 2D Topological Quantum Field Theories - London Mathematical Society Student Texts 59.* Cambridge University Press, 2003.

[La] Aaron D. Lauda. *Open-Closed Topological Quantum Field Theory and Tangle Homology.* Ph.D. Dissertation, University of Cambridge. 2006.

[LP] Aaron D. Lauda, Hendryk Pfeiffer. *Open-closed strings: Two-dimensional extended TQFT's and Frobenius algebras.* Topology and its Applications Vol. 155 No. 7, pp 623-666. 2008. http://arXiv.org/abs/math.AT/0510664. 2008.

[Lau] Gerd Laures. *On Cobordism of Manifolds with Corners. Transactions of the American Mathematical Society, vol. 252, no. 12, pp5667-5688.* 2000.

[Law] Ruth Lawrence. *Triangulation, Categories and Extended Field Theories,* in Quantum Topology. R. Baadhio and L. Kaufman, eds. World Scientific, Singapore, 1993, pp191-208.

[WL] F. William Lawvere. *Functorial Semantics of Algebraic Theories and Some Algebraic Problems in the Context of Functorial Semantics of Algebraic Theories. Reprints in Theory and Applications of Categories,* No. 5. 2004, pp 1-121. http://www.tac.mta.ca/tac/reprints/articles/5/tr5abs.html.

[Le] Tom Leinster. *Higher Operads, Higher Categories.* London Mathematical Society Lecture Notes Series, Cambridge University Press. 2003. Available as http://arXiv.org/abs/math.CT/0305049.

[Le2] Tom Leinster. *A Survey of Definitions of n-Category.* Theory and Applications of Categories **10** no. 1, pp1-70. 2002. http://www.tac.mta.ca/tac/volumes/10/1/10-01.ps.

[Mak] Michael Makkai. *Avoiding the axiom of choice in general category theory.* Journal of Pure and Applied Algebra, vol. 108, No. 2, pp. 109-173(65). 1996.

[Ma] N. Martins-Ferreira. *Pseudo-Categories.* Journal of Homotopy and Related Structures, Vol. 1 No. 1, pp. 47-78. 2006. .

[MV] S.S. Moskaliuk, A.T. Vlassov. *Double Categories in Mathematical Physics.* Preprint ESI 536. 1998.

[St] Ross Street. *The Algebra of Oriented Simplexes.* Journal of Pure and Applied Algebra, **49(3)** pp283-335. 1987.

[**Th**] René Thom. *Quelques propriétés des variétés differentiables.* Comm. Math. Helv. **28**, pp 17-86. 1954.

[**Tu**] Vladimir Turaev. *Cobordism of Knots in Surfaces.* Available as http://arXiv.org/abs/math.GT/0703055. 2007.

[**Ve**] Dominic Verity. *Enriched Categories, Internal Categories, and Change of Base.* Ph.D. Dissertation, University of Cambridge, 1992.

This article may be accessed via WWW at `http://www.rmi.acnet.ge/jhrs/`

Jeffrey C. Morton
`jeffrey.c.morton@gmail.com`

Mathematics Department
University of Western Ontario

www.ingramcontent.com/pod-product-compliance
Lightning Source LLC
Chambersburg PA
CBHW060949220326
41599CB00023B/3646